Understanding
Social Problems Third Edition

Linda A. Mooney

David Knox

Caroline Schacht

East Carolina University

WADSWORTH

THOMSON LEARNING

Australia • Canada • Mexico • Singapore • Spain • United Kingdom • United States

WADSWORTH

THOMSON LEARNING

Sociology Publisher: Eve Howard
Assistant Editor: Analie Barnett
Sr. Editorial Assistant: Stephanie M. Monzon
Marketing Manager: Matthew Wright
Marketing Assistant: Kasia Zagorski
Technology Project Manager: Dee Dee Zobian
Project Manager, Editorial Production: Lisa Weber
Print/Media Buyer: Mary Noel
Permissions Editor: Joohee Lee
Production Service: Graphic World Publishing Services

Text Designer: Ellen Pettengell
Photo Researcher: Sue C. Howard
Copy Editor: Lois Stagg
Illustrator: Graphic World Illustration Studio
Cover Designer: Cuttriss & Hambleton
Cover Image: PhotoDisc
Cover Printer: Phoenix Color Corporation
Compositor: Graphic World, Inc.
Printer: Quebecor/World, Taunton

Library of Congress Cataloging-in-Publication Data
Mooney, Linda A.
   Understanding social problems / Linda A. Mooney,
David Knox, Caroline Schacht.—3rd ed.
   p. cm.
Includes bibliographical references and index.
ISBN 0-534-58752-6
1. Social problems—United States. 2. United States—
Social conditions—1980-   I. Knox, David, 1943-
II. Schacht, Caroline.   III. Title.
HN59.2 .M66 2001
361.1′0973—dc21

Wadsworth/Thomson Learning
10 Davis Drive
Belmont, CA 94002-3098
USA

For more information about our products, contact us:
Thomson Learning Academic Resource Center
1-800-423-0563
http://www.wadsworth.com

International Headquarters
Thomson Learning
International Division
290 Harbor Drive, 2nd Floor
Stamford, CT 06902-7477
USA

UK/Europe/Middle East/South Africa
Thomson Learning
Berkshire House
168-173 High Holborn
London WC1V 7AA
United Kingdom

Asia
Thomson Learning
60 Albert Street, #15-01
Albert Complex
Singapore 189969

Canada
Nelson Thomson Learning
1120 Birchmount Road
Toronto, Ontario M1K 5G4
Canada

Brief Contents

Contents

SECTION 1   Problems of Well-Being   26

Chapter 7

Race and Ethnic Relations   184

Chapter 8

Gender Inequality   220

SECTION 3   Problems of Inequality and Power   280

Chapter 11

Work and Unemployment   318

Chapter 12

Problems in Education   354

SECTION 4   Problems of Modernization   410

Chapter 15

Science and Technology   453

Chapter 16

Conflict around the World   480

Preface

Violence in the home, school, and street; impoverished living conditions among millions of people throughout the world; increasing levels of environmental pollution and depletion of the earth's natural resources; persistent conflict between and within nations; ongoing oppression of minorities; and the widening gap between the "haves" and the "have-nots" paint a disturbing picture of our modern world. In *A Guide for the Perplexed*, E.F. Schumacher questions whether a "turning around will be accomplished by enough people quickly enough to save the modern world" (qtd in Safransky 1990, p.115). Schumacher notes that "this question is often asked, but whatever the answer given to it will mislead. The answer 'yes' would lead to complacency; the answer 'no' to despair. It is desirable to leave these perplexities behind us and get down to work."

In *Understanding Social Problems*, we "get down to work" by examining how the social structure and culture of society contribute to social problems and their consequences. Understanding the social forces that contribute to social problems is necessary for designing strategies for action—programs, policies and other interventions intended to ameliorate the social problem.

Academic Features of the New Edition

In response to feedback from teachers, reviewers, and students who have read the second edition of *Understanding Social Problems* (2000), we have retained several features and added several others:

Strong Integrative Theoretical Foundation. The three major sociological approaches—structural functionalism, symbolic interactionism, and conflict theory—are introduced in the first chapter and discussed and applied, where appropriate, to various social problems throughout the text. Other theories of social problems, as well as feminist approaches, are also presented where appropriate.

Emphasis on the Structure and Culture of Society. As noted above, the text emphasizes how the social structure and culture of society contribute to and maintain social problems, as well as provide the basis for alternative solutions.

Review of Basic Sociological Terms. An overview of basic sociological terms and concepts is presented in the first chapter. This overview is essential for students who have not taken an introductory course and is helpful, as a review, for those who have. Additionally, Appendix A details "Methods of Data Analysis."

Unique Organization. The order of the 16 chapters reflects a progression from micro to macro level of analysis, focusing first on problems of health care, drug use and crime and then broadening to the widening concerns of population growth and the environmental problems, science and technology and conflict around the world.

Two chapters merit special mention: "Sexual Orientation" (Chapter 9) and "Science and Technology" (Chapter 15). Whereas traditional texts discuss sexual orientation under the rubric of "deviance," this topic is examined in the section on problems of human diversity along with the related issues of age, gender, and racial and ethnic inequality. The chapter on science and technology includes such topics as biotechnology, the computer revolution, and the Internet. This chapter emphasizes the transformation of society through scientific and technological innovations, the societal costs of such innovations, and issues of social responsibility. This chapter is particularly relevant to college students, many of whom have never known a world without computers.

Expanded Coverage of Global Issues.   In the third edition, we place an even greater emphasis on examining social problems from a global perspective. Each chapter contains a heading entitled "The Global Context," and the number and scope of references to international issues have been expanded.

Consistent Chapter Format.   Each chapter follows a similar format: the social problem is defined, the theoretical explanations are discussed, the consequences of the social problem are explored, and the alternative solutions and policies are examined. A concluding section assesses the current state of knowledge for each social problem.

Increased Media Content.   New to the third edition is an emphasis on the media and the role it plays in defining, exacerbating, and ameliorating social problems. Examples include discussions of the media's role in stigmatizing mental illness (Chapter 2), minority representation on prime time television shows (Chapter 7), and media portrayals of the poor (Chapter 10).

Standard and Cutting Edge Topics.   In addition to problems that are typically addressed in social problems courses and texts, new and emerging topics are examined. Topics new to the third edition include the mailbox economy (Chapter 6), gender tourism (Chapter 8), genetically modified crops (Chapter 10), and eco-terrorism and environmental refugees (Chapter 14).

Pedagogical Features of the New Edition

Opening Vignettes.   New to the third edition, each chapter begins with a vignette designed to engage the student by illustrating the current relevance of the topic under discussion. Topics of opening vignettes include the New York City police shooting of Amadou Diallo (Chapter 4), the sex discrimination suit of Duke place kicker Heather Sue Mercer (Chapter 8), and life as Mr. Dot-ComGuy (Chapter 15).

Student-friendly Presentation.   To enhance the book's appeal to students, the third edition includes expanded information relevant to the college population. In Chapter 1, for example, we present data on the beliefs of college students about various social problems and Chapter 3 contains a section on binge drinking and other student alcohol-related problems. Further, Chapter 7 contains a new section on race and ethnic diversity on campus and in Chapter 12, "Problems in Education," students may complete a "Student Alienation Scale."

Self and Society. Each chapter includes a social survey designed to help students assess their own attitudes, beliefs, knowledge, or behavior regarding some aspect of a social problem. Examples include a Criminal Activities Survey (Chapter 4), a Beliefs about Women Scale (Chapter 8), and an Attitudes toward Economic Opportunity in the United States Inventory (Chapter 10).

The Human Side. To personalize the information being discussed, each chapter includes a feature entitled "The Human Side." These features describe personal experiences of individuals who have been affected by the social problem under discussion. Examples include a witness' description of an execution (Chapter 4), a college student's story of child sexual abuse (Chapter 5), the case study of a woman trying to change her sexual orientation (Chapter 9), and a Gulf War veteran's parents' description of their son's death (Chapter 16).

Social Problems Research Up Close. Now in every chapter, boxes called "Social Problems Research Up Close" present examples of social science research. These boxes demonstrate for students the sociological enterprise from theory and data collection, to findings and conclusions. Examples of research topics covered include "No Shame in My Game: The Working Poor in the Inner City" (Chapter 13), "The Social Construction of the Hacking Community" (Chapter 15), and "Family Adjustment to Military Deployment" (Chapter 16).

Focus on Technology. Boxes called "Focus on Technology" also now appear in every chapter. These boxes present information on how technology may contribute to social problems and their solutions. For example, in Chapter 4, Crime and Violence, the Focus on Technology feature highlights the use of DNA testing in criminal investigations. In Chapter 14, Population and the Environment, environmental and health hazards associated with computers are discussed.

Is It True? Each chapter begins with five true-false items to stimulate student interest and thinking.

Critical Thinking. Each chapter ends with a brief section called "Critical Thinking" that raises several questions related to the chapter topic. These questions invite the student to use critical thinking skills in applying the information discussed in the chapters.

InfoTrac College Edition. The third edition includes InfoTrac® College Edition—an online reference service that allows students to search for articles by subject, title, and author in hundreds of top academic journals and popular sources.

World Wide Web Home Page. http://sociology.wadsworth.com/. As an additional pedagogical tool, *Understanding Social Problems* has its own home page on the World Wide Web. Students and faculty can access relevant research studies, statistics, and theoretical links as suggested at the end of each chapter in a new Media Resources section. For example, in Chapter 12, Problems in Education, faculty and students are referred to Internet links on school vouchers, Title IX, distance learning and school violence.

In addition to chapter by chapter links, several additional features have been incorporated in the text's home page. For faculty, the *Understanding Social Problems* home page now contains an online instructor's manual and test bank, both password protected. For students, the home page now includes practice quizzes, flashcards, as well as the InfoTrac College Edition articles and questions listed at the end of each text chapter. Links to information about employment opportunities and sociology as a career are also listed. Students and faculty can send questions, comments, or suggestions directly to the authors and contact Wadsworth Publishing Company concerning book adoption.

New to This Edition—A Chapter by Chapter Look

In addition to the academic and pedagogical features noted above, *Understanding Social Problems'* content areas have been significantly revised. Over thirty new citations have been added to every chapter. Further, in addition to expanded coverage of important topics from the first and second editions, we have added new areas of research and theorizing. A partial list of new or expanded topics follows:

Chapter 1: Thinking About Social Problems.   The "sociological enterprise," journal article content, reading tables, and triangulation.

Chapter 2: Illness and the Health Care System.   New sections on mental illness, inadequacy of mental health care, strategies to improve mental health care, and expanded coverage of HIV/AIDS worldwide.

Chapter 3: Alcohol and Other Drugs.   Club drugs, mandatory drug testing, images of alcohol and tobacco in children's animated films, inpatient/out-patient treatment, and *Plan Colombia*.

Chapter 4: Crime and Violence.   Transnational crime, increase in prison population and decrease in crime rates, racial profiling, and restorative justice.

Chapter 5: Family Problems.   Parental alienation syndrome, interactive computer parenting, expanded global information on abuse of women, and expanded coverage of single parent families.

Chapter 6: The Young and the Old.   Dependency ratio, beliefs about the elderly, children and grandchildren as caretakers for the elderly, and an expanded section on children, violence and the media.

Chapter 7: Race and Ethnic Relations.   Hate on campus, new Census data on multiracial identification, expanded discussion of aversive and modern racism, and on racial/ethnic bias in the media.

Chapter 8: Gender Inequality.   Beliefs about gender equality, expanded section on media, language and cultural sexism, the "war on boys, " and a new section on international efforts toward gender equality.

Chapter 9: Sexual Orientation.   Updated terminology (lesbigay, transgender, LGBT), civil unions in Vermont, expanded global coverage of laws concerning sexual orientation, and a new section on how homophobia affects heterosexuals.

Chapter 10: The Haves and the Have-Nots.   New global qualitative research findings on poverty, expanded discussion of homelessness, and updated information on living wage laws.

Chapter 11: Work and Unemployment.   Limiting corporate power through campaign finance reform, job searching and recruitment through the Internet, and efforts to reform U.S. law on juvenile workers in agriculture.

Chapter 12: Problems in Education.   School shootings, the Early Head Start Program, President Bush's "no child left behind" proposal, and an expanded section on school choice.

Chapter 13: Cities in Crisis.   The effects of urban sprawl on wildlife, brownfields' redevelopment, and strategies of inner-city residents to avoid criminal victimization.

Chapter 14: Population and Environmental Problems.   Environmental education, intergovernmental report on global warming and its effects, on-line environmental activism, and corporate involvement in the environmental movement.

Chapter 15: Science and Technology.   Microsoft anti-trust lawsuit, completion of the Human Genome Project, the 2000 Discovery Innovation Awards, and cyberstalking.

Chapter 16: Conflict around the World.   Dual use technologies, heightened Israeli-Arab tensions, the bombing of the USS Cole, recent criticisms of the United Nations, and the Landmine Ban Treaty.

Supplements

The third edition of *Understanding Social Problems* comes with a full complement of supplements designed with both faculty and students in mind.

Web site for Students and Instructors

Virtual Society: The Wadsworth Sociology Resource Center
http://sociology.wadsworth.com

At *Virtual Society: The Wadsworth Sociology Resource Center,* you can find a Career Center, Sociology in the News, links to great sociology web sites and InfoTrac

College Edition, Virtual Tours in Sociology, an email link to the book's editor, instructor's resources, and many other selections. Through this site you can also access the *Understanding Social Problems* Web site, which offers the following online student study tools for each chapter of the text:

- Internet links and exercises
- Online practice quizzes
- InfoTrac College Edition exercises
- Flashcards

And much more!

Supplements for Instructors

Instructor's Edition. The Instructor's Edition of this text contains an eight-page Preview that provides a visual overview of the supplements that accompany the textbook as well as the features and themes of the textbook itself.

Instructor's Manual with Test Bank. Written by the main text authors, this supplement offers the instructor learning objectives, key terms, lecture outlines, student projects, classroom activities, and Internet and InfoTrac exercises. Test items include 60–100 multiple-choice and true/false questions with answers and page references, as well as short answer and essay questions for each chapter. Tips for using WebTutor and a concise user guide for *InfoTrac* are included as appendixes.

ExamView Computerized and Online Testing. Create, deliver, and customize tests and study guides (both print and online) in minutes with this easy-to-use assessment and tutorial system. ExamView offers both a Quick Test Wizard and an Online Test Wizard that guide you step-by-step through the process of creating tests, while its unique "WYSIWYG" capability allows you to see the test you are creating on the screen exactly as it will print or display online. Using ExamView's complete word processing capabilities, you can enter an unlimited number of new questions or edit existing questions.

Wadsworth's Introduction to Sociology Transparency Acetates. A selection of quality acetates from Wadsworth's introductory sociology texts. Free to qualified adopters.

SocLink 2002 CD-ROM: A Microsoft PowerPoint Presentation Tool. SocLink 2002 is an easy-to-use PowerPoint presentation tool that permits instructors to draw upon a digital library of hundreds of pieces of graphic art from Wadsworth sociology textbooks. In addition, photographs, CNN video segments, and pre-assembled lecture slides can be used to create customized lecture presentations.

***CNN Today Issues and Solutions* Video Series, Volumes I and II.** The *CNN Today Issues and Solutions* Video Series is an exclusive series jointly created by Wadsworth and CNN for the social problems course. Each video in the series consists of approximately 45-minutes of footage originally broadcast on CNN within in the last several years and selected specifically to illustrate important sociological concepts. The videos are broken into short two- to seven-minute segments, which are perfect for classroom use as lecture launchers, or to illustrate key sociological concepts. An annotated table of contents accompanies each video with descriptions of the segments. Special adoption conditions apply.

Wadsworth Sociology Video Library. Qualified adopters can also select from an extensive selection of videos from Films for the Humanities and Sciences. Ask your Wadsworth/Thomson Learning representative for more information.

Supplements for Students

Study Guide. Written by the authors of the main text, the Study Guide includes learning objectives, chapter outlines, key terms, completion exercises, and practice tests consisting of multiple-choice, true/false and short answer/essay questions to enhance and test student understanding of chapter concepts.

Researching Sociology on the Internet. Written by D.R. Wilson, Houston Baptist University, and David L. Carlson, Texas A&M University, this guide is designed to assist Sociology students with doing research on the Internet. Part One contains general information necessary to get started and answers questions about security, the type of sociology material available on the Internet, the information that is reliable and the sites that are not, the best ways to find research, and the best links to take students where they want to go. Part Two looks at each main discipline in the area of Sociology, and refers students to sites where the most enlightening research can be obtained.

Social Problems: A Reader with Four Questions. Written by Joel M. Charon, author of the best selling introductory sociology reader, *Ten Questions: A Sociological Perspective,* this NEW reader encourages students to take a critical, sociological look at today's diverse social problems. The four-question approach—What is the problem? What makes the problem a social problem? What causes the problem? What can be done?—provides a consistent framework through which students can objectively analyze and discuss our most complicated social issues.

Social Problems of the Modern World: A Reader. Written by Francis Moulder, Three Rivers Community Technical College, this FREE reader will appeal to Social Problems instructors that seek to internationalize the Social Problems course. The current trend in this course is to add a global focus, something that Moulder does in her reader without sacrificing focus on the United States.

Student Guide to *InfoTrac College Edition* for Sociology. This unique supplement, prepared by Michele Adams of the University of California–Riverside, consists of exercises based on 23 core subjects vital to the study of Sociology. These exercises utilize InfoTrac College Edition's huge database of articles. The exercises help students to narrow down the search of articles related to each subject and ask questions that enable students to see the ideas more clearly and pique students' interest.

Thomson Learning Web Tutor on Web CT and BlackBoard. WebTutor is a content-rich, Web-based teaching and learning tool that helps students succeed by taking the course beyond classroom boundaries to an anywhere, anytime environment. WebTutor is rich with study and mastery tools, communication tools, and course content. Professors can use WebTutor to provide virtual office hours, post syllabi, set up threaded discussions, track student progress with the quizzing material, and more.

Interactions CD-ROM. *Interactions* is a NEW free CD-ROM developed specifically for the third edition of *Understanding Social Problems*. Dynamic, colorful, and exciting, this all-new interactive tool for students contains

- Multimedia chapter summaries that include presentations of chapter topics with audio, photos, and selected videos
- Chapter-by-chapter quizzes with scoring and feedback
- Direct links to the Internet and InfoTrac College Edition
- Links to other online sociology resources and the *Understanding Social Problems* Web site

Acknowledgments

This text reflects the work of many people. We would like to thank the following for their contributions to the development of this text: Eve Howard, Stephanie Monzon, Analie Barnett, Dee Dee Zobian, Matthew Wright, Lisa Weber, Joy Westberg, and Marcia Craig.

We would also like to acknowledge the support and assistance of Blair Carr, Margaret and Thomas Mooney, Marieke Van Willigan, Bob Edwards, Ann Schehr, Christa Reiser, Richard Caston, Kristin Cmar, Brent Aspinwall, and Susann Mathews. To each you have our heartfelt thanks.

Additionally, we are indebted to those who read the manuscript in its various drafts and provided valuable insights and suggestions, many of which have been incorporated into the final manuscript:

Anna M. Cognetto
Dutchess Community College

Lynda D. Nyce
Bluffton College

Katherine Ann Dieetrich
Blinn College

Alice Van Ommeren
San Joaquin Delta College

Cooper Lansing
Erie Community College

Harry L. Vogel
Kansas State University

Tunga Lergo
Santa Fe Community College, Main Campus

We are also grateful to the reviewers of the first and second editions:

David Allen
University of New Orleans

Verghese Chirayath
John Carroll University

Patricia Atchison
Colorado State University

Kimberly Clark
DeKalb College–Central Campus

Walter Carroll
Bridgewater State College

Barbara Costello
Mississippi State University

Roland Chilton
University of Massachusetts

William Cross
Illinois College

Doug Degher
Northern Arizona University

Jane Ely
State University of New York Stony Brook

Joan Ferrante
Northern Kentucky University

Robert Gliner
San Jose State University

Julia Hall
Drexel University

Millie Harmon
Chemeketa Community College

Sylvia Jones
Jefferson Community College

Nancy Kleniewski
University of Massachusetts, Lowell

Daniel Klenow
North Dakota State University

Mary Ann Lamanna
University of Nebraska

Phyllis Langton
George Washington University

Lionel Maldonado
California State University, San Marcos

Judith Mayo
Arizona State University

Peter Meiksins
Cleveland State University

Madonna Harrington-Meyer
University of Illinois

Clifford Mottaz
University of Wisconsin–River Falls

Linda Nyce
Bluffton College

James Peacock
University of North Carolina

Ed Ponczek
William Rainey Harper College

Cynthia Reynaud
Louisiana State University

Rita Sakitt
Suffolk County Community College

Mareleyn Schneider
Yeshiva University

Paula Snyder
Columbus State Community College

Lawrence Stern
Collin County Community College

John Stratton
University of Iowa

Joseph Trumino
*St. Vincent's College of St. John's
University*

Joseph Vielbig
Arizona Western University

Rose Weitz
Arizona State University

Bob Weyer
County College of Morris

Oscar Williams
Diablo Valley College

Mark Winton
University of Central Florida

Diane Zablotsky
University of North Carolina

Finally, we are interested in ways to improve the text and invite your feedback and suggestions for new ideas and material to be included in subsequent editions.

Linda A. Mooney, David Knox, and
Caroline Schacht
Department of Sociology
East Carolina University
Greenville, NC 27858

E-mail addresses:

mooneyl@mail.ecu.edu
knoxd@mail.ecu.edu
schachtc@mail.ecu.edu

Thinking about Social Problems

Outline

What Is a Social Problem?

Elements of Social Structure and Culture

The Sociological Imagination

Theoretical Perspectives

Social Problems Research

Goals of the Text

Understanding Social Problems

Is It True?

1. An annual study of social well-being in America revealed that since 1970 social conditions have steadily improved in our society.

2. Prior to the nineteenth century, it was considered a husband's legal right and marital obligation to discipline and control his wife through the use of physical force.

3. In seventeenth- and eighteenth-century England, tea drinking was considered a social problem.

4. Questions involving values, religion, and morality can be answered only through scientific research.

5. Male high school students' use of condoms during sexual intercourse has decreased over the last five years.

Answers to "Is It True?": 1 = F; 2 = T; 3 = T; 4 = F; 5 = F

Unless someone like you cares a whole awful lot, nothing is going to get better. It's not.

DR. SEUSS
The Lorax

Researchers at Fordham University conduct an annual study called "The Index of Social Health." This study evaluates the cumulative effect on Americans of 16 major social problems including crime, unemployment, drug abuse, suicide rates, homicide rates, and child abuse. According to analyses of these 16 social indicators, the nation's social health has stagnated or declined throughout the 1980s and the 1990s. While some aspects of social life are improving, most are getting worse (Miringoff & Miringoff 1999).

A global perspective on social problems is even more troubling. In 1990 the United Nations Development Programme published its first annual Human Development Report, which measured the well-being of populations around the world according to a "human development index" (HDI). This index measures three basic dimensions of human development—longevity, knowledge (i.e., educational attainment), and a decent standard of living. The most recent report reveals that "globalization is increasing human insecurity by accelerating the spread of crime, disease, and financial volatility" (May 2000, 219). The results—100 million children live and work on the streets, 1.2 billion people live on less than a dollar a day, and 18 million people die every day from communicable diseases.

Social problems are fundamentally products of collective definition. . . . A social problem does not exist for society unless it is recognized by that society to exist.

HERBERT BLUMER
Sociologist

Problems related to poverty and malnutrition, inadequate education, acquired immunodeficiency syndrome (AIDS) and other sexually transmitted diseases (STDs), inadequate health care, crime, conflict, oppression of minorities, environmental destruction, and other social issues are both national and international concerns. Such problems present both a threat and a challenge to our national and global society.

The primary goal of this text is to facilitate increased awareness and understanding of problematic social conditions in U.S. society and throughout the world. Although the topics covered in this text vary widely, all chapters share common objectives: to explain how social problems are created and maintained; to indicate how they affect individuals, social groups, and societies as a whole; and to examine programs and policies for change. We begin by looking at the nature of social problems.

What Is a Social Problem?

There is no universal, constant, or absolute definition of what constitutes a social problem. Rather, social problems are defined by a combination of objective and subjective criteria that vary across societies, among individuals and groups within a society, and across historical time periods.

These woment were the victims of streaming, the assaulting of women in public places by multiple offenders. Prior to 1999, few streaming incidents were recorded. Others, however, have noted that the 1992 convention of Navy personnel that led to the "Tailhook" scandal is an earlier example of such behavior.

© AP/Wide World Photo

Objective and Subjective Elements of Social Problems

Although social problems take many forms, they all share two important elements: an objective social condition and a subjective interpretation of that social condition. The **objective element** of a social problem refers to the existence of a social condition. We become aware of social conditions through our own life experience, through the media, and through education. We see the homeless, hear gunfire in the streets, and see battered women in hospital emergency rooms. We read about employees losing their jobs as businesses downsize and factories close. In television news reports we see the anguished faces of parents whose children have been killed by violent youths.

The **subjective element** of a social problem refers to the belief that a particular social condition is harmful to society, or to a segment of society, and that it should and can be changed. We know that crime, drug addiction, poverty, racism, violence, and pollution exist. These social conditions are not considered social problems, however, unless at least a segment of society believes that these conditions diminish the quality of human life.

By combining these objective and subjective elements, we arrive at the following definition: A **social problem** is a social condition that a segment of society views as harmful to members of society and in need of remedy.

Variability in Definitions of Social Problems

Individuals and groups frequently disagree about what constitutes a social problem. For example, some Americans view the availability of abortion as a social problem, while others view restrictions on abortion as a social problem. Similarly, some Americans view homosexuality as a social problem, while others view prejudice and discrimination against homosexuals as a social problem. Such variations in what is considered a social problem are due to differences in values, beliefs, and life experiences.

It is easy to find things in the real world that are not perfect and to say that something should be done to correct the problems. However, for every ten people who see the same problem, there will be ten different ideal solutions.

RANDALL G. HOLCOMBE
Chairman of the Research Advisory Council of the James Madison Institute for Public Policy Studies

While some individuals view the availability of abortion as a social problem, others view restrictions on abortion as a social problem. The disagreement can lead to violence, destruction, and murder. In 1998, Dr. Barnett Slepian, who had performed abortions at a Buffalo, New York, clinic, was shot and killed by an antiabortion activist.

Sometimes a cigar is just a cigar.

SIGMUND FREUD
Founder of psychoanalysis

Definitions of social problems vary not only within societies, but across societies and historical time periods as well. For example, prior to the nineteenth century, it was a husband's legal right and marital obligation to discipline and control his wife through the use of physical force. Today, the use of physical force is regarded as a social problem rather than a marital right.

Tea drinking is another example of how what is considered a social problem can change over time. In seventeenth- and eighteenth-century England, tea drinking was regarded as a "base Indian practice" that was "pernicious to health, obscuring industry, and impoverishing the nation" (Ukers 1935, cited in Troyer & Markle 1984). Today, the English are known for their tradition of drinking tea in the afternoon.

Because social problems can be highly complex, it is helpful to have a framework within which to view them. Sociology provides such a framework. Using a sociological perspective to examine social problems requires a knowledge of the basic concepts and tools of sociology. In the remainder of this chapter, we discuss some of these concepts and tools: social structure, culture, the "sociological imagination," major theoretical perspectives, and types of research methods.

Elements of Social Structure and Culture

Although society surrounds us and permeates our lives, it is difficult to "see" society. By thinking of society in terms of a picture or image, however, we can visualize society and therefore better understand it. Imagine that society is a coin with two sides: on one side is the structure of society, and on the other is the culture of society. Although each "side" is distinct, both are inseparable from the whole. By looking at the various elements of social structure and culture, we can better understand the root causes of social problems.

Elements of Social Structure

The *structure* of a society refers to the way society is organized. Society is organized into different parts: institutions, social groups, statuses, and roles.

Institutions   An **institution** is an established and enduring pattern of social relationships. The five traditional institutions are family, religion, politics, economics, and education, but some sociologists argue that other social institutions, such as science and technology, mass media, medicine, sport, and the military, also play important roles in modern society.

Many social problems are generated by inadequacies in various institutions. For example, unemployment may be influenced by the educational institution's failure to prepare individuals for the job market and by alterations in the structure of the economic institution.

Social Groups Institutions are made up of social groups. A **social group** is defined as two or more people who have a common identity, interact, and form a social relationship. For example, the family in which you were reared is a social group that is part of the family institution. The religious association to which you may belong is a social group that is part of the religious institution.

Social groups may be categorized as primary or secondary. **Primary groups**, which tend to involve small numbers of individuals, are characterized by intimate and informal interaction. Families and friends are examples of primary groups. **Secondary groups**, which may involve small or large numbers of individuals, are task-oriented and characterized by impersonal and formal interaction. Examples of secondary groups include employers and their employees, and clerks and their customers.

Statuses Just as institutions consist of social groups, social groups consist of statuses. A **status** is a position a person occupies within a social group. The statuses we occupy largely define our social identity. The statuses in a family may consist of mother, father, stepmother, stepfather, wife, husband, child, and so on. Statuses may be either ascribed or achieved. An **ascribed status** is one that society assigns to an individual on the basis of factors over which the individual has no control. For example, we have no control over the sex, race, ethnic background, and socioeconomic status into which we are born. Similarly, we are assigned the status of "child," "teenager," "adult," or "senior citizen" on the basis of our age—something we do not choose or control.

An **achieved status** is assigned on the basis of some characteristic or behavior over which the individual has some control. Whether or not you achieve the status of college graduate, spouse, parent, bank president, or prison inmate depends largely on your own efforts, behavior, and choices. One's ascribed statuses may affect the likelihood of achieving other statuses, however. For example, if you are born into a poor socioeconomic status, you may find it more difficult to achieve the status of "college graduate" because of the high cost of a college education.

Every individual has numerous statuses simultaneously. You may be a student, parent, tutor, volunteer fund-raiser, female, and Hispanic. A person's **master status** is the status that is considered the most significant in a person's social identity. Typically, a person's occupational status is regarded as his or her master status. If you are a full-time student, your master status is likely to be "student."

Roles Every status is associated with many **roles,** or the set of rights, obligations, and expectations associated with a status. Roles guide our behavior and allow us to predict the behavior of others. As a student, you are expected to attend class, listen and take notes, study for tests, and complete assignments. Because you know what the role of teacher involves, you can predict that your teacher will lecture, give exams, and assign grades based on your performance on tests.

A single status involves more than one role. For example, the status of prison inmate includes one role for interacting with prison guards and another role for interacting with other prison inmates. Similarly, the status of nurse involves different roles for interacting with physicians and with patients.

> When I fulfill my obligations as a brother, husband, or citizen, when I execute contracts, I perform duties that are defined externally to myself. . . . Even if I conform in my own sentiments and feel their reality subjectively, such reality is still objective, for I did not create them; I merely inherited them.
>
> **EMILE DURKHEIM**
> *Sociologist*

Elements of Culture

Man is made by his belief. As he believes, so he is.

BHAGAVAD GITA

Whereas social structure refers to the organization of society, culture refers to the meanings and ways of life that characterize a society. The elements of culture include beliefs, values, norms, sanctions, and symbols.

Beliefs **Beliefs** refer to definitions and explanations about what is assumed to be true. The beliefs of an individual or group influence whether that individual or group views a particular social condition as a social problem. Does second-hand smoke harm nonsmokers? Are nuclear power plants safe? Does violence in movies and on television lead to increased aggression in children? Our beliefs regarding these issues influence whether we view the issues as social problems. Beliefs not only influence how a social condition is interpreted, they also influence the existence of the condition itself. For example, young women who believed that forced interactions with former boyfriends were romantic (i.e., he brought flowers, a card, or a gift) were less likely to define the relationship as frightening and, thus, less likely to terminate it (Dunn 2000). The *Self and Society* feature in this chapter allows you to assess your own beliefs about various social issues and compare your beliefs with a national sample of first-year college students.

Values **Values** are social agreements about what is considered good and bad, right and wrong, desirable and undesirable. Frequently, social conditions are viewed as social problems when the conditions are incompatible with or contradict closely held values. For example, poverty and homelessness violate the value of human welfare; crime contradicts the values of honesty, private property, and nonviolence; racism, sexism, and heterosexism violate the values of equality and fairness.

When people cherish some set of values and do not feel any threat to them, they experience well-being. When they cherish values but do feel them to be threatened, they experience a crisis—either as a personal trouble or as a public issue.

C. WRIGHT MILLS
Sociologist

Values play an important role not only in the interpretation of a condition as a social problem, but also in the development of the social condition itself. Sylvia Ann Hewlett (1992) explains how the American values of freedom and individualism are at the root of many of our social problems:

> There are two sides to the coin of freedom. On the one hand, there is enormous potential for prosperity and personal fulfillment; on the other are all the hazards of untrammeled opportunity and unfettered choice. Free markets can produce grinding poverty as well as spectacular wealth; unregulated industry can create dangerous levels of pollution as well as rapid rates of growth; and an unfettered drive for personal fulfillment can have disastrous effects on families and children. Rampant individualism does not bring with it sweet freedom; rather, it explodes in our faces and limits life's potential (pp. 350–51).

Absent or weak values may contribute to some social problems. For example, many industries do not value protection of the environment and thus contribute to environmental pollution.

Norms and Sanctions **Norms** are socially defined rules of behavior. Norms serve as guidelines for our behavior and for our expectations of the behavior of others.

There are three types of norms: folkways, laws, and mores. **Folkways** refer to the customs and manners of society. In many segments of our society, it is customary to shake hands when being introduced to a new acquaintance, to say "excuse me" after sneezing, and to give presents to family and friends on their

Personal Beliefs about Various Social Problems

Indicate whether you agree or disagree with each of the following statements:

Statement	Agree	Disagree
1. If two people really like each other, it's alright for them to have sex even if they have known each other for only a very short time.		
2. Colleges should prohibit racist/sexist speech on campus.		
3. There is too much concern in the courts for the rights of criminals.		
4. Abortion should be legal.		
5. The death penalty should be abolished.		
6. The activities of married women are best confined to the home and family.		
7. Marijuana should be legalized.		
8. It is important to have laws prohibiting homosexual relationships.		
9. Employers should be allowed to require drug testing of employees or job applicants.		
10. The federal government should do more to control the sale of handguns.		
11. Racial discrimination is no longer a major problem in America.		
12. Realistically, an individual can do little to bring about changes in our society.		
13. Wealthy people should pay a larger share of taxes than they do now.		
14. Affirmative action in college admissions should be abolished.		
15. Same-sex couples should have the right to legal marital status.		

Percentage* of First-Year College Students Agreeing with Belief Statements

Statement Number	Percentage Agreeing in 2000		
	Total	Women	Men
1. Have sex	42	31	55
2. Prohibit speech on campus	62	66	56
3. Too much concern for criminals' rights	67	66	68
4. Abortion rights	54	54	55
5. Abolishment of death penalty	31	34	27
6. Women's activities confined to home	22	17	29
7. Legalization of marijuana	34	29	40
8. Laws prohibiting gay relationships	27	20	36
9. Employers' right to drug test	77	79	73
10. Federal control of handgun sales	82	90	73
11. Racial discrimination not a problem	21	17	24
12. Individuals can't influence social change	27	24	32
13. Wealthy should pay higher taxes	52	52	53
14. Affirmative action abolished in college	50	45	56
15. Legal right of same-sex couples to marry	56	63	47

*Percentages are rounded.

Source: The American Freshman: *National Norms for Fall 2000.* Los Angeles: Higher Education Research Institute, UCLA. Copyright © 2000 by the Regents of the University of California. Used by permission.

birthdays. Although no laws require us to do these things, we are expected to do them because they are part of the cultural traditions, or folkways, of the society in which we live.

Laws are norms that are formalized and backed by political authority. A person who eats food out of a public garbage container is violating a folkway; no law prohibits this behavior. However, throwing trash onto a public street is considered littering and is against the law.

Some norms, called **mores**, have a moral basis. Violations of mores may produce shock, horror, and moral indignation. Both littering and child sexual abuse are violations of law, but child sexual abuse is also a violation of our mores because we view such behavior as immoral.

All norms are associated with **sanctions**, or social consequences for conforming to or violating norms. When we conform to a social norm, we may be rewarded by a positive sanction. These may range from an approving smile to a public ceremony in our honor. When we violate a social norm, we may be punished by a negative sanction, which may range from a disapproving look to the death penalty or life in prison. Most sanctions are spontaneous expressions of approval or disapproval by groups or individuals—these are referred to as informal sanctions. Sanctions that are carried out according to some recognized or formal procedure are referred to as formal sanctions. Types of sanctions, then, include positive informal sanctions, positive formal sanctions, negative informal sanctions, and negative formal sanctions (see Table 1.1).

Symbols A **symbol** is something that represents something else. Without symbols, we could not communicate with each other or live as social beings.

The symbols of a culture include language, gestures, and objects whose meaning is commonly understood by the members of a society. In our society, a red ribbon tied around a car antenna symbolizes Mothers Against Drunk Driving, a peace sign symbolizes the value of nonviolence, and a white hooded robe symbolizes the Ku Klux Klan. Sometimes people attach different meanings to the same symbol. The Confederate flag is a symbol of Southern pride to some, a symbol of racial bigotry to others.

The elements of the social structure and culture just discussed play a central role in the creation, maintenance, and social response to various social problems. One of the goals of taking a course in social problems is to develop an awareness of how the elements of social structure and culture contribute to social problems. Sociologists refer to this awareness as the "sociological imagination."

Table 1.1 *Types and Examples of Sanctions*

	Positive	**Negative**
Informal	Being praised by one's neighbors for organizing a neighborhood recycling program.	Being criticized by one's neighbors for refusing to participate in the neighborhood recycling program.
Formal	Being granted a citizen's award for organizing a neighborhood recycling program.	Being fined by the city for failing to dispose of trash properly.

The Sociological Imagination

The **sociological imagination**, a term developed by C. Wright Mills (1959), refers to the ability to see the connections between our personal lives and the social world in which we live. When we use our sociological imagination, we are able to distinguish between "private troubles" and "public issues" and to see connections between the events and conditions of our lives and the social and historical context in which we live.

For example, that one man is unemployed constitutes a private trouble. That millions of people are unemployed in the United States constitutes a public issue. Once we understand that personal troubles such as HIV infection, criminal victimization, and poverty are shared by other segments of society, we can look for the elements of social structure and culture that contribute to these public issues and private troubles. If the various elements of social structure and culture contribute to private troubles and public issues, then society's social structure and culture must be changed if these concerns are to be resolved.

Rather than viewing the private trouble of being unemployed as being due to an individual's faulty character or lack of job skills, we may understand unemployment as a public issue that results from the failure of the economic and political institutions of society to provide job opportunities to all citizens. Technological innovations emerging from the Industrial Revolution led to individual workers being replaced by machines. During the economic recession of the 1980s, employers fired employees so the firm could stay in business. Thus, in both these cases, social forces rather than individual skills largely determined whether a person was employed or not.

> Freedom is what you do with what's been done to you.
>
> JEAN-PAUL SARTRE
> *Philosopher*

Theoretical Perspectives

Theories in sociology provide us with different perspectives with which to view our social world. A perspective is simply a way of looking at the world. A theory is a set of interrelated propositions or principles designed to answer a question or explain a particular phenomenon; it provides us with a perspective. Sociological theories help us to explain and predict the social world in which we live.

Sociology includes three major theoretical perspectives: the structural-functionalist perspective, the conflict perspective, and the symbolic interactionist perspective. Each perspective offers a variety of explanations about the causes of and possible solutions for social problems.

> The most incomprehensible thing about the world is the fact that it is comprehensible.
>
> ALBERT EINSTEIN
> *Scientist*

> Some see the glass half-empty, some see the glass half-full. I see the glass as too big.
>
> GEORGE CARLIN
> *Comedian*

Structural-Functionalist Perspective

The structural-functionalist perspective is largely based on the works of Herbert Spencer, Emile Durkheim, Talcott Parsons, and Robert Merton. According to **structural-functionalism**, society is a system of interconnected parts that work together in harmony to maintain a state of balance and social equilibrium for the whole. For example, each of the social institutions contributes important functions for society: family provides a context for reproducing, nurturing, and socializing children; education offers a way to transmit a society's skills, knowledge, and culture to its youth; politics provides a means of governing members of society; economics provides for the production, distribution, and consumption of goods and services; and religion provides moral guidance and an outlet for worship of a higher power.

The structural-functionalist perspective emphasizes the interconnectedness of society by focusing on how each part influences and is influenced by other parts. For example, the increase in single-parent and dual-earner families has contributed to the number of children who are failing in school because parents have become less available to supervise their children's homework. As a result of changes in technology, colleges are offering more technical programs, and many adults are returning to school to learn new skills that are required in the workplace. The increasing number of women in the workforce has contributed to the formulation of policies against sexual harassment and job discrimination.

Structural-functionalists use the terms "functional" and "dysfunctional" to describe the effects of social elements on society. Elements of society are functional if they contribute to social stability and dysfunctional if they disrupt social stability. Some aspects of society may be both functional and dysfunctional for society. For example, crime is dysfunctional in that it is associated with physical violence, loss of property, and fear. But, according to Durkheim and other functionalists, crime is also functional for society because it leads to heightened awareness of shared moral bonds and increased social cohesion.

Sociologists have identified two types of functions: manifest and latent (Merton 1968). **Manifest functions** are consequences that are intended and commonly recognized. **Latent functions** are consequences that are unintended and often hidden. For example, the manifest function of education is to transmit knowledge and skills to society's youth. But public elementary schools also serve as baby-sitters for employed parents, and colleges offer a place for young adults to meet potential mates. The baby-sitting and mate selection functions are not the intended or commonly recognized functions of education—hence, they are latent functions.

Structural-Functionalist Theories of Social Problems

Two dominant theories of social problems grew out of the structural-functionalist perspective: social pathology and social disorganization.

Social Pathology According to the social pathology model, social problems result from some "sickness" in society. Just as the human body becomes ill when our systems, organs, and cells do not function normally, society becomes "ill" when its parts (i.e., elements of the structure and culture) no longer perform properly. For example, problems such as crime, violence, poverty, and juvenile delinquency are often attributed to the breakdown of the family institution, the decline of the religious institution, and inadequacies in our economic, educational, and political institutions.

Social "illness" also results when members of a society are not adequately socialized to adopt its norms and values. Persons who do not value honesty, for example, are prone to dishonesties of all sorts. Early theorists attributed the failure in socialization to "sick" people who could not be socialized. Later theorists recognized that failure in the socialization process stemmed from "sick" social

Everybody should live a good and productive life. When there are impediments to that, we as a society have a responsibility to help.

TIPPER GORE
Social activist

conditions, not "sick" people. To prevent or solve social problems, members of society must receive proper socialization and moral education, which may be accomplished in the family, schools, churches, workplace, and/or through the media.

Social Disorganization According to the social disorganization view of social problems, rapid social change disrupts the norms in a society. When norms become weak or are in conflict with each other, society is in a state of **anomie** or normlessness. Hence, people may steal, physically abuse their spouse or children, abuse drugs, commit rape, or engage in other deviant behavior because the norms regarding these behaviors are weak or conflicting. According to this view, the solution to social problems lies in slowing the pace of social change and strengthening social norms. For example, although the use of alcohol by teenagers is considered a violation of a social norm in our society, this norm is weak. The media portray young people drinking alcohol, teenagers teach each other to drink alcohol and buy fake identification cards (IDs) to purchase alcohol, and parents model drinking behavior by having a few drinks after work or at a social event. Solutions to teenage drinking may involve strengthening norms against it through public education, restricting media depictions of youth and alcohol, imposing stronger sanctions against the use of fake IDs to purchase alcohol, and educating parents to model moderate and responsible drinking behavior.

Conflict Perspective

Whereas the structural-functionalist perspective views society as comprising different parts working together, the **conflict perspective** views society as comprising different groups and interests competing for power and resources. The conflict perspective explains various aspects of our social world by looking at which groups have power and benefit from a particular social arrangement.

The origins of the conflict perspective can be traced to the classic works of Karl Marx. Marx suggested that all societies go through stages of economic development. As societies evolve from agricultural to industrial, concern over meeting survival needs is replaced by concern over making a profit, the hallmark of a capitalist system. Industrialization leads to the development of two classes of people: the bourgeoisie, or the owners of the means of production (e.g., factories, farms, businesses), and the proletariat, or the workers who earn wages.

The division of society into two broad classes of people—the "haves" and the "have-nots"—is beneficial to the owners of the means of production. The workers, who may earn only subsistence wages, are denied access to the many resources available to the wealthy owners. According to Marx, the bourgeoisie use their power to control the institutions of society to their advantage. For example, Marx suggested that religion serves as an "opiate of the masses" in that it soothes the distress and suffering associated with the working-class lifestyle and focuses the workers' attention on spirituality, God, and the afterlife rather than on such worldly concerns as living conditions. In essence, religion diverts the workers so that they concentrate on being rewarded in heaven for living a moral life rather than on questioning their exploitation.

Conflict Theories of Social Problems

Underlying virtually all social problems are conditions caused in whole or in part by social injustice.

PAMELA ANN ROBY
Sociologist, University of California, Santa Cruz

There are two general types of conflict theories of social problems: Marxist and non-Marxist. Marxist theories focus on social conflict that results from economic inequalities; non-Marxist theories focus on social conflict that results from competing values and interests among social groups.

Marxist Conflict Theories   According to contemporary Marxist theorists, social problems result from class inequality inherent in a capitalistic system. A system of "haves" and "have-nots" may be beneficial to the "haves" but often translates into poverty for the "have-nots." As we shall explore later in this text, many social problems, including physical and mental illness, low educational achievement, and crime, are linked to poverty.

In addition to creating an impoverished class of people, capitalism also encourages "corporate violence." Corporate violence may be defined as actual harm and/or risk of harm inflicted on consumers, workers, and the general public as a result of decisions by corporate executives or managers. Corporate violence may also result from corporate negligence, the quest for profits at any cost, and willful violations of health, safety, and environmental laws (Hills 1987). Our profit-motivated economy encourages individuals who are otherwise good, kind, and law-abiding to knowingly participate in the manufacturing and marketing of defective brakes on American jets, fuel tanks on automobiles, and contraceptive devices (i.e., intrauterine devices [IUDs]). The profit motive has also caused individuals to sell defective medical devices, toxic pesticides, and contaminated foods to developing countries. As Eitzen and Baca Zinn note, the "goal of profit is so central to capitalistic enterprises that many corporate decisions are made without consideration for the consequences. . ." (Eitzen & Baca Zinn 2000, 483).

Marxist conflict theories also focus on the problem of **alienation**, or powerlessness and meaninglessness in people's lives. In industrialized societies, workers often have little power or control over their jobs, which fosters a sense of powerlessness in their lives. The specialized nature of work requires workers to perform limited and repetitive tasks; as a result, the workers may come to feel that their lives are meaningless.

Alienation is bred not only in the workplace, but also in the classroom. Students have little power over their education and often find the curriculum is not meaningful to their lives. Like poverty, alienation is linked to other social problems, such as low educational achievement, violence, and suicide.

Marxist explanations of social problems imply that the solution lies in eliminating inequality among classes of people by creating a classless society. The nature of work must also change to avoid alienation. Finally, stronger controls must be applied to corporations to ensure that corporate decisions and practices are based on safety rather than profit considerations.

Non-Marxist Conflict Theories   Non-Marxist conflict theorists such as Ralf Dahrendorf are concerned with conflict that arises when groups have opposing values and interests. For example, anti-abortion activists value the life of unborn embryos and fetuses; pro-choice activists value the right of women to control their own body and reproductive decisions. These different value positions reflect different subjective interpretations of what constitutes a social problem. For anti-abortionists, the availability of abortion is the social problem; for pro-choice advocates, restrictions on abortion are the social problem. Sometimes the social problem is not the conflict itself, but rather the way that conflict is expressed. Even most pro-life advocates agree that shooting doctors who perform abortions and blowing up abortion clinics constitute unnecessary violence and lack of respect for life. Value conflicts may occur between diverse categories of people, including nonwhites versus whites, heterosexuals versus homosexuals, young versus old, Democrats versus Republicans, and environmentalists versus industrialists.

Solutions to the problems that are generated by competing values may involve ensuring that conflicting groups understand each other's views, resolving differences through negotiation or mediation, or agreeing to disagree. Ideally, solutions should be win-win; both conflicting groups are satisfied with the solution. However, outcomes of value conflicts are often influenced by power; the group with the most power may use its position to influence the outcome of value conflicts. For example, when Congress could not get all states to voluntarily increase the legal drinking age to 21, it threatened to withdraw federal highway funds from those that would not comply.

Symbolic Interactionist Perspective

Both the structural-functionalist and the conflict perspectives are concerned with how broad aspects of society, such as institutions and large social groups, influence the social world. This level of sociological analysis is called **macro sociology**: it looks at the "big picture" of society and suggests how social problems are affected at the institutional level.

Micro sociology, another level of sociological analysis, is concerned with the social psychological dynamics of individuals interacting in small groups. **Symbolic interactionism** reflects the micro sociological perspective and was largely influenced by the work of early sociologists and philosophers such as Max Weber, George Simmel, Charles Horton Cooley, G. H. Mead, W. I. Thomas, Erving Goffman, and Howard Becker. Symbolic interactionism emphasizes that human behavior is influenced by definitions and meanings that are created and maintained through symbolic interaction with others.

Sociologist W. I. Thomas ([1931] 1966) emphasized the importance of definitions and meanings in social behavior and its consequences. He suggested that humans respond to their definition of a situation rather than to the objective situation itself. Hence, Thomas noted that situations we define as real become real in their consequences.

Symbolic interactionism also suggests that our identity or sense of self is shaped by social interaction. We develop our self-concept by observing how others interact with us and label us. By observing how others view us, we see a reflection of ourselves that Cooley calls the "looking glass self."

Lastly, the symbolic interaction perspective has important implications for how social scientists conduct research. The German sociologist Max Weber (1864–1920) argued that in order to understand individual and group behavior, social scientists must see the world from the eyes of that individual or group. Weber called this approach *Verstehen*, which in German means "empathy." *Verstehen* implies that in conducting research, social scientists must try to understand others' view of reality and the subjective aspects of their experiences, including their symbols, values, attitudes, and beliefs.

> Each to each a looking glass, reflects the other that doth pass.
>
> CHARLES HORTON COOLEY
> *Sociologist*

Symbolic Interactionist Theories of Social Problems

A basic premise of symbolic interactionist theories of social problems is that a condition must be defined or recognized as a social problem in order for it to be a social problem. Based on this premise, Herbert Blumer (1971) suggested that social problems develop in stages. First, social problems pass through the stage of "societal recognition"—the process by which a social problem, for example, drunk driving, is "born." Second, "social legitimation" takes place when the social problem achieves recognition by the larger community, including the media, schools, and churches. As the visibility of traffic fatalities associated with alcohol increased, so did the legitimation of drunk driving as a social problem. The next stage in the development of a social problem involves "mobilization for action," which occurs when individuals and groups, such as Mothers Against Drunk Driving, become concerned about how to respond to the social condition. This mobilization leads to the "development and implementation of an official plan" for dealing with the problem, involving, for example, highway checkpoints, lower legal blood-alcohol levels, and tougher drunk driving regulations.

Blumer's stage development view of social problems is helpful in tracing the development of social problems. For example, although sexual harassment and date rape have occurred throughout this century, these issues did not begin to receive recognition as social problems until the 1970s. Social legitimation of these problems was achieved when high schools, colleges, churches, employers, and the media recognized their existence. Organized social groups mobilized to develop and implement plans to deal with these problems. For example, groups successfully lobbied for the enactment of laws against sexual harassment and the enforcement of sanctions against violators of these laws. Groups also mobilized to provide educational seminars on date rape for high school and college students and to offer support services to victims of date rape.

Some disagree with the symbolic interactionist view that social problems exist only if they are recognized. According to this view, individuals who were victims of date rape in the 1960s may be considered victims of a problem, even though date rape was not recognized at that time as a social problem.

Labeling theory, a major symbolic interactionist theory of social problems, suggests that a social condition or group is viewed as problematic if it is labeled as such. According to labeling theory, resolving social problems sometimes involves changing the meanings and definitions that are attributed to people and situations. For example, as long as teenagers define drinking alcohol as "cool" and "fun," they will continue to abuse alcohol. As long as our society defines providing sex education and contraceptives to teenagers as inappropriate or immoral, the teenage pregnancy rate in our country will continue to be higher than in other industrialized nations.

Table 1.2 summarizes and compares the major theoretical perspectives, their criticisms, and social policy recommendations as they relate to social problems. The study of social problems is based on research as well as theory, however. Indeed, research and theory are intricately related. As Wilson (1983) states,

> Most of us think of theorizing as quite divorced from the business of gathering facts. It seems to require an abstractness of thought remote from the practical activity of empirical research. But theory building is not a separate activity within sociology. Without theory, the empirical researcher would find it impossible to decide what to observe, how to observe it, or what to make of the observations . . . . (p. 1)

■ **Table 1.2** *Comparison of Theoretical Perspectives*

	Structural-Functionalism	Conflict Theory	Symbolic Interactionism
Representative Theorists	Emile Durkheim Talcott Parsons Robert Merton	Karl Marx Ralf Dahrendorf	George H. Mead Charles Cooley Erving Goffman
Society	Society is a set of interrelated parts; cultural consensus exists and leads to social order; natural state of society—balance and harmony.	Society is marked by power struggles over scarce resources; inequities result in conflict; social change is inevitable; natural state of society—imbalance.	Society is a network of interlocking roles; social order is constructed through interaction as individuals, through shared meaning, make sense out of their social world.
Individuals	Individuals are socialized by society's institutions; socialization is the process by which social control is exerted; people need society and its institutions.	People are inherently good but are corrupted by society and its economic structure; institutions are controlled by groups with power; "order" is part of the illusion.	Humans are interpretative and interactive; they are constantly changing as their "social beings" emerge and are molded by changing circumstances.
Cause of Social Problems?	Rapid social change: social disorganization that disrupts the harmony and balance; inadequate socialization and/or weak institutions.	Inequality; the dominance of groups of people over other groups of people; oppression and exploitation; competition between groups.	Different interpretations of roles; labeling of individuals, groups, or behaviors as deviant; definition of an objective condition as a social problem.
Social Policy/ Solutions	Repair weak institutions; assure proper socialization; cultivate a strong collective sense of right and wrong.	Minimize competition; create an equitable system for the distribution of resources.	Reduce impact of labeling and associated stigmatization; alter definitions of what is defined as a social problem.
Criticisms	Called "sunshine sociology"; supports the maintenance of the status quo; needs to ask "functional for whom?" Does not deal with issues of power and conflict; incorrectly assumes a consensus.	Utopian model; Marxist states have failed; denies existence of cooperation and equitable exchange. Can't explain cohesion and harmony.	Concentrates on micro issues only; fails to link micro issues to macro-level concerns; too psychological in its approach; assumes label amplifies problem.

Social Problems Research

Most students taking a course in social problems will not become researchers or conduct research on social problems. Nevertheless, we are all consumers of research that is reported in the media. Politicians, social activist groups, and organizations attempt to justify their decisions, actions, and positions by citing research results. As consumers of research, it is important to understand that our personal experiences and casual observations are less reliable than generalizations based on systematic research. One strength of scientific research is that it is subjected to critical examination by other researchers (see this chapter's *Social Problems Research Up Close* feature). The more you understand how research is done, the better able you will be to critically examine and question research, rather than to passively consume research findings. The remainder of this section discusses the stages of conducting a research study and the various methods of research used by sociologists.

■ In science (as in everyday life) things must be believed in order to be seen as well as seen in order to be believed.

WALTER L. WALLACE
Social scientist

The Sociological Enterprise

Each chapter in the book contains a *Social Problems Research Up Close* box that describes a research report or journal article that examines some sociologically significant topic. Some examples of the more prestigious journals in sociology include the *American Sociological Review,* the *American Journal of Sociology,* and *Social Forces.* Journal articles are the primary means by which sociologists, as well as other scientists, exchange ideas and information. Most journal articles begin with an *introduction and review of the literature.* It is here that the author examines previous research on the topic, identifies specific research areas, and otherwise "sets the stage" for the reader. It is often in this section that research hypotheses, if applicable, are set forth. A researcher, for example, might hypothesize that the sexual behavior of adolescents has changed over the years as a consequence of increased fear of sexually transmitted diseases, and that such changes vary on the basis of sex.

The next major section of a journal article is entitled *sample and methods.* In this section the author describes the characteristics of the sample, if any, and the details of the type of research conducted. The type of data analysis used is also presented in this section (see Appendix A). Using the above research question, a sociologist might obtain data from the Youth Risk Behavior Survey collected by the Centers for Disease Control. This self-administered questionnaire is distributed biennially to over 10,000 high school students across the United States.

The final section of a journal article includes the *findings and conclusions.* The findings of a study describe the results, that is, what the researcher found as a result of the investigation. Findings are then discussed within the context of the hypotheses and the conclusions that can be drawn. Often research results are presented in tabular form. Reading tables carefully is an important part of drawing accurate conclusions about the research hypotheses. In reading a table you should follow the steps below (see table on next page):

1. *Read the title of the table and make sure that you understand what the* *table contains.* The title of the table indicates the unit of analysis (high school students), the dependent variable (sexual risk behaviors), the independent variables (sex and year), and what the numbers represent (percentages).

2. *Read the information contained at the bottom of the table including the source and any other explanatory information.* For example, the information at the bottom of this table indicates that the data are from the Centers for Disease Control, that sexually active was defined as having intercourse in the last three months, and that data on condom use were only from those students who were defined as currently sexually active.

3. *Examine the row and column headings.* This table looks at the percentage of males and females, over four years, that reported ever having sexual intercourse, having four or more sex partners in a lifetime, being currently sexually active, and using condoms during the last sexual intercourse.

4. *Thoroughly examine the data contained within the table carefully*

Stages of Conducting a Research Study

Sociologists progress through various stages in conducting research on a social problem. This section describes the first four stages: formulating a research question, reviewing the literature, defining variables, and formulating a hypothesis.

Formulating a Research Question A research study usually begins with a research question. Where do research questions originate? How does a particular researcher come to ask a particular research question? In some cases, researchers have a personal interest in a specific topic because of their own life experience. For example, a researcher who has experienced spouse abuse may wish to do research on such questions as "What factors are associated with domestic violence?" and "How helpful are battered women's shelters in helping abused women break the cycle of abuse in their lives?" Other researchers may ask a particular research question because of their personal values—their concern for hu-

looking for patterns between variables. As indicated in the table, the first three columns indicate that "risky" sexual behaviors of both males and females, in general, have decreased between the years surveyed. There are several exceptions, however. For example, between 1997 and 1999 the percentage of males ever having sexual intercourse, having four or more sex partners, or currently being sexually active increased.

5. *Use the information you have gathered in step 4 to address the hypotheses.* Clearly sexual practices, as hypothesized, have changed over time. Not only have "risky" sexual practices, in general, declined over the time period study, there has been a general increase in condom use during sexual intercourse for both males and females.

6. *Draw conclusions consistent with the information presented.* From the table can we conclude that sexual practices have changed over time? The answer is probably yes although the limitations of the survey, the sample, and the measurement techniques used always should be considered. Can we conclude, however, that the observed changes are a consequence of the fear of sexually transmitted diseases? Although the data may imply it, having no measure of fear of sexually transmitted diseases over the time period studied, it would be premature to come to such a conclusion. More information, from a variety of sources, is needed. The use of multiple methods and approaches to study a social phenomenon is called **triangulation**.

Percentage of high school students reporting sexual risk behaviors, by sex and survey year

Survey year	Ever had sexual intercourse	Four or more sex partners during lifetime	Currently sexually active[a]	Condom used during last intercourse[b]
Male				
1993	55.6	22.3	37.5	59.2
1995	54.0	20.9	35.5	60.5
1997	48.8	17.6	33.4	62.5
1999	52.2	19.3	36.2	65.5
Female				
1993	50.2	15.0	37.5	46.0
1995	52.1	14.4	40.4	48.6
1997	47.7	14.1	36.5	50.8
1999	47.7	13.1	36.3	50.7

[a]Sexual intercourse during the three months preceding the survey.
[b]Among currently sexually active students.

Sources: Youth Risk Behavior Survey, Centers for Disease Control. 1999, Tables 30 and 32; "Trends in Sexual Risk Behaviors among High School Students—United States, 1991–1997." 1998. *Morbidity and Mortality Weekly Report* 47, September 18.

manity and the desire to improve human life. Researchers who are concerned about the spread of human immunodeficiency virus (HIV) infection and AIDS may conduct research on such questions as "How does the use of alcohol influence condom use?" and "What educational strategies are effective for increasing safer sex behavior?" Researchers may also want to test a particular sociological theory, or some aspect of it, in order to establish its validity or conduct studies to evaluate the effect of a social policy or program. Research questions may also be formulated by the concerns of community groups and social activist organizations in collaboration with academic researchers. Government and industry also hire researchers to answer questions such as "How many children are victimized by episodes of violence at school?" and "What types of computer technologies can protect children against being exposed to pornography on the Internet?"

Reviewing the Literature After a research question is formulated, the researcher reviews the published material on the topic to find out what is already known about it. Reviewing the literature also provides researchers with ideas about how to conduct their research and helps them formulate new research questions. A literature review also serves as an evaluation tool, allowing a comparison of research findings and other sources of information, such as expert opinions, political claims, and journalistic reports.

Defining Variables A **variable** is any measurable event, characteristic, or property that varies or is subject to change. Researchers must operationally define the variables they study. An **operational definition** specifies how a variable is to be measured. For example, an operational definition of the variable "religiosity" might be the number of times the respondent reports going to church or synagogue. Another operational definition of "religiosity" might be the respondent's answer to the question, "How important is religion in your life?" (1=not important, 2=somewhat important, 3=very important.)

Operational definitions are particularly important for defining variables that cannot be directly observed. For example, researchers cannot directly observe concepts such as "mental illness," "sexual harassment," "child neglect," "job satisfaction," and "drug abuse." Nor can researchers directly observe perceptions, values, and attitudes.

Formulating a Hypothesis After defining the research variables, researchers may formulate a **hypothesis**, which is a prediction or educated guess about how one variable is related to another variable. The **dependent variable** is the variable that the researcher wants to explain; that is, it is the variable of interest. The **independent variable** is the variable that is expected to explain change in the dependent variable. In formulating a hypothesis, the researcher predicts how the independent variable affects the dependent variable. For example, Mouw and Xie (1999) hypothesized that fluent bilingual children have higher levels of academic achievement than children who are English-only fluent. Analyzing data from first- and second-generation Asian-American eighth graders, the researchers found "no evidence that fluent bilinguals do better than students who are fluent only in English" (p.250). In this example, the independent variable is bilingualism and the dependent variable is school achievement.

In studying social problems, researchers often assess the effects of several independent variables on one or more dependent variables. For example, Jekielek (1998) examined the impact of parental conflict and marital disruption (two independent variables) on the emotional well-being of children (the dependent variable). Her research found that both parental conflict and marital disruption (separation or divorce) negatively affect children's emotional well-being. However, children in high-conflict intact families exhibit lower levels of well-being than children who have experienced high levels of parental conflict but whose parents divorce or separate.

Science is meaningless because it gives no answer to the question, the only question of importance for us: "What shall we do and how shall we live?"

LEO TOLSTOY
Novelist

Methods of Data Collection

After identifying a research topic, reviewing the literature, and developing hypotheses, researchers decide which method of data collection to use. Alternatives include experiments, surveys, field research, and secondary data.

Experiments **Experiments** involve manipulating the independent variable in order to determine how it affects the dependent variable. Experiments require one or more experimental groups that are exposed to the experimental treatment(s) and a control group that is not exposed. After the researcher randomly assigns participants to either an experimental or a control group, she or he measures the dependent variable. After the experimental groups are exposed to the treatment, the research measures the dependent variable again. If participants have been randomly assigned to the different groups, the researcher may conclude that any difference in the dependent variable among the groups is due to the effect of the independent variable.

An example of a "social problems" experiment on poverty would be to provide welfare payments to one group of unemployed single mothers (experimental group) and no such payments to another group of unemployed single mothers (control group). The independent variable would be welfare payments; the dependent variable would be employment. The researcher's hypothesis would be that mothers in the experimental group would be less likely to have a job after 12 months than mothers in the control group.

The major strength of the experimental method is that it provides evidence for causal relationships; that is, how one variable affects another. A primary weakness is that experiments are often conducted on small samples, usually in artificial laboratory settings; thus, the findings may not be generalized to other people in natural settings.

Surveys **Survey research** involves eliciting information from respondents through questions. An important part of survey research is selecting a sample of those to be questioned. A **sample** is a portion of the population, selected to be representative so that the information from the sample can be generalized to a larger population. For example, instead of asking all abused spouses about their experience, you could ask a representative sample of them and assume that those you did not question would give similar responses. After selecting a representative sample, survey researchers either interview people, ask them to complete written questionnaires, or elicit responses to research questions through computers.

1. *Interviews.* In interview survey research, trained interviewers ask respondents a series of questions and make written notes about or tape-record the respondents' answers. Interviews may be conducted over the telephone or face-to-face. A recent Gallup Poll (2000) involved telephone interviews with a randomly selected national sample of over 1,000 U.S. adults. One of the questions interviewers asked was, "What do you regard as the worst problem facing your community today?" The top three responses were crime (Chapter 4), education/schools (Chapter 12), and economic concerns (Chapter 11) (Gallup 2000).

   One advantage of interview research is that researchers are able to clarify questions for the respondent and follow up on answers to particular questions. Researchers often conduct face-to-face interviews with groups of individuals who might otherwise be inaccessible. For example, some AIDS-related research attempts to assess the degree to which individuals engage in behavior that places them at high risk for transmitting or contracting HIV. Street youth and intravenous drug users, both high-risk groups for HIV infection, may not have a telephone or address because of their transient lifestyle (Catania, Gibson, Chitwook, & Coates 1990). These groups may be accessible, however, if the researcher locates their hangouts and conducts face-to-face interviews. Research on homeless individuals may also require a face-to-face interview survey design.

My latest survey shows that people don't believe in surveys.

LAURENCE PETER
Humorist

The most serious disadvantages of interview research are cost and the lack of privacy and anonymity. Respondents may feel embarrassed or threatened when asked questions that relate to personal issues such as drug use, domestic violence, and sexual behavior. As a result, some respondents may choose not to participate in interview research on sensitive topics. Those who do participate may conceal or alter information or give socially desirable answers to the interviewer's questions (e.g., "No, I do not use drugs.").

2. *Questionnaires.* Instead of conducting personal or phone interviews, researchers may develop questionnaires that they either mail or give to a sample of respondents. Questionnaire research offers the advantages of being less expensive and time-consuming than face-to-face or telephone surveys. In addition, questionnaire research provides privacy and anonymity to the research participants. This reduces the likelihood that they will feel threatened or embarrassed when asked personal questions and increases the likelihood that they will provide answers that are not intentionally inaccurate or distorted.

   The major disadvantage of mail questionnaires is that it is difficult to obtain an adequate response rate. Many people do not want to take the time or make the effort to complete and mail a questionnaire. Others may be unable to read and understand the questionnaire.

3. *"Talking" Computers.* A new method of conducting survey research is asking respondents to provide answers to a computer that "talks." Romer et al. (1997) found that respondents rated computer interviews about sexual issues more favorably than face-to-face interviews and that the former were more reliable. Such increased reliability may be particularly valuable when conducting research on drug use, deviant sexual behavior, and sexual orientation as respondents reported the privacy of computers as a major advantage.

Field Research   **Field research** involves observing and studying social behavior in settings in which it occurs naturally. Two types of field research are participant observation and nonparticipant observation.

In participant observation research, the researcher participates in the phenomenon being studied in order to obtain an insider's perspective of the people and/or behavior being observed. Coleman (1990), a middle-class white male, changed clothes to live on the streets of New York as a homeless person for 10 days. In nonparticipant observation research, the researcher observes the phenomenon being studied without actively participating in the group or the activity. For example, Dordick (1997) studied homelessness by observing and talking with homeless individuals in a variety of settings, but she did not live as a homeless person as part of her research.

Sometimes sociologists conduct in-depth detailed analyses or case studies of an individual, group, or event. For example, Skeen (1991) conducted case studies of a prostitute and her adjustment to leaving the profession, an incest survivor, and a person with AIDS.

The main advantage of field research on social problems is that it provides detailed information about the values, rituals, norms, behaviors, symbols, beliefs, and emotions of those being studied. A potential problem with field research is that the researcher's observations may be biased (e.g., the researcher becomes too involved in the group to be objective). In addition, because field research is usually based on small samples, the findings may not be generalizable.

When I was younger I could remember anything—whether it happened or not.

MARK TWAIN
American humorist and writer

Feminists in all disciplines have demonstrated that objectivity has about as much substance as the emperor's new clothes.

CONNIE MILLER
Feminist scholar

Secondary Data Research Sometimes researchers analyze secondary data, which are data that have already been collected by other researchers or government agencies or that exist in forms such as historical documents, police reports, school records, and official records of marriages, births, and deaths. For example, Caldas and Bankston (1999) used information from Louisiana's 1990 Graduation Exit Examination to assess the relationship between school achievement and television viewing habits of over 40,000 tenth graders. The researchers found that, in general, television viewing is inversely related to academic achievement for whites but has little or no effect on school achievement for African-Americans. A major advantage of using secondary data in studying social problems is that the data are readily accessible, so researchers avoid the time and expense of collecting their own data. Secondary data are also often based on large representative samples. The disadvantage of secondary data is that the researcher is limited to the data already collected.

Goals of the Text

This text approaches the study of social problems with several goals in mind.

1. *Provide an integrated theoretical background.* This text reflects an integrative theoretical approach to the study of social problems. More than one theoretical perspective can be used to explain a social problem because social problems usually have multiple causes. For example, youth crime is linked to: (1) an increased number of youths living in inner-city neighborhoods with little or no parental supervision (social disorganization), (2) young people having no legitimate means of acquiring material wealth (anomie theory), (3) youths being angry and frustrated at the inequality and racism in our society (conflict theory), and (4) teachers regarding youths as "no good" and treating them accordingly (labeling theory).

2. *Encourage the development of a sociological imagination.* A survey study of over 2,000 Americans found that the majority believe that they are in control of their own lives (Mirowsky, Ross, & Van Willigen 1996). Ninety-three percent agreed with "I am responsible for my own success." More than two-thirds of the sample agreed with statements claiming responsibility for misfortunes and failures. However, a major insight of the sociological perspective is that various structural and cultural elements of society have far-reaching effects on individual lives and social well-being. This insight, known as the sociological imagination, enables us to understand how social forces underlie personal misfortunes and failures as well as contribute to personal successes and achievements. Each chapter in this text emphasizes how structural and cultural factors contribute to social problems. This emphasis encourages you to develop your sociological imagination by recognizing how structural and cultural factors influence private troubles and public issues.

3. *Provide global coverage of social problems.* The modern world is often referred to as a "global village." The Internet and fax machines connect individuals around the world, economies are interconnected, environmental destruction in one region of the world affects other regions of the world, and diseases cross national boundaries. Understanding social problems requires an awareness of how global trends and policies affect social problems. Many social problems call for collective action involving countries around the world; efforts to end poverty, protect the environment, control population growth,

The gulf between knowledge and truth is infinite.

HENRY MILLER
Novelist

For the first time in history, the 21st century should give each person the right to choose their government and enjoy the freedom to participate in decisions that affect their lives.

HUMAN DEVELOPMENT REPORT *2000*
United Nations Development Programme

In a certain sense, every single human soul has more meaning and value than the whole of history.

NICHOLAS BERDYAEV
Philosopher

and reduce the spread of HIV are some of the social problems that have been addressed at the global level. Each chapter in this text includes coverage of global aspects of social problems. We hope that attention to the global aspects of social problems broadens students' awareness of pressing world issues.

4. *Provide an opportunity to assess personal beliefs and attitudes.* Each chapter in this text contains a section called *Self and Society*, which offers you an opportunity to assess your attitudes and beliefs regarding some aspect of the social problem discussed. Earlier in this chapter, the *Self and Society* feature allowed you to assess your beliefs about a number of social problems and compare your beliefs with a national sample of first-year college students.

5. *Emphasize the human side of social problems.* Each chapter in this text contains a feature called *The Human Side*, which presents personal stories of how social problems have affected individual lives. By conveying the private pain and personal triumphs associated with social problems, we hope to elicit a level of understanding and compassion that may not be attained through the academic study of social problems alone. This chapter's *The Human Side* feature presents stories about how college students, disturbed by various social conditions, have participated in social activism.

6. *Encourage students to take pro-social action.* Individuals who understand the factors that contribute to social problems may be better able to formulate interventions to remedy those problems. Recognizing the personal pain and public costs associated with social problems encourages some to initiate social intervention.

Individuals can make a difference in society by the choices they make. Individuals may choose to vote for one candidate over another, demand the right to reproductive choice or protest government policies that permit it, drive drunk or stop a friend from driving drunk, repeat a racist or sexist joke or chastise the person who tells it, and practice safe sex or risk the transmission of sexually transmitted diseases. Individuals can also "make a difference" by addressing social concerns in their occupational role, as well as through volunteer work.

> Activism pays the rent on being alive and being here on the planet. . . . If I weren't active politically, I would feel as if I were sitting back eating at the banquet without washing the dishes or preparing the food. It wouldn't feel right.
>
> ALICE WALKER
> *Novelist*

> Most politicians will not stick their necks out unless they sense grass-roots support. . . Neither you nor I should expect someone else to take our responsibility.
>
> KATHARINE HEPBURN
> *Actress*

One way to affect social change is through demonstrations. A U.S. survey of first-year college students revealed that 45% reported having participated in organized demonstrations in the last year (Higher Education Research Institute, 2000). This photo depicts student demonstrators at The University of California, Berkeley, protesting the anti-affirmative action measure, Proposition 209.

© AP/Wide World Photos

College Student Activism

Some people believe that in order to promote social change one must be in a position of political power and/or have large financial resources. However, the most important prerequisite for becoming actively involved in improving levels of social well-being may be genuine concern and dedication to a social "cause." The following vignettes provide a sampler of college student activism—college students making a difference in the world:

- In May 1989, hundreds of Chinese college students protested in Tiananmen Square in Beijing, China, because Chinese government officials would not meet with them to hear their pleas for a democratic government. These students boycotted classes and started a hunger strike. On June 4, 1989, thousands of students and other protesters were massacred or arrested in Tiananmen Square.
- Students at the University of California-Berkeley recently came together to protest cutbacks in the campus' Ethnic Studies Department. The protest lasted over a month with over 100 arrests and six hunger strikes. As a consequence of the students activism, the Administration agreed to reopen a multi-cultural student center, hire eight tenure-track ethnic studies faculty over the next five years, and invest $100,000 in an Ethnic Research Center (Alvarado 2000).
- While a student at George Washington University, Ross Misher started an organization called Students Against Handgun Violence. When Ross was 13, his father was shot and killed by a coworker who had purchased a handgun during his lunch hour and returned to shoot Ross's father before killing himself (Lewis 1991).
- While a zoology major at the University of Colorado, Jeff Galus began the Animal Rights Student Group. This organization focuses on informing the public about how animals are treated in research and what corporations use animals in testing their products.
- Students from over 56 colleges and universities across the United States have convinced their schools to become members of the Workers' Rights Consortium (WRC). WRC is a student-run watchdog organization that inspects factories worldwide, monitoring the monitors, as part of the anti-sweatshop movement. Belonging to WRC requires that member schools agree to closely scrutinize manufacturers of collegiate apparel. Last year Nike invited students from 15 universities to monitor 32 of their facilities. Although some labor abuses were noted, the most significant finding of the students was that corporate monitors "would frequently announce themselves in advance, spend less than a day in each factory, and interview workers in locations accessible to management" (Gold 2000, 41). Nike has complained that the group is too strict in its regulations and has withdrawn funding from several supportive institutions (Aguilar 2000).

Students who are interested in becoming involved in student activism, or who are already involved, might explore the web site for the Center for Campus Organizing (2000)—a national organization that supports social justice activism and investigative journalism on campuses nationwide. The organization was founded on the premise that students and faculty have played critical roles in larger social movements for social justice in our society, including the Civil Rights movement, the anti-Vietnam War movement, the Anti-Apartheid movement, the women's rights movement, and the environmental movement. In a recent national poll of 18 to 21-year-olds, 58 percent of respondents reported that they were "more socially conscious" than members of their parents' generation (PRNewswire 2000).

Sources:
Alvarado, Diana. 2000. "Student Activism Today." *Diversity Digest*. http:www.inform.umd.edu/ DiversityWeb/Digest/sm99/activism.html
Center for Campus Organizing. 2000. http://www.cco.org/about.html
Charlton, Jacquie. 2000. "Higher Learning." *Adbusters* (August/September) 31:51.
Gold, Donna. 2000. "The Walls are Tumbling Down." *Hope* (Fall):39–41.
Lewis, Barbara A. 1991. *The Kid's Guide to Social Action*, pp. 110-11. Minneapolis, MN: Free Spirit Publishing, Inc.
PRNewswire. 2000. "Newsweek Poll: Young Voters."
http://www.news.excite.com/pr/00/08/ny-newsweek-poll
1998. http://www.colorado.edu/StudentGroups/animal rights/ Galus@UCSU.Colorado.edu
1998. http/www.compugraph.com/clr/alerts/alerts/campusactivism-q-a.html

Although individual choices make an important impact, collective social action often has a more pervasive effect. For example, while individual parents discourage their teenage children from driving under the influence of alcohol, Mothers Against Drunk Driving contributed to the enactment of national legislation that potentially will influence every U.S. citizen's decision about whether to use alcohol and drive.

Schwalbe (1998) reminds us that we don't have to join a group or organize a protest to make changes in the world.

> We can change a small part of the social world single-handedly. If we treat others with more respect and compassion, if we refuse to participate in re-creating inequalities even in little ways, if we raise questions about official representation of reality, if we refuse to work in destructive industries, then we are making change (p. 206).

Understanding *Social Problems*

Resolve to create a good future. It's where you'll spend the rest of your life.

CHARLES FRANKLIN KETTERING
American industrialist

At the end of each chapter to follow, we offer a section entitled *Understanding* in which we reemphasize the social origin of the problem being discussed, the consequences, and the alternative social solutions. It is our hope that the reader will end each chapter with a "sociological imagination" view of the problem and how, as a society, we might approach a solution.

Sociologists have been studying social problems since the Industrial Revolution at the turn of the twentieth century. Industrialization brought about massive social changes: the influence of religion declined; families became smaller and moved from traditional, rural communities to urban settings. These and other changes have been associated with increases in crime, pollution, divorce, and juvenile delinquency. As these social problems became more widespread, the need to understand their origins and possible solutions became more urgent. The field of sociology developed in response to this urgency. Social problems provided the initial impetus for the development of the field of sociology and continue to be a major focus of sociology.

Although the world is very full of suffering, it is also full of the overcoming of it.

HELEN KELLER
Social activist

There is no single agreed-upon definition of what constitutes a social problem. Most sociologists agree, however, that all social problems share two important elements: an objective social condition and a subjective interpretation of that condition. Each of the three major theoretical perspectives in sociology—structural-functionalist, conflict, and symbolic interactionist—has its own notion of the causes, consequences, and solutions of social problems.

Critical Thinking

1 People increasingly are using information technologies as a means of getting their daily news. As a matter of fact, some research indicates that news on the Internet is beginning to replace television news as the primary source of information among computer users (see Chapter 15). What role does the media play in our awareness of social problems, and will definitions of social problems change as sources of information change?

2 Each of you occupy several social statuses, each one carrying an expectation of role performance, that is, what you should and shouldn't do given your position. List five statuses you occupy, the expectations of their accompanying roles, and any role conflict that may result. What types of social problems are affected by role conflict?

3 Definitions of social problems change over time. Identify a social condition that is now widely accepted which might be viewed as a social problem in the future.

4 How would each of the three sociological perspectives analyze the recent decrease in violent crime?

Key Terms

achieved status	latent function	secondary group
alienation	law	social group
anomie	macro sociology	social problem
ascribed status	manifest function	sociological imagination
beliefs	master status	status
conflict perspective	micro sociology	structural-functionalism
dependent variable	mores	subjective element
experiment	norm	survey research
field research	objective element	symbol
folkway	operational definition	symbolic interactionism
hypothesis	primary group	values
independent variable	role	variable
institution	sanction	
labeling theory	sample	

Media Resources

The Wadsworth Sociology Resource Center: Virtual Society

http://sociology.wadsworth.com/

See the companion web site for this book to access general sociology resources and text-specific features that can further your understanding of this chapter. The site contains Internet links, Internet exercises, online practice quizzes, information on InfoTrac College Edition, and many more valuable materials designed to enrich your learning experience in social problems.

InfoTrac College Edition

You can access InfoTrac College Edition either from the Wadsworth Sociology Resource Center at **http://sociology.wadsworth.com** or directly from your web browser at **http://www.infotrac-college.com/wadsworth/**. InfoTrac College Edition is an online university library that includes over 700 popular and scholarly journals in which you can find articles related to the topics in this chapter such as information on the American Sociological Association, methods of research, and sociological theory.

Interactions CD-ROM

Go to the "Interactions" CD-ROM for *Understanding Social Problems,* Third Edition to access additional interactive learning tools, such as in-depth review materials, corresponding practice quizzes, and other engaging resources and activities to help you study the concepts in this chapter.

Problems of Well-Being

Section 1 deals with problems that are often regarded as private rather than public issues; that is, they are viewed as internally caused or as a function of individual free will. People often respond to these problems by assuming that the problem is the victims' fault—that in some way they have freely chosen their plight. In this set of problems, blame is most often attached to the individuals themselves. Thus, the physically and mentally ill (Chapter 2), the drunk and the drug addict (Chapter 3), the criminal and the delinquent (Chapter 4), and the divorced person and the child abuser (Chapter 5) are thought to be bad, or weak, or immoral, or somehow different from the average person. Consider the following scenarios:

1 A man on a fixed income without health insurance decides not to go to a doctor in order to save money to pay the rent and buy food for his children. When he becomes sick, his illness is blamed on his decision not to go to the doctor. As sociologists, we would say that the man did not want to be sick, but rather chose what he perceived as the best of several unfortunate alternatives. In this case, factors that underlie his illness include poverty, the structure of medical care in the United States, the rising cost of health insurance, and the value system that stresses parental responsibility and sacrifice.

2 A teenager from an urban lower-class neighborhood decides to sell drugs rather than stay in school or get a regular job. Such a teenager is generally viewed as being "weak" or having "low" morals. Sociologists view such a person as a lower-class, poorly educated individual with few alternatives in a society that values success. Raised in an environment where the most successful role models are often criminals, legitimate opportunities are few, traditional norms and values are weak, and peer pressure to use and sell drugs is strong, what are his choices? He can pump gas or serve fast food for minimum wage, or he can sell drugs for as much as $5,000 a week.

3 A mother comes home from work and finds her children playing and the house in disorder. She had told the children to clean the house while she was gone. She decides they need to be whipped with a belt because of their disobedience. The physical abuse she engages in is viewed as a reflection of her mental instability and her inability to control her temper. Research indicates, however, that fewer than 10 percent of identified child abusers are severely psychologically impaired.

If being mentally unstable does not explain the majority of child abuse cases, what does explain them? A history of being abused as a child is the strongest independent predictor of who will be a child abuser as an adult. Additionally, the culture of society includes a myriad of beliefs that contribute to child abuse: acceptance of corporal punishment of children and the ambiguity surrounding what constitutes appropriate discipline, the belief that parental control is an inalienable right, and the historical and lingering belief that children (as well as women) are property.

4 A college student drinks alcohol daily and often cuts classes. Although the general public views such behavior as a personal weakness, sociologists emphasize the role of the individual's socialization and society. For example, a disproportionate number of individuals with drinking problems were reared in homes where one or both parents drank heavily. In the general culture, media portrayals of drinking as desirable, fun, glamorous, and a source of status further promote drinking. College culture itself often emphasizes bars and drinking parties as primary sources of recreation and affiliation.

These examples illustrate that many behaviors result more from social factors than from individual choice. To the degree that individuals do make choices, these choices are socially determined in that the structure and culture of society limit and influence individual choices. For example, customers in a restaurant cannot choose anything they want to eat; they are limited to what is on the menu. Sociologically, one's social status—black, white, male, female, young, old, rich, poor—determines one's menu of life choices.

In each of the above examples, the alternatives were limited by the individual's position in the social structure of society and by the cultural and subcultural definitions of appropriate behavior. Although conflict theorists, structural-functionalists, and symbolic interactionists may disagree as to the relative importance and mechanisms of the shared structure and culture of society in determining the problems identified, all would agree that society, not the individual, is the primary source of the solutions. In this and the following sections, we emphasize the importance of the social structure and culture of society as both the sources of and solutions to social problems.

2

Illness and the Health Care Crisis

Is It True?

1. Individuals in the United States can expect to live longer than citizens of any other country.

2. Compared with other industrialized countries, the United States has the lowest infant death rate.

3. In the United States, major depression is the leading cause of disability.

4. According to the World Health Organization, the United States has the best system of health care in the world.

5. Half of the 1.3 million Americans who filed for bankruptcy in 1999 did so at least in part because of medical bills.

Answers to "Is It True?": 1 = F; 2 = F; 3 = T; 4 = F; 5 = T

It is ironic that in some parts of the world hundreds of millions of people suffer daily from a lack of basic health care while in other parts millions of people spend money on things that are not healthy. Think what a billion dollars could do to help immunize people against deadly diseases in developing countries. A billion dollars is not much money—it is what Americans spend on beer every 12 days and what Europeans spend on cigarettes every five days.

DAVID WRIGHT
Telemedicine and Developing Countries

In August 1997, Jeanne Calment, then the oldest woman in the world, died in France at the age of 122. When she was born in 1875, Thomas Edison had not yet discovered electricity; before she died, photographs from the planet Mars had been transmitted to Earth. During Jeanne Calment's lifetime, the world changed in unimaginable ways. One of the most profound changes over the last century has been the increase in the average length of life. Since the end of World War II, longevity of life in most developed and developing nations has increased by almost 25 years—the greatest increase seen in the history of humankind. (LaPorte 1997)

Despite overall improvements in living conditions and medical care, health problems and health care delivery are major concerns of individuals, families, communities, and nations. In this chapter, we review health concerns in the United States and throughout the world. The World Health Organization (1946) defines health as "a state of complete physical, mental, and social well-being" (p. 3). Sociologists are concerned with how social forces affect and are affected by health and illness, why some social groups suffer more illness than others, and how illness affects individuals' sense of identity and relationships with others. Sociologists also examine health care systems and explore how these systems can be improved.

The Global Context: Patterns of Health and Disease

The study of patterns of health and disease is called **epidemiology**. The field of epidemiology incorporates several disciplines, including public health, medicine, biology, and sociology. **Epidemiologists** are concerned with the social origins and distribution of health problems in a population and how patterns of health and disease vary between and within societies. Next, we look at global patterns of morbidity, longevity, mortality, and disease burden.

Patterns of Morbidity

Morbidity refers to acute and chronic illnesses and diseases and the symptoms and impairments they produce. **Acute conditions** are short-term; by definition they can last no more than three months. **Chronic conditions** are long-term health problems. The rate of serious morbidity in a population provides one measure of the health of that population. Morbidity may be measured according to the incidence and prevalence of specific illnesses and diseases. **Incidence** refers to the number of new cases of a specific health problem with a given population during a specified time period. **Prevalence** refers to the total number of cases of a specific health problem within a population that exist at a given time. For example, the incidence of HIV infection worldwide was 5.3 million in 2000; meaning that there were 5.3 million people newly infected with HIV in 2000. In the same year, the worldwide prevalence of HIV was 36.1 million, meaning that a total of 36.1 million people worldwide were living with HIV infection in 2000 (Global Summary of the HIV/AIDS Epidemic 2001).

As we discuss later in this chapter, patterns of morbidity vary according to social factors such as social class, education, sex, and race. Morbidity patterns also vary according to the level of development of a society and the age structure of the population. In the less developed countries, malnutrition, pneumonia, and infectious and parasitic diseases such as HIV disease, malaria (transmitted by mosquitoes), and measles are major health concerns. In the industrialized world, infectious and parasitic diseases have been largely controlled by advances in sanitation, immunizations, and antibiotics. Noninfectious diseases such as heart disease, cancer, mental disorders, and respiratory diseases pose the greatest health threat to the industrialized world. However, the widespread use of antibiotics in industrialized countries has contributed to a rise in infectious disease, as antibiotics kill the weaker disease-causing germs while allowing variants resistant to the drugs to flourish. Worldwide, the most alarming consequence of the development of drug-resistant germs is the reemergence of tuberculosis, which kills more people yearly than any other infectious disease. Tuberculosis is caused by bacilli that attack and destroy lung tissue and is spread when infected individuals cough or sneeze. The World Health Organization estimates that one-third of the world's population is infected, although only about 10 percent of infected persons ever develop symptoms (Weitz 2001).

The shift from a society characterized by low life expectancy and parasitic and infectious diseases to one characterized by high life expectancy and chronic and degenerative diseases is called the **epidemiological transition**. Declining birthrates and increased longevity have resulted in the aging of the world's population, which means that the major sources of morbidity are becoming those of adults rather than those of children. As societies make the epidemiological transition, diseases that need time to develop, such as cancer, heart disease, Alzheimer's disease, arthritis, and osteoporosis become more common, and childhood illnesses, typically caused by infectious and parasitic diseases, become less common.

Patterns of Longevity

One indicator of the health of a population is the average number of years individuals born in a given year can expect to live, referred to as **life expectancy**. Worldwide, life expectancy has increased dramatically over the last half century. However, wide disparities exist in life expectancy for different populations between and within societies. In 2000, Japan had the longest life expectancy: 81 years. In the same year, life expectancy was less than 50 in several countries (see Table 2.1). Several countries have life expectancies that are greater than the United States. Later, we discuss how U.S. life expectancy varies by sex and race.

The objective of good health is really twofold: the best attainable average level—goodness—and the smallest feasible differences among individuals and groups—fairness. A gain in either one of these, with no change in the other, constitutes an improvement.

WORLD HEALTH REPORT 2000

Table 2.1 *Countries* with the Longest and Shortest Life Expectancies***

Country	Life Expectancy
Lowest Life Expectancies	
Angola	35
Malawi	38
Mozambique	38
Zimbabwe	38
Niger	41
Uganda	43
Ethiopia	45
Cote d' Ivoire	45
Burkina Faso	47
Mali	47
Kenya	48
Congo	49
Highest Life Expectancies	
Japan	81
Australia	80
Canada	79
France	79
Italy	79
Spain	79
Belgium	78
Greece	78
Netherlands	78
United Kingdom	78
United States	77

*Includes countries with 10 million or more people in 2000.
**Life expectancies are rounded to nearest whole number.
Source: U.S. Census Bureau. 2000. *Statistical Abstract of the United States: 2000* (120th edition). Washington, D.C.

Patterns of Mortality

Rates of **mortality**, or death—especially those of infants, children, and women—provide sensitive indicators of the health of a population. Worldwide, the leading cause of death is infectious and parasitic diseases (World Health Organization 1998). In the United States, the three leading causes of death for both women and men are heart disease, cancer, and stroke (see Figure 2.1). Later, we discuss how patterns of mortality are related to social factors, such as social class, sex, race/ethnicity, and education. Mortality patterns also vary by age. For example, nearly three fourths of all deaths among U.S. youth and young adults aged 10 to 24 result from only four causes: motor-vehicle crashes (31 percent), other unintentional injuries (11 percent), homicide (18 percent), and suicide (12 percent) (Kann et al. 2000).

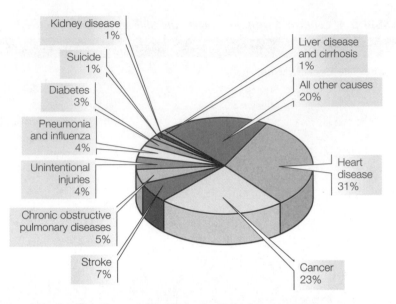

Figure 2.1 *Leading Causes of Death in the United States: 1998.*

SOURCE: Based on National Center for Health Statistics. 2000. *Health, United States, 2000 with Adolescent Health Chartbook.* Table 32. Hyattsville, MD: U.S. Government Printing Office.

Infant and Childhood Mortality Rates **Infant mortality rates**, the number of deaths of live-born infants under 1 year of age per 1,000 live births (in any given year), provide an important measure of the health of a population. In 1999, 25 countries had infant mortality rates over 100 (UNICEF 2001). That means that in 25 countries, one in every ten live-born babies died before they reached age 1. The African nation of Sierra Leone had the highest infant mortality rate in the world—an alarming 182 infants out of every 1,000 live births died before they reached their first birthday. The lowest rates of infant mortality in 1999 were in Sweden and Switzerland, where only 3 of every 1,000 live-born infants died in their first year of life. In 1999, the U.S. infant mortality rate was 7; 32 countries had infant mortality rates that were lower than that of the United States (UNICEF 2001).

Under-5 mortality rates, another useful measure of child health, refer to the rate of deaths of children under age 5. Approximately 12 million children younger than 5 years of age die every year; most of these children live in developing countries. More than half of these deaths are attributed to diarrhea, acute respiratory illness, malaria, or measles, conditions that are either preventable or treatable with low-cost interventions (Rice, Sacco, Hyder, & Black 2000). Malnutrition is associated with about half of all deaths among children (Rice et al. 2000). Mortality among infants and children has been declining in most developing countries from the mid-1980s through the 1990s. However, this decline has recently slowed, stopped, or reversed itself in some countries of sub-Saharan Africa, largely as a result of the rate of HIV infection among infants and children (Rustein 2000).

Maternal Mortality Rates **Maternal mortality rates**, a measure of deaths that result from complications associated with pregnancy, childbirth, and unsafe abortion, also provide a sensitive indicator of the health status of a population.

For a woman to die from pregnancy and childbirth is a social injustice. Such deaths are rooted in women's powerlessness and unequal access to employment, finances, education, basic health care, and other resources.

SAFE MOTHERHOOD INITIATIVE

Maternal deaths are the leading cause of death and disability for women ages 15 to 49 in developing countries (Family Care International 1999).

Of all the health statistics monitored by the World Health Organization, maternal mortality has the largest discrepancy between developed and developing countries. Women's lifetime risk of dying from pregnancy or childbirth is 1 in 48 in all developing countries compared to 1 in 1,800 in all developed countries (Family Care International 1999). In Africa, one in 16 women die from pregnancy or childbirth. The highest rate of maternal death is in Niger, where one woman in nine dies in pregnancy or childbirth (Lay 2000). The lowest maternal mortality rate is in Norway, where only one in 7,300 die in pregnancy or childbirth. In the United States, the maternal mortality rate is one in 3,500 (Lay 2000).

Several factors contribute to high maternal mortality rates in less developed countries. Poor quality and inaccessible health care, malnutrition, and poor sanitation contribute to adverse health effects of pregnancy and childbirth. In developing countries, only 53 percent of all births are attended to by professionals and nearly 30 percent of women who give birth in developing countries receive no care after the birth (United Nations Population Fund 2000). Also, women in less developed countries experience higher rates of pregnancy and childbearing and begin childbearing at earlier ages. Thus, they face the risk of maternal death more often and before their bodies are fully developed (see also Chapter 14). Women in many countries also lack access to family planning services and/or do not have the support of their male partners to use contraceptive methods such as condoms. Consequently, many women resort to abortion to limit their childbearing, even in countries where abortion is illegal.

Illegal abortions in less developed countries have an estimated mortality risk of 100 to 1,000 per 100,000 procedures (Miller & Rosenfield 1996). In contrast, the U.S. mortality risk for legal abortion is very low: 0.6 per 100,000. Unsafe abortion represents a serious threat to the health and lives of women. Each year, women undergo an estimated 50 million abortions, 20 million of which are unsafe, resulting in the deaths of 78,000 women (United Nations Population Fund 2000).

Patterns of Burden of Disease

Although infant and maternal mortality rates are sensitive indicators of the health of populations, researchers have developed a new approach to measuring the health status of a population that combines mortality and disability. This new approach provides an indicator of the overall burden of disease on a population through a single unit of measurement that combines not only the number of deaths but also the impact of premature death and disability on a population (Murray & Lopez 1996). This comprehensive unit of measurement, called the **disability-adjusted life year (DALY)**, reflects years of life lost to premature death and years lived with a disability. More simply, 1 DALY is equal to 1 lost year of healthy life. For example, The Global Burden of Disease Study (Murray & Lopez 1996) calculated the burden of disease for various diseases and injuries. The study concluded that worldwide, tobacco is a more serious threat to human health than any single disease, including HIV (see also Chapter 3). Table 2.2 lists the ten leading causes of DALYs worldwide in 1998.

It is not uncommon for women in Africa, when about to give birth, to bid their older children farewell.

UNITED NATIONS
POPULATION FUND
*The State of the World
Population 2000*

■ **Table 2.2** *Ten Leading Causes of Disability-Adjusted Life-Years (DALYs) Worldwide**

1. Lower respiratory infections
2. Perinatal conditions
3. Diarrhoeal diseases
4. HIV/AIDS
5. Major depression
6. Heart disease
7. Stroke
8. Malaria
9. Road traffic accidents
10. Measles

* Among all member states of the World Health Organization.

Source: World Health Organization. 1999. *The World Health Report 1999.* www.who.int

Some students have a hard time understanding that the consequences of one unprotected sexual encounter may not be reversible.

AMERICAN COLLEGE HEALTH ASSOCIATION

HIV/AIDS: A Global Health Concern

One of the most urgent public health concerns around the globe is the spread of the human immunodeficiency virus (HIV), which causes acquired immunodeficiency syndrome (AIDS). HIV is transmitted through sexual intercourse; through sharing unclean intravenous needles; perinatal transmission (from infected mother to fetus or newborn); through blood transfusions or blood products; and, rarely, through breast-milk. Worldwide, the predominant mode of HIV transmission is through heterosexual contact (Inciardi & Harrison 1997). The second most

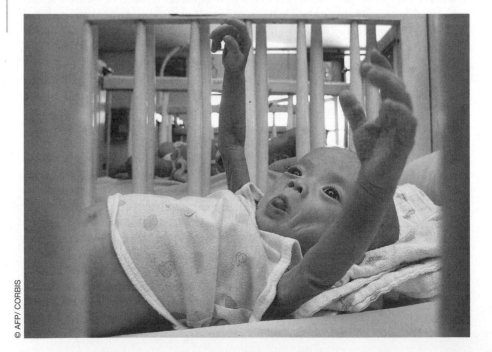

Although HIV/AIDS first emerged as an adult health problem, the disease is now a major killer of children in countries with high HIV prevalence rates.

© AFP/ CORBIS

common mode of transmission worldwide is **perinatal transmission**—the transmission of HIV from an infected mother to a fetus or newborn. An estimated 15 percent to 30 percent of babies born to HIV-infected mothers are HIV positive (Ward 1999). Homosexual activity accounts for less than 10 percent of new cases worldwide and less than half of new cases in the United States (Stine 1998).

Since the beginning of the epidemic, more than 18 million lives have been claimed by AIDS (Joint United Nations Programme on HIV/AIDS 2000a). HIV/AIDS is the fourth most common cause of death worldwide, and the leading cause of death in Africa (United Nations Population Fund 2000). In 16 countries in sub-Saharan Africa, more than one in ten persons ages 15 to 49 are HIV-infected. In seven countries in southern Africa, at least 20 percent of the adult population is living with HIV. In Botswana, where about one in three adults are HIV-infected—the highest prevalence rate in the world—at least two-thirds of today's 15-year-old boys will die prematurely of AIDS (Joint United Nations Programme on HIV/AIDS 2000b). Since the epidemic began, over 13 million children—95 percent of them in Africa— have lost their mother or both parents to AIDS (Joint United Nations Programme on HIV/AIDS 2000c). Although HIV/AIDS originally emerged as adult health problems, they have become a major killer of under-5-year-old children, especially in developing countries (Adetunji 2000).

The high rates of HIV in developing countries, particularly sub-Saharan Africa, are having alarming and devastating effects on societies. The HIV/AIDS epidemic creates an enormous burden on the limited health care resources of poor countries. Gains in life expectancy achieved in recent decades have been reversed in some countries. In Botswana, Zambia, and Zimbabwe—countries with high HIV prevalence—life expectancy at birth was lower in 2000 than it was in 1975 (Adetunji 2000). Economic development is threatened by the HIV epidemic, which diverts national funds to health-related needs and reduces the size of a nation's workforce.

HIV/AIDS in the United States

In 1999, the U.S. death rate from AIDS declined about 21 percent to the lowest level since 1987; AIDS no longer ranks in the top 15 leading causes of death in America. The decline reflects improvements in medical care and the development of new drug therapies to treat HIV (Ward 1999). However, among 25-to 44-year-olds, AIDS still ranks as the fifth leading cause of death and is the leading cause of death among African American men ages 25 to 44 (Centers for Disease Control and Prevention 2000a).

It is estimated that at least half of all new HIV infections in the United States are among people under age 25 and nearly half (48 percent) of HIV cases in this age group are among females (Centers for Disease Control and Prevention 1999). Among 13- to 24-year-olds, 51 percent of all AIDS cases reported among males in 1998 were among young men who have sex with men; 10 percent were among injection drug users; and 9 percent were among young men infected heterosexually. In 1998, among young women the same age, 47 percent were infected heterosexually and 14 percent were injection drug users.

Despite the widespread concern about HIV, many Americans—especially adolescents and young adults— engage in high risk behavior. (See "The Student Sexual Risks Scale" in this chapter's *Self and Society* feature). Although a national survey of U.S. teens found that most teens (81 percent) say that AIDS is a serious problem for people their age (Kaiser Family Foundation 2000) another

The Student Sexual Risks Scale

The following self-assessment allows you to evaluate the degree to which you may be at risk for engaging in behavior that exposes you to HIV. Safer sex means sexual activity that reduces the risk of transmitting the AIDs virus. Using condoms is an example of safer sex. Unsafe, risky, or unprotected sex refers to sex without a condom, or to other sexual activity that might increase the risk of AIDS virus transmission. For each of the following items, check the response that best characterizes your opinion.

A = Agree
U = Undecided
D = Disagree

	A	U	D
1. If my partner wanted me to have unprotected sex, I would probably give in.	___	___	___
2. The proper use of a condom could enhance sexual pleasure.	___	___	___
3. I may have had sex with someone who was at risk for HIV/AIDS.	___	___	___
4. If I were going to have sex, I would take precautions to reduce my risk of HIV/AIDS.	___	___	___
5. Condoms ruin the natural sex act.	___	___	___
6. When I think that one of my friends might have sex on a date, I ask him/her if he/she has a condom.	___	___	___
7. I am at risk for HIV/AIDS.	___	___	___
8. I would try to use a condom when I had sex.	___	___	___
9. Condoms interfere with romance.	___	___	___
10. My friends talk a lot about safer sex.	___	___	___
11. If my partner wanted me to participate in risky sex and I said we needed to be safer, we would still probably end up having unsafe sex.	___	___	___
12. Generally, I am in favor of using condoms.	___	___	___
13. I would avoid using condoms if at all possible.	___	___	___
14. If a friend knew that I might have sex on a date, he/she would ask me whether I was carrying a condom.	___	___	___
15. There is a possibility that I have HIV/AIDS.	___	___	___
16. If I had a date, I would probably not drink alcohol or use drugs.	___	___	___
17. Safer sex reduces the mental pleasure of sex.	___	___	___
18. If I thought that one of my friends had sex on a date, I would ask him/her if he/she used a condom.	___	___	___
19. The idea of using a condom doesn't appeal to me.	___	___	___
20. Safer sex is a habit for me.	___	___	___
21. If a friend knew that I had sex on a date, he/she wouldn't care whether I had used a condom or not.	___	___	___
22. If my partner wanted me to participate in risky sex and I suggested a lower-risk alternative, we would have the safer sex instead.	___	___	___
23. The sensory aspects (smell, touch, etc.) of condoms make them unpleasant.	___	___	___
24. I intend to follow "safer sex" guidelines within the next year.	___	___	___
25. With condoms, you can't really give yourself over to your partner.	___	___	___
26. I am determined to practice safer sex.	___	___	___
27. If my partner wanted me to have unprotected sex and I made some excuse to use a condom, we would still end up having unprotected sex.	___	___	___

28. If I had sex and I told my friends that I did not use a condom, they would be angry or disappointed.   _____ _____ _____

29. I think safer sex would get boring fast.   _____ _____ _____

30. My sexual experiences do not put me at risk for HIV/AIDS.   _____ _____ _____

31. Condoms are irritating.   _____ _____ _____

32. My friends and I encourage each other before dates to practice safer sex.   _____ _____ _____

33. When I socialize, I usually drink alcohol or use drugs.   _____ _____ _____

34. If I were going to have sex in the next year, I would use condoms.   _____ _____ _____

35. If a sexual partner didn't want to use condoms, we would have sex without using condoms.   _____ _____ _____

36. People can get the same pleasure from safer sex as from unprotected sex.   _____ _____ _____

37. Using condoms interrrupts sex play.   _____ _____ _____

38. It is a hassle to use condoms.   _____ _____ _____

(To be read after completing the scale.)

SCORING: Begin by giving yourself eighty points. Subtract one point of every undecided response. Subtract two points every time that you disagreed with odd-numbered items or with item number 38. Subtract two points every time you agreed with even-numbered items 2 through 36.

INTERPRETING YOUR SCORE: Research shows that students who make higher scores on the SSRS are more likely to engage in risky sexual activities, such as having multiple sex partners and failing to consistently use condoms during sex. In contrast, students who practice safer sex tend to endorse more positive attitudes toward safer sex, and tend to have peer networks that encourage safer sexual practices. These students usually plan on making sexual activity safer, and they feel confident in their ability to negotiate safer sex even when a dating partner may press for riskier sex. Students who practice safer sex often refrain from using alcohol or drugs, which may impede negotiation of safer sex, and often report having engaged in lower-risk activities in the past. How do you measure up?

(BELOW 15) LOWER RISK: (Of 200 students surveyed by DeHart and Birkimer, 16 percent were in this category.) Congratulations! Your score in the SSRS indicates that relative to other students your thoughts and behaviors are more supportive of safer sex. Is there any room for improvement in your score? If so, you may want to examine items for which you lost points and try to build safer sexual strengths in those areas. You can help protect others from HIV by educating your peers about making sexual activity safer.

(15 TO 37) AVERAGE RISK: (Of 200 students surveyed by DeHart and Birkimer, 68 percent were in this category.) Your score on the SSRS is about average in comparison with those of other college students. Although it is good that you don't fall into the higher-risk category, be aware that "average" people can get HIV, too. In fact, a recent survey indicated that the rate of HIV among college students is ten times that in the general heterosexual population. Thus, you may want to enhance your sexual safety by figuring out where you lost points and work toward safer sexual strengths in those areas.

(38 AND ABOVE) HIGHER RISK: (Of 200 students surveyed by DeHart and Birkimer, 16 percent were in this category.) Relative to other students, your score on the SSRS indicates that your thoughts and behaviors are less supportive of safer sex. Such high scores tend to be associated with greater HIV-risk behavior. Rather than simply giving in to riskier attitudes and behaviors, you may want to empower yourself and reduce your risk by critically examining areas for improvement. On which items did you lose points? Think about how you can strengthen your sexual safety in these areas. Reading more about safer sex can help, and sometimes colleges and health clinics offer courses or workshops on safer sex. You can get more information about resources in your area by contacting the CDC's HIV/AIDS Information Line at 1-800-342-2437.

Source: DeHart, D.D., and Birkimer, J.C. 1997. The Student Sexual Risks Scale (modification of SRS for popular use; facilitates student self-administration, scoring, and normative interpretation). Developed specifically for this text by Dana D. DeHart, College of Social Work at the University of South Carolina, John C. Birkimer, University of Louisville. Used by permission of Dana DeHart.

study of sexually active U.S. high school students found that only 58 percent reported that either they or their partner had used a condom during last sexual intercourse (Kann et al. 2000).

Mental Illness: The Invisible Epidemic

The concepts of mental illness and mental health are not easy to define. What it means to be mentally healthy varies across and within cultures. Furthermore, mental health and mental illness may be thought of as points on a continuum. **Mental health** has nevertheless been defined as "the successful performance of mental function, resulting in productive activities, fulfilling relationships with other people, and the ability to adapt to change and to cope with adversity" (U.S. Department of Health and Human Services 1999:ix). **Mental illness** refers collectively to all mental disorders. **Mental disorders** are health conditions that are characterized by alterations in thinking, mood, and/or behavior associated with distress and/or impaired functioning. Although we all experience problems in living, functioning, and emotional distress, such problems are not necessarily considered as mental illness, unless they meet specific criteria (such as level of intensity and duration) specified in the classification manual used to diagnose mental disorders: *The Diagnostic and Statistical Manual of Mental Disorders* (American Psychiatric Association 2000).

Some examples of mental disorders are presented in Table 2.3.

Extent and Impact of Mental Illness

Although cross-national estimates of the prevalence of mental disorders vary, one study found a 40 percent lifetime prevalence of any mental disorder in Netherlands and the United States; 12 percent lifetime prevalence in Turkey, and 20 percent in Mexico (WHO International Consortium in Psychiatric Epidemiology 2000). On an annual basis, more than 50 million U.S. adults—nearly 25 percent of the U.S. adult population—suffer from mental disorders or substance abuse disorders (U.S. Department of Health and Human Services 1999). About 20 percent of U.S. children under age 18 have mental disorders that result in mild functional impairment; between 5 and 9 percent experience severe functional impairments (Hyman 2000). In 1998, major depression was the leading cause of disability in developed nations, including the United States (World Health Organization 1999). Worldwide, mental disorders accounted for approximately 12 percent of all disability-adjusted life years lost in 1998 (Brundtland 2000). Five of the leading causes of disability worldwide are mental disorders: major depression, schizophrenia, bipolar disorders, alcohol use, and obsessive–compulsive disorders (Brundtland 2000). Mental disorders also contribute to mortality, with suicide representing one of the leading preventable causes of death in the United States and worldwide.

Causes of Mental Disorders

Mental illnesses are the result of a number of biological and social factors. A broad scope of research has linked many mental disorders with genetic or neurological causes involving some pathology of the brain. However, social and environmental influences, such as poverty, history of abuse or other se-

■ Table 2.3 *Mental Disorders Classified by the American Psychiatric Association*

Classification	Description
Anxiety Disorders	Disorders characterized by anxiety that is manifest in phobias, panic attacks, or obsessive–compulsive disorder
Dissociative Disorders	Problems involving a splitting or dissociation of normal consciousness such as amnesia and multiple personality
Disorders First Evident in Infancy, Childhood, or Adolescence	Including mental retardation, attention-deficit hyperactivity, anorexia nervosa, bulimia nervosa, and stuttering
Eating or Sleeping Disorders	Including such problems as anorexia and bulimia or insomnia and other problems associated with sleep
Impulse Control Disorders	Including the inability to control undesirable impulses such as kleptomania, pyromania, and pathological gambling
Mood Disorders	Emotional disorders such as major depression and bipolar (manic-depressive) disorder
Organic Mental Disorders	Psychological or behavioral disorders associated with dysfunctions of the brain caused by aging, disease, or brain damage (such as Alzheimer's disease)
Personality Disorders	Maladaptive personality traits that are generally resistant to treatment, such as paranoid and antisocial personality types
Schizophrenia and Other Psychotic Disorders	Disorders with symptoms such as delusions or hallucinations
Somatoform Disorders	Psychological problems that present themselves as symptoms of physical disease, such as hypochondria
Substance-Related Disorders	Disorders resulting from abuse of alcohol and/or drugs such as barbiturates, cocaine, or amphetamines

▮ The burdens of mental illnesses...have been seriously underestimated by traditional approaches that take account only of deaths and not disability. While psychiatric conditions are responsible for little more than one percent of deaths, they account for almost 11 percent of disease burden worldwide.

CHRISTOPHER MURRAY AND ALLEN LOPEZ
Harvard University

vere emotional trauma also affect individuals' vulnerability to mental illness and mental health problems. The global increase in life expectancy has contributed to mental illnesses that affect the elderly, such as Alzheimer's disease and other forms of dementia. War within and between countries may also contribute to mental illness (see Chapter 16 for a discussion of combat-related post-traumatic stress disorder). In addition, "many societies and communities that customarily offered support to their needier members through family and social bonds now find it much harder to do so" (Brundtland 2000, 411). Garfinkel and Goldbloom (2000) explain "the radical shifts in society towards technology, changes in family and societal supports and networks and the commercialization of existence...may account for the current epidemic of depression and other psychiatric disorders" (p. 503). It may be safe to conclude that, "the causes of most mental disorders lie in some combination of genetic and environmental factors, which may be biological or psychosocial" (U.S. Department of Health and Human Services 1999, xiv). See this chapter's *Human Side* feature for a discussion of a woman living with bipolar disorder.

▮ For too long the fear of mental illness has been profoundly destructive to people's lives...Mental illnesses are just as real as other illnesses...Yet fear and stigma persist, resulting in lost opportunities for individuals to seek treatment and improve or recover.

DONNA E. SHALALA
Secretary of Health and Human Services

Living with Bipolar Disorder: One Woman's Story

What is manic-depression or bipolar disorder? I will begin by telling you that it is NOT a term that defines a "CRAZY" person. It is a chemical imbalance in the brain that causes severe mood swings. The diagnosed person, before taking prescribed medications, lives a life on a roller coaster of emotions. Before treatment you would generally find us in one of two states. We are either dancing on top of the stars, or . . . digging our grave. There is rarely, if ever, an in-between mood. While in the manic state, the bipolar person is extremely hyperactive. We can go for days without sleep. We can clean what's already been clean. We talk so rapidly that conversing with others is always a one-sided conversation. We feel invincible. But what goes up . . . must come down . . . and that we call "crashing." When we crash, we can find not one reason why we should continue living another moment. We are not just sad, WE WANT TO DIE.

I fought depression my entire life. In public I would "fake it." My smiles, my laughter, my happiness. . . I had faked it all . . . in hopes that one day I would "make it." Make it past the depression, past the heartaches, and past the loneliness. But "faking it" just wasn't cutting it. So I sought help from a psychiatrist. I was diagnosed "severely depressed" and prescribed a new "wonder" drug . . . Prozac. That was the disaster that led to my current diagnosis. It seemed that Prozac took over my mind and my body. It made me feel more depressed. I became obsessed with the notion of dying. I searched my surroundings for the perfect suicide aide. I was scared . . . scared of myself.

My doctor admitted me into a local hospital and began testing. I was 26 years old, and I was consumed with the thought of ending my life. Then came the diagnosis; I was manic-depressive or in medical terminology, *bipolar*. The physician assured me that with proper medication I could live a perfectly normal life. But there was one small detail that was overlooked in the conversation with my psychiatrist. That was the fact that there was NO cure. My medications had one purpose and that was to aid in shifting moods. This did not mean I would never become manic or depressed again.

The toughest part of all is accepting the diagnosis. Who wants a physician to tell you in fancy words that you are "mentally" ill? No one. And then comes the daily reminder: the action of opening the pill bottles every single day of your life, sometimes several times a day . . . And then there's the fear of someone seeing you and questioning you about the medication. It ends many a friendship. Manic-depression is a frightening word.

Courtesy Lacy Hilliard

With the proper drug treatment and with the support of her family, Lacy Hilliard, who has bipolar disorder, is able to live a happy and productive life.

Then comes the day for all of us who are bipolar, that we decide, "I feel great . . . I must be healed," thus the nightmare begins again. No pills . . . get manic . . . crash.

I often think that the family of the manic-depressive person has it worse than the diagnosed person. This illness is a terrible strain on everyone who is a part of your life.

It has now been nearly eleven years since my first hospitalization. My medications have been changed so many times that I have literally lost track. Three suicide attempts and four hospitalizations later the doctors seem to have found what works. The combination of medications I currently take keep me very tired, but I can now function in society as well as the average person.

If one has cancer, that person generally seeks love, prayer, and support from family and friends. But if you are Bipolar, few must know of it. Although this disorder is not considered terminal, one-fifth of diagnosed manic depressives in America alone take their own lives each year. Two-thirds of Americans who are bipolar go without medical treatment. Why? Because most insurance companies in our country cover a minimal amount of treatment, if any at all, for mental conditions. It has been my experience that if a person is physically ill, society is extremely sympathetic and caring. On the other hand, if we have a mental illness, society shuns us and many who loved us before, forget us after.

My hope is that I have educated you on an illness rarely discussed. My desire is that you will realize that I may be different . . . BUT I AM STILL HUMAN.

Source: Adapted from Lacy Hilliard's Web Page at
http://www.geocities.com/heartland/oaks/3133/bipolar.html

Sociological Theories of Illness and Health Care

The sociological approach to the study of illness, health, and health care differs from medical, biological, and psychological approaches to these topics. Next, we discuss how three major sociological theories—structural-functionalism, conflict theory, and symbolic interactionism—contribute to our understanding of illness and health care.

Structural-Functionalist Perspective

The structural-functionalist perspective is concerned with how illness, health, and health care affect and are affected by changes in other aspects of social life. For example, the women's movement and changes in societal gender roles have led to more women smoking and drinking, and experiencing the negative health effects of these behaviors. Increased modernization and industrialization throughout the world has resulted in environmental pollution—a major health concern. Increasingly, patterns of health and disease are affected by **globalization**—the economic, political, and social interconnectedness among societies throughout the world. For example, increased business travel and tourism has encouraged the globalization of disease, such as the potentially fatal West Nile encephalitis which first appeared in the United States in 1999 (Weitz 2001).

Just as social change affects health, health concerns may lead to social change. The emergence of HIV and AIDS in the U.S. gay male population was a force that helped unite and mobilize gay rights activists. Concern over the hazards of using cellular phones while driving has led to policies banning the use of cellular phones while operating a vehicle (see this chapter's *Focus on Technology* feature).

According to the structural-functionalist perspective, health care is a social institution that functions to maintain the well-being of societal members and, consequently, of the social system as a whole. Illness is dysfunctional in that it interferes with people performing needed social roles. To cope with nonfunctioning members and to control the negative effects of illness, society assigns a temporary and unique role to those who are ill—the sick role (Parsons 1951). This role assures that societal members receive needed care and compassion, yet at the same time, it carries with it an expectation that the person who is ill will seek competent medical advice, adhere to the prescribed regimen, and return as soon as possible to normal role obligations.

Structural-functionalists explain the high cost of medical care by arguing that society must entice people into the medical profession by offering high salaries. Without such an incentive, individuals would not be motivated to endure the rigors of medical training or the stress of being a physician.

Conflict Perspective

The conflict perspective focuses on how wealth, status, and power, or the lack thereof, influence illness and health care. Worldwide, the have-nots not only experience the adverse health effects of poverty, they also have less access to medical insurance and quality medical care. In societies where women have little status and power, their life expectancy is lower than in industrialized

> Diseases respect no national borders. With rising globalization, diseases are spreading rapidly from developing to industrialized nations.
>
> ROSE WEITZ
> *Sociologist*

Health Hazards and Cellular Phones

Since they were introduced in 1983, cellular phones have become widespread with more than 400 million cell phones in use worldwide (Greenwald 2000). In the United States, 94 million Americans have cellular phone service; 27 percent of U.S. households report that at least one member owns a cellular phone (Lissy et al. 2000). Cell phones have been under scrutiny for possible contributions to two health problems: traffic accidents and cancer.

Do Cell Phones Contribute to Traffic Accidents?

Surveys have found that 80 to 90 percent of cellular phone owners use these devices while driving (Lissy et al. 2000). Some research suggests that cellular phone use is associated with a significantly increased rate of traffic accidents (Redelmeier & Tibshirani 1997; Violanti 1997). The Harvard Center for Risk Analysis concluded that "cellular phone use while driving poses a risk to the driver, to other motorists, and to pedestrians" (Lissy et al. 2000, 1). This report adds, however, "the risks appear to be small . . . but are uncertain because existing research is limited . . ." (p. 2).

Public perceptions of the dangers of driving while using cell phones are also mixed. A bumper sticker that reads "Hang Up and Drive" reflects the growing concern that driving while using a cell phone distracts drivers. In October 2000, Suffolk County New York became the nation's first county to adopt a ban on using a hand-held phone while driving, carrying a $150 fine (Kelley 2000). Despite research that found no safety advantages to hands-free as compared with hand-held phones (Redelmeier & Tibshirani 1997), drivers in Suffolk county are allowed to use a cell phone that is equipped with an earpiece or that can act like a speakerphone, leaving the driver's hands free. At least 11 local governments have enacted some type of ban or restriction on the use of cell phones (Clines 2001). Some restrictions are selective. For example, in New York City, taxi drivers are prohibited from using cell phones. The first statewide ban on cell phone use was one established in Massachusetts that prohibits school bus drivers from using cell phones while driving (Clines 2001). At least 22 nations have cell phone restrictions on drivers, ranging from a mandate against hand-held cell phones in Britain to an outright ban on the use of all types of cell phones by drivers in Japan (Clines 2001).

Critics of such bans argue that cellular phones improve safety on the road by enabling people to contact emergency services in the event of accidents and alerting police to drunk drivers. Unfortunately, this benefit of cell phone use is a mixed blessing: multiple calls for the same incident produces a significant burden on emergency response resources in some jurisdictions. Some localities have reported more than one hundred "911" calls for a single incident (National Highway Traffic Safety Administration 1998). Critics also argue that if cell phone use while driving is banned, so should other distracting activities such as eating and drinking, putting on makeup, and smoking. In response to research that finds even hands-free cell phone use to be distracting to drivers, critics ask if listening to radio talk shows or even having conversations with passengers is just as distracting, and if so, should those activities be banned as well?

countries because of several social factors: eating last and eating less, complications of frequent childbearing and sexually transmitted diseases (because they have no power to demand abstinence or condom use), infections and hemorrhages following genital mutilation (which is practiced in 29 countries), and restricted access to modern health care (World Health Organization 1997).

Medical research agendas are also shaped by wealth, status, and power. Although malaria kills twice as many people annually as does AIDS, malaria research receives less than one-tenth as much public funding as AIDS research (Morse 1998). Similarly, pneumonia and diarrheal diseases constitute 15.4 percent of the total global disease burden, but only 0.2 percent of the total global spending on research (Visschedijk & Simeant 1998). This is because northern developed countries (such as the United States), who provide most of the funding for world health-related research, do not feel threatened by malaria, pneu-

Can Using Cell Phones Cause Cancer?

Another public health issue concerning cell phones is the debate about whether they can cause cancer. Cell phones emit low levels of radiation, measured in "specific absorption rates," or SARS. An SAR measures the energy in watts per kilogram that one gram of body tissue absorbs from a cell phone. Although researchers have not been able to demonstrate a clear link between cell phone use and cancer, they have not been able to rule out the possibility either. An expert panel of scientists and physicians in Britain concluded, "it is not possible at present to say that exposure to [cell phone] radiation, even at levels below national guidelines, is totally without potential adverse health effects" (Raloff 2000, 326). This panel concludes that the available research findings on the cancer risks of cell phones justify a "precautionary approach." The British panel and other researchers and public health officials who are concerned about cancer risks and cell phone use recommend that consumers take the following precautions:

1. *Use hands-free cell phones.* Some research suggests that hands-free cell phones offer substantially reduced exposure to radiation (Dobson 2000).
2. *Choose cell phones with lower SAR levels.* In fall 2000 cell phone makers began including data on SAR levels in the packaging of the newest cell phone models (Greenwald 2000). Teral Communications Commission declares all phones that emit radiation below the SAR ceiling of 1.6 as being safe. Although there is no evidence that a cell phone with an SAR level of 0.24 is any safer than one with an SAR level of 1.49, concerned consumers may choose a phone with a lower SAR level.

3. *Limit the time one talks on cell phones.* U.S. cell phone users spend an average of 150 minutes a month using their cell phones (Greenwald 2000).
4. *Discourage children's use of cell phones.* Because children's developing brains absorb more radiation than adult brains, children should be discouraged from using cell phones. "Parents should permit children to use cell phones only for calls essential to safety" (Raloff 2000, 326).

Sources:

Clines, Francis X. 2001 (February 18). "Deaths Spur Laws Against Drivers on Cell Phones." *New York Times.* http://www.nytimes.com/2001/02/18/technology/18CELL.html?ex=9835240856ei=1

Dobson, Roger. 2000 (August 19). "'Hands-free' Mobile Phones May Be Safer than the Rest." *British Medical Journal* 321(7259):468.

Greenwald, John. 2000 (Oct. 9). "Do Cell Phones Need Warnings?" *Time* 66-67.

Kelley, Tina. 2000 (Oct. 4). "Phoning While Driving is Now Illegal in Suffolk." *New York Times.* www.nytimes.com/2000/10/04/nyregion/04CELL

Lissy, Karen; Joshua Cohen; Mary Park; and John D. Graham. 2000 (July). "Cellular Phones and Driving: Weighing the Risks and Benefits." *Risk in Perspective* 8(6):1–6. Harvard Center for Risk Analysis.

National Highway Traffic Safety Administration. 1998. "Cellular Telephone Use in America and Perceptions of Safety." www.nhtsa.dot.gov/

Raloff, J. 2000 (May 20). "Two Studies Offer Some Cell-Phone Cautions." *Science News* 157(21):326.

Redelmeier, Donald A. and Robert J. Tibshirani. 1997. "Association Between Cellular Telephone Calls and Motor Vehicle Collisions." *New England Journal of Medicine* 336(17):453–8.

Schultz, Stacey and Kenneth Terrell. 2000 (August 28). "Could Your Phone Cause Cancer? Don't Get Hung Up On It." *U.S. News & World Report* 129(8):54.

Senior, Kathryn. 2000 (May 20). "Mobile Phones: Are They Safe?" *The Lancet* 355(9217):1793.

Violanti, J.M. 1997. "Cellular Phones and Traffic Accidents." *Public Health* 111(6):423–8.

monia, and diarrheal diseases, which primarily affect less developed countries in Africa and Asia.

The male-dominated medical research community has also neglected women's health issues and has excluded women from medical research. When the male erectile dysfunction drug Viagra made its debut in 1998, women across the United States were outraged by the fact that some insurance policies covered Viagra (or were considering covering it), although female contraceptives were not covered. Women have also been excluded from participating in major health research studies, such as the Harvard Physicians Health Study (which looked at the relationship between aspirin use and heart disease) and the Multiple Risk Factor Intervention Trials (which examined how cholesterol levels, blood pressure, and smoking affect heart disease) (Johnson & Fee 1997).

The conflict perspective also focuses on how the profit motive influences health, illness, and health care. The profit motive underlies much of the illness, injury, and death that occurs from hazardous working conditions and dangerous consumer products. Corporations may influence health-related policies and laws through contributions to politicians and political candidates. In a study on the influence of tobacco industry campaign contributions on state legislators in six states, researchers found that legislators who received higher tobacco industry campaign contributions had more pro-tobacco policy positions (Monardi & Glantz 1998).

Conflict theorists argue that the high costs of medical care in the United States are a result of a capitalistic system in which health care is a commodity, rather than a right. The conflict perspective views power and concern for profits as the primary obstacles to U.S. health care reform. Insurance companies realize that health care reform translates into federal regulation of the insurance industry. In an effort to buy political influence to maintain profits, the insurance industry has contributed millions of dollars to congressional candidates.

Symbolic Interactionist Perspective

Symbolic interactionists focus on (1) how meanings, definitions, and labels influence health, illness, and health care and (2) how such meanings are learned through interaction with others and through media messages and portrayals. According to the symbolic interactionist perspective of illness, "there are no illness or diseases in nature. There are only conditions that society, or groups within it, have come to define as illness or disease" (Goldstein 1999, 31). Psychiatrist Thomas Szasz (1970) argued that what we call "mental illness" is no more than a label conferred on those individuals who are "different," that is, who don't conform to society's definitions of appropriate behavior.

Definitions of health and illness vary over time and from society to society. In some countries, being fat is a sign of health and wellness; in others it is an indication of mental illness or a lack of self-control. Before medical research documented the health hazards of tobacco, our society defined cigarette smoking as fashionable. Cigarette advertisements still attempt to associate positive meanings (such as youth, sex, and romance) with smoking to entice people to smoke. A study of top-grossing American films from 1985 to 1995 revealed that 98 percent had references that supported tobacco use and 96 percent had references that supported alcohol use (Everett, Schnuth, and Tribble 1998).

A growing number of behaviors and conditions are being defined as medical problems—a trend known as **medicalization.** Hyperactivity, insomnia, anxiety, and learning disabilities are examples of phenomena that some view as medical conditions in need of medical intervention. Increasingly, "normal" aspects of life, such as birth, aging, sexual development, menopause, and death, have come to be seen as medical events (Goldstein 1999).

Symbolic interactionists also focus on the stigmatizing effects of being labeled "ill." A **stigma** refers to any personal characteristic associated with social disgrace, rejection, or discrediting. (Originally, the word *stigma* referred to a mark burned into the skin of a criminal or slave.) Individuals with mental illnesses, drug addictions, physical deformities and impairments, and HIV and AIDS are particularly prone to being stigmatized. Stigmatization may lead to prejudice and discrimination and even violence against individuals with illnesses or impairments.

Having a stigmatized illness or condition often becomes a master status, obscuring other aspects of a person's social identity. One wheelchair-bound individual commented: "When I am in my chair I am invisible to some people; they see only the chair" (Ostrof 1998, 36).

Social Factors Associated with Health and Illness

Public health education campaigns, articles in popular magazines, college-level health courses, and health professionals emphasize that to be healthy, we must adopt a healthy lifestyle. In response, many people have at least attempted to quit smoking, eat a healthier diet, and include exercise in their daily or weekly routine. However, health and illness are affected by more than personal lifestyle choices. In the following section, we examine how social factors such as social class, poverty, education, race, and gender affect health and illness. Health problems related to environmental problems are discussed in Chapter 14.

Recent years have seen a tendency to blame individuals for their own health problems . . . Yet . . . patterns of disease reflect social conditions as much as, if not more than, individual behaviors or biological characteristics.

ROSE WEITZ
Sociologist

Social Class and Poverty

Poverty has been identified as the world's leading health problem by an international group of physicians ("Poverty Threatens Crisis" 1998). Poverty is associated with unsanitary living conditions, hazardous working conditions, lack of access to medical care, and inadequate nutrition (see also Chapter 10).

In the United States, socioeconomic status is related to numerous aspects of health and illness. People with lower family income tend to die at younger ages than those with higher income. Children from low-income families are less likely to be fully vaccinated against childhood diseases and are less likely to have medical insurance. Poor persons are also more likely to report an unmet need for health care and the percentage of Americans reporting fair or poor health is nearly 4 times as high for persons living below the poverty line as for those with family income at least twice the poverty threshold (National Center for Health Statistics 2000).

Low socioeconomic status is also associated with increased risk of a broad range of psychiatric conditions (Williams & Collins 1999). Rates of depression and substance abuse, for example, are higher in the lower socioeconomic classes (Kessler et al. 1994). Why do poor people have higher rates of mental illness? One explanation suggests that lower class individuals experience greater stress as a result of their deprived and difficult living conditions. Others argue that members of the lower class are simply more likely to have their behaviors identified and treated as mental illness.

Lower socioeconomic groups have higher rates of mortality, in part, because they have higher rates of health risk behaviors such as smoking, alcohol drinking, being overweight, and being physically inactive. Other factors that explain the relationship between socioeconomic status and mortality include exposure to environmental health hazards and inequalities in access to and use of preventive and therapeutic medical care (Lantz et al. 1998). In addition, the lower class tends to experience high levels of stress, while having few resources to cope with it (Cockerham 1998). Stress has been linked to a variety of physical and mental health problems, including high blood pressure, cancer, chronic fatigue, and substance abuse.

Education

In general, lower levels of education are associated with higher rates of health problems and mortality (National Center for Health Statistics 2000) (see Table 2.4). For example, less educated women and men have higher rates of suicide.

■ **Table 2.4** *Death Rates* for Persons Ages 25–64 According to Educational Attainment: 1998*

Cause of Death	Years of Educational Attainment (Highest Grade Completed)		
	Less than 12	12	13 or More
All causes	583	446	213
Chronic and noncommunicable diseases	402	343	171
Injury	96	75	31
Communicable diseases	40	27	11
HIV infection	17	12	4

*Per million population (percentages are rounded)

Source: National Center for Health Statistics. 2000. *Health, United States, 2000 with Adolescent Health Chartbook*. Table 35. Hyattsville, MD: National Center for Health Statistics.

Low birthweight and high infant mortality are also more common among the children of less educated mothers than among children of more educated mothers. Low-birthweight babies (weighing less than 5.5 lbs.) are more likely to die in infancy, and those who survive are more likely to suffer illness, stunted growth, and other health problems into adult life. One in 13 children in the United States is born with low birthweight (Children's Defense Fund 2000).

One reason that lower education levels are associated with higher rates of health problems and mortality is that individuals with low levels of education are more likely to engage in health risk behaviors such as smoking and heavy drinking. The well-educated, in contrast, are less likely to smoke and drink heavily and are more likely to exercise. Women with less education are less likely to seek prenatal care and are more likely to smoke during pregnancy. However, research findings suggest that educational differences in health and mortality are best explained by the strong association between education and income (Lantz et al. 1998). Despite the association between education and better health, many college students engage in high-risk health behaviors (see the *Social Problems Research Up Close* feature in this chapter).

Gender

Gender issues affect the health of both women and men. Gender discrimination and violence against women produce adverse health effects in girls and women worldwide. Violence against women is a major public health concern: at least one in three women has been beaten, coerced into sex, or abused in some way—most often by someone she knows (United Nations Population Fund 2000). "Although neither health care workers nor the general public typically thinks of battering as a health problem, woman battering is a major cause of injury, disability, and death among American women, as among women worldwide" (Weitz 2001, 56). In Africa, where the leading cause of death is HIV/AIDS, HIV positive women outnumber men by 2 million, in part because African women do not have the social power to refuse sexual intercourse and/or to demand that their male partners use condoms (United Nations Population Fund 2000). As noted earlier, women in developing countries suffer high rates of mortality and mor-

Female infants and young children, like the child on the left in the photo, are more likely than males to be denied food and medical care.

bidity due to the high rates of complications associated with pregnancy and childbirth. The low status of women in many less developed countries results in their being nutritionally deprived and having less access to medical care than do men. For example, in some countries in Asia and Africa, boys receive more medicine and medical treatment than girls. In Latin America and India girls are often immunized later than boys or not at all (United Nations Population Fund 2000).

In the United States before the twentieth century, the life expectancy of U.S. women was shorter than that of men because of the high rate of maternal mortality that resulted from complications of pregnancy and childbirth. Today, although U.S. women experience adverse health effects of battering and poverty (see also chapter 10), they live longer than U.S. men (see Table 2.5).

Although U.S. women tend to live longer than men, they have higher rates of illness and disability than do U.S. men (Verbrugge 1999). Prevalence rates for nonfatal chronic conditions (such as arthritis, thyroid disease, and migraine headache) are typically higher for women. However, men tend to have higher rates of fatal chronic conditions (such as high blood pressure, heart disease, and diabetes). Women also tend to experience a higher incidence of acute conditions, such as colds and influenza, infections, and digestive conditions. "In sum, women live longer than men but experience more illness, whereas men experience relatively little illness but die quickly when illness strikes" (Weitz 2001, 55). Regarding mental health, men are more likely to abuse drugs and have

■ **Table 2.5** *Life Expectancy of U.S. Individuals Born in 1998: By Sex:*

Both Sexes	Male	Female
76.7	73.8	79.5

Source: National Center for Health Statistics. 2000. *Health, United States, 2000 with Adolescent Health Chartbook*. Hyattsville, MD: U.S. Government Printing Office. Table 28.

The National College Health Risk Behavior Survey

In 1995, the first national college-based survey was conducted to measure a broad range of health-risk behaviors among U.S. college students: behaviors that contribute to unintentional and intentional injury; tobacco, alcohol, and other drug use; sexual behaviors that contribute to sexually transmitted infections; unhealthy dietary behaviors; and physical inactivity. A brief description of the methods and selected findings from the National College Health Risk Behavior Survey follows.

Methods and Sample

A nationally representative sample of 2- and 4-year colleges and universities were selected, resulting in 136 institutions participating in the survey. A random sample of full- and part-time undergraduate students aged 18 and older were selected from the 136 participating colleges and universities. A total of 7,442 students were selected and eligible for the study, of which 4,838 (65 percent) completed the questionnaire. Students aged 18–24 represented 64 percent of the sample. The survey questionnaire, developed by the Centers for Disease Control and Prevention, was sent by mail to students in the sample. The questionnaire, available in both English and Spanish, consisted of a booklet that could be scanned by a computer and contained 96 multiple-choice questions. Responses to the survey were voluntary and confidential.

Findings and Conclusions*

- *Use of safety belts*: Nationwide, 10 percent of college students rarely or never used safety belts when riding in a car driven by someone else. Of those students who had driven a car, 9 percent said they rarely or never used safety belts when driving a car.

- *Riding with a driver who had been drinking alcohol and driving after drinking alcohol*: During the 30 days preceding the survey, more than one third (35 percent) of college students nationwide had ridden with a driver who had been drinking alcohol and 27 percent of students had driven a vehicle after drinking alcohol. Male students were more likely than female students to report these behaviors, and white students were more likely than black students to report these behaviors.

- *Suicide*: Nationwide, 10 percent of college students had seriously considered attempting suicide during the 12 months preceding the survey; 7 percent had made a specific plan to attempt suicide, and 2 percent had attempted suicide.

- *Tobacco use*: Nearly one-third (32 percent) of college students nationwide reported either current cigarette use or current smokeless-tobacco use. Male students

higher rates of personality disorders, whereas women are more likely to suffer from mood disorders such as depression and anxiety (Cockerham 1998).

Men are more prone to chronic and life-threatening diseases, such as coronary disease, because they are more likely than women to smoke, use alcohol and illegal drugs, and work in hazardous occupations such as agriculture or commercial fishing. U.S. culture socializes men to be aggressive and competitive and to engage in risky behaviors (such as dangerous sports, driving fast, and violence) which contributes to their higher risk of death from injuries and accidents. Although women are more likely to attempt suicide, men are more likely to succeed at it because they use deadlier methods. HIV infections and AIDS deaths in men outnumber those in women on every continent except sub-Saharan Africa (Joint United Nations Programme on HIV/AIDS 2000d), in part because of socialization men receive which permits and even encourages them to engage in sexually promiscuous behavior. Men are also less likely than women to seek medical care. Boys who are brought up to believe that 'real men don't get sick' often see themselves as invulnerable to illness or risk. This is reflected in the under-use of health services by men (Joint United Nations Programme on HIV/AIDS 2000d).

(37 percent) were significantly more likely than female students (29 percent) to report current tobacco use. White students (36 percent) were significantly more likely than black (15 percent) and Hispanic (26 percent) students to report this behavior.

- *Sexual behavior and HIV testing*: More than one third (35 percent) of college students nationwide had had sexual intercourse with six or more sex partners during their lifetime. Students aged 25 and older (50 percent) were significantly more likely to report this behavior than students ages 18 to 24 years. Among currently sexually active college students nationwide, 30 percent reported that either they or their partner had used a condom during last intercourse. Nationwide, 39 percent of college students had ever had their blood tested for HIV infection.
- *Dietary behaviors*: About one-fourth (26 percent) of college students had eaten five or more servings of fruits and vegetables during the day preceding the survey.
- *Weight*: Although 21 percent of college students were classified as being overweight based on body mass index calculations, 42 percent of college students believed themselves to be overweight. Black students (34 percent) were significantly more likely than white (20 percent) and Hispanic (21 percent) students to be overweight. Female students (49 percent) were significantly more likely than male students (32 percent) to perceive themselves as overweight. Nationwide, 4 percent of female students and less than 1 percent of male students had either vomited or taken laxatives to lose weight or to keep from gaining weight. Female students (7 percent) were significantly more likely than male students (1 percent) to have taken diet pills to either lose weight or keep from gaining weight.
- *Physical activity*: More than one third (38 percent) of college students had participated in activities that had made them sweat and breathe hard for at least 20 minutes on at least 3 of the 7 days preceding the survey. Male students (44 percent) were significantly more likely than female students (33 percent) to report vigorous physical activity.

The results of the National College Health Risk Behavior Survey indicate that many U.S. college students engage in behaviors that place them at risk for serious health problems. The survey results provide important baseline data for college leaders and health officials to use in reducing health-risk behaviors among college students.

*Percentages are rounded.

Source: Centers for Disease Control and Prevention. 1997. "Youth Risk Behavior Surveillance: National College Health Risk Behavior Survey—United States, 1995." www.cdc.gov/nccdphp/dash/MMWRFile/ss4606.htm

Racial and Ethnic Minority Status

In the United States, Asian Americans typically enjoy high levels of health, while other racial and ethnic minorities tend to suffer higher rates of mortality and morbidity. U.S. blacks, especially black men, have a lower life expectancy compared with whites and also have higher rates of infant mortality (see Figures 2.2 and 2.3).

Black Americans are more likely than white Americans to die from stroke, heart disease, cancer, HIV infection, unintentional injuries, diabetes, cirrhosis, and homicide. Compared with white Americans, American Indians have higher death rates from motor vehicle injuries, diabetes, and cirrhosis of the liver (caused by alcoholism). Compared with non-Hispanic whites, Hispanics have more diabetes, high blood pressure, tuberculosis, lung cancer, sexually transmitted infections, alcoholism, and homicide (National Center for Health Statistics 2000). Because Asian Americans have the highest levels of income and education of any racial/ethnic U.S. minority group, they typically have high levels of health.

Socioeconomic differences between racial and ethnic groups are largely responsible for racial and ethnic differences in health status (Weitz 2001; Williams

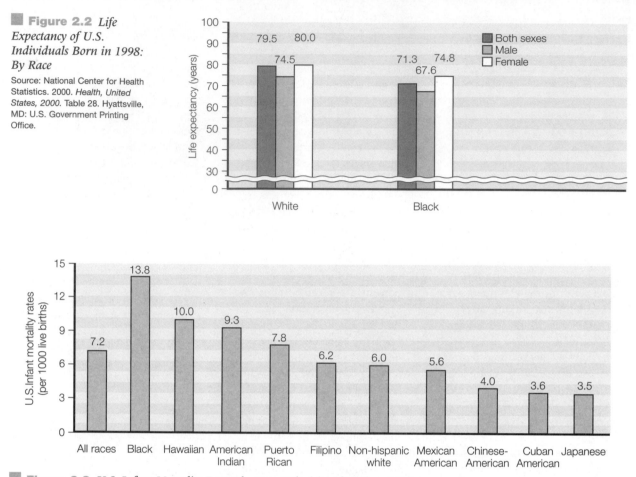

Figure 2.2 *Life Expectancy of U.S. Individuals Born in 1998: By Race*

Source: National Center for Health Statistics. 2000. *Health, United States, 2000.* Table 28. Hyattsville, MD: U.S. Government Printing Office.

Figure 2.3 *U.S. Infant Mortality Rates* by Race/Ethnicity of Mother: 1998.*

Source: Mathews, T.J., Sally C. Curtin, and Marian F. MacDorman. 2000 (July 20). "Infant Mortality Statistics from the 1998 Period Linked Birth/Infant Death Data Set." *National Vital Statistics Reports* 48(12):1–28.

> Of all the forms of inequality, injustice in health care is the most shocking and inhumane.
>
> MARTIN LUTHER KING, JR.

& Collins 1999). In addition, discrimination contributes to poorer health among oppressed racial and ethnic populations by restricting access to the quantity and quality of public education, housing, and health care. Racial and ethnic minorities are less likely to have insurance coverage for health care and are more likely than whites to live and work in environments where they are exposed to hazards such as toxic chemicals, dust, and fumes. However, "it is also important to note that African Americans have lower rates of suicide, substance use, and mental illness than whites, possibly because of the higher levels of social support many African Americans receive from their families and churches" (Weitz 2001, 69).

Problems in U.S. Health Care

The United States boasts of having the best physicians, hospitals, and advanced medical technology in the world, yet problems in U.S. health care remain a major concern on the national agenda. In 2000, the World Health Organization released its report on the first ever analysis of the world's health systems (World Health Organization 2000). This report notes that although the United States spends a higher portion of its gross domestic product on health care than any other country, it ranks 37 out of 191 countries according to its performance. The report concluded that France provides the best overall health care among major countries followed by Italy, Spain, Oman, Austria, and Japan. After presenting a brief overview of U.S. health care, we address some of the major health care problems in the United States—the high cost of medical care and insurance, unequal access to medical care, inadequate mental health care, and the managed care crisis.

U.S. Health Care: An Overview

Various types of health insurance exist in the United States. In traditional health insurance plans, the insured choose their health care provider, who is reimbursed by the insurance company on a fee-for-service basis. The insured individual typically must pay out-of-pocket a certain amount called a "deductible" (perhaps $250.00 per year per person or per family) and then is often required to pay a percentage of medical expenses (e.g., 20 percent) until a maximum out-of-pocket expense amount is reached (after which insurance will cover 100 percent of medical costs).

Health maintenance organizations (HMOs) are prepaid group plans in which a person pays a monthly premium for comprehensive health care services. HMOs attempt to minimize hospitalization costs by emphasizing preventive health care. **Preferred provider organizations** (PPOs) are health care organizations in which employers who purchase group health insurance agree to send their employees to certain health care providers or hospitals in return for cost discounts. In this arrangement, health care providers obtain more patients, but charge lower fees to buyers of group insurance.

The two major publicly-funded health programs are Medicare and Medicaid. **Medicare** is funded by the Federal government and reimburses the elderly and the disabled for their health care (see also Chapter 6). Medicare consists of two separate programs: a hospital insurance program and a supplementary medical insurance program. The hospital insurance program is free, but enrollees may pay a deductible and a copayment, and there is limited coverage of home health nursing and hospice care. Medicare's medical insurance program care is not free; enrollees must pay a monthly premium as well as a copayment for services. Medicare does not cover prescription drugs, long-term nursing home care, and other types of services, which is why many individuals who receive Medicare also purchase supplementary private insurance known as medigap policies. **Medicaid**, which provides health care coverage for the poor, is jointly funded by the federal and state governments (see also Chapter 10). Eligibility rules and benefits vary from state to state and in many states Medicaid provides health care only for the very poor who are well below the federal poverty level.

Perhaps the most dramatic change in the U.S. health care system is the rise of managed care. **Managed care** refers to any medical insurance plan that controls costs through monitoring and controlling the decisions of health care providers. In many plans, doctors must call a utilization review office to receive approval before they can hospitalize a patient, perform surgery, or order an expensive diagnostic test. Although the terms HMO and managed care are often used interchangeably, HMOs are only one form of managed care. About three-fourths of all Americans with private insurance belong to some form of managed care plan (Iglehart 1999). Recipients of Medicaid and Medicare may also belong to a managed care plan.

Inadequate Health Insurance Coverage

Healthcare is one of our most basic needs. Making it available is perhaps our greatest test of humanity.

LINDA PEENO, M.D.

Virtually all elderly Americans have Medicare coverage and most non-elderly Americans receive health insurance coverage through their employers. But millions of U.S. children and adults lack health insurance or have inadequate health insurance benefits. In 1999, 15.5 percent of the U.S. population, or 42.6 million people, were without health insurance coverage during the entire year (Mills 2000). The percentage of uninsured Americans dropped from 16.3 in 1998 to 15.5 in 1999, representing the first decline in percent of the population without health insurance since 1987 when health insurance statistics were first available. The increase in health insurance coverage is largely a result of the increase in employment-based health insurance associated with the low unemployment rate in 1999 (Mills 2000). As insurance premiums are expected to rise in the coming years, the percentage of employers offering health insurance may decrease.

The poor are least likely to have health insurance. Although Medicaid insured 12.9 million poor people during at least a portion of 1999, 10.4 million poor, or 32.4 percent, had no insurance of any kind during the year (Mills 2000).

More than 100 million Americans lack dental coverage ((PNHP Data Update 2000). Tooth decay—the most common chronic U.S. childhood disease—goes untreated among nearly half of all poor African American and Latino children.

Individuals who lack health insurance are much more likely than individuals with insurance to delay or forego needed medical care. A survey of U.S. adults found that of those who were uninsured, 20 percent needed but did not get care for a serious medical problem, 39 percent skipped a recommended test or treatment, 30 percent did not get a prescription filled, and 13 percent had problems getting needed mental health care (Kaiser Commission on Medicaid and the Uninsured 2000). Decisions made by the uninsured to delay or forego needed care because of its cost can lead to poorer health outcomes, as reflected in the following examples (Kaiser Commission on Medicaid and the Uninsured 2000):

- Uninsured children are at least 70 percent more likely than insured children not to have received medical care for ear infections, which can lead to more serious health problems when left untreated.
- Uninsured adults are over 30 percent less likely to have had a check-up in the past year; uninsured men 40 percent less likely to have had a prostate exam and uninsured women 60 percent less likely to have had a mammogram compared with the insured.
- The uninsured are more likely than those with insurance to be hospitalized for conditions that could have been avoided if they had sought medical treatment earlier (e.g., pneumonia and uncontrolled diabetes).
- The uninsured with various forms of cancer are more likely to be diagnosed with late stage cancer. Death rates for uninsured women with breast cancer are significantly higher compared with women with insurance.

For individuals who have health insurance, the benefits are often limited. For example, nearly one-third (31 percent) of Medicare beneficiaries lack prescription drug coverage, and most seniors with private coverage have limited benefits (such as paying only 50 percent of drug costs up to a yearly limit of $1,000).

The High Cost of Health Insurance and Health Care

The United States spends more on health care per person than any other country in the world. In 1998, health care costs in the United States totaled 14 percent of gross domestic product (GDP), or $4,270 per person, compared to an average of 8 percent of GDP, or $2,000 per person among the 23 countries of the Organization for Economic Cooperation and Development (PNHP Data Update 2000). Health care expenditures are expected to increase to 16.2 percent of GDP by 2008 (Hoffman, Klees, & Curtis 2000).

Almost 9 million U.S. families with health insurance spent more than 10 percent of their annual income for health care in 1997. On average, Medicare beneficiaries spend 19 percent of their annual income on medical care, and the 2 million elderly who are poor but not covered by Medicaid spend an average of 54 percent of income on out-of-pocket health care costs (PNHP Data Updates 1998). Half of these health care costs were spent on insurance. The high cost of health care creates an enormous economic burden on many U.S. individuals and families. Indeed, almost half of the 1.3 million Americans who filed for bankruptcy in 1999 did so at least in part because of medical bills (PNHP Data Update 2000).

Several factors have contributed to escalating medical costs. First, the U.S. population is aging. Because people over age 65 use medical services more than younger individuals, the growing segment of the elderly population means more money is spent on medical care. Conrad and Brown (1999) forewarn that "we have yet to feel the full effect of medical costs on our aging population; as the so called 'baby boom' generation reaches older ages in the next century, health costs are expected to rise even more steeply" (p. 582).

Health care administrative expenses, which are higher in the United States than in any other nation, are another reason for high medical care costs. About 26¢ of every health care dollar is spent on administrative overhead (Health Care Financing Administration 1998).

High costs of Medicaid and Medicare are partly to the result of fraud and abuse. Fraud and abuse are committed by individuals, hospitals, clinics, nursing homes, physicians, pharmacies, medical equipment companies, dentists, and other providers of health services and products (PNHP Data Update 1997). "Errors" include billing for medically unnecessary treatment and billing for products and services that were not provided. Fraud and abuse costs Medicare an estimated $23 billion annually (PNHP Data Update 1997), and costs Medicaid at least $365 million a year (Brown 2000).

The high salaries and benefits paid to executives in the pharmaceutical industry and for-profit managed care corporations also contribute to health care costs to the consumer. In 1998, the CEOs of the top ten drug companies averaged $290 million in annual compensation including stock options. Executives at 17 for-profit HMOs received $42 million in pay and $87 million in stock options in 1999 (PNHP Data Update 2000).

High doctors' fees and hospital costs are also factors in the rising costs of health care. The use of expensive medical technology, unavailable just decades ago, also contributes to high medical bills. Increasingly, hospitals are purchasing more expensive equipment than is needed. And new expensive hospitals are being built in areas where there is already a surplus of hospital beds (PNHP Data Updates 1998).

High costs of public and private insurance have also contributed to escalating health care expenditures. In February 2000, the average monthly insurance premium for a family was more than $500.00 (Insure.com 2000). Businesses that provide health insurance benefits bear much of the expense of the insurance for their employees. With the rising cost of medical insurance, fewer companies today fully finance medical care coverage for their employees than in the past. The percentage of U.S. workers covered by employer-provided health insurance declined from 70 percent in 1979 to 63 percent in 1998 (Michel, Bernstein, & Schmitt 2001).

Inadequate Mental Health Care

Recent tragic episodes of violence in our schools remind us that inadequately treated emotional and behavioral disorders among our children can literally have lethal consequences in terms of suicide and murder.

DR. STEVEN E. HYMAN
Director, National Institute of Mental Health

Nearly half of all Americans who have a severe mental disorder do not seek treatment (U.S. Department of Health and Human Services 1999). In part, this is because of the stigma associated with mental illness (and thus with treatment for mental illness) and the discriminatory policies in mental health care. Nearly 98 percent of private health insurance plans discriminate against patients seeking treatment for mental illness by requiring higher copayments, allowing fewer doctor visits or days in the hospital, or imposing higher deductibles than those on other medical illnesses (American Psychiatric Association 1999a). Dr. Steven Hyman, Director of the National Institute of Mental Health, makes the point that health insurance discrimination against individuals with mental health disorders is senseless.

> Try to explain to the family member of a person with schizophrenia why Parkinson's disease—a chronic and not yet curable disease that affects dopamine systems in the brain—might be fully covered by insurance while schizophrenia—another chronic and not yet curable disease that affects dopamine systems in the brain—is not. Think of how you might explain to the parent of an autistic child why chronic genetic diseases that cause childhood seizures might be fully covered but why a highly genetic brain disease such as autism is not. (Hyman 2000, 3)

Another glaring inadequacy in U.S mental health care is under-identification of children needing mental health care. Between 15 and 25 percent of children evaluated in primary care settings have significant psychosocial disorders requiring some type of intervention; yet fewer than one in five of these at risk children are identified as needing help (American Psychiatric Association 1999b). Almost half the students with untreated mental disorders drop out of high school. Of those drop-outs, 73 percent are arrested within 5 years of leaving school (American Psychiatric Association 1999b).

In adults, untreated mental disorders can lead to lost productivity, unsuccessful relationships, significant distress, and abuse or neglect of any children they may have. The high societal costs of untreated and under-treated mental illnesses are well-documented. The National Institutes of Health estimates that the yearly cost of untreated mental illnesses exceeds $300 billion, primarily as a result of loss of productivity because of lost days at work and premature death, health care costs, and increased use of the criminal justice system and social welfare benefits (American Psychiatric Association 1999a).

The Managed Care Crisis

Increasingly, Americans are concerned about the reduced quality of health care resulting from the emphasis on cost-containment in managed care. A 2000 Harris Poll phone survey of 1,000 adults found that 59 percent believed that HMOs compromise the quality of medical care, up from 39 percent in 1995 (PNHP Data Update 2000). In a survey of physicians' views on the effects of managed care, the majority responded that managed care has negative effects on the quality of patient care because of limitations on diagnostic tests, length of hospital stay, and choice of specialists (Feldman, Novack, & Gracely 1998). One former director for a health maintenance organization describes:

> I've seen from the inside how managed care works. I've been pressured to deny care, even when it was necessary. I have seen the bonus checks given to nurses and doctors for their denials. I have seen the medical policies that keep patients from getting care they need . . . and the inadequate appeal procedures. (Peeno 2000, 20)

Strategies for Action: Improving Health and Health Care

Because poverty underlies many of the world's health problems, improving the world's health requires strategies that alleviate poverty (see Chapter 10). Other chapters in this text discuss strategies to alleviate problems associated with tobacco and illegal drugs (Chapter 3), health hazards in the workplace (Chapter 11), and environmental health problems (Chapter 14). Here, we discuss other strategies for improving health, including global strategies to improve maternal and infant health, HIV/AIDS prevention and alleviation strategies, the use of computer technology in health care, and U.S. health care reform.

Improving Maternal and Infant Health

As discussed earlier, maternal deaths are a major cause of death among women of reproductive age in the developing world. In 1987, the Safe Motherhood Initiative was launched. This global initiative involves a partnership of governments, nongovernmental organizations, agencies, donors, and women's health advocates working to protect women's health and lives, especially during pregnancy and childbirth. Improving women's health also improves the health of infants; 30 to 40 percent of infant deaths are the result of poor care during labor and delivery (Safe Motherhood Initiative 1998). The cost of ensuring that women in low-income countries get health care during pregnancy, delivery, and after birth; family planning services; and newborn care is estimated at only $3 (U.S.) per person per year (Family Care International 1999).

The Safe Motherhood Initiative advocates improving maternal and infant health by first identifying the powerlessness that women face as an injustice that countries must remedy through political, health, and legal systems. In many developing countries, men make the decisions about whether or when their wives (or partners) will have sexual relations, use contraception, or bear children. Improving the status and power of women involves ensuring that they have the right to make decisions about their health and reproductive lives.

A report published by the Save the Children Organization entitled "State of the World's Mothers 2000" found that access to family planning and female education are the two most important determinants of the well being of mothers and their children (Lay 2000). Women must have access to family planning services, affordable methods of contraception, and safe abortion services where legal. The Safe Motherhood Initiative recommends reforming laws and policies to support women's reproductive health and improve access to family planning services. This implies removing legal barriers to abortion—a highly controversial issue in many countries. Promoting women's education increases the status and power of women to control their reproductive lives, exposes women to information about health issues, and also delays marriage and childbearing.

HIV/AIDS Prevention and Alleviation Strategies

> If true control of the HIV epidemic is to be achieved, development of an anti-HIV vaccine is an absolute must.
>
> DARRELL E. WARD
> *The AmFAR AIDS Handbook*

In some regions, injection drug use accounts for more than half of HIV infections (U.S. Department of Health and Human Services 1998). To reduce transmission of HIV among injection drug users, their sex partners, and their children, some countries and U.S. communities have established needle exchange programs. These programs provide new, sterile syringes in exchange for used, contaminated syringes. Many needle exchange programs also provide drug users with a referral to drug counseling and treatment. Research evidence clearly suggests that needle exchange programs result in lower rates of HIV transmission (U.S. Department of Health and Human Services 1998). In Canada, sterile injection equipment is available to drug users in pharmacies and through numerous needle-exchange programs. In contrast, 45 of the 50 states in the United States prohibit the sale or possession of sterile needles or syringes without a medical prescription, and only a small number of communities have legal needle exchange programs (Centers for Disease Control and Prevention 2000b).

Another strategy to curb the spread of HIV involves encouraging individuals to get tested for HIV so they can modify their behavior (to avoid transmitting the virus to others) and so they can receive early medical intervention which can slow or prevent the onset of AIDS. Millions of HIV-infected people throughout the world are not aware they are infected. Many are hesitant to find out if they have HIV because of the shame and blame that can be associated with HIV/AIDS. Facilities for HIV testing are also inadequate in many developing countries. From 1987 to 1995, the percentage of U.S. adults ever tested for HIV increased from 16 percent to 40 percent (Anderson, Carey, & Taveras 2000). However, individuals who have been diagnosed with HIV often continue to engage in risky behaviors such as unprotected anal, genital, or oral sex and needle sharing without bleach. One study compared risky behaviors in HIV-infected youths (younger than 25 years) and HIV-infected adults (25 years or older) and found that 66 percent of young women and 46 percent of adult women engaged in risky behaviors after HIV infection. Twenty-eight percent of young men with HIV infection and 16 percent of infected adult men engaged in risky behavior (Diamond & Buskin 2000).

Alleviating HIV/AIDS also requires educating populations about how to protect against HIV and providing access to condoms. Because about half of all people who acquire HIV become infected before they reach age 25, it is crucial that HIV-prevention education be targeted to young people. Yet, there continues to be widespread concern that sex education and access to condoms will encourage young people to become prematurely sexually active. Consequently, many sex education programs in the United States have focused solely upon abstinence. Several studies have concluded, however, that sex education programs that

Many HIV/AIDS public educational campaigns target young adults—the age group that is most at risk for acquiring HIV.

combine messages about abstinence and safer sex practices (e.g., condom use) may delay the initiation of sexual behavior as well as increase preventive behaviors among young people who are already sexually active (Joint United Nations Programme on HIV/AIDS 2000e).

Alleviating HIV/AIDS requires making medical interventions accessible and affordable, especially to the poor populations in developing countries. In an effort to combat AIDS in Africa, five of the world's leading pharmaceutical companies have agreed to decrease the price of drugs used to treat HIV. But even at discounted prices, the drugs will be beyond what many Africans can afford.

There is currently no "cure" for HIV or AIDS. Although various prevention and alleviation strategies may help reduce the spread of HIV and AIDS deaths, the HIV pandemic "will ultimately be controlled only by immunization against HIV using a protective, cheap, simple, and widely available vaccine" (Ward 1999, 199). As of this writing, no such effective vaccine exists, although at least 29 AIDS vaccines have been clinically tested in the United States and around the world without success (Gottlieb 2000).

U.S. Health Care Reform

The United States is the only country in the industrialized world that does not have any mechanism for guaranteeing health care to its citizens. Other countries such as Canada, Great Britain, Sweden, Germany, and Italy, have national health insurance systems, also referred to as **socialized medicine** and **universal health care**. Despite differences in how socialized medicine works in various countries, what is common to all systems of socialized medicine is that the government (1) directly controls the financing and organization of health services, (2) directly pays providers, (3) owns most of the medical facilities (Canada is an exception), (4) guarantees equal access to health care, and (5) allows some private care for individuals who are willing to pay for their medical expenses (Cockerham 1998).

Advocates of universal health care continue to lobby for health care reform that would provide health care benefits to all Americans. The primary goal of Physicians for a National Health Program, an organization with over 8,000 U.S. physicians, is to promote universal health care, eliminating for-profit health insurance and substituting a single-payer system funded through a payroll deduction. In a single-payer system, a single tax-financed public insurance program replaces private insurance companies. Massive numbers of administrative personnel needed to handle itemized billing to 1,500 private insurance companies would no longer be needed (Single Payer Fact Sheet 1999). In testimony before the National Medicare Commission, Douglas Robins (1998) stated:

> As a result of the enormous savings that could be realized by eliminating the present insurance system . . . we could cover all of the 43 million Americans who now lack health care coverage. . . The GAO (Government Accounting Office) estimated a 10 percent administrative savings with a single payer system which would amount to more than 100 billion dollars a year in savings. (p. 1)

The insurance industry, not surprisingly, opposes the adoption of such a system because the private health insurance industry would be virtually eliminated. The health insurance industry's opposition to a single-payer universal health plan is matched only by the persistent efforts of those who advocate such a plan. As these opposing forces continue their battle, other health care reform

If we view obtaining health care as an individual responsibility, we are likely to oppose any attempts to extend government sponsorship of health care. However, if we view health care as a basic human right, we are likely to support extending health care to all.

ROSE WEITZ
Sociologist

measures have taken place. In Massachusetts, a Coalition for Health Care is trying to establish Massachusetts as the first state to offer comprehensive universal health care coverage in July 2002. In 2000, Representative John Tierney (D-MA) introduced a bill (States Right to Innovate in Health Care Act 2000), making states eligible for grants to assist them in developing plans for universal coverage (PNHP Data Update 2000). Other reform measures are in the next few pages.

Strategies to Improve Mental Health Care   In 1999 the first White House Conference on Mental Health and the first Surgeon General's report on mental health established mental health care as a national health priority. Two areas for improving mental health care in the United States are eliminating the stigma associated with mental illness and improving health insurance coverage for treating mental disorders. Efforts in these areas will, hopefully, promote the delivery of treatment to individuals suffering from mental illness. As noted earlier, nearly half of all Americans who have a severe mental illness do not seek treatment. Yet, most mental disorders may be successfully treated with medications and/or psychotherapy or counseling (U.S. Department of Health and Human Services 1999).

The first White House Conference on Mental Health called for a national campaign to eliminate the stigma associated with mental illness. Fearing the negative label of "mental illness" and the social rejection and stigmatization associated with mental illness, individuals are reluctant to seek care. In addition, "stigma deters the public from wanting to pay for care and, thus, reduces consumers' access to resources and opportunities for treatment and social services" (U.S. Department of Health and Human Services 1999, viii). Reducing stigma associated with mental illness may be achieved through encouraging individuals to seek treatment and making treatment accessible and affordable. The Surgeon General's Report on Mental Health explains,

> Effective treatment for mental disorders promises to be the most effective antidote to stigma. Effective interventions help people to understand that mental disorders are not character flaws but are legitimate illnesses that respond to specific treatments, just as other health conditions respond to medical interventions. (U.S. Department of Health and Human Services 1999, viii)

Americans assign high priority to preventing disease and promoting personal well-being and public health; so too must we assign priority to the task of promoting mental health and preventing mental disorders.

MENTAL HEALTH: A REPORT OF THE SURGEON GENERAL

The National Alliance for the Mentally Ill has a StigmaBusters campaign, whereby the public submits instances of media content that stigmatize individuals with mental illness to StigmaBusters, who then investigate and take action (www.nami.org/). For example, StigmaBusters became aware of an episode of "The Simpsons" that featured Marge Simpson being ostracized by the whole town after being in an insane asylum. Marge Simpson was referred to on the show as "the woman who just flew in from the cuckoo's nest" and the "dancing Marge Simpsons" were depicted as dancing in straitjackets. StigmaBusters telephoned the executive producer of "The Simpsons" and expressed concerns about the negative and stereotypical portrayal of mental illness on the show and urged him not to repeat this particular episode or to deal with mental illness in this fashion in the future. StigmaBusters has also discouraged 20th Century Fox from using the phrase "From Gentle to Mental" in the advertising of the Jim Carrey movie "Me, Myself & Irene."

Another priority on the agenda to improve the nation's mental health care system involves eliminating the inequalities in health care coverage for mental

A 1999 U.S. Department of Justice Report revealed that 16 percent of all inmates in state and federal prisons and jails suffer from severe mental illness.

disorders. The Mental Health Parity Act of 1996 was an important step in ending health care discrimination against individuals with mental illnesses by requiring equality between mental health care insurance coverage and other health care coverage—a concept known as **parity**. However, this federal law only applies to annual or lifetime cost limits, but not to substance abuse, copayments, deductibles, or inpatient/outpatient treatment limits. Some states have enacted mental health parity laws, which vary in their scope and application: many do not address substance abuse, are limited to the more serious mental illnesses, or apply only to government employees. In 1999, 28 states had enacted some type of mental health parity laws (National Alliance for the Mentally Ill 2000a).

As of this writing Congress is considering the Mental Health Equitable Treatment Act that was introduced in 1999 by Senators Domenici and Wellstone. This act calls for full parity for the most severe and disabling mental illness and partial parity (limits on duration of treatment only) for all mental disorders.

Another bill being considered by Congress as of this writing is the Mental Health Early Intervention, Treatment and Prevention Act of 2000. This bill provides a series of programs to raise awareness about mental illness; to increase resources for the screening, diagnosis, and treatment of mental illness; and to increase resources to enable the criminal justice system to respond more effectively to persons with mental illness (National Alliance for the Mentally Ill 2000b).

The State Children's Health Insurance Program (SCHIP) In 2000, nearly 11 million American children (1 in 7) had no health insurance. Many of these children come from families with incomes too high to qualify for Medicaid but too low to afford private health insurance. In 1997, SCHIP was created to expand health coverage to uninsured children (Health Care Financing Administration 2000). Under this initiative between the federal and state governments, states receive matching federal funds to provide medical insurance to uninsured

> Mental illness is just like any other medical illness, but treating it differently in health care plans is unconscionable.
>
> AMERICAN PSYCHIATRIC ASSOCIATION

children. As of September 1999, nearly 2 million children were covered by SCHIP (Health Care Financing Administration 2000).

Nearly two-thirds of the nation's 11 million uninsured children are eligible for either SCHIP or Medicaid, but their parents mistakenly believe they don't qualify (Associated Press 2000). Outreach efforts are underway to identify uninsured children who are eligible for SCHIP or Medicaid and to encourage parents to enroll their children.

Telemedicine: Computer Technology in Health Care

Computer technology offers numerous ways to reduce costs associated with health care delivery and to improve patient care. **Telemedicine** involves using information and communication technologies to deliver a wide range of health care services, including diagnosis, treatment, prevention, health support and information, and education of health care workers. Telemedicine can involve the transmission of three main types of information: data, audio, and images. A patient's medical records or vital signs (such as heart rate and blood pressure) can be transmitted from one location to another. Many hospitals and clinics store their medical records electronically, allowing doctors to access information about their patients very quickly and to update patient data from a distance. Specialized medical databases, such as MEDLINE, can be accessed via the Internet and offer a valuable resource for health care practitioners and researchers. The public may also utilize the Internet to gain health information and support. The transmission of radiological images (such as x-ray and ultrasound images) from one location to another for the purpose of interpretation or consultation, has become one of the most commonly used telemedicine services. Images of tissue samples may also be transmitted to a pathologist in another location, who can look at the image on a monitor and offer an interpretation.

Benefits of Telemedicine Telemedicine has the potential to improve public health by making health care available in rural and remote areas and by providing health information to health care workers and to the general population. In addition, "telemedicine allows the scarce resources of specialists and expensive equipment to be shared by a much greater number of patients. Doctors are no longer restricted by geographical boundaries; international specialists are able to spread their skills across continents, without leaving their own hospitals" (LaPorte 1997, 38).

Telemedicine can also be used in training and educating health care professionals and providing health care workers with up-to-date health information. For example, East Carolina University School of Medicine has a family practice training program in which medical trainees live and practice in rural areas and are supervised over the state telemedicine network.

Another benefit of telemedicine is the provision of health information and support services on the Internet, which helps empower individuals in managing their health concerns. Through e-mail, bulletin boards, and chat rooms, individuals with specific health problems can network with other similarly affected individuals. This social support assists in patient recovery, reduces the number of visits to physicians and clinics, and "provides disabled individuals with an opportunity to achieve levels of social integration that were simply not possible before" (LaPorte 1997, 33).

In the twentieth century, advances in public health have been due largely to improvements in sanitation and immunization. Advocates of telemedicine have forecasted that in the twenty-first century, improvements in public health will result from the increased uses of information technology (LaPorte 1997). Telemedicine holds the promise of improving the health of individuals, families, communities, and nations. But whether or not telemedicine achieves its promise depends, in part, on whether resources are allocated to provide the technology and the training to use it.

Understanding *Illness and Health Care*

Human health has probably improved more over the past half century than over the previous three millennia (Feachem 2000). Yet the gap in health between rich and poor remains very wide and the very poor suffer appallingly. Health problems are affected not only by economic resources, but also by other social factors such as aging of the population, gender, education, and race/ethnicity.

> The health of each person affects the health of our families, our workplaces, our communities, our economy, and our society.
>
> LINDA PEENO, M.D.

U.S. cultural values and beliefs emphasize the ability of individuals to control their lives through the choices they make. Thus, Americans and other westerners view health and illness as a result of individual behavior and lifestyle choices, rather than as a result of social, economic, and political forces. We agree that an individual's health is affected by the choices he or she makes—choices such as whether or not to smoke, exercise, engage in sexual activity, use condoms, wear a seatbelt, etc. However, the choices individuals make are influenced by social, economic, and political forces that must be taken into account if the goal is to improve the health of not only individuals, but also entire populations. Further, by focusing on individual behaviors that affect health and illness, we often overlook not only social causes of health problems, but social solutions as well. For example, at an individual level, the public has been advised to rinse and cook meat, poultry, and eggs thoroughly and to carefully wash hands, knives, cutting boards, and so on in order to avoid illness caused by *Escherichia coli* and *Salmonella* bacteria. But whether or not one becomes ill from contaminated meat, eggs, or poultry is affected by more than individual behaviors in the kitchen. Governmental action in the 1980s reduced the number of government food inspectors and deregulated the meat-processing industry (Link & Phelan 1998). Just as governmental actions created the need for individuals to use caution in food preparation, governmental actions can also offer solutions by providing for more food inspectors and stricter regulations on food industries.

Although certain changes in medical practices and policies may help to improve world health, "the health sector should be seen as an important, but not the sole, force in the movement toward global health" (Lerer, Lopez, Kjellstrom, & Yach 1998, 18). Improving the health of a society requires addressing diverse issues, including poverty and economic inequality, gender inequality, population growth, environmental issues, education, housing, energy, water and sanitation, agriculture, and workplace safety. Health promotion is important not only in the hospital, clinic, or doctor's office—it must also occur in the various settings where people live, work, play, and learn (Antezana, Chollat-Traquet, & Yach 1998).

Perhaps the most critical public health agenda today is reducing the gap between the health of advantaged and disadvantaged populations. Feachem (2000) suggests that "addressing this problem, both between countries and within countries, constitutes one of the greatest challenges of the new century. Failure to do so properly will have dire consequences for the global economy, for social order and justice, and for civilization as a whole" (p. 1).

Critical Thinking

1 An analysis of 161 countries found that, in general, countries with high levels of literacy have low levels of HIV (World Health Organization and United Nations Joint Programme on HIV/AIDS 1998). However, in the region of the world worst affected by HIV, sub-Saharan Africa, there is also a relationship between literacy rates and HIV, but the direction of the relationship is reversed. In this region, the countries with the highest levels of HIV infection are also those whose men and women are most literate. What are some possible explanations for this?

2 The Centers for Disease Control and Prevention (CDC) and the American College of Sports Medicine (ACSM) recommend that people aged 6 and older engage regularly, preferably daily, in light to moderate physical activity for at least 30 minutes per day. Experts agree that "if Americans who lead sedentary lives would adopt a more active lifestyle, there would be enormous benefit to the public's health and to individual well-being" (Pate et al. 1995, 406). Yet in a telephone survey of over 87,000 U.S. adults, only about 22 percent reported being active at the recommended level; 24 percent reported that they led a completely sedentary lifestyle (i.e., they reported no leisure-time physical activity in the past month) (Pate et al. 1995). What social and cultural factors contribute to the sedentary lifestyle of many Americans?

3 Why do you think The American Psychiatric Association (2000) avoids the use of such expressions as "a schizophrenic" or "an alcoholic" and instead uses the expressions "an individual with Schizophrenia" or "an individual with Alcohol Dependence?"

Key Terms

acute conditions

burden of disease

chronic conditions

disability-adjusted life year (DALY)

epidemiological transition

epidemiologists

epidemiology

globalization

health maintenance organizations (HMOs)

incidence

infant mortality rates

life expectancy

managed care

maternal mortality rates

Medicaid

Medicare

mental disorder

mental health

mental illness

morbidity

mortality

needle exchange programs

parity

perinatal transmission

preferred provider organizations (PPOs)

prevalence

socialized medicine

stigma

telemedicine

under-5 mortality rate

universal health care

Media Resources

The Wadsworth Sociology Resource Center: Virtual Society

http://sociology.wadsworth.com/

See the companion web site for this book to access general sociology resources and text-specific features that can further your understanding of this chapter. The site contains Internet links, Internet exercises, online practice quizzes, information on InfoTrac College Edition, and many more valuable materials designed to enrich your learning experience in social problems.

InfoTrac College Edition

You can access InfoTrac College Edition either from the Wadsworth Sociology Resource Center at **http://sociology.wadsworth.com** or directly from your web browser at **http://www.infotrac-college.com/wadsworth/**. InfoTrac College Edition is an online university library that includes over 700 popular and scholarly journals in which you can find articles related to the topics in this chapter such as information on the American Sociological Association, methods of research, and sociological theory.

Interactions CD-ROM

Go to the "Interactions" CD-ROM for *Understanding Social Problems*, Third Edition to access additional interactive learning tools, such as in-depth review materials, corresponding practice quizzes, and other engaging resources and activities to help you study the concepts in this chapter.

3

Alcohol and Other Drugs

Is It True?

1. Use of illicit drugs reached record levels in 2000.

2. In 1995, the Supreme Court ruled that random drug testing of student athletes in public schools is unconstitutional.

3. Alcoholics are seven times more likely to separate or divorce than non-alcoholics.

4. The most commonly used illicit drug in the United States is marijuana.

5. The number of ecstacy users is quite large, comprising more than 17 percent of the population.

Answers to "Is It True?": 1 = T; 2 = F; 3 = T; 4 = T; 5 = F

Substance abuse, the nation's number-one preventable health problem, places an enormous burden on American society, harming health, family life, the economy, and public safety, and threatening many other aspects of life.

ROBERT WOOD JOHNSON FOUNDATION
Institute for Health Policy, Brandeis University

Scott Krueger was athletic, intelligent, handsome, and what you'd call an all-around "nice guy." A freshman at Massachusetts Institute of Technology, he was a three-letter athlete and one of the top 10 students in his high school graduating class of over 300. He was a "giver" not a "taker," tutoring other students in math after school while studying second-year calculus so he could pursue his own career in engineering. While at MIT he rushed a fraternity and celebrated his official acceptance into the brotherhood. The night he celebrated he was found in his room, unconscious, and after 3 days in an alcoholic coma he died. He was 18 years old (Moore 1997). In September of 2000, MIT agreed to pay Scott's parents, Bob and Darlene Krueger, $4.75 million in a settlement over the death of their son and to establish a scholarship in his name. As a consequence of Scott's death, as of August 2002 all first year students at MIT must live in "University-owned, -operated and -supervised housing" (AP 2000,11).

Drug-induced death is just one of many negative consequences that can result from alcohol and drug abuse. The abuse of alcohol and other drugs is a social problem when it interferes with the well-being of individuals and/or the societies in which they live—when it jeopardizes health, safety, work and academic success, family, and friends. But managing the drug problem is a difficult undertaking. In dealing with drugs, a society must balance individual rights and civil liberties against the personal and social harm that drugs promote—crack babies, suicide, drunk driving, industrial accidents, mental illness, unemployment, and teenage addiction. When to regulate, what to regulate, and who should regulate are complex social issues. Our discussion begins by looking at how drugs are used and regulated in other societies.

The Global Context: Drug Use and Abuse

Pharmacologically, a **drug** is any substance other than food that alters the structure or functioning of a living organism when it enters the bloodstream. Using this definition, everything from vitamins to aspirin constitutes a drug. Sociologically, the term drug refers to any chemical substance that (1) has a direct effect on the user's physical, psychological, and/or intellectual functioning, (2) has the potential to be abused, and (3) has adverse consequences for the individual and/or society. Societies vary in how they define and respond to drug use. Thus, drug use is influenced by the social context of the particular society in which it occurs.

Drug Use and Abuse around the World

According to estimates by the European Information Network on Drugs and Drug Addiction, the prevalence of drug use around the world varies dramatically. For example, the proportion of adults who report ever using marijuana ranges from 5 to 8 percent in Finland, Belgium, and Sweden, to 11 to 16 percent in France, Spain and the United States, to 30 percent in Denmark (Reitox 2000). Use by adolescents varies as well. In Greece, Finland, Portugal, and Sweden, approximately 5 percent of the 15- and 16-year-old population admit to ever using marijuana. However, in Belgium, France, and Denmark that number is 30 percent.

In England, illegal drug use continues to spread particularly among those under 25 (ISDD 1999). One-third of the adult population of the UK (Great Britain, Scotland, Wales, and Northern Ireland) is estimated to have used drugs at some time in their life—49 percent of those under 30. Statistics also indicate that drug users in the UK are more likely to be male, unemployed, and living in or around London when compared with non-users.

Some of the differences in international drug use may be attributed to variations in drug policies. The Netherlands, for example, has had an official government policy of treating the use of such drugs as marijuana, hashish, and heroin as a health issue rather than a crime issue since the mid-1970s. In the first decade of the policy, drug use did not appear to increase. However, increases in marijuana use were reported in the early 1990s with the advent of "cannabis cafes." These coffee shops sell small amounts of marijuana for personal use and, presumably, prevent casual marijuana users from coming into contact with drug dealers (MacCoun & Reuter 2001). More recent evidence suggests that marijuana use among Dutch youth is decreasing (Sheldon 2000).

Great Britain has also adopted a "medical model," particularly in regard to heroin and cocaine. As early as the 1960s, English doctors prescribed opiates and cocaine for their patients who were unlikely to quit using drugs on their own and for the treatment of withdrawal symptoms. By the 1970s, however, British laws had become more restrictive making it difficult for either physicians or users to obtain drugs legally. Today, British government policy provides for limited distribution of drugs by licensed drug treatment specialists to addicts who might otherwise resort to crime to support their habits. Recent increased

U.S. citizens visiting the Netherlands may be shocked or surprised to find people smoking marijuana and hashish openly in public.

© Georges Merillon/Liaison Agency

drug use, however, has led to discussions of a British "zero-tolerance" policy (Francis 2000).

In stark contrast to such health-based policies, other countries execute drug users and/or dealers, or subject them to corporal punishment. The latter may include whipping, stoning, beating, and torture. Such policies are found primarily in less developed nations such as Malaysia, where religious and cultural prohibitions condemn any type of drug use, including alcohol and tobacco.

Drug Use and Abuse in the United States

According to government officials there is a drug crisis in the United States—a crisis so serious that it warrants a multibillion-dollar-a-year "war on drugs." Americans' concern with drugs, however, has varied over the years. Ironically, in the 1970s when drug use was at its highest, concern over drugs was relatively low. Today, when a sample of Americans were asked, "What would you say is the most urgent health problem facing this country at the present time . . . ?", alcohol abuse and drugs ranked fifth behind healthcare costs, cancer, AIDS, and heart disease (Gallup Poll 2000a).

As Table 3.1 indicates, use of alcohol and tobacco are much more widespread than use of illicit drugs, such as marijuana and cocaine. In the United States, cultural definitions of drug use are contradictory—condemning it on the one hand (e.g., heroin), yet encouraging and tolerating it on the other (e.g., alcohol). At various times in U.S. history, many drugs that are illegal today were legal and readily available. In the 1800s and the early 1900s, opium was routinely used in medicines as a pain reliever, and morphine was taken as a treatment for

■ **Table 3.1** *Percentages Reporting Lifetime, Past Year, and Past Month Use of Illicit and Licit Drugs among Persons Aged 12 and Older: 1999.*

Drug	Time Period		
	Lifetime	Past Year	Past Month
Marijuana/hashish	34.6	8.9	5.1
Cocaine	11.5	1.7	0.7
Crack	2.7	0.5	0.2
Heroin	1.4	0.2	0.1
Hallucinogens	11.3	1.4	0.4
LSD	8.7	0.9	0.2
PCP	2.6	0.1	0.0
Inhalants	7.8	1.1	0.5
Methamphetamine	4.3	0.5	0.2
Non-medical use of any psychotherapeutic drug[1]	15.4	4.2	1.8
Any tobacco[2]	72.0	36.1	30.2
Cigarettes	68.2	30.1	25.8
Alcohol	81.3	62.6	47.3

[1]Non-medical use of any prescription pain reliever, stimulant, sedative, or tranquilizer.

[2]Use of any tobacco product indicates using at least once cigarettes, smokeless tobacco, cigars, or pipe tobacco except in the "past year" which does not include use of pipe tobacco.

Source: HHS (U.S. Department of Health and Human Services). 2000. "1999 National Household Survey on Drug Abuse." Substance Abuse and Mental Health Service Administration. Washington, D.C.: U.S. Government Printing Office.

dysentery and fatigue. Amphetamine-based inhalers were legally available until 1949, and cocaine was an active ingredient in Coca-Cola until 1906 when it was replaced with another drug—caffeine (Witters, Venturelli, & Hanson 1992).

Sociological Theories of Drug Use and Abuse

Most theories of drug use and abuse concentrate on what are called psychoactive drugs. These drugs alter the functioning of the brain, affecting the moods, emotions, and perceptions of the user. Such drugs include alcohol, cocaine, heroin, and marijuana. **Drug abuse** occurs when acceptable social standards of drug use are violated resulting in adverse physiological, psychological, and/or social consequences. For example, when an individual's drug use leads to hospitalization, arrest, or divorce, such use is usually considered abusive. Drug abuse, however, does not always entail drug addiction. **Drug addiction**, or **chemical dependency,** refers to a condition in which drug use is compulsive—users are unable to stop because of their dependency. The dependency may be psychological, in that the individual needs the drug to achieve a feeling of well-being, and/or physical, in that withdrawal symptoms occur when the individual stops taking the drug.

Various theories provide explanations for why some people use and abuse drugs. Drug use is not simply a matter of individual choice. Theories of drug use explain how structural and cultural forces, as well as biological factors, influence drug use and society's responses to it.

Structural-Functionalist Perspective

Functionalists argue that drug abuse is a response to the weakening of norms in society. As society becomes more complex and rapid social change occurs, norms and values become unclear and ambiguous, resulting in **anomie**—a state of normlessness. Anomie may exist at the societal level, resulting in social strains and inconsistencies that lead to drug use. For example, research indicates that increased alcohol consumption in the 1830s and the 1960s was a response to rapid social change and the resulting stress (Rorabaugh 1979). Anomie produces inconsistencies in cultural norms regarding drug use. For example, while public health officials and health care professionals warn of the dangers of alcohol and tobacco use, advertisers glorify the use of alcohol and tobacco and the U.S. government subsidizes alcohol and tobacco industries. Further, cultural traditions, such as giving away cigars to celebrate the birth of a child and toasting a bride and groom with champagne, persist.

There are but three ways for the populace to escape its wretched lot. The first two are by route of the wine-shop or the church; the third is by that of the social revolution.

MIKHAIL A. BAKUNIN
Anarchist and revolutionary

Anomie may also exist at the individual level as when a person suffers feelings of estrangement, isolation, and turmoil over appropriate and inappropriate behavior. An adolescent whose parents are experiencing a divorce, who is separated from friends and family as a consequence of moving, or who lacks parental supervision and discipline may be more vulnerable to drug use because of such conditions. Thus, from a structural-functionalist perspective, drug use is a response to the absence of a perceived bond between the individual and society, and to the weakening of a consensus regarding what is considered acceptable. Consistent with this perspective, Nylander, Tung, and Xu (1996) found that adolescents who reported that religion was important in their lives were less likely to use drugs than those who didn't.

Conflict Perspective

Conflict perspectives emphasize the importance of power differentials in influencing drug use behavior and societal values concerning drug use. From a conflict perspective, drug use occurs as a response to the inequality perpetuated by a capitalist system. Societal members, alienated from work, friends, and family, as well as from society and its institutions, turn to drugs as a means of escaping the oppression and frustration caused by the inequality they experience. Further, conflict theorists emphasize that the most powerful members of society influence the definitions of which drugs are illegal and the penalties associated with illegal drug production, sales, and use.

For example, alcohol is legal because it is often consumed by those who have the power and influence to define its acceptability—white males (HHS 2000). This group also disproportionately profits from the sale and distribution of alcohol and can afford powerful lobbying groups in Washington to guard the alcohol industry's interests. Since tobacco and caffeine are also commonly used by this group, societal definitions of these substances are also relatively accepting.

Conversely, crack cocaine and heroin are disproportionately used by minority group members, specifically, blacks and Hispanics (HHS 2000). Consequently, the stigma and criminal consequences associated with the use of these drugs are severe. The use of opium by Chinese immigrants in the 1800s provides a historic example. The Chinese, who had been brought to the United States to work on the railroads, regularly smoked opium as part of their cultural tradition. As unemployment among white workers increased, however, so did resentment of Chinese laborers. Attacking the use of opium became a convenient means of attacking the Chinese, and in 1877 Nevada became the first of many states to prohibit opium use. As Morgan (1978) observes:

> The first opium laws in California were not the result of a moral crusade against the drug itself. Instead, it represented a coercive action directed against a vice that was merely an appendage of the real menace—the Chinese—and not the Chinese per se, but the laboring "Chinamen" who threatened the economic security of the white working class. (p. 59)

The criminalization of other drugs, including cocaine, heroin, and marijuana, follows similar patterns of social control of the powerless, political opponents, and/or minorities. In the 1940s, marijuana was used primarily by minority group members and carried with it severe criminal penalties. But after white middle-class college students began to use marijuana in the 1970s, the government reduced the penalties associated with its use. Though the nature and pharmacological properties of the drug had not changed, the population of users was now connected to power and influence. Thus, conflict theorists regard the regulation of certain drugs, as well as drug use itself, as a reflection of differences in the political, economic, and social power of various interest groups.

Symbolic Interactionist Perspective

Symbolic interactionism, emphasizing the importance of definitions and labeling, concentrates on the social meanings associated with drug use. If the initial drug use experience is defined as pleasurable, it is likely to recur, and over time, the individual may earn the label of "drug user." If this definition is internalized so that the individual assumes an identity of a drug user, the behavior will likely continue and may even escalate.

Drug use by friends is consistently the strongest predictor of a person's involvement in drug use.

U.S. DEPARTMENT OF JUSTICE

Drug use is also learned through symbolic interaction in small groups. Friends, for example, are the most common source of drugs for teenagers (Leinwand 2000). First-time users learn not only the motivations for drug use and its techniques, but also what to experience. Becker (1966) explains how marijuana users learn to ingest the drug. A novice being coached by a regular user reports the experience:

> I was smoking like I did an ordinary cigarette. He said, "No, don't do it like that." He said, "Suck it, you know, draw in and hold it in your lungs . . . for a period of time." I said, "Is there any limit of time to hold it?" He said, "No, just till you feel that you want to let it out, let it out." So I did that three or four times. (Becker 1966, 47)

Marijuana users not only learn how to ingest the smoke, but also learn to label the experience positively. When certain drugs, behaviors, and experiences are defined by peers as not only acceptable but pleasurable, drug use is likely to continue.

> Because they (first-time users) think they're going to keep going up, up, up till they lose their minds or begin doing weird things or something. You have to like reassure them, explain to them that they're not really flipping or anything, that they're gonna be all right. You have to just talk them out of being afraid. (Becker 1966, 55)

Interactionists also emphasize that symbols may be manipulated and used for political and economic agendas. The popular DARE (Drug Abuse Resistance Education) program, with its anti-drug emphasis fostered by local schools and police, carries a powerful symbolic value that politicians want the public to identify with. "Thus, ameliorative programs which are imbued with these potent symbolic qualities (like DARE's links to schools and police) are virtually assured wide-spread public acceptance (regardless of actual effectiveness) which in turn advances the interests of political leaders who benefit from being associated with highly visible, popular symbolic programs" (Wysong, Aniskiewicz, & Wright 1994, 461).

Biological and Psychological Theories

Drug use and addiction are likely the result of a complex interplay of social, psychological, and biological forces. Biological research has primarily concentrated on the role of genetics in predisposing an individual to drug use. According to a recent report by the National Institute on Alcohol Abuse and Alcoholism (2000, xiii), "50 to 60 percent of the risk for developing alcoholism is genetic." Research also indicates that by examining inherited traits science ". . . can predict [in] childhood with 80 percent accuracy who is going to develop alcoholism later in life" (AAP 1998). At the same time, many alcoholics do not have parents who abuse alcohol, and many alcoholic parents have offspring who do not abuse alcohol.

Biological theories of drug use hypothesize that some individuals are physiologically predisposed to experience more pleasure from drugs than others and, consequently, are more likely to be drug users. According to these theories, the central nervous system, which is composed primarily of the brain and spinal cord, processes drugs through neurotransmitters in a way that produces an unusually euphoric experience. Individuals not so physiologically inclined report less pleasant experiences and are less likely to continue use (Jarvik 1990; Alcohol Alert 2000).

Psychological explanations focus on the tendency of certain personality types to be more susceptible to drug use. Individuals who are particularly prone to anxiety may be more likely to use drugs as a way to relax, gain self-confidence, or ease tension. For example, research indicates that female adolescents who have been sexually abused or who have poor relationships with their parents are more likely to have severe drug problems (NIDA 2000a). Psychological theories of drug abuse also emphasize that drug use may be maintained by positive and negative reinforcement.

Frequently Used Legal and Illegal Drugs

Over 14 million people in the United States are illicit drug users, which represents 6.7 percent of the 12-and-older population. Users of illegal drugs, although varying by type of drug used, are more likely to live in metropolitan areas, to be male, young, and minority group members (HHS 2000). Social definitions regarding which drugs are legal or illegal, however, have varied over time, circumstance, and societal forces. In the United States, two of the most dangerous and widely abused drugs, alcohol and tobacco, are legal.

Alcohol

Americans' attitudes toward alcohol have had a long and varied history (this chapter's *Self and Society* feature deals with attitudes toward alcohol). Although alcohol was a common beverage in early America, by 1920 the federal government had prohibited its manufacture, sale, and distribution through the passage of the Eighteenth Amendment to the Constitution. Many have argued that Prohibition, like the opium regulations of the late 1800s, was in fact a "moral crusade" (Gusfield 1963) against immigrant groups who were more likely to use alcohol. The amendment had little popular support and was repealed in 1933. Today, the United States is experiencing a resurgence of concern about alcohol. What has been called a "new temperance" has manifested itself in federally mandated 21-year-old drinking age laws, warning labels on alcohol bottles, increased concern over fetal alcohol syndrome and teenage drinking, and stricter enforcement of drinking and driving regulations.

> Drunkenness is the ruin of reason. It is premature old age. It is temporary death.
>
> St. Basil
> *Bishop of Caesarea*

Alcohol is the most widely used and abused drug in America. Although most people who drink alcohol do so moderately and experience few negative effects, alcoholics are psychologically and physically addicted to alcohol and suffer various degrees of physical, economic, psychological, and personal harm.

The National Household Survey on Drug Abuse conducted by the Department of Health and Human Services reports that 105 million Americans 12 and older consumed alcohol at least once in the month preceding the survey, that is, were current users. Of these, 12.4 million were heavy drinkers (defined as drinking 5 or more drinks per occasion on 5 or more days in the survey month) and 45 million were binge drinkers (defined as drinking 5 or more drinks on at least one occasion during the survey month). Even more troubling were the 10 million current users of alcohol who were 12 to 20 years old, over half of whom were binge drinkers (HHS 2000). For many students drinking began before high school, with almost one-third having their first drink before age 13. Research indicates that the younger the age of onset, the higher the probability that an individual will develop a drinking dis-

Alcohol Attitude Test

If you strongly agree with the following statements, write in 1. If you agree, but not strongly, write in 2. If you neither agree nor disagree, write in 3. If you disagree, but not strongly, write in 4. If you strongly disagree, write in 5.

SET 1

_____ 1. If I tried to stop someone from driving after drinking, the person would probably think I was butting in where I shouldn't.

_____ 2. Even if I wanted to, I would probably not be able to stop someone from driving after drinking.

_____ 3. If people want to kill themselves, that's their business.

_____ 4. I wouldn't like someone to try to stop me from driving after drinking.

_____ 5. Usually, if you try to help someone else out of a dangerous situation, you risk getting yourself into one.

_____ Total score for questions 1 through 5

SET 2

_____ 6. My friends would not disapprove of me for driving after drinking.

_____ 7. Getting into trouble with my parents would not keep me from driving after drinking.

_____ 8. The thought that I might get into trouble with the police would not keep me from driving after drinking.

_____ 9. I am not scared by the thought that I might seriously injure myself or someone else by driving after drinking.

_____ 10. The fear of damaging the car would not keep me from driving after drinking.

_____ Total score for questions 6 through 10

SET 3

_____ 11. The 55 mph speed limit on the open roads spoils the pleasure of driving for most teenagers.

_____ 12. Many teenagers use driving to let off steam.

_____ 13. Being able to drive a car makes teenagers feel more confident in their relations with others their age.

_____ 14. An evening with friends is not much fun unless one of them has a car.

_____ 15. There is something about being behind the wheel of a car that makes one feel more adult.

_____ Total score for questions 11 through 15

SCORING

Set 1. 15–25 points: takes responsibility to keep others from driving when drunk; 5–9 points: wouldn't take steps to stop a drunk friend from driving.

Set 2. 12–25 points: hesitates to drive after drinking; 5–7 points: is not deterred by the consequences of drinking and driving.

Set 3. 19–25 points: perceives auto as means of transportation; 5–14 points: uses car to satisfy psychological needs, not just transportation.

Source: Courtesy of National Highway Traffic Safety Administration. National Center for Statistics and Analysis, from *Drunk Driving Facts*. Washington, D.C.: NHTSA, 1988.

order at some time in their life. The likelihood of alcohol dependence, as defined by the National Household Survey, peaks at age 21 and declines with each successive age group. Heavy teenage drinkers, as with their adult counterparts, are more likely to be white, non-Hispanic, and male (HHS 2000; CDC 2000).

A recent study by the Harvard School of Public Health (Wechsler, Lee, Kuo, & Lee 2000) of over 14,000 college students at 119 nationally representative 4-year colleges in 39 states indicates that:

* In the 2 weeks preceding the survey, 44 percent of college students had engaged in binge drinking—51 percent of men and 40 percent of women.
* Students who were white, 23 or younger, lived in a fraternity or sorority house, and who had binged in high school were the most likely to binge drink in college.
* The prevalence of binge drinking has slightly decreased since 1993; however, it has increased since1997.
* Even though alcohol consumption is prohibited under the age of 21, drinking rates varied little between college freshmen, sophomores, juniors, and seniors.
* Frequent binge drinkers were significantly more likely than non-binge drinkers to have missed a class, damaged property, been hurt or injured, engaged in unplanned sex, or driven a car after drinking.

Not only are binge drinkers more likely to report using other controlled substances, but the more frequently a student binges, the higher the probability of reporting other drug use. Some evidence suggests that certain combinations of drugs, for example alcohol and cocaine, may heighten the negative effects of either drug separately, that is, there is a negative drug interaction (NIDA 2000b).

> Although technology was developed years ago to remove nicotine from cigarettes and to control with precision the amount of nicotine in cigarettes, they are still marketing cigarettes with levels of nicotine that are sufficient to produce and sustain addiction.
>
> DAVID KESSLER
> *Former Head of Federal Drug Administration*

Tobacco

Although nicotine is an addictive psychoactive drug and tobacco smoke has been classified by the Environmental Protection Agency as a Group A carcinogen, tobacco continues to be among the most widely used drugs in the United States. According to the U.S. Department of Health and Human Services survey, 67 million Americans continue to smoke cigarettes—30 percent of the 12-and-older population and 15 percent of the 12- to 17-year-old population. Interestingly, among 12- to 17-year-olds, three brands account for over 50 percent of the tobacco market—Marlboro, Newport, and Camel. Use of all tobacco products including smokeless tobacco, cigars, and cigarettes is higher for high school compared with college graduates, males, and Native Americans/Alaska Natives (HHS 2000).

Much of the concern about smoking surrounds the use of tobacco by young people. Between 1991 and 1999, the number of adolescent smokers increased 27 percent—56 percent among African Americans students (CDC 2000). Advertising campaigns that appeal to youth such as the now defunct "Joe Camel" and the placement of billboards advertising cigarettes at a higher rate in black compared with white communities has often been blamed for these increases (American Heart Association 2000). Advertising is but one venue criticized for their positive portrayal of tobacco use. In this chapter's *Social*

© John Kobal Foundation/Hulton/Archive

Up until the 1960s, the harmful effects of tobacco were not part of the American collective consciousness. Tobacco was glamorized in the movies, and on television, and it was not unusual to use "stars" to advertise tobacco products. Countless numbers of television and movie personalities have died from tobacco-related illnesses including actor Humphrey Bogart.

Images of Alcohol and Tobacco Use in Children's Animated Films

The impact of media on drug and alcohol use is likely to be recursive—media images affect drug use while, alternatively, societal drug use helps define media presentations. Previous research has documented the rate of tobacco and alcohol use in print media, advertising, and Hollywood movies. In the present research, Goldstein, Sobel, and Newman (1999) use content analysis to investigate the prevalence of tobacco and alcohol use is children's animated films as one step in assessing the growing concern with media influence on children's smoking and drinking behavior.

Sample and Methods

The researcher's examined all G-rated animated films released between 1937 (*Snow White and the Seven Dwarfs*) and 1997 (*Hercules, Anastasia, Pippi Longstocking,* and *Cats Don't Dance*). Criteria for sample inclusion included that the film be at least 60 minutes in length and, before video distribution, was released to theaters. The resulting sample included all of Disney's animated children's films produced during the target years with the exception of three that were unavailable on videocassette. The remaining films included all children's animated films produced by MGM/United Artists, Universal, 20th Century Fox, and Warner Brothers since 1982. Variables coded included the: (1) presence of alcohol or tobacco use, (2) length of time of use on screen, (3) number of characters using alcohol or tobacco, (4) value of the character using tobacco or alcohol (i.e., good, neutral, or bad), (5) any implied messages about the drug use, and (6) the type of tobacco or alcohol being used.

Findings and Conclusions

Of the 50 films analyzed, at least one episode of alcohol and/or tobacco use was portrayed in 34 (68%) with tobacco use (N=28) slightly exceeding portrayals of alcohol use (N=25). Tobacco was used by 76 different characters with an onscreen time of 45 minutes—an average of 1.62 minutes per movie. Characters were most likely to use cigars followed by cigarettes, and pipes. Of the 76 characters using tobacco, 28 (37%) were classified as good. Surprisingly, the use of tobacco products by "good" characters has increased rather than decreased over time.

Everybody knew it's addictive. Everybody knew it causes cancer. We were all in it for the money.

VICTOR CRAWFORD
Former tobacco lobbyist and smoker who developed lung cancer

Problems Research Up Close feature, images of tobacco and alcohol use in children's animated films are examined.

Tobacco was first cultivated by Native Americans, who introduced it to the European settlers in the 1500s. The Europeans believed it had medicinal properties, and its use spread throughout Europe, assuring the economic success of the colonies in the New World. Tobacco was initially used primarily through chewing and snuffing, but in time smoking became more popular even though scientific evidence that linked tobacco smoking to lung cancer existed as early as 1859 (Feagin and Feagin 1994). However, it was not until 1989 that the U.S. Surgeon General concluded that tobacco products are addictive and that it is nicotine that causes the dependency. Today, the health hazards of tobacco use are well documented and have resulted in the passage of federal laws that require warning labels on cigarette packages and prohibit cigarette advertising on radio and television. By the year 2030, tobacco-related diseases will be the number one cause of death worldwide, killing one of every six people. Eighty percent of the deaths will take place in poor nations where many smokers are unaware of the health hazards associated with their behavior (Mayell 1999).

Marijuana

Although drug abuse ranks among the top concerns of many Americans, surveys indicate that illegal drug use of all kinds is far less common than alcohol and tobacco abuse. Marijuana remains the most commonly used and most

Sixty-two characters, averaging 2.5 per film, were shown using alcohol with a total duration of 27 minutes across all films. Characters were most likely to consume wine followed by beer, spirits, and champagne. The number of good characters using alcohol was similar to the number of characters classified as bad. In 19 of the 25 films in which alcohol use was portrayed, tobacco use was also pictured. Although several films portrayed the physical consequences of smoking (N=10) (e.g., coughing) or drinking (N=7) (e.g., passing out), no film verbally referred to the health hazards of either drug.

One particularly interesting finding of the research concerned the use of alcohol and tobacco as a visual prop in character development. For example, although cigar smokers were portrayed as tough and powerful (e.g., Sykes in *Oliver and Company*), pipe smokers were most often older, kindly, and wise (e.g., Geppetto in Pinocchio), and cigarette smokers independent, witty, and intelligent (e.g., the Genie in *Aladdin*). There was also a tendency for alcohol and tobacco use to be portrayed together. When one, the other, or both are associated with positively defined characters the impact may be detrimental to the lifestyle choices of viewers.

Although this study cannot assess the "impact question," advertising campaigns such as Joe Camel and the Budweiser frogs have been linked to detrimental results. Although in each of these cases the motivation for the use of such appealing characters is clear, the presentation of "good" characters using alcohol and tobacco products in children's animated films remains unexplained. Interpretation of the results is further complicated by the lack of change over time, that is, as our knowledge of the harmful effects of these products increased, their presence in children's films did not, as expected, decrease. In light of these results, the researchers call for an end to the portrayal of alcohol and tobacco use in all children's animated films and associated products (e.g., posters, books, games).

Source: Adam Goldstein, Rachel Sobel, and Glen Newman. 1999. "Tobacco and Alcohol Use in G-rated Children's Animated Films." *Journal of the American Medical Association* 281:1121–1136.

heavily trafficked illicit drug in the world. It is estimated that over 40 million Europeans have tried cannabis at least once in their lifetime and that there are 200 to 250 million marijuana users worldwide, predominantly in Africa and Asia. Recently, several European countries including Switzerland, Portugal, and Luxembourg have considered softening penalties for marijuana use (Francis 2000).

Marijuana's active ingredient is THC (delta-9-tetrahydrocannabinol) which, in varying amounts, may act as a sedative or as a hallucinogen. Marijuana use dates back to 2737 B.C. in China and has a long tradition of use in India, the Middle East, and Europe. In North America, hemp, as it was then called, was used for making rope and as a treatment for various ailments. Nevertheless, in 1937 Congress passed the Marijuana Tax Act, which restricted its use; the law was passed as a result of a media campaign that portrayed marijuana users as "dope fiends" and, as conflict theorists note, was enacted at a time of growing sentiment against Mexican immigrants (Witters et al 1992, 357–59).

There are more than 14 million illicit drug users in the United States. Of that number, 75 percent used marijuana only or used it in conjunction with some other illicit drug (see Figure 3.1). Marijuana use is particularly high among the young, being the illegal drug of choice of 14-year-olds. Although recently marijuana use by 12- to 17-year-olds decreased significantly (26 percent), use by 18- to 25-year-olds increased by 28 percent (HHS 2000). Lifetime rates of marijuana use by high school seniors was higher than any year since 1987 though remaining below peak 1970s levels. Not surprisingly, 89 percent of 12th graders

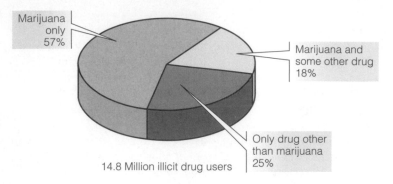

Marijuana
only
57%

Marijuana and
some other drug
18%

Only drug other
than marijuana
25%

14.8 Million illicit drug users

Figure 3.1. *Types of Drugs Used in Past Month by Illicit Drug Users, Age 12 and Older, 1999.*

Source: HHS (U.S. Department of Health and Human Services). 2000. "1999 National Household Survey on Drug Abuse." Substance Abuse and Mental Health Service Administration. Washington, D.C.: U.S. Government Printing Office.

report that it would be "fairly easy" or "very easy" to get marijuana if so desired (MTF 2000).

Although the effects of alcohol and tobacco are, in large part, indisputable, there is less agreement about the effects of marijuana. Although the carcinogenic effects of marijuana are as lethal as nicotine's, other long-term physiological effects are unknown. An important concern is that marijuana may be a **gateway drug** that causes progression to other drugs such as cocaine and heroin. More likely, however, is that persons who experiment with one drug are more likely to experiment with another. Indeed, most drug users are polydrug users with the most common combination being alcohol, tobacco, and marijuana.

Cocaine

Crack is a drug peddler's dream: it is cheap, easily concealed and provides a short-duration high that invariably leaves the user craving more.

TOM MORGANTHAU
Journalist

Cocaine is classified as a stimulant and, as such, produces feelings of excitation, alertness, and euphoria. Although such prescription stimulants as methamphetamine and dextroamphetamine are commonly abused, over the last 10 to 20 years societal concern over drug abuse has focused on cocaine. Such concerns have been fueled by its increased use, addictive qualities, physiological effects, and worldwide distribution. More than any other single substance, cocaine has led to the present "war on drugs."

Cocaine, which is made from the coca plant, has been used for thousands of years, but anti-cocaine sentiment in the United States did not emerge until the early 1900s, when it was primarily a response to cocaine's heavy use among urban blacks (Witters et al. 1992, 260). Cocaine was outlawed in 1914 by the Harrison Narcotics Act, but its use and effects continued to be misunderstood. For example, a 1982 Scientific American article suggested that cocaine was no more habit forming than potato chips (Van Dyck & Byck 1982). As demand and then supply increased, prices fell from $100 a dose to $10 a dose, and "from 1978 to 1987 the United States experienced the largest cocaine epidemic in history" (Witters et al. 1992, 256, 261).

In 1999, there were an estimated 1.5 million cocaine users, a decrease from the previous year (HHS 2000). Whether the decrease is a function of increased government surveillance, anti-cocaine media messages and celebrity deaths, demographic changes, or a greater awareness of the substance's dangers is un-

An Excerpt from The Cocaine Kids

The door opens a crack before I can knock, a tall African-American man brusquely thrusts his palm toward me and asks, "You got three dollars?" He motions excitedly, "If you ain't got three dollars you can't come in here." The entrance fee. I pay and walk in.

The establishment is desolate, uninviting, dank and smoky. The carpet in the first room is shit-brown and heavily stained, pockmarked by so many smoke burns that it looks like an abstract design. In the dim light, all the people on the scene seem to be in repose, almost inanimate, for a moment.

As my eyes adjust to the smoke, several bodies emerge. I see jaws moving, hear voices barking hoarsely into walkie-talkies—something about money; their talk is jagged, nasal, and female. One woman takes out an aluminum foil packet, snorts some of its contents, passes it to her partner then disappears into another room. In a corner near the window, a shadowy figure moans. One woman sits with her skirt over her head, while a bobbing head writhes underneath her. In an adjacent alcove, I see another couple copulating. Somewhere in the corridor a man and woman argue loudly in Spanish. Staccato rap music sneaks over the grunts and hollers.

The smell is a nauseating mix of semen, crack, sweat, other human body odors, funk and filth. Two men dicker about who took the last "hit" (puff); two others are on their hands and knees looking for crack particles they claim they have lost in the carpet.

In the crack houses, the sharing rituals associated with snorting are being supplanted by more individualistic, detached arrangements where people come together for erotic stimulation, sexual activity, and cocaine smoking. They may be total strangers, seeking only brief and superficial physical contact, encounters designed to heighten sensations; the smoking act is a narcissistic fix—there is little thought for the other person. The emotional content is largely due to the momentary excitation of the setting and the cocaine. Much of the sexual behavior is performed to acquire more cocaine.

Nothing better exemplifies the new attitude than the act of *Sancocho* (a word meaning to cut up in little pieces and stew). To sancocho is to steal crack, drugs or money from a friend or other person who is not alert, a regular practice in the crack houses. Another example is the "hit kiss" ritual: after inhaling deeply, basers literally "kiss"—put their lips together and exhale the smoke into each other's mouths. This not only saves all the valuable smoke, but also stimulates the other sexually. Other versions of the kiss extend to other orifices.

Source: Terry Williams. 1990. From *The Cocaine Kids: The Inside Story of a Teenage Drug Ring*, by Terry Williams. Copyright © 1990, by Terry Williams. Reprinted by permission of Perseus Books Publishers, a member of Perseus Books, L.L.C.

known. About half of current cocaine users are under the age of 26, and some evidence suggests that teenage use in rural areas is higher than in urban areas (HHS 2000; *The Economist* 2000).

Crack is a crystallized product made by boiling a mixture of baking soda, water, and cocaine. The result, also called rock, base, and gravel, is relatively inexpensive and was not popular until the mid-1980s. Crack is one of the most dangerous drugs to surface in recent years. Crack dealers often give drug users their first few "hits" free, knowing the drug's intense high and addictive qualities are likely to lead to returning customers. Recent data, however, suggest that the number of new users may be decreasing as young people begin to associate crack use with "burnouts" and "junkies." In 1999 there were an estimated 413,000 crack users (HHS 2000). This chapter's *The Human Side* feature graphically describes conditions in a crack house and associated criminal behaviors through the eyes of sociologist-ethnographer Terry Williams.

Other Drugs

Other drugs abused in the United States include "club drugs" (e.g., LSD, ecstacy), heroin, prescription drugs (e.g., tranquilizers, amphetamines), and inhalants (e.g., glue).

Club Drugs **Club drugs** is a general term used to refer to illicit, often synthetic drugs commonly used at nightclubs or all night dances called raves. Club drugs include ecstacy (MDMA), ketamine ("Special K"), LSD ("acid"), GHB ("liquid ecstacy"), and Rohypnol ("roofies"). Ecstacy, manufactured in and trafficked from Europe, is the most popular of the club drugs ranging in price from $20 to $30 a dose (DEA 2000). Use of ecstacy is small compared with other drug use—less than 1 percent of the population—but is growing in numbers. Ecstacy is associated with feelings of euphoria and inner peace, yet critics argue that as the "new cocaine" both long- (e.g., permanent brain damage) and short-term (e.g., hyperthermia) negative effects are possible (Cloud 2000; DEA 2000).

Ketamine and LSD (lysergic acid diethylamide) both produce visual effects when ingested. Use of ketamine can also cause loss of long-term memory, respiratory problems, and cognitive difficulties. LSD is a synthetic hallucinogen, although many other hallucinogens are produced naturally (e.g., peyote). Hallucinogens, in general, have recently increased in use with 1.2 million new users in 1999. Specifically, use of LSD by 8th, 10th, and 12th graders has increased substantially since the early 1990s (MTF 2000).

GHB (gamma hydroxybutyrate) and Rohypnol (flunitrazepam) are often called **date-rape drugs** because of their use in rendering victims incapable of resisting sexual assaults. GHB, a central nervous system depressant, was banned by the Food and Drug Administration in 1990 although kits containing all the necessary ingredients to manufacture the drug continued to be available on the Internet. On February 18, 2000, President Clinton signed a bill which made GHB a controlled substance and, thus, illegal to manufacture, possess, or sell. Rohypnol, presently illegal in the United States, is lawfully sold in Europe and Latin America. It belongs to a class of drugs known as benzodiazepines, which also includes such common prescription drugs as Valium, Halcion, and Xanax. Rohypnol is tasteless and odorless; 1 mg of the drug can incapacitate a victim for up to 12 hours (NIDA 2000c; DEA 2000).

The only livin' thing that counts is the fix . . . Like I would steal off anybody—anybody, at all, my own mother gladly included.

HEROIN ADDICT

Heroin Although heroin use has decreased in the last couple of years, rates remain higher than early 1990s levels with estimates ranging from 900,000 to 1.2 million current heroin users (International Narcotics Control Strategy Report 2000). Heroin is a highly addictive drug that is increasingly popular among school-aged youth. While crack cocaine has become less fashionable among youthful offenders, heroin, an opium-based narcotic, has increased in acceptability to the point of being glamorized in recent motion pictures and song lyrics (Heroin Drug Conference 1997: NIDA 2000d). In 1999, the rate of heroin use for 12- to 17-year-olds increased. In addition to the negative repercussions of all other drugs, heroin users are subjected to the risks of HIV/AIDS if using intravenously (HHS 2000; NIDA 2000d).

Psychotherapeutic Drugs Use of psychotherapeutic drugs, that is, nonmedical use of any prescription pain reliever (2.6 million users), stimulant (.9 million users), sedative (.2 million users) or tranquilizer (1.1 million users) has increased in recent years. There were 6.4 million users of psychotherapeutic drugs in 1999 with half of all users being under the age of 26. Unlike many drugs, use of psychotherapeutic drugs is equally distributed between males and females (HSS 2000).

Methamphetamine, a stimulant, is one example of a popular psychotherapeutic drug. Although occasionally prescribed for legitimate medical reasons, "meth" is often made in clandestine laboratories in Mexico and the United States. Recent increases in the use of methamphetamine have alarmed international authorities with use in some areas of the world rivaling that of cocaine. It is notable that such increases are not just in industrialized nations but in many developing countries as well. For example, in Thailand methamphetamine has replaced heroin as the most commonly used drug. Methamphetamine is linked to violent behavior and is often used in combination with other drugs (Final Report 2000; International Narcotics Control Strategy Report 2000).

Inhalants Common inhalants include lighter fluid, air fresheners, hair spray, glue, paint, and correction fluid, although over 1,000 other household products are currently abused. Inhalants are the most common illegal substance used by 12-year-olds although by age 14 they are the third most common preceded by psychotherapeutic drugs and marijuana (HHS 2000). Young people often use inhalants believing they are harmless, or that any harm caused requires prolonged use. In fact, inhalants are very dangerous because of their toxicity and may result in what is called Sudden Sniff Death Syndrome.

Societal Consequences of Drug Use and Abuse

Drugs are a social problem, not only because of their adverse effects on individuals, but also as a result of the negative consequences their use has for society as a whole. Everyone is a victim of drug abuse. Drugs contribute to problems within the family and to crime rates, and the economic costs of drug abuse are enormous. Drug abuse also has serious consequences for health at both the individual and societal level.

Family Costs

The cost to families of drug use is incalculable. When one or both parents use or abuse drugs, needed family funds may be diverted to purchasing drugs rather than necessities. Children raised in such homes have a higher probability of neglect, behavioral disorders, and absenteeism from school as well as lower self-concepts and increased risk of drug abuse (Tubman 1993; Easley and Epstein 1991; AP 1999; ONDCP 2000a). Drug abuse is also associated with family disintegration. For example, alcoholics are seven times more likely to separate or divorce than nonalcoholics, and as much as 40 percent of family court problems are alcohol related (Sullivan and Thompson 1994, 347). In a recent Gallup Poll (2000b), 22 percent of Americans reported that drug abuse was or had been a source of trouble in their family.

Abuse between intimates is also linked to drug use. Research indicates that between 25 and 50 percent of men who are involved in domestic violence have substance abuse problems, and drug use contributes to 7 out of 10 cases of child maltreatment (ONDCP 2000a). In a study of 320 men who were married or living with someone, twice as many reported hitting their partner only after they had been drinking compared with those who reported the same behavior while sober (Leonard and Blane 1992). Additionally, Straus and Sweet (1992) found

This poster from the Office of National Drug Control Policy National Youth Anti-Drug Media Campaign emphasizes the importance of a close relationship between parent and child in the fight against youthful drug use.

that alcohol consumption and drug use were associated with higher levels of verbal abuse among spouses.

Crime Costs

> . . . every $1 invested in addiction treatment programs yields a return of between $4 and $7 in reduced drug-related crime . . .
>
> PRINCIPLES OF DRUG ADDICTION TREATMENT
> *NIDA*

The drug behavior of persons arrested, those incarcerated, and persons in drug treatment programs provides evidence of a link between drugs and crime. Drug users commit a disproportionate number of crimes. For example, in 45 to 75 percent of date rape cases either the victim, the offender, or both had used alcohol. Further, 25 percent of all drug offenders in state prisons have been previously sentenced for a violent crime (ONDCP 2000a).

The relationship between crime and drug use, however, is a complex one. Sociologists disagree as to whether drugs actually "cause" crime or whether, instead, criminal activity leads to drug involvement. Further, because both crime and drug use are associated with low socioeconomic status, poverty may actually be the more powerful explanatory variable. After extensive study of the assumed drug–crime link, Gentry (1995) concludes that "the assumption that drugs and crime are causally related weakens when more representative or affluent subjects are considered" (p. 491).

In addition to the hypothesized crime–drug use link, some criminal offenses are drug defined: possession, cultivation, production, and sale of controlled substances; public intoxication; drunk and disorderly conduct, and driving while intoxicated. Driving while intoxicated is one of the most common drug-related crimes.

Nationwide, drunk drivers account for 14 percent of all probationers, 7 percent of local jail inmates, and 2 percent of state prisoners. Of drunk driving offenders, most

(89 percent) were on probation; only 11 percent were sentenced to jail time. The average length of incarceration for those who did serve time was 11 months, and nearly half were sentenced to at least six months in jail. About two-thirds of those incarcerated for drunk driving were repeat offenders (State Legislatures, 2000).

In 1999, 38 percent of all traffic crashes were alcohol-related and 15,786 Americans were killed in drunk driving accidents (MADD 2000; Hunt 2000). A recent federal law requires states to adopt a .08 blood alcohol content limit by 2004 under threat of loss of highway funds (also, see Collective Action).

Economic Costs

The economic costs of drug use are high, over $65 billion, 70 percent of which is attributable to the cost of drug-related crime. Federal, state, and local governments spend an estimated $25 billion on drug enforcement alone—50¢ for every dollar spent on illicit drugs (ONDCP 1998). Also lost are billions of corporate dollars as a result of reduced worker productivity, absenteeism, premature deaths, and insurance and health care costs. On the average, a pack of cigarettes sold in the United States costs Americans $3.80 in smoking-related expenses (ACS 2000a). Concern that on-the-job drug use may impair performance and/or cause fatal accidents has led to drug testing. For many employees, such tests are routine both as a condition for employment and as a requirement for keeping their job. This chapter's *Focus on Technology* feature reviews some of the issues related to privacy rights and drug testing.

Other economic costs of drug abuse include the cost of homelessness, the cost of implementing and maintaining educational and rehabilitation programs, and the cost of health care. Also, the cost of fighting the "war on drugs" is likely to increase as organized crime develops new patterns of involvement in the illicit drug trade. What is the total economic cost of substance abuse in this country? Over $1,000 annually, for every man, woman, and child (Healthy People 2000 Review).

Health Costs

The physical health consequences of drug use for the *individual* are tremendous: shortened life expectancy; higher morbidity (e.g., cirrhosis of the liver, lung cancer); exposure to HIV infection, hepatitis, and other diseases through shared needles; a weakened immune system; birth defects such as fetal alcohol syndrome; drug addiction in children; and higher death rates. Over 400,000 people in the United States die every year from cigarette smoking alone including "21 percent of all coronary heart disease deaths, 87 percent of all lung cancer deaths, and 82 percent of all deaths from chronic obstructive pulmonary disease" (Healthy People 2000 Review). Worldwide, it is estimated that by the year 2020 over 10 million tobacco-related deaths will occur annually (Mayell 1999).

Heavy alcohol and drug use are also associated with negative consequences for an individual's mental health. Longitudinal data on both male and female adults have shown that drug users are more likely to suffer from anxiety disorders (e.g., phobias), depression, and antisocial personalities (White & Labouvie 1994). Marijuana, the drug most commonly used by adolescents, is also linked to short-term memory loss, learning disabilities, motivational deficits, and retarded emotional development.

The Question of Drug Testing

The technology available to detect whether a person has taken drugs was used during the 1970s by crime laboratories, drug treatment centers, and the military. Today, employers in private industry have turned to chemical laboratories for help in making decisions on employment and retention, and parents and school officials use commercial testing devices to detect the presence of drugs. An individual's drug use can be assessed through the analysis of hair, blood, or urine. New technologies include portable breath (or saliva) alcohol testers, THC detection strips, passive alcohol sensors, interlock vehicle ignition systems, and fingerprint screening devices. Counter-technologies have even been developed, for example, shampoos that rid hair of toxins and "Urine Luck," a urine additive that is advertised to speed the breakdown of unwanted chemicals.

In 1986, the President's Commission on Organized Crime recommended that all employees of private companies contracting with the federal government be regularly subjected to urine testing for drugs as a condition of employment. This recommendation was based on the belief that if employees such as air traffic controllers, airline pilots, and railroad operators are using drugs, human lives may be in jeopardy as a result of impaired job performance. In 1987, an Amtrak passenger train crashed outside Baltimore, killing 16 and injuring hundreds. There was evidence of drug use by those responsible for the train's safety. As a result, the Supreme Court ruled in 1989 (by a vote of 7–2), that it is constitutional for the Federal Railroad Administration to administer a drug test to railroad crews if they are involved in an accident. Testing those in "sensitive" jobs for drug use may save lives.

An alternative perspective is that drug testing may be harmful. One concern is the accuracy of the tests and the possible impact of false positives. An innocent person, for example, could lose his or her job. Concern with accuracy of drug tests has led the U.S. Department of Health and Human Services to begin an investigation of all federally certified drug testing laboratories (Brannigan 2000).

A second issue concerns the constitutionality of drug testing. A recent Indiana Court of Appeals decision held that schools could not "require students to submit to random drug tests as a condition of driving to school, playing sports or joining other extracurricular activities" ("Indiana ruling . . ." 2000). At the heart of the debate is the Fourth Amendment which states that "the right of the people to be secure in their persons . . . against unreasonable searches and seizures, shall not be violated." Specifically at issue is the definition of "special needs" — an exception to the fourth amendment. The special needs exception argues that when a circumstance arises (e.g., drug use among student athletes) that requires action

The *societal* costs of drug-related health concerns are also extraordinary—an estimated $12 billion annually (ONDCP 2000b). Health costs include medical services for drug users, the cost of disability insurance, the effects of secondhand smoke, the spread of AIDS, and the medical costs of accident and crime victims, as well as unhealthy infants and children. For example, cocaine use in pregnant women may lead to low birth weight babies, increased risk of spontaneous abortions, and abnormal placental functioning (Klutt 2000).

Treatment Alternatives

Drug treatment reduces drug use by approximately 40 to 60 percent and is as effective as treating many other chronic diseases (e.g., diabetes, asthma) (NIDA 1999, 15). Helping others to overcome chemical dependency is, however, expensive. For example, the average annual cost of methadone treatment, a synthetic opiate used to block the euphoric effect of heroin and thus its motivation for use, is $4,700 per patient (NIDA 1999). Persons who are interested in overcoming chemical dependency have a number of treatment alternatives from which to choose. Some options include family therapy, coun-

(e.g., controlling drug use), an exception to the fourth amendment based on "special needs" (e.g., requiring drug testing as a condition of eligibility) may be made. Thus in 1995, the Supreme Court ruled that random drug testing of student athletes in public schools, where a pattern of drug use had been established and athletes voluntarily submitted to physical examinations, is not unconstitutional. It is the same principle that allows a police officer to conduct a warrantless search of an apartment building in the case of a bomb threat. What the Indiana Court of Appeals held is that in the absence of a federally defined "special needs" exception, and in light of Indiana's requirement of suspicion as a prerequisite for searches, the school's policy of random drug testing is unconstitutional ("Indiana ruling . . ." 2000).

The issue continues to grow in complexity as drug testing spreads to other venues. Michigan, for example, became the first state in the United States to pass a law that requires that new welfare applicants submit to drug testing. Those who refuse to take the test will be denied benefits and those who fail the test will be required to participate in a drug treatment program or have their benefits cut by 25 percent (*Alcoholism and Drug Abuse Weekly* 2000). However, a recent class action suit filed on behalf of welfare recipients has resulted in an injunction preventing implementation of the state policy. Not surprisingly, the suit claims that such a stipulation,

that is, requiring drug tests for welfare eligibility, is a fourth amendment violation.

Of equal significance is a case involving the drug testing of pregnant women, during routine pregnancy examinations, who exhibited certain warning signs of drug use (Kahn 2000). The case, now before the Supreme Court, was brought by 10 women who had tested positive for cocaine and were arrested after giving birth. In *Ferguson v. City of Charleston* the legal issue is whether or not the hospital violated the plaintiffs' fourth amendment rights by disclosing to nonmedical personnel the results of the urine analyses. The hospital argues that it is an effective and non-intrusive means of protecting the "special needs" of the fetus (CNN 2000). The question in a complex and increasingly technologically dependent society is how to balance the rights of an individual with the needs of society as a whole.

Sources:
Alcoholism and Drug Abuse Weekly. 2000. September 18, 6.
Brannigan, Martha. 2000. "Labs that Test Transportation Workers for Drugs Face Inquiry over Samples." *Wall Street Journal,* October 2, A4.
CNN.com. 2000. "Attorney Jennifer Granick Discusses U.S. Supreme Court Case about Drug Testing of Pregnant Women." October 11.
"Indiana Ruling Has Chilling Effect on School." 2000. *Drug Detection Report* 10:133.
Kahn, Jeffery. 2000. "Criminally Pregnant." CNN.com. October 30.
Vernonia School District v Wayne Acton et ux, Guardians Ad Litem for James Acton, 115 S. Ct. 2386, 132 L. ed. 2d 564 (1995).

seling, private and state treatment facilities, community care programs, pharmacotherapy (i.e., use of treatment medications), behavior modification, drug maintenance programs, and employee assistance programs. Two commonly used techniques are inpatient/outpatient treatment and supportive communities.

Inpatient/Outpatient Treatment

Inpatient treatment refers "to the treatment of drug dependence in a hospital and includes medical supervision of detoxification" (McCaffrey 1998, 2). Most inpatient programs last between 30 and 90 days and target individuals whose withdrawal symptoms require close monitoring (e.g., alcoholics, cocaine addicts). Some drug-dependent patients, however, can be safely treated as outpatients. Outpatient treatment allows the individual to remain in their home and work environments and is often less expensive. In outpatient treatment the patient is under the care of a physician who evaluates the patient's progress regularly, prescribes needed medication, and watches for signs of a relapse.

The longer a patient stays in treatment, the greater the likelihood of a successful recovery. Variables that predict success include the user's motivation to

People who smoke are more likely to be heavy drinkers and current illicit drug users. Some evidence suggests that giving up smoking leads to a reduction in alcohol consumption.

© AP/Wide World Photo

change, support of family and friends, criminal justice or employer intervention, a positive relationship with therapeutic staff, and a program of recovery that addresses many of the needs of the patient.

Peer Support Groups

AA members essentially trade addiction to the bottle to a network of friends who share a common bond.

BARRY LUBETKIN
Psychologist

12-Step Programs Both Alcoholics Anonymous (AA) and Narcotics Anonymous (NA) are voluntary associations whose only membership requirement is the desire to stop drinking or taking drugs. AA and NA are self-help groups in that they are operated by nonprofessionals, offer "sponsors" to each new member, and proceed along a continuum of 12 steps to recovery. Members are immediately immersed in a fellowship of caring individuals with whom they meet daily or weekly to affirm their commitment. Some have argued that AA and NA members trade their addiction to drugs for feelings of interpersonal connectedness by bonding with other group members.

Symbolic interactionists emphasize that AA and NA provide social contexts in which people develop new meanings. Abusers are surrounded by others who convey positive labels, encouragement, and social support for sobriety. Sponsors tell the new members that they can be successful in controlling alcohol and/or drugs "one day at a time" and provide regular interpersonal reinforcement for doing so. Although thought of as a "crutch" by some, AA members may also take medications to help prevent relapses. In a study of 222 AA members, Rychtarik and colleagues (2000) found that although over half of those surveyed thought the use of relapse-preventing medication was or might be a good idea, 29 percent reported pressures from others to stop taking the medication.

Therapeutic Communities In **therapeutic communities**, which house between 35 and 500 people for up to 15 months, participants abstain from drugs, develop marketable skills, and receive counseling. Synanon, which was established in 1958, was the first therapeutic community for alcoholics and was later expanded to include other drug users. More than 400 residential treatment centers are now in existence, including Daytop Village and Phoenix House. The longer a person stays at such a facility, the greater the chance of overcoming his or her dependency. Symbolic interactionists argue that behavioral changes appear to be a consequence of revised self-definition and the positive expectations of others.

The goal of a drug-free America is an unrealistic one; it is realistic, however, to strive to reduce use to the pre-1960 era when drug use was a small problem in America.

HERBERT D. KLEBER, M.D.
Former deputy director, White House Office of National Drug Control Policy

Strategies for Action: America Responds

Drug use is a complex social issue exacerbated by the structural and cultural forces of society that contribute to its existence. While the structure of society perpetuates a system of inequality creating in some the need to escape, the culture of society, through the media and normative contradictions, sends mixed messages about the acceptability of drug use. Thus, developing programs, laws, or initiatives that are likely to end drug use may be unrealistic. Nevertheless, numerous social policies have been implemented or proposed to help control drug use and its negative consequences (see Table 3.2).

Table 3.2. *Government Initiatives in the Fight Against Drugs*

Drug Courts—divert drug offenders out of the criminal justice system and into treatment facilities; those who complete treatment may have their sentences reduced.

National Youth Anti-Drug Campaign—5-year bipartisan program to combat teenage drug use through an aggressive media campaign.

Special Operations Division (SOD)—multi-agency effort, including the Drug Enforcement Administration (DEA) and the Federal Bureau of Investigation (FBI), to investigate major drug trafficking organizations.

After School Initiatives—funding for after-school programs (e.g., mentoring programs) designed to deal with the antecedents of drug abuse including lack of attachment to family and aggressive behavior.

Law Enforcement Partnerships—law enforcement and drug prevention program partnerships including the Bureau of Justice Statistics and D.A.R.E. (Drug Abuse Resistance Education), Bureau of Alcohol, Tobacco and Firearms and G.R.E.A.T. (Gang Reduction Education and Training), and the Department of Justice and D.E.F.Y. (Drug Education For Youth).

High Intensity Drug Trafficking Area—31 problem areas identified by federal officials, including the U.S. Attorney General, in the hopes of fostering interagency (local, state, and federal) law enforcement cooperation.

Drug-Free Prison Zone Demonstration Project—a project jointly sponsored by the Department of Justice, the Office of National Drug Control Policy, and the Bureau of Prisons, this initiative is designed to reduce the availability of drugs in prison through the use of detection devices and drug testing.

Domestic Cannabis Eradication and Suppression Program—coordinated by the DEA, this initiative provides support to local and state law enforcement agencies for the detection and eradication of domestically grown marijuana.

Plan Colombia and the Andean Region—U.S. military aid and financial support of *Plan Colombia*, a strategy to fight corruption, the drug trade, and coca cultivation in that country.

Source: Office of National Drug Control Policy. 2000c. "Report on Programs and Initiatives." Chapter III. *Annual Report and National Drug Strategy*. http://whitehousedrugpolicy.gov

Government Regulations

The largest social policy attempt to control drug use in the United States was Prohibition. Although this effort was a failure by most indicators, the government continues to develop programs and initiatives designed to combat drug use. In the 1980s the federal government declared a "war on drugs" based on the belief that controlling drug availability would limit drug use and, in turn, drug-related problems. In contrast to a **harm reduction** position which focuses on minimizing the costs of drug use for both user and society (e.g., distributing clean syringes to decrease the risk of HIV infection), this "zero-tolerance" approach advocates get-tough law enforcement policies. However, Yale law professor Steven Duke and co-author Albert C. Gross, in their book *America's Longest War* (1994), argue that the war on drugs, much like Prohibition, has only intensified other social problems: drug-related gang violence and turf wars, the creation of syndicate-controlled black markets, unemployment, the spread of AIDS, overcrowded prisons, corrupt law enforcement officials, and the diversion of police from other serious crimes.

> The nation working together has made substantial progress in confronting illegal drug abuse and trafficking.
>
> BARRY MCCAFFREY
> *Director, ONDCP*

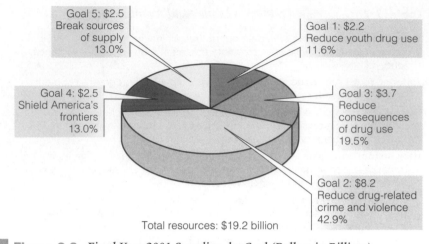

Figure 3.2. *Fiscal Year 2001 Spending, by Goal (Dollars in Billions).*
Source: ONDCP 2000. "The National Drug Control Budget." Chapter III. *The National Drug Control Strategy 2000 Annual Report.* http://www.whitehousedrugpolicy.gov.

Consistent with conflict theory, still others argue that the war on drugs unfairly targets minorities (see Chapter 4). U.S. Department of Justice data analyzed by Human Rights Watch, an international human rights organization, indicates that, in general, black men are 13 times more likely to be incarcerated in state prisons for drug charges than white men (Fletcher 2000, A10).

Despite concerns, the war on drugs continues at an astronomical societal cost. The U.S. drug control strategy focuses on five general areas as identified in Figure 3.2. Two of these areas, shielding U.S. frontiers from the drug threat and breaking sources of supply, point to the international quality of the war on drugs. U.S. drug policies have implications that extend beyond domestic concerns and affect international relations and the economies of foreign countries.

Many of the countries in which drug trafficking occurs are characterized by government corruption and crime, military coups, and political instability. Such is the case with Colombia, supplier of over 80 percent of cocaine in the United States. Congress recently approved a multi-billion dollar aid package entitled *Plan Colombia* to not only fight drug traffickers but to "help foster democracy in Colombia over the long run by stabilizing civil society and putting it [back] in the hands of the government" (Ebinger 2000, 2). Rather than foreign aid and military assistance, many argue that trade sanctions should be imposed in addition to crop eradication programs and interdiction efforts. Others, however, noting the relative failure of such programs in reducing the supply of illegal drugs entering the United States, argue that the "war on drugs" should be abandoned and that deregulation is preferable to the side effects of regulation.

Deregulation or Legalization: The Debate

Although Americans believe that drugs are a serious problem in the United States, little consensus exists about what progress is being made in the "war on drugs." In a recent Gallup Poll, 47 percent of the respondents reported that "much" or "some" progress "has been made over the last year or two in coping with the problem of illegal drugs." The remainder, however, reported that "some" or "much" ground has been lost, or that we have "stood still" in the fight against illegal drugs (Gallup Poll 2000b). Given such ambivalence, it is not surprising that some advocate alternatives to the punitive emphasis of the last several decades.

Deregulation is the reduction of government control over certain drugs. For example, although individuals must be 21 years old to purchase alcohol and 18 to purchase cigarettes, both substances are legal and purchased freely. Further, in some states, possession of marijuana in small amounts is now a misdemeanor rather than a felony, and in other states marijuana is lawfully used for medical purposes. In 1996 both Arizona and California passed acts known as "marijuana medical bills," that made the use and cultivation of marijuana, under a physician's orders, legal. Medical use of marijuana has also been approved in Alaska, Hawaii, Maine, Nevada, Oregon, and Washington (*Alcoholism and Drug Abuse Weekly* 2000).

Proponents for the **legalization** of drugs affirm the right of adults to make an informed choice. They also argue that the tremendous revenues realized from drug taxes could be used to benefit all citizens, that purity and safety controls could be implemented, and that legalization would expand the number of distributors, thereby increasing competition and reducing prices. Drugs would thus be safer, drug-related crimes would be reduced, and production and distribution of previously controlled substances would be taken out of the hands of the underworld.

Those in favor of legalization also suggest that the greater availability of drugs would not increase demand, pointing to countries where some drugs have already been decriminalized. **Decriminalization**, or the removing of penalties for certain drugs, would promote a medical rather than criminal approach to drug use that would encourage users to seek treatment and adopt preventive practices. For example, making it a criminal offense to sell or possess hypodermic needles without a prescription encourages the use of non-sterile needles that spread infections such as HIV and hepatitis.

Opponents of legalization argue that it would be construed as government approval of drug use and, as a consequence, drug experimentation and abuse would increase. Further, while the legalization of drugs would result in substantial revenues for the government, because all drugs would not be decriminalized (e.g., crack), drug trafficking and black markets would still flourish. Legalization would also require an extensive and costly bureaucracy to regulate the manufacture, sale, and distribution of drugs. Finally, the position that drug use is an individual's right cannot guarantee that others will not be harmed. It is illogical to assume that a greater availability of drugs will translate into a safer society.

Collective Action

Social action groups such as Mothers Against Drunk Driving **(MADD)** have successfully lobbied legislators to raise the drinking age to 21 and to provide harsher penalties for driving while impaired. MADD, with 3.5 million members and 600 chapters, has also put pressure on alcohol establishments to stop "two for one" offers and has pushed for laws that hold the bartender personally liable if a served person is later involved in an alcohol-related accident. Even hosts in private homes can now be held liable if they allow a guest to drive who became impaired while drinking at their house. Recently MADD successfully lobbied for the passage of a new drunk driving standard that established "a uniform blood-alcohol limit of .08 percent—the equivalent of an average size male drinking four beers in one hour on an empty stomach" (Worden 2000, 1).

If some drunk gets out and kills your kid, you'd probably be a little crazy about it, too.

KATHY PRESCOTT
Former MADD President

> You can't litigate a case against an organized and extremely well-financed defendant without minimal levels of funding.
>
> JANET RENO
> *U.S. Attorney General*

Collective action is also being taken against tobacco companies by smokers, ex-smokers, and the families of smoking victims. They charge that tobacco executives knew over 30 years ago that tobacco was addictive and concealed this fact from both the public and the government. Furthermore, they charge that tobacco companies manipulate nicotine levels in cigarettes with the intention of causing addiction. Recently, in a class-action suit of over 300,000 Florida smokers, a jury ordered the top five cigarette producers to pay $145 billion—the largest settlement to date—to the plaintiffs (Wilson 2000). Additionally, the federal government has brought action against tobacco companies for relief of the over $20 billion spent annually on behalf of Americans with smoking-related illnesses. In response to legal assaults on the tobacco industry, cigarette exports and foreign production have increased dramatically in recent years (ACS 2000b)

Finally, several recent initiatives have resulted in state-wide referendums concerning the cost-effectiveness of government policies. For example, as a result of the passage of Proposition 36, California, as well as many other states, will now require that nonviolent first- and second-time minor drug offenders receive treatment including job training, therapy, literacy education and family counseling, rather than jail time. The initiative provides over $100 million a year over the next 5 1/2 years for community-based treatment facilities. The program is predicted to divert over 30,000 state and county prisoners to treatment programs resulting in a net savings to California taxpayers of $1.5 billion over the course of the program. Evaluation of a similar strategy in Arizona boasts a treatment success rate of 71 percent (Mann 2000a; Thompson 2000).

Understanding *Alcohol and Other Drugs*

In summarizing what we know about substance abuse, drugs and their use are socially defined. As the structure of society changes, the acceptability of one drug or another changes as well. As conflict theorists assert, the status of a drug as legal or illegal is intricately linked to those who have the power to define acceptable and unacceptable drug use. There is also little doubt that rapid social change, anomie, alienation, and inequality further drug use and abuse. Symbolic interactionism also plays a significant role in the process—if people are labeled as "drug users" and expected to behave accordingly, drug use is likely to continue. If there is positive reinforcement of such behaviors and/or a biological predisposition to use drugs, the probability of drug involvement is even higher. Thus, the theories of drug use complement rather than contradict one another.

Drug use must also be conceptualized within the social context in which it occurs. In a study of high-risk youths who had become drug involved, Dembo and coworkers (1994) suggest that many youths in their study had been "failed by society":

> Many of them were born into economically-strained circumstances, often raised by families who neglected or abused them, or in other ways did not provide for their nurturance and wholesome development . . . few youths in our sample received the mental health and substance abuse treatment services they needed. (p. 25)

However, many treatment alternatives, emanating from a clinical model of drug use, assume that the origin of the problem lies within the individual rather than in the structure and culture of society. Although admittedly the problem may lie within the individual at the time treatment occurs, policies that address the social causes of drug abuse provide a better means of dealing with the drug problem in the United States.

Prevention is preferable to intervention, and given the social portrait of hard drug users—young, male, minority—prevention must entail dealing with the social conditions that foster drug use. Some data suggest that inner city adolescents are particularly vulnerable to drug involvement because of their lack of legitimate alternatives (Van Kammen & Loeber 1994).

> Illegal drug use may be a way to escape the strains of the severe urban conditions and dealing illegal drugs may be one of the few, if not the only, way to provide for material needs. Intervention and treatment programs, therefore, should include efforts to find alternate ways to deal with the limiting circumstances of inner-city life, as well as create opportunities for youngsters to find more conventional ways of earning a living. (p. 22)

Social policies dealing with drug use have been predominantly punitive rather than preventive. Recently, however, there appears to be some movement toward educating the public on the dangers of alcohol and drug use. For example, a new national campaign, "Keep your Brain Healthy, Don't Use Drugs" will use public service announcements (PSA) to inform consumers of the dangers of drug use, and the physiological impact of drugs on the structure of the brain. The National Institute on Drug Use has already begun producing PSAs in English and Spanish to be aired on television and the radio (NIDA 2000e).

In the United States and throughout the world, millions of people depend on legal drugs for the treatment of a variety of conditions, including pain, anxiety and nervousness, insomnia, overeating, and fatigue. Although drugs for these purposes are relatively harmless, the cultural message "better living through chemistry" contributes to alcohol and drug use and its consequences. But these and other drugs are embedded in a political and economic context that determines who defines what drugs, in what amounts, as licit or illicit and what programs are developed in reference to them.

> A child who reaches 21 without smoking, abusing alcohol or using drugs is virtually certain never to do so.
>
> JOESPH A. CALIFANO, JR.
> *President, Center on Addiction and Substance Abuse*

Critical Thinking

1 Are alcoholism and other drug addictions a consequence of nature or nurture? If nurture, what environmental factors contribute to such problems? Which of the three sociological theories best explains drug addiction?

2 Measuring alcohol and drug use is often very difficult. This is particularly true given the tendency for respondents to acquiesce, that is, respond in a way they believe is socially desirable. Consider this and other problems in doing research on alcohol and other drugs, and how such problems would be remedied.

3 If, as symbolic interactionists argue, social problems are those conditions so defined, how might the manipulation of social definitions virtually eliminate many "drug" problems?

4 Growing hemp, the plant from which marijuana is cultivated, has many non-drug uses and is illegal in the United States although legal in Canada. A recent survey of farmers in 5 midwestern and western states indicates that the majority favor the legalization of hemp (Mann 2000b). Argue for or against this position.

Key Terms

anomie	decriminalization	gateway drug
chemical dependency	deregulation	harm reduction
club drugs	drug	legalization
crack	drug abuse	MADD
date-rape drugs	drug addiction	therapeutic communities

Media Resources

 ### The Wadsworth Sociology Resource Center: Virtual Society

http://sociology.wadsworth.com

See the companion Web site for this book to access general sociology resources and text-specific features that can further your understanding of this chapter. The site contains Internet links, Internet exercises, online practice quizzes, information on InfoTrac College Edition, and many more valuable materials designed to enrich your learning experience in social problems.

InfoTrac College Edition

You can access InfoTrac College Edition either from the Wadsworth Sociology Resource Center at **http://sociology.wadsworth.com** or directly from your web browser at **http://www.infotrac-college.com/wadsworth/**. InfoTrac College Edition is an online university library that includes over 700 popular and scholarly journals in which you can find articles related to the topics in this chapter such as information on the effects of drugs, drug testing in the workplace, and arguments for and against the legalization of marijuana.

 ### Interactions CD-ROM

Go to the "Interactions" CD-ROM for *Understanding Social Problems,* Third Edition to access additional interactive learning tools, such as in-depth review materials, corresponding practice quizzes, and other engaging resources and activities to help you study the concepts in this chapter.

Crime and Violence

Is it True?

1. U.S. crime rates have decreased dramatically in recent years.

2. Most persons arrested for violent crime are between the ages of 35 and 45.

3. Although sociologists have different theories about the causes of crime, they agree that crime is always harmful to society.

4. The majority of female homicide victims had some type of relationship with their murderer.

5. The average cost of a vehicle at the time it was stolen is over $10,000.

Answers to "Is It True?": 1 = T; 2 = F; 3 = F; 4 = T; 5 = F

Amadou Diallo was a Western African immigrant who lived in the Bronx, a borough of New York City. Late one February evening, he was standing in the hallway of his apartment building when he was approached by several members of the NYPD Street Crime Unit who were investigating a rape case. Fitting the description of the rape suspect, Diallo was told to "freeze," but with limited understanding of English he continued to reach for his wallet apparently believing that the officers were asking for his identification. One police officer, thinking he saw a weapon, yelled "gun" and in seconds Diallo was dead. The four police officers fired a total of forty-one shots—19 hitting the 22-year-old. Diallo was unarmed and innocent of the suspected crime (PBS 2000; Human Rights Watch 2000).

One year later, in February of 2000, all four officers were acquitted of second-degree murder after jurors found that the officers were in reasonable fear for their safety. The decision spurred countless protests and, in the wake of accusations of racial profiling, contributed to a growing distrust of the criminal justice system.

Clearly what happened in the Diallo case is not representative of the millions of police–citizen interactions that take place every year. Nonetheless, in a recent Gallup poll when respondents where asked "How much confidence do you have in the criminal justice system?" 33 percent said "very little" or "none," 42 percent said "some," and 24 percent said "a great deal" or "quite a lot" (BJS 2000a). This chapter examines the criminal justice system as well as theories, types, and demographic patterns of criminal behavior. The economic, social, and psychological costs of crime and violence are also examined. The chapter concludes with a discussion of social policies and prevention programs designed to reduce crime and violence in America.

. . . crime is no longer bound by the constraints of borders. Such offenses as terrorism, nuclear smuggling, organized crime, computer crime and drug trafficking can spill over from other countries into the United States. Regardless of origin, these and other overseas crimes impact directly on our citizens and our economy.

LOUIS FREEH
Director, FBI

The Global Context: International Crime and Violence

According to the *United Nations Global Report on Crime and Justice*, several facts about crime are true throughout the world. First, crime is ubiquitous; that is, there is no country where crime does not exist. Second, most countries have the same components in their criminal justice systems—police, courts, and prisons. Third, worldwide, adult males comprise the largest category of crime suspects and, fourth, there is no general trend toward increased use of incarceration. Finally, in all countries, theft is the most common crime committed with violent crime being a relatively rare event—about 10 to 15 percent of all reported crime (Global Report 1999).

Dramatic differences do exist, however, in international crime and violence rates. In general, industrialized countries have higher rates of reported crime than nonindustrialized countries, with Arab states having among the lowest crime rates worldwide. There are also variations within industrialized countries. The U.S. homicide rate is four times higher than western Europe's, six times higher than Great Britain's, and seven times higher than Japan's (Doyle 2000; Siegel 2000). It is important to note, however, that the incidences of violent but non-deadly crimes (e.g., robbery, burglary) in the United States are not higher than in other industrialized nations. Zimring and Hawkins (1997) have thus concluded that "crime is not the problem" in the United States—the problem is lethal violence.

Recent concerns have focused on **transnational crime** defined by the United Nations as "offenses whose inception, prevention, and/or direct or indirect effects involve more than one country"(Finckenauer 2000, 3). For example, Russian rubles, arms, and precious metals are smuggled out of that country daily; Chinese Triads operate in large cities worldwide netting billions of dollars a year from prostitution, drugs, and other organized crime activities; children are trafficked through Canada and Mexico for use in pornography rings; and Colombian cocaine cartels flourish and spread to sub-Saharan countries with needy economies (United Nations 1997; INTERPOL 1998; Finckenauer 2000). Transnational crime is facilitated by recent trends in globalization including enhanced transportation and communication technologies. For example, the U.S. Customs Service estimates that, worldwide, there are over 100,000 Web sites involved in child pornography (ABCNews 2001).

Sources of Crime Statistics

The U.S. government spends millions of dollars annually compiling and analyzing crime statistics. A **crime** is a violation of a federal, state, or local criminal law. For a violation to be a crime, however, the offender must have acted voluntarily and with intent and have no legally acceptable excuse (e.g., insanity) or justification (e.g., self-defense) for the behavior. The three major types of statistics used to measure crime are official statistics, victimization surveys, and self-report offender surveys.

Official Statistics

Local sheriffs' departments and police departments throughout the United States collect information on the number of reported crimes and arrests and voluntarily report them to the Federal Bureau of Investigation (FBI). The FBI then compiles these statistics annually and publishes them, in summary form, in the Uniform Crime Reports (UCR). The UCR lists **crime rates** or the number of crimes committed per 100,000 population, the actual number of crimes and the percentage of change over time, as well as clearance rates. **Clearance rates** measure the percentage of cases in which an arrest and official charge have been made and the case turned over to the courts.

These statistics have several shortcomings (DiIulio 1999). Not only do many incidents of crime go unreported, not all crimes reported to the police are recorded. Alternatively, some rates may be exaggerated. Motivation for such distortions may come from the public (e.g., demanding that something be done) or from political e.g., election of a sheriff) and/or organizational pressures (e.g., budget requests). For example, a police department may "crack down" on drug-

> Ultimately, any crime statistic is only as useful as the reader's understanding of the processes that generated it.
>
> ROBERT M. O'BRIEN
> *Sociologist, University of Oregon*

related crimes in an election year. The result is an increase in the recorded number of these offenses. Such an increase reflects a change in the behavior of law enforcement personnel, not a change in the number of drug violations. Thus, official crime statistics may be a better indicator of what police are doing than what criminals are doing.

Victimization Surveys

Victimization surveys ask people if they have been victims of crime. The Department of Justice's National Crime Victimization Survey (NCVS), conducted annually, interviews nearly 100,000 people about their experiences as victims of crime. Interviewers collect a variety of information, including the victim's background (e.g., age, race and ethnicity, sex, marital status, education, and area of residence), relationship to offender (stranger or non-stranger), and the extent to which the victim was harmed. Although victimization surveys provide detailed information about crime victims, they provide less reliable data on offenders.

Self-Report Offender Surveys

Self-report surveys ask offenders about their criminal behavior. The sample may consist of a population with known police records, such as a prison population, or it may include respondents from the general population, such as college students. Self-report data compensate for many of the problems associated with official statistics but are still subject to exaggerations and concealment. The Criminal Activities Survey in this chapter's *Self and Society* feature asks you to indicate whether you have engaged in a variety of illegal activities.

   Self-report surveys reveal that virtually every adult has engaged in some type of criminal activity. Why then is only a fraction of the population labeled as criminal? Like a funnel, which is large at one end and small at the other, only a small proportion of the total population of law violators is ever convicted of a crime. For an individual to be officially labeled as a criminal, his or her behavior (1) must become known to have occurred, (2) must come to the attention of the police who then file a report, conduct an investigation, and make an arrest and, finally, (3) the arrestee must go through a preliminary hearing, an arraignment, and a trial and may or may not be convicted. At every stage of the process, an offender may be "funneled" out. As Figure 4.1 indicates, the measures of crime used at various points in time lead to different results.

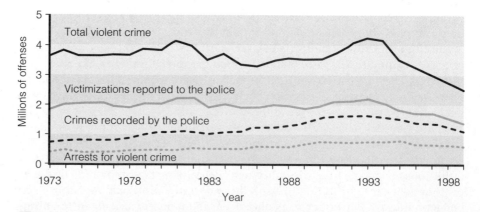

■ **Figure 4.1** *Four Measures of Serious Violent Crime.*

Criminal Activities Survey

Read each of the following questions. If, since the age of 16, you have ever engaged in the behavior described, place a "1" in the space provided. If you have not engaged in the behavior, put a "0" in the space provided. After completing the survey, read the section on interpretation to see what your answers mean.

Questions	1 (Yes)	0 (No)
1. Have you ever been in possession of drug paraphernalia?	_____	_____
2. Have you ever lied about your age, or about anything else when making application to rent an automobile?	_____	_____
3. Have you ever intentionally destroyed or erased someone else's phone messages?	_____	_____
4. Have you ever tampered with a coin-operated vending machine or parking meter?	_____	_____
5. Have you ever loaned or given away lewd or obscene materials?	_____	_____
6. Have you ever begun and/or participated in an office basketball or football pool?	_____	_____
7. Have you ever used "filthy, obscene, annoying, or offensive" language while on the telephone?	_____	_____
8. Have you ever given or sold a beer to someone under the age of 21?	_____	_____
9. Have you ever been on someone else's property (land, house, boat, structure, etc.) without their permission?	_____	_____
10. Have you ever forwarded a chain letter with the intent to profit from it?	_____	_____
11. Have you ever improperly gained access to someone else's e-mail or other computer account?	_____	_____
12. Have you ever written a check when you knew it was bad?	_____	_____

INTERPRETATION

Each of the activities described in these questions represents criminal behavior that was subject to fines, imprisonment, or both under the laws of Florida in 2000. For each activity, the following table lists the maximum prison sentence and/or fine for a first-time offender. To calculate your "prison time" and/or fines, sum the numbers corresponding to each activity you have engaged in.

Maximum Prison Sentence	Maximum Fine	Offense
1. One year	$1000	Possession of drug paraphernalia
2. Five years	$5000	Fraud
3. Two months	$500	Unlawful interference with telecommunications
4. Two months	$500	Fraud
5. One year	$1000	Unlawful dissemination of obscene material
6. Two months	$500	Illegal gambling
7. Two months	$500	Harassing/obscene telecommunications
8. Two months	$500	Illegal distribution of alcohol
9. One year	$1000	Trespassing
10. One year	$1000	Illegal gambling
11. Five years	$5000	Illegal misappropriation of cybercommunication
12. One year	$1000	Worthless check

Source: Florida Criminal Code. 2000. http://www.leg.state.fl./statutes/index

Sociological Theories of Crime and Violence

Some explanations of crime and violence focus on psychological aspects of the offender such as psychopathic personalities, unhealthy relationships with parents, and mental illness. Other crime theories focus on the role of biological variables such as central nervous system malfunctioning, stress hormones, vitamin or mineral deficiencies, chromosomal abnormalities, and a genetic predisposition toward aggression. Sociological theories of crime and violence emphasize the role of social factors in criminal behavior and societal responses to it.

Structural-Functionalist Perspective

> Poverty is the mother of crime.
>
> **MAGNUS AURELIUS CASSIODORUS**
> *Roman historian*

According to Durkheim and other structural-functionalists, crime is functional for society. One of the functions of crime and other deviant behavior is that it strengthens group cohesion:

> The deviant individual violates rules of conduct that the rest of the community holds in high respect; and when these people come together to express their outrage over the offense . . . they develop a tighter bond of solidarity than existed earlier. (Erikson 1966, 4)

Crime may also lead to social change. For example, an episode of local violence may "achieve broad improvements in city services . . . [and] be a catalyst for making public agencies more effective and responsive, for strengthening families and social institutions, and for creating public-private partnerships" (National Research Council 1994, 9–10).

> Every time you stop a school, you will have to build a jail. What you gain at one end you lose at the other. It's like feeding a dog on its own tail. It won't fatten the dog.
>
> **MARK TWAIN**
> *American author*

Although functionalism as a theoretical perspective deals directly with some aspects of crime and violence, it is not a theory of crime *per se*. Three major theories of crime and violence have developed from functionalism, however. The first, called **strain theory**, was developed by Robert Merton (1957) using Durkheim's concept of anomie, or normlessness. Merton argues that when legitimate means (e.g., a job) of acquiring culturally defined goals (e.g., money) are limited by the structure of society, the resulting strain may lead to crime.

Individuals, then, must adapt to the inconsistency between means and goals in a society that socializes everyone into wanting the same thing but only provides opportunities for some (see Table 4.1). Conformity occurs when individuals accept the culturally defined goals and the socially legitimate means of

Table 4.1 *Merton's Strain Theory*

Mode of Adaption	Seeks Culturally Defined Goals?	Uses Structurally Defined Means to Achieve Them?
1. Conformity	Yes	Yes
2. Innovation	Yes	No
3. Ritualism	No	Yes
4. Retreatism	No	No
5. Rebellion	No—seeks to replace	No—seeks to replace

Source: Adapted from Robert K. Merton's *Social Theory and Social Structure* (1957).

achieving them. Merton suggests that most individuals, even those who do not have easy access to the means and goals, remain conformists. Innovation occurs when an individual accepts the goals of society, but rejects or lacks the socially legitimate means of achieving them. Innovation, the mode of adaptation most associated with criminal behavior, explains the high rate of crime committed by uneducated and poor individuals who do not have access to legitimate means of achieving the social goals of wealth and power.

Another adaptation is ritualism, in which the individual accepts a lifestyle of hard work, but rejects the cultural goal of monetary rewards. The ritualist goes through the motions of getting an education and working hard, yet is not committed to the goal of accumulating wealth or power. Retreatism involves rejecting both the cultural goal of success and the socially legitimate means of achieving it. The retreatist withdraws or retreats from society and may become an alcoholic, drug addict, or vagrant. Finally, rebellion occurs when an individual rejects both culturally defined goals and means and substitutes new goals and means. For example, rebels may use social or political activism to replace the goal of personal wealth with the goal of social justice and equality.

Whereas strain theory explains criminal behavior as a result of blocked opportunities, **subcultural theories** argue that certain groups or subcultures in society have values and attitudes that are conducive to crime and violence. Members of these groups and subcultures, as well as other individuals who interact with them, may adopt the crime-promoting attitudes and values of the group. For example, subcultural norms and values contribute to street crime. Sociologist Elijah Anderson (1994) explains that many inner-city African-American youths live by a survival code on the streets that emphasizes gaining the respect of others through violence—the tougher you are and the more others fear you, the more respect you have in the community.

But, if blocked opportunities and subcultural values are responsible for crime, why don't all members of the affected groups become criminals? **Control theory** may answer that question. Hirschi (1969), consistent with Durkheim's emphasis on social solidarity, suggests that a strong social bond between individuals and the social order constrains some individuals from violating social norms. Hirschi identified four elements of the social bond: attachment to significant others, commitment to conventional goals, involvement in conventional activities, and belief in the moral standards of society. Several empirical tests of Hirschi's theory support the notion that the higher the attachment, commitment, involvement, and belief, the higher the social bond and the lower the probability of criminal behavior. For example, Laub, Nagan, and Sampson (1998) found that a good marriage contributes to the cessation of a criminal career. Further, Warner and Rountree (1997) report that local community ties, although varying by neighborhood and offense, decrease the probability of crimes occurring.

Conflict Perspective

Conflict theories of crime suggest that deviance is inevitable whenever two groups have differing degrees of power; in addition, the more inequality in a society, the greater the crime rate in that society. Social inequality leads individuals to commit crimes such as larceny and burglary as a means of economic survival. Other individuals, who are angry and frustrated by their low position in the socioeconomic hierarchy, express their rage and frustration through crimes such as drug use, assault, and homicide. In Argentina, for example, the souring violent crime

There are two criminal justice systems in this country. There is a whole different system for poor people. It's the same courthouse—it's not separate—but it's not equal.

PAUL PETTERSON
Public defender

To Marxists, the cultural definition of women as property contributes to the high rates of female criminality and, specifically, involvement in prostitution, drug abuse, and petty theft.

rate is hypothesized to be "a product of the enormous imbalance in income distribution . . . between the rich and the poor" (Pertossi 2000).

According to the conflict perspective, those in power define what is criminal and what is not, and these definitions reflect the interests of the ruling class. Laws against vagrancy, for example, penalize individuals who do not contribute to the capitalist system of work and consumerism. Rather than viewing law as a mechanism that protects all members of society, conflict theorists focus on how laws are created by those in power to protect the ruling class. For example, wealthy corporations contribute money to campaigns to influence politicians to enact tax laws that serve corporate interests (Jacobs 1988), and the "criminal justice system grows increasingly punitive as labor surplus increases," that is, as greater social control is needed (Hochstetler & Shover 1997).

Furthermore, conflict theorists argue that law enforcement is applied differentially, penalizing those without power and benefiting those with power. For example, female prostitutes are more likely to be arrested than are the men who seek their services. Unlike street criminals, corporate criminals are often punished by fines rather than by lengthy prison terms, and rape laws originated to serve the interests of husbands and fathers who wanted to protect their property—wives and unmarried daughters.

Societal beliefs also reflect power differentials. For example, "rape myths" are perpetuated by the male-dominated culture to foster the belief that women are to blame for their own victimization, thereby, in the minds of many, exonerating the offender. Such myths include the notion that when a woman says "no" she means "yes," that "good girls" don't get raped, that appearance indicates willingness, and that women secretly want to be raped. Not surprisingly, in societies where women and men have greater equality, there is less rape (Sanday 1981).

Symbolic Interactionist Perspective

Two important theories of crime and violence emanate from the symbolic interactionist perspective. The first, **labeling theory**, focuses on two questions: How do crime and deviance come to be defined as such, and what are the effects of being labeled as criminal or deviant? According to Howard Becker (1963):

> Social groups create deviance by making rules whose infractions constitute deviance, and by applying those rules to particular people and labeling them as outsiders. From this point of view, deviance is not a quality of the act a person commits, but rather a consequence of the application by others of rules and sanctions to an "offender." The deviant is one to whom the label has successfully been applied; deviant behavior is behavior that people so label. (p. 238)

© Stephen Shames/Matrix

Labeling theorists make a distinction between primary deviance, which is deviant behavior committed before a person is caught and labeled as an offender, and secondary deviance, which is deviance that results from being caught and labeled. After a person violates the law and is apprehended, that person is stigmatized as a criminal. This deviant label often dominates the social identity of the person to whom it is applied and becomes the person's "master status," that is, the primary basis on which the person is defined by others.

Being labeled as deviant often leads to further deviant behavior because (1) the person who is labeled as deviant is often denied opportunities for engaging in nondeviant behavior, and (2) the labeled person internalizes the deviant label, adopts a deviant self-concept, and acts accordingly. For example, the teenager who is caught selling drugs at school may be expelled and thus denied opportunities to participate in nondeviant school activities (e.g., sports, clubs) and associate with nondeviant peer groups. The labeled and stigmatized teenager may also adopt the self-concept of a "druggie" or "pusher" and continue to pursue drug-related activities and membership in the drug culture.

The assignment of meaning and definitions learned from others is also central to the second symbolic interactionist theory of crime, **differential association**. Edwin Sutherland (1939) proposed that, through interaction with others, individuals learn the values and attitudes associated with crime as well as the techniques and motivations for criminal behavior. Individuals who are exposed to more definitions favorable to law violation (e.g., "crime pays") than unfavorable (e.g., "do the crime, you'll do the time") are more likely to engage in criminal behavior. Thus children who see their parents benefit from crime, or who live in high crime neighborhoods where success is associated with illegal behavior, are more likely to engage in criminal behavior.

Types of Crime

The FBI identifies eight **index offenses** as the most serious crimes in the United States. The index offenses, or street crimes as they are often called, may be against a person (called violent or personal crimes) or against property (see Table 4.2). Other types of crime include vice crime, such as drug use, gambling, and prostitution, as well as organized crime, white-collar crime, computer crime, and juvenile delinquency.

Street Crime: Violent Offenses

Statistics from the Department of Justice indicate that 1999 levels of crime were the lowest in over 20 years and represent the largest single-year decline since 1973. It should be remembered, however, that crime statistics represent only those crimes *reported* to the police—1.4 million violent crimes in 1999. Victim surveys indicate that less than half of all violent crime is reported to the police (Dorning 2000).

Violent crime includes homicide, assault, rape, and robbery. Homicide refers to the willful or nonnegligent killing of one human being by another individual or group of individuals. Although homicide is the most serious of the violent crimes, it is also the least common, accounting for less than 1 percent of all index crimes in 1999. A typical homicide scenario includes a male killing a male with a handgun after a heated argument. The victim and offender are dispro-

The continuing drop in violent crime is good news for all Americans. It demonstrates that the innovative and collaborative policies and programs among federal and state and local law enforcement work.

JANET RENO
Former U.S. Attorney General

Table 4.2 *Index Crime Rates, Percent Change, and Clearance Rates, 1999*

	Rate per 100,000, 1999	Percentage Change in Rate, (1995–1999)	Percent Cleared, 1999
Total Index Crimes	**4,481.7**	**−19.1**	**21.4**
Murder	6.1	−30.5	69.1
Forcible Rape	33.7	−11.9	49.5
Robbery	166.7	−32.0	28.5
Aggravated Assault	357.7	−19.7	59.2
Violent Crime Total	**563.4**	**−23.4**	**50.0**
Burglary	805.4	−22.0	13.7
Larceny/Theft	2,653.3	−16.2	19.1
Motor Vehicle Theft	459.6	−24.9	14.9
Arson	37.1[a]	n/a	17.2
Property Crime Total[b]	**3,918.3**	**−18.5**	**17.5**

[a]Arson rates per 100,000 are calculated independently because population coverage for arson is lower than for the other index offenses.
[b]Property Crime totals do not include arson.
n/a = not available

Source: Federal Bureau of Investigation. 2000. *Uniform Crime Reports, 1999*. Washington, D.C.: U.S. Department of Justice.

portionately young and of minority status. When a woman is murdered and the victim–offender relationship is known, she is most likely to have been killed by her husband or boyfriend; men are more likely to be killed by a stranger (FBI 2000; BJS 2000a).

Another form of violent crime, aggravated assault, involves the attacking of another with the intent to cause serious bodily injury. Like homicide, aggravated assault occurs most often between members of the same race and, as with violent crime in general, is more likely to occur in warm weather months. In 1999, the assault rate was over 50 times greater than the murder rate, assaults comprising 7.9 percent of all index crime (FBI 2000).

Rape is also classified as a violent crime and is also intra-racial, that is, the victim and offender are from the same racial group. The FBI definition of rape contains three elements: sexual penetration, force or the threat of force, and non-consent of the victim. In 1999, more than 89,000 forcible rapes were reported in the United States, a 4 percent decline from the previous year (FBI 2000). However, data from the 1999 NCVS suggests that the actual number of rapes has not decreased in that fewer than half of all rapes are reported to the police (Siegel 2000).

Perhaps as many as 80 percent of all rapes are **acquaintance rapes**—rapes committed by someone the victim knows. Although acquaintance rapes are the most likely to occur, they are the least likely to be reported and the most difficult to prosecute. Unless the rape is what Williams (1984) calls a **classic rape**—that is, the rapist was a stranger who used a weapon, and the attack resulted in serious bodily injury—women hesitate to report the crime out of fear of not being believed. The increased use of "rape drugs" such as Rohypnol may lower reporting levels even further (see Chapter 3).

Robbery, unlike simple theft, also involves force or the threat of force, or putting a victim in fear, and is thus considered a violent crime. Officially, in 1999 nearly 400,000 robberies took place in the United States. Robberies are committed by young people, with over 42 percent of all robberies entailing "strong-arm" tactics (FBI 2000). As with rape, victims who resist a robbery are more likely to stop the crime, but are also more likely to be physically harmed.

> The rich rob the poor and the poor rob one another.
>
> **SOJOURNER TRUTH**
> *Abolitimist*

Street Crime: Property Offenses

Property crimes are those in which someone's property is damaged, destroyed, or stolen; they include larceny, motor vehicle theft, burglary, and arson. Property crimes have declined in recent years, decreasing 18.5 percent between 1995 and 1999. Larceny, or simple theft, is the most common property crime, comprising over half of all property arrests.

Larcenies involving automobiles and auto accessories are the largest single category of thefts. However, because of the cost involved, motor vehicle theft is considered a separate index offense. Although numbering over 1 million, 1999 motor vehicle thefts were at their lowest recorded level. Because of insurance requirements, vehicle theft is one of the most highly reported index crimes and, consequently, estimates between the URC and the NCVS are fairly compatible (Siegel 2000).

Burglary, which is the second most common index offense, entails entering a structure, usually a house, with the intent to commit a crime while inside. Official statistics indicate that, in 1999, 1,804,200 burglaries occurred, for a rate of 805.4 per 100,000 population. Interestingly, whether or not the burglary is reported to the police depends on the type of entry. When forced entry occurs (breaking a window or a door), 77 percent of the burglaries are reported; when unlawful entry takes place (going through an open window or door), only 43 percent are reported (Barkan 1997).

Arson involves the malicious burning of the property of another. Estimating the frequency and nature of arson is difficult given the legal requirement of "maliciousness." Of the reported cases of arson, 45 percent involved structures (the majority of which were residential), 30 percent movable property (e.g., boat, car), and the remainder were miscellaneous (e.g., crops, timber) personal items (FBI 2000).

Vice Crimes

Vice crimes are illegal activities that have no complaining party and are therefore often called **victimless crimes**. Vice crimes include using illegal drugs, engaging in or soliciting prostitution (except for legalized prostitution, which exists in Nevada), illegal gambling, and pornography.

Most Americans view drug use as socially disruptive (see Chapter 3). In a recent survey, drugs were listed as one of the most significant problems facing America today (Gallup Poll 2000a). Less consensus exists, nationally or internationally, that gambling and prostitution are problematic. In the Netherlands the Prostitution Information Centre in Amsterdam offers a 6-day course on "prostitution as a career-option," and the Australia Council of Trade Unions recently recognized women in prostitution as a labor sector (CATW 1997). In the United States, many states have legalized gambling, including casinos in Nevada, New Jersey, Connecticut, and many other states, as well as state lotteries, bingo parlors, horse and dog racing, and jai alai. Further, some have argued that there is little difference, other than societal definitions of acceptable and unacceptable behavior, between gambling and other risky ventures such as investing in the stock market. Conflict theorists are quick to note that the difference is who's making the wager.

Organized crime refers to criminal activity conducted by members of a hierarchically arranged structure devoted primarily to making money through illegal means. Although often discussed under victimless crimes because of its association with prostitution, drugs, and gambling, organized crime often uses coercive techniques. For example, organized crime groups may force legitimate businesses to pay "protection money" by threatening vandalism or violence.

The traditional notion of organized crime is the Mafia—a national band of interlocked Italian families—but members of many ethnic groups engage in organized crime. In recent years Asian organized crime groups have emerged as one of the most violent and economically successful groups in the United States dealing primarily in kidnapping, gambling, drug trafficking, counterfeiting, and prostitution (Lindberg. Petrenko, Gladden, & Johnson 1997).

> . . . organized crime in Russia threatens not only the safety of this country but the safety of the U.S.A.
>
> LOUIS FREEH
> *Director, FBI*

Organized crime also occurs at the international level, such as smuggling illegal drugs and arms. Since the fall of the Soviet Union, it is estimated that as much as 25 percent of the Russian gross national income is from organized crime activities generated by 5,600 separate crime groups. Today, organized crime is considered ". . . one the most important political problems in Russia . . ." whereby a ". . . Russian citizen's personal safety" can no longer be guaranteed (Shabalin, Alibini, & Rogers 1995).

White-Collar Crime

White-collar crime includes both occupational crime, in which individuals commit crimes in the course of their employment, and corporate crime, in which corporations violate the law in the interest of maximizing profit. Occupational

Residents of Love Canal were victimized by corporate violence when toxic waste, dumped by Hooker Chemical Company, began to seep into the basements of homes and schools. Many residents of Love Canal moved out of the area; others complained of high rates of miscarriages, birth defects, and cancer. Although Love Canal was allegedly cleaned up, residents protested against resettling the area. After 20 years of litigation, the last suit was settled in 1998 with the City of Love Canal receiving $250,000 and the construction of a park near the spill.

© William Campbell/Sygma

■ **Table 4.3** *Types of White-Collar Crime*

Crimes against Consumers	Crimes against Employees
Deceptive advertising	Health and safety violations
Antitrust violations	Wage and hour violations
Dangerous products	Discriminatory hiring practices
Manufacturer kickbacks	Illegal labor practices
Physician insurance fraud	Unlawful surveillance practices
Crimes against the Public	**Crimes against Employers**
Toxic waste disposal	Embezzlement
Pollution violations	Pilferage
Tax fraud	Misappropriation of government funds
Security violations	Counterfeit production of goods
Police brutality	Business credit fraud

crime is motivated by individual gain. Employee theft of merchandise, or pilferage, is one of the most common types of occupational crime. Other examples include embezzlement, forgery and counterfeiting, and insurance fraud. Price-fixing, anti-trust violations, and "churning" are all examples of corporate crime, that is, crime that benefits the organization. As Sherrill notes, "churning" is:

> a racket in which as many as 10 million customers were sweet-talked into using the cash value of their old insurance policies to pay the premiums on new, more expensive policies. They were not warned that the upgrading could be so costly that it would eat up their equity, leaving them with premiums they couldn't afford—and therefore no coverage (2000, 304).

Corporate violence, another form of corporate crime, refers to the production of unsafe products and the failure of corporations to provide a safe working environment for their employees. Corporate violence is the result of negligence, the pursuit of profit at any cost, and intentional violations of health, safety, and environmental regulations. For example, after more than a year of recalls in 16 countries, in August of 2000 Bridgestone/Firestone began a U.S. recall of over 6.5 million tires. The tires, many of which were standard equipment on the popular Ford Explorer, had a 10-year history of tread separation. It was only after 88 U.S. traffic deaths were linked to the defective tires, prompting a Congressional investigation, that Ford and Bridgestone/Firestone acknowledged the overseas recalls and the tires' questionable safety history (Pickler 2000). Table 4.3 summarizes some of the major categories of white-collar crime.

■ I come before you to apologize to you, the American people and especially to the families who lost loved ones in these terrible rollover accidents.

MASATOSHI ONO
Chief Executive,
Bridgestone/Firestone, Inc.

Computer Crime

Computer crime refers to any violation of the law in which a computer is the target or means of criminal activity. It is one of the fastest growing crimes in the United States, costing an estimated $226 million dollars in 2000—over twice the previous year's estimates. Hacking, or unauthorized computer intrusion, is one type of computer crime. In just one month, hackers successfully attacked the computer systems of Walt Disney World, Yahoo, eBay, and Amazon.com through "denial of service" invasions (Kong & Swartz 2000).

The increase in computer break-ins has also led to an increase in identity theft—the use of someone else's identification (e.g., social security number, birth date) to obtain credit. In 1999, 40,000 cases of identity theft occurred, costing an estimated $2 to $3 billion in credit card fraud alone (Miller 1999; Fields 2000). Although mail theft is one of the most common modes of obtaining the needed information, new technologies have contributed to the increased rate of this offense. For example, government officials "are in Russia to look into the theft of 300,000 credit card numbers from CD Universe, an online music retailer" (Fields 2000, 6a).

Conklin (1998) has identified other examples of computer crime:

- Two individuals were charged with theft of 80,000 cellular phone numbers. Using a device purchased from a catalogue, the thieves picked up radio waves from passing cars, determined private cellular codes, reprogrammed computer chips with the stolen codes and then, by inserting the new chips into their own cellular phones, charged calls to the original owners.
- A programmer made $300 a week by programming a computer to round off each employee's paycheck down to the nearest 10¢ and then to deposit the extra few pennies in the offender's account.
- An oil company illegally tapped into another oil company's computer to get information that allowed the offending company to underbid the other company for leasing rights.

Juvenile Delinquency

In America today, no population poses a greater threat to public safety than juvenile criminals.

BILL McCOLLUM
U.S. Representative

In general, children under the age of 18 are handled by the juvenile court either as status offenders or as delinquent offenders. A status offense is a violation that can only be committed by a juvenile, such as running away from home, truancy, and underage drinking. A delinquent offense is an offense that would be a crime if committed by an adult, such as the eight index offenses. The most common status offenses handled in juvenile court are underage drinking, truancy, and running away. In 1999, 17.4 percent of all arrests (excluding traffic violations) were of offenders under the age of 18 and 6 percent were under the age of 15. Those under 18 accounted for 28 percent of all arrests for index crimes (FBI 2000).

As is the case with adults, juveniles commit more property than violent offenses and the number of violent offenses has dropped in recent years (OJJDP 2000). Nonetheless, Americans are concerned about the high rate of juvenile violence including violence in schools (see Chapter 12) and gang-related violence. Gang-related crime is, in part, a function of two interrelated social forces—the increased availability of guns in the 1980s and the lucrative and expanding drug trade. It is estimated that the United States has 30,000 street gangs with over 800,000 members, three-quarters of whom are racial minorities (Bartollas 2000).

Demographic Patterns of Crime

Although virtually everyone violates a law at some time, persons with certain demographic characteristics are disproportionately represented in the crime statistics. Victims, for example, are disproportionately young, lower-class, minority males from urban areas. Similarly, the probability of being an offender varies by gender, age, race, social class, and region (see Figure 4.2).

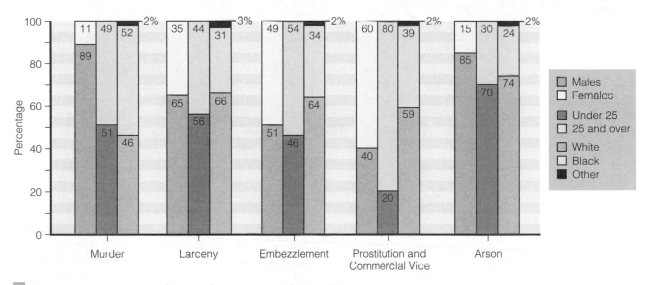

Figure 4.2 *Percentage of Arrests by Sex, Age, and Race, 1999.*
Source: U.S. Department of Justice. 2000. *Crime in the United States, 1999.* Washington, D.C.

Gender and Crime

Both official statistics and self-report data indicate that males commit more crime than females (see Figure 4.2). Why are males more likely to commit crime than females? One explanation is that society views female lawbreaking as less acceptable and thus places more constraints on female behavior: "women may need a higher level of provocation before turning to crime—especially serious crime. Females who choose criminality must traverse a greater moral and psychological distance than males making the same choice" (Steffensmeier & Allan 1995, 88).

Further, data suggest that males and females tend to commit different types of crimes. Men, partly because of more aggressive socialization experiences, are more likely than women to commit violent crimes. In 1999, males accounted

Females who join gangs often do so to win approval from their boyfriends who are gang members; increasingly, however, females are forming independent "girl gangs."

Race and Ethnicity in Sentencing Outcomes

Certainly one of the more pressing issues of recent times is the question of whether legally irrelevant information (e.g., race, sex of offender) affects criminal justice outcomes. Given recent accounts of alleged police brutality, accusations of racial profiling, and concern over race, class, and ethnic bias in death penalty cases, Steffensmeier and Demuth's (2000) investigation of the impact of extralegal variables in federal court sentencing is particularly timely.

Sample and Methods

Steffensmeier and Demuth (2000) investigate whether the criminal justice system discriminates on the basis of race and ethnicity. Although considerable research supports the contention that minorities compared with other members of society are more likely to receive harsher sentences (having fewer resources, perceived as "dangerous," threatening to the status quo, etc.), Steffensmeier's and Demuth's (2000) analysis examines both inter- *and* intragroup variations. The researchers argue that court decisions take place within a powerful cultural context—a "perceptual shorthand" of sorts—where the definitions of an offender's blameworthiness, the need to protect society, and the practical limitations of the sentence imposed affect the decision-making process.

Given that federal sentencing guidelines allow for some discretion and that "Hispanic defendants may seem even more culturally dissimilar and be even more disadvantaged than their black counterparts" (Steffensmeier and Demuth 2000, 709), the researchers hypothesize that: (1) Hispanics will receive more severe sentences than whites or blacks, (2) in drug cases, Hispanics specifically, but minorities in general, will receive harsher sentences than whites, and (3) Hispanics identified as black will receive harsher sentences than Hispanics identified as white. The dependent variable, severity of sentence, was measured by whether the court imposed a prison sentence and, if so, the length of the sentence. The independent

for 83 percent of all arrests for violent crime (FBI 2000). Females are less likely than males to commit serious crimes, and the monetary value of female involvement in theft, property damage, and illegal drugs is typically less than that for similar offenses committed by males. Nevertheless, a growing number of women have become involved in characteristically male criminal activities such as gang-related crime and drug use.

The recent increase in crimes committed by females has led to the development of a feminist criminology. Feminist criminology focuses on how the subordinate position of women in the social structure affects the criminal behavior of women. For example, Chesney-Lind and Shelden (1998) report that arrest rates for runaway juvenile females are higher than for males not only because they are more likely to run away as a consequence of sexual abuse in the home, but also because police with paternalistic attitudes are more likely to arrest female runaways than male runaways. Feminist criminology thus adds insights into understanding crime and violence often neglected by conventional theories by concentrating on gender inequality in society.

Age and Crime

In general, criminal activity is more prevalent among younger persons than older persons. The highest arrest rates are for individuals under the age of 25. Crimes committed by people in their teens or early 20s tend to be property crimes like burglary, larceny, arson, and vandalism. The median age of people who commit more serious crimes such as aggravated assault and homicide is in the late 20s. However, given the increase of young people in the next decade, it is projected that by the year 2005, arrests of teenagers for homicide will increase 25 percent (Heubusch 1997, 1).

variables are race and ethnicity. The data includes all convictions of U.S. citizens in federal courts between 1993 and 1996 (N= 89,637).

Findings and Conclusions

The authors note the "considerable consistency" of sentencing of federal criminal defendants across racial and ethnic categories for *similar cases*.

> *. . . judges, on balance, prescribe similar sentences for similar defendants convicted of the same offense. Whether they were white, black, or Hispanic—defendants who committed more serious crimes, had more extensive criminal histories, or were convicted at trial (as opposed to a guilty plea), were much more likely to be incarcerated and receive longer prison sentences (p. 724).*

They also note, however, small to moderate effects of race and ethnicity in terms of the *overall* sentencing with Hispanics receiving harsher sentences than whites and blacks. These differences are likely the result of what is called "substantial assistance departures." When a defendant, for example, "substantially assists" the State in the prosecution or investigation of another offender, the judge may "depart" from the sentencing guidelines. Because, Hispanics may be more likely to be involved in drug trafficking circles who seek retribution against those who cooperate with authorities, they may be less likely to provide "substantial assistance" and, thus, are less likely to have their sentences reduced.

Finally, the researchers conclude that their results "provide[s] strong evidence for the continuing significance of race and ethnicity in the larger society in general and in organizational–decision-making processes in particular"(p. 726). While acknowledging that the results are less an indication of discrimination than the unintended consequence of seemingly neutral policies, the authors call for more research in the area of race and ethnic relations.

Source: Steffensmeier, Darryl and Stephen Demuth. 2000. "Ethnicity and Sentencing Outcomes in U.S. Federal Courts: Who is Punished More Harshly?" *American Sociological Review* 65:705-729.

Why is criminal activity more prevalent among individuals in their teens and early 20s? One reason is that juveniles are insulated from many of the legal penalties for criminal behavior. Younger individuals are also more likely to be unemployed or employed in low-wage jobs. Thus, as strain theorists argue, they have less access to legitimate means for acquiring material goods.

Some research suggests, however, that high school students who have jobs become more, rather than less, involved in crime (Felson 1998, 120). In earlier generations, teenagers who worked did so to support themselves and/or their families. Today, teenagers who work typically spend their earnings on recreation and "extras," including car payments and gasoline. The increased mobility associated with having a vehicle also increases opportunity for criminal behavior and reduces parental control.

Race, Social Class, and Crime

Race is a factor in who gets arrested. Minorities are disproportionately represented in official statistics. For example, although African Americans represent about 12 percent of the population, they account for more than 38 percent of all violent index offenses, 31 percent of all property index offenses, and 33 percent of the crime index total.

Nevertheless, it is inaccurate to conclude that race and crime are causally related. First, official statistics reflect the behaviors and policies of criminal justice actors. Thus, the high rate of arrests, conviction, and incarceration of minorities may be a consequence of individual and institutional bias against minorities (see this chapter's *Social Problems Research Up Close* feature). For example, blacks are sent to prison for drug offenses at a rate 8.2 times higher than the rate for whites. If present trends continue, by 2020 two out of every three black men between the ages of 18 and 34 will be in prison (Fletcher 2000; Dickerson 2000). These disturbing statistics have led to concerns over **racial profiling**—the practice of targeting suspects based upon race status. Proponents of the practice argue that because race, as gender, is a significant predictor of who commits crime, the practice should be allowed. Opponents hold that racial profiling is little more than discrimination and should therefore be abolished.

Second, race and social class are closely related in that nonwhites are over-represented in the lower classes. Because lower-class members lack legitimate means to acquire material goods, they may turn to instrumental, or economically motivated, crimes. Further, while the "haves" typically earn social respect through their socioeconomic status, educational achievement, and occupational role, the "have-nots" more often live in communities where respect is based on physical strength and violence, as subcultural theorists argue.

Thus the apparent relationship between race and crime may, in part, be a consequence of the relationship between these variables and social class. Research indicates, however, that even when social class backgrounds of blacks and whites are comparable, blacks have higher rates of criminality, particularly for violent crime (Wolfgang, Figlio, & Sellen 1972; Elliot & Ageton 1980). Further, to avoid the bias inherent in official statistics, researchers have compared race, class, and criminality by examining self-report data and victim studies. Their findings indicate that although racial and class differences in criminal offenses exist, the differences are not as great as official data would indicate (U.S. Department of Justice 1993; Walker, Spohn, & Delone 1996; Hagan & Peterson 1995).

Region and Crime

In general, crime rates, and particularly violent crime rates, are higher in urban than suburban areas, and higher in suburban areas than rural areas. Higher crime rates in urban areas result from several factors. First, social control is a function of small intimate groups socializing their members to engage in law-abiding behavior, expressing approval for their doing so and disapproval for their noncompliance. In large urban areas, people are less likely to know each other and thus are not influenced by the approval or disapproval of strangers. Demographic factors also explain why crime rates are higher in urban areas: cities have large concentrations of poor, unemployed, and minority individuals.

Crime rates also vary by region of the country. In 1999, both violent and property crimes were highest in southern states followed by western, midwestern, and northeastern states. The murder rate is particularly high in the South with 43 percent of all murders recorded in Southern states (FBI 2000). The high rate of southern lethal violence has been linked to high rates of poverty and minority populations in the South, a southern "subculture of violence," higher rates of gun ownership, and a warmer climate that facilitates victimization by increasing the frequency of social interaction.

> Obviously crime pays, or there'd be no crime.
>
> G. GORDON LIDDY
> *Radio personality*

Costs of Crime and Violence

Crime and violence often result in physical injury and loss of life. In 1997, the most recent year in which data are available, one person out of every 240 people in the United States was likely to be a victim of murder in their lifetime (FBI 2000).

Not surprisingly, in a recent Gallup survey Americans ranked crime and violence the third most pressing issue in the country preceded only by education, and ethics and morals (Gallup Poll 2000b). Moreover, the U.S. Public Health Service now defines "violence" as one of the top health concerns facing Americans. In addition to death, physical injury, and loss of property, crime also has economic, social, and psychological costs.

Economic Costs of Crime and Violence

Conklin (1998, 71–72) suggests that the financial costs of crime can be classified into at least six categories. First are direct losses from crime such as the destruction of buildings through arson, of private property through vandalism, and of the environment by polluters. Second are costs associated with the transferring of property. Bank robbers, car thieves, and embezzlers have all taken property from its rightful owner at tremendous expense to the victim and society. For example, in 1999 it is estimated that $7 billion was lost to motor vehicle theft; the average value per vehicle at the time of the theft was $6,105 (FBI 2000).

> We cannot expect to rein in the costs of our health-care system if emergency rooms are overflowing with victims of gun violence.
>
> MAJOR OWENS
> *U.S. Congressman*

A third major cost of crime is that associated with criminal violence, for example, the medical cost of treating crime victims—$9.9 billion annually—or the loss of productivity of injured workers (Anderson 1999). Fourth are the costs associated with the production and sale of illegal goods and services, that is, illegal expenditures. The expenditure of money on drugs, gambling, and prostitution diverts funds away from the legitimate economy and enterprises and lowers property values in high crime neighborhoods. Fifth is the cost of prevention and protection, that is, the billions of dollars spent on locks and safes ($4 billion), surveillance cameras ($1.4 billion), guard dogs ($49 million), and the like (Anderson 1999).

Finally, there is the cost of the criminal justice system—law enforcement, litigative and judicial activities, corrections, and victim's assistance. In 2000, the cost of the federal criminal justice system alone was $27 billion estimated to grow to $33 billion in 2005 (BJS 2000a). Reasons for such growth include increases in the rates of arrest and conviction, changes in the sentencing structure, public attitudes toward criminals, the growing number of young males, and the war on drugs. Regardless of the cause, however, the staggering cost of public institutions has led to the "privatization" of prisons whereby the private sector increasingly supplies needed prison services.

What is the total economic cost of crime? One estimate suggests that the total cost of crime and violence in the United States is more than $1.7 trillion a year (Anderson 1999). Although costs from "street crimes" are staggering, the costs from "crimes in the suites" such as tax evasion, fraud, false advertising, and antitrust violations may be as high as 50 times the cost of property crime (Rosoff, Pontell, & Tillman 1998).

Social and Psychological Costs of Crime and Violence

Crime and violence entail social and psychological, as well as economic, costs. When asked, in general, "How safe do you feel in each of these locations?," 67 percent of respondents reported feeling "very safe" in their homes, 49 percent "very safe" walking in their neighborhoods after dark, and 22 percent "very safe" when at a shopping mall at night (Pew Research Center 2000). Fear of crime may be fueled by media presentations which may not accurately reflect the crime picture. For example, in a content analysis of local television portray-

> The possibility of being a victim of a crime is ever present on my mind. . . . thinking about it is as natural as . . . breathing. . . .
>
> 40-YEAR-OLD WOMAN IN
> NEW YORK CITY

The eight blunders that lead to violence in society:

-Wealth without work

-Pleasure without conscience

-Knowledge without character

-Commerce without morality

-Science without humanity

-Worship without sacrifice

-Politics without principle

-Rights without responsibilities

MOHANDAS K. GANDHI
Indian nationalist leader and peace activist

als of crime victims and offenders, whites were overrepresented as victims and blacks as offenders when compared with official statistics (Dixon & Linz 2000). Fear of crime and violence also affects community life:

> If frightened citizens remain locked in their homes instead of enjoying public spaces, there is a loss of public and community life, as well as a loss of "social capital"—the family and neighborhood channels that transmit positive social values from one generation to the next. (National Research Council 1994, 5–6)

This is particularly true of women and the elderly who restrict their activities, living "limited lives," as a consequence of fear (Madriz 2000).

White-collar crimes also take a social and psychological toll at both the individual and the societal level. Moore and Mills (1990, 414) state that the effects of white-collar crime include "(a) diminished faith in a free economy and in business leaders, (b) loss of confidence in political institutions, processes and leaders, and (c) erosion of public morality." Crime also causes personal pain and suffering, the destruction of families, lowered self-esteem, and shortened life expectancy and disease.

Strategies for Action: Responding to Crime and Violence

In addition to economic policies designed to reduce unemployment and poverty, numerous social policies and programs have been initiated to alleviate the problem of crime and violence. These policies and programs include local initiatives, criminal justice policies, and legislative action.

Local Initiatives

Youth Programs Early intervention programs acknowledge that it is better to prevent crime than to "cure" it once it has occurred. Preschool enrichment programs, such as the Perry Preschool Project, have been successful in reducing rates of aggression in young children. After random assignment of children to either a control or experimental group, experimental group members received academically oriented interventions for 1 to 2 years, frequent home visits, and weekly parent–teacher conferences. When control and experimental groups were compared, the experimental group had better grades, higher rates of high school graduation, lower rates of unemployment, and fewer arrests (Murray, Guerre, & Williams 1997).

Although we seem to go to extraordinarily expensive lengths to punish criminals, we are reluctant to spend money on programs that have proven effective at preventing youngsters from slipping into a life of crime in the first place.

SYLVIA ANNE HEWLETT
President of National Parenting Association

Recognizing the link between juvenile delinquency and adult criminality, many anti-crime programs are directed toward at-risk youths. These prevention strategies, including youth programs such as Boys and Girls Clubs, are designed to keep young people "off the streets," provide a safe and supportive environment, and offer activities that promote skill development and self-esteem. According to Gest and Friedman (1994), housing projects with such clubs report 13 percent fewer juvenile crimes and a 25 percent decrease in the use of crack.

Finally, many youth programs are designed to engage juveniles in noncriminal activities and integrate them into the community. Weed and Seed, a program under the Department of Justice, "aims to prevent, control, and reduce violent crime, drug abuse and gang activity in targeted high crime neighborhoods across the country" by 'weeding' out the bad (e.g., violent criminals, drug traffickers) and 'seeding' the good (e.g., neighborhood restoration, crime prevention strategies). As part of the program, Safe Havens are established in, for example, schools where multi-agency services are provided for youth (Weed and Seed 2000).

Community Programs Neighborhood watch programs involve local residents in crime-prevention strategies. For example, MAD DADS in Omaha, Nebraska, patrol the streets in high crime areas of the city on weekend nights, providing positive adult role models for troubled children. Members also report crime and drug sales to police, paint over gang graffiti, organize gun buy-back programs, and counsel incarcerated fathers. In 1999, 9,500 communities participated in "National Night Out," a crime prevention event in which citizens, businesses, neighborhood organizations, and local officials joined together in outdoor activities to heighten awareness of neighborhood problems, promote anti-crime messages, and strengthen community ties (NNO 2000).

Mediation and victim-offender dispute resolution programs are also increasing with over 300 such programs now in the United States. The growth of these programs is a reflection of their success rate: two-thirds of cases referred result in face-to-face meetings, over 90 percent of these cases result in a written restitution agreement, and 90 percent of the written restitution agreements are completed within 1 year (VORP 1998).

Criminal Justice Policy

The criminal justice system is based on the principle of **deterrence**, that is, the use of harm or the threat of harm to prevent unwanted behaviors. It assumes that people rationally choose to commit crime, weighing the rewards and consequences of their actions. Thus, the recent emphasis on "get tough" measures holds that maximizing punishment will increase deterrence and cause crime rates to decrease. Research indicates, however, that the effectiveness of deterrence is a function of not only the severity of the punishment, but the certainty and swiftness of the punishment as well. Further, "get tough" policies create other criminal justice problems, including overcrowded prisons and, consequently, the need for plea bargaining and early release programs.

Law Enforcement Agencies In 1999, the United States had 420,544 law enforcement officers with an average of 2.5 officers for every 1,000 people (FBI 2000). Ironically, despite recent increases in the number of law enforcement personnel, public opinion that "more police on the streets" will reduce violent crime has decreased (Pew Research Center 2000). Further, accusations of racial profiling, police brutality, and discriminatory arrest practices have shaken public confidence in the police, particularly among non-whites and those with low incomes (Gallup Poll 1999; BJS 2000a).

In response to such trends, the Crime Control Act of 1994 established the Office of Community Oriented Policing Services (COPS). Community-oriented policing involves collaborative efforts among the police, the citizens of a community, and local leaders. As part of community policing efforts, officers speak to citizen groups, consult with social agencies, and enlist the aid of corporate and political leaders in the fight against neighborhood crime (COPS 1998; Lehrur 1999; Worden 2000a).

Officers using community policing techniques often employ "practical approaches" to crime intervention. Such solutions may include what Felson (1998) calls "situational crime prevention." Felson argues that much of crime could be prevented simply by minimizing the opportunity for its occurrence. For example, cars could be outfitted with unbreakable glass, flush-sill lock buttons, an audible reminder to remove keys, and a high-security lock for steering columns (Felson 1998, 168). These techniques, and community oriented policing in general, have been fairly successful. After COPS was implemented in New York City and New Orleans, homicide rates dropped dramatically—40 and 18 percent, respectively (Sileo 2000).

Rehabilitation versus Incapacitation   An important debate concerns the primary purpose of the criminal justice system: Is it to rehabilitate offenders or to incapacitate them through incarceration? Both **rehabilitation** and **incapacitation** are concerned with recidivism rates, or the extent to which criminals commit another crime. Advocates of rehabilitation believe recidivism can be reduced by changing the criminal, whereas proponents of incapacitation think it can best be reduced by placing the offender in prison so that he or she is unable to commit further crimes against the general public.

Societal fear of crime has led to a public emphasis on incapacitation and a demand for tougher mandatory sentences, a reduction in the use of probation and parole, support of a "three strikes and you're out" policy, and truth-in-sentencing laws (DiIulio 1999; Human Rights Watch 2000). Although incapacitation is clearly enhanced by longer prison sentences, rehabilitation may not be. Rehabilitation assumes that criminal behavior is caused by sociological, psychological, and/or biological forces rather than being solely a product of free will. If such forces can be identified, the necessary change can be instituted. Rehabilitation programs include education and job training, individual and group therapy, substance abuse counseling, and behavior modification.

Capital Punishment   With capital punishment, the State (the federal government or a state) takes the life of a person as punishment for a crime. Thirty-eight states have capital punishment statutes and in 1999, 98 persons in 20 states were executed—94 carried out by lethal injection, 3 by electrocution, and 1 by lethal gas (BJS 2001). Over 100 countries have banned state executions; America, India, and Japan are the only remaining democracies that have capital punishment. Further, in 1999, 85 percent of the world's executions took place in the

Salaries and other benefits for custodial staff constitute the largest portion of prison expenses. In 1999, over 1,500 correctional facilities employed some 400,000 personnel.

© Billy E. Barnes/PhotoEdit

Witness to an Execution

At 10:58 the prisoner entered the death chamber. He was, I knew from my research, a man with a checkered, tragic past. He had been grossly abused as a child, and went on to become grossly abusive of others. I was told he could not describe his life, from childhood on, without talking about confrontations in defense of a precarious sense of self—at home, in school, on the streets, in the prison yard. Belittled by life and choking with rage, he was hungry to be noticed. Paradoxically, he had found his moment in the spotlight. . . .

En route to the chair, the prisoner stumbled slightly, as if the momentum of the event had overtaken him. Were he not held securely by two officers, one at each elbow, he might have fallen. . . . Once the prisoner was seated, again with help, the officers strapped him into the chair.

Arms, legs, stomach, chest, and head were secured in a matter of seconds. Electrodes were attached to the cap holding his head and to the strap holding his exposed right leg. A leather mask was placed over his face. The last officer mopped the prisoner's brow, then touched his hand in a gesture of farewell. . . .

The strapped and masked figure sat before us, utterly alone, waiting to be killed . . . waiting for a blast of electricity that would extinguish his life. Endless seconds passed. His last act was to swallow, nervously, pathetically, with his Adam's apple bobbing. I was struck by the simple movement then, and can't forget it even now. It told me, as nothing else did, that in the prisoner's restrained body, behind that mask, lurked a fellow human being who, at some level, however primitive, knew or sensed himself to be moments from death.

. . . Finally, the electricity hit him. His body stiffened spasmodically, though only briefly. A thin swirl of smoke trailed away from his head and then dissipated quickly. The body remained taut, with the right foot raised slightly at the heel, seemingly frozen there. A brief pause, then another minute of shock. When it was over, the body was flaccid and inert.

Three minutes passed while the officials let the body cool. (Immediately after the execution, I'm told, the body would be too hot to touch and would blister anyone who did.) All eyes were riveted to the chair; I felt trapped in my witness seat, at once transfixed and yet eager for release. I can't recall any clear thoughts from that moment. One of the death watch officers later volunteered that he shared this experience of staring blankly at the execution scene. Had the prisoner's mind been mercifully blank before the end? I hope so.

The physician listened for a heartbeat. Hearing none, he turned to the warden and said, "This man has expired." The warden, speaking to the Director, solemnly intoned: "Mr. Director, the court order has been fulfilled." . . .

Source: Robert Johnson. 1989. "'This Man Has Expired': Witness to an Execution." *Commonwealth*, January 13, 9–15. Copyright by the Commonwealth Foundation. Reprinted with permission.

United States, China, Congo, Iran, and Saudi Arabia (*The Economist* 2000, 21). In this chapter's *The Human Side* feature, Robert Johnson, a professor at American University, describes his reaction to being witness to an execution.

Proponents of capital punishment argue that executions of convicted murderers are necessary to convey public disapproval and intolerance for such heinous crimes. Those against capital punishment believe that no one, including the State, has the right to take another person's life and that putting convicted murderers behind bars for life is a "social death" that conveys the necessary societal disapproval.

Proponents of capital punishment also argue that it deters individuals from committing murder. Critics of capital punishment hold, however, that because most homicides are situational, and are not planned, offenders do not consider the consequences of their actions before they commit the offense. Critics also point out that the United States has a much higher murder rate than Western European nations that do not practice capital punishment and that death sentences are racially discriminatory. For example, a recent study on federal capital cases found that blacks compared with whites were less likely to have their sentences reduced through plea bargaining (Worden 2000b).

Capital punishment advocates suggest that executing a convicted murderer relieves the taxpayer of the costs involved in housing, feeding, guarding, and providing medical care for inmates. Opponents of capital punishment argue that

■ The legal system is greatly overestimated in its ability to sort out innocent from guilty.

STEPHEN BRIGHT
Visiting lecturer, Yale Law School

DNA Evidence

In 1954, Sam Sheppard, a prominent Cleveland physician, was accused of killing his wife, Marilyn Sheppard. Over 40 years later this case was the basis for the movie *The Fugitive* and the new television series of the same name. The real-life drama carries on as Sam Sheppard, Jr. continues to try to clear his father's name. This time, however, he is armed with DNA evidence—evidence that suggests that someone other than his father may have killed Marilyn Sheppard (Parker 1998; Hagan and Ewinger 2000).

Increasingly, law enforcement officers both in the United States and in Europe are using DNA evidence in the identification of criminal suspects. DNA stands for deoxyribonucleic acid, which is found in the nucleus of every cell and contains an individual's complete and unique genetic makeup. Developed in the mid-1980s, DNA fingerprinting is a general term used to describe the process of analyzing and comparing DNA from different sources, including evidence found at a crime scene, for example, blood, semen, hair, saliva, fibers, and skin tissue, and the DNA of a suspect.

According to Barry Scheck, co-director of the Innocence Project at Benjamin N. Cardozo School of Law, to date 82 post-conviction exonerations have occurred in Canada and the United States, eight of which pertained to death row prisoners. In 16 of the 82 cases DNA evidence not only vindicated the falsely accused, it helped identify the real perpetrator of the crime (Milloy 2000). Although primarily used in rape and homicide cases in the United States, DNA evidence is routinely used in burglary cases in Great Britain (NIJ, 2000a).

Concern over the number of post-conviction exonerations has led to the establishment of a *National Commission on the Future of DNA Evidence*, an independent panel under the National Institute of Justice. The commission will provide recommendations to the U.S. Attorney General on the "use of current and future DNA methods, applications, and technologies in the operation of the criminal justice system, from the crime scene to the courtroom" (NIJ 2000b, 1). Five essential policy areas will be addressed: (1) use of DNA in post-conviction cases, (2) legal issues including privacy concerns, (3) training and technical assistance in the use of DNA evidence, (4) laboratory needs, and (5) the impact of DNA on the future of the criminal justice system (NIJ 2000b).

> I feel morally and intellectually obligated to concede that the death penalty experiment has failed.
>
> **JUSTICE HARRY BLACKMUN**
> *U.S. Supreme Court*

the principles that decide life and death issues should not be determined by financial considerations. In addition, taking care of convicted murderers for life may actually be less costly than sentencing them to death, because of the lengthy and costly appeal process for capital punishment cases (Garey 1985; Myths1998).

Nevertheless, those in favor of capital punishment argue that it protects society by preventing convicted individuals from committing another crime, including the murder of another inmate or prison official. Opponents contend, however, that capital punishment may result in innocent people being sentenced to death. Since 1973, 87 death row inmates have been released when new evidence supported their innocence (*The Economist* 2000). Further, a recent report by the Justice Project entitled, "A Broken System" found "serious, reversible error in nearly 7 of every 10 of the thousands of capital sentences" that were reviewed over the 23-year study period (Liebman, Fagan, & West 2000).

Legislative Action

Gun Control Fueled by the Columbine school shooting and other recent images of children as both gun victims and offenders, in May of 2000 tens of thousands of women descended on Washington, D.C. demanding that something be done about gun violence. Although the impact of the Million Mom March is still unknown, most Americans, and particularly women, "support new restrictions on handguns, including child safety locks, mandatory safety courses, and registration" (Hinds 2000,1). However, when asked whether or not handguns should be banned, as in some cities, 62 percent of Americans responded "no" ("Guns" 2000).

Some concern exists, however, about the use of DNA evidence and, specifically, questions about how donors will be selected, what methods of data collection will be used, and potential abuses of analysis results. For example, DNA evidence might be used to simply *imply* guilt as in the case of DNA information that indicates "proclivities for aggression" (McCullagh 1999, 1). That is why, until recently, England had the only nationwide DNA data bank in the world. However, in 1998 the FBI initiated the Combined DNA Index System (CODIS). Today, every state in the nation is in the process of indexing DNA information on offenders convicted of certain crimes—information that will enable police officers to link offenders to crimes across jurisdictions (NIJ 1999). The extent of the collection effort has led some to ask what will happen to the tens of thousands of samples which, among other things, reveal a donor's genetic disposition for certain diseases (McCullagh 1999).

It is likely that use of DNA will face many legal battles but, nonetheless, the future of DNA evidence is bright. It is less expensive than ever before, predicted to be as low as $10 per test within a few years compared with earlier costs of $200 to $300. And as technology has become more and more sophisticated, the time it takes to conduct the analysis has decreased from weeks to days and portable DNA analysis units are in the making (NIJ 2000a). Further, even if it doesn't survive the legal scrutiny that is likely, it remains a valuable identification technique used in biology, archeology, medical diagnosis, paleontology, and forensics.

Sources:

Hagan, John and James Ewinger. 2000. "Jurors for Sheppard Trial Selected." *The Plain Dealer*. February 11. http://www.cleveland.com/news/sreports/sam

National Institute of Justice. 2000a. "National Commission on the Future of DNA Evidence." Washington D.C.: U.S. Department of Justice. http://www.ojp.usdoj.gov/nij/dna.

National Institute of Justice. 2000b. "Annual Report to Congress." Washington D.C.: U.S. Department of Justice.

National Institute of Justice. 1999. "What Every Law Enforcement Officer Should Know." BC000614. Washington D.C.: U.S. Department of Justice.

McCullagh, Declan. 1999. "The Debate of DNA Evidence." *Wired News*. July 12, 1. http://www.wirednews.com/news

Milloy, Ross E. 2000. "Some Prosecutors Willing to Review DNA Evidence." *The New York Times on the Web*. October 20.

Parker, Laura. 1998. "Tests Point to Sheppard's Innocence." *USA Today*, March 5, A3.

Those against gun control argue that not only do citizens have a constitutional right to own guns, but that more guns may actually lead to less crime as would-be offenders retreat in self-defense (Lott 2000). Advocates of gun control, however, insist that the 200 million privately owned guns in this country, 70 million of which are handguns (Albanese 2000), significantly contribute to the violent crime rate in the United States as well as distinguishing it from other industrialized nations.

After a 7-year battle with the National Rifle Association (NRA), gun control advocates achieved a small victory in 1993 when Congress passed the **Brady Bill**. This law requires a 5-day waiting period on handgun purchases so that sellers can screen buyers for criminal records or mental instability. Today, the law requires background checks of not just handgun users, but for those who purchase rifles and shotguns as well (Siegel 2000).

Other Legislation Major legislative initiatives have been passed in recent years including the 1994 Crime Control Act that created community policing, "three strikes and you're out," and truth-in-sentencing laws. Significant crime-related legislation presently before Congress include (Worden 2000c):

- *The Violence Against Women Act*—designed to prevent domestic violence and protect the elderly and the disabled from crime; provides $200 million in victim assistance.
- *The Innocence Protection Act*—intended to reduce the number of wrongful executions by providing enhanced legal services and DNA testing for the accused (see this chapter's *Focus on Technology* feature).

> Each day, 10 children and teens are killed by firearms, and that is 10 too many.
>
> **DONNA SHALALA**
> *Secretary, Department of Health and Human Services*

- *The Traffic Stop Statistics Study Act*—requires police departments to keep records of the race, ethnicity, and gender of drivers stopped, searched, and/or issued a citation.
- *The Law Enforcement Trust and Integrity Act*—defines police use of excessive force and establishes guidelines for reporting in-custody deaths; designed to discourage police misconduct.

Understanding *Crime and Violence*

> To do justice, to break the cycle of violence, to make Americans safer, prisons need to offer inmates a chance to heal like a human, not merely heel like a dog.
>
> RICHARD STRATTON
> *Former prison inmate*

What can we conclude from the information presented in this chapter? Research on crime and violence supports the contentions of both functionalists and conflict theorists. Inequality in society, along with the emphasis on material well being and corporate profit, produces societal strains and individual frustrations. Poverty, unemployment, urban decay, and substandard schools, the symptoms of social inequality, in turn lead to the development of criminal subcultures and definitions favorable to law violation. Further, the continued weakening of social bonds between members of society and between individuals and society as a whole, the labeling of some acts and actors as "deviant," and the differential treatment of minority groups at the hands of the criminal justice system encourage criminal behavior.

With the recent decrease in crime it is tempting to conclude that "get tough" criminal justice policies are responsible for the reductions. Other valid explanations exist and are likely to have contributed to the falling rates—a healthy economy, changing demographics, community policing, stricter gun control, and a reduction in the use of crack cocaine (Hinds 2000; Meek 2000). Nonetheless, "nail'em and jail'em" policies have been embraced by citizens and politicians alike. Get-tough measures include building more prisons and imposing lengthier mandatory prison sentences on criminal offenders. Advocates of harsher prison sentences argue that "getting tough on crime" makes society safer by keeping criminals off the streets and deterring potential criminals from committing crime. Yet, for example, an analysis of over 200 studies comparing institutionalized and non-institutionalized serious juvenile offenders found that community-based treatment was superior to incarceration in lowering recidivism rates (Lipsey & Wilson 1998).

Prison sentences may not always be effective in preventing crime. In fact, they may promote it by creating an environment in which prisoners learn criminal behavior, values, and attitudes from each other. Two-thirds of all parolees are re-arrested within 3 years of release. With over a million men and women in prison in 2000—over 600,000 re-entering communities annually—punitive policies may be a short-sighted temporary fix (Petersilia 2000; BJS 2000b, Gullo 2001).

Rather than getting tough on crime after the fact, some advocate getting serious about prevention. Reemphasizing the values of honesty and, most importantly, taking responsibility for one's actions is a basic line of prevention with which most agree. The recent movement toward **restorative justice,** a philosophy primarily concerned with repairing the victim-offender-community relation, is in direct response to the concerns of an adversarial criminal justice system that encourages offenders to deny, justify, or otherwise avoid taking responsibility for their actions.

Restorative justice holds that the justice system should be a "healing process rather than a distributor of retribution and revenge" (Siegel 2000, 278). Key components of restorative justice include restitution to the victim, remedying the harm to the community, and mediation. Restorative justice programs have been instituted in several states including Vermont and New York. At a recent meeting of the United Nations' Congress on Crime Prevention and Treatment of Offenders, a summary resolution called the *Vienna Declaration* was approved. The resolution calls for the worldwide use of restorative justice interventions (United Nations 2000).

Critical Thinking

1 Crime statistics are sensitive to demographic changes. Explain why crime rates in the United States began to rise in the 1960s as baby-boom teenagers entered high school, and why they may increase again as we move into the 21st century.

2 Some countries have high gun ownership and low crime rates. Others have low gun ownership and low crime rates. What do you think accounts for the differences between these countries?

3 One of the criticisms of crime theories is that they don't explain all crime, all of the time. Identify the theories of crime that are most useful in explaining categories of crime, for example, white-collar crime, violent crime, sex crimes, etc. Explain your choices.

4 The use of technology in crime-related matters is likely to increase dramatically over the next several decades. DNA testing, and the use of heat sensors, blood detecting chemicals, and computer surveillance are just some of the ways science will help fight crime. As with all technological innovations, however, there is the question, "Who benefits?" Do these new technologies have gender, race, and/or class implications?

Key Terms

acquaintance rape	crime rate	rehabilitation
Brady Bill	deterrence	restorative justice
classic rape	differential association	strain theory
clearance rate	incapacitation	subcultural theory
computer crime	index offenses	transnational crime
control theory	labeling theory	victimless crimes
corporate violence	organized crime	white-collar crime
crime	racial profiling	

Media Resources

 The Wadsworth Sociology Resource Center: Virtual Society

http://sociology.wadsworth.com

See the companion Web site for this book to access general sociology resources and text-specific features that can further your understanding of this chapter. The site contains Internet links, Internet exercises, online practice quizzes, information on InfoTrac College Edition, and many more valuable materials designed to enrich your learning experience in social problems.

 InfoTrac College Edition

You can access InfoTrac College Edition either from the Wadsworth Sociology Resource Center at **http://sociology.wadsworth.com** or directly from your web browser at **http://www.infotrac-college.com/wadsworth/**. InfoTrac College Edition is an online university library that includes over 700 popular and scholarly journals in which you can find articles related to the topics in this chapter such as information on crime statistics and the measurement of crime, DNA testing, gun control, and capital punishment.

 Interactions CD-ROM

Go to the Interactions CD-ROM for *Understanding Social Problems*, Third Edition to access additional interactive learning tools, such as in-depth review materials, corresponding practice quizzes, and other engaging resources and activities to help you study the concepts in this chapter.

Family Problems

Is It True?

1. In 1999, less than one-third of U.S. families included an employed Dad and a stay-at-home Mom.

2. In 1999, more than one in four U.S. children lived with one parent.

3. In the United States, people are more likely to be killed and physically assaulted, abused, and neglected, and sexually assaulted and molested in their own homes and by other family members than anywhere else or by anyone else.

4. About one-third of all U.S. births are to unmarried women.

5. Most out-of-wedlock births in the United States are to teenagers.

Answers to "Is It True?": 1 = T; 2 = T; 3 = T; 4 = T; 5 = F

An unhealthy culture cannot have healthy families.

MARY PIPHER
from The Shelter of Each Other: Rebuilding Our Families

Carol, a 20-year-old college sophomore, usually sat in the front of her sociology class. She was one of those students who always had something to say about whatever the topic was for that day. One day mid-semester, Carol came to class 10 minutes late. Although it was a cloudy day, she wore sunglasses. Rather than sit in her usual seat, she took a seat at the back. On this day, Carol said nothing during the entire class—highly unusual for her. Concerned by Carol's atypical behavior, her teacher approached her after class. "Are you OK?" she asked. Carol's voice quivered as she replied, "I had a fight with my boyfriend last night and . . . well . . . things got out of hand. I'm sorry I was late." "Did he hurt you?" the teacher asked. "He didn't mean to . . . he's real sorry." Underneath those sunglasses, Carol's eye was bruised and swollen.

Patty, an education major, had maintained a 3.2 GPA and had a high B average in her sociology class. Her attendance had been perfect all semester, so when she missed 3 consecutive classes, her teacher e-mailed her to inquire about her absences. In Patty's reply, she said she was having some personal problems. Her teacher asked Patty to meet with her during office hours. When Patty showed up, her face was pale and drawn and she looked exhausted. "What's going on?" the teacher asked. Patty broke down and cried as she explained that she recently found out that her parents are separated and are planning to divorce. The teacher encouraged Patty to make an appointment with a counselor on campus. A week later, just 3 weeks before the end of the Spring semester, the teacher received an official notice that Patty is taking a medical leave of absence. Patty's depression surrounding her parents' divorce required medication and ongoing counseling through the summer. By fall, she was well enough to return to school.

Lamar, a high-school senior, was looking forward to going to college on a basketball scholarship, although it would mean leaving his girlfriend Felicia, a high-school junior, behind. After saying his goodbyes to Felicia, Lamar left home to begin his journey of what he expected to be 4 years of college. During his separation from Felicia, Lamar kept in touch with her through letters, phone calls, and e-mails. After about a month into the semester, Felicia called Lamar to tell him she was pregnant. Although Lamar loved Felicia, he was not sure he was ready to commit to one woman, and he knew he was not ready for the responsibilities of fatherhood. But for Felicia, abortion was out of the question. "How could this have happened?" Lamar agonized. "What will happen to my future? My hopes and dreams? My basketball scholarship?"

Each of the above scenarios depicts a real situation that is based on the authors' experiences with college students. These scenarios exemplify the major problems that are discussed in this chapter: violence and abuse in intimate partner and family relationships, problems of divorce, and problems associated with nonmarital and teenage childbearing. Other problems facing families are discussed in other chapters in this text. For example, families are affected by health problems (Chapter 2), substance abuse (Chapter 3), problems of youth and the elderly (Chapter 6), problems with employment and poverty (Chapters 10 and 11), and so on. Before discussing these problems, we look briefly at diversity in families worldwide, and we provide an overview of the changing patterns and structures of households and families in the United States.

The Global Context: Families of the World

Family is a central aspect of every society throughout the world. But family forms are diverse. In the United States, the only legal form of marriage is **monogamy**—a marriage between two partners. A common marriage pattern in the United States is **serial monogamy**—a succession of marriages in which a person has more than one spouse over a lifetime but is legally married to only one person at a time. Although Vermont allows same-sex couples to apply for a "civil union" certificate that entitles them to the same legal rights and responsibilities of marriage, same-sex couples may not be legally married in the United States (see Chapter 9). At least six European countries as well as Canada, Brazil, and Portugal give legal recognition to same-sex couples, and two countries (Denmark and The Netherlands) permit same-sex marriages.

In France, Parliament passed a new law creating a type of marital arrangement called *Pacte civil de solidarite* (civil solidarity pact), or PACS. PACS are similar to marriage in that each party is responsible for financially supporting the other and sharing in each others' debts accumulated during the pact, and partners are eligible for the others' work benefits. But unlike traditional marriages, PACS can be broken in a short period of time and can be entered into by same-sex couples, although nearly 40 percent of couples opting for PACS in France are heterosexual couples (Daley 2000).

Some societies practice **polygamy**—a form of marriage in which one person may have two or more spouses. The most common form of polygamy is **polygyny**—the concurrent marriage of one man with two or more women. Polygyny may be practiced in some Islamic societies, including some regions of Africa. The second form of polygamy is **polyandry**—the concurrent marriage of one woman with two or more men. Polyandry is very rare, but does occur in some groups including Tibetans and various African groups.

Family values, roles, and norms are also highly variable across societies. For example, unlike in Western societies, in Asian countries some marriages are arranged by the parents, who select mates for their children. Another traditional Asian practice is for the eldest son and his wife to move in with the son's parents, where his wife takes care of her husband's parents. In some societies, it is normative for married couples to view each other as equal partners in the marriage, whereas in other societies social values dictate that wives be subservient to their husbands. The role of children also varies across societies. In some societies, children as young as 5 years old are expected to work full-time to help support the family. Social acceptance of unmarried couples having children also varies widely throughout the world. Acceptance of this lifestyle ranges from 90 percent or more in parts of Western Europe to fewer than 15 percent in Singapore and India (Global Study of Family Values 1998).

Finally, the ease of obtaining a divorce varies by country. Ireland did not allow divorce under any conditions until 1995, when voters narrowly approved a public referendum allowing divorce for the first time (Thompson & Wyatt

1999). In 2000, the Egyptian Parliament voted to allow women to file for divorce on grounds of incompatibility (Eltahawy 2000). Prior to this new law, Egyptian women could file for divorce only in cases of proven physical or psychological abuse. By contrast, a man could simply say, "I divorce you" three times or get a divorce by filing a paper with the marriage registrar without even notifying his wife. Under the new law, a woman who wants a divorce must return her husband's dowry and relinquish all financial claims, including alimony.

It is clear from the previous discussion that families are shaped by the social and cultural context in which they exist. As we discuss the family problems addressed in this chapter—violence and abuse, divorce, and nonmarital and teenage childbearing—we refer to social and cultural forces that contribute to these problematic events. Next we look at changing patterns and structures of U.S. families and households.

Changing Patterns and Structures in U.S. Families and Households

Over the last century, dramatic changes have occurred in the patterns and structures of U.S. families and households. A **family household**, as defined by the U.S. Census Bureau, consists of two or more persons related by birth, marriage, or adoption who reside together. **Nonfamily households** may consist of one person who lives alone, two or more people as roommates, or cohabiting heterosexual or homosexual couples involved in a committed relationship.

The **traditional family** in which the husband is a breadwinner and the wife is a homemaker has declined significantly since the 1940s (see Figure 5.1). In the 1940s, about two-thirds of U.S. families fit the traditional model. In 1999, two-thirds (64 percent) of U.S. married-couple families with children under 18 were dual-earner marriages; less than one-third (29 percent) of these families involved an employed Dad and a stay-at-home Mom (*Statistical Abstract 2000*).

In addition to the rise in dual-earner marriages, singlehood and cohabitation have also become more common in U.S. society. Although most households today (69 percent) are family households, nonfamily households have increased significantly, from 19 percent of households in 1970 to 31 percent in 1999 (*Statistical Abstract 2000*). The majority of nonfamily households consist of one person living alone. However, many households considered "nonfamily" by the Census Bureau include cohabiting heterosexual couples and same-sex couples

■ **Figure 5.1** *The Decline of Traditional Families in the United States: 1940–2000. Traditional families refer to families in which the husband is the breadwinner and the wife is a homemaker; other families include single-mother families, single-father families, married couples with only the wife working, married couples with no workers, and other related people living together, with or without workers.*

Source: Scott Coltrane and Randall Collins. 2001. *Sociology of Marriage and the Family: Gender, Love, and Property,* 5th ed, p. 12. Belmont, CA: Wadsworth.

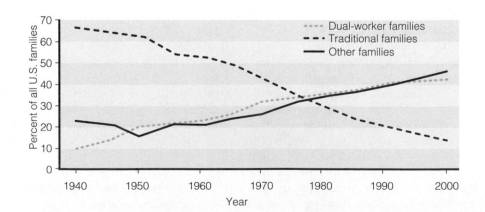

who function economically and emotionally as a family. About half of all women in the United States ages 25 to 40 have cohabited (*Statistical Abstract 2000*). For some, cohabitation is a part of the modern courtship process; for others, cohabitation is an informal type of marriage or alternative to marriage.

Increasingly, social and legal policy and court decisions are expanding the concept of family to include unmarried living-together couples and same-sex couples who live together and view themselves as a family. In some U.S. cities and counties, heterosexual cohabiting couples as well as gay and lesbian couples who apply for a **domestic partnership** designation are granted legal entitlements such as health insurance benefits and inheritance rights that have traditionally been reserved for married couples. In Vermont, same-sex couples may obtain a "civil union" license which carries many of the same rights and responsibilities as a marriage license. Some employers offer health insurance benefits to cohabiting partners of employees and grant family leave to an employee whose cohabiting partner becomes ill, dies, or gives birth or adopts.

Single-parent family households have also increased in recent decades, although their growth is slowing. As shown in Figure 5.2, one in ten U.S. households consists of a single parent with children under 18. In 1999, over half (58 percent) of black families with children under 18 were single-parent families, compared with 30 percent of Hispanic and 23 percent of white families (*Statistical Abstract 2000*). The percentage of U.S. children living with one parent increased from 20 percent in 1980 to 27 percent in 1999; 23 percent live with their mother only and 4 percent live with their father only (Forum on Child and Family Statistics 2000). Some children live with a single parent who has a cohabiting partner: 16 percent of children living with single fathers and 9 percent of children living with single mothers also live with their parents' partners (Forum on Child and Family Statistics 2000).

Divorce, remarriage, and stepfamilies have also become more common in the last several decades. As we discuss later, about half of marriages today are expected to end in divorce. More than half of persons who divorce get remarried. Nearly half (46 percent) of all marriages in 1998 involved a remarriage for either the bride, groom, or both (up from 31 percent in 1970) (*Statistical Abstract 1999*). About one-third of children will live in a remarried or cohabiting stepfamily household before they reach adulthood (Bumpass, Raley, & Sweet 1995).

The trend toward diversification of family forms is viewed by some as signaling the disappearance of marriage and the breakdown of the family. But al-

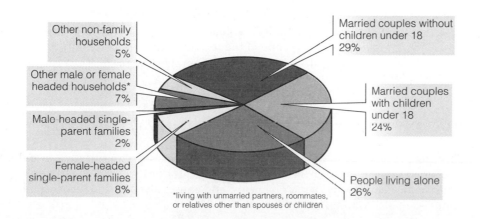

Other non-family households 5%

Other male or female headed households* 7%

Male headed single-parent families 2%

Female-headed single-parent families 8%

*living with unmarried partners, roommates, or relatives other than spouses or children

Married couples without children under 18 29%

Married couples with children under 18 24%

People living alone 26%

■ **Figure 5.2** *The Many Types of U.S. Households, 1999.*

Source: *Statistical Abstract of the United States: 2000.* 2000. U.S. Bureau of the Census. Table 60.

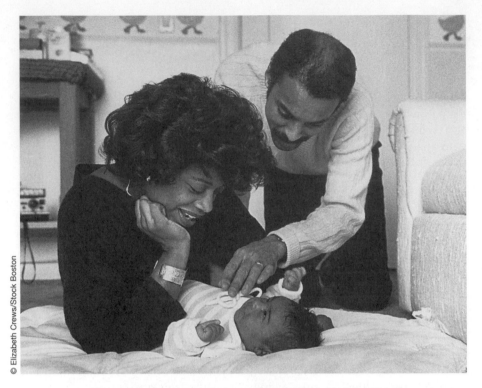

Two parents are present in just under half of U.S. black families.

though traditionalists may view cohabitation (both heterosexual and homosexual), divorce, and single parenthood as threats to the institution of marriage, it is unlikely that marriage will disappear. Most people who cohabit eventually marry, although not necessarily to the person they cohabited with. Very few Americans remain single; less than 4 percent of Americans 65 years old and over have never married (*Statistical Abstract 2000*). Although the high rate of marital dissolution seems to suggest a weakening of marriage, divorce may also be viewed as resulting from placing a high value on marriage, such that a less-than-satisfactory marriage is unacceptable. The high rate of out-of-wedlock childbirth and single parenting is also not necessarily indicative of a decline in the value of marriage. Edin (2000) found that in a sample of low-income single women with children, most said they would like to be married, but just haven't found "Mr. Right". The persistent value placed on family is also evident in a national survey of first-year students in U.S. colleges and universities in which students listed "raising a family" as one of the top two values (second only to "being well-off financially") (American Council on Education & University of California 2000).

Sociological Theories of Family Problems

Three major sociological theories—structural-functionalism, conflict theory, and symbolic interactionism—help to explain different aspects of the family problems we focus on in this chapter: violence and abuse, divorce, and teenage and nonmarital childbearing.

Structural-Functionalist Perspective

The structural-functionalist perspective views the family as a social institution that performs important functions for society, including reproducing new members, regulating sexual activity and procreation, socializing the young, and providing physical and emotional care for family members. According to the structural-functionalist perspective, traditional gender roles contribute to family functioning. Thus, heterosexual two-parent families in which women perform the "expressive" role of managing household tasks and providing emotional care and nurturing to family members and men perform the "instrumental" role of earning income and making major family decisions are viewed as "normal" or "functional." Other family forms (e.g., single-parent families, same-sex unions) are viewed as abnormal or dysfunctional.

According to the structural-functionalist perspective, the high rate of divorce and the rising numbers of single-parent households constitute a "breakdown" of the family. This breakdown is the result of rapid social change and social disorganization. The functionalist perspective views the breakdown of the family as one of the primary social problems in the world today—a problem of such magnitude that it leads to such secondary social problems as crime, poverty, and substance abuse.

Functionalist explanations of family problems examine how changes in other social institutions contribute to family problems. For example, a structural-functionalist view of divorce examines how changes in the economy (more dual-earner marriages) and in the legal system (such as the adoption of "no-fault" divorce) contribute to high rates of divorce. Changes in the economic institution, specifically falling wages among unskilled and semi-skilled men, also contribute to both intimate partner abuse and the rise in female-headed single parent households (Edin 2000).

Conflict Perspective

Conflict theory focuses on how social class and power influence marriages and families. Within families, the unequal distribution of power among women and men may contribute to domestic violence. The traditional male domination of families—a system known as **patriarchy**—includes the attitude that married women are essentially the property of husbands. When wives violate or challenge the male head-of-household's authority, the male may react by "disciplining" his wife or using anger and violence to reassert his position of power in the family. The unequal distribution of wealth between men and women, with men traditionally earning more money than women, contributes to inequities in power and fosters economic dependence of wives on husbands.

The conflict perspective also helps to explain many racial and ethnic differences in families. Higher rates of teenage and single parenting among African Americans, for example, are related to the lower socioeconomic status of this population.

Conflict theorists emphasize that social programs and policies that affect families are largely shaped by powerful and wealthy segments of society. The interests of corporations and businesses are often in conflict with the needs of families. Corporations and businesses strenuously fought the passage of the 1993 Family and Medical Leave Act, giving people employed full-time for at least 12 months in companies with at least 50 employees up to 12 weeks of unpaid time

Our leaders talk as though they value families, but act as though families were a last priority.

SYLVIA HEWLETT AND
CORNELL WEST
Family advocates

off for parenting leave, illness or death of a family member, and elder care. Hewlett and West (1998) note that corporate interests undermine family life "by exerting enormous downward pressure on wage levels for young, child-raising adults" (p. 32). Government, which is largely influenced by corporate interests through lobbying and political financial contributions, enacts policies and laws that serve the interests of for-profit corporations, rather than families.

Symbolic Interactionist Perspective

Symbolic interactionism emphasizes that human behavior is largely dependent upon the meanings and definitions that emerge from small group interaction. Divorce, for example, was once highly stigmatized and informally sanctioned through the criticism and rejection of divorced friends and relatives. As societal definitions of divorce became less negative, however, the divorce rate increased. The social meanings surrounding single parenthood, cohabitation, and delayed childbearing and marriage have changed in similar ways. As the definitions of each of these family variations became less negative, the behaviors became more common.

The culture we have does not make people feel good about themselves. And you have to be strong enough to say if the culture doesn't work, don't buy it.

MORRIE SCHWARTZ
Sociologist
From **Tuesdays with Morrie**

Symbolic interactionists also point to the effects of labeling on one's self-concept and the way the self-fulfilling prophecy can affect family members' behavior toward one another. The **self-fulfilling prophecy** implies that we behave according to the expectations of others. For example, when a non-custodial divorced parent (usually a father) is awarded visitation rights, he may view himself as a visitor in his children's lives. The meaning attached to the visitor status can be an obstacle to the father's involvement as the label "visitor" minimizes the importance of the non-custodial parent's role and results in conflict and emotional turmoil for fathers (Pasley & Minton 2001). Fathers' rights advocates suggest replacing the term "visitation" with such terms as "parenting plan" or "time-sharing arrangement," as these latter terms do not minimize either parent's role.

The symbolic interactionist perspective is useful in understanding the dynamics of domestic violence and abuse. For example, some abusers and their victims learn to define intimate partner violence as an expression of love (Lloyd 2000). **Emotional abuse** often involves using negative labels (e.g., "stupid," "whore," "bad,") to define a partner or family member. Such labels negatively affect the self-concept of abuse victims, often convincing them that they deserve the abuse. Next, we discuss violence and abuse in intimate and family relationships, noting the scope, causes, and consequences of this troubling social problem.

Violence and Abuse in Intimate and Family Relationships

Although intimate and family relationships provide many individuals with a sense of intimacy and well-being, for others, these relationships involve physical violence, verbal and emotional abuse, sexual abuse, and/or neglect. Indeed, in U.S. society, people are more likely to be physically assaulted, abused and neglected, sexually assaulted and molested, and killed in their own homes and by other family members than anywhere else or by anyone else (Gelles 2000, 183). Before reading further, you may want to take the Abusive Behavior Inventory in this chapter's *Self and Society* feature.

Abusive Behavior Inventory

Circle the number that best represents your closest estimate of how often each of the behaviors happened in your relationship with your partner or former partner during the previous six months.

1. **Never**
2. **Rarely**
3. **Occasonally**
4. **Frequently**
5. **Very frequently**

1. Called you a name and/or criticized you.   1 2 3 4 5
2. Tried to keep you from doing something you wanted to do (e.g., going out with friends, going to meetings).   1 2 3 4 5
3. Gave you angry stares or looks.   1 2 3 4 5
4. Prevented you from having money for your own use.   1 2 3 4 5
5. Ended a discussion with you and made the decision himself/herself.   1 2 3 4 5
6. Threatened to hit or throw something at you.   1 2 3 4 5
7. Pushed, grabbed, or shoved you.   1 2 3 4 5
8. Put down your family and friends.   1 2 3 4 5
9. Accused you of paying too much attention to someone or something else.   1 2 3 4 5
10. Put you on an allowance.   1 2 3 4 5
11. Used your children to threaten you (e.g., told you that you would lose custody, said he/she would leave town with the children).   1 2 3 4 5
12. Became very upset with you because dinner, housework, or laundry was not done when he/she wanted it done or done the way he/she thought it should be.   1 2 3 4 5
13. Said things to scare you (e.g., told you something "bad" would happen, threatened to commit suicide.)   1 2 3 4 5
14. Slapped, hit, or punched you.   1 2 3 4 5
15. Made you do something humiliating or degrading (e.g., begging for forgiveness, having to ask his/her permission to use the car or to do something).   1 2 3 4 5
16. Checked up on you (e.g., listened to your phone calls, checked the mileage on your car, called you repeatedly at work).   1 2 3 4 5
17. Drove recklessly when you were in the car.   1 2 3 4 5
18. Pressured you to have sex in a way you didn't like or want.   1 2 3 4 5
19. Refused to do housework or child care.   1 2 3 4 5
20. Threatened you with a knife, gun, or other weapon.   1 2 3 4 5
21. Spanked you.   1 2 3 4 5
22. Told you that you were a bad parent.   1 2 3 4 5
23. Stopped you or tried to stop you from going to work or school.   1 2 3 4 5
24. Threw, hit, kicked, or smashed something.   1 2 3 4 5
25. Kicked you.   1 2 3 4 5
26. Physically forced you to have sex.   1 2 3 4 5
27. Threw you around.   1 2 3 4 5
28. Physically attacked the sexual parts of your body.   1 2 3 4 5
29. Choked or strangled you.   1 2 3 4 5
30. Used a knife, gun, or other weapon against you.   1 2 3 4 5

SCORING: Add the numbers you circled and divide the total by 30 points to find your score. The higher your score, the more abusive your relationship.

The inventory was given to 100 men and 78 women equally divided into groups of abusers/abused and nonabusers/nonabused. The men were members of a chemical dependency treatment program in a veterans' hospital and the women were partners of these men. Abusing or abused men earned an average score of 1.8; abusing or abused women earned an average score of 2.3. Nonabusing/abused men and women earned scores of 1.3 and 1.6, respectively.

Source: Melanie F. Shepard and James A. Campbell. The abusive behavior inventory: A measure of psychological and physical abuse. *Journal of Interpersonal Violence*, September 1992, 7, no. 3, 291–305. Inventory is on pages 303–304. Used by permission of Sage Publications, 2455 Teller Road, Newbury Park, CA 91320.

Breast bruised, brains
battered.
Skin scarred, soul
shattered.
Can't scream—neighbors
stare,
Cry for help—no one's
there.

STANZA FROM A POEM BY
NENNA NEHRU
a battered Indian woman

■ Intimate Partner Violence and Abuse

Globally, one woman in every three has been subjected to violence in an intimate relationship (United Nations Development Programme 2000). **Intimate partner violence** refers to actual or threatened violent crimes committed against persons by their current or former spouses, boyfriends, or girlfriends. Data from the National Crime Victimization Survey indicate that most (85 percent) acts of intimate partner violence are committed against women and one out of four U.S. women say they've been victims of violence or stalking by a spouse, partner, or date at some point in their lives (Rennison & Welchans 2000). Although women also assault their male partners, these assaults tend to be acts of retaliation or self-defense (Johnson 2001).

Factors associated with higher rates of intimate partner victimization against women include being young (ages 16 to 24), black, divorced or separated, and earning lower incomes (see Figures 5.3, 5.4, 5.5, and 5.6). The rate and severity of physical violence tend to be higher among cohabiting couples compared with dating and marital partners (Johnson & Ferraro 2001; Magdol, Moffitt, Caspi, & Silva 1998; Stets & Straus 1989).

Four patterns of partner violence have been identified: common couple violence, intimate terrorism, violent resistance, and mutual violent control (Johnson & Ferraro 2001). **Common couple violence** refers to occasional acts of violence arising from arguments that get "out of hand." Common couple violence

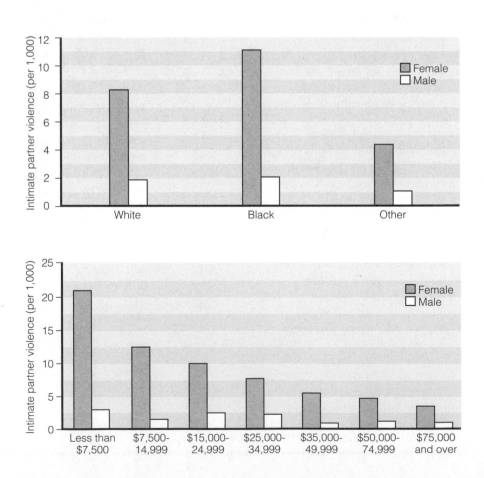

■ **Figure 5.3** *Rate of Intimate Partner Violence, by Victim's Race, 1993–1998.*

Source: Rennison, C. M. and S. Welchans. 2000. "Intimate Partner Violence." U.S. Department of Justice. Office of Justice Programs, Washington, D.C.: Bureau of Justice Statistics.

■ **Figure 5.4** *Rate of Intimate Partner Violence, by Annual Household Income, 1993–1998.*

Source: Rennison, C. M. and S. Welchans. 2000. "Intimate Partner Violence." U.S. Department of Justice. Office of Justice Programs, Washington, D.C.: Bureau of Justice Statistics.

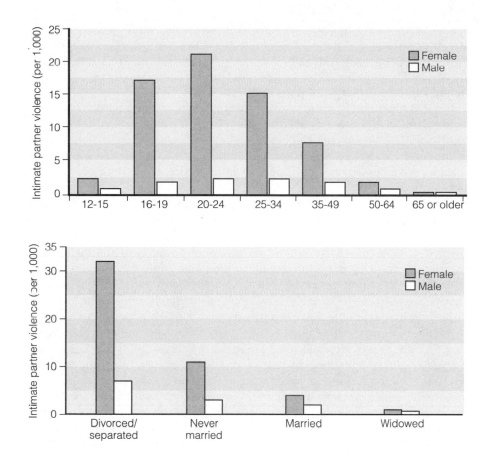

■ **Figure 5.5** *Rate of Intimate Partner Violence, by Age, 1993–1998.*

Source: Rennison, C. M. and S. Welchans. 2000. "Intimate Partner Violence." U.S. Department of Justice. Office of Justice Programs, Washington, D.C.: Bureau of Justice Statistics.

■ **Figure 5.6** *Rate of Intimate Partner Violence, by Marital Status, 1993–1998.*

Source: Rennison, C. M. and S. Welchans. 2000. "Intimate Partner Violence." U.S. Department of Justice. Office of Justice Programs, Washington, D.C.: Bureau of Justice Statistics.

usually does not escalate into serious or life-threatening violence. **Intimate terrorism** is violence that is motivated by a wish to control one's partner and involves the systematic use of not only violence, but economic subordination, threats, isolation, verbal and emotional abuse, and other control tactics. Intimate terrorism is almost entirely perpetrated by men and is more likely to escalate over time and to involve serious injury. **Violent resistance** refers to acts of violence that are committed in self-defense. Violent resistance is almost exclusively perpetrated by women against a male partner. **Mutual violent control** is a rare pattern of abuse "that could be viewed as two intimate terrorists battling for control" (Johnson & Ferraro 2001, 169).

Intimate partner abuse also takes the form of sexual aggression, including rape. **Sexual aggression** refers to sexual interaction that occurs against one's will through use of physical force, threat of force, pressure, use of alcohol/drugs, or use of position of authority. An estimated 7 to 14 percent of married women have been raped by their husbands (Monson, Byrd, & Langhinrichsen-Rohling 1996). A national study found that about 3 percent of college women experience a completed or attempted rape during a college year (Fisher, Cullen, & Turner 2000). Nearly 90 percent of the sexually assaulted college women knew the person who assaulted them.

Battering is the single major cause of injury to women in the United States ("Domestic Violence Fact Sheet" 1999). Data from the National Crime Victimization Survey (Rennison & Welchans 2000) reveal that, between 1993 and 1998, most (about two-thirds) of the male and female victims of intimate partner violence were physically attacked; the remaining third were victims of threats or attempted violence. Half of female victims were physically injured by their abuser versus one-third of male victims. Most injuries are minor, involving cuts and bruises. However, 5 percent of female victims and 4 percent of male victims reported sustaining serious injuries.

In 1998 about 1,830 murders were attributed to intimate partner violence; nearly 3 out of 4 of these murder victims were women (Rennison & Welchans 2000). In 1996, 30 percent of all female murders were perpetrated by husbands, ex-husbands, or boyfriends. Three percent of all male murder victims were killed by wives, ex-wives, or girlfriends (National Center for Injury Prevention and Control 2000).

Many battered women are abused during pregnancy, resulting in a high rate of miscarriage and birth defects. Psychological consequences for victims of intimate partner violence can include depression, suicidal thoughts and attempts, lowered self-esteem, alcohol and other drug abuse, and post-traumatic stress disorder (National Center for Injury Prevention and Control 2000).

Battering also interferes with women's employment. Some abusers prohibit their partners from working. Other abusers "deliberately undermine women's employment by depriving them of transportation, harassing them at work, turning off alarm clocks, beating them before job interviews, and disappearing when they promise to provide child care" (Johnson & Ferraro 2000, 177). Battering also undermines employment by causing repeated absences, impairing women's ability to concentrate, and lowering their self-esteem and aspirations.

Abuse, whether physical or emotional, is no doubt a factor in many divorces and is also a primary cause of homelessness. One study of homeless parents found that 22 percent were fleeing from abuse (National Coalition for the Homeless 1999).

In 1998, about 4 of 10 female victims of intimate partner violence lived in households with children under age 12 (Rennison & Welchans 2000). Children who witness marital violence may experience emotional, behavioral, and academic problems and may commit violence in their own relationships later (Parker, Bergmark, Attridge, & Miller-Burke 2000). Children may also commit violent acts against a parent's abusing partner.

Why do abused adults stay in abusive relationships? Reasons include love, emotional dependency, commitment to the relationship, hope that things will get better, the view that violence is legitimate because they "deserve" it, guilt, fear, economic dependency, and feeling stuck. Abuse in relationships is usually not ongoing and constant, but rather occurs in cycles. The **cycle of abuse** involves a violent or abusive episode, followed by a makeup period where the abuser expresses sorrow and asks for forgiveness and "one more chance." The honeymoon period may last for days, weeks, or even months before the next outburst of violence occurs.

Child Abuse

Child abuse refers to the physical or mental injury, sexual abuse, negligent treatment, or maltreatment of a child under the age of 18 by a person who is responsible for the child's welfare (Willis, Holden, & Rosenberg 1992). In 1998, nearly 12 percent of victims of child maltreatment were sexually abused, almost a quarter (23 percent) suffered physical abuse, and more than half (54 percent) suffered **neglect**—the child caregiver's failure to provide adequate attention and supervision, food and nutrition, hygiene, medical care, and a safe and clean living environment. In 1998, over 1,000 U.S. children died of abuse and neglect ("Child Abuse and Neglect National Statistics" 2000). In a national survey of U.S. adults, 27 percent of the women and 16 percent of the men reported being victims of child sexual abuse (Finkelhor, Hotaling, Lewis, & Smith 1990).

Children ages 0 to 3 have the highest rates of victimization. Asian/Pacific Islander children have the lowest rates of victimization: 3.8 per 1,000 children. ("Child Abuse and Neglect National Statistics" 2000). The highest rate of child victimization rates are for African Americans (20.7 victims per 1,000 children). Child victimization rates for American Indians/Alaska Natives are 19.8, for Hispanics 10.6, and for Caucasians 8.5. Children with developmental disabilities, emotional problems, or hyperactivity are especially likely to be abused, probably because caring for such children is more demanding and stressful for parents and caregivers.

The effects of child abuse and neglect vary according to the frequency and intensity of the abuse or neglect. Research on the effects of child abuse suggest that abused children are at higher risk for aggressive behavior, low self-esteem, depression, and low academic achievement (Gelles & Conte 1991; Lloyd & Emery 1993). Adolescents and adults who were abused as children are more vulnerable to low self-esteem, depression, unhappiness, anxiety, increased risk of substance abuse, juvenile delinquency, and suicide. Physical injuries sustained by child abuse cause pain, disfigurement, scarring, physical disability, and death.

© Carol Geddes/Sonoma Image Studios

Although many adult survivors of child abuse suffer long-term negative consequences, survivors can also heal and overcome their painful past. David Pelzer, author of the autobiographical bestsellers A Child Called "It," The Lost Boy, *and* A Man Named Dave *describes his mother holding his arm in the flame of the gas stove, requiring him to sleep in the cold garage, denying him food for days at a time, slamming his face into a soiled baby's diaper, forcing him to vomit and then eat it, forcing ammonia and dishwashing soap down his throat, throwing a knife into his stomach, and repeatedly beating him until he was black and blue, among other horrendous acts. Although David's experience of child abuse was one of the most severe ever discovered in California, he has gone on to a fulfilling and successful life as a husband, father, inspirational speaker, volunteer, child abuse prevention activist, and bestselling author.*

Child Sexual Abuse: A College Student Tells Her Story

Jennifer Doyle is one of thousands of women who have experienced child sexual abuse. As a way to heal from this traumatic experience, help others who have had similar experiences, and educate the public at large, she has told her story on the Internet. The following excerpts represent bits and pieces of Jennifer's story. . . .

Remembering the Abuse

I didn't admit to the abuse for a long time. I couldn't let myself believe it. I have a tendency to "split" my father into two people. It's very hard for my mind to connect the man who came into my bed at night with the man who used to throw me up in the air and catch me until I laughed like crazy, with the man who bought me my first puppy, with the man who used to throw money in the front lawn some mornings so my sister and I could have a treasure hunt. There is a side of that man that I loathe with every cell in my body, but there is also a side of me that loves him simply because he was my father, and because he was not always "bad."

I've always known what happened . . . I just couldn't let myself know or think about it because I could not handle the pain and couldn't deal with it. I believe this defense mechanism is called repression. I always knew something was wrong. I always knew what had happened, but I didn't want to think about it, so I didn't. I forced it to the back of my mind and, from everyone else's point of view, was a fairly normal child.

I didn't have specific memories for a while, but it's not because they weren't there—it's because I didn't allow myself access to them. I didn't want to remember.

And they didn't come to me "all at once," or simply appear one day when I woke up. They started with a vague suspicion I tried to deny that became a horrible certainty I didn't know how to express. It was more of an awakening. What set it in motion, I think, was when I was in junior high, my father would brush across my knee repeatedly or rub my back incessantly, despite my constant pleas for him to stop and my attempts to move away from those hands that never seemed to go away. I felt then just so utterly dirty and used and destroyed. Once I admitted my feelings to myself, that I had felt this way before and I knew why I did—because it had happened before—it was like opening Pandora's box. I was totally unprepared for what spewed forth, and what continues to.

Sometimes when I think I've remembered all there is to know, another memory comes to mind and I think, "Oh, yes, I'd forgotten about that. Why did I forget that? Why did I remember it now? What does it mean?" Other times I'm just scared and angry and I think, "How could he do that to me? Why didn't anyone stop him? Why didn't I tell someone sooner?"

The Abuse

I think that the abuse must have started before I was five, maybe at two or three—possibly back into infancy. It started out with just fondling, and then it turned into oral sex. And after that, I'm not ready to talk about at this point. Not in detail. But I lost my virginity before I was in the second grade. I have this one memory, when I was five or six, of climbing into my father's bed, and I know he's naked, and the sheets are pale pink, and I ask him what game we're going to play this time. I know that when the abuse started, I didn't say no. I

Among females, early forced sex is associated with lower self-esteem, higher levels of depression, running away from home, alcohol and drug use, and more sexual partners (Jasinski, Williams, & Siegel 2000, Lanz 1995; Whiffen, Thompson, & Aube 2000). Sexually abused girls are also more likely to experience teenage pregnancy, have higher numbers of sexual partners in adulthood, and are more likely to acquire sexually transmitted infections and experience forced sex (Browning & Laumann 1997; Stock, Bell, Boyer, & Connell 1997). Women who were sexually abused as children also report a higher frequency of post-traumatic stress disorder (Spiegel 2000) and suicide ideation (Thakkar, Gutierrez, Kuczen, & McCanne 2000). Spouses who were physically and sexually abused as children report lower marital satisfaction, higher stress, and lower family cohesion than spouses with no abuse history (Nelson & Wampler 2000).

liked it; I thought it was a game. And I didn't think there was anything wrong with it. I thought everyone played with their fathers that way.

I'm not exactly sure how long the sexual abuse went on . . . it had stopped by the time I was in sixth grade, I'm pretty sure. What I remember most is always feeling dirty and not knowing, at the time, exactly why. I remember that by seventh grade I was continually warding off my father's advances and removing his hands from my body. He also came into the bathroom a few times when I was taking a bath, assumingly by accident. I think other people (like my sister and mother) noticed but never really said anything.

In addition to the sexual abuse, my father also abused me emotionally and physically. He accused me of being a drug addict, a lesbian, and a slut. He begged me to forgive him and made me feel like it was my fault. He would pretend like he was going to hit me, and then hit the wall above my head at the last second, just to scare me. He hit me with boards and with a belt. He would sit on me for long periods of time and refuse to move until I begged him. He threw my body against walls. He grabbed my arms, pushed me into furniture, and slapped me across the face. He hit me in the back of my head and knocked me unconscious. He held a gun to my head.

Talking About the Abuse; Telling People About It

One of the hardest things for me to do is talk about my past. In the past few years, I've done a lot of talking though. I've told most of my friends that I was sexually abused by my father, and through my web page I've told a lot of strangers as well.

As for my family . . . I haven't talked to them about it much at all. I brought the subject up to my older sister one time. She says she was molested by the son of one of our parents' friends. I asked her if maybe it had been our father, and she emphatically said no and that nothing had happened to me either. I don't know if she either doesn't remember or is in denial. Since that time, she has conceded that she believes me that our father abused me, but at the same time she says she will deny it because she doesn't want to think about it. She still sees him regularly.

My mother found out over the summer of 1998, when she saw a book on sexual abuse survivors I had beside my bed. She asked me if my father had ever sexually abused me, and I said yes, and made it clear I didn't wish to discuss it further. She assumed it happened while I was in high school, and for years I didn't correct her that the abuse occurred way before then.

Once I told my sister about my web site, she told my mother, father, and stepmother—along with various other relatives. Most of them no longer talk to me. My mother asserts that sexual abuse survivors have to be careful who they tell. She also believes me, and feels that the abuse is not her fault because when she asked me if anything was going on, I would never answer her. It's true—I never did answer her, because I knew that she would do nothing to stop it and, while I was pretty sure she had to know what was going on, to admit that she knew concretely and chose not to help me would have been unbearable.

I read that most survivors go through a "tell everyone" stage, and I know I did. It's starting to come to a close now, because I've learned there's more to me than the abuse.

Source: http://www.geocities.com/midsecret78. Used by permission of Jennifer Doyle.

Adult males who were sexually abused as children tend to exhibit depression, substance abuse, and difficulty establishing intimate relationships (Krug 1989). Sexually abused males also have a higher risk of anxiety disorders, sleep and eating disorders, and sexual dysfunctions (Elliott & Briere 1992).

Effects of child sexual abuse are likely to be severe when the sexual abuse is forceful, is prolonged, and involves intercourse, and when the abuse was perpetrated by a father or stepfather (Beitchman et al. 1992). Not only has the child been violated physically, she or he has lost an important social support. One woman who had been sexually abused by her father described feeling that she had lost her father; he was no longer a person to love and protect her (Spiegel 2000). This chapter's *The Human Side* feature describes one young woman's experiences of being sexually abused by her father.

Elder Abuse

Community-based surveys suggest that 3 to 6 percent of elders in the United States (over age 65) report having experienced elder abuse (Lachs, Williams, O'Brien, Hurst, & Horwitz 1997). **Elder abuse** includes physical abuse, psychological abuse, financial abuse (such as improper use of the elder's financial resources), and neglect. Elder neglect includes failure to provide basic health and hygiene needs such as clean clothes, doctor visits, medication, and adequate nutrition. Neglect also involves unreasonable confinement, isolation of elderly family members, lack of supervision, and abandonment. Most abuses of the elderly are committed by adult children, followed by spouses (Lachs et al. 1997).

Factors Contributing to Intimate Partner and Family Violence and Abuse

Research suggests that cultural, community, and individual and family factors contribute to domestic violence and abuse (Willis et al. 1992).

Cultural Factors Violence in the family stems from our society's acceptance of violence as a legitimate means of enforcing compliance and solving conflicts at personal, national, and international levels (Viano 1992). Violence and abuse in the family may be linked to cultural factors such as violence in the media (see Chapter 4), acceptance of corporal punishment, gender inequality, and the view of women and children as property:

1. *Acceptance of corporal punishment.* **Corporal punishment** involves the use of physical force with the intention of causing a child to experience pain, but not injury, for the purpose of correction or control of a child's behavior (Straus 2000). Many mental health professionals and child development specialists argue that it is ineffective and damaging to children. Children who experience corporal punishment display more antisocial behavior, are more violent, and have an increased incidence of depression as adults (Straus 2000). Yet many parents accept the cultural tradition of spanking as an appropriate form of child discipline. More than 90 percent of parents of toddlers reported using corporal punishment (Straus 2000). Eighty-three percent of more than 11,000 undergraduate students at the University of Iowa reported that they had experienced some form of physical punishment during their childhood (Knutson & Selner 1994). In a national study of mothers of children aged 6 to 9, 44 percent reported having spanked their children during the week prior to the study and that they had spanked them an average of 2.1 times that week (Straus, Sugarman, & Giles-Sims 1997). A national survey of parental discipline practices found that in 1998, 45 percent of parents reported that they had spanked or hit their child in the past month (compared to 62 percent in 1988) (Daro 1998). Although not everyone agrees that all instances of corporal punishment constitute abuse, undoubtedly, some episodes of parental "discipline" are abusive.
2. *Gender role socialization.* Traditional male gender roles have taught men to be aggressive and to be dominant in male-female relationships. Male abusers are likely to hold traditional attitudes toward women and male-female roles (Lloyd & Emery 2000). Anderson (1997) found that men who earn less money than their partners are more likely to be violent toward them. "Disenfranchised men then must rely on other social practices to construct a

masculine image. Because it is so clearly associated with masculinity in American culture, violence is a social practice that enables men to express a masculine identity" (p. 667). Traditional female gender roles have also taught women to be submissive to their male partner's control.

3. *View of women and children as property.* Before the late nineteenth century, a married woman was considered to be the property of her husband. A husband had a legal right and marital obligation to discipline and control his wife through the use of physical force. The expression "rule of thumb" can be traced to an old English law that permitted a husband to beat his wife with a rod no thicker than his thumb; it was originally intended as a humane measure to limit how harshly men could beat their wives. This traditional view of women as property may contribute to men doing with their "property" as they wish.

The view of women and children as property also explains marital rape and father-daughter incest. Historically, the penalties for rape were based on property rights laws designed to protect a man's property—his wife or daughter—from rape by other men; a husband or father "taking" his own property was not considered rape (Russell 1990). In the past, a married woman who was raped by her husband could not have her husband arrested because marital rape was not considered a crime. Today, marital rape is considered a crime in all 50 states.

> Our culture promotes sexual victimization of women and children when it encourages males to believe that they have overpowering sexual needs that must be met by whatever means available.
>
> EDWIN SCHUR
> *Sociologist*

Community Factors Community factors that contribute to violence and abuse in the family include social isolation and inaccessible or unaffordable health care, day care, elder care, and respite care facilities:

1. *Social isolation.* Living in social isolation from extended family and community members increases a family's risk for abuse. Isolated families are removed from material benefits, care-giving assistance, and emotional support from extended family and community members.

2. *Inaccessible or unaffordable community services.* Failure to provide medical care to children and elderly family members (a form of neglect) is sometimes a result of the lack of accessible or affordable health care services in the community. Failure to provide supervision for children and adults may result from inaccessible day care and elder care services. Without elder care and respite care facilities, socially isolated families may not have any help with the stresses of caring for elderly family members and children with special needs.

Individual and Family Factors Individual and family factors associated with intimate partner and family violence and abuse include a history of violence, drug and alcohol abuse, poverty, and fatherless homes:

1. *History of abuse.* Mothers who have been sexually abused as children are more likely to physically abuse their own children (DiLillo, Tremblay, & Peterson 2000). Although a history of abuse is associated with an *increased likelihood* of being abusive as an adult, *most* adults who were abused as children do not continue the pattern of abuse in their own relationships (Gelles 2000).

2. *Drug and alcohol abuse.* More than half of prison and jail inmates serving time for violence against an intimate had been using alcohol, drugs, or both at the time of the incident for which they were incarcerated (U.S. Depart-

ment of Justice 1998). Alcohol use is reported as a factor in 50 to 75 percent of incidents of physical and sexual aggression in intimate relationships (Lloyd & Emery 2000). Alcohol and other drugs increase aggression in some individuals and enable the offender to avoid responsibility by blaming his or her violent behavior on drugs/alcohol.

3. *Poverty.* Abuse in adult relationships occurs among all socioeconomic groups. However, Kaufman and Zigler (1992) point to a relationship between poverty and child abuse:

> Although most poor people do not maltreat their children, and poverty, per se, does not cause abuse and neglect, the correlates of poverty, including stress, drug abuse, and inadequate resources for food and medical care, increase the likelihood of maltreatment (p. 284).

Strategies for Action: Preventing and Responding to Violence and Abuse in Intimate and Family Relationships

Ameliorating and preventing intimate violence against women will ultimately require a deep commitment to ending both the glorification of violence and the many facets of patriarchy that are still embedded in our culture, our laws, and our intimate lives.

SALLY A. LLOYD
Director of Women's Studies, Miami University

Strategies to prevent family violence and abuse include **primary prevention** strategies that target the general population, **secondary prevention** strategies that target groups at high risk for family violence and abuse, and **tertiary prevention** strategies that target families who have experienced abuse (Gelles 1993; Harrington & Dubowitz 1993).

Primary Prevention Strategies

Preventing violence and abuse may require broad, sweeping social changes such as eliminating the norms that legitimize and glorify violence in society and changing the sexist character of society (Gelles 2000). Specific abuse prevention strategies include public education and media campaigns that may help reduce domestic violence by conveying the criminal nature of domestic assault and offering ways to prevent abuse ("When you are angry at your child, count to 10 and call a friend . . ."). Other prevention efforts focus on parent education to teach parents realistic expectations about child behavior and methods of child discipline that do not involve corporal punishment. Sweden passed a law in 1979 that bans the use of corporal punishment on children. Parent training classes are a high school graduation requirement in some states (Shapiro & Schrof 1995).

Another strategy involves reducing violence-provoking stress by reducing poverty and unemployment and providing adequate housing, child care programs and facilities, nutrition, medical care, and educational opportunities. However, rather than strengthening the supports for poor families with children, welfare reform legislation enacted in 1996 limits cash assistance to poor single parents to 2 consecutive years with a 5-year lifetime limit (some exceptions are granted) and forces women into the labor force. Many women going from welfare to work will experience greater hardships as a result of a loss of food stamp benefits, increases in federal housing rent, loss of Medicaid benefits, cost of transportation to work, and child care costs (Edin & Lein 1997). Some women forced to go to work and unable to afford childcare will leave their chil-

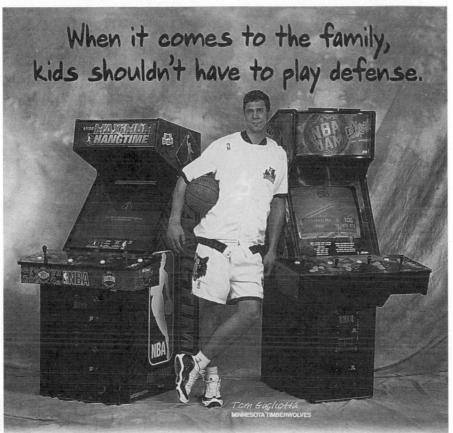

Public service campaigns against child abuse use sports figures such as Tom Gugliotta to promote their message of non-violence.

dren unattended, increasing child neglect. The cumulative stresses and hardships that the welfare-to-work legislation will have on single parents may very well contribute to increases in child neglect and abuse.

Secondary Prevention Strategies

Families at risk of experiencing violence and abuse include low-income families, parents with a history of depression or psychiatric care, single parents, teenage mothers, parents with few social and family contacts, individuals who experienced abuse in their own childhood, and parents or spouses who abuse drugs or alcohol. Secondary prevention strategies, designed to prevent abuse from occurring in high-risk families, include parent education programs, parent support groups, individual counseling, substance abuse treatment, and home visiting programs.

A nationwide program called Healthy Families America (HFA) offers one or two home visits to all new parents. During these visits, service providers share information about baby care and development and parenting skills, and explain how parents can utilize other organizations and agencies that provide family support. Families most at risk of child maltreatment are offered intensive home visitation services (at least once a week) for 3 to 5 years. There are more than 3000 HFA sites around the country.

Tertiary Prevention Strategies

The National Domestic Violence Hotline (1-800-799-SAFE) is a 24- hour, toll-free service that provides crisis assistance and local domestic violence shelter and safe house referrals for callers across the country. Shelters provide abused women and their children with housing, food, and counseling services. Safe houses are private homes of individuals who volunteer to provide temporary housing to abused women who decide to leave their violent home. Some communities have abuse shelters for victims of elder abuse. The fact that more shelters exist for abandoned and abused animals in the United States than shelters for battered women is commonly cited as reflecting the inadequacy of services available to abused women. The passage of the Violence Against Women Act in 2000 promises more federal funds for more battered women's services and transitional housing.

Abused or neglected children may be removed from the home. All states have laws that permit abused or neglected children to be placed in out-of-home care, such as foster care (Stein 1993). However, federal law requires that states have programs to prevent family breakup when desirable and possible without jeopardizing the welfare of children in the home. **Family preservation programs** are in-home interventions for families who are at risk of having a child removed from the home because of abuse or neglect.

Alternatively, a court may order an abusing spouse or parent to leave the home. Abused spouses or cohabiting partners may obtain a restraining order prohibiting the perpetrator from going near the abused partner. About half of the states and Washington, D.C., now have mandatory arrest policies that require police to arrest abusers, even if the victim does not want to press charges. However, between 1993 and 1998 only about half of all victims of intimate partner violence reported the violence to law enforcement authorities (Rennison & Welchans 2000). Reasons for not reporting intimate partner violence to the police include (1) such violence is a private or personal matter, (2) fear of retaliation, (3) viewing the violence as a "minor" crime, (4) to protect the offender, and (5) the belief that the police will not help or will be ineffective (Rennison & Welchans 2000). Unfortunately, legal action is not always effective in preventing subsequent violence. Nearly 4 in 100 violent offenders sentenced to local jail for intimate violence were on probation or under a restraining order at the time they committed the offense (U.S. Department of Justice 1998).

Treatment for abusers—which may be voluntary or mandated by the court—typically involves group and/or individual counseling; substance abuse counseling; and/or training in communication, conflict resolution, and anger management. Men who stop abusing their partners learn to take responsibility for their abusive behavior, develop empathy for their partner's victimization, reduce their dependency on their partners, and improve their communication skills (Scott & Wolfe 2000).

Divorce

The United States has the highest divorce rate among Western nations (Thompson & Wyatt 1999). U.S. divorce rates peaked around 1980 and have remained relatively stable (with a slight decline) since that time (Emery 1999). Whereas marriages made in 1890 had only a one in ten chance of ending in divorce, by 1930, the chances rose to one in four, and by the 1970s, the chances were one in two. This rate has more or less stabilized; thus half of first marriages today are estimated to end in divorce (Coltrane & Collins 2001). Nearly one-fourth (23.3 percent) of U.S. first-year college students have parents who are divorced (American Council on Education & University of California 2000).

Individual and relationship factors that contribute to divorce include incompatibility in values or goals, poor communication, lack of conflict resolution skills, sexual incompatibility, extramarital relationships, substance abuse, emotional or physical abuse or neglect, boredom, jealousy, and difficulty coping with change or stress related to parenting, employment, finances, in-laws, and illness. Other demographic and life-course factors that are predictive of divorce include marriage order (second and subsequent marriages are more prone to divorce than first marriages), cohabitation (couples who live together before marriage are more prone to divorce), teenage marriage, premarital pregnancy, and low socioeconomic status.

Various social factors also contribute to the high rate of divorce in U.S. society. These include the following structural and cultural forces.

1. *Changing family functions.* Before the Industrial Revolution, the family constituted a unit of economic production and consumption, provided care and protection to its members, and was responsible for socializing and educating children. During industrialization, other institutions took over these functions. For example, the educational institution has virtually taken over the systematic teaching and socialization of children. Today, the primary function of marriage and the family is the provision of emotional support, intimacy, affection, and love. When marital partners no longer derive these emotional benefits from their marriage, they may consider divorce with the hope of finding a new marriage partner to fulfill these affectional needs.

2. *Increased economic autonomy of women.* As noted earlier, before 1940 most wives were not employed outside the home and depended on their husband's income. Today, nearly two-thirds of married women are in the labor force (*Statistical Abstract 2000*). Wives who are unhappy in their marriage are more likely to leave the marriage if they have the economic means to support themselves. Unhappy husbands may also be more likely to leave a marriage if their wives are self-sufficient and can contribute to the support of the children.

3. *Increased work demands and dissatisfaction with marital division of labor.* Another factor influencing divorce is increased work demands and the stresses of balancing work and family roles. Workers are putting in longer hours, often working overtime or taking second jobs. Many employed parents, particularly mothers, come home to work a **second shift**—the work involved in caring for children and household chores (Hochschild 1989). Wives are more likely than husbands to perceive the marital division of labor—household chores and child care—as unfair (Nock 1995). This perception of un-

> Women simply expect different things from a marriage than men do, and if they don't get them, they prefer to live alone.
>
> VIBEKE NISSEN
> *Danish psychologist*

fairness can lead to marital tension and resentment, as reflected in the following excerpt:

> My husband's a great help watching our baby. But as far as doing housework or even taking the baby when I'm at home, no. He figures he works five days a week; he's not going to come home and clean. But he doesn't stop to think that I work seven days a week. Why should I have to come home and do the housework without help from anybody else? My husband and I have been through this over and over again. Even if he would just pick up from the kitchen table and stack the dishes for me, that would make a big difference. He does nothing. On his weekends off, I have to provide a sitter for the baby so he can go fishing. When I have a day off, I have the baby all day long without a break. He'll help out if I'm not here, but the minute I am, all the work at home is mine (quoted in Hochschild 1997, 37–38).

4. *Liberalized divorce laws.* Before 1970, the law required a couple who wanted a divorce to prove that one of the spouses was at fault and had committed an act defined by the state as grounds for divorce—adultery, cruelty, or desertion. In 1969, California became the first state to initiate **no-fault divorce**, which permitted a divorce based on the claim that there were "irreconcilable differences" in the marriage. No-fault divorce law has contributed to the U.S. divorce rate by making divorce easier to obtain. Although U.S. divorce rates started climbing before California instituted the first no-fault divorce law, the widespread adoption of such laws has probably contributed to their continued escalation. Today, all 50 states recognize some form of no-fault divorce.

5. *Changing cultural values.* U.S. society is characterized by **individualism**— the tendency to focus on one's individual self-interests rather than on the interests of one's family and community. The high divorce rate has been linked to the continued rise of individualism. "For many, concerns with self-fulfillment and careerism diminished their commitment to family, rendering marriage and other intimate relationships vulnerable" (Demo, Fine, & Ganong 2000, 281). **Familism**—the view that the family unit is more important than individual interests— is still prevalent among Asian Americans and Mexican Americans, which helps to explain why the divorce rate is lower among these groups than among whites and African Americans (Mindel, Habenstein, & Wright 1998).

   The value of marriage has also changed in American culture. The increased social acceptance of nonmarital sexuality, nonmarital childbearing, cohabitation, and singlehood reflect the view that marriage is an option, rather than an inevitability. The view of marriage as an option, rather than as an imperative, is mirrored by the view that divorce is also an acceptable option. Divorce today has less social stigma than in previous generations. "Marital dissolution, once considered to be a rare, unfortunate, and somewhat shameful deviation from normal family life, has become part of mainstream America" (Thompson & Amato 1999, xi).

6. *Increased life expectancy.* Finally, more marriages today end in divorce, in part, because people live longer than they did in previous generations. Americans born in 1900 could expect to live an average of only 47 years. This figure increased to 59 for those born in 1925, 68 for those born in 1950, and 77 for those born in 1997 (*Statistical Abstract 2000*). Because people live longer today than in previous generations, "till death do us part" involves a longer commitment than it once did. Indeed, one can argue that "marriage once was as unstable as it is today, but it was cut short by death not divorce" (Emery 2000, 7).

In this new model of family and work life, a tired parent flees a world of unresolved quarrels and unwashed laundry for the reliable orderliness, harmony, and managed cheer of work.

ARLIE HOCHSCHILD
Sociologist

If it's not fun don't do it.

INDIVIDUALISTIC SLOGAN IN BEN AND JERRY'S ICE CREAM STORE

It is now widely accepted that men and women have the right to expect a happy marriage, and that if a marriage does not work out, no one has to stay trapped.

SYLVIA ANN HEWLETT
Family advocate

Consequences of Divorce

When parents have bitter and unresolvable conflict, and/or if one parent is abusing the child or the other parent, divorce may offer a solution to family problems. But divorce often has negative effects for ex-spouses and their children and also contributes to problems that affect society as a whole.

Health Consequences A large number of studies have found that divorced individuals, compared to married individuals, experience lower levels of psychological well-being, including more unhappiness, depression, anxiety, and poorer self-concepts. Divorced individuals also have more health problems and a higher risk of mortality (Amato 2001; Waite & Gallagher 2000). Both divorced and never-married individuals are, on average, more distressed than married people because unmarried people are more likely than married people to have low social attachment, low emotional support, and increased economic hardship (Walker 2001).

Economic Consequences In families at the lowest income levels, divorce can improve women's income because men who earn little income can be a drain on the family's finances. However, "there is a clear pattern that the economic well-being of divorced women and their children plunges in comparison to predivorce levels " (Demo, Fine, & Ganong 2000, 281). The economic costs of divorce are greater for women and children because women tend to earn less than men (see Chapter 8) and mothers devote substantially more time to household and child care tasks than fathers do. The time women invest in this unpaid labor restricts their educational and job opportunities as well as their income.

Men are less likely than women to be economically disadvantaged after divorce. Men often enjoy a better financial situation after divorce (Peterson 1996) as they continue to profit from earlier investments in education and career. However, men with low and unstable earnings often experience financial strain following divorce, which often underlies failure to pay child support. When custodial mothers who did not receive child support were asked the reasons why they did not receive it, 66 percent gave the reason as "father unable to pay" (Henry 1999). The Fragile Families Coalition estimates that over 3 million noncustodial fathers are eligible for food stamps (Henry 1999). If these fathers are so poor that they need assistance simply to put food on the table for themselves, is it fair to characterize them as "deadbeats" when they do not pay child support? Instead of being deadbeats, some Dads are simply dead broke (Henry 1999).

Effects on Children Although divorce, following high conflict, may actually improve the emotional well-being of children relative to staying in a conflicted home environment (Jekielek 1998), for many children parental divorce has detrimental effects. With only one parent in the home, children of divorce, as well as children of never-married mothers, tend to have less adult supervision compared with children in two-parent homes. Lack of adult supervision is related to higher rates of juvenile delinquency, school failure, and teenage pregnancy (Popenoe 1993). A survey of 90,000 teenagers found that the mere physical presence of a parent in the home after school, at dinner, and at bedtime significantly reduces the risk of teenage suicide, violence, and drug use (Resnick et al. 1997). Many of the negative effects of divorce on children are related to the economic hardship associated with divorce. Economic hardship is associated with less effective and less supportive parenting, inconsistent and harsh discipline, and emotional distress in children (Demo et al. 2000).

In a review of the literature on the effects of parental divorce on college-age students, Nielson (1999) found that when their parents divorce and their mother remarries within a few years, most children do not suffer serious long-term consequences in terms of self-confidence, mental health, or academic achievements. However, in those cases where the mother does not remarry (about 15 percent), the consequences are more likely to be negative. Children of divorced women who do not remarry tend to experience a reduced standard of living and a lack of parental discipline. These children "fail to develop as much self-control, self-motivation, self-reliance, and self-direction as people their own age whose mothers have remarried" (Nielsen 1999, 547).

Children who live with their mothers may suffer from a damaged relationship with their nonresidential father, especially if he becomes disengaged from their lives. On the other hand, children may benefit from having more quality time with their fathers after parental divorce. Some fathers report that they became more active in the role of father after divorce. One father commented:

> In the last 4½ years, I have developed an incredibly strong and loving bond with my two sons. I am actively involved in all aspects of their lives. I have even coached their soccer and basketball teams. . . . The time I spend with them is very quality time—if anything, the divorce has made me a better and more caring father . . . not to say this would not have happened if my marriage had worked out (quoted in Pasley & Minton 2001, 248).

Divorce often results in decreased involvement of fathers in their children's lives. However, many noncustodial fathers report that although their time with their children is limited (for example, to every other weekend), they are more focused on and involved with their children than before the divorce.

Divorced fathers and their children usually remain close only if the mother actively encourages and facilitates their relationship (Nielsen 1999). However, some divorced mothers not only fail to encourage their children's relationships with their fathers, but they actively attempt to alienate the children from their father. (We note that some divorced fathers do likewise.) Thus, some children of divorce suffer from **parental alienation syndrome** (PAS), which is defined as an emotional and psychological disturbance in which children engage in exaggerated and unjustified denigration and criticism of a parent (Gardner 1998). "Children of PAS show negative parental reactions and perceptions which can be grossly exaggerated. . . . Put simply, they profess rejection and hatred of a previously loved parent, most often in the context of divorce and child custody conflicts" (Family Court Reform Council of America 2000). Parental alienation syndrome has been described as a form of "psychological kidnapping" whereby one parent manipulates children's psyches to make them hate and reject the other parent. Children who suffer from PAS are victims of a form of child abuse in which one parent essentially brain-

© Myrleen Ferguson/PhotoEdit

washes the child to hate the other parent. A parent may alienate his or her child from the other parent by engaging in the following behaviors (Schacht 2000):

- Minimizing the importance of contact and relationship with the other parent.
- Being rude to the other parent; refusing to speak to or tolerate the presence of the other parent, even at events important to the child; refusing to allow the other parent near the home for drop-off or pick-up visitations.
- Failing to express concern for missed visits with the other parent.
- Failing to display any positive interest in the child's activities or experiences during visits.
- Expressing disapproval or dislike of the child's spending time with the other parent.
- Refusing to discuss anything about the other parent ("I don't want to hear about . . .") or selective willingness to discuss only negative matters.
- Making innuendoes and accusations against the other parent, including statements that are false.
- Portraying the child as an actual or potential victim of the other parents' behavior.
- Demanding that the child keep secrets from the other parent.
- Destruction of gifts or memorabilia from the other parent.
- Promoting loyalty conflicts (for example, offering an opportunity for a desired activity that conflicts with scheduled visitation).

Long-term effects of PAS on children are extremely serious and can include long-term depression, inability to function, guilt, hostility, alcoholism and other drug abuse, and other symptoms of internal distress (Family Court Reform Council of America 2000). Indeed, the effects on the rejected parent are equally devastating.

Some noncustodial divorced fathers discontinue contact with their children as a coping strategy for managing emotional pain (Pasley & Minton 2001). Many divorced fathers are overwhelmed with feelings of failure, guilt, anger, and sadness over the separation from their children (Knox 1998). Hewlett and West (1998) explain that "visiting their children only serves to remind these men of their painful loss, and they respond to this feeling by withdrawing completely" (p. 69). Divorced fathers commonly experience the legal system as favoring the mother in child-related matters. One divorced father commented:

> I believe that the system [judges, attorneys, etc.] have [sic] little or no consideration for the father. At some point the system creates an environment where the father loses any natural desire to see his children because it becomes so difficult, both financially and emotionally. At that point, he convinces himself that the best thing to do is wait until they are older. (quoted in Pasley & Minton 2001, 242)

In conclusion, we note that "while some children emerge from divorce with a strong sense of loss and feelings of anger or despair, others are capable of acknowledging that divorce has pain but also benefits for themselves and other family members" (Thompson & Amato 1999, xix). Factors that determine whether children are scathed or strengthened by parental divorce include the quality of the child's relationship with both parents, the extent of ongoing conflict versus cooperation between the parents, and the economic circumstances of the child's household. In most circumstances, children adapt to divorce, "showing resiliency, not dysfunction" (Thompson & Amato 1999, xix).

It seems clear that the majority of children from divorced families "pass" rather than "fail" the test of healthy child development as measured by key indicators of their academic, social, and psychological adjustment.

ROBERT E. EMERY
University of Virginia

Strategies for Action: Responding to the Problems of Divorce

Two general strategies for responding to the problems of divorce are (1) strategies to prevent divorce and strengthen marriages and (2) strategies to strengthen post-divorce families.

Divorce Prevention Strategies

One strategy for preventing divorce is to prepare individuals and couples for marriage through family life education courses (taught in high schools and colleges) and premarital education (also known as premarital counseling) which usually takes place at churches or with private counselors. Premarital education involves evaluating the relationship, exploring potential problems, and learning positive ways of communicating and resolving conflict. Several states have proposed legislation requiring premarital education or encouraging it by lowering marriage license fees for those who participate, and imposing delays on issuing marriage licenses for those who refuse premarital education (Clark 1996; Peterson 1997). Researchers have found that couples who participated in a widely used couples' education program called PREP (the Prevention and Relationship Enhancement Program), had a lower divorce/separation rate 5 years after completing the program compared with couples who did not participate (Stanley, Markman, St. Peters, & Leber 1995). Marriage enrichment programs for newlyweds have also been found to be effective in improving communication and conflict resolution and overall marital satisfaction (Cole, Cole, & Gandolfo 2000).

Several states are considering legislation that is intended to decrease the number of divorces by extending the waiting period required before a divorce is granted or requiring proof of fault (e.g., adultery, abuse) (Amato 1999). Opponents argue that **divorce law reform** measures would increase acrimony between divorcing spouses (which harms the children as well as the adults involved), increase the legal costs of getting a divorce (which leaves less money to support any children), and delay court decisions on child support and custody and distribution of assets. Efforts in many state legislatures to repeal no-fault divorce laws have largely failed.

However, in 1997, the Louisiana legislature became the first in the nation to pass a law creating a new kind of marriage contract. Under the new law, couples can voluntarily choose between the standard marriage contract that allows a no-fault divorce (after a 6-month separation) or a **covenant marriage** that permits divorce only under condition of fault or after a marital separation of more than 2 years. Couples who choose a covenant marriage must also get premarital counseling. The covenant marriage law was designed to strengthen marriages and discourage divorce. However, less than 1 percent of Louisiana couples have opted for a covenant marriage license (Flory 2000). As of this writing, several states are considering marriage laws similar to the covenant marriage, but only one other state—Arizona—has passed a similar marriage law.

Strengthening marriages may be achieved through policies and services that provide greater resources to couples whose marriages are at risk. These supports include marriage counseling, flexible workplace policies that decrease work/family conflict, affordable child care, and economic support such as the

Earned Income Tax Credit (Amato 1999). In a nationwide poll of parents' political priorities (National Parenting Association 1996) parents indicated that they want economic help in raising their children and work policies that help them balance their work and family roles.

Strengthening Postdivorce Families

Negative consequences of divorce for children may be minimized if both parents continue to spend time with their children on a regular and consistent basis and communicate to their children that they love them and are interested in their lives. Parental conflict, either in intact families or divorced families, negatively influences the psychological well-being of children (Demo 1992; Demo 1993). Ongoing conflict between divorced parents also tends to result in decreased involvement of nonresidential fathers with their children (Leite & McKenry 2000). By maintaining a civil coparenting relationship during a separation and after divorce, parents can minimize the negative effects of divorce on their children.

What can society do to promote cooperative parenting by ex-spouses? One answer is to encourage, or even mandate, divorcing couples to participate in divorce mediation. In **divorce mediation** divorcing couples meet with a neutral third party, a mediator, who helps them resolve issues of property division, child custody, child support, and spousal support in a way that minimizes conflict and encourages cooperation. Children of mediated divorces adjust better to the divorce than children of litigated divorces (Marlow & Sauber 1990). An increasing number of jurisdictions and states have mandatory child custody mediation programs, whereby parents in a custody or visitation dispute must attempt to resolve their dispute through mediation before a court will hear the case.

Another trend aimed at strengthening postdivorce families is the establishment of parenting programs for divorcing parents. Programs such as "Sensible Approach to Divorce" (Wyandotte County, Kansas) "Parenting after Divorce" (Orange County, North Carolina) and "Focus on Children in Separation" (Jackson County, Missouri) emphasize the importance of cooperative co-parenting for the well-being of children. Parents are taught about children's reactions to divorce, nonconflictual co-parenting skills, and how to avoid negative behavior toward their ex-spouse. In some programs (such as "Focus on Children in Separation"), children participate and are taught that they are not the cause of the divorce, how to deal with grief reactions to divorce, and techniques for talking to parents about their concerns. A few states require all divorcing parents to attend a parenting program, as do more than 100 courts throughout the United States.

> Although divorce ends a marriage, it does not end the family.
>
> ROSS A. THOMPSON AND PAUL R. AMATO
> *The Postdivorce Family*

> My toughest fight was with my first wife.
>
> MUHAMMAD ALI

Nonmarital and Teenage Childbearing

In addition to domestic violence and the problems associated with divorce, another concern for families is the set of problems related to the high rate of unmarried and teenage childbearing. More than a million births occur to unmarried women annually in the United States, representing about a third of all births (Curtin & Martin 2000). The percent of U.S. births to unmarried women rose from only 3.8 percent in 1940 to 33 percent in 1999 and seems to have leveled off in the 1990s (Ventura & Bachrach 2000). Compared with some countries, the United States has a high proportion of nonmarital births. In Germany,

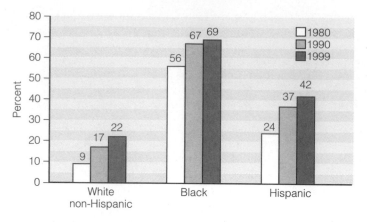

Figure 5.7 *Percent of all Births to Unmarried Women: United States, 1980, 1990, and 1999.*

Source: Stephanie J. Ventura and Christine A. Bachrach. 2000 (Oct. 18). "Nonmarital Childbearing in the United States, 1940–99." U.S. Department of Health and Human Services. Figure 9, p. 6.

Italy, Greece, and Japan, less than 15 percent of births occur out of wedlock. Some countries, however, have rates of out of wedlock childbearing that are much higher than that of the United States. In 1998, two-thirds of births in Iceland were to unmarried women and half or more of births in Norway and Sweden were out of wedlock (Ventura & Bachrach 2000).

Rates of nonmarital childbearing vary by age, race, and ethnicity. In 1970, half of all nonmarital births were to teens. However, in 1999, less than 3 in 10 (29 percent) of nonmarital births were to teenagers; the remaining 70 percent of unmarried births were to women ages 20 and older (Ventura & Bachrach 2000). Although the *number* of out-of-wedlock births to non-Hispanic whites is higher, the *rates* of such births are higher for blacks and Hispanics (see Figure 5.7). The highest rate of nonmarital childbirth is among blacks. As shown in Figure 5.7 more than two-thirds of all births to non-Hispanic black women in the United States are to unmarried women.

Social Factors Related to Nonmarital and Teenage Childbearing

To understand the rise in the number of children born to teenage and unmarried women, we must consider several social factors. First, having a baby outside of marriage has become more socially acceptable and does not carry the stigma it once did. In the 1950s and 60s, more than half of U.S. women who gave birth to a baby conceived out of wedlock married before the birth of the baby. In the 1990s, less than one in four (23 percent) of such women married before the birth of the baby (Ventura & Bachrach 2000). About half of pregnant teens carry their babies to term, 35 percent have an abortion, and the remainder miscarry. Most teens today who carry their babies to term (95 percent) keep the baby rather than place it for adoption (Jorgensen 2000). When single women in a Time/CNN poll were asked if they would consider raising a child on their own, 61 percent said "yes" (Edwards 2000). This chapter's *Social Problems Research Up Close* feature examines attitudes of low-income single mothers toward marriage that help to explain why they are not married.

Perceptions of Marriage among Low-Income Single Mothers

Single mothers with low incomes face both economic hardships and parenting challenges. These hardships and challenges could potentially be alleviated by marriage to a partner who contributes income to the family, and who shares the responsibilities of housework and child care and supervision. Yet, low-income single women have low rates of marriage and remarriage. Sociologist Kathryn Edin (2000) conducted research on the perceptions of marriage among low-income single mothers that helps explain why these mothers are not married. Here we summarize her research and its findings.

Sample and Methods

This study focused on transcripts and field notes collected from multiple qualitative interviews with 292 low-income single mothers. Initially, interviewers utilized the resources of community groups and local institutions to recruit eligible mothers. They then relied on references from the mothers themselves to gain access to others who might also participate in the interviews. The sample was evenly divided between African Americans and whites in three U.S. cities. Approximately half of the participants in each city depended on welfare, and about half worked at low-wage jobs earning less than $7.50/hour.

Researchers scheduled a minimum of two interviews with each mother to maximize rapport, and to encourage them to expand and clarify responses that may have been initially unclear. Interviewers asked re-

spondents to describe the circumstances surrounding the births of their children, their current family situations and relationship with each child's father, their views about how their family situations might change over time, and their views of marriage in general. Interviewers focused on two issues in their conversations with mothers: why women with few economic resources choose to have children and why these same women fail to marry or remarry.

Findings and Conclusions

Edin found four primary reasons why low-income single mothers do not marry. These include the low economic status of available men; the desire for parental, economic, and relationship control; mistrust; and domestic violence.

1. *Low economic status of men.* Edin found that men's income is a significant factor in poor single mothers' willingness to marry, as "they simply could not afford to keep an unproductive man around the house" (p. 118). Nearly every mother in Edin's sample said that it was important for any man they would marry to have a "good job." Although total earnings is the most important aspect for single mothers, also important are the regularity of those earnings, the effort men make to find and keep employment, and the source of the income. "Women whose male partners couldn't, or wouldn't find work, often lost respect for them and 'just couldn't stand' to keep them around" (p. 119). Mothers did not consider earnings from crime (e.g., drug dealing) as legitimate earnings and reported that

they wouldn't marry such a man no matter how much money he made from crime.

Respectability was also an issue for poor single mothers:

> Mothers said that they could not achieve respectability by marrying someone who was frequently out of work, otherwise underemployed, supplemented his income through criminal activity, and had little chance of improving his situation over time. Mothers believed that marriage to such a man would diminish their respectability, rather than enhance it. (p. 120)

Mothers also believed that marriage to a lower-class man would be unlikely to last because the economic pressures on the relationship would be too great. Mothers talked about the "sacred" nature of marriage, and believed that no "respectable" woman would marry a man whose economic situation would likely lead to a future divorce. Edin notes that:

> In interview after interview, mothers stressed the seriousness of the marriage commitment and their belief that "it should last forever." Thus, it is not that mothers held marriage in low esteem, but rather the fact that they held it in such high esteem that convinced them to forego marriage, at least until their prospective marriage partner could prove himself worthy economically or they could find another partner who could. (pp. 120–121)

2. *Desire for parental, economic, and relationship control.* When asked what they liked best about being a single parent, the most common response was "I am in charge," or "I am in control."

Continued

Most mothers felt that the presence of fathers often interfered with their parental control. Single mothers in Edin's study felt that a husband might be "too demanding" of them and thus impede their efforts to spend time with their children.

Mothers were also concerned about losing control of the family's economic situation. One mother said, "[I won't marry because] the men take over the money. I'm too afraid to lose control of my money again" (p. 121).

Regardless of whether or not the prospective wife worked, mothers feared that prospective husbands would expect to be the "head of the house," who, being "in charge," would want to make the "final" decisions about child rearing, finances, and other matters. Mothers also expressed the view that if they married, their husbands would expect them to do all of the household chores. Some women described their relationships with their ex-partners as "like having one more kid to take care of" (p. 122).

Most single mothers in Edin's sample wanted to marry eventu-

ally and thought that the best way to maximize their control in a marriage was by working and earning an income. One African-American mother explained,

> One thing my mom did teach me is that you must work some and bring some money into the household so you can have a say in what happens. If you completely live off a man, you are helpless. That is why I don't want to get married until I get my own [career] and get off of welfare. (p. 122)

Mothers also wanted to develop their labor market skills before marriage to ensure against destitution in the event of a divorce and to increase their bargaining power in their relationships. Mothers believed that women who were economically dependent on men had to "put up with all kinds of behavior" because they could not legitimately threaten to leave.

> Mothers felt that if they became more economically independent . . . they could legitimately threaten to leave their husbands if certain conditions (i.e., sexual fidelity) weren't met. These threats would, in turn they be-

lieved, keep a husband on his best behavior. (p. 122)

Many of the single mothers in Edin's study had held traditional gender role attitudes when they were younger and still in a relationship with their children's fathers. When the men for whom they sacrificed so much gave them nothing but pain and anguish, they felt they had been 'duped' and were no longer willing to be dependent or subservient to men.

3. *Mistrust.* Many of the single mothers in Edin's sample did not marry because they did not trust their partners or men in general, and/or because their partners did not trust them. Some fathers claimed that the child was not theirs because the mother was "a whore." One partner of a pregnant woman in the sample told the interviewer, "how do I know the baby's mine? Who knows if she hasn't been stepping out on me with some other man and now she wants me to support another man's child!" (p. 124).

Mothers tended to mistrust men because their previous boyfriends and partners had

Singlehood has also become more acceptable and more common. The dictionary once defined a *spinster* as an unmarried woman above age 30. In previous generations, being a spinster meant the fear of isolation, living alone, and being somewhat of a social outcast. Today, women are "more confident, more self-sufficient, and more choosy than ever [and] no longer see marriage as a matter of survival and acceptance" (Edwards 2000, 48). The women's movement created both new opportunities for women and expectations of egalitarian relationships with men. A Time/CNN poll of single women found that 61 percent felt that men were "too controlling" (Edwards 2000, 50). Increasingly, college-educated women in their 30s and 40s are making "the conscious decision to have a child on their own because they haven't found Mr. Adequate, let alone Mr. Right" (Drummond 2000, 54).

Increased acceptance of cohabitation and same-sex relationships has also contributed to the high rate of nonmarital births. In the 1990s, about 40 percent

been unfaithful. This experience was so common among respondents that many simply did not believe men could be faithful to only one woman. Most said they would rather never marry than to "let him make a fool out of me" (p. 124). "Nonetheless, many of *these same women* often held out hope of finding a man who was 'different', one who could be trusted" (p. 125).

4. *Domestic violence.* For some mothers in Edin's sample, domestic violence in either their childhood or adult lives played a role in their negative attitudes toward marriage. Many mothers reported physical abuse during pregnancy and several mothers had miscarriages because of such abuse. One woman who had been in the hospital three times because of abuse said, "I was terrified to leave because I knew it would mean going on welfare. . . . But that is okay. I can handle that. The thing I couldn't deal with is being beat up" (p.126). Edin commented,

> The fact that women tended to experience repeated abuse from their children's fathers before

they decided to leave attests to their strong desire to make things work with their children's fathers. Many women finally left when they saw the abuse beginning to affect their children's well-being. (p. 126)

In conclusion, low-income single mothers in Edin's study were reluctant to marry the father of their children because these men had low economic status, traditional notions of male domination in household and parental decisions, and patterns of untrustworthy and even violent behavior. Given the low level of trust these mothers have of men and given their view that husbands want more control than the women are willing to give them, women realize that a marriage that is also economically strained is likely to be conflictual and short-lived. "Interestingly, mothers say they reject entering into economically risky marital unions out of respect for the institution of marriage, rather than because of a rejection of the marriage norm" (Edin 2000, 130). Low-income single mothers in Edin's study "say they are willing, even eager, to marry if the marriage represents an increase in their class

standing and if . . . their prospective husbands' behavior indicates he won't beat them, abuse their children, refuse to share household tasks, insist on making all decisions, be sexually unfaithful, or abuse alcohol or drugs" (p. 113):

> In short, the mothers interviewed here believe that marriage will probably make their lives more difficult than they are currently. . . . If they are to marry, they want to get something out of it. If they cannot enjoy economic stability and gain upward mobility from marriage, they see little reason to risk the loss of control and other costs they fear marriage might exact from them. Unless low-skilled men's economic situations improve and they begin to change their behaviors toward women, it is quite likely that most low-income women will continue to resist marriage. (p. 130)

Source: Edin, Kathryn. 2000. "What Do Low-Income Single Mothers Say about Marriage?" *Social Problems* 47(1):112133. Used by permission of University of California Press.

of out-of-wedlock births were to women who were cohabiting, presumably with the child's father (Bachu 1999), but in some cases with the mother's same-sex partner.

Teenage pregnancy has been related to a variety of factors, including low self-esteem and hopelessness, low parental supervision, and perceived lack of future occupational opportunities (Jorgensen 2000; Luker 1996). Although lack of information about and access to contraceptives contributes to unintended teenage pregnancy, 30 to 40 percent of adolescent pregnancies are intended (Jorgensen 2000). Teenage females who do poorly in school may have little hope of success and achievement in pursuing educational and occupational goals. They may think that their only remaining option for a meaningful role in life is to become a parent. In addition, some teenagers feel lonely and unloved and have a baby to create a sense of feeling needed and wanted.

For many disadvantaged teenagers, childbearing reflects—rather than causes—the limitations of their lives.

ELLEN W. FREEMAN AND KARL RICKELS

Social Problems Related to Nonmarital and Teenage Childbearing

Teenage and unmarried childbirth are considered social problems because of the adverse consequences for women and children that are associated with such births.

1. *Poverty for single mothers and children*. Many unmarried mothers, especially teenagers, have no means of economic support or have limited earning capacity. Single mothers and their children often live in substandard housing and have inadequate nutrition and medical care. Even with public assistance, many unwed parents struggle to survive economically. By the late 1990s, more than half of the families below the official poverty level were female-headed single-parent families. As shown in Figure 5.8, median household income for female-headed single-parent families is less than half of household income for married couples, and is significantly less than male-headed single-parent families (U.S. Bureau of the Census 2000).

2. *Poor health outcomes*. Compared with older pregnant women, pregnant teenagers are less likely to receive timely prenatal care and to gain adequate weight and are more likely to smoke and use alcohol and drugs during pregnancy (Jorgensen 2000; Ventura, Curtin, & Mathews 2000). As a consequence of these and other factors, infants born to teenagers are at higher risk of low birth weight, of premature birth, and of dying in the first year of life.

3. *Low academic achievement*. Low academic achievement is both a contributing factor and a potential outcome of teenage parenthood. Teens whose parents have not graduated from high school are at higher risk for becoming pregnant (Hogan, Sun, & Cornwell 2000). Three-fifths of teenage mothers drop out of school and, as a consequence, have a much higher probability of remaining poor throughout their life. Because poverty is linked to unmarried parenthood, a cycle of successive generations of teenage pregnancy may develop.

   Some research has found that children of single mothers are more prone to academic problems. However, Henry Ricciuti, professor of human development, found that the mother's educational level and ability, rather than the absence of a father, have the most influence on a child's school readiness (Drummond 2000).

■ **Figure 5.8** *Median Family Income by Family Type, 1999.*

Source: U.S. Bureau of the Census. 2000 (Sept.). "Money Income in the U.S." *Current Population Reports.* Figure 1, p. x.

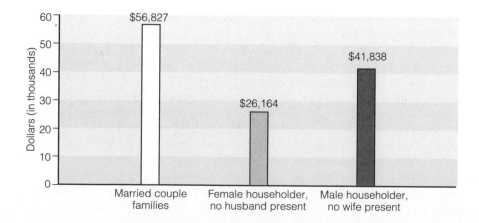

4. *Children without fathers.* About one-third of U.S. children who live in households without their fathers are the products of unmarried childbirth (Hewlett & West 1998). Shapiro and Schrof (1995) report that children who grow up without fathers are more likely to drop out of school, be unemployed, abuse drugs, experience mental illness, and be a target of child sexual abuse. They also note that "a missing father is a better predictor of criminal activity than race or poverty" (p. 39). Popenoe (1996) believes that fatherlessness is a major cause of the degenerating conditions of our young.

However, others argue that the absence of a father is not, in and of itself, damaging to children. Rather, other conditions associated with female-headed single-parent families, such as low educational attainment of the mother, poverty, and lack of child supervision contribute to negative outcomes for children.

Strategies for Action: Interventions in Teenage and Nonmarital Childbearing

Interventions in teenage and nonmarital childbearing include efforts to prevent such births and strategies to minimize the negative effects of such births to women and children. Although most nonmarital births are to women who are not on welfare, the 1996 welfare reform law contained measures designed to discourage nonmarital childbearing. The welfare reform law of 1996 included an "illegitimacy bonus," which rewards states that reduce out-of-wedlock births and also decrease abortions. From 1999 to 2003, the federal government will provide generous financial bonuses to up to five states each year that achieve the greatest declines in out-of-wedlock births and reduce their abortion rate to below its 1995 level (Donovan 1999). Substantial progress was made in the 1990s in reducing teenage pregnancy and childbearing. In 1996 the teenage pregnancy rate was lower than in any year since 1976 (Ventura et al. 2000). The teenage birthrate declined between 1991 and 1998 for all races and Hispanic origin groups, with the steepest decline reported for black teenagers. Reductions in teenage childbearing by state (1991 to 1998) varied from 10 to 38 percent (Ventura et al. 2000).

The European approach to teenage sexual activity is to provide widespread confidential and accessible contraceptive services to adolescents. The provision of contraceptive services to European teens is believed to be a central factor in explaining the more rapid declines in teenage childbearing in northern and western European countries, in contrast to slower declines in the United States (Singh & Darroch 2000). Although sex education is provided in schools throughout the United States, most programs emphasize abstinence and do not provide students with access to contraception.

Other programs aim at both preventing teenage and unmarried childbearing and minimizing its negative effects by increasing the life options of teenagers and unmarried mothers. Such programs include educational programs, job training, and skill-building programs. Other programs designed to help teenage and unwed mothers and their children include prenatal programs to help ensure the health of the mother and baby, and public welfare such as WIC (Women, Infants, and Children) and TANF (Temporary Assistance to Needy Families). However, public assistance to low-income single mothers has been

Parenting Wisely: An Interactive Computer Parenting Education Program

Parenting Wisely (FamilyWorks, Inc. 2000) is an interactive computer parenting education program that teaches communication skills (such as active listening), assertive discipline (such as using praise and setting consequences), and child supervision techniques (such as how to monitor homework and friends). The program contains nine case studies depicting nine problems common in families. After a video of a family problem is shown, three possible responses are presented. Some responses result in a worsening of the situation, whereas others improve the situation. Parents choose a response, see a video of how their choice would work, and get feedback on the pros and cons of their choice. Parents who have difficulty reading can choose to have the computer text read aloud. It usually takes 2 to 3 hours working through the nine case studies.

The nine scenarios include two-parent, single-parent, and stepfamilies from diverse racial and ethnic backgrounds (Caucasian, African American, Hispanic, and Asian). Preteens and teenagers appear in the scenarios, but *Parenting Wisely* has been shown to be equally successful with parents of younger children.

The program can be used by parents who are unfamiliar with computers, as well as those with computer experience. It is designed to be entirely self-administered, eliminating the need for an instructor to guide parents through the program. Poorly educated parents with no computer experience have no difficulty using the program without assistance, and commonly report that the program is fun and highly engaging to use (FamilyWorks, Inc. 2000).

How effective is the *Parenting Wisely* program? A summary of research evaluating *Parenting Wisely* reveals that parents with pre-teens and teens showing significant behavior problems showed increased knowledge in parenting principles and skills, increased use of the parenting skills taught in the program, and reductions in problem behaviors of their children ("Parenting Wisely Evaluation Results" 2000). In one study (Lagges & Gordon 1999), 62 pregnant or parenting teens were randomly assigned to either the *Parenting Wisely* program or to a control group. Both groups attended a teen parenting class in their high schools. Compared with the control group, the intervention group scored significantly higher at 2 months follow-up in parenting knowledge, belief in the effectiveness of positive parenting practices over coercive practices (yelling and spanking), and application of positive parenting skills to hypothetical problem situations. In two studies, children's problem behavior showed at least a 50 percent reduction one month after parents used the program ("Parenting Wisely Evaluation Results" 2000). A matched control group showed no improvement. In a study with teen mothers, use of the *Parenting Wisely* program resulted in improved knowledge of parenting principles and skills and problem-solving for toddler misbehavior.

The effectiveness of *Parenting Wisely* may be attributed in part to the self-administered aspect of the computer program. "Parental resistance to change is minimal with this learning format. Parents are not being judged by a person as they are in therapy. Parents also move through the program at their own pace, and can repeat a section when they wish" ("Parenting Wisely Evaluation Results" 2000, 1). Other benefits of using interactive computer technology in parenting education include that it can be disseminated fairly quickly and inexpensively. The convenience and lack of stigma can increase parent participation in computer-based parent education.

Evaluation results of the *Parenting Wisely* program suggest at least short-term improvements in children's behavior after their parents complete the program. Ongoing evaluation is currently assessing longer-term effects.

Sources:
FamilyWorks, Inc. 2000. "Parenting Wisely." 20 East Circle Drive, Suite 190. Athens, OH 45701-3751.

"Parenting Wisely Evaluation Results." 2000. http://familyworksinc.com/index2.html

A. Lagges and D. A. Gordon. 1999. "Interactive Videodisk Parent Training for Teen Mothers." *Child and Family Behavior Therapy* 21: 1937.

blamed for contributing to the problem by discouraging the poor from marrying. Up until 1996, single mothers with no or little income could receive welfare as long as they had a dependent child living with them. If single mothers on welfare married, they could lose welfare benefits. In 1996, the Personal Responsibility and Work Opportunity Reconciliation Act (PRWORA) reformed the welfare program (see Chapter 10) by limiting cash benefits to adults. For example, adults can receive welfare benefits for no more than 2 consecutive years (unless they work at least 20 hours per week) and are limited to 5 years of welfare cash benefits over their lifetime. One goal of welfare reform is to encourage marriage by decreasing the benefits an unmarried mother can claim. Despite the new time limitations on welfare benefits, marriage rates among low-income single mothers continue to be low.

Strategies to increase and support fathers' involvement with their children are relevant to both children of unwed mothers and children of divorce. At the federal and state levels, Fatherhood Initiative programs encourage fathers' involvement with children through a variety of means (U.S. Department of Health and Human Services 2000). These include promoting responsible fatherhood by improving work opportunities for low-income fathers, increasing child support collections, providing parent education training for men, supporting access and visitation by noncustodial parents, and involving boys and young men in preventing teenage pregnancy and early parenting.

Because teenage parents are less likely than older parents to use positive and effective child-rearing techniques, parent education programs for teen mothers and fathers are an important component of improving the lives of young parents and their children. One such program that utilizes interactive computer technology is presented in this chapter's *Focus on Technology* feature.

Understanding *Family Problems*

Family problems can best be understood within the context of the society and culture in which they occur. Although domestic violence, divorce, and teenage pregnancy and unmarried parenthood may appear to result from individual decisions, these decisions are influenced by a myriad of social and cultural forces.

The impact of family problems, including divorce, abuse, and nonmarital childbearing, is felt not only by family members, but by the larger society as well. Family members experience such life difficulties as poverty, school failure, low self-esteem, and mental and physical health problems. Each of these difficulties contributes to a cycle of family problems in the next generation. The impact on society includes public expenditures to assist single-parent families and victims of domestic violence and neglect, increased rates of juvenile delinquency, and lower worker productivity.

For some, the solution to family problems implies encouraging marriage and discouraging other family forms, such as single parenting, cohabitation, and same-sex unions. But many family scholars argue that the fundamental issue is making sure that children are well-cared for, whether or not their parents are married. Some even suggest that marriage is part of the problem, not part of the solution. Professor Martha Fineman of Cornell Law School says, "This obsession with marriage prevents us from looking at our social problems and addressing them . . . . Marriage is nothing more than a piece of paper, and yet we rely on marriage to do a lot of work in this society: It becomes our family policy, our police in regard to welfare and children, the cure for poverty" (quoted in Lewin 2000, 2). Strengthening marriage is a worthy goal because strong marriages offer many benefits to individuals and their children. However, "strengthening marriage does not have to mean a return to the patriarchal family of an earlier era . . . . Indeed, greater marital stability will only come about when men are willing to share power, as well as housework and childcare, equally with women" (Amato 1999, 184). And strengthening marriage does not mean that

If we are to achieve a richer culture, rich in contrasting values, we must recognize the whole gamut of human potentialities, and so weave a less arbitrary social fabric, one in which each diverse human will find a fitting place.

MARGARET MEAD
Anthropologist

We recognize today that children can be effectively raised in many different family systems and that it is the emotional climate of the family, rather than its kinship structure, that primarily determines a child's emotional well-being and healthy development.

DAVID ELKIND
Child development specialist

other family forms should not also be supported. The reality is that the post-modern family comes in many forms, each with its strengths, needs, and challenges. Given the diversity of families today, social historian Stephanie Coontz (2000) suggests that, "The only way forward at this point in history is to find better ways to make *both* marriage and its alternatives work" (p. 15).

Critical Thinking

1 Some scholars and politicians argue that "stable families are the bedrock of stable communities." Others argue that "stable communities and economies are the bedrock of stable families." Which of these two positions would you take and why?

2 Lloyd and Emery (2000) note that "one of the primary ways that power disguises itself in courtship and marriage is through the 'myth of equality between the sexes'. . . . The widespread discourse on 'marriage between equals' serves as a cover for the presence of male domination in intimate relationships . . . and allows couples to create an illusion of equality that masks the inequities in their relationships" (pp. 25–26). Do you agree that the modern view of marriage between equal partners is an illusion? Why or why not?

3 Research has suggested that secondhand smoke from cigarettes represents a health hazard for those who are exposed to it. Consequently, smoking is now banned in many public places and workplaces. Do you think that parents should be banned from smoking in enclosed areas (home or car) to protect their children from secondhand smoke? Do you think that smoking in enclosed areas with one's children present should be considered a form of child abuse? Why or why not?

4 Some judges are imposing "shame sentences" on convicted abusers. For example, Texas Judge Ted Poe ordered an abusive husband to publicly apologize to his wife on the steps of City Hall and ordered another batterer to go to a local mall carrying a sign that read, "I went to jail for assaulting my wife. This could be you" ("In the News" 1998). What do you think about these types of "shame sentences" for batterers?

5 In the United States, women are more likely than men to initiate divorce. Why do you think this is so?

Key Terms

child abuse	elder abuse	intimate terrorism
common couple violence	emotional abuse	monogamy
corporal punishment	familism	mutual violent control
covenant marriage	family household	neglect
cycle of abuse	family preservation program	no-fault divorce
divorce law reform	individualism	nonfamily household
divorce mediation	intimate partner violence	parental alienation syndrome (PAS)
domestic partnership		

patriarchy

polyandry

polygamy

polygyny

primary prevention

second shift

secondary prevention

self-fulfilling prophecy

serial monogamy

sexual aggression

tertiary prevention

traditional family

violent resistance

Media Resources

The Wadsworth Sociology Resource Center: Virtual Society

http://sociology.wadsworth.com/

See the companion Web site for this book to access general sociology resources and text-specific features that can further your understanding of this chapter. The site contains Internet links, Internet exercises, online practice quizzes, information on InfoTrac College edition, and many more valuable materials designed to enrich your learning experience in social problems.

InfoTrac College Edition

You can access InfoTrac College Edition either from the Wadsworth Sociology Resource Center at **http://sociology.wadsworth.com** or directly from your web browser at **http://www.infotrac-college.com/wadsworth/**. InfoTrac College Edition is an online university library that includes over 700 popular and scholarly journals in which you can find articles related to the topics in this chapter such as information on divorce, domestic violence and abuse, child abuse, the Fatherhood Initiative, teenage pregnancy and parenting, and single parenting.

Interactions CD-ROM

Go to the "Interactions" CD-ROM for *Understanding Social Problems*, Third Edition to access additional interactive learning tools, such as in-depth review materials, corresponding practice quizzes, and other engaging resources and activities to help you study the concepts in this chapter.

Problems of Human Diversity

People are diverse. They vary on a number of dimensions, including age, gender, sexual orientation, and race and ethnicity. In most societies, including the United States, these characteristics are imbued with social significances and are used to make judgments about an individual's worth, intelligence, skills, and personality. Such labeling creates categories of people who are perceived as "different" by others as well as themselves and, as a result, are often treated differently.

A **minority** is defined as a category of people who have unequal access to positions of power, prestige, and wealth in a society. In effect, minorities have unequal opportunities and are disadvantaged in their attempt to gain societal resources. Even though they may be a majority in terms of numbers, they may still be a minority sociologically. Before Nelson Mandela was elected president of South Africa, South African blacks suffered the disadvantages of a minority, even though they were a numerical majority of the population.

Terms that apply to all minorities include stereotyping, prejudice, and discrimination. A **stereotype** is an exaggerated and overgeneralized truth (e.g., "all Puerto Ricans like spicy food"). **Prejudice** is an attitude, often negative, that prejudges an individual. **Discrimination** is differential treatment by members of the majority group against members of the minority that has a harmful impact on members of the subordinate group. The groups we will discuss in this section are all victims of stereotyping, prejudice, and discrimination.

Minority groups usually have certain characteristics in common. In general, members of a minority group know that they are members of a minority, stay within their own group, have relatively low levels of self-esteem, are disproportionately in the lower socioeconomic strata, and are viewed as having negative traits. Other characteristics of specific minority groups are identified in the accompanying table.

In the following chapters, we discuss categories of minorities based upon age (Chapter 6), race and ethnicity (Chapter 7), gender (Chapter 8), and sexual orientation (Chapter 9). Although other categories of minorities exist (e.g., disabled/handicapped, religious minorities), we have chosen to concentrate on these four because each is surrounded by issues and policies that have far-reaching social, political, and economic implications.

Nine Characteristics of Four Minorities

	Old/Young	Racial and Ethnic Minorities	Women	Homosexuals
1. Status ascribed based on:	Age	Race/ethnicity	Sex	Sexual orientation
2. Visibility:	High	High	High	Low
3. Attribution of minority status (correctly or incorrectly) based on:	Hair Skin elasticity/Size Posture	Skin color Facial features Hair	Anatomy Shape	Mannerisms Style of dress
4. Summary image:	Dependent	Inferior	Weak	Sick or immoral
5. Derogatory and offensive terms:	Old codger/Brats Battle-Ax/Punks	Nigger Spic	Bitch Whore	Faggot Dyke
6. Control through feigning various characteristics:*	Frailty/Helplessness	Ignorance	Weakness	Heterosexuality
7. Discrimination:	Yes	Yes	Yes	Yes
8. Victims of violence:	Yes	Yes	Yes	Yes
9. Segregation:	Yes	Yes	Yes	Yes

*Being aware of their lack of power, minority group members may try to exert control by feigning certain characteristics. For example, a slave couldn't tell his or her owner, "I'm not going to plow the fields—do it yourself!" But by acting incompetent, the slave may avoid the work. Gays pass as straight by bragging about heterosexual conquests. The elderly in nursing homes whose family members might otherwise not visit act ill in the hope of eliciting a visit. The problem with these acts is that they contribute to and stabilize the summary images.

6

The Young and the Old

Is It True?

1. The more primitive a society, the more likely its members are to practice senilicide—the killing of the elderly.

2. The concepts of "middle-age" and "adolescence" have always existed.

3. Social Security is not a major source of income for most people 65 years of age and older.

4. In the United States, individuals age 65 and older are more likely to vote than any other age group.

5. The U.S. poverty rate for children is more than double that of every other major industrialized country.

"Who am I?" asked Nathan Grieco. "That's a question I haven't asked myself for quite some time. Many things (mostly bad), have happened in my life. There are so many it would take me two whole lifetimes to type about it..." (quoted in Carpenter & Kopas 1999, 1).

There is no doubt that Nathan Grieco was depressed. Things had not been easy for him. His girlfriend had broken up with him, he was socially awkward and not very popular at school, and although his grades were good, he had been diagnosed with attention deficit disorder and was on medication.

The greatest "torture" in his life, however, was the ongoing custody battle between his feuding parents. After 8 years there was no end in sight for him and his two younger brothers. After several occasions of being forced to see their father, a man they described as abusive and overbearing, the most recent court order held that failure to visit their father would result in contempt of court charges against their mother, and her possible incarceration. Having no legal rights of their own, they had few choices. Within a year of the Order, at age 16, Nathan Grieco was dead—found by his mother, kneeling next to his bed, with a leather belt around his neck. The coroner ruled that there was "insufficient evidence of either a suicide or an accident" (Carpenter & Kopas 1999, 1).

Like too many other youths, Nathan Grieco defined his life as one of "endless torment." Interestingly, some research suggests that depression is "curvilinear" with age, that is, highest at the extremes of the age continuum (DeAngelis 1997). Depression is not the only characteristic shared by the young and the old. Both groups are often the victims of stereotyping, physical abuse, age discrimination, and poverty; both groups are also major population segments of American society. In this chapter, we examine the problems and potential solutions associated with youth and aging. We begin by looking at age in a cross-cultural context.

The Global Context: The Young and the Old around the World

The young and the old receive different treatment in different societies. Differences in the treatment of the dependent young and old have traditionally been associated with whether the country is developed or less developed. Although proportionately more elderly live in developed countries than in less developed ones, these societies have fewer statuses for the elderly to occupy. Their positions as caretakers, homeowners, employees, and producers are often usurped by those aged 18 to 64. Paradoxically, the more primitive the society the more likely that society is to practice senilicide—the killing of the elderly. In some societies the elderly are considered a burden and left to die or, in some cases, actually killed.

Not all societies treat the elderly as a burden. Scandinavian countries provide government support for in-home care workers for elderly who can no longer perform such tasks as cooking and cleaning. Eastern cultures such as Japan revere the elderly, in part, because of their presumed proximity to honored ancestors. By 2005, if present trends continue, Japan will have the highest proportion of elderly of any country in the world—19.6 percent of their population (Yamaguchi 2000).

Societies also differ in the way they treat children. In less developed societies, children work as adults, marry at a young age, and pass from childhood directly to adulthood with no recognized period of adolescence. In contrast, in industrialized nations, children are often expected to attend school for 12 to 16 years and, during this time, to remain financially and emotionally dependent on their families.

Because of this extended period of dependence, the United States treats "minors" differently than adults. There is a separate justice system for juveniles and age limits for driving, drinking alcohol, joining the military, entering into a contract, marrying, dropping out of school, and voting. These limitations would not be tolerated if placed on individuals on the basis of sex or race. Hence **ageism**, the belief that age is associated with certain psychological, behavioral and/or intellectual traits, at least in reference to children, is significantly more tolerated than sexism or racism in the United States.

> I . . . remember when young people had respect for their elders and for the law. Those were the good old days.
>
> MESSAGES TO THE NEXT GENERATION
> *AARP Online*

Despite this differential treatment, people in the United States are fascinated with youth and being young. This was not always the case. The elderly were once highly valued in the United States—particularly older men who headed families and businesses. Younger men even powdered their hair, wore wigs, and dressed in a way that made them look older. It should be remembered, however, that in 1900 the average life expectancy in the United States was 47 and over half the population was under the age of 16 (AOA 2000a). Being old was rare and respected; to some it was a sign that God looked upon the individual favorably.

One theory argues that the shift from valuing the old to valuing the young took place during the transition from an agriculturally based society to an industrial one. Land, which was often owned by elders, became less important as did their knowledge and skills about land-based economies. With industrialization, technological skills, training, and education became more important than land ownership. Called **modernization theory**, this position argues that as a society becomes more technologically advanced, the position of the elderly declines (Cowgill & Holmes 1972).

Youth and Aging

> There were only two periods in the life of a Chinese male when he possessed maximum security and minimal responsibility—infancy and old age. Of the two, old age was the better because one was conscious of the pleasure to be derived from such an almost perfect period.
>
> PAUL T. WELTY
> *Historian*

Age is largely socially defined. Cultural definitions of "old" and "young" vary from society to society, from time to time, and from person to person. For example, in ancient Greece or Rome where the average life expectancy was 20 years, one was old at 18; similarly, one was old at 30 in medieval Europe and at 40 in the United States in 1850.

Age is also a variable that has a dramatic impact on one's life (Matras 1990, identified 1–4):

1. Age determines one's life experiences because the date of birth determines the historical period in which a person lives. Twenty years ago cell phones and palm-sized computers couldn't have been imagined.
2. Different ages are associated with different developmental stages (physiological, psychological, and social) and abilities. Ben Franklin observed, "[A]t 20 years of age the will reigns; at 30 the wit; at 40 judgment."

3. Age defines roles and expectations of behavior. The expression "act your age" implies that some behaviors are not considered appropriate for people of certain ages.
4. Age influences the social groups to which one belongs. Whether one is part of a sixth-grade class, a labor union, or a senior's bridge club depends on one's age.
5. Age defines one's legal status. Sixteen-year-olds can get a driver's license, 18-year-olds can vote and get married without their parents' permission, and 65-year-olds are eligible for Social Security benefits.

Childhood, Adulthood, and Elderhood

Every society assigns different social roles to different age groups. **Age grading** is the assignment of social roles to given chronological ages (Matras 1990). Although the number of age grades varies by society, most societies make at least three distinctions—childhood, adulthood, and elderhood.

Childhood The period of childhood in our society is from birth through age 17 and is often subdivided into infancy, childhood, and adolescence. Infancy has always been recognized as a stage of life, but the social category of childhood only developed after industrialization, urbanization, and modernization took place. Before industrialization, infant mortality was high because of the lack of adequate health care and proper nutrition. Once infants could be expected to survive infancy, the concept of childhood emerged, and society began to develop norms in reference to children. In the United States, child labor laws prohibit children from being used as cheap labor, educational mandates require that children attend school until the age of 16, and federal child pornography laws impose severe penalties for the sexual exploitation of children.

Adulthood The period from age 18 through 64 is generally subdivided into young adulthood, adulthood, and middle age. Each of these statuses involves dramatic role changes related to entering the workforce, getting married, and having children. The concept of "middle age" is a relatively recent one that has developed as life expectancy has been extended. Some people in this phase are known as members of the **sandwich generation** because they are often emotionally and economically responsible for both their young children and their aging parents.

Elderhood At age 65 one is likely to be considered elderly, a category that is often subdivided into the young-old, old, and old-old. Membership in one of these categories does not necessarily depend on chronological age. The growing number of healthy, active, independent, elderly are often considered to be the young-old, whereas the old-old are less healthy, less active, and more dependent.

Sociological Theories of Age Inequality

Three sociological theories help explain age inequality and the continued existence of ageism in the United States. These theories—structural-functionalism, conflict theory, and symbolic interactionism—are discussed in the following sections.

> That's exactly how it goes. You're young and the whole world is ahead of you, and then one day you're not young anymore, and the whole world is behind you, and it's all over.
>
> **DAVID LETTERMAN**
> *Comedian*

> I'm a little older now. I've moved from the sandwich generation to the club sandwich generation.
>
> **ELLEN GOODMAN**
> *Syndicated columnist*

Structural-Functionalist Perspective

Structural-functionalism emphasizes the interdependence of society—how one part of a social system interacts with other parts to benefit the whole. From a functionalist perspective, the elderly must gradually relinquish their roles to younger members of society. This transition is viewed as natural and necessary to maintain the integrity of the social system. The elderly gradually withdraw as they prepare for death, and society withdraws from the elderly by segregating them in housing such as retirement villages and nursing homes. In the interim, the young have learned through the educational institution how to function in the roles surrendered by the elderly. In essence, a balance in society is achieved whereby the various age groups perform their respective functions: the young go to school, adults fill occupational roles, and the elderly, with obsolete skills and knowledge, disengage. As this process continues, each new group moves up and replaces another, benefiting society and all of its members.

This theory is known as **disengagement theory** (Cummings & Henry 1961). Some researchers no longer accept this position as valid, however, given the increased number of elderly who remain active throughout life (Riley 1987). In contrast to disengagement theory, **activity theory** emphasizes that the elderly disengage in part because they are structurally segregated and isolated, not because they have a natural tendency to do so. For those elderly who remain active, role loss may be minimal. In studying 1,720 respondents who reported using a senior center in the previous year, Miner, Logan, and Spitze (1993) found that those who attended were less disengaged and more socially active than those who did not.

Conflict Perspective

The conflict perspective focuses on age grading as another form of inequality as both the young and the old occupy subordinate statuses. Some conflict theorists emphasize that individuals at both ends of the age continuum are superfluous to a capitalist economy. Children are untrained, inexperienced, and neither actively producing nor consuming in an economy that requires both. Similarly, the elderly, although once working, are no longer productive and often lack required skills and levels of education. Both young and old are considered part of what is called the dependent population; that is, they are an economic drain on society. Hence, children are required to go to school in preparation for entry into a capitalist economy, and the elderly are forced to retire.

Other conflict theorists focus on how different age strata represent different interest groups that compete with one another for scarce resources. Debates about funding for public schools, child

© Mark Allan/Alpha/Globe Photos

Despite age grading, many singers such as Stevie Nicks, Mick Jagger, and Cher continue to perform, and are commercially successful. Tina Turner, almost 60 in this picture, defies age stereotypes.

health programs, Social Security, and Medicare largely represent conflicting interests of the young versus the old.

Symbolic Interactionist Perspective

The symbolic interactionist perspective emphasizes the importance of examining the social meaning and definitions associated with age. Teenagers are often portrayed as lazy, aimless, and awkward. The elderly are also defined in a number of stereotypical ways contributing to a host of myths surrounding the inevitability of physical and mental decline. Table 6.1 identifies some of these myths.

Media portrayals of the elderly contribute to the negative image of the elderly. The young are typically portrayed in active, vital roles and are often overrepresented in commercials. In contrast, the elderly are portrayed as difficult, complaining, and burdensome and are often underrepresented in commercials. A

Table 6.1 *Myths and Facts about the Elderly*

Health

Myth The elderly are always sick; most are in nursing homes.

Fact Over 85 percent of the elderly are healthy enough to engage in their normal activities. Only 6 percent are confined to a nursing home.

Automobile Accidents

Myth The elderly are dangerous drivers and have a lot of accidents.

Fact Drivers ages 45 to 74 have only 12 accidents per 100 drivers per year. Drivers under age 20 have 37 accidents per 100 drivers per year. Up to age 75, older drivers tend to drive fewer miles than younger drivers and compensate for their reduced reaction time by driving more carefully. However, drivers 75 and older have 25 accidents per 100 drivers per year. The increased incidence is "rooted in the normal processes of aging: diminishing vision and hearing and decreasing attention spans" (Carney 1989).

Mental Status

Myth The elderly are senile.

Fact Although some of the elderly learn more slowly and forget more quickly, most remain oriented and mentally intact. Only 20–25 percent develop Alzheimer's disease or some other incurable form of brain disease. Senility is not inevitable as one ages.

Employment

Myth The elderly are inefficient employees.

Fact Although only about 17 percent of men and 9 percent of women 65 years old and over are still employed, those who continue to work are efficient workers. When compared with younger workers, the elderly have lower job turnover, fewer accidents, and less absenteeism. Older workers also report higher satisfaction in their work.

Politics

Myth The elderly are not politically active.

Fact In 1986, 1988, 1990, 1992, 1994, 1996, and 1998 individuals age 65 and older were more likely to be registered to vote and/or to vote than any other age group.

Sexuality

Myth Sexual satisfaction disappears with age.

Fact Many elderly people report active and satisfying sex lives. For example, of couples 75 years old and older, over 25 percent report having sexual intercourse once a week (Toner 1999).

Adaptability

Myth The elderly cannot adapt to new working conditions.

Fact A high proportion of the elderly are flexible in accepting change in their occupations and earnings. Adaptability depends on the individual: many young are set in their ways, and many older people adapt to change readily.

Sources: Robert H. Binstock. 1986. "Public Policy and the Elderly." *Journal of Geriatric Psychiatry* 19: 115–43; Teresa E. Seeman and Nancy Adler. 1998. "Older Americans: Who Will They Be?" National Forum, Spring, 22–25; James Carney. 1989. "Can a Driver Be Too Old?" *Time*, January 16, 28; Administration on Aging. 2000. "Profile of Older Americans: 1999." http://aoa.dhhs.gov/aoa/stats/profile/default.htm; Erdman B. Palmore. 1984. "The Retired." In *Handbook on the Aged in the United States*, ed. Erdman B. Palmore, pp. 63–76, Westport, CT: Greenwood Press; Federal Elections Commission. "Voter Registration and Turnout by Age, Gender and Race." http://www.fec.gov; Robin Toner. 1999. "A Majority Over 45 Say Sex Lives are Just Fine." *New York Times*, August 4, A10; and *Statistical Abstract 2000*, 120th ed. U.S. Bureau of the Census. Washington, D.C.: U.S. Government Printing Offices.

recent study of the elderly in popular 1940s through 1980s films concluded that "[O]lder individuals of both genders were portrayed as less friendly, having less romantic activity, and enjoying fewer positive outcomes than younger characters at a movie's conclusion" (Brazzini, McIntosh, Smith, Cook, & Harris 1997, 541).

The elderly are also portrayed as childlike in terms of clothes, facial expressions, temperament, and activities—a phenomenon known as **infantilizing elders** (Arluke & Levin 1990). For example, young and old are often paired together. A promotional advertisement for the movie *Just You and Me, Kid* with Brooke Shields and George Burns described it as "the story of two juvenile delinquents." Jack Lemmon and Walter Matthau in *Grumpy Old Men* get "cranky" when they get tired, and the media focus on images of Santa visiting nursing homes and local elementary school children teaching residents arts and crafts. Finally, the elderly are often depicted in role reversal, cared for by their adult children as in the situation comedies *Golden Girls, Frasier,* and *King of Queens.*

Negative stereotypes and media images of the elderly engender **gerontophobia**—a shared fear or dread of the elderly, which may create a self-fulfilling prophecy. In a recent study, seniors received one of two types of subliminal messages—those negatively stereotyping the elderly (e.g., senile) and those positively stereotyping the elderly (e.g., wise). Compared with those who received negative messages, subjects who received the positive messages scored better on memory tests, and were more likely to respond that they would accept life-prolonging health interventions (Begley 2000).

> Since it is the Other within us who is old, it is natural that the revelation of our age should come to us from outside—from others.
>
> SIIMONE DE BEAUVOIR
> *Feminist author*

Problems of Youth in America

> For in its innermost depths, youth is lonelier than old age.
>
> UNKNOWN

The number of people under the age of 18, 71.4 million in 2000, is the largest in history and will grow to 80 million by the year 2020. Although recent evidence indicates that conditions for children are improving, numerous problems remain. Indeed, some of our most pressing social problems can be traced to early childhood experiences and adolescent behavioral problems (Weissberg & Kuster 1997).

A Children's Defense Fund report documents the condition of children in America (see Table 6.2). Not surprisingly, what happens to children is increasingly defined as a social problem.

Table 6.2 *Key Facts about American Children*

1 in 2 never completes a single year of college

1 in 3 is born to unmarried parents

1 in 5 is living in poverty

1 in 6 has no health insurance

1 in 8 will never graduate from high school

1 in 8 is born to a teenage mother

1 in 12 has a disability

1 in 13 is born with low birth weight

1 in 24 lives with neither parent

1 in 138 will die before their first birthday

1 in 910 will be killed by a gun by age 20

Source: Children's Defense Fund. 2000. "The State of America's Children Yearbook 2000". http://www.childrensdefense.org/keyfacts.html

Children and the Law

Historically, children have had little control over their lives. They have been "double dependent" on both their parents and the state. Indeed, colonists in America regarded children as property. Beginning in the 1950s, however, the view that children should have greater autonomy became popular and was codified in several legal decisions and international treaties. In 1959, the United Nations General Assembly approved the *Declaration on the Rights of the Child,* which held that health care, housing, and education, as well as freedom from abuse, neglect, and exploitation, are fundamental children's rights.

A second measure, the *Convention on the Rights of the Child,* further articulated the rights of children and was adopted by the United Nations in 1989. Countries ratifying the document have made significant improvements in the lives of children. For example, the Democratic Republic of the Congo's draft constitution now prohibits conscription into the army before age 18 (UNICEF 2000). Only two countries, the United States and Somalia, have failed to ratify the *Convention.* One reason the United States has not signed the pact may be Article 11, which requires that children be assured "the highest attainable standard of health." Some people are concerned that accepting such a position would result in cases being brought to court that they are simply unwilling, at present, to hear.

Children are both discriminated against and granted special protections under the law. Although legal mandates require that children go to school until age 16, other laws provide a separate justice system whereby children have limited legal responsibility based upon their age status. Concern with youth violence has many Americans questioning the wisdom of a separate legal structure for minors. The laws are changing: most states have lowered the age of accountability, capital punishment of 16-year-olds is allowed in a number of states, and the number of parent-liability laws has increased. Although children's rights may expand in some areas such as self-determination, recent legal changes may cost minors their protected status in the courts. In 2001, 13-year-old Lionel Tate was convicted of first-degree murder and sentenced to life in prison without parole (CNN 2001).

> American children now stand in a transitional state between chattel and constitutionally protected child-citizen.
>
> CHARLES GILL
> *Attorney*

Poverty and Economic Discrimination

Based on a recent worldwide study, the U.S. poverty rate for children is more than double that of every other major industrialized country. In 1999, 22 percent of U.S. children lived below the poverty line, up from 15 percent in 1960 (UNICEF 1994; Bergmann 1999; Leeman 2000). Childhood poverty is related to school failure (Fields & Smith 1998; NIH 2000), negative involvement with parents (Harris & Marmer 1996), stunted growth, reduced cognitive abilities, limited emotional development (Brooks-Gunn & Duncan 1997), and a higher likelihood of dropping out of school (Duncan, Yeung, Brooks-Gunn, & Smith 1998).

Three decades ago the elderly were the poorest age group in the United States; today it is children (Snapshots 2000). Although Social Security, Medicare, housing subsidies, and Supplemental Security Income (SSI) keep millions of elderly out of poverty, the United States has no universal government policy to protect children. Further, recent welfare reform has led to cutbacks in what few programs do benefit children. For example, more than half of food stamp recipients are under the age of 18 (CDF 2000a).

Children are also the victims of discrimination in terms of employment, age restrictions, wages, training programs, and health benefits. Traditionally, children worked on farms and in factories but were displaced by the Industrial

Growing Up in the Other America

Author Alex Kotlowitz conducted a 2-year participant observation research study of children in a Chicago housing project. The following description captures the horrific living conditions endured by Lafeyette, Pharoah, and Dede—three children living in the "jects."

The children called home "Hornets" or, more frequently, "the projects" or, simply, the "jects" (pronounced jets). Pharoah called it "the graveyard." But they never referred to it by its full name: the Governor Henry Horner Homes.

Nothing here, the children would tell you, was as it should be. Lafeyette and Pharoah lived at 1920 West Washington Boulevard, even though their high-rise sat on Lake Street. Their building had no enclosed lobby; a dark tunnel cut through the middle of the building, and the wind and strangers passed freely along it. Those tenants who received public aid had their checks sent to the local currency exchange, since the building's first-floor mailboxes had all been broken into. And since darkness engulfed the building's corridors, even in the daytime, the residents always carried flashlights, some of which had been handed out by a local politician during her campaign.

Summer, too, was never as it should be. It had become a season of duplicity.

On June 13, a couple of weeks after their peaceful afternoon on the railroad tracks, Lafeyette celebrated his twelfth birthday. Under the gentle afternoon sun, yellow daisies poked through the cracks in the sidewalk as children's bright faces peered out from behind their windows. Green leaves clothed the cottonwoods, and pastel cotton shirts and shorts, which had sat for months in layaway, clothed the children. And like the fresh buds on the crabapple trees, the children's spirits blossomed with the onset of summer.

Lafeyette and his nine-year-old cousin Dede danced across the worn lawn outside their building, singing the lyrics of an L.L. Cool J rap, their small hips and spindly legs moving in rhythm. The boy and girl were on their way to a nearby shopping strip, where Lafeyette planned to buy radio headphones with $8.00 he had received as a birthday gift.

Suddenly, gunfire erupted. The frightened children fell to the ground. "Hold your head down!" Lafeyette snapped, as he covered Dede's head with her pink nylon jacket. If he hadn't physically restrained her, she might have sprinted for home, a dangerous action when the gangs started warring. "Stay down," he ordered the trembling girl.

The two lay pressed to the beaten grass for half a minute, until the shooting subsided. Lafeyette held Dede's hand as they cautiously crawled through the dirt toward home. When they finally made it inside, all but fifty cents of Lafeyette's birthday money had trickled from his pockets.

Source: Alec Kotlowitz. 1991. *There Are No Children Here*. Copyright © 1991 by Alex Kotlowitz. Used by permission of Doubleday, a division of Random House, Inc.

By the end of the 20th century, a child in America is almost twice as likely to be poor as an adult. This is a condition that has never before existed in our history.

DANIEL PATRICK MOYNIHAN
Retired U.S. Senator

Revolution. In 1938, Congress passed the Fair Standards Labor Act, which required factory workers to be at least 16. Although the law was designed to protect children, it was also discriminatory in that it prohibited minors from having free access to jobs and economic independence. Today, 14- and 15-year-olds are restricted in the number of hours per day and hours per week they can work; those under 14 are, in general, prohibited from working.

Children, Violence, and the Media

Gang violence, suicide, child abuse, and crime are all-too-common childhood experiences (see this chapter's *The Human Side* feature). When a sample of 12- to 17-year-olds were asked about the most important problems facing their age group, "drugs" was the number one response; "crime and violence in school" ranked third, and "other crime and violence" ranked fifth (CASA 2000).

According to the Centers for Disease Control and Prevention, compared with children in other industrialized nations, U.S. children are 9 times more likely to die in a firearms accident, 16 times more likely to be murdered by a gun, and 11 times more likely to commit suicide with a firearm (CDF 2000c). Child abuse and neglect remain at epidemic levels with an estimated 3 million children victimized annually, and although juvenile crime has gone down in recent years (see Chapter 4), more than 1.5 million arrests of juveniles occurred in 1999 (FBI 2000).

School violence, and particularly the tragic deaths of 12 students and one teacher at Columbine High School, has focused attention on the relationship between youth violence, guns, and the media. Forty-three percent of respondents in a Gallup survey reported fearing for their child's safety at school—almost twice the number than in 1977. When asked about the causes of violence, the majority of respondents reported that movies; video or computer games; song lyrics on CDs, tapes, or the radio; and television were each an "extremely serious" or "serious" problem (Gallup Poll 2000). Despite denials, a recent report by the Federal Trade Commission concludes that the "motion picture, music recording, and electronic games industry routinely targets children under 17" and "makes little effort to restrict access to violent material" (FTC 2000). Before completing elementary school, the average child views more than 100,000 acts of violence on television (CDF 2000b). A national report on children, violence, and the media concluded that television alone is responsible for 10 percent of all youth violence (U.S. Senate 1999).

Kids in Crisis

Childhood is a stage of life that is socially constructed by structural and cultural forces of the past and present. The old roles for children as laborers and farm helpers are disappearing, yet no new roles have emerged. While being bombarded by the media, children must face the challenges of an uncertain future, peer culture, music videos, divorce, poverty, and crime. Parents and public alike fear children are becoming increasingly involved with sex, drugs, alcohol, and violence. Some even argue that childhood as a stage of life is disappearing.

Recent statistics, however, indicate that children's lives are getting better. A 2000 government report that examines key indicators of children's well-being concluded that "American children are less likely to die during childhood, less likely to live in poverty, less likely to be at risk for hunger, and less likely to give birth in adolescence" than in recent years (NIH 2000a). Critics, however, are quick to note that given today's economic prosperity, achieving 1980 child poverty levels is hardly a success (Russakoff 2000).

Despite some gains, conditions remain intolerable for millions of children in the United States and around the world—homelessness, sexual exploitation, low birth weight, hunger, childhood depression, absent or inadequate health care, and dangerous living conditions (e.g., foster care, child abuse). A report from the National Research Council recently called for a "new national dialogue focused on rethinking the meaning of both shared responsibility for children and a strategic investment in their future" (Jacobson 2000, 1). Such an investment, particularly in preventive initiatives, is not only humane but would save millions of dollars spent on solving child-related problems after they occur.

> So long as little children are allowed to suffer, there is no true love in this world.
>
> **ISADORA DUNCAN**
> *Dancer*

> Knowledge by definition puts an end to innocence.
>
> **PHILIP ADLER**
> *Writer*

Demographics: The "Graying of America"

In recent years, the ratio of children to old people has changed dramatically (see Figure 6.1). In 1900, the ratio of those under 18 to those over 65 was 10:1. Today that ratio is 2:1, and by 2030 is expected to be 1.2:1 (Uhlenberg 2000). In

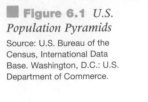

Figure 6.1 *U.S. Population Pyramids*

Source: U.S. Bureau of the Census, International Data Base. Washington, D.C.: U.S. Department of Commerce.

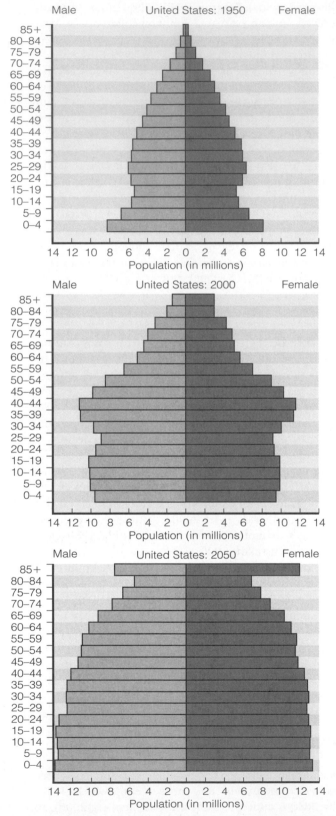

1999, one in eight Americans was over the age of 65; in 2030 one in five people will be over the age of 65 (AOA 2000a). These statistics reflect three significant demographic changes. First, 76 million baby boomers born between 1946 and 1964 are getting older. Second, life expectancy has increased as a result of a general trend toward modernization including better medical care, sanitation, and nutrition. Finally, lowered birth rates mean fewer children and a higher percentage of the elderly. For example, in Japan, low birth rates and rising life expectancies have contributed to the highest proportion of elderly in the world—19.6 percent of the population by 2005 (Yamaguchi 2000).

Age and Race and Ethnicity

In 2000, about 16 percent of the elderly were racial/ethnic minorities (NIH 2000b). Minority populations are projected to represent 25 percent of the U.S. elderly population by 2030, up from 13 percent in 1990. Although the elderly white population is projected to increase 79 percent between 1997 and 2030, the elderly minority population is expected to increase 238 percent in the same time period. The growth of the elderly minority population is a consequence of higher fertility rates and immigration patterns, particularly among Hispanics (AOA 2000c). Given their higher rates of diabetes, arthritis, and cardiovascular disease, the increased numbers of older minorities presents a unique challenge to health care providers (AOA 2000d; Newman 2000).

Age and Gender

In the United States, elderly women outnumber elderly men. The sex ratio for persons 65 and older is 70 men for every 100 women. (*Statistical Abstract* 2000). Men die at an earlier age than women for both biological and sociological reasons—heart disease, stress, and occupational risk (see Chapter 2). The fact that women live longer results in a sizable number of elderly women who are poor. Not only do women, in general, make less money than men, but older women may have spent their savings on their husband's illness, and as homemakers, they are not eligible for Social Security benefits. Further, retirement benefits and other major sources of income may be lost with a husband's death. Almost three-quarters of the elderly living in poverty are women (AOA 2000e).

Age and Social Class

How long a person lives is influenced by his or her social class. In general, the higher the social class, the longer the person lives, the fewer the debilitating illnesses, the greater the number of social contacts and friends, the less likely to define oneself as "old," and the greater the likelihood of success in adapting to retirement. Higher social class is also related to fewer residential moves, higher life satisfaction, more leisure time, and more positive self-rated health. Of

© Tony Freeman/Photo Edit

Women, as they get older, are often portrayed as sexually unattractive, something men are rarely subjected to. Advertising contributes to this image by idealizing youth in advertisements which advocate the use of wrinkle creams, hair dyes, and getting one's "girlish figure" back.

those 65 and older, 26 percent of those with annual incomes of $35,000 or higher reported that their health was "excellent," whereas only 10 percent of those with incomes under $10,000 rated their health as such. Functional limitations such as problems with walking, dressing, and bathing were also less common among higher income groups (Seeman & Adler 1998). In short, the higher one's socioeconomic status, the longer, happier, and healthier one's life.

Problems of the Elderly

This is the century of old age, or, as it has been called, the "Age of Aging."

Robert Butler
Gerontologist

The increase in the number of the elderly, worldwide, presents a number of institutional problems. The **dependency ratio**—the number of societal members that are under 18 or 65 and over compared with the number of people who are between 18 and 64—is increasing. In 2000, there were 62 "dependents" for every 100 persons between 18 and 64. By 2050, the estimated ratio will be 80 to 100 (AOA 2000c). This dramatic increase, and the general movement toward global aging, may lead to a shortage of workers and military personnel, foundering pension plans, and declining consumer markets. It may also lead to increased taxes as governments struggle to finance elder care programs and services, heightening intergenerational tensions as societal members compete for scarce resources (Peterson 2000; Schieber 2000; Goldberg 2000). In addition to these macro level concerns, the elderly face a number of challenges of their own. This chapter's *Self and Society* feature tests your knowledge of the aged and some of these concerns.

Work and Retirement

What one does (occupation), for how long (work history), and for how much (wages), are important determinants of retirement income. Indeed, employment is important because it provides the foundation for economic resources later in life. Yet, for the elderly who want to work, entering and remaining in the labor force may be difficult because of negative stereotypes, lower levels of education, reduced geographical mobility, fewer employable skills, and discrimination. The likelihood of being employed has decreased for elderly men and women over the last few decades.

In 1967, Congress passed the Age Discrimination in Employment Act (ADEA), which was designed to ensure continued employment for people between the ages of 40 and 65. In 1986, the upper limit was removed, making mandatory retirement illegal in most occupations. Nevertheless, thousands of cases of age discrimination occur annually. Although employers cannot advertise a position by age, they can state that the position is an "entry level" one or that "2 to 3 years' experience" is required. Despite strong evidence to the contrary, a study conducted by the National Council of Aging revealed that "50 percent of employers surveyed believed that older workers cannot perform as well as younger workers" (Reio & Sanders-Reto 1999). Further, if displaced, older workers remain unemployed longer than younger workers, often have to accept lower salaries, and may be more likely to give up looking for work (Goldberg 2000).

Retirement is a relatively recent phenomenon. Before Social Security, individuals continued to work into old age. Today, social security payments are limited to those over 65 years of age—age 67 by the year 2022. Most people, however, retire before age 65, with 60 percent of the labor force retiring at age 62 (Goldberg 2000).

Facts on Aging Quiz

Answer the following questions about the elderly and assess your knowledge of the world's fastest-growing age group.

	True	False
1. Lung capacity tends to decline in old age.	_____	_____
2. The majority of old people say they are seldom bored.	_____	_____
3. Old people tend to become more religious as they age.	_____	_____
4. The health and economic status of old people will be about the same or worse in the year 2010 (compared with younger people).	_____	_____
5. A person's height tends to decline in old age.	_____	_____
6. The aged are more fearful of crime than are younger persons.	_____	_____
7. The proportion of blacks among the aged is growing.	_____	_____
8. The majority of old people live alone.	_____	_____
9. The five senses all tend to weaken in old age.	_____	_____
10. Medicare pays over half of the medical expenses of the elderly.	_____	_____
11. Older persons who reduce their activity tend to be happier than those who do not.	_____	_____
12. Older persons have more injuries in the home than younger persons.	_____	_____
13. Physical strength tends to decline with age.	_____	_____
14. The aged are the most law abiding of all adult age groups.	_____	_____
15. Older persons have more acute illnesses than do younger persons.	_____	_____

Answers: 1, 2, 5–7, 9, 13, and 14 are true. The remainder are false.

Source: Erdman B. Palmore. 1999. *Ageism: Negative and Positive*. New York: Springer Publishing Co.

Retirement is difficult in that in the United States "work" is often equated with "worth." A job structures one's life and provides an identity; retirement often culturally signifies the end of one's productivity. Retirement may also involve a dramatic decrease in personal income. The desire to remain financially independent, a lack of confidence in the Social Security system, increased educational levels and technological skills, and the desire to continue working have led to a recent reduction in early retirements (AOA 2000f). Additionally, the number of people who stop working once they retire is also decreasing as retirees open their own businesses, work part-time, continue their educations, volunteer in the community, and begin second careers (Ennis 2000; Costa 2000; Goldberg 2000; AARP 2000).

Poverty

Poverty among the elderly varies dramatically by gender, race, ethnicity, marital status, and age: women, minorities, those who are single or widowed, and the "old-old" are most likely to be poor (NCSC 2000a). Seventy percent of all elderly poor are women, half of whom were not poor before the death of their

■ **Table 6.3** *Persons 65 years Old and Older Below Poverty Level*

	Percent Below Poverty Level		
	1970	**1990**	**1999**
Black	48.0	33.8	22.7
Hispanic	n/a	22.5	20.4
White	22.6	10.1	7.6
All	24.6	12.2	9.7

Sources: U.S. Bureau of the Census. 2000. "Historical Tables: Poverty", Table 3. *Current Population Survey*. Washington, D.C.: U.S. Department of Commerce.

husbands. Older minority women are particularly vulnerable. The poverty rate of white elderly women is 11 percent; among elderly African American women it is 30 percent, and among elderly Hispanic American women, 25 percent (AOA 2000e). Although the number of elderly poor has declined in recent years, over 3 million older persons live in poverty, and an additional 2 million are classified as "nearly poor" (AOA 2000g) (see Table 6.3).

Actually titled "Old Age, Survivors, Disability, and Health Insurance," Social Security is a major source of income for the elderly. When Social Security was established in 1935, it was never intended to be a person's sole economic support in old age; rather, it was to supplement other savings and assets. However, a recent Social Security Administration survey revealed that when asked "major sources of income," 90 percent of the elderly reported social security, 62 percent income from assets, 44 percent public and private pensions, and 21 percent earnings (AOA 2000g). Spending on the elderly has increased over time as have Social Security benefits. Today, 92 percent of Americans aged 65 and over receive Social Security, and another 3 percent will be eligible once they retire.

Although Social Security may keep as many as 4 out of 10 elderly from being poor (NCSC 2000a), the Social Security system has been criticized for being based on number of years of paid work and pre-retirement earnings. Hence women and minorities, who often earn less during their employment years, receive less in retirement benefits. Another concern is whether funding for the Social Security system will be adequate to provide benefits for the increased numbers of aged in the next decades. Fear of its demise has led some economists and government officials to recommend privatization. Similar to an Individual Retirement Account, privatization would allow workers to put the money they now have deducted from their payroll for social security (FICA) into a personally owned and invested retirement account.

Health Issues

The biology of aging is called **senescence.** It follows a universal pattern but does not have universal consequences. "Massive research evidence demonstrates that the aging process is neither fixed nor immutable. Biologists are now showing that many symptoms that were formerly attributed to aging—for example, certain disturbances in cardiac function or in glucose metabolism in the brain—are instead produced by disease" (Riley & Riley 1992, 221). Biological functioning is also intricately related to social variables. Altering lifestyles, activities, and social contacts affects mortality and morbidity. For example, a longitudinal study of men and women between 70 and 79 found that regular physical activity, higher levels of ongoing positive social relationships, and a sense of self-efficacy enhanced physical and cognitive functioning (Seeman & Adler 1998).

Biological changes are consequences of either **primary aging** caused by physiological variables such as cellular and/or molecular variation (e.g., gray hair) or secondary aging. **Secondary aging** entails changes attributable to poor diet, lack of exercise, increased stress, and the like. Secondary aging exacerbates and accelerates primary aging.

Alzheimer's disease received national attention when former President Ronald Reagan announced that he had the disease. Named for German neurologist Alois Alzheimer, the debilitating disease affects both the mental and physical condition of some 4 million Americans—a projected 22 million by 2025 (AP 2000).

Recent data indicate that elder health is improving, and that most older Americans rate their health as good or excellent (NIH 2000b). Nonetheless, with age, health tends to decline. Older people account for 36 percent of all hospital stays, averaging 6.8 days compared to 5.5 days for people under 65 (AOA 2000b). Health is a major quality-of-life issue for the elderly, especially because they face higher medical bills with reduced incomes. Older Americans spend three times as much on health care as their younger counterparts, over half of which is spent on insurance. The poor elderly, often women and/or minorities, spend an even higher proportion of their resources on health care.

Medicare was established in 1966 to provide medical coverage for those over the age of 65 and today insures approximately 39 million people (HCFA 2000). Although it is widely assumed that the medical bills of the elderly are paid by the government, the elderly are responsible for as much as 25 percent of their total health costs (AOA 2000e). Medicare, for example, pays about half the cost of a visit to the doctor, but does not pay for prescriptions, most long-term care, dental care, glasses, and hearing aids. The difference between Medicare benefits and the actual cost of medical care is called the **medigap**.

Because health is associated with income, the poorest old are often the most ill: they receive less preventive medicine, have less knowledge about health care issues, and have limited access to health care delivery systems. Medicaid is a federally and state-funded program for those who cannot afford to pay for medical care. However, eligibility requirements often disqualify many of the aged poor, often minorities and women.

Living Arrangements

The elderly live in a variety of contexts, depending on their health and financial status. Most elderly do not want to be institutionalized but prefer to remain in their own homes or in other private households with friends and relatives (see Table 6.4). Of the noninstitutionalized elderly population, 67 percent live in a family setting although that number decreases with age (AOA 2000b). Homes of the elderly, however, are usually older, located in inner-city neighborhoods, in need of repair, and often too large to be cared for easily. Six percent of the elderly, nearly one-and-a-half million households, live in homes that need serious repairs or modifications (NCSC 2000b).

If exercise could be put in a pill it would be the number one anti-aging medicine...

ROBERT BUTLER
Gerontologist

It's just a very hard process to watch yourself age. It's sad sometimes.

JESSICA LANGE
Actress

■ **Table 6.4** *Living Arrangements of Noninstitutionalized Women and Men 65 Years and Older, 1998*

Women

Living with spouse 41%

Living with other relative 17%

Living alone or with nonrelatives 42%

Men

Living with spouse 73%

Living with other relative 7%

Living alone or with nonrelatives 20%

Source: AOA (Administration on Aging). 2000. "A Profile of Older Americans" Washington, D.C.: Department of Health and Human Services.

Although many of the elderly poor live in government housing or apartments with subsidized monthly payments, the wealthier aged often live in retirement communities. These are often planned communities, located in states with warmer climates, and are often very expensive. These communities offer various amenities and activities, have special security, and are restricted by age. One criticism of these communities is that they segregate the elderly from the young and discriminate against younger people by prohibiting them from living in certain areas.

Those who cannot afford retirement communities or may not be eligible for subsidized housing often live with relatives in their own home or in the homes of others. It is estimated that more than 7 million people provide care for aging family members. Over three-quarters are women, many of whom are daughters who care not only for their elderly parents but for their own children as well (see this chapter's *Social Problems Research Up Close* feature).

Other living arrangements include shared housing, modified independent living arrangements, and nursing homes. With shared housing, people of different

Little did members of this age cohort know that they were to become the middle-aged "sandwich generation," emotionally and economically responsible for both their children and their aging parents. The stress and pressures from these caregiver roles are often part of what's called the "mid-life crisis."

© Tom Miner/The Image Works

ages live together in the same house or apartment; they have separate bedrooms but share a common kitchen and dining area. They share chores and financial responsibilities. In modified independent living arrangements, the elderly live in their own house, apartment, or condominium within a planned community where special services such as meals, transportation, and home repairs are provided. Skilled or semiskilled health care professionals are available on the premises, and call buttons are installed so help can be summoned in case of an emergency.

Nursing homes are residential facilities that provide full-time nursing care for residents. Nursing homes may be private or public. Private facilities are very expensive and are operated for profit by an individual or a corporation. Public facilities are nonprofit and are operated by a governmental agency, religious organization, or the like. The probability of being in such an extended care facility is associated with race, age, and sex: whites, the old-old, and women are more likely to be in residence. The elderly with chronic health problems are also more likely to be admitted to nursing homes. Nursing homes vary dramatically in cost, services provided, and quality of care.

Victimization and Abuse

Elder abuse refers to physical or psychological abuse, financial exploitation, or medical abuse or neglect of the elderly (Hooyman & Kiyak 1999). Although it can take place in private homes by family members, the elderly, like children, are particularly vulnerable to abuse when they are institutionalized. In 1990 the Nursing Home Reform Act was passed, establishing various rights for nursing care residents: the right to be free of mental and physical abuse, the right not to be restrained unless necessary as a safety precaution, the right to choose one's physician, and the right to receive mail and telephone communication (Harris 1990, 362).

In 1998, in response to continued reports of nursing home abuses, President Clinton instructed state agencies to make random and frequent inspections of such facilities, and to immediately institute fines where violations of federal law were found. Today, almost all states have mandatory reporting laws and some type of legal protection for nursing home residents including elder-abuse statutes, consumer protection laws, and adult protective services.

Whether the abuse occurs within the home or in an institution, the victim is most likely to be female, widowed, white, frail, and over 75. The abuser tends to be an adult child or spouse of the victim, who misuses alcohol (Anetzberger, Korbin, & Austin 1994). Some research suggests that the perpetrator of the abuse is more often an adult child who is financially dependent on the elderly victim (Boudreau 1993). Whether the abuser is an adult child or a spouse may simply depend on whom the elder victim lives with.

Many of the problems of the elderly are compounded by their lack of interaction with others, loneliness, and inactivity. This is particularly true for the old-old. The elderly are also segregated in nursing homes and retirement communities, separated from family and friends, and isolated from the flow of work and school. As with most problems of the elderly, the problems of isolation, loneliness, and inactivity are not randomly distributed. They are higher among the elderly poor, women, and minorities. A cycle is perpetuated—being poor and old results in being isolated and engaging in fewer activities. Such withdrawal affects health, which makes the individual less able to establish relationships or participate in activities.

Children and Grandchildren as Primary Caregivers

The single most significant demographic change of the next several decades will be the dramatic increase in the number of elderly worldwide. The increase in the elderly is associated with a variety of concerns including who will care for the growing number of the old. The present study by Dellman-Jenkins, Blankemeyer, and Pinkard (2000) examines a recent trend in caregiving—young adult children and grandchildren as primary caregivers.

Sample and Methods

The present research focuses on three areas: (1) characteristics of young caregivers and the assistance they provide, (2) the consequences of being a primary caregiver for the individual, and (3) caregivers' social support. Specifically, the researchers "compared the caregiving motivations, experiences and support

needs of young adult grandchildren providing assistance to grandparents with those of young adult children caring for their older parents" (p. 178). Participants were recruited from a variety of sources including social service agencies, hospitals, and assisted living facilities. An open-ended interview was used in conjunction with a 65-item questionnaire. The final sample was composed of 43 caregivers—20 daughters, 2 sons, 19 granddaughters, and 2 grandsons (N = 43). Eighty-one percent of the sample was white, 53 percent married, and 54 percent lived with the care-recipient; the remainder living in a separate residence.

Findings and Conclusions

Caregiver Role and Assistance Given. Children, compared with grandchildren, were significantly more likely

to respond that they were caring for their parents because no one else would. Grandchildren were most likely to respond that they were caring for their grandparent(s) out of a sense of duty and the desire to avoid nursing home care. Grandchildren also reported volunteering to care for grandparent(s) to help their parents.

> When Alzheimer's became apparent, my parents weren't coping well with it and were mean to her (not physically). I couldn't stand it...I told them I would take her to my house. My dad told me it would be the biggest mistake I ever made. Three years later...I still disagree.

The modal category for length of caregiving for both grandchildren and children was 1 to 5 years with approximately equal proportions of both providing a round-the-clock (47 percent) versus daily assistance (53 percent). Both sets of caregivers provided transportation, compan-

Quality of Life

Although some elderly do suffer from declining mental and physical functioning, many others do not. Being old does not mean being depressed, poor, and sick. Less than 6 percent of those 65 and over suffer from any type of depression (NIMH 1999). Interestingly, Blumenthal and colleagues report that depression in the elderly is better treated by exercise than medication alone, or by medication and exercise combined (Livni 2000).

Among the elderly who are depressed, two social factors tend to be in operation. One is society's negative attitude toward the elderly. Words and phrases such as "old," "useless," and "a has-been" reflect cultural connotations of the aged that influence feelings of self-worth. The roles of the elderly also lose their clarity. How is a retiree supposed to feel or act? What does a retiree do? As a result, the elderly become dependent on external validation that may be weak or absent.

The second factor contributing to depression among the elderly is the process of "growing old." This process carries with it a barrage of stressful life events all converging in a relatively short time period. These include health concerns, retirement, economic instability, loss of significant other(s), physical isolation, job displacement, and increased salience of the inevitability of death as a result of physiological decline. All of these events converge on the elderly and increase the incidence of depression and anxiety, and may affect the decision not to prolong life (see the *Focus on Technology* feature in this chapter).

All would live long, but none would be old.

BEN FRANKLIN
Inventor

ionship and emotional support, personal care, and household support (e.g., cleaning house, paying bills, making appointments).

Consequences of Role. Caretakers in general, and single caretakers in particular, reported that caretaking activities interfered with their social life. All caretakers reported spending at least 3 hours a day in the caretaker role. One 20-year-old granddaughter commented that, "I sometimes get mad because I think I'm doing too much for my grandpa and I don't have a life of my own... then I feel selfish for having such thoughts" (p. 184). Caretakers also reported strained family relations, particularly with spouse and children.

> I would like to go camping with my husband once in a while, but I can't just get up and go away, because of taking care of my grandparents. Even though they have the medical alert, I'm afraid

they won't use it if something goes wrong.

Further, careers were negatively affected as relocating, job performance, and achievement of long-term goals became difficult because of the demands of caregiving. Increases in stress were also reported by both groups although significantly higher for grandchildren (81 percent) than children (59 percent). Positive outcomes were also expressed with over 97 percent of respondents identifying some benefit or reward. Grandchildren were most likely to note maintaining a strong relationship with the care-recipient, whereas children more often listed prevention of nursing home placement as the primary benefit.

Sources of Support. All respondents reported seeking informal assistance from some source. Children most often turned to their siblings for help, whereas grandchildren most often turned to their spouse or dating part-

ner. Formal support, although seldom sought, most often came from nursing/home health care providers or community support groups.

The two generations of caregivers, children and grandchildren, differ little in their caregiving activities, behaviors, and strains. They do, however, differ in their motivation to assist, grandchildren more often noting attachment rather than need or obligation as the primary reason for becoming a caregiver. Further, the authors conclude, grandchildren are "more likely to report personal rewards (e.g., greater closeness, positive memories), while children were more apt to report instrumental rewards (e.g., providing quality home-care and avoiding nursing home placement)" (p. 185).

Source: Dellman-Jenkins, Mary, Maureen Blankemeyer, and Odessa Pinkard. 2000. "Young Adult Children and Grandchildren in Primary Caregiver Roles to Older Relatives and their Service Needs." *Family Relations* 49:177–187.

Strategies for Action: Growing Up and Growing Old

Activism by or on behalf of children or the elderly has been increasing in recent years and, as their numbers grow, such activism is likely to escalate and to be increasingly successful. For example, global attention to the elderly led to 1999 being declared the "International Year of Older Persons," and "the first nearly universally ratified human rights treaty in history" deals with children's rights (UNICEF 1998). Such activism takes several forms, including collective action through established organizations and the exercise of political and economic power.

Collective Action

Countless organizations are working on behalf of children, including the Children's Defense Fund, UNICEF, Children's Partnership, Children Now, and the Children's Action Network. Many successes take place at the local level where parents, teachers, corporate officials, politicians, and citizens join together in the interest of children. In Kentucky, AmeriCorp volunteers raised the reading competency of underachieving youths 116 percent in just 6 months; in Dayton, Ohio, behavioral problems of students at an elementary school were signifi-

> For age is opportunity, no less than youth itself; although in another dress.
>
> And as the evening twilight fades away
>
> The sky is filled with stars Invisible by day.
>
> HENRY WADSWORTH LONGFELLOW
> *Poet*

Physician-Assisted Suicide and Social Policy

Given the dramatic increase in the number of elderly and the technological ability to extend life, the debate over physician-assisted suicide (PAS) is likely to continue. When 2,000 doctors of terminally ill patients were surveyed, 6 percent said they had assisted in patient suicides and 33 percent said they would prescribe lethal amounts of drugs if permitted by law (Finsterbusch 2001). Further, in a study of pharmacists in Great Britain, 70 percent agreed that it was the patient's right to choose to die, and 57 percent responded that it was also the patient's right to involve the physician in the decision-making process (Hanlon, Weiss, & Rees 2000).

One of the most important factors influencing a physician's decision to restrict technological interventions is "family preference" (Randolph, Zollo, Wigton, & Yeh 1997). The family's decision, however, is most often based on the physician's recommendations to limit care whether it be withdrawal of life support (food, water, or mechanical ventilation), administering medications to end life (intravenous vasopressors), or withholding certain procedures that would prolong life (cardiopulmonary resuscitation) (Luce 1997).

In 1997, the Supreme Court ruled that the right to die is not constitutionally protected. However, the court also validated the concept of **double effect**. Double effect refers to the use of medical interventions to relieve pain and suffering but which may hasten death. Although such practices are not prohibited, a recent bill before Congress would criminalize such behaviors making it difficult for any state to uphold or to enact a physician-assisted suicide law (Orentlicher & Caplan 2000; Death with Dignity 2000).

As of 2000, only Oregon recognizes the right of PAS with its Death with Dignity Act. Two physicians must agree that the patient is terminally ill and is expected to die within 6 months, the patient must ask three times for death both orally and in writing, and the patients must swallow the barbiturates themselves rather than be injected with a drug by the physician. In 1999, 27 physician-assisted suicides occurred in Oregon—59 percent were male, 96 percent were white, and 44 percent were married. The majority were suffering from end-stage cancer (63 percent) and the median age was 71. The median time between ingestion and unconsciousness was 10 minutes; between ingestion and death 30 minutes (Annual Report 2000).

The national debate over PAS was fueled by images of Dr. Jack Kevorkian administering a deadly dose of drugs, at the request of a terminally ill patient, on the CBS evening show "60 Minutes." Although Kevorkian was convicted of second-degree murder for the "60 Minutes" death, advocates of PAS have tried to get Oregon-like provisions passed in other states. The most recent attempt was in Maine

---

cantly reduced once "character education" was introduced; and in Los Angeles' Crenshaw High School, student entrepreneurs established "Food from the Hood," a garden project that is projected to earn $50,000 in profit this year.

Some programs combine the interests of both the young and the old. The Adopt a Grandparent Program arranges for children to visit nursing home residents, and the Foster Grandparent Program pairs children with special needs with low-income elderly. Further, the Children's Defense Fund, the Child Welfare League of America, the American Association of Retired Persons (AARP) and the National Council on Aging joined together in 1986 to create Generations United, a "national advocacy organization to promote age integration" (Uhlenberg 2000, 279).

More than a thousand organizations are directed toward realizing political power, economic security, and better living conditions for the elderly. One of the earliest and most radical groups is the Gray Panthers, founded in 1970 by Margaret Kuhn. The Gray Panthers were responsible for revealing the unscrupulous practices of the hearing aid industry, persuading the National Association of Broadcasters to add "age" to "sex" and "race" in the Television Code of Ethics statement on media images, and eliminating the mandatory retirement age. In view of these successes, it is interesting to note that the Gray Panthers, with only 50,000 to 70,000 members, is a relatively small organization when compared with the AARP.

No man or woman stands as tall as one who stoops to help a child.

TEDDY ROOSEVELT
Former U.S. President

where, in November of 2000, the voters defeated the measure by a slim 51 percent margin. At the same time, The Netherlands passed a law that legalized PAS. The law requires that three conditions be met, that: (1) there be unbearable suffering, (2) the case be clinically hopeless, and (3) a voluntary request for assistance be made (Weber 2000).

Despite what some say is a movement toward a greater acceptance of PAS, the official position of the American Medical Association is that physicians must respect the patient's decision to forgo life-sustaining treatment but should not participate in physician-assisted suicide. One argument against PAS is that the practice is subject to abuses—a spouse with self-serving interests, a depressed patient making a hasty decision, or an overburdened family pressuring a vulnerable loved one. Concern also exists that legalizing PAS may disproportionately end the lives of minority, ethnic, or psychiatrically disturbed individuals (Allen 1998).

Some would argue, however, that ultimately the decision should reside with the patient. As one elderly person said:

I came into this world as a human-being and I wish to leave in the same manner. Being able to walk, to communicate, to take care of my own needs, to think, to feel....There is no need for me to reexperience my first few months of life though my last months in this world...I do not wish to be once again in a diaper. Just the thought of me losing control over my body frightens me. (Leichtenritt & Rettig 2000, 3)

Sources:

C. L. Allen. 1998. Euthanasia: Why Torture Dying People When We Have Sick Animals Put Down? *Australian Psychologist* 33:12–15.

Annual Report. 2000. "Oregon's Death with Dignity Act: The Second Year Experience." Oregon's Death with Dignity Act Annual Report, 1999. Center for Heath Statistics. Oregon Health Division.

Death with Dignity. 2000. "Chronology of the Issue." http://web.lwc.ewdu/administrative/library/death.html

Kurt Finsterbusch, 2001. *Clashing Views on Controversial Social Issues.* Guilford, CN: Dushkin Publishing.

Timothy Hanlon, Marjorie Weiss and Judith Rees. 2000. "British Community Pharmacists' Views of Physicians-assisted Suicide." *Journal of Medical Ethics* 26:363–370.

R. D. Leichtenritt and K. D. Rettig. 2000. "Conflicting value considerations for end-of-life decisions." Poster session at 62nd Annual Meeting of National Council on Family Relations, Minneapolis, Minnesota. November 12.

J. M. Luce. 1997. "Withholding and Withdrawal of Life Support: Ethical, Legal, and Clinical Aspects." *New Horizons* 5: 30–37.

David Orentlicher and Arthur Caplan. 2000. "The Pain Relief Promotion Act of 1999." *Journal of the American Medical Association* 283:255–258.

A. G. Randolph, M. B. Zollo, R. S. Wigton, and T. S. Yeh. 1997. "Factors Explaining Variability among Caregivers in the Intent to Restrict Life-Support Interventions in a Pediatric Intensive Care Unit." *Critical Care Medicine* 25:435–39.

Kim Weber. 2000. "Dutch Euthanasia Law passed by Parliament." *Lancet* 356:1911.

Kim Weber. 2000. "Netherlands Proposal for Comprehensive Euthanasia Legislation." *Lancet* 356:1666.

The AARP has more than 33 million members age 50 and above, 90 percent of whom are white. It has over 1,700 paid employees and ten times as many volunteers with a budget of over $5 billion a year (Peterson 2000). Services of the AARP include discounted mail-order drugs, investment opportunities, travel information, volunteer opportunities, and health insurance. The AARP is the largest volunteer organization in the United States with the exception of the Roman Catholic Church. Not surprisingly, it is one of the most powerful lobbying groups in Washington.

Political Power

Children are unable to hold office, to vote, or to lobby political leaders. Child advocates, however, acting on behalf of children, have wielded considerable political influence in such areas as child care, education, health care reform, and crime prevention. Further, funding of such programs is supported by most Americans. A recent Children's Defense Fund study found that the majority of those surveyed favored federally funded after-school programs and subsidized child care even if it meant raising taxes (CDF 2000d).

As conflict theorists emphasize, the elderly compete with the young for limited resources. They have more political power than the young and more polit-

Protests are not limited to the young. The elderly have already been active in protesting changes to Medicare and other government policies that they believe are detrimental. As the number of the elderly grows, social activism is likely to increase.

© Frank Fournier/Contact Press Images

ical power in some states than in others. In Florida there is concern that the elderly may eventually wield too much political power and act as a voting bloc, demanding excessive services at the cost of other needy groups. For example, if the elderly were concentrated in a particular district, they could block tax increases for local schools. To the extent that future political issues are age-based and the elderly are able to band together, their political power may increase as their numbers grow over time (Matras 1990; Thurow 1996). By 2030, almost half of all adults in developed countries and two-thirds of all voters will be near or at retirement age. The growing political power of the aged is already becoming evident; The Netherlands has a new political party called the Pension Party.

Economic Power

Working to make a better life for the next generation.

SOUTH KOREAN COMPANY MOTTO

Although children have little economic power, the economic power of the elderly has grown considerably in recent years leading one economist to refer to the elderly as a "revolutionary class" (Thurow 1996). The 1999 median income for males 65 and over was $19,079; for females 65 and over, $10,943. Households headed by persons 65 and over had a median income of $31,568 (AOA 2000b, 2000g). Although these incomes are significantly lower than for men and women between 45 and 54 years of age, income is only one source of economic power. The fact that many elderly own their homes, have substantial savings and investments, enjoy high levels of disposable income, and are growing in number contribute to their increased importance as consumers.

Finally, in many parts of the country the elderly have become a major economic power as part of what is called the **"mailbox economy."** The mailbox economy refers to the tendency for a substantial portion of local economies to be dependent on pension and social security checks received in the mail by older residents (Atchley 2000). States with the highest percent of the elderly include Florida (18.3 percent), Pennsylvania (15.9 percent), Rhode Island (15.6 percent), and West Virginia (15.2 percent) (AOA 2000b).

Government Policy

In a recent nationwide survey, registered voters were asked what they thought was the single most important issue facing the government. Three of the four top answers dealt with issues confronting the elderly—health care (discussed in Chapter 2), social security, and Medicare/ Medicaid (CBS/New York Times Poll 2000). As mentioned earlier, Medicaid provides health care for the poor; Medicare is a national health care insurance program designed for people over the age of 65. In 1988, Congress passed the Medicare Catastrophic Coverage Act (MCCA), which was the most significant change in Medicare since its establishment in 1966. The new benefits included unlimited hospitalization, an upper limit on the amount of money recipients would pay for physicians' services, home health care and nursing home services, and unlimited hospice care. These changes were particularly significant because many of the illnesses of the elderly are chronic in nature.

The reforms were financed by increasing monthly medical premiums $4 a month and imposing an annual fee based on a person's federal income tax bracket. The maximum premium paid was $800 per person and $1,600 per couple (Harris 1990; Torres-Gil 1990). The AARP initially supported the reforms but later withdrew its support as did many other organizations. The additional monies paid were simply not worth the new benefits, they contended. The AARP also argued that the elderly should not have to bear the burden of reforms necessary for the general public. In 1989, under pressure from the AARP and other organizations of the elderly, Congress repealed the MCCA. Pressured by public demands and fear of bankruptcy, Congress is currently in the process of reforming health care policy (see Chapter 2) including Medicare, which is projected to triple in cost over the next several years (Weiss & Lonnquist 2000).

Other recent government initiatives for the elderly include the Elderly Nutrition Program and the Older Americans Act Amendments of 2000 (AOA 2001). The Elderly Nutrition Program provides delivered meals to locations where elderly congregate—homes, senior centers, and schools, and provides nutritional training, counseling, and education (AOA 2000h). The Older Americans Act Amendments 2000 created the National Family Caregivers Support Program. This program provides "critical information, training and counseling, as well as much-needed quality respite care for those caregivers who are juggling jobs and other family responsibilities while meeting the special needs of loved ones in their care" (Clinton 2000, 1).

Finally, in 1997 Congress passed the State Children's Health Care Insurance Program (SCHIP) which "is the largest single expansion of health insurance coverage for children in more than 30 years" (HHS 1999, 1). The SCHIP was designed for families who cannot afford health insurance but whose incomes are too high to qualify for Medicaid. The program, at a cost of $24 billion, will provide health care to some of the 10 million children who do not have health care coverage (see Chapter 2).

Understanding The Young and the Old

What can we conclude about youth and aging in American society? Age is an ascribed status and, as such, is culturally defined by role expectations and implied personality traits. Society regards both the young and the old as dependent and in need of the care and protection of others. Society also defines the young and old as physically, emotionally, and intellectually inferior. As a consequence of these and other attributions, both age groups are sociologically a minority with limited opportunity to obtain some or all of society's resources.

Although both the young and the old are treated as minority groups, different meanings are assigned to each group. In the United States, in general, the young are more highly valued than the old. Functionalists argue that this priority on youth reflects the fact that the young are preparing to take over important statuses while the elderly are relinquishing them. Conflict theorists emphasize that in a capitalist society, both the young and the old are less valued than more productive members of society. Conflict theorists also point out the importance of propagation, that is, the reproduction of workers, which may account for the greater value placed on the young than the old. Finally, symbolic interactionists describe the way images of the young and the old intersect and are socially constructed.

The collective concerns for the elderly and the significance of defining ageism as a social problem have resulted in improved economic conditions for the elderly. Currently, they are one of society's more powerful minorities. Research indicates, however, that despite their increased economic status, the elderly are still subject to discrimination in such areas as housing, employment, and medical care and are victimized by systematic patterns of stereotyping, abuse, and prejudice.

In contrast, the position of children in the United States, although improving, remains tragic with one in five children living in poverty. Wherever there are poor families, there are poor children who are educated in inner city schools, live in dangerous environments, and lack basic nutrition and medical care. Further, age-based restrictions limit their entry into certain roles (e.g., employee) and demand others (e.g., student). Although most of society's members would agree that children require special protections, concerns regarding quality-of-life issues and rights of self-determination are only recently being debated.

Age-based decisions are potentially harmful. For example, "by 2003 expenditures on the elderly (plus interest payments) will take 75 percent, and by 2013…100 percent" of tax revenues if present laws remain unchanged (Thurow 1996). If budget allocations were based on indigence rather than age, more resources would be available for those truly in need. Further, age-based decisions may encourage intergenerational conflict. Government assistance is a zero-sum relationship—the more resources one group gets, the fewer resources another group receives.

Social policies that allocate resources on the basis of need rather than age would shift the attention of policy makers to remedying social problems rather than serving the needs of special interest groups. Age should not be used to cause negative effects on an individual's life any more than race, ethnicity, gender, or sexual orientation. Although eliminating all age barriers or requirements is unrealistic, a movement toward assessing the needs of individuals and their abilities would be more consistent with the American ideal of equal opportunity for all.

> One of the first priorities of any civilized society is to take care of its children, to prevent needless suffering in the ranks of the vulnerable and the blameless.
>
> SYLVIA ANN HEWLETT
> *President, National Parenting Association*

Critical Thinking

1 In many ways, American society discriminates against children. Children are segregated in schools, in a separate justice system, and in the workplace. Identify everyday examples of the ways in which children are treated like "second-class" citizens in the United States.

2 **Age pyramids** pictorially display the distribution of people by age (see Figure 6.1). How do different age pyramids influence the treatment of the elderly?

3 Regarding children and the elderly, what public policies or programs from other countries might be beneficial to incorporate in the United States? Do you think policies from other countries would necessarily be successful here?

4 Argue for or against the privatization of the social security system.

Key Terms

activity theory	double effect	modernization theory
age grading	elder abuse	prejudice
age pyramids	gerontophobia	primary aging
ageism	infantilizing elders	sandwich generation
dependency ratio	mailbox economy	secondary aging
discrimination	medigap	senescence
disengagement theory	minority	stereotype

Media Resources

The Wadsworth Sociology Resource Center: Virtual

http://sociology.wadsworth.com/

See the companion Web site for this book to access general sociology resources and text-specific features that can further your understanding of this chapter. The site contains Internet links, Internet exercises, online practice quizzes, information on InfoTrac College Edition, and many more valuable materials designed to enrich your learning experience in social problems.

InfoTrac College Edition

You can access InfoTrac College Edition either from the Wadsworth Sociology Resource Center at **http://sociology.wadsworth.com** or directly from your web browser at **http://www.infotrac-college.com/wadsworth/**. InfoTrac College Edition is an online university library that includes over 700 popular and scholarly journals in which you can find articles related to the topics in this chapter such as information on youth suicide, teenage employment, social security, and Alzheimer's Disease.

Interactions CD-ROM

Go to the "Interactions" CD-ROM for *Understanding Social Problems*, Third Edition to access additional interactive learning tools, such as in-depth review materials, corresponding practice quizzes, and other engaging resources and activities to help you study the concepts in this chapter.

7

Race and Ethnic Relations

Is It True?

1. Many anthropologists and other scientists have concluded that "races" do not really exist.

2. In the 2000 elections, 41 percent of voters in Alabama voted *against* removing the state constitution's ban on interracial marriage.

3. In California, minority group members outnumber non-Hispanic whites.

4. The largest category of individuals targeted to benefit from affirmative action is African-Americans.

5. The third most common place for hate crimes to occur is on college campuses.

Answers to "Is it True?" 1 = T; 2 = T; 3 = T; 4 = F; 5 = T

The 21st century will be the century in which we redefine ourselves as the first country in world history which is literally made up of every part of the world.

KENNETH PREWITT
Census Bureau director

Karl Nichols, a white residence hall director at the University of Mississippi, learned about racism in college, but not in the classroom and not from a textbook. Two chunks of asphalt were hurled through Karl's dormroom window along with a note warning, "You're going to get it, you Godforsaken nigger-lover" ("Hate on Campus" 2000, 10). The next night, someone attempted to set Karl's door on fire. According to the university's investigation of these incidents, Karl Nichols "may have violated racist taboos . . . by openly displaying his affinity for African-American individuals and black culture, by dating black women, by playing [black] music . . . and by promoting diversity" in his dormitory (p. 10).

Karl Nichols is not alone. At Brown University in Rhode Island, a black student is beaten by three white students who tell her she is a "quota" who doesn't belong at a university. At the State University of New York at Binghamton, an Asian-American student is left with a fractured skull after a racially motivated assault by three students. Two students at the University of Kentucky—one white, one black— were crossing the street just off campus when they were attacked by 10 white men. The attackers yelled racist slurs at the black student and choked him until he couldn't speak or move. The assailants called the white student a "nigger-lover" as they broke his hand and nose. "I definitely thought I was going to lose my life," the black student said later. The white student was shocked by the incident, commenting, "I didn't know that much hate existed." ("Hate on Campus" 2000, 7)

The United States is becoming increasingly diversified in the racial and ethnic characteristics of its population. Indeed, 2000 census data revealed that for the first time in modern U.S. history, non-Hispanic whites comprise less than half of the population of California (Purdum 2001). The majority of Californians are minorities, with Hispanic residents making up nearly one-third of the state's population. Racial and ethnic group relations continue to be a major social concern in the United States and throughout the world. Despite significant improvements in U.S. race and ethnic relations over the last two centuries, a great deal of work remains to be done. And much pessimism about the future of race relations in the United States is evident. In response to a question asking whether relations between blacks and whites will always be a problem for the United States, or whether a solution will eventually be worked out, 51 percent of whites and 59 percent of blacks say that race relations will always be a problem (Ludwig 2000).

In this chapter, we discuss the nature and origins of prejudice and examine the extent of discrimination and its consequences for racial and ethnic minorities. We also discuss strategies designed to reduce prejudice and discrimination. We begin by examining racial and ethnic diversity worldwide and in the United States, emphasizing first that the concept of race is based on social rather than biological definitions.

The Global Context: Diversity Worldwide

A first-grade teacher asked the class, "What is the color of apples?" Most of the children answered red. A few said green. One boy raised his hand and said "white." The teacher tried to explain that apples could be red, green, or sometimes golden, but never white. The boy insisted his answer was right and finally said, "Look inside" (Goldstein 1999). Like apples, human beings may be similar on the "inside," but are often classified into categories according to external appearance. After examining the social construction of racial categories, we review patterns of interaction among racial and ethnic groups and examine racial and ethnic diversity in the United States.

The Social Construction of Race

> Race and ethnicity are in the eye of the beholder.
>
> HAROLD L. HODGKINSON
> *Director of the Center for Demographic Policy*

The concept **race** refers to a category of people who are believed to share distinct physical characteristics that are deemed socially significant. Racial groups are sometimes distinguished on the basis of such physical characteristics as skin color, hair texture, facial features, and body shape and size. Some physical variations among people are the result of living for thousands of years in different geographical regions (Molnar 1983). For example, humans living in regions with hotter climates developed darker skin from the natural skin pigment, melanin, which protects the skin from the sun's rays. In regions with moderate or colder climates, people had no need for protection from the sun and thus developed lighter skin.

Cultural definitions of race have taught us to view race as a scientific categorization of people based on biological differences between groups of individuals. Yet, racial categories are based more on social definitions than on biological differences.

Anthropologist Mark Cohen (1998) explains that distinctions among human populations are graded, not abrupt. Skin color is not black or white, but rather ranges from dark to light with many gradations of shades. Noses are not either broad or narrow, but come in a range of shapes. Physical traits such as these, as well as hair color and other characteristics, come in an infinite number of combinations. For example, a person with dark skin can have any blood type and can have a broad nose (a common combination in West Africa), a narrow nose (a common trait in East Africa), or even blond hair (a combination found in Australia and New Guinea) (Cohen 1998).

The science of genetics also challenges the notion of race. Geneticists have discovered that "the genes of black and white Americans probably are 99.9 percent alike" (Cohen 1998, B4). Furthermore, genetic studies indicate that genetic variation is greater *within* racially classified populations than *between* racial groups (Keita & Kittles 1997). Classifying people into different races fails to recognize that over the course of human history, migration and intermarriage have resulted in the blending of genetically transmitted traits. Thus, there are no "pure" races; people in virtually all societies have genetically mixed backgrounds.

The American Anthropological Association has passed a resolution stating that "differentiating species into biologically defined 'race' has proven meaningless and unscientific" (Etzioni 1997, 39). Scientists who reject the race concept now speak of **populations** when referring to groups that most people would call races (Zack 1998).

Clear evidence that race is a social, rather than biological concept is the fact that different societies construct different systems of racial classification, and that these systems change over time (Niemonen 1999). Leggon (1999) explains, "The major significance of race is not biological but social and political, insofar as race is used as the primary line of demarcation separating 'we' from 'they' and, consequently, becomes a basis for distinctive treatment of a group by another" (p. 382). Despite the increasing acceptance that race is not a valid biological categorization, its social significance continues to be evident throughout the world.

Patterns of Racial and Ethnic Group Interaction

When two or more racial or ethnic groups come into contact, one of several patterns of interaction may occur, including genocide, expulsion or population transfer, slavery, colonialism, segregation, acculturation, assimilation, pluralism, and amalgamation. These patterns of interaction may occur when two or more groups exist in the same society or when different groups from different societies come into contact. Although not all patterns of interaction between racial and ethnic groups are destructive, author and Mayan shaman Martin Prechtel reminds us that, "Every human on this earth, whether from Africa, Asia, Europe, or the Americas, has ancestors whose stories, rituals, ingenuity, language, and life ways were taken away, enslaved, banned, exploited, twisted, or destroyed..." (quoted in Jensen 2001, 13).

- **Genocide** refers to the deliberate, systematic annihilation of an entire nation or people. The European invasion of the Americas, beginning in the sixteenth century, resulted in the decimation of most of the original inhabitants of North and South America. Some native groups were intentionally killed, whereas others fell victim to diseases brought by the Europeans. In the twentieth century, Hitler led the Nazi extermination of more than 12 million people, including over 6 million Jews, in what has become known as the Holocaust. More recently, ethnic Serbs have attempted to eliminate Muslims from parts of Bosnia—a process they call "ethnic cleansing."

tol•er•ance••n.: the capacity for or the practice of recognizing and respecting the beliefs or practices of others.

THE AMERICAN HERITAGE DICTIONARY

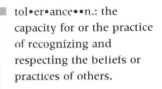

The most cited twentieth-century example of genocide is the "extermination" of over 12 million Jews and others considered by Hitler not to be members of his "superior race."

AP/Wide World Photos

- **Expulsion** or **population transfer** occurs when a dominant group forces a subordinate group to leave the country or to live only in designated areas of the country. The 1830 Indian Removal Act called for the relocation of eastern tribes to land west of the Mississippi River. The movement, lasting more than a decade, has been called the Trail of Tears because tribes were forced to leave their ancestral lands and endure harsh conditions of inadequate supplies and epidemics that caused illness and death. After Japan's attack on Pearl Harbor in 1941, President Franklin Roosevelt authorized the removal of any people considered threats to national security. All people on the West Coast of at least one-eighth Japanese ancestry were transferred to evacuation camps surrounded by barbed wire, where 120,000 Japanese Americans experienced economic and psychological devastation. In 1979, Vietnam expelled nearly 1 million Chinese from the country as a result of long-standing hostilities between China and Vietnam.

- **Slavery** exists when one group treats another group as property to exploit for financial gain. The dominant group forces the enslaved group to live a life of servitude, without the basic rights and privileges enjoyed by the dominant group. In early American history, slavery was tolerated and legal for three centuries. Enslavement of Africans also occurred in Canada from 1689 to 1833 (Schaefer 1998).

- **Colonialism** occurs when a racial or ethnic group from one society takes over and dominates the racial or ethnic group(s) of another society. The European invasion of North America, the British occupation of India, and the Dutch presence in South Africa before the end of apartheid are examples of outsiders taking over a country and controlling the native population. As a territory of the United States, Puerto Rico is essentially a colony whose residents are U.S. citizens but cannot vote in presidential elections unless they move to the mainland.

- **Segregation** refers to the physical separation of two groups in residence, workplace, and social functions. Segregation may be **de jure** (Latin meaning "by law") or **de facto** ("in fact"). Between 1890 and 1910, a series of U.S. laws that came to be known as **Jim Crow laws** were enacted that separated blacks from whites by prohibiting blacks from using "white" buses, hotels, restaurants, and drinking fountains. In 1896, the U.S. Supreme Court (in *Plessy v. Ferguson*) supported de jure segregation of blacks and whites by declaring that "separate but equal" facilities were constitutional. Blacks were forced to live in separate neighborhoods and attend separate schools. Beginning in the 1950s, various rulings overturned these Jim Crow laws, making it illegal to enforce racial segregation. Although de jure segregation is illegal in the United States, de facto segregation still exists in the tendency for racial and ethnic groups to live and go to school in segregated neighborhoods.

- **Acculturation** refers to learning the culture of a group different from the one in which a person was originally raised. Acculturation may involve learning the dominant language, adopting new values and behaviors, and changing the spelling of the family name. In some instances, acculturation may be forced, as in the California decision to discontinue bilingual education and force students to learn English in school.

- **Pluralism** refers to a state in which racial and ethnic groups maintain their distinctness, but respect each other and have equal access to social resources. In Switzerland, for example, four ethnic groups—French, Italians,

Germans, and Swiss Germans—maintain their distinct cultural heritage and group identity in an atmosphere of mutual respect and social equality. In the United States, the political and educational recognition of multiculturalism reflects efforts to promote pluralism. Given the extent of prejudice and discrimination toward racial and ethnic minorities (discussed later in this chapter), however, the United States is far from pluralistic.

- **Assimilation** is the process by which formerly distinct and separate groups merge and become integrated as one. One form of assimilation is referred to as the **melting pot**, whereby different groups come together and contribute equally to a new, common culture. Although the United States has been referred to as a "melting pot," in reality, many minorities have been excluded or limited in their cultural contributions to the predominant white Anglo-Saxon Protestant tradition.

  Assimilation may be of two types: secondary and primary. **Secondary assimilation** occurs when different groups become integrated in public areas and in social institutions, such as neighborhoods, schools, the workplace, and in government. **Primary assimilation** occurs when members of different groups are integrated in personal, intimate associations such as friends, family, and spouses.

- **Amalgamation**, also known as **marital assimilation**, occurs when different ethnic or racial groups become married or pair-bonded and produce children. Nineteen states had **antimiscegenation laws** banning interracial marriage until 1967 when the Supreme Court (in *Loving v. Virginia*) declared these laws unconstitutional and required all states to recognize interracial marriage (Reid 1995). In the United States, marriages between individuals with different ethnic backgrounds are not unusual, but interracial marriages are relatively rare, with less than 3 percent of U.S. married couples being interracial (*Statistical Abstract* 2000). In the 2000 elections, the citizens of Alabama voted to remove the state constitution's ban on interracial marriage. Even though the ban had been deemed unconstitutional by the 1967 Supreme Court ruling and could not be enforced, 41 percent of Alabama voters voted *against* removing the ban from the state constitution in the 2000 election.

  The degree of acculturation and assimilation that occurs between majority and minority groups depends in part on (1) whether minority group members have voluntary or involuntary contact with the majority group and (2) whether majority group members accept or reject newcomers or minority group members. Groups that voluntarily immigrate and who are accepted by "host" society members will experience an easier time acculturating and assimilating than those who are forced (through slavery, frontier expansion, or military conquest) into contact with and are rejected by the majority group.

Racial Diversity in the United States

The first census in 1790 divided the U.S. population into four groups: free white males, free white females, slaves, and other persons (including free blacks and Indians). In order to increase the size of the slave population, the **one drop of blood rule** appeared, which specified that even one drop of Negroid blood defined a person as black and, therefore, eligible for slavery. In 1960, the census recognized only two categories: white and nonwhite. In 1970, the census cate-

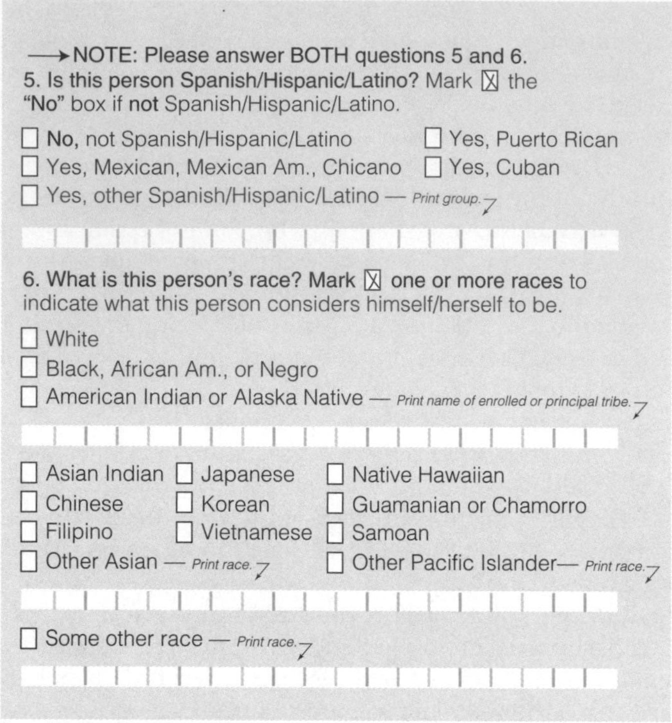

> → NOTE: Please answer BOTH questions 5 and 6.
>
> **5. Is this person Spanish/Hispanic/Latino?** Mark ☒ the "No" box if **not** Spanish/Hispanic/Latino.
>
> ☐ **No,** not Spanish/Hispanic/Latino   ☐ Yes, Puerto Rican
> ☐ Yes, Mexican, Mexican Am., Chicano   ☐ Yes, Cuban
> ☐ Yes, other Spanish/Hispanic/Latino — *Print group.* ⌐
>
> [write-in boxes]
>
> **6. What is this person's race?** Mark ☒ one or more races to indicate what this person considers himself/herself to be.
>
> ☐ White
> ☐ Black, African Am., or Negro
> ☐ American Indian or Alaska Native — *Print name of enrolled or principal tribe.* ⌐
>
> [write-in boxes]
>
> ☐ Asian Indian   ☐ Japanese   ☐ Native Hawaiian
> ☐ Chinese   ☐ Korean   ☐ Guamanian or Chamorro
> ☐ Filipino   ☐ Vietnamese ☐ Samoan
> ☐ Other Asian — *Print race.* ⌐   ☐ Other Pacific Islander— *Print race.* ⌐
>
> [write-in boxes]
>
> ☐ Some other race — *Print race.* ⌐
>
> [write-in boxes]
>
> Source: U.S. Census Bureau, Census 2000 questionnaire.

■ **Figure 7.1** *Reproduction of Questions on Race and Hispanic Origin from Census 2000*

gories consisted of white, black, and "other" (Hodgkinson 1995). In 1990, the U.S. Bureau of the Census recognized four racial classifications: (1) white, (2) black, (3) American Indian, Aleut, or Eskimo, and (4) Asian or Pacific Islander. The 1990 census also included the category of "other." Beginning with the 2000 census, the Office of Management and Budget requires federal agencies to use a minimum of five race categories: (1) white, (2) black or African American, (3) American Indian or Alaska Native, (4) Asian, and (5) Native Hawaiian or other Pacific Islander (Grieco & Cassidy 2001). In addition, respondents to federal surveys and the census now have the option of officially identifying themselves as being more than one race, rather than checking only one racial category (see Figure 7.1). Figure 7.2 presents the newest Census 2000 data on the racial composition of the United States.

A mixed-race option for self-identification avoids putting children of mixed-race parents in the difficult position of choosing the race of one parent over the other when filling out race data on school and other forms. It also avoids impairment of children's self-esteem and social functioning that comes from choosing the racial category of "other." Such a category implies that the society does not recognize and respect mixed-race individuals, and thus "children growing up within mixed families may feel ashamed of their 'irregular' racial makeup and may experience rejection and alienation in the wider social community" (Zack 1998, 23).

Some critics of the new mixed-race option are concerned that the wide-scale recognition of mixed-race identity will decrease the numbers within minority

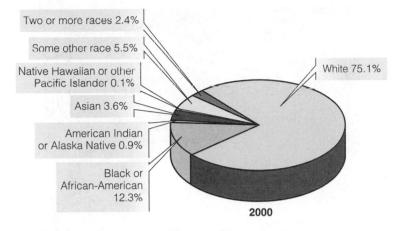

Two or more races 2.4%

Some other race 5.5%

Native Hawaiian or other Pacific Islander 0.1%

Asian 3.6%

American Indian or Alaska Native 0.9%

Black or African-American 12.3%

White 75.1%

2000

Figure 7.2 *Race Composition of the United States, 2000.*
Source: Grieco, Elizabeth M. & Rachel C. Cassidy. 2001 (March). "Overview of Race and Hispanic Origin: Census 2000 Brief." U.S. Census Bureau.

groups and disrupt the solidarity and loyalty based on racial identification. What will happen, for example, to organizations and movements devoted to equal rights for blacks if much of the "black" population acquires a new mixed-racial identity? However, 2000 census data suggest that the mixed-race option will not have the large national impact that critics fear. As shown in Figure 7.2 only 2.4 percent of the U.S. population identified themselves as being of more than one race in the 2000 Census. About 5 percent of U.S. black people said they were multiracial (Schmitt 2001).

The recognition of mixed race is also criticized as perpetuating scientifically unfounded classifications by race. This criticism suggests that mixed-race categorization is just as unscientific as the concept of race itself. However, mixed-race classification may encourage the realization that racial categorization is meaningless:

> Recognized mixed-race identity would undo the assumption that everyone is racially pure, and may, in the process, undo the assumption that everyone belongs to a race or that race is a meaningful way to type people. (Zack 1998, 27)

Ethnic Diversity in the United States

Ethnicity refers to a shared cultural heritage or nationality. Ethnic groups may be distinguished on the basis of language, forms of family structures and roles of family members, religious beliefs and practices, dietary customs, forms of artistic expression such as music and dance, and national origin.

Two individuals with the same racial identity may have different ethnicities. For example, a black American and a black Jamaican have different cultural, or ethnic, backgrounds. Conversely, two individuals with the same ethnic background may identify with different races. Hispanics, for example, may be white or black. The current Census Bureau classification system does not allow people of mixed Hispanic/Latino ethnicity to identify themselves as such. Individuals with one Hispanic and one non-Hispanic parent still must say they are either Hispanic or not Hispanic. And Hispanics must select one country of origin, even if their parents are from different countries.

Golf pro Tiger Woods has referred to himself as "Cabrinasian"—reflecting his mixed heritage that includes Caucasian, Black, Asian, and Indian.

U.S. citizens come from a variety of ethnic backgrounds. The largest ethnic population in the United States is of Hispanic origin. (The terms "Hispanic" and "Latino" are used interchangeably.) Nearly 13 percent of people in the United States are Hispanic or Latino and about two-thirds of all U.S. Hispanics/Latinos are of Mexican origin (Grieco & Cassidy 2001; Ramirez 2000). Other Hispanic Americans have origins in Central and South America (14.3 percent), Puerto Rico (9.6 percent), and Cuba (4.3 percent). The fastest growing racial/ethnic population in the United States is Asian/Pacific Islanders, followed by Hispanics. Estimates project that the percentage of Asian/Pacific Islanders in the United States will increase 36.1 percent from 2000 to 2010; the percentage of Hispanics will increase 31.2 percent over the same time period (Gardyn & Fetto 2000).

> If you look at me, I'm black. But if you're asking my ethnicity, it's Haitian. The term "black" just doesn't cut it.
>
> **DENISE BERNARD**
> *President of Queens College student government*

The use of racial and ethnic labels is often misleading. The ethnic classification of "Hispanic/Latino," for example, lumps together such disparate groups as Puerto Ricans, Mexicans, Cubans, Venezuelans, Colombians, and others from Latin American countries. The racial term "American Indian" includes more than 300 separate tribal groups that differ enormously in language, tradition, and social structure. The racial label "Asian American" includes individuals with ancestors from China, Japan, Korea, India, the Philippines, or one of the countries of Southeast Asia.

U.S. Immigration

The growing racial and ethnic diversity of the United States is largely the result of immigration (as well as the higher average birth rates among many minority groups). The many hardships experienced by poor people throughout the world "push" them to leave their home countries, while the economic opportunities that exist in more affluent countries "pull" them to those countries. In the 1960s, most immigrants were from Europe, but now they are from Central America (predominantly Mexico) or Asia (see Figure 7.3). In 2000, an estimated

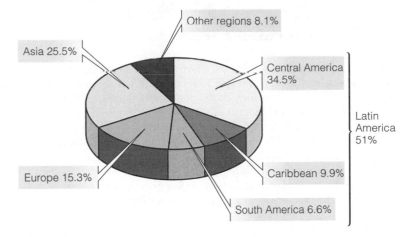

Other regions 8.1%

Asia 25.5%

Central America 34.5%

Latin America 51%

Europe 15.3%

Caribbean 9.9%

South America 6.6%

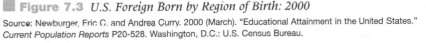

Figure 7.3 *U.S. Foreign Born by Region of Birth: 2000*

Source: Newburger, Eric C. and Andrea Curry. 2000 (March). "Educational Attainment in the United States." *Current Population Reports* P20-528. Washington, D.C.: U.S. Census Bureau.

one in ten U.S. residents was born in a foreign country (Lollock 2001) (see Table 7.1), and one in five U.S. children 18 and under is the child of an immigrant (Urban Institute 2000).

For the first 100 years of U.S. history, all immigrants were allowed to enter and become permanent residents. The continuing influx of immigrants, especially those coming from non-white, non-European countries, created fear and resentment among native-born Americans who competed with immigrants for jobs and who held racist views toward some racial and ethnic immigrant populations. Increasing pressures from U.S. citizens to restrict, or halt entirely, the immigration of various national groups led to legislation that did just that. America's open door policy on immigration ended in 1882 with the Chinese Exclusion Act which suspended for 10 years the entrance of the Chinese to the United States and declared them ineligible for U.S. citizenship. The Immigration Act of 1917 included the requirement that all immigrants must pass a literacy test before entering the United States. And in 1921, the Johnson Act for the first time introduced a limit on the number of immigrants who could enter the country in a single year, with stricter limitations for certain countries (including Africa and the Near East). The 1924

Ethnic diversity is an opportunity rather than a problem.

ANDREW GREELEY
Sociologist/Author

Table 7.1 *Foreign-Born Population and Percent of Total Population for the United States: 1890 to 2000*

Year	Number (in millions)	Percent of total
2000	28.4	10.4
1990	19.8	7.9
1970	9.6	4.7
1950	10.3	6.9
1930	14.2	11.6
1910	13.5	14.7
1890	9.2	14.8

Source: Lisa Lollock. 2001. "The Foreign-Born Population in the United States: March 2000." *Current Population Reports* P20-534. Washington, D.C.: U.S. Bureau of the Census.

Becoming an American Citizen: Could You Pass the Test?

To become a U.S. citizen, immigrants must have been lawfully admitted for permanent residence; have resided continuously as a lawful permanent U.S. resident for at least 5 years; be able to read, write, speak, and understand basic English (certain exemptions apply); and they must show that they have "good moral character" (Immigration and Naturalization Service 2001). Applicants who have been convicted of murder or an aggravated felony are permanently denied U.S. citizenship. In addition, applicants are denied if in the last 5 years they have engaged in any one of a variety of offenses, including prostitution, illegal gambling, controlled substance law violation (except for a single offense of possession of 30 grams or less of marijuana), habitual drunkardness, willful failure or refusal to support dependents, and criminal behavior involving "moral turpitude."

To become a U.S. citizen, one must take the oath of allegiance and swear to support the Constitution and obey U.S. laws, renounce any foreign allegiance, and bear arms for the U.S. military or perform services for the U.S. government when required.

Finally, applicants for U.S. citizenship must pass an examination administered by the U.S. Immigration and Naturalization Service. The following questions are typical of those on the examination given to immigrants seeking U.S. citizenship. Applicants may choose between an oral and a written test. On the oral test, they must answer all the questions correctly. On the written test, they must correctly answer 12 of 20 questions. Based on your answers to these questions, would you be granted U.S. citizenship if you were an immigrant (that is, could you correctly answer 6 out of the following 10 questions)? After selecting an answer for each of the following questions, check your answers using the answer key provided.

SAMPLE QUESTIONS FOR BECOMING A U.S. CITIZEN

1. Who becomes the president of the United States if the president and vice president should die?
   a. The speaker of the House of Representatives
   b. The Senate majority leader
   c. The chairman of the Joint Chiefs of Staff
   d. The chief justice of the Supreme Court

2. Who said, "Give me liberty or give me death"?
   a. George Washington   c. Patrick Henry
   b. Benjamin Franklin   d. Thomas Jefferson

3. How many branches are there in our government?
   a. 2   c. 4
   b. 3   d. 6

Immigration Act further limited the number of immigrants allowed into the United States, and completely excluded the Japanese. Other federal immigration laws include the 1943 Repeal of the Chinese Exclusion Act, the 1948 Displaced Persons Act (which permitted **refugees** from Europe), and the 1952 Immigration and Naturalization Act (which permitted a quota of Japanese immigrants).

A major concern regarding immigration is the number of people entering the United States illegally. Each day, an estimated 5,000 foreigners enter the United States illegally; 4,000 illegal immigrants are apprehended, and about 1,000 escape detection (Population Reference Bureau 1999). In 1986, Congress approved the Immigration Reform and Control Act, which made hiring illegal aliens an illegal act punishable by fines and even prison sentences. This act also prohibits employers from discriminating against legal aliens who are not U.S. citizens. The more recent 1996 Illegal Immigrant Reform and Immigrant Responsibility Act increased border enforcement and penalties for illegal entry into the U.S. and placed new restrictions on federal public benefits for immigrants (McLemore, Romo, & Baker 2001).

Many immigrants struggle to adjust to life in the United States. But despite the prejudice, discrimination, and lack of social support they experience, many foreign-born U.S. residents work hard to succeed educationally and occupationally. Foreign born residents of the United States may or may not apply for, and be granted, U.S. citizenship. In 2000, of all foreign-born U.S. residents, 35

4. Which countries were our enemies during World War II?
   a. Iraq, Libya, and Turkey
   b. Germany, Japan, and the Soviet Union
   c. Japan, Italy, and Germany
   d. Italy, Germany, and France

5. What are the duties of Congress?
   a. To execute laws
   b. To naturalize citizens
   c. To sign bills into law
   d. To make laws

6. Which list contains three rights or freedoms guaranteed by the Bill of Rights?
   a. Right to life, right to liberty, right to the pursuit of happiness
   b. Freedom of speech, freedom of press, freedom of religion
   c. Right to protest, right to protection under the law, freedom of religion
   d. Freedom of religion, right to elect representatives, human rights

7. How many times may a senator be reelected?
   a. There is no limit
   b. Once
   c. Twice
   d. 4 times

8. Who signs bills into law?
   a. The Supreme Court
   b. The president
   c. Congress
   d. The Senate

9. How many changes or amendments are there to the Constitution?
   a. 5
   b. 9
   c. 13
   d. 27

10. Who has the power to declare war?
   a. Congress
   b. The president
   c. Chief justice of the Supreme Court
   d. Chairman of the Joint Chiefs of Staff

ANSWER KEY

1 = a; 2 = c; 3 = b; 4 = c; 5 = d; 6 = b; 7 = a; 8 = b; 9 =d; 10 = a

Sources:

Immigration and Naturalization Service, U.S. Department of Justice. 1989. *By the People . . . U.S. Government Structure.* Washington, D.C.: U.S. Government Printing Office.

Immigration and Naturalization Service, 2001. "General Naturalization Requirements." http://www.ins.usdoj.gov/graphics/services/natz/general.htm

percent were **naturalized citizens** (immigrants who applied and met the requirements for U.S. citizenship); 65 percent were not U.S. citizens (Lollock 2001). Requirements for becoming a U.S. citizen are discussed in this chapter's *Self and Society* feature.

According to a study by the National Academy of Sciences, immigration produces economic benefits for the United States as a whole ("Study Finds Benefits from Immigration" 1997). Economist James Smith explained: "It's true that some Americans are now paying more taxes because of immigration, and native-born Americans without a high school education have seen their wages fall slightly because of the competition sparked by lower-skilled, newly arrived immigrants. But the vast majority of Americans are enjoying a healthier economy as a result of the increased supply of labor and lower prices that result from immigration" (p. 4A).

Sociological Theories of Race and Ethnic Relations

Some theories of race and ethnic relations suggest that individuals with certain personality types are more likely to be prejudiced or to direct hostility toward minority group members. Sociologists, however, concentrate on the impact of the structure and culture of society on race and ethnic relations. Three major sociological theories lend insight into the continued subordination of minorities.

Structural-Functionalist Perspective

Functionalists emphasize that each component of society contributes to the stability of the whole. In the past, inequality between majority and minority groups was functional for some groups in society. For example, the belief in the superiority of one group over another provided moral justification for slavery, supplying the South with the means to develop an agricultural economy based on cotton. Further, southern whites perpetuated the belief that emancipation would be detrimental for blacks, who were highly dependent upon their "white masters" for survival (Nash 1962).

Functionalists recognize, however, that racial and ethnic inequality is also dysfunctional for society (Schaefer 1998; Williams & Morris 1993). A society that practices discrimination fails to develop and utilize the resources of minority members. Prejudice and discrimination aggravate social problems such as crime and violence, war, poverty, health problems, urban decay, and drug use—problems that cause human suffering as well as financial burdens on individuals and society.

The structural-functionalist analysis of manifest and latent functions also sheds light on issues of race and ethnic relations. For example, the manifest function of the civil rights legislation in the 1960s was to improve conditions for racial minorities. However, civil rights legislation produced an unexpected consequence, or latent function. Because civil rights legislation supposedly ended racial discrimination, whites were more likely to blame blacks for their social disadvantages and thus, perpetuate negative stereotypes such as "blacks lack motivation" and "blacks have less ability" (Schuman & Krysan 1999).

Conflict Perspective

Conflict theorists emphasize the role of competition over wealth, power, and prestige in creating and maintaining racial and ethnic group tensions. Majority group subordination of racial and ethnic minorities reflects perceived or actual economic threats by the minority. For example, between 1840 and 1870, large numbers of Chinese immigrants came to the United States to work in mining (California Gold Rush of 1848), railroads (transcontinental railroad completed in 1860), and construction. As Chinese workers displaced whites, anti-Chinese sentiment rose, resulting in increased prejudice and discrimination and the eventual passage of the Chinese Exclusion Act of 1882, which restricted Chinese immigration until 1924. More recently, ethnic conflict among Albanians and Serbs in Kosovo has been explained as resulting not only from differences in religious beliefs, but from competition over scarce resources (Kumovich 1999).

Further, conflict theorists suggest that capitalists profit by maintaining a surplus labor force, that is, having more workers than are needed. A surplus labor force assures that wages remain low, because someone is always available to take a disgruntled worker's place. Minorities who are disproportionately unemployed serve the interests of the business owners by providing surplus labor, keeping wages low, and, consequently, enabling them to maximize profits.

Conflict theorists also argue that the wealthy and powerful elite fosters negative attitudes toward minorities in order to maintain racial and ethnic tensions

Racism . . . has been the most persistent and divisive element in this society and one that has limited our growth and happiness as a nation.

JAMES E. JONES JR.
Director, Center for the Study of Affirmative Action

among workers. As long as workers are divided along racial and ethnic lines, they are less likely to join forces to advance their own interests at the expense of the capitalists. In addition, the "haves" perpetuate racial and ethnic tensions among the "have-nots" to deflect attention away from their own greed and exploitation of workers.

Struggles over political power also affect race and ethnic relations. When the first free elections were held in South Africa, the Black African National Congress (ANC) led by Nelson Mandela had campaigned on a platform that promised several plans of affirmative action to reverse the four decades of apartheid that barred Black South Africans from political participation and denied them many basic human rights. The White National Party, which was trying to maintain the position of political power it had held during apartheid, also promised affirmative action in the 1994 election campaign, but it did not announce this change in platform until 2 months before the elections, when polls began to predict a landslide victory for the African National Congress (Guillebeau 1999).

Symbolic Interactionist Perspective

The symbolic interactionist perspective focuses on how meanings, labels, and definitions affect racial and ethnic groups. The different connotations of the colors white and black, for example, may contribute to negative attitudes toward people of color. The white knight is good, and the black knight is evil; angel food cake is white, devil's food cake is black. Other negative terms associated with black include black sheep, black plague, black magic, black mass, blackballed, and blacklisted. The continued use of such derogatory terms as Jap, Gook, Spic, Frog, Kraut, Coon, Chink, Wop, and Mick also confirms the power of language in perpetuating negative attitudes toward minority group members.

The labeling perspective directs us to consider the role that negative stereotypes play in race and ethnicity. **Stereotypes** are exaggerations or generalizations about the characteristics and behavior of a particular group. Negative stereotyping of minorities leads to a self-fulfilling prophecy. As Schaefer (1998, 17) explains:

> Self-fulfilling prophecies can be devastating for minority groups. Such groups often find that they are allowed to hold only low-paying jobs with little prestige or opportunity for advancement. The rationale of the dominant society is that these minority individuals lack the ability to perform in more important and lucrative positions. Training to become scientists, executives, or physicians is denied to many subordinate group individuals, who are then locked into society's inferior jobs. As a result, the false definition becomes real. The subordinate group has become inferior because it was defined at the start as inferior and was therefore prevented from achieving the levels attained by the majority.

Prejudice and Racism

Prejudice refers to negative attitudes and feelings toward or about an entire category of people. Prejudice may be directed toward individuals of a particular religion, sexual orientation, political affiliation, age, social class, sex, race, or

That both black and white in our country can today say we are to one another brother and sister, a united rainbow nation that derives its strength from the bonding of its many races and colours, constitutes a celebration of the oneness of the human race.

NELSON MANDELA
President, Republic of South Africa

I have a dream that my four little children will one day live in a nation where they will not be judged by the color of their skin, but by the content of their character.

MARTIN LUTHER KING JR.
Civil rights leader

ethnicity. **Racism** is a belief system, or ideology, that includes three basic ideas (Marger 2000):

1. Humans are divided naturally into different physical types.
2. The physical traits associated with each human type are innately related to the culture, personality, and intelligence of each type.
3. On the basis of their genetic inheritance, some groups are innately superior to others.

It is never too late to give up your prejudices.

HENRY DAVID THOREAU
Writer/civil activist

Surveys of racial attitudes continue to reveal disturbing levels of prejudice and racism. In a study of adults living in the Detroit area, 13 percent of white respondents expressed the view that whites have more inborn ability than blacks; 31 percent endorsed the view that blacks do not work as hard as whites, 73 percent indicated that they are personally opposed to romantic relations with a black person, and over half indicated a preference for living in an all-white neighborhood (Williams et al. 1999).

Aversive and Modern Racism

Compared with traditional, "old-fashioned" prejudice which is blatant, direct, and conscious, contemporary forms of prejudice are often subtle, indirect, and unconscious. Two variants of these more subtle forms of prejudice include aversive racism and modern racism.

Aversive Racism **Aversive racism** represents a subtle, often unintentional form of prejudice exhibited by many well-intentioned white Americans who possess strong egalitarian values and who view themselves as nonprejudiced. The negative feelings that aversive racists have toward blacks and other minority groups are not feelings of hostility or hate, but rather, feelings of discomfort, uneasiness, disgust, and sometimes fear (Gaertner & Dovidio 2000). Aversive racists may not be fully aware that they harbor these negative racial feelings; indeed, they disapprove of individuals who are prejudiced and would feel falsely accused if they were labeled as prejudiced. "Aversive racists find Blacks 'aversive,' while at the same time find any suggestion that they might be prejudiced 'aversive' as well" (Gaertner & Dovidio 2000, 14).

Another aspect of aversive racism is the presence of pro-white attitudes, as opposed to anti-black attitudes. In several studies, respondents did not indicate that blacks were worse than whites, only that whites were better than blacks (Gaertner & Dovidio 2000). For example, blacks were not rated as being lazier than whites, but whites were rated as being more ambitious than blacks. Gaertner and Dovidio (2000) explain that "aversive racists would not characterize blacks more negatively than whites because that response could readily be interpreted by others or oneself, to reflect racial prejudice" (p. 27). Compared with anti-black attitudes, pro-white attitudes reflect a more subtle prejudice 'that, although less overtly negative, is still racial bias.

In a national survey of first-year college students, one in five agreed that "racial discrimination is no longer a major problem in America."

"THIS YEAR'S FRESHMEN: A STATISTICAL PROFILE" 2000

Modern Racism Like aversive racism, **modern racism** involves the rejection of traditional racist beliefs, but a modern racist displaces negative racial feelings onto more abstract social and political issues. The modern racist believes that serious discrimination in America no longer exists, that any continuing racial inequality is the fault of minority group members, and that demands for affirmative action for minorities are unfair and unjustified. "Modern racism tends to 'blame the victim' and place the responsibility for change and improvements on the minority groups, not on the larger society" (Healey 1997, 55). Like the aversive racist, modern racists tend to be unaware of their negative racial feelings and do not view themselves as prejudiced.

Learning to be Prejudiced: The Role of Socialization, Stereotypes, and the Media

Psychological theories of prejudice focus on forces within the individual that give rise to prejudice. For example, the **frustration-aggression theory** of prejudice (also known as the **scapegoating theory**), suggests that prejudice is a form of hostility that results from frustration. According to this theory, minority groups serve as convenient targets of displaced aggression. The **authoritarian-personality theory** of prejudice suggests that prejudice arises in people with a certain personality type. According to this theory, people with an authoritarian personality—who are highly conformist, intolerant, cynical, and preoccupied with power—are prone to being prejudiced.

Rather than focus on the individual, sociologists focus on social forces that contribute to prejudice. Earlier we explained how intergroup conflict over wealth, power, and prestige give rise to negative feelings and attitudes that serve to protect and enhance dominant group interests. In the following discussion, we explain how prejudice is learned through socialization, stereotypes, and the media.

Learning Prejudice through Socialization   In the socialization process, individuals adopt the values, beliefs, and perceptions of their family, peers, culture, and social groups. Prejudice is taught and learned through socialization, although it need not be taught directly and intentionally. Parents who teach their children to not be prejudiced, yet live in an all-white neighborhood, attend an all-white church, and have only white friends may be indirectly teaching negative racial attitudes to their children. Socialization may also be direct, as in the case of a parent who uses racial slurs in the presence of her children, or forbids her children from playing with children from a certain racial or ethnic background. Children may also learn prejudicial attitudes from their peers. The telling of racial and ethnic jokes among friends, for example, perpetuates stereotypes that foster negative racial and ethnic attitudes.

> We are not born with hatred; we learn to hate.
>
> A HOLOCAUST SURVIVOR

Stereotypes   Prejudicial attitudes toward racial and ethnic groups are based on false or inadequate group images known as stereotypes. As noted earlier, stereotypes are exaggerations or generalizations about the characteristics and behavior of a particular group.

When Americans in a 1990 National Opinion Research Center poll were asked to evaluate various racial and ethnic groups, blacks were rated least favorably (Shipler 1998). Most of the respondents labeled blacks as less intelligent than whites (53 percent), lazier than whites (62 percent), and more likely than whites to prefer being on welfare over being self-supporting (78 percent).

> Why are blacks stereotyped as lazy? Wasn't it whites who enslaved blacks to do the hard work on the plantations? If you ask me, whites were too lazy to do the work themselves.
>
> STUDENT IN A SOCIOLOGY CLASS

Prejudice and the Media   The media contribute to prejudice by portraying minorities in negative and stereotypical ways, or by not portraying them at all. In the 1999–2000 television season, a majority (61 percent) of prime time shows had more than one minority character when the entire casts of characters were considered. However, when only main characters were considered, nearly half (48 percent) of the shows had all-white casts (Children Now 2000a). Another analysis of the 1999–2000 prime time television season revealed evidence of progress in the ways minority characters were portrayed: characters of color were more likely than white characters to be shown as "good," competent at work, and law-abiding (Children Now 2000b). The report found that although negative stereotyping of African-Americans was virtually absent, Asian Americans and Latino characters were vulnerable to stereotyping. Stereotypes of Asians on prime time television programs included the nerdy student, the martial arts master, the seductive "Dragon Lady," and the "clueless immigrant." Another analysis of 1999–2000 prime time television programs found that Latino characters were virtually absent, accounting for only 3 percent of total prime time characters (Children Now and the National Hispanic Foundation for the Arts 2000). Latinas were especially underrepresented in prime time shows; two-thirds of Latino characters were male. The few Latina women on prime time were typically portrayed in service roles, such as nanny, nurse, and maid. Latino men were often stereotyped as "Latin lovers."

Negative, stereotypical views of minorities are also found on the Internet. "Mr. Wong," an Internet cartoon series appearing on Icebox.com, features a buck-toothed, yellow-faced Chinese servant and his white socialite boss, Miss Pam, who is always insulting him. Finding the portrayal of Mr. Wong offensive, many Asian Americans have demanded that the cartoon be discontinued (Liu 2000). The Internet also spreads messages of hate toward minority groups through the web sites of various white supremacist and hate group organizations (see this chapter's *Focus on Technology* feature).

Another media form that contributes to hatred of minority groups is "white power music": music with racist lyrics and titles such as *Coon Hunt*, *Race Riot*, and *White Revolution*. Consider the following music lyrics:

> . . . Niggers just hit this side of town, watch my property values go down. Bang, gang, watch them die, watch those niggers drop like flies . . . .—Berserkr.

Resistance Records, a company that sells "white power" music sells 50,000 compact discs a year in Europe, South Africa, South America, Canada, and the United States (*Intelligence Report* 1998). "Skinhead" music, which contains anti-Semitic, racist, and homophobic lyrics, has become a leading recruitment tool for white supremacist groups ("Intelligence Briefs" 2000).

> A frightening reality is that every child in America is now just a mouseclick away from hate.
>
> Morris Dees
> *Co-founder of the Southern Poverty Law Center*

Discrimination against Racial and Ethnic Minorities

Whereas prejudice refers to attitudes, **discrimination** refers to actions or practices that result in differential treatment of categories of individuals. Although prejudicial attitudes often accompany discriminatory behavior or practices, one may be evident without the other.

Individual versus Institutional Discrimination

Individual discrimination occurs when individuals treat persons unfairly or unequally because of their group membership. Individual discrimination may be overt or adaptive. In **overt discrimination** the individual discriminates because of his or her own prejudicial attitudes. For example, a white landlord may refuse to rent to a Mexican-American family because of her own prejudice against Mexican Americans. Or, a Taiwanese-American college student who shares a dorm room with an African-American student may request a room-

Hate on the Web

The Internet provides access to a wide range of information for the over 250 million people worldwide using the World Wide Web. Unfortunately, many websites promote hate and intolerance toward various minority groups. In Chapter 9, for example, we refer to a web site that promotes antigay sentiments: www.godhatesfags.com. Here, we focus on the use of Internet technology to promote hatred toward racial and ethnic groups.

The Southern Poverty Law Center, a non-profit organization that combats hate, intolerance, and discrimination, has tracked over 400 hate websites, many of which are designed to lure children and teenagers into the ideologies and organizations that promote hate (Dees 2000). According to the Center, "the gospel of hate is being projected worldwide, more cheaply and effectively than ever before, and it is attracting a new demographic of youthful followers . . ." (*SPLC Report* 2000).

Many German neo-Nazi white supremacist groups use U.S.-based Internet servers to avoid prosecution under German law. German intelligence officials report that 70 percent of the nearly 400 German neo-Nazi sites are on U.S. servers, and about a third of those would be illegal under German law (Kaplan & Kim 2000). One U.S.-based Web page posted in German offered a $7,500 reward for the murder of a young, left-wing activist, giving his home address, job, and phone number.

Although the Internet is used as a vehicle for spreading messages of hate, it is also used to combat such messages. Although it is now offline, for six years the website of "HateWatch" (http://www.hatewatch.org) was devoted to educating the public about the proliferation of hate on the Internet. The Southern Poverty Law Center helped make a documentary called "HATE.COM: Extremists on the Internet," which premiered on Home Box Office in October 2000. In conjunction with the re-

White supremacist groups such as Stormfront spread their message of racial hate through their Web site.

lease of HATE.COM, HBO developed on its Web site (www.hbo.com) a special cyber-campaign called "Hate Hurts," about the impact of hate on individuals, families, and communities.

The corporate sector can also play a role in the fight against hate on the Web. In January 2001, Yahoo announced it would actively try to keep hateful material out of its auctions, classified sections, and shopping areas. This policy came shortly after a French court ordered Yahoo to pay fines of about $13,000 a day if the company did not install technology that would shield French Web users from seeing Nazi-related memorabilia in its auction site (French law prohibits the display of such material).

Sources: Dees, Morris. 2000 (Dec. 28). Personal correspondence. Morris Dees, co-founder of the Southern Poverty Law Center. 400 Washington Avenue, Montgomery, AL 36104; SPLC Report 2000 (September). "HBO, Center Document Hate on Net." 30(3):1. Southern Poverty Law Center. 400 Washington Ave. Montgomery, AL 36104.

---

mate reassignment from the student housing office because he is prejudiced against blacks.

Suppose a Cuban-American family wants to rent an apartment in a predominantly non-Hispanic neighborhood. If the landlord is prejudiced against Cubans and does not allow the family to rent the apartment, that landlord has engaged in overt discrimination. But what if the landlord is not prejudiced against Cubans but still refuses to rent to a Cuban family? Perhaps that landlord is engaging in **adaptive discrimination**, or discrimination that is based on the prejudice of others. In this example, the landlord may fear that if he or she rents to a Cuban-American family, other renters who are prejudiced against Cubans may move out of the building or neighborhood and leave the landlord with unrented apartments. Overt and adaptive individual discrimination may coexist. For example, a landlord may not rent an apartment to a Cuban family because of his or her own prejudices *and* the fear that other tenants may move out.

Institutional discrimination occurs when normal operations and procedures of social institutions result in unequal treatment of minorities. Institutional discrimination is covert and insidious and maintains the subordinate position of minorities in society. When businesses move out of inner-city areas, they are removing employment opportunities for America's highly urbanized minority groups. When schools use standard intelligence tests to decide which children will be placed in college preparatory tracks, they are limiting the educational advancement of minorities whose intelligence is not fairly measured by culturally biased tests developed from white middle-class experiences. Institutional discrimination is also found in the criminal justice system, which more heavily penalizes crimes that are more likely to be committed by minorities. For example, the penalties for crack cocaine, more often used by minorities, have traditionally been higher than those for other forms of cocaine use even though the same prohibited chemical substance is involved. As conflict theorists emphasize, majority group members make rules that favor their own group.

Racial and ethnic minorities experience discrimination and its effects in almost every sphere of social life. Next, we look at discrimination in education, employment and income, housing, and politics. Finally, we expose the extent and brutality of physical and verbal violence against minorities.

Educational Discrimination

Both institutional and individual discrimination in education negatively affect racial and ethnic minorities and help to explain why minorities (with the exception of Asian Americans) tend to achieve lower levels of academic attainment and success (see also Chapter 12). Institutional discrimination is evidenced by inequalities in school funding—a practice that disproportionately hurts minority students (Kozol 1991). Inner-city schools, which primarily serve minority students, receive less funding per student than do schools in more affluent, primarily white areas. For example, inequalities in school funding resulted in New York City receiving $2,000 less per pupil than Buffalo, Rochester, Syracuse, and Yonkers. New York State Supreme Court Judge Leland DeGrasse found that New York City's system of school funding violated federal civil rights laws because it disproportionately hurt minority students (more than 70 percent of the state's Asian, black, and Hispanic students live in New York City) (Goodnough 2001).

Minorities also experience individual discrimination in the schools, as a result of continuing prejudice among teachers. One college student completing a teaching practicum reported that some of the teachers in her school often spoke about children of color as "wild kids who slam doors in your face" (Lawrence 1997, 111). In a survey conducted by the Southern Poverty Law Center, 1,100 educators were asked if they had heard racist comments from their colleagues in the past year. More than one-quarter of survey respondents answered, "Yes" (*Teaching Tolerance* 2000). It is likely that teachers who are prejudiced against minorities discriminate against them, giving them less teaching attention and less encouragement.

Racial and ethnic minorities are also treated unfairly in educational materials, such as textbooks, which often distort the history and heritages of people of

Demography is not destiny. The amount of melanin in a student's skin, the home country of her antecedents, the amount of money in the family bank account, are not the inexorable determinants of academic success.

LELAND DEGRASSE
New York State Supreme Court Judge

How much do we really value the lives of children? . . . On the basis of what I've seen in the South Bronx, it's hard to believe we love these children. The physical degradation of their neighborhood, the chaotic and dysfunctional medical care we give them, the apartheid school system we provide them, the poisons we pump into their neighborhood by putting every kind of toxic-waste incinerator in their neighborhood, all that together does not give me the impression we value their lives.

JONATHON KOZOL
Author and children's advocate

color (King 2000). For example, Zinn (1993) observes, "[T]o emphasize the heroism of Columbus and his successors as navigators and discoverers, and to de-emphasize their genocide, is not a technical necessity but an ideological choice. It serves, unwittingly—to justify what was done" (p. 355).

Employment Discrimination and the Racial/Ethnic Wealth Gap

In comparison with U.S. whites, Hispanics and blacks are more likely to be unemployed, live in poverty, and earn lower incomes. The 1998 black median household income ($25,351) is only 60 percent of the median white household income ($42,439), and the Hispanic median income ($28,330) is only 67 percent of white median income (United for a Fair Economy 1999). Lower levels of educational attainment among these minority groups account for some, but not all, of the disadvantages they experience in employment and income. As shown in Figure 7.4, average earnings of whites is higher than for blacks and Hispanics at the same level of educational attainment. A study by Ryan Smith (School of Management and Labor Relations at Rutgers University) found that for whites in large cities, each added year of education raises the chance of being employed in a position of authority by 9 percent (reported in *Race Relations Reporter* 1999). For blacks, each year of education only increases the likelihood that they will be employed in a position of authority by 1 percent.

Despite laws against it, discrimination against minorities occurs today at all phases of the employment process, from recruitment to interview, job offer, salary, and promotion. In an investigation of employment agencies in the San Francisco Bay area, "testers" were sent to 17 employment agencies and found that seven of the firms favored white applicants over black applicants for entry-level positions (*Race Relations Reporter* 1999).

> At one employment agency, a black applicant was told that interviews were not being conducted and that he should come back a week later. Hours later, a white applicant with identical educational and employment qualifications was interviewed at the agency and was offered a job within 30 minutes of completing the interview. (p. 1)

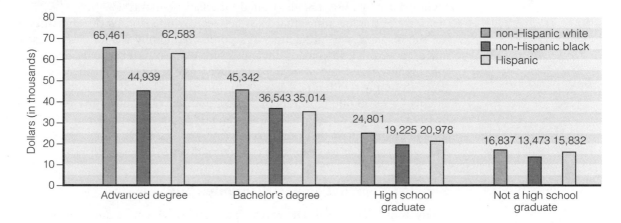

Figure 7.4 *Average 1998 Earnings by Educational Attainment, Race, and Hispanic Origin, for All Workers Ages 18 and Over*

Source: Newburger, Eric C. and Andrea Curry. 2000 (March). "Educational Attainment in the United States." *Current Population Reports* P20-528. Washington, D.C.: U.S. Census Bureau.

Even when workplaces seem integrated, they are often segregated within, as whites—particularly white males—tend to occupy the positions of power and authority. Social segregation contributes to minority disadvantage in the workplace, as illustrated in the following account (Shipler 1998):

> A black man worked for IBM for three years before learning that every evening a happy hour was taking place in a nearby bar. Only white men from the office were involved—no women, no minorities. Had it been strictly social it would have been merely offensive, but it was also professionally damaging, for business was being done over drinks, plans were being designed, connections made. Excluded from that network, the black man was excluded from opportunity for advancement, and he left the job. (p. 2)

Darity (2000) cited research that concluded that current discrimination in the labor market causes black men to earn 12 to 15 percent less than white men. In addition to the effects of current discrimination, historical discrimination against minorities in previous generations continues to affect minorities today. Darity (2000) cited research that shows strong correlations between the occupational status of U.S. men and women in the 1980 and 1990 censuses and whether or not their ethnic or racial group had experienced discrimination a century ago.

Because of past discrimination, minorities have not achieved the accumulation of wealth across the generations that whites have. Wealth refers to the total assets of an individual or household (savings, home, car, stocks, business, etc.) minus the liabilities (debt). A major source of wealth today is inheritance; however, current generations of minorities have not had the same advantage of inheriting wealth from their parents, grandparents, and great-grandparents and in turn, have less wealth to transmit to their children. Darity (2000) notes that "the sharpest economic gap between blacks and whites in the United States today is the gap in wealth" (p. B18). In 1995 (the latest year for which complete data are available), the typical white household had $18,000 in wealth, the typical black household had just $200 and the typical Hispanic household had zero wealth (United for a Fair Economy 1999). The percentage of black or Hispanic households with zero or negative wealth (greater debt than assets) is twice as high as for white households (United for a Fair Economy 1999).

Many of the inequities that exist between whites and blacks and Hispanics are the result of wide differences in accumulated family wealth. Racial discrepancies in wealth lead to many advantages for whites: better schools, better housing, higher wages, and more opportunities to save, invest, and thereby further their economic position (Conley 1999). Minorities' lack of access to wealth influences a multitude of factors that affect both current and future generations, including how well they perform on standardized tests, whether or not they go to college, how well they can survive emergencies (such as the loss of a job), and what they can leave to their children (Oliver & Shapiro 1997).

Housing Segregation and Discrimination

Analysis of 2000 census data revealed that although blacks and whites lived in neighborhoods that were slightly more integrated than they were in 1990, people still lived in largely segregated neighborhoods. In 2000, an average white person living in a metropolitan area (which includes city dwellers and suburban residents) lived in a neighborhood that is about 80 percent white and 7 percent black (Schmitt 2001). In contrast, the average black person lived in a neighborhood that is 33 percent white and 51 percent black.

U.S. minorities, who are disproportionately represented among the poor, tend to be segregated in concentrated areas of low-income housing, often in inner-city areas of concentrated poverty (Massey & Denton 1993). Zoning regulations in affluent suburbs restrict development of affordable housing in order to keep out "undesirables" and maintain high property values. Suburban zoning regulations that require large lot sizes, minimum room sizes, and single family dwellings serve as barriers to low-income development in suburban areas.

Segregation also results from discriminatory practices such as redlining, racial steering, and restrictive home covenants. **Redlining** occurs when mortgage companies deny loans for the purchase of houses in minority neighborhoods, arguing that the financial risk is too great. **Racial steering** occurs when realtors discourage minorities from moving into certain areas by showing them homes only in minority neighborhoods. **Restrictive home covenants**, illegal since the 1950s, involve a pact between property owners that they will not sell or rent their property to minority group members.

Although housing discrimination is illegal, it is not uncommon. In a study in Jacksonville, Florida, white and African-American volunteer applicants, or "testers," with similar backgrounds tried to rent the same apartments. More than half of the apartment owners tested in the study broke laws against racial discrimination (Halton 1998). Black testers were quoted higher rents and security deposits, told that units were not available, or were not granted appointments or applications. In other research on mortgage lending discrimination, minorities were less likely to receive information about loan products, they received less time and information from loan officers, they were often quoted higher interest rates, and had higher loan denial rates compared with whites, other things being equal (Turner & Skidmore 1999). This chapter's *Social Problems Research Up Close* feature presents research on housing discrimination conducted by college students.

Political Discrimination

Historically, African Americans have been discouraged from political involvement by segregated primaries, poll taxes, literacy tests, and threats of violence. However, tremendous strides have been made since the passage of the 1965 Voting Rights Act, which prohibited literacy tests and provided for poll observers. Blacks have won mayoral elections in Atlanta, Cleveland, Washington, D.C., and New York City, and the governorship in Virginia. However, racial minorities and Hispanics continue to be underrepresented in political positions and voting participation.

Discrimination against racial and ethnic minorities in the U.S. political process persists. After "voting irregularities" in the 2000 national elections, the National Association for the Advancement of Colored People (NAACP) and several civil rights groups filed a lawsuit in Florida to eliminate unfair voting practices (NAACP Press Release 2001). In the 2000 election, thousands of black voters complained that they were wrongfully turned away from the polls or had trouble casting their ballots. Complaints were not limited to Florida; black voters in about a dozen other states reported similar unfair treatment (NAACP Press Release 2000).

An Undergraduate Sociology Class Uncovers Racial Discrimination in Housing

Previous research indicates that Americans can infer the race of a speaker through the speaker's accent, grammar, and diction, thus offering rental agents an opportunity to discriminate over the phone. Under the guidance of their sociology professor (Douglas Massey) and a postdoctoral fellow (Garvey Lundy), students in an undergraduate sociology research methods class at the University of Pennsylvania designed a study to assess if rental agents discriminated over the phone.

Sample and Methods

Students involved in this study developed a data collection instrument which consisted of a standard script that auditors (students posing as renters) followed in their telephone interactions with rental agents. For example, if a machine answered the call, the auditor said, "Hello. My name is _____. I'm interested in the apartment you advertised in _____. Please call me back at _____." If a rental agent an-

swered the phone, the auditor said, "Hello. My name is _____. I'm interested in the apartment you advertised in _____. Are any apartments still available?" Other questions auditors asked included "How much do I have to put down?" and "Are there any other fees?"

After devising the data collection instrument, students chose 79 rental listings from three sources: *Apartments for Rent* magazine, *The Apartment Hunter,* published by the *Philadelphia Inquirer,* and the Sunday Real Estate Section of the *Philadelphia Inquirer* itself. The listings covered all zones of the Philadelphia metropolitan area. Four male and nine female auditors telephoned the selected rental agents, following the script, and presenting a profile of a recent college graduate in his or her early to mid-20s with an annual income of $25,000 to $30,000. The auditors spoke White Middle-Class English (WME), Black Accented English (BAE), or Black English Vernacular (BEV). Black Accented English and Black English

Vernacular are widely spoken by African Americans in the United States. Massey and Lundy (2001) explain that "although BEV and BAE may both be identified as 'black sounding,' we suspect that most listeners can tell the difference between the two dialects and that they attach different class labels to each style of speech" (p. 456). Specifically, when an African American speaks Standard English with a black pronunciation of certain words (BAE), listeners conclude that the speaker is a middle-class black person, whereas the combination of nonstandard grammar with a black accent (BEV) signals lower-class status.

Based on this assumption, the researchers were able to employ six independent variables to test for a three-way interaction between race (black-white), gender (male-female), and class (lower-middle). The six independent variables were: (1) male BEV, (2) female BEV, (3) male BAE, (4) female BEV, (5) male WME, and (6) female WME. The dependent variables included the various responses of the rental agents, such as whether or not the rental agent returned the

Racial and Ethnic Harassment

Harassment is illegal in the workplace and most schools have policies prohibiting harassment. Nevertheless, racial and ethnic minorities are often targets for harassment in places of employment and education. Charges of racial harassment filed with the Equal Employment Opportunity Commission (EEOC) have more than doubled in the 1990s, from 2,849 charges filed in fiscal year 1991 to about 6,550 in 2000 (EEOC Press Release 2000). For example, Sun Ag, Inc., a major citrus grower in Florida, settled a quarter-million-dollar lawsuit which charged that the company subjected African-American employees to a racially hostile work environment, which included threatening remarks, displaying of a hangman's noose, and racial slurs by supervisors and co-workers (EEOC Press Release 2000).

Hate Crimes

Hate crimes have become a growing threat to the well-being of our society—on college campuses, in the workplace, and in our neighborhoods.

JACK LEVIN
Sociologist
JACK MCDEVITT
Criminologist

In June 1998, James Byrd Jr., a 49-year-old father of three, was walking home from a niece's bridal shower in the small town of Jasper, Texas. According to police reports, three white men riding in a gray pickup truck saw Byrd, a black

auditor's phone call, whether or not the rental agent indicated that a housing unit was still available, and whether or not the agent indicated that an application fee was required (and if so, how much).

Findings and Conclusions

The researchers found "clear and often dramatic evidence of phone-based racial discrimination" (p. 466). Compared with whites, African Americans were less likely to speak to a rental agent (their calls were less likely to be returned). African Americans were also less likely to be told that a unit was available, more likely to pay application fees, and more likely to have credit mentioned as a potential problem in qualifying for a lease.

These racial effects were exacerbated by gender and class. Lower-class blacks experienced less access to rental housing than middle-class blacks, and black females experienced less access than black males. Lower-class black females were the most disadvantaged group. They experienced the lowest probability of

contacting and speaking to a rental agent and, even if they did make contact, they faced the lowest probability of being told of a housing unit's availability. Lower-class black females also faced the highest chance of paying an application fee. On average, lower-class black females were assessed $32 more per application than white middle-class males.

The share of auditors reaching an agent and the share being told a unit was available was combined to indicate an overall measure of access to rental units in the Philadelphia housing market. Whereas more than three-quarters (76 percent) of white middle-class males gained access to a potential rental unit, the figure dropped to 63 percent for middle-class black men (those speaking BAE), 60 percent for white middle-class females (those speaking WME), 57 percent for black middle-class females (those speaking BAE), 44 percent for lower-class black men (those speaking BEV), and only 38 percent for lower-class black women (those speaking BEV). "In other words, for every call a white male makes to find out about a rental unit in the

Philadelphia housing market, a poor black female must make two calls to achieve the same level of access, roughly doubling her time and effort compared with his" (p. 461). In sum, "being identified as black on the basis of one's speech pattern clearly reduces access to rental housing, but being black and female lowers it further, and being black, female, and poor lowers it further still" (p. 467).

These findings suggest that much housing discrimination probably occurs over the phone. "Through technology, a racist landlord may discriminate without actually having to experience the inconvenience or discomfort of personal contact with his or her victim" (Massey & Lundy 2001, 454-455). The authors also conclude that "telephone audit studies offer social scientists a cheap, effective, and timely way to measure the incidence and severity of racial discrimination in urban housing markets" (p. 455).

Source: Massey, Douglas S. and Garvey Lundy. 2001. "Use of Black English and Racial Discrimination in Urban Housing Markets: New Methods and Findings." *Urban Affairs Review* 36(4):252-469.

man, walking down the road and offered him a ride. The men reportedly drove down a dirt lane and, after beating Byrd, chained him to the back of the pickup truck and dragged him for 2 miles down a winding, narrow road. The next day, police found Byrd's mangled and dismembered body. The three men who were arrested had ties to white supremacist groups. James Byrd had been brutally murdered simply because he was black.

The murder of James Byrd exemplifies a **hate crime**—an act of violence motivated by prejudice or bias. Examples of hate crimes, also known as **bias-motivated crimes**, include intimidation (e.g., threats), destruction/damage of property, physical assault, and murder. In 1999, 17 hate crime murders were reported to the FBI, nine of which were motivated by racial bias (Federal Bureau of Investigation 2001). The majority of hate crimes are based on racial bias (see Figure 7.5).

Levin and McDevitt (1995) found that the motivations for hate crimes were of three distinct types: thrill, defensive, and mission. Thrill hate crimes are committed by offenders who are looking for excitement and attack victims for the "fun of it." Defensive hate crimes involve offenders who view their attacks as necessary to protect their community, workplace, or college

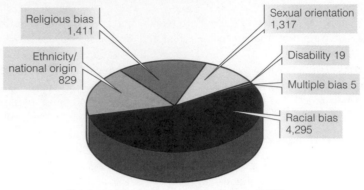

Religious bias
1,411

Ethnicity/
national origin
829

Sexual orientation
1,317

Disability 19

Multiple bias 5

Racial bias
4,295

Total number of hate crime incidences: 7,876

■ **Figure 7.5** *Hate Crime Incidences, by Category of Bias: 1999*
Source: Federal Bureau of Investigation 2001. *Hate Crime Statistics 1999.* http://www.fbi.gov

campus from "outsiders." Perpetrators of defensive hate crimes are trying to send a message that their victims do not belong in a particular community, workplace, or campus and that anyone in the victim's group who dares "intrude" could be the next victim. Mission hate crimes are perpetrated by white supremacist group members or other offenders who have dedicated their lives to bigotry.

The Ku Klux Klan, the first major racist, white supremacist group in the United States, began in Tennessee shortly after the Civil War. Klansmen have threatened, beat, mutilated, and lynched blacks as well as whites who dared to oppose them. White Aryan Resistance (WAR), another white supremacist group, fosters hatred that breeds violence. The following message is typical of one received by calling a White Aryan Resistance telephone number in one of many states (Kleg 1993, 205):

> This is WAR hotline. How long, White men, are you going to sit around while these non-White mud races breed you out of existence? They have your jobs, your homes, and your country. Have you stepped outside lately and looked around while these niggers and Mexicans hep and jive to this Africanized rap music? While these Gooks and Flips are buying up the businesses around you? . . . This racial melting pot is more like a garbage pail. Just look at your liquor stores. Most of them are owned by Sand niggers from Iraq, Egypt, or Iran. Most of the apartments are owned by the scum from India, or some other kind of raghead . . . [Jews] are like maggots eating off a dead carcass. When you see what these Jews and their white lackeys have done, the gas chambers don't sound like such a bad idea after all. For more information write us at. . . .

Other racist groups known to engage in hate crimes are the Identity Church Movement, neo-Nazis, and the skinheads. Indeed, the growing diversity of the U.S. population is matched by a growing number of hate groups, which increased from 457 in 1999 to 602 in 2000 ("The Year in Hate" 2001).

In this chapter's *The Human Side* feature, a former member of a racist hate group describes how he became involved in the skinhead movement, how he recruited others into the movement, and what led him to leave the movement and denounce racial hatred.

I'm not against Blacks. I'm against all non-whites.

KU KLUX KLAN MEMBER

An Interview with a Former Racist Skinhead

After spending 15 years in the skinhead movement, Thomas (T.J.) Leyden renounced racism and went to work for the Simon Wiesenthal Center in Los Angeles. Since joining the human rights organization in June 1996, Leyden has given speeches at more than 100 high schools, the Pentagon, FBI headquarters, and police agencies. In the following interview Leyden talks about his life in a racist hate group, his views on why youth join hate groups, and why he left the racist skinhead movement.

INTELLIGENCE REPORT: What brought you into the Skinhead movement?

T.J. LEYDEN: I was hanging out in the punk rock scene in the late '70s and early '80s . . . In 1980, my parents got a divorce, and I started to hang out in the street. I was venting a lot of my frustration and anger over the divorce. I went around attacking kids, punching them and beating them up. A group of older kids who were known as Skinheads saw this, and I got in with them . . . In 1981, four big-time racist bands came into the Skinhead movement . . . We started to listen to their music, and that broke the Skinhead movement into two factions, SHARPs [Skinheads Against Racial Prejudice] and the neo-Nazi Skinheads. Since I lived in a very upper-middle class, white neighborhood, we decided to establish one of the first neo-Nazi Skinhead gangs in Southern California. If we caught somebody black, Hispanic, or Asian, we'd attack them, beat them for sure . . .

INTELLIGENCE REPORT: Did your racism come partly from your parents?

T.J. LEYDEN: My mom was nonracist and my dad was a stereotypical man. I mean, if somebody cut him off on the freeway, if they were black, he'd use the word 'nigger' . . . But the racism I really learned came from my grandfather, a staunch Irish Catholic. He would say, "You don't bring darkies home" and "Jews killed Christ."

INTELLIGENCE REPORT: What are the circumstances that lead teenagers to join neo-Nazi gangs?

T.J. LEYDEN: We were middle-class to rich, bored white kids. We had a lot of time on our hands so we decided to become gang members. When a kid doesn't have something else constructive to do, he's going to find something, whether it's football, baseball or hanging with neo-Nazi Skinheads . . .

INTELLIGENCE REPORT: When did you start to really learn the ideology of racism?

T.J. LEYDEN: After I joined the Marine Corps in 1988. They teach a philosophy that if you do something, you do it all the way, not half-assed. So since I was a racist, I started reading everything I could about Nazism . . . I was recruiting, organizing Marines to join the racist movement . . . Eventually, I was kicked out for alcohol-related incidents—not for being a racist. If you look at my military packet you're not going to find anything about me being a racist. And I had two-inch high Nazi SS bolts tattooed on my neck! Once I got cut, I decided to be a [Skinhead] recruiter. I was going to get younger kids to be street soldiers.

INTELLIGENCE REPORT: How did recruitment work?

T.J. LEYDEN: We incited violence on high school campuses. We'd put out literature that got black kids to think the white kids were racist. Then the black kids would attack the white kids and the white kids would say, "I'm not going to get beat up by these black guys anymore." They'd start fighting back, and we'd go and fight with them. They'd say, "God, these guys are really cool" . . . That put my foot in the door. Then I could start talking to them, giving them comic books with racist overtones or CDs of racist music. And I would just keep talking to them, giving them literature, indoctrinating them over a period of time . . . .

INTELLIGENCE REPORT: What finally brought you to leave the racist movement?

T.J. LEYDEN: It was an incident with my son that woke me up more than anything. We were watching a Caribbean-style show. My 3-year-old walked over to the TV, turned it off and said, "Daddy, we don't watch shows with niggers." My first impression was, "Wow, this kid's pretty cool." Then I started seeing something different. I started seeing my son acting like someone 10 times tougher than I was, 10 times more loyal, and I thought he'd end up actually doing something and going to prison. Or he was going to get hurt or killed. I started looking at the hypocrisy. A white guy, even if he does crystal meth and sells crack to kids, if he's a Nazi he's okay. And yet this black gentleman here, who's got a Ph.D. and is helping out white kids, he's still a "scummy nigger." In 1996, when I was at the Aryan Nations Congress [in Hayden Lake, Idaho], I started listening to everybody and I felt like, "God, this is pathetic." I asked the guy sitting next to me, "If we wake up tomorrow and the race war is over and we've won, what are we going to do next?" And he

Continued

said, "Oh, come on, T.J., you know we're going to start with hair color next, dude." I laughed at it, but when I drove home, 800 miles, that question and answer kept popping into my head. I thought that kid was so right. Next it'll be you have black hair so you can't be white, or you have brown eyes so somebody in your past must have been black, or you wear glasses so you have a genetic defect.

A little over 2 years after my son said the thing about the "niggers" on TV, I left the racist movement . . .

INTELLIGENCE REPORT: What has been the personal cost of your involvement in the movement?

T.J. LEYDEN: A little bit of my dignity. I look at myself as two people, who I am now and who I was then. I see the destruction I did to people by bringing them into the movement, the families I hurt. I ruined a lot of lives. That's the biggest thing I have to pay back. I don't forgive myself. Only my victims can forgive me.

Source: Excerpted from "A Skinhead's Story: An Interview with a Former Racist." From 1998 *Intelligence Report*. Winter, Issue no. 89, pp. 21–23. Montgomery, AL: Southern Poverty Law Center. Used by permission of the Southern Poverty Law Center.

Hate on Campus According to FBI hate crime statistics, 250 incidents of hate crime occurred on campuses in 1998, making campuses the third most common site for hate crimes ("Hate on Campus" 2000). Far more common than hate crimes are "bias incidents," which are "events that do not rise to the level of prosecutable offenses but that may nevertheless poison the atmosphere at a college and lead to more serious trouble" (p. 8). About 10 percent of bias-motivated acts on campus are committed by faculty, although faculty can also be targets of hate ("Hate on Campus" 2000).

Strategies for Action: Responding to Prejudice, Racism, and Discrimination

Strategies that address problems of prejudice, racism, and discrimination can be categorized as legal/political, educational, and religion-based.

Legal/Political Strategies

Legal and political strategies to reduce prejudice, racism, and discrimination include increasing minority participation and representation in government, eliminating inequalities in school funding, and creating equal opportunities through affirmative action. (Because affirmative action includes voluntary programs and policies in addition to legally mandated ones, we consider affirmative action under a separate heading.) Although it is readily apparent that such strategies can prevent or reduce discriminatory practices, the effects of legal and political strategies on prejudice are more complex. Legal policies that prohibit discrimination can actually increase modern forms of prejudice, as in the case of individuals who conclude that because laws and policies prohibit discrimination, any social disadvantages of minorities must be their own fault. On the other hand, any improvement in the socioeconomic status of minorities that results from legal/political policies may help to replace negative images of minorities with positive images.

Increasing Minority Representation and Participation in Government Increasing minority representation and participation in government promises to increase minorities' voices in influencing public policy that addresses their interests. In addition, minority representation in government affects race relations. One study found a high degree of mistrust of U.S. government among African Americans; however, those who believed that blacks could influence the political process were less likely to distrust the government (Parsons, Simmons, Shinhoster, & Kilburn 1999). The researchers concluded that African Americans' "distrust of government will not be reduced until African Americans perceive that they have more of a role to play in their government" (p. 218).

Various national, state, and local minority groups and organizations encourage minorities to register to vote and to vote in governmental elections. Such efforts contributed to an upsurge in black voter participation in the 2000 presidential election. For example, in 1996, the black share of the vote in Florida was 10 percent. In 2000, the black share of the vote was 16 percent. Several states reported similar increases (National Coalition on Black Civic Participation 2000). Efforts to increase voting participation have also targeted Asian Americans and Hispanic populations.

Representation of racial and ethnic minorities in elected government positions has also increased. Members of the 107th U.S. Congress include 38 African Americans, 21 Hispanics, 8 Asians/Pacific Islanders, and 2 Native Americans; 6 members were foreign-born.

Reducing Disparities in Education Legal remedies have also sought to address institutional discrimination in education by reducing or eliminating disparities in school funding. As noted earlier, schools in poor districts—which predominantly serve minority students—have traditionally received less funding per pupil than do schools in middle and upper class districts (which predominantly serve white students). In recent years, more than two dozen states have

> If we are to move the society toward freedom, away from racism, we must choose leaders deeply committed to that objective.
>
> EDDIE WILLIAM AND MILTON D. MORRIS
> *Joint Center for Political and Economic Studies*

> Voting has proven to be the most effective weapon against racism in America. At a time when sweeping changes in public policy are being decided in the White House, the Congress, and the states, it is imperative that African Americans make their voices heard in the political process.
>
> KWEISI MUFUME
> *President and CEO of NAACP*

Posters such as the one depicted in this photo are part of the efforts to increase political participation of minorities.

© Mark Richards/PhotoEdit

been forced by the courts to come up with a new system of financing schools to increase inadequate funding of schools in poor districts (Goodnough 2001).

Affirmative Action

The term **affirmative action** refers to a broad range of policies and practices in the workplace and educational institutions to promote equal opportunity and to eliminate discrimination. Affirmative action represents an attempt to compensate for the effects of past discrimination and prevent current discrimination against women and racial and ethnic minorities. Affirmative action policies developed in the 1960s from federal legislation requiring that any employer (universities as well as businesses) receiving contracts from the federal government must take steps to increase the numbers of women and minorities in their organization.

Various myths and misunderstandings surround the concept of affirmative action. Beeman, Chowdhry, and Todd (2000) explain that confusion exists about affirmative action in the United States because "unlike India, which specifically defined affirmative action policy in its constitution, the U.S. policy is a collection of executive orders, bureaucratic decisions, court cases, and . . . state legislation" (p. 99). Table 7.2 presents myths and realities about affirmative action.

A 2000 public opinion poll of U.S. adults found that more than half (58 percent) favor affirmative action programs, whereas 33 percent oppose them (Gallup/CNN/USA Today Poll 2000). Among first-year college students, 45 percent of women and 56 percent of men agreed that "affirmative action in college admissions should be abolished" (American Council on Education and University of California 2000). Williams and coworkers (1999) found that various measures of negative racial attitudes are strongly associated with whites' lack of support for affirmative action in employment and other government programs to assist blacks. Women are more likely than men to support affirmative action (Williams et al. 1999)—not a surprising finding given that women are the largest category targeted to benefit from affirmative action.

Critics of affirmative action often base their opposition on distorted views of affirmative action programs and policies (see Table 7.2). Some African Americans are also critical of affirmative action, arguing that it perpetuates feelings of inferiority among minorities and fails to help the most impoverished of minorities (Shipler 1998; Wilson 1987).

Supporters of affirmative action suggest that affirmative action has many social benefits. In a review of over 200 scientific studies of affirmative action, Holzer and Neumark (2000) concluded that affirmative action produces benefits for women, minorities, and the overall economy. Holzer and Neumark (2000) found that employers adopting affirmative action increase the relative number of women and minorities by an average of 10 to 15 percent. Since the early 1960s, affirmative action in education has contributed to an increase in the percentage of blacks attending college by a factor of three and the percentage of blacks enrolled in medical school by a factor of four. Black doctors choose more often than their white medical school classmates to practice medicine in inner cities and rural areas serving poor or minority patients (Holzer & Neumark 2000). Increasing the numbers of minorities in educational and professional positions also provides positive role models for other, especially younger, minorities "who can identify with them and form realistic goals to occupy the same roles themselves" (Zack 1998, 51).

The great enemy of truth is very often not a lie—deliberate, continued, and dishonest—but the myth, persistent, persuasive, and unrealistic.

JOHN F. KENNEDY
Former U.S. President

You do not take a person who for years has been hobbled by chains, and liberate him, bring him up to the starting line, and then say, "You are free to compete with all others."

LYNDON BAINES JOHNSON
Former U.S. president

I am colored and America is colorblind. Do you see me?

JAMES M. JONES
University of Delaware

■ **Table 7.2** *Affirmative Action: Myths and Realities*

Myth: Affirmative action is a program based on quotas.

Reality: Affirmative action is *not* a quota system. In 1978, the Supreme Court (in *Regents of the University of California v. Bakke*) ruled that racial quotas were unconstitutional. (Some exceptions to the prohibition on quotas have occurred where court orders have been imposed on employers who have engaged in flagrant discrimination or in cases where private employers established their own policies to address past discrimination.)

Myth: Affirmative action is a program for racial minorities. As evidence of the acceptance of this myth, the majority of students in two sociology classes did not know that women were covered by affirmative action (Beeman et al., 2000).

Reality: The largest category of affirmative action beneficiaries is women, most of whom are white. White male Hispanics also qualify for affirmative action and in some cases, Vietnam veterans and people with disabilities may also qualify under affirmative action policies. Although affirmative action *does* provide opportunities and protection for racial minorities, "referring to affirmative action as a policy just for 'racial' minorities distorts the policy and misleads the public" (Beeman et al., 2000, 100).

The myth that affirmative action is a race-based program is perpetuated by textbooks (Beeman et al. 2000) and by news media (television, newspapers, and news weeklies). A survey of news media in 1998 found that only 7 of 314 news stories (2 percent) concerning affirmative action mainly focused on the program's impact on women (Jackson 1999). Only 19 percent of news articles on affirmative action mentioned its impact on women at all.

Myth: Affirmative action is a system of preferential treatment that gives jobs to unqualified minorities.

Reality: A U.S. Department of Labor (1999) Fact Sheet on affirmative action explains that:

> Affirmative action is *not* preferential treatment. It does *not* mean that unqualified persons should be hired or promoted over other people. What affirmative action *does* mean is that positive steps must be taken to ensure equal employment opportunity for traditionally disadvantaged groups.

Although blacks and Hispanics hired under affirmative action may have fewer credentials, such as education, they typically perform their jobs as well as non-minority employees (Holzer & Neumark 2000).

Myth: Affirmative action creates "**reverse discrimination,**" or the unfair treatment of white males as a result of efforts to provide women and minorities with educational and employment opportunities.

Reality: Of the 3,000 reported discrimination cases filed in Federal court between 1990 and 1994, less than 2 percent alleged reverse discrimination against white males and very few of these cases were upheld (U.S. Department of Labor, undated). Further evidence that contradicts the view that affirmative action creates reverse discrimination lies in the continuing disparity between the economic well being of white males compared with women and racial and ethnic minorities. White males in the United States have more wealth and income, and lower rates of poverty and unemployment, compared with women, blacks, and Hispanics.

Myth: Affirmative action gives women and racial and ethnic minorities special advantages which they no longer need, because laws prohibiting discrimination already exist.

Reality: Despite laws prohibiting discrimination, women (see chapter 8) and minorities continue to experience discrimination. Even if minorities today are given equal opportunities in education and employment, they do not enjoy the advantages that accumulated, inherited wealth from prior generations provides to many whites.

The view that affirmative action is not needed because discrimination is not a problem is perpetuated by news media that omit discrimination from their discussions of affirmative action. A study of news media stories on affirmative action found that only 15 percent made any reference to inequity or bias against women or racial/ethnic minorities (Jackson 1999).

Sources: Beeman, Mark, Geeta Chowdhry, and Karmen Todd. 2000. "Educating Students About Affirmative Action: An Analysis of University Sociology Texts." *Teaching Sociology* 28(2):98–115.

Holzer, Harry and David Neumark. 2000. "Assessing Affirmative Action." *Journal of Economic Literature* 38(3) 483–568.

Jackson, Janine. 1999. "Affirmative Action Coverage Ignores Women—and Discrimination." *Extra!* (January/February). http://www.fair.org/extra/9901/affirmative-action.html

Rivera, Ismael. 2000 (August 17). "The Affirmative Action Debate." American Association for Affirmative Action. http://www.affirmativeaction.org/alerts/aadebate.html

U.S. Department of Labor. 1999 (December). "Facts on Executive Order 11246 Affirmative Action." http://www.dol.gov/dol/esa/public/ofcp_org.htm

U.S. Department of Labor. Undated. "Affirmative Action at OFCCP." http://www.dol.gov/dol/esa/public/regs/compliance/ofccp/what_is.htm

While we react to those wearing white sheets, it is those who wear black robes who take away our protection.

JESSE JACKSON
Civil rights leader

Numerous legal battles have been fought over affirmative action. In 1990, a Hispanic student sued the University of Maryland for denying him a scholarship that was limited to black students. A lower court ruled the black scholarship program was unconstitutional and constituted **reverse discrimination**. In 1995, the U.S. Supreme Court refused to hear the appeal, which means that any race-based scholarship in the United States can now be challenged. In Texas, judges ruled that the University of Texas Law School could no longer use race as a factor in its admissions decisions. Voters in California and Washington State passed ballot initiatives prohibiting the consideration of race by public agencies and universities in hiring and admissions. These measures foreshadow the growing trend toward limiting or perhaps even abolishing altogether affirmative action in the United States.

Opponents of affirmative action would view the elimination of affirmative action programs and policies as progress. Others would agree with the comments of Ismael Rivera (2000): "Eliminating affirmative action would halt the progress that the nation has made to end discrimination and would be a giant leap backwards in our journey toward equal opportunity" (p. 2).

Educational Strategies

Teach tolerance. Because open minds open doors for all our children.

RADIO PUBLIC SERVICE
ANNOUNCEMENT

Educational strategies to reduce prejudice, racism, and discrimination have been implemented in schools (primary, secondary, and higher education levels), in communities and community organizations, and in the workplace.

Multicultural Education in Schools and Communities   In schools across the nation, **multicultural education**, which encompasses a broad range of programs and strategies, works to dispel myths, stereotypes, and ignorance about minorities, promotes tolerance and appreciation of diversity, and includes minority groups in the school curriculum (see also Chapter 12). With multicultural education, the school curriculum reflects the diversity of American society and fosters an awareness and appreciation of the contributions of different racial and ethnic groups to American culture. The Southern Poverty Law Center's program "Teaching Tolerance" has published and distributed materials and videos designed to promote better human relations among diverse groups to schools, colleges, religious organizations, and a variety of community groups across the nation.

Today, hundreds of colleges and universities recognize the educational value of diversity and view student and faculty diversity as an essential resource for optimizing teaching and learning.

AMERICAN COUNCIL ON
EDUCATION AND AMERICAN
ASSOCIATION OF UNIVERSITY
PROFESSORS

Many colleges and universities have made efforts to promote awareness and appreciation of diversity by offering courses and degree programs in racial and ethnic studies, and multicultural events and student organizations. A national survey by the Association of American Colleges and Universities found that 54 percent of colleges and universities required students to take at least one course that emphasizes diversity and another 8 percent were in the process of developing such a requirement (Humphreys 2000). Examples of diversity courses include "Intergroup Relations, Conflict and Community" (University of Michigan), "American Pluralism and the Search for Equality" (SUNY-Buffalo), "Sociology and Culture of American Ethnicity" (North Seattle Community College), and "Comparative Race Relations: A History of Race Relations in South Africa, Brazil and the United States" (Rowan University, NJ), to name a few. Evidence suggests a number of positive outcomes for both minority and majority students who take college diversity courses, including increased racial understanding and cultural awareness, increased social interaction with students who have back-

More than 600,000 teachers receive Teaching Tolerance *magazine—a free resource for teaching tolerance education in the classroom (Dees 2000). Other materials available free from the Southern Poverty Law Center (www.spl-center.org) include* Responding to Hate at School, Ten Ways to Fight Hate, *and* 101 Tools for Tolerance.

Southern Poverty Law Center

grounds different from their own, improved cognitive development, increased support for efforts to achieve educational equity, and higher satisfaction with their college experience (Humphreys 1999).

Diversification of College Student Populations Efforts to recruit and admit racial and ethnic minorities in institutions of higher education have also been found to foster positive relationships among diverse groups and enrich the educational experience of all students. In the United States, about one in five undergraduates at 4-year colleges is a minority (American Council on Education and the American Association of University Professors 2000). Research indicates that racial and ethnic diversity on campus provides educational benefits for *all* students—minority and nonminority alike (American Council on Education and the American Association of University Professors 2000). Gurin (1999) found that students with the most exposure to diverse populations during college had the most cross-racial interactions 5 years after leaving college. A poll of law students at Harvard Law School and the University of Michigan found that nearly 90 percent of the students said that diversity in the classroom provided them with a better educational experience (*Race Relations Reporter* 1999). Nearly 90 percent of the law students said that the contact they had with students of different racial or ethnic backgrounds influenced them to change their view on some aspect of civil rights.

Diversity Training in the Workplace Increasingly, corporations have begun to implement efforts to reduce prejudice and discrimination in the workplace through an educational approach known as diversity training. Broadly defined, **diversity training** involves "raising personal awareness about individual 'differences' in the workplace and how those differences inhibit or enhance the way people work together and get work done" (Wheeler 1994, 10). Diversity training may address such issues as stereotyping and cross-cultural insensitivity, as well as provide workers with specific information on cultural norms of different groups and how these norms affect work behavior and social interactions.

In a survey of 45 organizations that provide diversity training, Wheeler (1994) found that for 85 percent of the respondents, the primary motive for offering diversity training was to enhance productivity and profits. In the words of one survey respondent, "The company's philosophy is that a diverse work force that recognizes and respects differing opinions and ideas adds to the creativity, productivity, and profitability of the company" (p. 12). Only 4 percent of respondents said they offered diversity training out of a sense of social responsibility.

Religion's Role in Promoting Racial Harmony

Politics strives to transform people by altering the structure of society; religion strives to change society by transforming individuals.

Patrick Glynn
*Institute for Communitarian Policy Studies,
George Washington University*

Beginning in the early 1990s, a racial reconciliation movement emerged among various religious organizations. Religious organizations that have launched initiatives advocating racial harmony include the Southern Baptist Convention, the National Association of Evangelicals, the Promise Keepers, and the United Methodist Church. Efforts by religious organizations to promote racial harmony include the following (Glynn 1998; Snyder 2000):

- In 1995, the Southern Baptist Convention passed a resolution apologizing for past support of slavery and racism.
- In the same year, the president of the National Association of Evangelicals publicly confessed and repented past racism by white evangelicals.
- In 1996, the Promise Keepers organization sponsored a gathering of more than 39,000 male pastors of diverse racial, ethnic, and denominational backgrounds under the theme "Breaking Down the Walls." One of the seven promises that members of Promise Keepers make is a promise to overcome racial and denominational differences. "Promise Keepers materials encourage members to go out of their way to engage with those of different races and ethnic groups for purposes of advancing reconciliation" (Glynn 1998, 840–41).

- In 2000, the United Methodist Church announced plans to change its constitution to include a commitment to racial justice. United Methodist leaders declared that the denomination is committed to creating "one America"
for all people, regardless of race (Snyder 2000).

Unlike political strategies to promote racial harmony, religious-based racial
reconciliation efforts provide a spiritual imperative. The religious reconciliation
movement promotes belief in God's ability to provide healing and forgiveness
for past racism. Glynn (1998) suggests that "the mere belief in the possibility of
divine forgiveness and in divine aid at arriving at reconciliation could provide a
strong psychological impetus for positive group interaction" (p. 840).

Given the recentness of the religious racial reconciliation movement, the extent and duration of the effects of this movement are not known. Glynn (1998)
predicts that social change will come increasingly from grassroots community
and religious activists, "who strive to change the nature of society one community, and one soul, at a time" (p. 841).

Understanding *Race and Ethnic Relations*

After considering the material presented in this chapter, what understanding
about race and ethnic relations are we left with? First, we have seen that racial
and ethnic categories are socially constructed; they are largely arbitrary, im-
precise, and misleading. Although some scholars suggest we abandon racial and
ethnic labels, others advocate adding new categories—multiethnic and multiracial—to reflect the identities of a growing segment of the U.S. and world
population.

Conflict theorists and functionalists agree that prejudice, discrimination, and
racism have benefited certain groups in society. But racial and ethnic dishar-
mony has created tensions that disrupt social equilibrium. Symbolic interactionists note that negative labeling of minority group members, which is
learned through interaction with others, contributes to the subordinate position of minorities.

Prejudice, racism, and discrimination are debilitating forces in the lives of
minorities. In spite of these negative forces, many minority group members
succeed in living productive, meaningful, and prosperous lives. But many others cannot overcome the social disadvantages associated with their minority
status and become victims of a cycle of poverty (see Chapter 10). Minorities
are disproportionately poor, receive inferior education, and with continued
discrimination in the workplace, have difficulty improving their standard of
living.

Alterations in the structure of society that increase opportunities for minorities—in education, employment and income, and political participation—are
crucial to achieving racial and ethnic equality. In addition, policy makers concerned with racial and ethnic equality must find ways to reduce the racial/
ethnic wealth gap and foster wealth accumulation among minorities (Conley
1999). As noted earlier, access to wealth affects many dimensions of well being.

Civil rights activist Lani Guinier (1998) suggests that "the real challenge is to
. . . use race as a window on issues of class, issues of gender, and issues of fundamental fairness, not just to talk about race as if it's a question of individual

> The problem is . . . we
> don't believe we are as
> much alike as we are.
> Whites and blacks,
> Catholics and Protestants,
> men and women. If we
> saw each other as more
> alike, we might be very
> eager to join in one big
> human family in this
> world, and to care about
> that family the way we
> care about our own.
>
> **MORRIE SCHWARTZ**
> *Sociologist (from Tuesdays
> with Morrie)*

> One of the great divides in
> this country is between
> those Americans who see
> only blatant racism and
> those who see the subtle
> forms as well.
>
> **DAVID K. SHIPLER**
> *Author, researcher*

bigotry or individual prejudice. The issue is more than about making friends—it's about making change." But, as Shipler (1998) alludes to, making change requires that members of society recognize that change is necessary, that there is a problem that needs rectifying.

> One has to perceive the problem to embrace the solutions. If you think racism isn't harmful unless it wears sheets or burns crosses or bars blacks from motels and restaurants, you will support only the crudest anti-discrimination laws and not the more refined methods of affirmative action and diversity training. (p. 2).

Finally, it is important to consider the role of class in race and ethnic relations. bell hooks (2000) warns that focusing on issues of race and gender can deflect attention away from the larger issue of class division that increasingly separates the "haves" from the "have-nots." Addressing class inequality must, suggests hooks, be part of any meaningful strategy to reduce inequalities suffered by minority groups.

Critical Thinking

1 At colleges and universities around North America, a number of professors are endorsing Holocaust denial, race-based theories of intelligence, and other racist ideas. For example, Associate Professor Arthur Butz of Northwestern University publically rejects the claim that millions of Jews were exterminated in the Holocaust ("Hate on Campus" 2000). Professor Edward M. Miller of the University of New Orleans has concluded that blacks are "small-headed, over-equipped in genitalia, oversexed, hyper-violent and . . . unintelligent" (p. 9). Professor Glayde Whitney of Florida State and Professor J. Philippe Rushton of the University of Western Ontario have both described blacks as having smaller brains. How should institutions of higher learning respond to such racist claims made by faculty members? What role does the right to free speech and academic freedom play?

2 Women, most of whom are white, are the largest category designated to benefit from affirmative action. Yet, a survey of 35 introductory sociology texts published in the 1990s found that nearly 90 percent of the texts did not mention affirmative action in their sections on gender inequality, and only 20 percent of texts included women in their definitions of affirmative action (Beeman et al. 2000). Why do you think many textbooks overlook or minimize the benefits women may receive from affirmative action?

3 Should race be a factor in adoption placements? Should people be discouraged from adopting a child that is of a different race than the adoptive parents? Why or why not?

4 Under Swedish law, giving Nazi salutes is a crime (Lofthus 1998). Do you think that the social benefits of outlawing racist expressions outweigh the impingement of free speech? Do you think such a law should be proposed in the United States? Why or why not?

5 Do you think that the time will ever come when a racial classification system will no longer be used? Why or why not? What arguments can be made for discontinuing racial classification? What arguments can be made for continuing it?

Key Terms

acculturation

adaptive discrimination

affirmative action

amalgamation

antimiscegenation laws

assimilation (primary and secondary)

authoritarian-personality theory

aversive racism

bias-motivated crimes

colonialism

de facto segregation

de jure segregation

discrimination

diversity training

ethnicity

expulsion

frustration-aggression theory

genocide

hate crime

individual discrimination

institutional discrimination

Jim Crow laws

marital assimilation

melting pot

modern racism

multicultural education

naturalized citizen

one drop of blood rule

overt discrimination

pluralism

population transfer

populations

prejudice

primary assimilation

race

racial steering

racism

redlining

refugees

restrictive home covenants

reverse discrimination

scapegoating theory

secondary assimilation

segregation

slavery

stereotype

Media Resources

The Wadsworth Sociology Resource Center: Virtual Society

http://sociology.wadsworth.com/

See the companion Web site for this book to access general sociology resources and text-specific features that can further your understanding of this chapter. The site contains Internet links, Internet exercises, online practice quizzes, information in InfoTrac College Edition, and many more valuable materials designed to enrich your learning experience in social problems.

InfoTrac College Edition

You can access InfoTrac College Edition either from the Wadsworth Sociology Resource Center at **http://sociology.wadsworth.com** or directly from your web browser at **http://www.infotrac-college.com/wadsworth/**. InfoTrac College Edition is an online university library that includes over 700 popular and scholarly journals in which you can find articles related to the topics in this chapter such as immigration, interracial relationships, hate crimes, affirmative action, and multicultural education.

Interactions CD-ROM

Go to the "Interactions" CD-ROM for *Understanding Social Problems*, Third Edition to access additional interactive learning tools, such as in-depth review materials, corresponding practice quizzes, and other engaging resources and activities to help you study the concepts in this chapter.

8

Gender Inequality

Is It True?

1. Worldwide, one in three women has been abused, beaten, or coerced into sex.

2. Although women receive 15 percent of all psychology doctorates, they earned 67 percent of all doctorates in engineering.

3. The gender gap between male and female incomes decreases as education increases.

4. Men have higher rates of suicide, homicide, and accidental deaths than women.

5. In a recent "report card," the U.S. received a "B" in efforts to reduce poverty among women and an "F" for placing women in decision-making positions.

Answers to "Is It True?": 1 = T; 2 = F; 3 = F; 4 = T; 5 = F

O nly a radical transformation of the relationship between women and men to one of full and equal partnership will enable the world to meet the challenges of the 21st century.

BEIJING DECLARATION AND PLATFORM FOR ACTION

In high school Heather Sue Mercer was an all-state place kicker, so that she wanted to play collegiate football wasn't much of a surprise. In the spring of 1995, things were looking good. She had emerged as the 2nd ranked kicker for the Duke Blue Devils and even scored the winning points in an intra-squad scrimmage. Said Coach Goldsmith, "She's very accurate from short [range], and she has very good technique" (Cohen 1995, 1). But that coming fall the Coach "refused to let her even sit on the sidelines...and dismissed her from the team before spring practice began in 1996" (Suggs 2000, 1).

In 1997, Heather Sue Mercer sued Duke University under Title IX, which prohibits sex discrimination in schools that receive federal funds. Although colleges and universities may decide not to let women try out for contact sports, if they are permitted to and make the team, they cannot be treated differently because they're female. School officials argued that Ms. Mercer's dismissal was based solely on her inability to play Division I football but the coach's advice to forget about "boy's games" and consider entering a beauty pageant suggested otherwise, at least to the jury. In October of 2000, Heather Sue Mercer was awarded $2 million in damages, a portion of which she has set aside as a scholarship fund for female place kickers. (Suggs 2000)

The Mercer decision was based on a finding of gender inequality. The term "gender inequality," however, begs the question. Unequal in what way? Depending on the issue, both women and men are victims of inequality. When income, career advancement, and sexual harassment are the focus, women are most often disadvantaged. But when life expectancy, mental and physical illness, and access to one's children following divorce is considered, it is often men who are disadvantaged. In this chapter, we seek to understand inequalities for both genders.

In the previous two chapters, we discussed the social consequences of youth and aging, and race and ethnicity. This chapter looks at **sexism**—the belief that innate psychological, behavioral, and/or intellectual differences exist between women and men and that these differences connote the superiority of one group and the inferiority of the other. As with age, race, and ethnicity, such attitudes often result in prejudice and discrimination at both the individual and institutional levels. Individual discrimination is reflected by the physician who will not hire a male nurse because he or she believes that women are more nurturing and empathetic and are, therefore, better nurses. Institutional discrimination, that is, discrimination built into the fabric of society, is exemplified by the difficulty many women experience in finding employment; they may have no work history and few job skills as a consequence of living in traditionally defined marriage roles.

Discerning the basis for discrimination is often difficult because gender, age, sexual orientation, and race intersect. For example, elderly African-American and Hispanic women are more likely to receive lower wages and work in fewer prestigious jobs than younger white women. They may also experience discrimination if they are "out" as homosexuals. Such **double** or **triple jeopardy** occurs when a person is a member of two or more minority groups. In this chapter, however, we emphasize the impact of gender inequality. **Gender** refers to the social definitions and expectations associated with being female or male and should be distinguished from **sex**, which refers to one's biological identity.

The Global Context: The Status of Women and Men

A recent United Nations Report finds that despite some progress, millions of women around the world remain victims of violence, discrimination, and abuse (Leeman 2000; Austin 2000). For example, worldwide:

- Over 60 million young girls, predominantly in Asia, are listed as "missing" and are likely the victims of infanticide or neglect.
- 2 million girls between the ages of 5 and 15 are forced into the sex trade each year.
- 500,000 women each year die from complications from childbirth
- Two-thirds of the world's 876 million women are illiterate.
- One in three women has been abused, beaten, or coerced into sex.

> In childhood, a woman must be subject to her father; in youth, to her husband; when her husband is dead, to her sons. A woman must never be free of subjugation.
>
> HINDU CODE

One specific type of violence suffered by millions of women is female genital mutilation (FGM). Clitoridectomy and infibulation are two forms of FGM. In a clitoridectomy, the entire glans and shaft of the clitoris and the labia minora are removed or excised. With infibulation the two sides of the vulva are stitched together shortly after birth, leaving only a small opening for the passage of urine and menstrual blood. After marriage, the sealed opening is reopened to permit intercourse and delivery. After childbirth, the woman is often reinfibulated. Although some progress is being made, worldwide, about 100 to 140 million women and girls have undergone genital mutilation (WHO 2001).

The societies that practice clitoridectomy and infibulation do so for a variety of economic, social, and religious reasons. A virgin bride can inherit from her father, thus making her an economic asset to her husband. A clitoridectomy increases a woman's worth because a woman whose clitoris is removed is thought to have less sexual desire and therefore to be less likely to be tempted to have sex before marriage. Older women in the community also generate income by performing the surgery so its perpetuation has an economic function (Kopelman 1994, 62). Various cultural beliefs also justify FGM. In Muslim cultures, for example, female circumcision is justified on both social and religious grounds. Muslim women are regarded as inferior to men: they cannot divorce their husbands, but their husbands can divorce them; they are restricted from buying and inheriting property; and they are not allowed to have custody of their children in the event of divorce. Female circumcision is merely an expression of the inequality and low social status women have in Muslim society.

Inequality in the United States

Although attitudes toward gender equality are becoming increasingly liberal, the United States has a long history of gender inequality. (You can assess your own beliefs about gender equality in this chapter's *Self and Society* feature.)

The Beliefs About Women Scale (BAWS)

INSTRUCTIONS: The statements listed below describe different attitudes toward men and women. There are no right or wrong answers, only opinions. Indicate how much you agree or disagree with each statement, using the following scale:

A = Strongly Disagree; B = Slightly Disagree; C = Neither agree nor disagree; D = Slightly Agree; E = Strongly Agree

_____ 1. Women are more passive than men.

_____ 2. Women are less career-motivated than men.

_____ 3. Women don't generally like to be active in their sexual relationships.

_____ 4. Women are more concerned about their physical appearance than men are.

_____ 5. Women comply more often than men do.

_____ 6. Women care as much as men do about developing a job career.

_____ 7. Most women don't like to express their sexuality.

_____ 8. Men are as conceited about their appearance as women are.

_____ 9. Men are as submissive as women are.

_____ 10. Women are as skillful in business-related activities as men are.

_____ 11. Most women want their partner to take the initiative in their sexual relationships.

_____ 12. Women spend more time attending to their physical appearance than men do.

_____ 13. Women tend to give up more easily than men do.

_____ 14. Women dislike being in leadership positions more than men do.

_____ 15. Women are as interested in sex as men are.

_____ 16. Women pay more attention to their looks than most men do.

_____ 17. Women are more easily influenced than men are.

_____ 18. Women don't like responsibility as much as men do.

_____ 19. Women's sexual desires are less intense than men's.

_____ 20. Women gain more status from their physical appearance than men do.

The Beliefs About Women Scale (BAWS) consists of fifteen (15) separate sub-scales; only four are used here. The items for these four sub-scales and coding instructions are as follows:

SUB-SCALES:

___1. Women are more passive than men. (Items 1, 5, 9, 13, 17)

___2. Women are interested in careers less than men. (Items 2, 6, 10, 14, 18)

___3. Women are less sexual than men. (Items 3, 7, 11, 15, 19)

___4. Women are more appearance conscious than men (Items 4, 8, 12, 16, 20)

Items 1–5, 7, 11–14, and 16–20 should be scored as followed: Strongly Agree = +2, Slightly Agree = +1, Neither Agree Nor Disagree = 0, Slightly Disagree = −1, and Strongly Disagree = −2.
Items 6, 8, 9, 10, and 15 are scored so that: Strongly Agree = −2, Slightly Agree = −1, Neither Agree Nor Disagree = 0, Slightly Disagree = +1, and Strongly Disagree = +2. Scores range from −40 to +40; sub-scale scores range from −10 to +10. The higher your score the more traditional your gender beliefs about men and women.

Source: William E. Snell, Jr., Ph.D. (1997). College of Liberal Arts, Department of Psychology, Southeast Missouri State University. Reprinted with permission.

Women have had to fight for equality: the right to vote, equal pay for comparable work, quality education, entrance into male-dominated occupations, and legal equality. Even today most U.S. citizens agree that American society does not treat women and men equally—women have lower incomes, hold fewer prestigious jobs, earn fewer academic degrees, and are more likely than men to live in poverty.

Increasingly, however, society recognizes that men are also the victims of gender inequality. When U.S. college students were asked to list the best and worst things about being the opposite sex, the same qualities, although in opposite categories, emerged (Cohen 2001). For example, what males list as the best thing about being female (e.g., free to be emotional), females list as the worst thing about being male (e.g., not free to be emotional). Similarly, what females list as the best thing about being male (e.g., higher pay), males listed as the worst thing about being female (e.g., lower pay). As Cohen notes (2001, 3), although "some differences are exaggerated or oversimplified, . . . we identif[ied] a host of ways in which we 'win' or 'lose' simply because we are male or female."

Sociological Theories of Gender Inequality

> The history of mankind is a history of repeated injuries and usurpations on the part of man toward woman, having in direct object the establishment of a tyranny over her.
>
> Manifesto, Seneca Falls, New York, 1848

Both structural-functionalism and conflict theory concentrate on how the structure of society and, specifically, its institutions contribute to gender inequality. However, these two theoretical perspectives offer opposing views of the development and maintenance of gender inequality. Symbolic interactionism, on the other hand, focuses on the culture of society and how gender roles are learned through the socialization process.

Structural-Functionalist Perspective

Structural-functionalists argue that pre-industrial society required a division of labor based on gender. Women, out of biological necessity, remained in the home performing such functions as bearing, nursing, and caring for children. Men, who were physically stronger and could be away from home for long periods of time, were responsible for providing food, clothing, and shelter for their families. This division of labor was functional for society and, over time, became defined as both normal and natural.

Industrialization rendered the traditional division of labor less functional, although remnants of the supporting belief system still persist. Today, because of day care facilities, lower fertility rates, and the less physically demanding and dangerous nature of jobs, the traditional division of labor is no longer as functional. Thus modern conceptions of the family have, to some extent, replaced traditional ones—families have evolved from extended to nuclear, authority is more egalitarian, more women work outside the home, and greater role variation exists in the division of labor. Functionalists argue, therefore, that as the needs of society change, the associated institutional arrangements also change.

Conflict Perspective

Many conflict theorists hold that male dominance and female subordination are shaped by the relationship men and women have to the production process. During the hunting and gathering stage of development, males and females were economic equals, each controlling their own labor and producing needed subsistence. As society evolved to agricultural and industrial modes of production, private property developed and men gained control of the modes of production, while women remained in the home to bear and care for children. Male domination was furthered by inheritance laws that ensured that ownership would remain in their hands. Laws that regarded women as property ensured that women would remain confined to the home.

As industrialization continued and the production of goods and services moved away from the home, the male-female gaps continued to grow—women had less education, lower incomes, and fewer occupational skills and were rarely owners. World War II necessitated the entry of large numbers of women into the labor force, but in contrast to previous periods, many did not return home at the end of the war. They had established their own place in the workforce and, facilitated by the changing nature of work and technological advances, now competed directly with men for jobs and wages.

Conflict theorists also argue that continued domination by males requires a belief system that supports gender inequality. Two such beliefs are that: (1) women are inferior outside the home (e.g., they are less intelligent, less reliable, and less rational) and (2) women are more valuable in the home (e.g., they have maternal instincts and are naturally nurturing). Thus, unlike functionalists, conflict theorists hold that the subordinate position of women in society is a consequence of social inducement rather than biological differences that led to the traditional division of labor.

Symbolic Interactionist Perspective

Although some scientists argue that gender differences are innate, symbolic interactionists emphasize that through the socialization process both females and males are taught the meanings associated with being feminine and masculine. Gender assignment begins at birth as a child is classified as either female or male. However, the learning of gender roles is a lifelong process whereby individuals acquire society's definitions of appropriate and inappropriate gender behavior.

> We live in a state of gender warfare....there is a growing public awareness that many women are abused, discriminated against, and hindered in their personal development.
>
> —DANIEL J. LEVINSON
> *Author*

© Rick Berkowitz/Index Stock Imagery

Women as well as girls are often portrayed provocatively as a means of selling a product or service. This billboard is a good example of the cultural emphasis placed on women's physical appearance.

Gender roles are taught in the family, the school, and in peer groups, and by media presentations of girls and boys, and women and men (see the section "The Social Construction of Gender Roles" later in this chapter). Most importantly, however, gender roles are learned through symbolic interaction as the messages others send us reaffirm or challenge our gender performances. As Lorber (1998, 213) notes:

> Gender is so pervasive that in our society we assume it is bred into our genes. Most people find it hard to believe that gender is constantly created and recreated out of human interaction, out of social life, and is the texture and order of social life. Yet gender, like culture, is a human production that depends on everyone constantly "doing gender". . .

Conceptions of gender are, thus, socially constructed as societal expectations dictate what it means to be female or what it means to be male. Although race and class variations exist, in general, women are socialized into **expressive** or nurturing and emotionally supportive roles, and males are more often socialized into **instrumental** or task-oriented roles. These roles are then acted out in countless daily interactions as boss and secretary, doctor and nurse, football player and cheerleader "do gender."

Gender Stratification: Structural Sexism

As structural-functionalists and conflict theorists agree, the social structure underlies and perpetuates much of the sexism in society. **Structural sexism**, also known as "institutional sexism," refers to the ways in which the organization of society, and specifically its institutions, subordinate individuals and groups based on their sex classification. Structural sexism has resulted in significant differences between the education and income levels, occupational and political involvement, and civil rights of women and men.

Education and Structural Sexism

Literacy rates worldwide indicate that women are less likely to be able to read and write than men, with millions of women being denied access to even the most basic education (Population Reference Bureau 1999; Leeman 2000). For example, on the average women in South Asia have only half as many years of education as men (World Bank 2001). In some areas of the world, however, conditions are improving. In Iran, for example, between 1990 and 2000, the number of women entering universities tripled. Today, more than 60 percent of all university entrants in Iran are women (Sachs 2000).

In 1999, little difference existed between U.S. men and women in their completion rates of high school and college degrees (U.S. Census 2000a; U.S. Census 2000b). However, as Figure 8.1 indicates, dramatic differences appear in the type of advanced degrees men and women earned. For example, although women received 67 percent of all psychology doctorates, they earned only 15 percent of all doctorates in engineering. Note that, with the exception of psychology, men are more likely to earn a doctorate than women in every degree listed. Although the general trend is for men to have higher levels of education than women, in recent years African-American women's educational levels have increased at a faster rate than African-American men's educational levels (Shepard 2000).

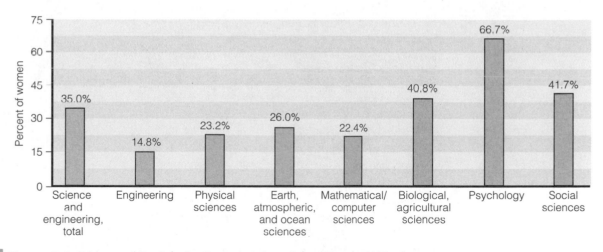

Figure 8.1 *Science and Engineering Doctorates Awarded to Women, 1999.*
Source: National Science Foundation, Division of Science Resource Studies. *Science and Engineering Doctorate Awards:* 1999.

One explanation for why women earn fewer advanced degrees than men is that women are socialized to choose marriage and motherhood over long-term career preparation (Olson, Frieze, & Detlefsen 1990). From an early age, women are exposed to images and models of femininity that stress the importance of domestic family life. When 821 undergraduate women were asked to identify their lifestyle preference, less than 1 percent selected being unmarried and working full-time. In contrast, 53 percent selected "graduation, full-time work, marriage, children, stop working at least until youngest child is in school, then pursue a full-time job" as their preferred lifestyle sequence (Schroeder, Blood, & Maluso 1993, 243). Only 6 percent of 535 undergraduate men selected this same pattern.

Structural limitations also discourage women from advancing in the educational profession itself. For example, women seeking academic careers may find that promotion in higher education is more difficult than for men. Long, Allison, and McGinnis (1993) examined the promotions of 556 men and 450 women with Ph.D.s in biochemistry. They found that women were less likely to be promoted to associate or full professor, were held to a higher standard than men, and were particularly disadvantaged in more prestigious departments. Even in public schools, where women comprise 74 percent of all classroom teachers, only 45 percent of principals and assistant principals are women (*Statistical Abstract 2000*).

Income and Structural Sexism

In general, the higher one's education, the higher one's income. Yet, even when men and women have identical levels of educational achievement and both work full-time, women, on the average, earn significantly less than men (see Table 8.1). Racial differences also exist. For example, although in 1998 white women earned 68 percent as much as white men, African-American and Hispanic-American women earned just 48 and 39 percent, respectively, of white men's salaries (U.S. Census 1999). Further, minority women earn more compared with minority men than white women earn compared with white men.

Tomaskovic-Devey (1993) examined the income differences between males and females and found that the percentage of females in an occupation was the

■ **Table 8.1** *Effects of Education and Sex on Income, 1999*

Level of Educational Attainment	Average Annual Income		
	Women	Men	Women's Income as a Percentage of Men's
Some high school, no diploma	$11,353	$19,155	59
High school diploma	17,898	28,742	62
Some college, no degree	19,327	32,005	60
Associate's degree	25,390	40,082	63
Bachelor's degree or more	31,452	55,057	57
Master's degree	40,429	64,533	63
Professional degree	65,351	108,926	60
Doctorate	54,552	82,619	66
Total persons	22,818	38,134	60

Source: *Statistical Abstract 2000*, 120th ed. U.S. Bureau of the Census. Washington, D.C.: U.S. Government Printing Office, Table 252.

best predictor of an income gender gap—the higher the percentage of females, the lower the pay. Supporting this observation, a team of researchers (Kilbourne, Farkas, Beron, Weir, & England 1994) analyzed data from the National Longitudinal Survey that included more than 5,000 women and 5,000 men. They concluded that occupational pay is gendered and that "occupations lose pay if they have a higher percentage of female workers or require nurturant skills" (p. 708).

Two hypotheses are frequently cited in the literature as to why the income gender gap continues to exist. One is called the **devaluation hypothesis**. It argues that women are paid less because the work they perform is socially defined as less valuable than the work performed by men. The other hypothesis, the **human capital hypothesis,** argues that female-male pay differences are a function of differences in women's and men's levels of education, skills, training, and work experience.

Tam (1997), in testing these hypotheses, concludes that human capital differences are more important in explaining the income gender gap than the devaluation hypothesis. Marini and Fan (1997) also found support for the human capital hypothesis, although their research supports a third category of variables as well. They found that organizational variables (characteristics of the business, corporation, or industry) explain, in part, the gender income gap. For example, women and men upon career entry are channeled by employers into gender-specific jobs that carry different wage rates.

Work and Structural Sexism

... everyone assumed I was someone's wife ... It took years before my mail was addressed to Ms. instead of Mr.

LYNN KIMMEL

Owner of Indianapolis-based, multimillion-dollar business

Women now make up one third of the world's labor force. Worldwide, women tend to work in jobs that have little prestige and low or no pay, where no product is produced, and where women are the facilitators for others. Women are also more likely to hold positions of little or no authority within the work environment and to have more frequent and longer periods of unemployment (United Nations 2000a). Women of color may be even less likely to hold positions of power. In an investigation of female and male African-American and white firefighters, black women were the most subordinated group, as black males and white females relied on their superordinate gender and race statuses, respectively (Yoder & Aniakudo 1997).

No matter what the job, if a woman does it, it is likely to be valued less than if a man does it. For example, in the early 1800s, 90 percent of all clerks were men, and being a clerk was a very prestigious profession. As the job became more routine, in part because of the advent of the typewriter, the pay and prestige of the job declined and the number of female clerks increased. Today, female clerks predominate, and the position is one of relatively low pay and prestige.

The concentration of women in certain occupations and men in other occupations is referred to as **occupational sex segregation** (see Table 8.2). For example, women are over-represented in semiskilled and unskilled occupations, and men are disproportionately concentrated in professional, administrative, and managerial positions.

Table 8.2 *Highly Sex-Segregated Occupations, 1999*

Female-Dominated Occupations	Percentage of Female Workers
Child care workers	97
Cleaners and servants	94
Dental hygienists	99
Dietitians	84
Elementary school teachers	84
Librarians	84
Prekindergarten and kindergarten teachers	98
Receptionists	95
Registered nurses	93
Secretaries	98
Speech therapists	93
Teachers' aides	96
Typists	94
Male-Dominated Occupations	**Percentage of Male Workers**
Announcers	79
Airplane pilots and navigators	97
Architects	84
Automobile mechanics	99
Clergy	86
Construction	97
Dentists	83
Engineers	89
Firefighters	97
Lawyer	71
Mechanics and repairers	95
Physicians	75
Police and detectives	83

Source. *Statistical Abstract 2000*, 120th ed. U.S. Bureau of the Census. Washington, D.C.: U.S. Government Printing Office, Table 669.

In some occupations, sex segregation has decreased in recent years. For example, the percentage of female physicians increased from 16 percent to 25 percent between 1983 and 1999, female dentists increased from 7 to 17 percent, female engineers increased from 6 to 11 percent, and female clergy increased from 6 to 14 percent (*Statistical Abstract 2000*).

Nevertheless, despite these and other changes, women are still heavily represented in low-prestige, low-wage **"pink-collar" jobs** that offer few benefits. Even those women in higher-paying jobs are often victimized by a **glass ceiling**—an invisible barrier that prevents women and other minorities from moving into top corporate positions. A study of Fortune 500 companies found that less than 11 percent of all seats on the Fortune 500 company boards are held by women (Klein 1998). Interestingly, Cianni and Romberger's investigation (1997) of Asian, African-American, and Hispanic women and men in Fortune 500 companies indicates that gender has more of a role in "organizational treatment" than race.

Sex segregation in occupations continues for several reasons (Martin 1992; Williams 1995). First, cultural beliefs about what is an "appropriate" job for a man or a woman still exist. Cejka and Eagly (1999) report that the more college students believed that an occupation was male- or female-dominated, the more they attributed success in that occupation to masculine or feminine characteristics. Further, socialization experiences in which males and females learn different skills and acquire different aspirations remain. A recent government report documents that work at age 12 is sex-segregated: girls baby-sit and boys do lawn work (BLS 2000).

Opportunity structures differ as well. Women have fewer opportunities in the more prestigious and higher-paying male-dominated professions, resulting in women comprising more than 70 percent of all minimum-wage earners. Women may also be excluded by male employers and employees who fear the prestige of their profession will be lessened with the entrance of women, or who simply believe that "the ideal worker is normatively masculine" (Martin 1992, 220).

Finally, because family responsibilities primarily remain with women, working mothers may feel pressure to choose professions that permit flexible hours and career paths, sometimes known as "mommy tracks" (Moen and Yu 2000). Thus, for example, women dominate the field of elementary education, which permits them to be home when their children are not in school. Nursing, also dominated by women, often offers flexible hours.

Politics and Structural Sexism

Women are barely tokens in the decision-making bodies of our nation, so the laws that govern us are made by men.

National Organization for Women

Women received the right to vote in 1920 with the passage of the Nineteenth Amendment. Even though this amendment went into effect almost 80 years ago, women still play a rather minor role in the political arena. In general, the more important the political office, the lower the probability a woman will hold it. Although women comprise 52 percent of the population, the United States has never had a woman president or vice president and, until 1993 when a second woman was appointed, had only one female U.S. Supreme Court justice. The highest-ranking woman ever to serve in U.S. government was Madeleine Albright, who became the U.S. Secretary of State in 1997. In 2001, women comprised only 10 percent of all governors and held only 13.5 percent of all U.S. Congressional seats (see Table 8.3). Worldwide, the percentage of legislative seats held by women ranges from 30 to 40 percent in Scandinavian countries to

Table 8.3 *Percentage of Women Elected by Level and Type of Government Position, 2001*

Level of Government/Position	No. Seats	No. Women	Percent Held by Women
U.S. President	1	0	0.0
U.S. Vice President	1	0	0.0
U.S. Congress	535	72	13.5
House	435	59	13.5
Senate	100	13	13.0
Governors	50	5	10.0
State Legislators	7, 426	1,656	22.3

Source: Center for American Women in Politics. "Election 2000." http://rci.rutgers.edu/~cawp/facts/Summary 2000.

less than 1 percent in several Middle Eastern and African countries (Kenworthy & Malami 1999). Further, in no developing country do women hold more than 8 percent of ministerial positions (World Bank 2001, 5).

In response to the under-representation of women in the political arena, some countries have instituted quotas. In India, a 1993 amendment held one-third of all seats in local contests for women. The result? Eight hundred thousand women were elected. A 1996 law in Brazil requires a minimum of 20 percent of each party's candidates be women. Countries with similar policies include Finland, Germany, Mexico, South Africa, and Spain (Sheehan 2000).

The relative absence of women in politics, as in higher education and in high-paying, high-prestige jobs in general, is a consequence of structural limitations. Running for office requires large sums of money, the political backing of powerful individuals and interest groups, and a willingness of the voting public to elect women. Thus, minority women have even greater structural barriers to election and, not surprisingly, represent an even smaller percentage of elected officials. Nonetheless, 80 percent of U.S. women believe that by 2024, a woman will be in the White House (Thomas 1999).

Civil Rights, the Law, and Structural Sexism

The 1963 Equal Pay Act and Title VII of the 1964 Civil Rights Act made it illegal for employers to discriminate in wages or employment on the basis of sex. Nevertheless, such discrimination still occurs as evidenced by the thousands of grievances filed each year with the Equal Employment Opportunity Commission (EEOC). One technique used to justify differences in pay is the use of different job titles for the same work. The courts have ruled, however, that jobs that are "substantially equal," regardless of title, must result in equal pay. In 2000, President Clinton proposed increased federal funding for initiatives directed toward reducing the gender pay gap, a proposal supported by 79 percent of the American people (Saad 2000).

Women are also discriminated against in employment. Discrimination, although illegal, takes place at both the institutional and the individual level (see Chapter 7). Institutional discrimination includes male-dominated recruiting networks (Reskin & McBrier 2000), employment screening devices designed for

> The price of inequality is just too high.
>
> **NASFIS SADIK**
> *United Nations Population Fund*

men, hiring preferences for veterans, and the practice of promoting from within an organization, based on seniority. One of the most blatant forms of individual discrimination is sexual harassment.

Discrimination takes place in other forms as well. In the United States, having lower incomes, shorter work histories, and less collateral, women often have difficulty obtaining home mortgages or rental property. Until fairly recently, husbands who raped their wives were exempt from prosecution. Even today, some states require a legal separation agreement and/or separate residences for a raped wife to receive full protection under the law. Women in the military have traditionally been restricted in the duties they can perform and, finally, since the U.S. Supreme Court's 1973 Roe v. Wade decision, the right of a woman to obtain an abortion has been limited. The debate continues with several recent court decisions weakening the self-determination doctrine (see Chapter 15).

The Social Construction of Gender Roles: Cultural Sexism

As symbolic interactionists note, structural sexism is supported by a system of cultural sexism that perpetuates beliefs about the differences between women and men. **Cultural sexism** refers to the ways in which the culture of society—its norms, values, beliefs, and symbols—perpetuates the subordination of an individual or group because of the sex classification of that individual or group.

For example, the *belief* that females are less valuable than males has serious consequences. In one study in Bombay, India, of 8,000 abortions performed after amniocentesis, 7,900 were of female fetuses (Anderson & Moore 1998). Cultural sexism takes place in a variety of settings including the family, the school, and media, as well as in everyday interactions.

Family Relations and Cultural Sexism

From birth, males and females are treated differently. For example, the toys male and female children receive convey different messages about appropriate gender behavior. Recently, retail giant Toys-R-Us, after much criticism, removed store directories labeled "Boy's World" and "Girl's World." Similarly, toy manufacturer Mattel came under fire after producing a pink, flowered Barbie computer for girls, and a blue Hot Wheels computer for boys. The social significance of the gender-specific computers and the public criticism came after it was revealed that the accompanying software packages were different—the boys' package had more educational titles (Bannon 2000). This chapter's *Focus on Technology* feature documents the negative consequences of such seemingly harmless differences.

Household Division of Labor
Little girls and boys work within the home in approximately equal amounts until the age of 18 when the female-to-male ratio begins to change (Robinson & Bianchi 1997). In a recent study of household labor in ten Western countries, Bittman and Wajcman (2000) report that, "women continue to be responsible for the majority of hours of unpaid labor" ranging from a low of 70 percent in Sweden to a high of 88 percent in Italy (p.173). The fact that women, even when working full time, contribute signifi-

Women, Men, and Computers

Technology has changed the world in which we live. The technological revolution has brought the possibility of greater gender equality for, unlike tasks dominating industrialization, sex differences in size, weight, and strength are less relevant. Although feminists have long decried the gendering of technology (see Chapter 15), surely computers and other information technologies are gender neutral—or are they?

Girls spend less time playing video games (to wit, "Game Boy") and, consequently, software that appeals to girls is less likely to be manufactured (O'Neal 1998; Children Now 2001). In a recent study of the ten top-selling video games for each of three major systems (Sony PlayStation, Sega Dreamcast, and Nintendo 64), 54 percent of the games contained female characters whereas 92 percent contained male characters. Of the female characters displayed, over one third had significantly exposed breasts, thighs, stomachs, midriffs, or bottoms, and 46 percent had "unusually small" waists. Further, despite evidence that girls, contrary to boys, prefer video games that are nonviolent, over half of the female characters were portrayed engaging in violent behavior (Children Now 2000).

In adolescence and beyond, girls are not as interested in computers as boys. When they are interested, unlike boys who define it as a toy to explore, something that's fun, girls define computers as a tool to accomplish a task, a kind of homework helper. Says Jane Margolis, a researcher at Carnegie Mellon (quoted in Breidenbach 1997, 69):

> Girls want to do something constructive with computers, while boys get into hacking and using computers for their own sake...Computers are just one interest of many for girls, while they become an object of love and fascination for boys.

If women do not pursue computer-based information technologies, they will have an "intellectual and workplace handicap that can only get worse as technology grows more prevalent" (Currid 1996, 114). For example, whereas social work positions, held primarily by women, are projected to grow by 36 percent, systems analyst and computer engineer positions—two of the four fastest-growing occupations between 1998 and 2008—are projected to grow 94 and 108 percent, respectively (*Statistical Abstract 2000*).

The culture and structure of high-tech occupations, however, are often not conducive to female employment. Rapidly changing knowledge bases make taking time off for motherhood almost impossible; long hours make child care arrangements and time away from home difficult; and lingering stereotypes as "women can't handle stress" and women don't "understand technology as well as men" (*Informationweek* 1996) make advancement difficult. A Massachusetts Institute of Technology report on the gender gap in computing found that "women are often judged as less qualified than men even when their performance is identical" (Breidenbach 1997, 69).

There are some signs of optimism. The number of women getting computer science degrees—although in recent years declining—has increased over the last several decades (*Statistical Abstract 2000*). Some companies such as Sun Microsystems and Hewlett-Packard have begun aggressive recruitment and hiring programs for women and people of color (Wilde 1997). However, for significant changes to take place in reference to women, men, and computers we must recognize that computers specifically, and technology in general, are not gender neutral and, in fact, have emerged and flourished within the context of a male-dominated industry.

Sources:

Susan Breidenbach. 1997. "Where Are All the Women?" *Network World* 14(41):68–69.

Cheryl Currid. 1996. "Bridging the Gender Gap: Women Will Lose Out Unless They Catch Up with Men in Technology Use." *Informationweek*, April 1, 114.

Children Now. 2001. "Girls and Gaming: Gender and Video Game Marketing." *Media Now* (Winter). http://www. childrennow.org/media/medianow

Children Now. 2000. "Top Selling Video Games "unhealthy" for Girls, Research Show." News Release, December 12. http://www.childrennow.org/newsroom

Informationweek. 1996. "Women Gain in IS Ranks," September 16, 158.

Glenn O'Neal. 1998. "Girls Often Dropped from Computer Equation." *USA Today*, March 10, D4.

Statistical Abstract of the United States: 1999, 119th ed. U.S. Bureau of the Census.

Washington, D.C.: U.S. Government Printing Office, Tables 325 and 326.

Candee Wilde. 1997. "Women Cut through IT's Glass Ceiling." *Informationweek*, January 20, 83–86.

cantly more hours to home care than men is known as the "second shift" (Hochschild 1989).

Three explanations for the continued traditional division of labor emerge from the literature. The first explanation is the "time-availability approach." Consistent with the structural-functionalist perspective, this position emphasizes that role performance is a function of who has the time to accomplish certain tasks. Because women are more likely to be at home, they are more likely to perform domestic chores.

A second explanation is the "relative resources approach." This explanation, consistent with a conflict perspective, suggests that the spouse with the least power is relegated the most unrewarding tasks. Because men have more education, higher incomes, and more prestigious occupations, they are less responsible for domestic labor.

"Gender role ideology," the final explanation, is consistent with a symbolic interactionist perspective. It argues that the division of labor is a consequence of traditional socialization and the accompanying attitudes and beliefs. Females and males have been socialized to perform various roles and to expect their partners to perform other complementary roles. Women typically take care of the house, men the yard. This division of labor is learned in the socialization process through the media, schools, books, and toys. A recent test of the three positions found that although all three had some support, gender role ideology was the weakest of the three in predicting work allocation (Bianchi, Milkie, Sayer, & Robinson 2000).

The School Experience and Cultural Sexism

Sexism is also evident in the schools. It can be found in the books students read, the curricula and tests they are exposed to, and the different ways teachers interact with students.

Textbooks The bulk of research on gender images in textbooks and other instructional materials documents the way males and females are portrayed stereotypically. For example, Purcell and Stewart (1990) analyzed 1,883 storybooks used in schools and found that they tended to depict males as clever, brave, adventurous, and income-producing and females as passive and as victims. Females were more likely to be in need of rescue, and were also depicted in fewer occupational roles than males. Witt (1996), in a study of third grade textbooks from six publishers, reports that little girls were more likely to be portrayed as having both traditionally masculine *and* feminine traits whereas little boys were more likely pictured as having masculine characteristics only. These results are consistent with research that suggests that boys are much less free to explore gender differences than females, and with Purcell and Stewart's conclusion that boys are often depicted as having "to deny their feelings to show their manhood" (1990, 184). Although some recent evidence suggests that the frequency of male and female textbook characters is increasingly equal, portrayals of girls and boys largely remain stereotypical (Evans & Davies 2000).

Curricula and Testing Encouragement to participate in sports, academic programs, and extracurricular activities is gender-biased despite Title IX of the 1972 Educational Amendments Act, which prohibits officials from "tracking" students by sex. Although women's and girls' participation is now at an all-time high

Men kinda have to choose between marriage and death. I guess they figure at least with marriage they get meals. Then they get married and find out we don't cook anymore.

RITA RUDNER
Comedian

In the school room, more than any other place, does the difference of sex, if there is any, need to be forgotten.

SUSAN B. ANTHONY
Feminist

with, for example, over 120,000 women participating in NCAA sporting events, differences remain in the sports males and females play (U.S. Census 2000c). Males are more likely to participate in competitive sports that emphasize traditional male characteristics such as winning, aggression, physical strength, and dominance. Women are more likely to participate in sports that emphasize individual achievement (e.g., figure skating) or cooperation (e.g., synchronized swimming).

The differing expectations and/or encouragement that females and males receive also contribute to their varying abilities, as measured by standardized tests, in such disciplines as math and science. Boys and girls have the same mathematics and science proficiency at age 9; by age 13, males outperform females in science, although not math. By age 17, males outperform females in both math and science. Are such differences a matter of aptitude? In an experiment at the University of Waterloo, male and female college students, all of whom said they were good in math, were shown either gender-stereotyped or gender-neutral advertisements. When, subsequently, female students who had seen the female stereotyped advertisements took a math test, they performed not only lower than women who had seen the gender-neutral advertisements but lower than their male counterparts (Begley 2000). Research also indicates that standardized tests themselves are biased—almost exclusively being timed, multiple-choice tests—a format favoring males according to some advocates (Smolken 2000).

Teacher-Student Interactions Sexism is also reflected in the way teachers treat their students. After interviewing 800 adolescents, parents, and teachers in three school districts in Kenya, Mensch and Lloyd (1997) report that teachers were more likely to describe girls as lazy and unintelligent. "And when the girls do badly," the researchers remark, "it undoubtedly reinforces teachers' prejudices, becoming a vicious cycle." Similarly, in the United States Sadker and Sadker (1990) observed that elementary and secondary school teachers pay more attention to boys than to girls. Teachers talk to boys more, ask them more questions, listen to them more, counsel them more, give them more extended directions, and criticize and reward them more frequently. However, a recent book by philosopher Christina Sommers (2000) entitled, *The War Against Boys*, argues that it is "boys, not girls, on the weak side of the educational gender gap" (p. 14). Noting that boys are at a higher risk for learning disabilities and lag behind in reading and writing scores, Sommers argues that the belief that females are educationally shortchanged is untrue (see Chapter 12).

Media, Language, and Cultural Sexism

One concern voiced by social scientists in reference to cultural sexism is the extent to which the media portrays females and males in a limited and stereotypical fashion, and the impact of such portrayals.

Signorielli (1998) analyzed gender images presented in six media: television, movies, magazines, music videos, TV commercials, and print media advertisements. The specific items selected from each medium were those most often consumed by 12- to 17-year-old girls, for example, the 25 most-watched television shows. The results indicate:

- In general, media content stresses the importance of appearance and relationships for girls/women and of careers and work for boys/men.

- Across the six media, 26 to 46 percent of women are portrayed as "thin or very thin" in contrast to 4 to 16 percent of men; 70 percent of girls wanted to look like, fix their hair like, or dress like a character on television, compared with 40 percent of boys.
- In a survey of boys and girls, both agreed that "worrying about weight, crying or whining, weakness, and flirting" are characteristics associated more with girls than with boys, and that "playing sports, being a leader, and wanting to kiss or have sex. . . ." are more often characteristic of male characters.

Like media images, both the words we use and the way we use them can reflect gender inequality. The term nurse carries the meaning of "a woman who . . ." and the term engineer suggests "a man who. . . ." Terms like "broad," "old maid," and "spinster" have no male counterpart. Language is so gender-stereotyped that the placement of male or female before titles is sometimes necessary as in the case of "female police officer" or "male prostitute." Further, as symbolic interactionists note, the embedded meanings of words carry expectations of behavior.

Virginia Sapiro (1994) has shown how male-female differences in communication style reflect the structure of power and authority relations between men and women. For example, women are more likely to use disclaimers ("I could be wrong but. . . .") and self-qualifying tags ("That was a good movie, wasn't it?"), reflecting less certainty about their opinions. Communication differences between women and men also reflect different socialization experiences. Women are more often passive and polite in conversation; men are less polite, interrupt more often, and talk more (Tannen 1990).

Social Problems and Traditional Gender Role Socialization

Cultural sexism, transmitted through the family, school, media, and language, perpetuates traditional gender role socialization. Gender roles, however slowly, are changing. As one commentator observed (Fitzpatrick 2000, 1), "...the hard lines that once helped to define masculine [and feminine] identity are blurring. Women serve in the military, play pro basketball, run corporations and govern. Men diet, undergo cosmetic surgery, bare their souls in support groups and cook . . . ."

Despite this "**gender tourism**" (Fitzpatrick 2000), most research indicates that traditional gender roles remain dominant particularly for males who, in general, have less freedom to explore the gender continuum. Social problems that result from traditional gender socialization include the feminization of poverty, social-psychological and health costs, and conflict in relationships.

The Feminization of Poverty

The poor of America are women: the poor of the world are women.

MARILYN FRENCH
Novelist/author

Globally, the percentage of female households is increasing dramatically with approximately 25 percent of all households in Africa, North America, the Caribbean, and parts of Europe headed by women (Population Reference Bureau 1999). Often living in poverty, many of these households are headed by young women with dependent children and older women who have outlived their spouses. More than 900 million women, worldwide, live on less than $1 a day (United Nations 2000b).

Recently a "report card" of U.S. efforts to reduce poverty among women was released by U.S. Women Connect, a nonprofit activist group. Although the United States received a "B" for placing women in decision-making positions, it received an "F" for efforts to reduce female poverty. Citing federal statistics, the group reports that although the overall poverty rate in the United States has decreased, female poverty has increased over the last 5 years (Winfield 2000). As noted earlier, both individual and institutional discrimination contribute to the economic plight of women.

Traditional gender role socialization also contributes to poverty among women. Women are often socialized to put family ahead of their education and careers. Women are expected to take primary responsibility for child care, which contributes to the alarming rate of single-parent poor families in the United States. Hispanic and black female-headed households are the poorest of all families headed by a single woman. Further, a study of the relationship between marital status, gender, and poverty in the United States, Australia, Canada, and France indicates that never-married women compared with ever-married women in all four countries are more likely to live in poverty (Nichols-Casebolt & Krysik 1997).

Social-Psychological and Other Health Costs

Many of the costs of traditional gender socialization are social-psychological in nature. Reid and Comas-Diaz (1990) noted that the cultural subordination of women results in women having low self-esteem and being dissatisfied with their roles as spouses, homemakers/workers, mothers, and friends. In a study of self-esteem among more than 1,160 students in grades 6 through 10, girls were significantly more likely to have "steadily decreasing self-esteem," whereas boys were more likely to fall into the "moderate and rising" self-esteem group (Zimmerman, Copeland, Shope, & Dielman 1997).

Not all researchers have found that women have a more negative self-concept than men. Summarizing their research on the self-concepts of women and men in the United States, Williams and Best (1990) found "no evidence of an appreciable difference" (p. 153). They also found no consistency in the self-concepts of women and men in 14 countries: "[I]n some of the countries the men's perceived self was noticeably more favorable than the women's, whereas in others the reverse was found" (p. 152). More recent research also documents that women are becoming more assertive and desirous of controlling their own lives rather than merely responding to the wishes of others or the limitations of the social structure (Burger & Solano 1994).

Men also suffer from traditional gender socialization. Men experience enormous cultural pressure to be successful in their work and earn a high income. Research indicates that men who have higher incomes feel more "masculine" than those with lower incomes (Rubenstein 1990). Not surprisingly, males are more likely than females to value materialism and competition over compassion and self-actualization (Beutel & Marini 1995; McCammon, Knox, & Schacht 1998; Cohen 2001). Traditional male socialization also discourages males from expressing emotion—part of what Pollock (2000a) calls the "boy code." This chapter's *The Human Side* feature describes the problems and pressures of being male in American society.

On the average, men in the United States die about 6 years earlier than women, although gender differences in life expectancy have been shrinking. Tra-

> I learned from my father how to work. I learned from him that work is life and life is work, and work is hard.
>
> PHILIP ROTH
> *Author*

Real Boys' Voices

Psychologist and author William Pollock traveled from coast to coast talking to boys about the "boy code."

Brad, 14, from a suburb in the Northwest
Guys aren't supposed to be weak or vulnerable. Guys aren't supposed to be sweet. A friend of mine died in the hospital...I knew that, as a guy, I was supposed to be strong and I wasn't supposed to show any emotion...I was supposed to be tough...when I went home, I just sat by myself and let myself cry.(p.17)

Sam, 16, from a city in New England
I think most of the macho stuff guys do is stupid...like the kids who do wrestling moves in the hall. At the same time, there are things that I wouldn't do because I'm a guy. I've never gone to a guy friend, for example, and said, "I'm feeling hurt right now and let's talk about it." (p. 31)

Gordon, 18, from a small town in the South
...all the men in my family...have been the epitome of negativity. Some have become wrapped up in infidelity, some abuse, some alcoholism. I don't want to become a man, because I don't want to become this. (p. 53)

Jeff, 16, from a small town in New England
Your virginity is what determines whether you're a man or a boy in the eyes of every teenage male. Teenage men see sex as a race: the first one to the finish line wins.(p. 69)

Brett, 17, from a city in the South
I think most guys are kind of isolated because it's thought of as weird if you have any really close guy friends. To get around it, guys will go fishing or hunting or bowling or something else "masculine," and then talk about personal or serious things while they're doing that activity. (p. 116)

Jesse, 17, from a suburb in New England
From the girls I've spoken to about relationships, one of their biggest complaints is that they're doing all the giving and the guy is doing all the taking. Girls also tend to be better able to understand social situations. Girls can look at someone and tell what they're feeling. They have more social intuitiveness, more than we clueless guys do. I think that makes them more aware of what's happening in a relationship than we are.(p. 253)

Graham, 17, from a suburb in the West
We would live in a better society if guys could share their feelings more easily. But guys still hear mixed messages from our society. On the one hand they hear that it's OK now to talk about their feelings, but on the other hand they still hear that they have to be tough and that only girls get emotional. My friend who talked to me and cried about his girlfriend was on the football team. His teammates would laugh at him if he tried to talk to them about that sort of stuff. (p. 272)

Jake, 16, from a suburb in southern New England
Ever since I've played Little League the word "win" has been forced into my mind. When I was eight years old, the coach would tell us at the beginning of the season that we were just out there for fun, but I knew that it wasn't true. Every day that there was a game, my day would be ruined. (p. 280–281)

Dylan, 17, from a suburb of Chicago
If I get in shape, if I develop a more attractive body, I'd be more popular. It's like the way life is around here, what society shows you. It's a problem to be naturally skinny like me. You're not as athletic or muscular or attractive; you're not as good as the other kids are. (p.302)

Kirk, 18, from a suburb in the Northwest
I think it's hard growing up in the year 2000. It's definitely hard for a guy. Going through high school is tough. I have pressures in sports, school, life all rolled into one. My parents pressure me to do well in school, do well in sports, and I pressure myself to do well in life....I worry about life a lot. I feel that everything is going to work out for everybody except me, that I'll be left in the dust. (p. 341)

Source: Excerpted from *Real Boys' Voices* by William Pollock (2000). New York: Random House. Used by permission.

ditional male gender socialization is linked to males' higher rates of cirrhosis of the liver, most cancers, AIDS, homicide, drug- and alcohol- induced deaths, suicide, and firearm and motor vehicle accidents (*Statistical Abstract 2000*). Females, however, after the age of 15, are twice as likely to suffer from depression than males (Kalb 2000) and much more likely to suffer from anorexia or bulimia. Seven million women and girls suffer from eating disorders (Wilmot 1999).

Are gender differences in morbidity and mortality a consequence of socialization differentials or physiological differences? Although both nature and nurture are likely to be involved, social rather than biological factors may be dominant. As part of the "masculine mystique," men tend to engage in self-destructive behaviors—heavy drinking and smoking, poor diets, lack of exercise, stress-related activities, higher drug use, and a refusal to ask for help. Men are also more likely to work in hazardous occupations than women. Women's higher rates of depression are also likely to be rooted in traditional gender roles. The heavy burden of childcare and household responsibilities, the gender pay and occupation gap, and fewer socially acceptable reactions to stress (e.g., it is more acceptable for males than females to drink alcohol), contribute to gender differences in depression (Klein 1997).

Conflict in Relationships

Gender inequality also has an impact on relationships. For example, negotiating work and home life can be a source of relationship problems. Whereas men in traditional versus dual-income relationships are more likely to report being satisfied with household task arrangements, women in dual-income families are the most likely to be dissatisfied with household task arrangements (Baker, Kriger, & Riley 1996). Further, the belief that one's partner is not performing an equitable portion of the housework is associated with a reduction in the perception of spousal social support (Van Willigen & Drentea 1997).

> I was more of a man with you as a woman than I have ever been with a woman as a man. I just need to learn to do it without the dress.
>
> DUSTIN HOFFMAN
> *To Jessica Lange in the film "Tootsie"*

We must consider also, of course, the practical difficulties of raising a family, having a career, and maintaining a happy and healthy relationship with a significant other. In a recent survey, over 80 percent of both men and women responded that changing gender roles make it more difficult to have a successful marriage (Morin & Rosenfeld 2000). Successfully balancing work, marriage, and children may require a number of strategies, including: (1) a mutually satisfying distribution of household labor, (2) rejection of such stereotypical roles as "super-mom" and "breadwinner dad" (see this chapter's *Social Problems Research Up Close* feature), (3) seeking outside help from others (e.g., childcare providers, domestic workers), and (4) a strong commitment to the family unit.

Finally, violence in relationships is gender-specific (see Chapters 4 and 5). Although men are more likely to be victims of violent crime, women are more likely to be victims of rape and domestic violence. Violence against women reflects male socialization that emphasizes aggression and dominance over women. Male violence is a consequence of gender socialization and a definition of masculinity which holds that, "[A]s long as nobody is seriously hurt, no lethal weapons are employed, and especially within the framework of sports and games—football, soccer, boxing, wrestling—aggression and violence are widely accepted and even encouraged in boys" (Pollock 2000b, 40).

Strategies for Action: Toward Gender Equality

Efforts to achieve gender equality have been largely fueled by the feminist movement. Despite a conservative backlash, feminists, and to a lesser extent men's activists groups, have made some gains in reducing structural and cultural sexism in the workplace and in the political arena.

Family, Gender Ideology, and Social Change

One of the most important questions concerning gender is the extent to which gender ideologies affect family roles. An investigation by Zuo and Tang (2000) addresses this issue by focusing on two research questions: (1) are men less likely than women to believe in the equality of roles and, (2) is the male "breadwinner" status predictive of beliefs about gender ideology?

Sample and Methods

Data for this investigation came from a randomly selected national sample of married persons collected by the Bureau of Sociological Research at the University of Nebraska. As part of a larger longitudinal study, respondents (N= 400 married men and 640 married women) were interviewed in 1980, 1983, and 1992. All were between the ages of 18 and 55, 95 percent were white, and 67 percent had 1992 annual incomes between $25,000 and $45,000. The independent variable, *breadwinner status*, was measured by the proportion of a husband's income to the total family income. For example, a husband who provided 90 percent of the total family income received a higher score than a husband who provided 50 percent of the total family income. The higher a respondent's score, the higher their breadwinner status and the lower the breadwinner status of their spouse. *Gender ideology*, the dependent variable, was measured by the extent to which a respondent agreed or disagreed with statements concerning: (1) the wife's economic role (e.g., ". . . a woman should not be employed if jobs are scarce"), (2) the provider role (e.g., ". . . a husband should be the main breadwinner even if his wife works"), and (3) the women's maternal role (e.g., ". . . a woman's most important task in life is being a mother"). In combination, these three variables indicated the extent to which respondents adhered to a traditional or egalitarian (i.e., equal partners) gender ideology.

Findings and Conclusions

The results signify that over the years studied, both men and

Grassroots Movements

Feminism and the Women's Movement **Feminism** is the belief that women and men should have equal rights and responsibilities. The American feminist movement began in Seneca Falls, New York, in 1848 when a group of women wrote and adopted a women's rights manifesto modeled after the Declaration of Independence. Although many of the early feminists were primarily concerned with suffrage, feminism has its "political origins...in the abolitionist movement of the 1830s...," when women learned to question the assumption of "natural superiority" (Andersen 1997, 305). Early feminists were also involved in the temperance movement, which advocated restricting the sale and consumption of alcohol, although their greatest success was the passing of the Nineteenth Amendment in 1920, which guaranteed women the right to vote.

The rebirth of feminism almost 50 years later was facilitated by a number of interacting forces: an increase in the number of women in the labor force, an escalating divorce rate, the socially and politically liberal climate of the 1960s, student activism, and the establishment of the Commission on the Status of Women by John F. Kennedy. The National Organization for Women (NOW) was established in 1966 and remains the largest feminist organization in the United States with more than 100,000 members. One of NOW's hardest-fought battles was the struggle to win ratification of the Equal Rights Amendment (ERA), which states that "[E]quality of rights under the law shall not be denied or abridged by the United States, or by any state, on account of sex." The proposed amendment passed both the House of Representatives and the Senate in 1972 but failed to be ratified by the required 38 states by the 1978 deadline, later ex-

> Be ready when the hour comes, to show that women are human and have the pride and dignity of human beings.
>
> CHRISTABEL PANKHURST
> *American Feminist, 1911*

women have shifted toward a more egalitarian gender role ideology. The shift, however, is greater for women than for men. One notable exception is in reference to beliefs about a woman's maternal role. Here, men held more egalitarian beliefs than women. The author's caution, however, that this result does not necessarily indicate that women hold more traditional beliefs about motherhood than men. It may be, for example, that women's stress over the lack of childcare facilities outside of the home is responsible for gender differences on this indicator.

Results also indicate that the higher a husband's breadwinner status, that is, the more he contributes to household finances, the more likely he is to hold traditional gender beliefs. Conversely, the lower a husband's breadwinner status, the more likely he is to hold egalitarian gender beliefs. Similarly, the higher a wife's breadwinner status, that is, the more she contributes to household finances, the more likely she is to hold egalitarian beliefs, and the lower her breadwinner status, the more likely she is to hold traditional beliefs.

The authors conclude that the results of their study support what is called the *benefits hypothesis*. The benefits hypothesis holds that men whose wives earn high wages, that is, men who have lower breadwinner statuses, are likely to embrace rather than be threatened by role equality. Given the empirical support for this hypothesis, the authors predict a continued narrowing in male–female differences in gender role ideology.

The present trend is that men's breadwinner status continues to decline; more and more individuals perform non-gendered family roles. Based on these facts, it may be predicted that the movement toward egalitarianism for both men and women will continue and a further decrease in the gender gap in gender ideology is down the road. (2000, 36)

Source: Jiping Zuo and Shengming Tang. 2000. "Breadwinner Status and Gender Ideologies of Men and Women Regarding Family Roles." *Sociological Perspectives* 43:29–44.

tended to 1982. Thirty-five states have ratified the ERA, and it is presently in several state legislatures awaiting action.

Supporters of the ERA argue that its opponents used scare tactics—saying the ERA would lead to unisex bathrooms, mothers losing custody of their children, and mandatory military service for women—to create a conservative backlash.

Brown Brothers

The United States has a long history of gender inequality; women have had to fight for the right to vote, equal pay for comparable work, and other rights.

Susan Faludi in *Backlash: The Undeclared War against American Women* (1991) contends that contemporary arguments against feminism are the same as those levied against the movement a hundred years ago and that the negative consequences opponents of feminism predict (e.g., women unfulfilled and children suffering) have no empirical support.

Today, a new wave of feminism is being led by young women and men who grew up with the benefits won by their mothers but shocked by the realities of the Tailhook scandal, Paula Jones's accusations of sexual harassment against a sitting president, and packs of men roaming Central Park openly assaulting women. These young feminists are more inclusive than their predecessors, welcoming all who champion the cause of global equality. Not surprisingly, the new feminists are likely to attract a more diverse group of supporters than their predecessors as future feminist efforts focus on "gender equality" over "gender sameness" (Parker 2000).

As I talked to boys across America, I'm struck by how trapped they feel. Our culture puts boys in a gender straightjacket.

WILLIAM POLLOCK
Psychologist

The Men's Movement As a consequence of the women's rights movement, men began to reevaluate their own gender status. In *Unlocking the Iron Cage: The Men's Movement, Gender Politics, and American Culture,* Michael Schwalbe (1996) examines the men's movement as both participant and researcher. For 3 years, he attended meetings and interviewed active members. His research indicates that participants, in general, are white middle-class men who feel they have little emotional support, question relationships with their fathers and sons, and are overburdened by responsibilities, unsatisfactory careers, and what is perceived as an overly competitive society.

As with any grassroots movement, the men's movement has a variety of factions. Some men's organizations advocate gender equality, that is, they are pro-feminists; others developed to oppose "feminism" and what was perceived as male-bashing. For example, the Promise Keepers are part of a Christian men's movement that has often been criticized as racially intolerant, patriarchal, and anti-feminist. However, one woman researcher/author who attended meetings incognito (i.e., as a man) reports: "I'm struck with how close it all sounds like feminism" (Leo 1997).

Today, issues of custody and fathers' rights, led by such groups as Dads against Discrimination, Texas Father's Alliance, and the National Coalition of Free Men (NCFM), headline the men's rights movement and have led to increased visibility. Many members of such groups argue that society portrays men as "disposable," and that as fathers and husbands, workers and soldiers, they feel that they can simply be replaced by other men willing to do the "job." They also hold that nothing in society is male-affirming and that the social reform of the last 30 years has "been the deliberate degradation and disempowerment of men economically, legally, and socially" (NCFM 1998, 7). Still other men's advocates concentrate less on men's rights and more on personal growth, advocating "the restoration of earlier versions of masculinity" (Cohen 2001, 393).

Public Policy

The world needs a man's heart.

JOSEPH JASTRAB
Author

A number of statutes have been passed to help reduce gender inequality. They include the 1963 Equal Pay Act, Title VII of the Civil Rights Act of 1964, Title IX of the Educational Amendments Act of 1972, the Family Leave Act of 1993, the 1994 Violence Against Women Act, and the Victims of Trafficking and Violence Protection Act of 2000. The National Organization for Women encourages women to be politically active, to run for political office, and to participate in the decision-making processes of the nation. Recently, public policy has focused on two issues—sexual harassment and affirmative action.

Sexual Harassment During the 1980s and 1990s, the courts held that Title VII of the 1964 Civil Rights Act prohibited **sexual harassment** involving members of the opposite sex. In 1998, the U.S. Supreme Court extended protection to victims of same-sex harassment. According to the U.S. Equal Employment Opportunity Commission, more than 15,000 cases of sexual harassment were reported in 1999, a 50 percent increase since 1991 (Civil Rights Monitor 2000). It is likely that the dramatic increase in cases over the last decade is, at least in part, a response to the publicity surrounding the Anita Hill-Clarence Thomas controversy.

Sexual harassment can be of two types: (1) *quid pro quo*, in which an employer requires sexual favors in exchange for a promotion, salary increase, or any other employee benefit, and (2) the existence of a hostile environment that unreasonably interferes with job performance, as in the case of sexually explicit comments or insults being made to an employee. According to a 1993 Supreme Court decision, a person no longer has to demonstrate "severe psychological damage" in order to win damages. Sexual harassment occurs at all occupational levels, and some research suggests that the number of incidents of sexual harassment is inversely proportional to the number of women in an occupational category (Fitzgerald & Shullman 1993; *Civil Rights Monitor* 2000). For example, female doctors (Schneider & Phillips 1997) and lawyers (Rosenberg, Perlstadt, & Phillips 1997) report high rates of sexual harassment, in the first case by male patients and in the second by male colleagues. Sexual harassment is also a worldwide phenomenon. Seventy percent of female government employees in Japan report being sexually harassed at work (Yamaguchi 2000).

Affirmative Action The 1964 Civil Rights Act provided for **affirmative action** to end employment discrimination based on sex and race (see Chapter 7). Such programs require employers to make a "good faith effort" to provide equal opportunity to women and other minorities. However, in response to the growing sentiment that affirmative action programs constitute "reverse discrimination," recent court decisions have begun to dismantle affirmative action programs despite former President Clinton's plea to "mend it, but don't end it." In 1996, a California ballot initiative abolished racial and sexual preferences in government programs which included state colleges and universities (Chavez 2000). Washington state voters passed a similar initiative in 1998, and in 1999 the governor of Florida signed an executive order ending that state's affirmative action program (see Chapter 7). Several other states have anti-affirmative action initiatives pending including Colorado, Michigan, and Oregon (ACE 2001).

International Efforts

The Convention to Eliminate All Forms of Discrimination against Women (CEDAW), also known as the International Women's Bill of Rights, was adopted by the United Nations in 1979. CEDAW establishes rights for women not previously recognized internationally in a variety of areas, including education, politics, work, law, and family life. The United States signed the document on July 17, 1980, although it has yet to be ratified by the required two-thirds vote of the U.S. Senate. Over 166 countries have ratified the treaty, including every country in the Western Hemisphere and every industrialized nation in the world with the exception of Switzerland and the United States. Contrary to what some critics argue, provisions in CEDAW would not supersede existing U.S. law (United Nations 2000c; Rabin 2000).

> Give to every other human being every right that you claim for yourself.
>
> THOMAS PAINE
> *Political and social activist*

In April of 2000, heads of state and other high ranking officials from South and North America met to consider a proposed plan entitled the *Inter-American Program on the Promotion of Women 's Human Rights and Gender Equality* (*Americas* 2000). The proposal, if implemented, would "incorporate women more fully in every aspect of life, including economic development opportunities, equal access to education and health services, and full participation in political decision-making" (p. 52).

In addition to these and many other global efforts, individual countries have instituted programs or policies designed to combat sexism and gender inequality. For example, Japan has implemented the Basic Law for a Gender-Equal Society, a "blueprint for gender equality in the home and workplace" (Yumiko 2000, 41). The new South African Bill of Rights prohibits discrimination on the basis of, among other things, gender, pregnancy, and marital status (IWRP 2000), and China has recently established a Programme for the Development of Chinese Women which focuses on empowering women in the areas of education, human rights, health, child care, employment, and political power (*WIN News* 2000).

Understanding *Gender Inequality*

Gender roles and the social inequality they create are ingrained in our social and cultural ideologies and institutions and are, therefore, difficult to alter. For example, in almost all societies women are primarily responsible for child care and men for military service and national defense (World Bank 2001). Nevertheless, as we have seen in this chapter, growing attention to gender issues in social life has spurred some change. Women who have traditionally been expected to give

Definitions of appropriate gender roles change over time. Fifty years ago, women playing hockey, and winning the Olympics would have been unimaginable! Men's roles, although more slowly, are also changing.

domestic life first priority are now finding it acceptable to be more ambitious in seeking a career outside the home. Men who have traditionally been expected to be aggressive and task-oriented are now expected to be more caring and nurturing. Women seem to value gender equality more than men, however, perhaps because women have more to gain. For instance, 84 percent of 600 adult women said that the ideal man is caring and nurturing; only 52 percent of 601 adult men said that the ideal woman is ambitious (Rubenstein 1990, 160).

But men also have much to gain by gender equality. Eliminating gender stereotypes and redefining gender in terms of equality does not mean simply liberating women, but liberating men and our society as well. "What we have been talking about is allowing people to be more fully human and creating a society that will reflect that humanity. Surely that is a goal worth striving for" (Basow 1992, 359). Regardless of whether traditional gender roles emerged out of biological necessity as the functionalists argue or economic oppression as the conflict theorists hold, or both, it is clear today that gender inequality carries a high price: poverty, loss of human capital, feelings of worthlessness, violence, physical and mental illness, and death. Surely, the costs are too high to continue to pay.

Critical Thinking

1 Some research suggests that "[Men] and women with more androgynous gender orientations—that is to say, those having a balance of masculine and feminine personality characteristics—show signs of greater mental health and more positive self-images" (Anderson 1997, 34). Do you agree or disagree? Why or why not?

2 Recent evidence suggests that a "gender gap" exists in the number of men and women entering college, particularly among African Americans, with women attending at higher rates than men. Although the number of females in the population is slightly higher, the difference does not explain the projected gap in enrollments. Why are black women entering college at a higher rate than black men?

3 What have been the interpersonal costs, if any, of sensitizing U.S. society to the "political correctness" of female–male interactions?

4 Why are women more likely to work in traditionally male occupations than men are to work in traditionally female occupations? Are the barriers that prevent men from doing "women's work" cultural, structural, or both? Explain.

Key Terms

affirmative action	gender	pink-collar jobs
cultural sexism	gender tourism	sex
devaluation hypothesis	glass ceiling	sexism
double or triple (multiple) jeopardy	human capital hypothesis	sexual harassment
	instrumental roles	structural sexism
expressive roles	occupational sex	
feminism	segregation	

Media Resources

 The Wadsworth Sociology Resource Center: Virtual Society

http://sociology.wadsworth.com/

See the companion Web site for this book to access general sociology resources and text-specific features that can further your understanding of this chapter. The site contains Internet links, Internet exercises, online practice quizzes, information on InfoTrac College Edition, and many more valuable materials designed to enrich your learning experience in social problems.

 InfoTrac College Edition

You can access InfoTrac College Edition either from the Wadsworth Sociology Resource Center at **http://sociology.wadsworth.com** or directly from your web browser at **http://www.infotrac-college.com/wadsworth/**. InfoTrac College Edition is an online university library that includes over 700 popular and scholarly journals in which you can find articles related to the topics in this chapter such as men's parental rights, feminism, and the gender pay gap.

Interactions CD-ROM

Go to the "Interactions" CD-ROM for *Understanding Social Problems*, Third Edition to access additional interactive learning tools, such as in-depth review materials, corresponding practice quizzes, and other engaging resources and activities to help you study the concepts in this chapter.

Sexual Orientation

9

Is It True?

1. In some countries, homosexual behavior is punishable by the death penalty.

2. The majority of Americans say that gays should have equal rights in the workplace.

3. People who believe that gay individuals are "born that way" tend to be more tolerant of gays than people who believe that gay individuals choose their sexual orientation.

4. The American Psychiatric Association currently classifies homosexuality as a mental disorder.

5. The majority of Americans support same-sex marriage.

Answers to "Is It True?": 1 = T; 2 = T; 3 = T; 4 = F; 5 = F

have friends. Some of them are straight . . . Year after year I continue to realize that the facts of my life are irrelevant to them and that I am only half listened to, that I am an appendage to the doings of a greater world, a world of power and privilege, . . . a world of exclusion. "That's not true," argue my straight friends. There is only one certainty in the politics of power; those left out beg for inclusion, while the insiders claim that they already are. Men do it to women, whites do it to blacks, and everyone does it to queers.

GAY PRIDE PARADE FLIER

James Dale was a highly decorated scoutmaster, until he was removed from his position because he was gay. Dale sued under a New Jersey state law prohibiting sexual orientation discrimination in public accommodations, and the New Jersey Supreme Court entered a unanimous decision in his favor in 1999. However, in June 2000, in a 5 to 4 decision, the Supreme Court overruled New Jersey's decision asserting that the Boy Scouts of America (BSA) had a First Amendment right to expel a gay scoutmaster (LAWbriefs 2000).

The James Dale case focused national attention on sexual orientation minorities —gay men, lesbians, and bisexuals—and their treatment in society. While conservative groups applauded the decision to remove James Dale from his scoutmaster position, other groups protested the decision. As of this writing, the backlash against the Supreme Court's ruling is sweeping the country: Ten members of Congress signed a letter asking President Clinton to resign as honorary head of Boy Scouts of America and Rep. Lynn Woolsey (D-Calif.) introduced legislation in the House to revoke BSA's federal charter (LAWbriefs 2000). The Connecticut Commission on Human Rights and Opportunities ruled that BSA would not be included in the State Employee Campaign for charitable giving because of BSA's exclusionary policy toward gays. Several chapters of United Way have ended or reduced their donations to the Scouts. School districts in California, Massachusetts, Minnesota, and New York have severed ties with the Scouts. City councils, mayors, school districts, corporations, and charities across the country are considering proposals to revoke funds from groups that discriminate against gays. Gregg Shields, spokesman for the Boy Scouts of America asserts:

> *The Boy Scouts of America since 1910 have taught traditional family values. We feel that an avowed homosexual isn't a role model for those values. (Zernike 2000, 2)*

Thomas Jager, director of the United Way of Evanston, Illinois, points out that "in local schools, gay people are allowed to teach. I would say that perhaps the Scouts are slightly behind the times." (Parker & Garcia 2000, 2A)

In this chapter we examine prejudice and discrimination toward homosexual (or gay) women (also known as lesbians), homosexual (or gay) men, and bisexual individuals. It is beyond the scope of this chapter to explore how sexual diversity and its cultural meanings vary throughout the world. Rather, this chapter focuses on Western conceptions of diversity in sexual orientation. The term **sexual orientation** refers to the classification of individuals as heterosexual, bisexual, or homosexual, based on their emotional and sexual attractions, relationships, self-identity, and lifestyle. **Heterosexuality** refers to the predominance of emotional and sexual attraction to persons of the other sex. **Homosexuality** refers to the predominance of emotional and sexual attraction to persons of the same sex, and **bisexuality** is emotional and sexual attraction to members of both sexes. Lesbians, gays, and bisexuals, sometimes referred to collectively as **lesbigays**, are considered to be part of a larger population referred to as the transgendered community. **Transgendered individuals** include persons who do not fit neatly into either the male or female category, or their behavior is not congruent with the rules and expectations for their sex in the society in which they live (Bullough 2000). Transgendered individuals include not only homosexuals and bisexuals, but also cross-dressers (individuals who occasionally dress in the clothing of the opposite sex) and transsexuals (individuals who have undergone hormone treatment and sex reassignment surgery to achieve a new identity as a member of the biologically opposite sex). Much of the current literature on the treatment and political and social agendas of the lesbigay population includes other members of the transgendered community; hence the term **LGBT** is often used to refer collectively to lesbians, gays, bisexuals, and transgendered individuals.

We begin by summarizing the legal status of lesbians and gay men around the world. Then we discuss the prevalence of homosexuality, heterosexuality, and bisexuality in the United States, review biological and environmental explanations for sexual orientation diversity, and apply sociological theories to better understand societal reactions to sexual diversity. The chapter ends with a discussion of strategies to reduce antigay prejudice and discrimination.

The Global Context: A World View of Laws Pertaining to Homosexuality

Homosexual behavior has existed throughout human history and in most, perhaps all, human societies (Kirkpatrick 2000). A global perspective on laws and social attitudes regarding homosexuality reveals that countries vary tremendously

In July 2000, an estimated 200,000 people participated in World Pride 2000 held in Rome, Italy, the first global gay pride event.

© AP/Wide World Photos

■ **Table 9.1** *Countries in Which Homosexual Acts are Subject to the Death Penalty*

Mauritania
Sudan
Afghanistan
Pakistan
Chechen Republic
Iran
Saudi Arabia
United Arab Emirates
Yemen
Somalia

Sources: The International Lesbian and Gay Association. 1999. "World Legal Survey 1999." www.ilga.org.

"Jail, Death Sentences in Africa." 2001 (February 21). PlanetOut.com. http://www.planetout.com/news/article-print.html?2001/02/21/2

in their treatment of homosexuals—from intolerance and criminalization to acceptance and legal protection. A global overview of laws that criminalize sexual behavior between consenting adults indicates that such behavior is illegal in 85 countries ("Sodomy Fact Sheet: A Global Overview" 2000). In 52 of these countries, laws criminalizing same-sex sexual behavior apply to both female and male homosexuality. In 33 of these countries, laws criminalizing same-sex sexual behavior apply to male homosexuality only. Legal penalties for violating laws that prohibit homosexual sexual acts vary. In ten countries, individuals found guilty of engaging in same-sex sexual behavior may receive the death penalty (see Table 9.1). For example, a Somali lesbian couple were sentenced to death for "exercising unnatural behavior" ("Jail, Death Sentences in Africa," 2001). Although executions in this region are performed by firing squads, religious tradition dictates that those convicted of homosexuality should either have a wall pushed over onto them or be thrown off a roof or other high place.

In general, countries throughout the world are moving toward increased legal protection of sexual orientation minorities. Between 1984 and 1995, 86 countries changed their policies regarding sex between men, sex between women, or both, and nearly every change was toward increased liberalization of policies regarding same-sex sexual behavior (Frank & McEneaney 1999). According to the International Gay and Lesbian Human Rights Commission (1999), 22 countries have national laws that ban various forms of discrimination against gays, lesbians, and bisexuals. In 1996 South Africa became the first country in the world to include in its constitution a clause banning discrimination based on sexual orientation. Fiji, Canada, and Ecuador also have constitutions that ban discrimination based on sexual orientation ("Constitutional Protection" 1999).

In Brazil, a gay, lesbian, bisexual, or transgendered individual is murdered on the average of every two days. However, the brutal gay-bashing murder of Edson Neris da Silva by a gang of about 30 people resulted in what some believe is Brazil's first trial and convictions in an antigay hate crime ("Brazilian Killers Sentenced," 2001) The first two gang members tried for this murder were sentenced to 21 years in prison.

In recent years legal recognition of same-sex relationships has become more widespread. Six countries that recognize gay and lesbian partnerships include

Denmark, Norway, Sweden, Iceland, Greenland, France, Portugal, and the Netherlands (Alsdorf 2001; Gay and Lesbian International Lobby 2000). One U.S. state—Vermont—gives legal recognition to same-sex "civil unions" (discussed further later in this chapter). As of this writing, seven other European countries considering same-sex partnership laws include Finland, Switzerland, Germany, Luxembourg, Belgium, and Spain. In June 2000, Canada enacted a bill that extends to same-sex couples and unmarried heterosexual couples who have lived together for at least a year all the benefits and obligations of married couples (*LAWbriefs* 2000). In the same month, Brazil extended to same-sex couples the right to inherit each other's pension and social security benefits. The law represents the first time a Latin American country has legally recognized gay relationships (*LAWbriefs* 2000). In 2000, the Netherlands enacted a law allowing same-sex marriages. Just after the stroke of midnight on the day the Dutch law went into effect (April 1, 2001), the world's first fully legal same-sex civil marriages took place in Amsterdam (Drinkwater 2001). Same-sex married couples and opposite-sex married couples in the Netherlands will be treated identically, with two exceptions. Unlike opposite-sex marriages,

Gert Kasteel (left) *and his partner Dolf Pasker, are one of four same-sex couples who were the world's first to get legally married under a new Dutch law.*

same-sex couples married in the Netherlands are unlikely to have their marriages recognized as fully legal abroad. Regarding children, parental rights will not automatically be granted to the non-biological spouse in gay couples. To become a fully legal parent, the spouse of the biological parent must adopt the child.

Clearly, public legitimization of same-sex relations is occurring in the global society. Human rights treaties and transnational social movement organizations have increasingly asserted the rights of persons to engage in same-sex relations. International organizations such as Amnesty International, which resolved in 1991 to defend those imprisoned for homosexuality, the International Lesbian and Gay Association (founded in 1978), and the International Gay and Lesbian Human Rights Commission (founded in 1990) continue to fight prejudice and discrimination against lesbians and gays. Despite the worldwide movement toward increased acceptance and protection of homosexual individuals, the status and rights of lesbians and gays in the United States continues to be one of the most divisive issues in American society.

Homosexuality and Bisexuality in the United States: Prevalence and Explanations

What percent of the U.S. population is gay, lesbian, or bisexual? After addressing this question, we briefly view explanations regarding the "causes" of homosexuality.

Prevalence of Homosexuality, Heterosexuality, and Bisexuality in the United States

In early research on sexual behavior, Kinsey and his colleagues (1948, 1953) found that a substantial proportion of respondents reported having had same-sex sexual experiences. The data revealed that 37 percent of men and 13 percent of women had at least one homosexual experience since adolescence. Yet, very few of the individuals in Kinsey's research reported exclusive homosexual behavior. These data led Kinsey to conclude that most people are not exclusively heterosexual or homosexual. Rather, Kinsey suggested an individual's sexual orientation may have both heterosexual and homosexual elements. In other words, Kinsey suggested that heterosexuality and homosexuality represent two ends of a sexual orientation continuum, and that most individuals are neither entirely homosexual nor entirely heterosexual, but fall somewhere within this continuum.

Kinsey's early sex research demonstrated the difficulty of classifying individuals as heterosexual, homosexual, or bisexual, as the distinctions between these classifications are not as clear-cut as some people believe. Current research has confirmed Kinsey's finding that sexual behavior, desire, and sexual orientation identity do not always match. In a national study of U.S. adults aged 18 to 59, researchers focused on three aspects of homosexuality: sexual attraction to persons of the same sex, sexual behavior with people of the same sex, and homosexual self-identification (Michael, Gagnon, Laumann, & Kolata 1994). This survey found that 4 percent of women and 6 percent of men said they are sexually attracted to individuals of the same sex, and 4 percent of women and 5 percent of men reported that they had sexual relations with a same-sex partner after age 18. Yet less than 3 percent of U.S. men and less than 2 percent of U.S. women identified themselves as homosexual or bisexual (Michael et al. 1994). What these data tell us is that first, "those who acknowledge homosexual desires may be far more numerous than those who actually act on those desires" (Black, Gates, Sanders, & Taylor 2000, 140). Second, not all people who are sexually attracted to or have had sexual relations with individuals of the same sex view themselves as homosexual or bisexual.

Nationally representative General Social Survey data reveal that the percentage of women age 59 and younger reporting same-sex sexual partnering in the previous five years increased from 0.2 percent in 1988 to 2.8 percent in 1998 (Butler 2000). The percentage of men reporting same-sex sexual partnering in the previous five years increased from 1.7 percent in 1988 to 4.1 percent in 1998 (Butler 2000). According to Butler, these increases may have resulted from declining social and legal constraints against same-sex sexual behavior, as well as more positive images of gay men and lesbians in the media, which have made it easier for people to recognize and act on their sexual attraction to others of their same sex. The researcher also notes that:

> These estimates of same-gender sex partnering should not be taken as estimates of the proportion of the population that is gay or lesbian. Some people may engage in same-gender sexual activity and yet identify as heterosexual, whereas other people may identify as gay or lesbian but may not have been sexually active in recent years. (Butler 2000, 342)

Exit polling data provide additional information concerning the prevalence of gay, lesbian, and bisexual voters in the United States. Five percent of voters in the 1996 Congressional House elections self-identified as gay, lesbian, or bisexual (Bailey 1999). In the same year 6.4 percent of voters under age 40 self-identified as gay, lesbian, or bisexual. Exit polling data also reveal a higher percentage of sexual orientation minorities in large cities. In urban areas during the 1996–98 Congressional elections nearly 9 percent of voters identified as gay, lesbian, or bisexual (Bailey 1999). In the 2000 presidential election, self-identified gays, lesbians, and bisexuals represented 4 percent of voters nationwide ("Post-Election Analysis" 2000).

Origins of Sexual Orientation Diversity: Nature or Nurture?

One of the prevailing questions raised regarding sexual orientation centers on its origin or "cause." However, questions about the "causes" of sexual orientation are typically concerned with the origins of homosexuality and bisexuality. Because heterosexuality is considered normative and "natural," causes of heterosexuality are rarely considered.

Despite the growing research on this topic, a concrete cause of sexual orientation diversity has yet to be discovered. Many researchers believe that an interaction of biological and environmental forces is involved in the development of one's sexual orientation (De Cecco & Parker 1995).

Environmental Explanations of Sexual Orientation According to Doell (1995), ". . . we all probably develop, from infancy, the capacity to have heterosexual, homosexual, or bisexual relationships" (p. 352). Environmental theories propose that such factors as availability of sexual partners, early sexual experiences, and sexual reinforcement influence subsequent sexual orientation. The degree to which early sexual experiences have been negative or positive has been hypothesized as influencing sexual orientation. Having pleasurable same-sex experiences would be likely to increase the probability of a homosexual orientation. By the same token, early traumatic heterosexual sexual experiences have been suggested as causing fear of heterosexual activity. However, a study that compared sexual histories of lesbian and heterosexual women found no differences in the incidence of traumatic experiences with men (Brannock & Chapman 1990).

Biological Origins of Sexual Orientation Biological explanations of sexual orientation diversity usually focus on genetic or hormonal differences between heterosexuals and homosexuals. In an overview of genetics research on homosexual and heterosexual orientations, Pillard and Bailey (1998) conclude that genes account for at least half of the variance in sexual orientation. Their review of family, twin, and adoptee studies indicates that homosexuality (and thus heterosexuality) runs in families. However, science has yet to identify a "homosexuality" gene. Even though the evidence for any link among humans between genes, hormones (such as testosterone), and sexual orientation is very unclear, researchers continue in the quest to isolate a biological cause for human sexual orientation.

Most gays and gay rights advocates believe that homosexuality is an inherited, inborn trait. In a national study of homosexual men, 90 percent believe that they were born with their homosexual orientation; only 4 percent believe that environmental factors are the sole cause (Lever 1994). The percentage of Americans who believe that homosexuality is something a person is born with has increased in recent decades, from 13 percent in 1977 to 34 percent in 1999 (Gallup Organization 2000).

> I knew in my bones that my own sexuality was not a decision but a natural part of who I am.
>
> JONATHON TOLINS

Can Homosexuals Change Their Sexual Orientation? A Case Study

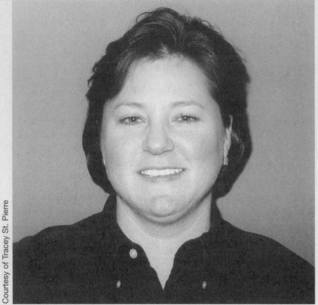

Courtesy of Tracey St. Pierre

I sat peacefully in the prayer circle, nodding in agreement or whispering my "amen," as different women prayed. As I peeked up, my eyes met those of my church counselor. She was watching me. With a nod and a quick wave of her hand, she instructed me to sit in a more "feminine manner." This is something we had talked about before, my needing to become more feminine.

I knew I was different at an early age. When my hormones started raging in high school, I didn't go boy crazy. I fell in love with another girl. When we met, sparks flew. We were both in love for the first time. But this was a small town in the buckle of the Bible Belt during the late 70s. Society, organized religion, and Anita Bryant, all with uncompromising certainty, declared that our kind of love was wrong. As a result, we kept our 2-year relationship a secret. We lived in constant fear that someone would find out and label us as lesbians.

By the time I left for college, I was desperate to talk to someone about my relationship and my sexual orientation. As it happened, that someone came in the form of the church counselor. When I told her about my relationship, she said that God did not create me gay and the love I had shared was sinful. She assured me that God could heal me and make be "whole again"—a real woman. She prayed for me, laid her hands on me and rebuked the demonic "spirit of homosexuality" that I had "allowed to control me." She gave me Bible verses to memorize, told me to avoid temptation and to break off

my 2-year relationship. She encouraged me to develop "godly" relationships with women—but not too close.

We scheduled time together to work on my femininity. The first afternoon we spent in front of a mirror. I learned how to apply make-up—eyeliner, mascara, eye shadow, lipstick, the works. Another day, she criticized my short, popular Dorothy Hamill haircut that all the girls had. "Let it grow longer," she said. She told me to rid my closet of my old jeans, sweat pants, and gym shorts and replace them with skirts and dresses. My fam-

Can Homosexuals Change Their Sexual Orientation? Individuals who believe that homosexuality is biologically based tend to be more accepting of homosexuality. In contrast, "those who believe homosexuals choose their sexual orientation are far less tolerant of gays and lesbians and more likely to conclude homosexuality should be illegal than those who think sexual orientation is not a matter of personal choice" (Rosin & Morin 1999, 8). Individuals who believe that homosexuals choose their sexual orientation also tend to think that homosexuals can and should change their sexual orientation. Various forms of **reparative therapy** or **conversion therapy** are dedicated to changing homosexuals' sexual orientation. Some religious organizations sponsor **ex-gay ministries,** which claim to "cure" homosexuals and transform them into heterosexuals through prayer and other forms of "therapy" (see this chapter's *The Human Side* feature). Ex-gay ministries attract new recruits by claiming that there is a "cure" for people who are unhappy being gay. Critics of ex-gay ministries take a different approach:

ily, particularly my mother and tomboy sister (heterosexual and married now for 20 years), noticed the changes in me. I was zealous in my pursuit of heterosexuality. God would give me a husband, the counselor said.

My church counselor explained that my same-sex attraction was because of either sexual abuse, a flaw in my upbringing, or a deficient parent-child relationship. But try as I might, I could never pinpoint what "made me gay." I have wonderful, loving parents (married for 45 years now), four brothers and a sister (all heterosexual), and no history of abuse. In the meantime, I prayed, fasted (sometimes for days), and literally begged God to give me an attraction to men.

For me, the journey of self-acceptance began after almost 15 years of celibacy and intense personal struggle. After years of trying to suppress my sexual orientation, I could not shake a crush on a fellow female congressional staffer. So I decided to seek the help of a mental health professional. It didn't take long for me to realize that I am a lesbian, and happily so. My therapist said that the church did no more than brainwash me, and I now know that "reparative therapy"—which purports to be able to change one's sexual orientation from homosexual to heterosexual—is a lie.

My story is not unique. The reality is that societal prejudice and religious intolerance can drive people to take drastic actions, often with devastating consequences. In my work at the Human Rights Campaign, I have heard horror stories about parents who coerce their children to try to change. One 21-year-old man struggling with his orientation and his church's "gay deprogramming" recently wrote me. He was raised in a strict religious family who, upon learning of his orientation, threatened to disown him and have him excommunicated unless he changed. He went into the program voluntarily because he couldn't bear to lose his family, but other young men, he said, were forced against their will after being kidnapped. They strapped him to a chair with electrodes and sensors and showed him pictures of nude men, shocking him when he became aroused, he said. They continued this until he didn't respond. He finally fled the program after being sexually abused by a male orderly. The experience left him traumatized with incredible feelings of self-hatred, fear and thoughts of suicide. He knows that he is still gay.

I hope that the American people will see that these groups who call gay people sinners, telling us we can and should change when we cannot, are simply masking their message of prejudice in religious terms for political and monetary gain. It is unconscionable to me that people use God, religion or "Christian love" as an excuse for name-calling and discrimination. It is simply wrong. As for me, I have never felt as complete or peaceful as I have since coming out. And, I have never felt closer to God.

—*Tracey St. Pierre*

Source: Human Rights Campaign. 2000. *Finally Free: Personal Stories: How Love and Self-Acceptance Saved Us from "Ex-Gay" Ministries*. Human Rights Campaign Foundation. 919 18th St. N.W. Suite 800. Washington, D.C. 20006. Used by permission.

The cure for unhappiness is not the 'ex-gay' ministries—but coming out with dignity and self-respect. It is not gay men and lesbians who need to change . . . but negative attitudes and discrimination against gay people that need to be abolished. (Besen 2000, 7)

The American Psychiatric Association, American Psychological Association, American Academy of Pediatrics, and the American Medical Association agree that sexual orientation cannot be changed, and that efforts to change sexual orientation do not work and may, in fact, be harmful (Human Rights Campaign 2000a). According to the American Psychological Association, close scrutiny of reports of "successful" reparative therapy reveal that (1) many claims come from organizations with an ideological perspective on sexual orientation, rather than from unbiased researchers; (2) the treatments and their outcomes are poorly documented; and (3) the length of time that clients are followed after treatment is too short (Human Rights Campaign 2000a). In addition, at least 13 ministries of one reparative therapy group, Exodus, have closed because their directors reverted to homosexuality (Fone 2000).

Sociological Theories of Sexual Orientation

Although sociological theories do not explain the origin or "cause" of sexual orientation diversity, they help explain societal reactions to homosexuality and bisexuality. In addition, the symbolic interaction perspective sheds light on the process of adopting a gay, lesbian, or bisexual identity.

Structural-Functionalist Perspective

Structural-functionalists, consistent with their emphasis on institutions and the functions they fulfill, emphasize the importance of monogamous heterosexual relationships for the reproduction, nurturance, and socialization of children. From a functionalist perspective, homosexual relations, as well as heterosexual nonmarital relations, are defined as "deviant" because they do not fulfill the family institution's main function of producing and rearing children. Clearly, however, this argument is less salient in a society in which (1) other institutions, most notably schools, have supplemented the traditional functions of the family; (2) reducing (rather than increasing) population is a societal goal; and (3) same-sex couples can and do raise children.

Some functionalists argue that antagonisms between heterosexuals and homosexuals may disrupt the natural state, or equilibrium, of society. Durkheim, however, recognized that deviation from society's norms may also be functional. As Durkheim observed, deviation ". . . may be useful as a prelude to reforms which daily become more necessary" (Durkheim [1938] 1993, 66). Specifically, the gay rights movement has motivated many people to reexamine their treatment of sexual orientation minorities and has produced a sense of cohesion and solidarity among members of the gay population (although bisexuals have often been excluded from gay and lesbian communities and organizations). Gay activism has been instrumental in advocating for more research on HIV and AIDS, more and better health services for HIV and AIDS patients, protection of the rights of HIV-infected individuals, and HIV/AIDS public education. Such HIV/AIDS prevention strategies and health services benefit the society as a whole.

Finally, the structural-functionalist perspective is concerned with how changes in one part of society affect other aspects. With this focus on the interconnectedness of society, we note that urbanization has contributed to the formation of strong social networks of gays and bisexuals. Cities "acted as magnets, drawing in gay migrants who felt isolated and threatened in smaller towns and rural areas" (Button, Rienzo, & Wald 1997, 15). Given the formation of gay communities in large cities, it is not surprising that the gay rights movement first emerged in large urban centers.

Other research has demonstrated that the worldwide rise in liberalized national policies on same-sex relations and the lesbian and gay rights social movement has been influenced by three cultural changes: the rise of individualism, increasing gender equality, and the emergence of a global society in which nations are influenced by international pressures (Frank & McEneaney 1999). Individualism "appears to loosen the tie between sex and procreation, allowing more personal modes of sexual expression" (p. 930).

> Whereas once sex was approved strictly for the purpose of family reproduction, sex increasingly serves to pleasure individualized men and women in society. This shift has involved the casting off of many traditional regulations on sexual behavior, including prohibitions of male-male and female-female sex. (Frank & McEneaney 2000, 936)

Gender equality involves the breakdown of sharply differentiated sex roles, thereby supporting the varied expressions of male and female sexuality. Globalization permits the international community to influence individual nations. For example, when Zimbabwe president Robert Mugabe pursued antihomosexual policies in 1995, 70 members of the U.S. Congress signed a letter asking him to halt his antihomosexual campaign. Many international organizations and human rights associations joined the protest. The pressure of international opinion led Zimbabwe's Supreme Court to rule in favor of lesbian and gay groups' right to organize.

Conflict Perspective

Conflict theorists, particularly those who do not emphasize a purely economic perspective, note that the antagonisms between heterosexuals and nonheterosexuals represent a basic division in society between those with power and those without power. When one group has control of society's institutions and resources, as in the case of heterosexuals, they have the authority to dominate other groups. The recent battle over gay rights is just one example of the political struggle between those with power and those without it.

A classic example of the power struggle between gays and straights took place in 1973 when the American Psychiatric Association (APA) met to revise its classification scheme of mental disorders. Homosexual activists had been appealing to the APA for years to remove homosexuality from its list of mental illnesses but with little success. The view of homosexuals as mentally sick contributed to their low social prestige in the eyes of the heterosexual majority. In 1973, the APA's board of directors voted to remove homosexuality from its official list of mental disorders. The board's move encountered a great deal of resistance from conservative APA members and was put to a referendum, which reaffirmed the board's decision (Bayer 1987).

More currently, gays and lesbians are waging a political battle to win civil rights protections in the form of laws prohibiting discrimination on the basis of sexual orientation (discussed later in this chapter). Conflict theory helps to explain why many business owners and corporate leaders oppose civil rights protection for gays and lesbians. Employers fear that such protection would result in costly lawsuits if they refused to hire homosexuals, regardless of the reason for their decision. Business owners also fear that granting civil rights protections to homosexual employees would undermine the economic health of a community by discouraging the development of new businesses and even driving out some established firms (Button et al. 1997).

However, some companies are recognizing that implementing antidiscrimination policies that include sexual orientation is good for the "bottom line." Over half (51 percent) of Fortune 500 companies and 82 percent of Fortune 100 companies have included sexual orientation in their nondiscrimination policies and employers are increasingly offering benefits to domestic partners of LGBT employees (Human Rights Campaign 2000b). Gay-friendly work policies help employers maintain a competitive edge in recruiting and maintaining a talented and productive work force.

> We want to attract the brightest talent that values and embraces diversity in the workplace. The competition for talent is tough out there and we view this move as an important step to keep attracting people.
>
> **KATHLEEN OSWALD**
> *Senior vice-president for human resources at Daimler-Chrysler commenting on the automaker's decision to extend health coverage to partners of lesbian and gay employees.*

In summary, conflict theory frames the gay rights movement and the opposition to it as a struggle over power, prestige, and economic resources. Recent trends toward increased social acceptance of homosexuality may, in part, reflect the corporate world's competition over the gay and lesbian consumer dollar.

Symbolic Interactionist Perspective

I've made the choice that my personal liberty and the emancipation that I felt about coming out was definitely more important than my career at that point.

kd LANG
Music artist

Symbolic interactionism focuses on the meanings of heterosexuality, homosexuality, and bisexuality and how these meanings are socially constructed. The meanings we associate with same-sex relations are learned from society—from family, peers, religion, and the media. Freedman and D'Emilio (1990) observed that ". . . sexual meanings are subject to the forces of culture. Human beings learn how to express themselves sexually, and the content and outcome of that learning vary widely across cultures and across time" (p. 485).

Historical and cross-cultural research on homosexuality reveals the socially constructed nature of homosexuality and its meaning. Although many Americans assume that same-sex romantic relationships have always been taboo in our society, during the nineteenth century, "romantic friendships" between women were encouraged and regarded as preparation for a successful marriage. The nature of these friendships bordered on lesbianism. President Grover Cleveland's sister Rose wrote to her friend Evangeline Whipple in 1890: ". . . It makes me heavy with emotion . . . all my whole being leans out to you . . . I dare not think of your arms" (Goode & Wagner 1993, 49).

The symbolic interactionist perspective also points to the effects of labeling on individuals. Once individuals become identified or labeled as lesbian, gay, or bisexual, that label tends to become their **master status.** In other words, the dominant heterosexual community tends to view "gay," "lesbian," and "bisexual" as the most socially significant statuses of individuals who are identified as such. Esterberg (1997) notes that "unlike heterosexuals, who are defined by their family structures, communities, occupations, or other aspects of their lives, lesbians, gay men, and bisexuals are often defined primarily by what they do in bed. Many lesbians, gay men, and bisexuals, however, view their identity as social and political as well as sexual" (p. 377).

Heterosexism, Homophobia, and Biphobia

Homophobia alienates mothers and fathers from sons and daughters, friend from friend, neighbor from neighbor, Americans from one another. So long as it is legitimated by society, religion, and politics, homophobia will spawn hatred, contempt, and violence, and it will remain our last acceptable prejudice.

BYRNE FONE

The United States, along with many other countries throughout the world, is predominantly heterosexist. **Heterosexism** refers to "the institutional and societal reinforcement of heterosexuality as the privileged and powerful norm" (SIECUS 2000). Heterosexism is based on the belief that heterosexuality is superior to homosexuality and results in prejudice and discrimination against homosexuals and bisexuals. Prejudice refers to negative attitudes, whereas discrimination refers to behavior that denies individuals or groups equality of treatment. Before reading further, you may wish to complete this chapter's *Self and Society* feature "The Homophobia Scale."

Homophobia

SIECUS, the Sex Information and Education Council of the United States, "strongly supports the right of each individual to accept, acknowledge, and live in accordance with his or her orientation . . . and deplores all forms of prejudice

The Homophobia Scale

Directions: Indicate the extent to which you agree or disagree with each statement by placing a check mark on the appropriate line.

	Strongly Agree	Agree	Undecided	Disagree	Strongly Disagree
1. Homosexuals contribute positively to society.	_____	_____	_____	_____	_____
2. Homosexuality is disgusting.	_____	_____	_____	_____	_____
3. Homosexuals are just as moral as heterosexuals.	_____	_____	_____	_____	_____
4. Homosexuals should have equal civil rights.	_____	_____	_____	_____	_____
5. Homosexuals corrupt young people.	_____	_____	_____	_____	_____
6. Homosexuality is a sin.	_____	_____	_____	_____	_____
7. Homosexuality should be against the law.	_____	_____	_____	_____	_____

SCORING

Assign scores of 0, 1, 2, 3, and 4 to the five choices respectively ("strongly agree" through "strongly disagree"). Reverse-score items 2, 5, 6, and 7 (0 = 4; 1 = 3; 2 = 2; 3 = 1; 4 = 0). All items are summed for the total score. The possible range is 0 to 28; high scores indicate greater homophobia.

COMPARISON DATA

The Homophobia Scale was administered to 524 students enrolled in introductory psychology courses at the University of Texas. The mean score for men was 15.8; for women it was 13.8. The difference was statistically significant.

Source: From Richard A. Bouton, P.E. Gallagher, P.A. Garlinghouse, T. Leal, et al. 1987. "Scales for Measuring Fear of AIDS and Homophobia." *Journal of Personality Assessment* 67(1):609. Copyright © 1987 by Lawrence Erlbaum Associates, Inc. Reprinted by permission.

---

and discrimination against people based on sexual orientation" (SIECUS 2000, 1). Nevertheless, negative attitudes toward homosexuality are reflected in the high percentage of the U.S. population who disapprove of homosexuality. In 1999, nearly half (46 percent) of respondents in a Gallup Poll survey felt that homosexuality should not be considered an acceptable alternative lifestyle (Gallup Organization 2000). It is important to note, however, that the number of people who disapprove has decreased in recent years. Whereas roughly half of Americans disapproved of homosexuality in 2000, three-quarters disapproved in the late 1980s (Yang 1999).

The term **homophobia** is commonly used to refer to negative attitudes and emotions toward homosexuality and those who engage in it. Homophobia is not necessarily a clinical phobia (that is, one involving a compelling desire to avoid the feared object in spite of recognizing that the fear is unreasonable). Other terms that refer to negative attitudes and emotions toward homosexuality include "homonegativity" and "antigay bias."

In general, certain categories of people are more likely to have negative attitudes toward homosexuals. Persons who are older, less educated, living in the South or Midwest, living in lightly populated rural areas, and Protestant are the most likely to have negative attitudes. In contrast, people who are younger, more educated, never married, living in the West, living in heavily populated urban areas, and Jewish are least likely to have antigay attitudes (Klassen, Williams, & Levitt 1989). Also, positive contact with homosexuals or having homosexuals as friends is associated with less homophobia (Simon 1995). Public opinion surveys also indicate that men are more likely than women to have negative attitudes toward gays (Moore 1993). But many studies on attitudes toward homosexuality do not distinguish between attitudes toward gay men and attitudes toward lesbians (Kite & Whitley 1996). Research that has assessed attitudes toward male versus female homosexuality has found that heterosexual women and men hold similar attitudes toward lesbians, but men are more negative toward gay men (Louderback & Whitley 1997; Price & Dalecki 1998).

A study of undergraduates (110 men, 98 women) attending a Canadian university found that attitudes toward gay men were more negative than toward lesbians (Schellenberg, Hirt, & Sears 1999). When compared with science or business majors, students in the arts and social sciences had more positive attitudes toward gay men, and women were more positive than men.

Cultural Origins of Homophobia

Why do many Americans disapprove of homosexuality? Antigay bias has its roots in various aspects of U.S. culture.

1. *Religion.* Most Americans who view homosexuality as unacceptable say they object on religious grounds (Rosin & Morin 1999). Many religious leaders teach that homosexuality is sinful and prohibited by God. The Roman Catholic Church rejects all homosexual expression and resists any attempt to validate or sanction the homosexual orientation. Some fundamentalist churches have endorsed the death penalty for homosexual people and teach the view that AIDS is God's punishment for engaging in homosexual

In January 1999, 95 United Methodist Church ministers violated the Church's ban on "ceremonies that celebrate homosexual unions" as they officiated at a same-sex union church service in Sacramento, California.

© AP/Wide World Photos

sex (Nugent & Gramick 1989). The Westboro Baptist Church (Topeka, Kansas), headed by the antigay Rev. Fred Phelps, maintains a Web site called godhatesfags.com. Members of this church have held antigay demonstrations near the funerals of people who have died from AIDS carrying signs reading, "Gays Deserve to Die."

Some religious groups, such as the Quakers, are accepting of homosexuality and other groups have made reforms toward increased acceptance of lesbians and gays. Some Episcopal priests perform "ceremonies of union" between same-sex couples; some Reform Jewish groups sponsor gay synagogues; and the United Church of Christ allows homosexuals to be ordained (Fone 2000). Although the official position of the United Methodist Church is one that condemns homosexuality, some Methodist ministers advocate acceptance of and equal rights for lesbians and gay men. The United Methodist Church has considered splitting into two organizations: one in favor of equal rights for gays and lesbians and the other opposed (The United Methodist Church and Homosexuality 1999).

2. *Marital and procreative bias.* Many societies have traditionally condoned sex only when it occurs in a marital context that provides for the possibility of producing and rearing children. Although Vermont has granted legal recognition of same-sex "civil unions" (which we discuss later in this chapter), as of this writing, same-sex marriages are not legally recognized in the United States. Further, even though assisted reproductive technologies make it possible for gay individuals and couples to have children, many people believe that these advances should only be used by heterosexual married couples (Franklin 1993).

3. *Concern about HIV and AIDS.* Although most cases of HIV and AIDS worldwide are attributed to heterosexual transmission, the rates of HIV and AIDS in the United States are much higher among gay and bisexual men than among other groups. Because of this, many people associate HIV and AIDS with homosexuality and bisexuality. This association between AIDS and homosexuality has fueled antigay sentiments. During a 1985 mayoral campaign in Houston, the challenging candidate was asked what he would do about the AIDS problem. His answer: he would "shoot the queers" (Button et al. 1997, 70). Lesbians, incidentally, have a very low risk for sexually transmitted HIV—a lower risk than heterosexual women.

4. *Threat to the power of the majority.* Like other minority groups, the gay minority threatens the power of the majority. Fearing loss of power, the majority group stigmatizes homosexuals as a way of limiting their power.

5. *Rigid gender roles.* Antigay sentiments also stem from rigid gender roles. When Cooper Thompson (1995) was asked to give a guest presentation on male roles at a suburban high school, male students told him that the most humiliating put-down was being called a "fag." The boys in this school gave Thompson the impression they were expected to conform to rigid, narrow standards of masculinity in order to avoid being labeled in this way.

From a conflict perspective, heterosexual men's subordination and devaluation of gay men reinforces gender inequality. "By devaluing gay men . . . heterosexual men devalue the feminine and anything associated with it" (Price & Dalecki 1998, 155–56). Negative views toward lesbians also reinforce the patriarchal system of male dominance. Social disapproval of lesbians is a form of punishment for women who relinquish traditional female sexual and economic dependence on men. Not surprisingly, research findings suggest that individuals with traditional gender role attitudes tend to hold more negative views toward homosexuality (Louderback & Whitley 1997).

6. *Psychiatric labeling.* As noted earlier, before 1973 the APA defined homosexuality as a mental disorder. Treatments for the "illness" of homosexuality included lobotomies, aversive conditioning, and, in some cases, castration. The social label of mental illness by such a powerful labeling group as the APA contributed to heterosexuals' negative reactions to gays. Further, it created feelings of guilt, low self-esteem, anger, and depression for many homosexuals. Thus, the psychiatric care system is now busily treating the very conditions it, in part, created.

7. *Myths and negative stereotypes.* Prejudice toward homosexuals may also stem from some of the unsupported beliefs and negative stereotypes regarding homosexuality. One negative myth about homosexuals is that they are sexually promiscuous and lack "family values" such as monogamy and commitment to relationships. Although some homosexuals do engage in casual sex, as do some heterosexuals, many homosexual couples develop and maintain long-term committed relationships.

Another myth that is not supported by data is that homosexuals, as a group, are child molesters. However, 95 percent of all reported incidents of child sexual abuse are committed by heterosexual men (SIECUS 2000). Most often the abuser is a father, stepfather, or heterosexual relative of the family. When a father rapes his daughter, the media do not report that something is wrong with heterosexuality or with traditional families. But when a homosexual child molestation is reported, it is viewed as confirmation of "the way homosexuals are" (Mohr 1995, 404).

Excerpts from "The Heterosexual Questionnaire" in Table 9.2 parody the different attitudes many Americans have toward heterosexuals and homosexuals in our society.

■ **Table 9.2** *Excerpts from "The Heterosexual Questionnaire"*

1. What do you think caused your heterosexuality?
2. When and how did you decide you were a heterosexual?
3. Is it possible that your heterosexuality is just a phase you may grow out of?
4. Is it possible that your heterosexuality stems from a neurotic fear of others of the same sex?
5. If you have never slept with a person of the same sex, is it possible that all you need is a good gay lover?
6. Do your parents know that you are straight? Do your friends and/or roommate(s) know? How did they react?
7. Why do heterosexuals place so much emphasis on sex?
8. A disproportionate majority of child molesters are heterosexual. Do you consider it safe to expose children to heterosexual teachers?
9. Just what do men and women do in bed together? How can they truly know how to please each other, being so anatomically different?
10. There seem to be very few happy heterosexuals. Techniques have been developed that might enable you to change if you really want to. Have you considered trying aversion therapy?

Source: Rochlin, M. 1982. "The Heterosexual Questionnaire." *Changing Men* (Spring).

Biphobia

Just as the term "homophobia" is used to refer to negative attitudes toward gay men and lesbians, **biphobia** refers to "the parallel set of negative beliefs about and stigmatization of bisexuality and those identified as bisexual" (Paul 1996, 449). Although both homosexual- and bisexual-identified individuals are often rejected by heterosexuals, bisexual-identified women and men also face rejection from many homosexual individuals. Thus, bisexuals experience "double discrimination."

Biphobia includes negative stereotyping of bisexuals; the exclusion of bisexuals from social and political organizations of lesbians and gay men; and fear and distrust of, as well as anger and hostility toward, people who identify themselves as bisexual (Firestein 1996). Individuals who are biphobic often believe that bisexuals are really homosexuals afraid to acknowledge their real identity or homosexuals maintaining heterosexual relationships to avoid rejection by the heterosexual mainstream. Bisexual individuals are sometimes viewed as heterosexuals who are looking for exotic sexual experiences.

Lesbians seem to exhibit greater levels of biphobia than do gay men. This is because many lesbian women associate their identity with a political stance against sexism and patriarchy. Some lesbians view heterosexual and bisexual women who "sleep with the enemy" as traitors to the feminist movement.

One negative stereotype that encourages biphobia is the belief that bisexuals are, by definition, nonmonogamous. However, many bisexual women and men prefer and have long-term committed monogamous relationships.

Effects of Homophobia and Heterosexism on Heterosexuals

The homophobic and heterosexist social climate of our society is often viewed in terms of how it victimizes the gay population. However, heterosexuals are also victimized by homophobia and heterosexism. "Hatred, fear, and ignorance are bad for the bigot as well as the victim" (*Homophobia 101*, 2000).

Because of the antigay climate, heterosexuals, especially males, are hindered in their own self-expression and intimacy in same-sex relationships. "The threat of victimization (i.e., antigay violence) probably also causes many heterosexuals to conform to gender roles and to restrict their expressions of (nonsexual) physical affection for members of their own sex" (Garnets, Herek, & Levy 1990, 380). Homophobic epithets frighten youth who do not conform to gender role expectations, leading some youth to avoid activities that they might otherwise enjoy and benefit from (arts for boys, athletics for girls, for example) (*Homophobia 101*, 2000).

Some cases of rape and sexual assault are related to homophobia and compulsory heterosexuality. For example, college men who participate in gang rape, also known as "pulling train," entice each other into the act "by implying that those who do not participate are unmanly or homosexual" (Sanday 1995, 399). Homonegativity also encourages early sexual activity among adolescent men. Adolescent male virgins are often teased by their male peers, who say things like "You mean you don't do it with girls yet? What are you, a fag or something?" Not wanting to be labeled and stigmatized as a "fag," some adolescent boys "prove" their heterosexuality by having sex with girls.

Antigay cultural attitudes also affect family members and friends of homosexuals, who often fear that their lesbian or gay friend or family member will be

> Injustice anywhere is a threat to justice everywhere. We are caught in an inescapable network of mutuality, tied in a single garment of destiny. Whatever affects one directly affects all indirectly.
>
> MARTIN LUTHER KING, JR.

victimized by antigay prejudice and discrimination. Youth with gay and lesbian family members are often taunted by their peers.

As we discuss later in this chapter, extreme homophobia contributes to instances of physical violence against homosexuals—acts known as hate crimes. But hate crimes are crimes of perception, meaning that victims of antigay hate crimes may not be homosexual; they may just be perceived as being homosexual. Many heterosexuals have been victims of antigay physical violence because the attacker(s) perceived the victim to be gay.

Antigay harassment has also been a factor in many of the school shootings in recent years. In March 2001, 15-year-old Charles Andrew Williams fired more than 30 rounds in a San Diego suburban high school, killing two and injuring 13 others. A woman who knew Williams reported that the students had teased Williams, and called him gay (Dozetos 2001). According to the Gay, Lesbian and Straight Education Network (GLSEN), Williams' story is not unusual. Referring to a study of harassment of U.S. students commissioned by the American Association of University Women, a GLSEN report concluded, "For boys, no other type of harassment provoked as strong a reaction on average; boys in this study would be less upset about physical abuse than they would be if someone called them gay" (Dozetos 2001).

Discrimination against Sexual Orientation Minorities

The Constitution does not allow government to subordinate a class of persons simply because others do not like them.

CHIEF JUDGE ABNER MIKAVA

Like other minority groups in American society, homosexuals and bisexuals experience various forms of discrimination. Next, we look at sexual orientation discrimination in the workplace, in family matters, in the application and enforcement of sodomy laws, and in violent expressions of hate. This chapter's *Focus on Technology* feature discusses the discriminatory effects of Internet filtering and monitoring technology on sexual orientation minorities.

Discrimination in the Workplace

The majority of Americans agree that homosexuals should have equal rights in the workplace. Gallup polls reveal that the percentage of U.S. adults who believe that homosexuals should have equal rights in terms of job opportunities has increased from 56 percent in 1977 to 83 percent in 1999 (Gallup Organization 2000). However, when asked if homosexuals should or should not be hired for various specific occupations, Americans reveal biases against homosexuals. For example, although 90 percent of a national sample of U.S. adults agree that homosexuals should be hired in sales, only 54 percent believe that homosexuals should be hired in the clergy or as elementary school teachers (Gallup Organization 2000).

I know now that I had more than one teacher who happened to be gay. I also know I wouldn't be a member of Congress today had it not been for those great teachers.

REPRESENTATIVE CHRISTOPHER SHAYS
R–Conn.

As of this writing, it is legal in 38 states to fire, decline to hire or promote, or otherwise discriminate against an employee because of his or her sexual orientation. Between 1994 and 2000, the Human Rights Campaign has documented more than 800 cases of sexual orientation discrimination in U.S. workplaces (Human Rights Campaign 2000b). The Cracker Barrel restaurant chain fired at least 11 openly gay and lesbian employees in 1991. Cracker Barrel stated that it would refuse employment to people "whose sexual preferences fail to demonstrate normal heterosexual values which have been the foundation of families

The Impact of Internet Filtering and Monitoring on Sexual Orientation Minorities

Joan Garry is the Executive Director of the Gay & Lesbian Alliance Against Defamation (GLAAD) and a mother of three children. When her 10-year-old daughter Sarah approached her for information about COLAGE (Children of Lesbians and Gays Everywhere), a support organization for the children of lesbian and gay parents, they signed on to America Online to look up the web site. The COLAGE site was "Web Restricted": Sarah could not access the site because her computer was equipped with AOL's filtering software called Kids Only. In fact, when Joan tried to look up various family, youth, and national organization Web sites with lesbian and gay content, most sites came up "Web Restricted" as well (Garry 1999). For example, the Kids Only software blocked access to several youth-oriented gay and lesbian resource sites, including PFLAG (Parents, Family and Friends of Lesbians and Gays), Family Pride, !OutProud!, GLSEN (Gay, Lesbian & Straight Education Network), and Oasis Magazine, a gay and lesbian youth Webzine (Javier 1999).

An explosion of filtering technologies to help maintain "decency" and "community standards" on the Internet has occurred. Examples of filtering technologies include those that block key words or URLs. For example, the search engine Jayde.com, which marketed itself as being able to filter out pornographic material, decided to block the word "lesbian" in its search engine. The filtering software "CYBERsitter" automatically filters out words and phrases like "gay," "lesbian," "gay rights," and "gay community" (Javier 1999). Another filtering software called "Bess" is used in libraries and schools across the country. Filtering technologies are promoted as tools to help parents, schools, libraries, and communities prevent children's access to sexually explicit and pornographic material on the Internet. However, some filtering software also denies users access to several lesbian, gay, and bisexual youth resource sites.

Schneider (1999) presents the following scenario:

Imagine a teenager seeking information about his or her sexual orientation whose attempts to go to gay-related Web sites are met with a warning that tells him or her "Bess doesn't want you to go there." In an unfiltered environment . . . the teen could be expected to privately retrieve all kinds of information in a discrete, nonjudgmental environment. But place a filter

on that computer, and . . . the teen gets the message that he or she is somehow . . . "inappropriate". (p. 13)

First Amendment challenges to Internet filtering have become increasingly common. In Mainstream Loudoun v. Loudoun County Library, the use of X-STOP filtering software in the Loudoun County (Va.) public library system was challenged. In November 1998, a U.S. District Judge found that the mandatory use of the software by all library patrons was unconstitutional. The library now offers filtered and unfiltered Internet access (Schneider 1999). Nevertheless, efforts to legislate mandatory filtering in public libraries and schools have not slowed. The Children's Internet Protection Act (CHIPA), passed by Congress in 2000, requires libraries that participate in certain federal programs to install filtering technology on all computer terminals with Internet access. The purpose of the law is to protect children from "indecent" online content. In March 2001, the American Civil Liberties Union filed a lawsuit against the federal government in an attempt to overturn CHIPA (Dozetos 2001). As of this writing the case is pending.

Given the impact of Internet filtering on the gay community, the Gay & Lesbian Alliance Against Defamation does not support the use of filtering software. Instead, they advocate parental oversight, school supervision, and training of young Internet users (Bowes 1999).

Another concern of sexual orientation minorities is the use of monitoring software that allows parents, teachers, and other authority figures to track sites a Web surfer tries to access. The monitoring software industry markets its products by claiming that they allow for parental awareness without censorship, and that parental use of monitoring software encourages open family communication. But for youth who are not ready to reveal their sexual orientation to their family, "such software could potentially 'out' them before they are ready, leading to strained family relations and deeper isolation" (Javier 1999, 8).

The best-known story of a gay person affected by an "online outing" is that of a naval officer who was outed in 1998 after an America Online employee divulged his personal data and online history to the Navy. This invasion of privacy cost this officer his career.

Internet service providers may record every click of your mouse, including personal information you exchange in chat rooms, emails, and instant messages.

Say you visit a Web magazine: the time and date of your visit are recorded. You conduct a search for all

Continued

articles containing the word "gay." Your inquiry is recorded. You click on an article about new AIDS drugs. Your request for that article is recorded. While reading the article, you click on an ad about online dating services, and that choice is recorded . . . .You decide to enter a gay-oriented chat-room and see if you can make any new friends, or even get a date. You chat with a number of people, and finally connect with one special person, with whom you share instant messages, pictures, and perhaps even a virtual date. Depending on . . . the Internet provider in question, that date could have just been recorded. (Aravosis 1999, 30)

Web sites also record information about your Internet usage. When you visit a free news Web site, you may be required to "subscribe" by giving them your name and email address. After subscribing, you read a few articles. But unbeknownst to you, the news Web site has put a "cookie," or piece of computer code, on the hard drive of your computer. This cookie contains a unique identifier permitting the news site to recognize you when you return to that site. Cookies can do helpful things like remember your password for accessing that site, and tailoring Web pages for preset preferences. But they can also contain a record of what you did on that site, including what searches you made and what articles you read (Aravosis 1999). According to the Georgetown Internet Privacy Policy Study, of the 7,500 busiest servers on the Web, 93 percent of the sites collect personal information from consumers, yet only 66 percent post any disclosure about their information practices (Bowes 1999).

Although some web sites have strict privacy policies, others sell information about site visitors to corporate America. One company boasts on its Web site that it is "the world's oldest and largest mailing list manager and broker for Gay, Lesbian, and HIV-related names, currently managing almost two million names, which we estimate to be about 65 percent of all those commercially available in this segment" (Aravosis 1999, 33). Databases containing the names of gay consumers are valuable because gays are perceived as a "wealthy and wired market." A 1998 study found that the average household income of lesbians and gay men on the Internet was $57,300, slightly higher than the $52,000 for the general Internet population. Another study by Computer Economics predicts that within the next few years, the

worldwide gay and lesbian Internet population will increase from 9.2 million to 17.1 million (Aravosis 1999).

The Internet has been a useful tool for gays, lesbians, and bisexuals. Going online has allowed the gay community to create safe places for support and information. However, Internet filtering and monitoring software that has been installed on computers in homes, schools, libraries, and workplaces throughout the country, represent a threat to the gay community. Filtering and monitoring technologies make it impossible or dangerous for a closeted gay or lesbian to seek out support and information about their community.

We live in an age in which the Internet has become an extremely important part of the coming out process for many gay and lesbian youth. In many cases, it can be a lifeline to those in geographically isolated areas. To deny basic educational and support resources to lesbian and gay youth could seriously endanger their physical and emotional well being. (Appendix A 1999, 46)

Sources:

Appendix A. 1999. "Frequently Asked Questions." In *Access Denied Version 2.0, The Continuing Threat Against Internet Access and Privacy and its Impact on the Lesbian, Gay, Bisexual and Transgender Community.* New York: Gay & Lesbian Alliance Against Defamation, pp. 45–48.

Aravosis, John. 1999. "Privacy: The Impact on Lesbian, Gay, Bisexual and Transgender Community. In *Access Denied Version 2.0, The Continuing Threat Against Internet Access and Privacy and its Impact on the Lesbian, Gay, Bisexual and Trangender Community.* New York: Gay & Lesbian Alliance Against Defamation, pp. 30–33.

Bowes, John. 1999. "Conclusions." In *Access Denied Version 2.0, The Continuing Threat Against Internet Access and Privacy and its Impact on the Lesbian, Gay, Bisexual and Trangender Community.* New York: Gay & Lesbian Alliance Against Defamation, pp. 38–44.

Dozetos, Barbara. 2001 (March 20). "ACLU Sues Over Censorship of GLBT Sites." PlanetOut.com. http://www.planetout.com/news/article-print.html?2001/03/20/2

Garry, Joan M. 1999. "Introduction: How Access and Privacy Impact the Lesbian, Gay, Bisexual and Transgender Community." In *Access Denied Version 2.0, The Continuing Threat Against Internet Access and Privacy and its Impact on the Lesbian, Gay, Bisexual and Trangender Community.* New York: Gay & Lesbian Alliance Against Defamation, pp. 3–5.

Javier, Loren. 1999. "The World Since Access Denied." In *Access Denied Version 2.0 The Continuing Threat Against Internet Access and Privacy and its Impact on the Lesbian, Gay, Bisexual and Trangender Community.* New York: Gay & Lesbian Alliance Against Defamation, pp. 6–9.

Schneider, Karen G. 1999. "Access: The Impact on the Lesbian, Gay, Bisexual and Transgender Community." In *Access Denied Version 2.0, The Continuing Threat Against Internet Access and Privacy and its Impact on the Lesbian, Gay, Bisexual and Trangender Community.* New York: Gay & Lesbian Alliance Against Defamation, pp. 10–15.

in our society" (cited in Button et al. 1997, 126). Another publicized example of employment discrimination based on sexual orientation involves the case of a woman who was offered a job as a lawyer in the Georgia attorney general's office but lost the job offer after the state attorney general learned that she was planning a commitment ceremony with her lesbian partner. The Supreme Court decided not to hear her appeal (Greenhouse 1998). In 1998, Oklahoma passed a measure to ban homosexuals from working in public schools ("1998 in Review" 1999).

Occupational discrimination also occurs in the military. In 1992, a gay Navy serviceman was fired after revealing his homosexuality but was reinstated by orders of a federal judge. In 1993, President Clinton instituted a "Don't ask, don't tell" policy in which recruiting officers are not allowed to ask about sexual orientation, and homosexuals are encouraged not to volunteer such information. Although the "Don't ask, don't tell" policy was intended to protect homosexuals from discrimination in the military, the number of service members discharged from the armed forces under the policy doubled between 1994 to 1998 (Human Rights Watch 2001). More than 5,400 service members were discharged under the policy from 1994 to 1999. Women were discharged at a disproportionately high rate. The "Don't ask, don't tell" policy also provided an additional means for servicemen to harass lesbian service members by threatening to "out" those who refused their advances or threatened to report them, thus ending their careers (Human Rights Watch 2001). Discrimination toward gays in the military is not surprising; in the largest survey ever conducted on military attitudes toward homosexuality, 80 percent of those questioned had heard offensive comments about gays within the previous year (Ricks 2000).

Americans are divided on the issue of gays serving in the military. A 2000 Gallup poll revealed that 41 percent of U.S. adults believe that gays should be able to serve openly in the military; 38 percent agree with the current "Don't ask, don't tell" policy; and 17 percent believe that gays should not serve in the military under any circumstances (Gallup Organization 2000).

Discrimination in Family Relationships

In addition to discrimination in the workplace, sexual orientation minorities experience discrimination in policies concerning marriage. States and judges have also used a person's sexual orientation to deny custody and visitation, adoption, and foster care.

Same-Sex Marriage   As of this writing, no U.S. state grants marriage licenses to same-sex couples. In December 1999, the Hawaii Supreme Court dropped a case that could have legalized same-sex marriage. Recognizing the possibility that states may one day grant marriage licenses to same-sex couples, opponents of same-sex marriage have prompted antigay marriage legislation. In 1996, Congress passed and President Clinton signed the **Defense of Marriage Act**, which states that marriage is a "legal union between one man and one woman" and denies federal recognition of same-sex marriage. In effect, this law allows states to either recognize or not recognize same-sex marriages performed in other states. As of January 2001, 35 states had antigay-marriage laws, declaring that they will not recognize same-sex marriages (National Gay and Lesbian Task Force 2001a).

The prejudice against gays and lesbians today is meaner, nastier—a vindictiveness that's rooted in hatred, not ignorance.

BETTY DEGENERES
Comedienne Ellen DeGeneres' Mom

Antigay-marriage laws are consistent with public opinion on the issue: a 2000 Gallup poll of U.S. adults found that 62 percent of Americans think that marriages between homosexuals should not be legally recognized; 34 percent said that gay marriages should be recognized by the law as valid (Gallup Organization 2000). Forty percent of a national sample of U.S. adults believe that gay partners who make a legal commitment to each other should be entitled to the same rights and benefits as couples in traditional marriages (Gallup Organization 2000).

Advocates of same-sex marriage argue that banning same-sex marriages, or refusing to recognize same-sex marriages granted in other states, denies same-sex couples the many legal and financial benefits that are granted to heterosexual married couples. For example, married couples have the right to inherit from a spouse who dies without a will; to avoid inheritance taxes between spouses; to make crucial medical decisions for a partner and take family leave in order to care for a partner in the event of the partner's critical injury or illness; to receive Social Security survivor benefits; and to include a partner in his or her health insurance coverage. Other rights bestowed to married (or once married) partners include assumption of spouse's pension, bereavement leave, burial determination, domestic violence protection, reduced rate memberships, divorce protections (such as equitable division of assets and visitation of partner's children), automatic housing lease transfer, and immunity from testifying against a spouse.

Hate is not a family value.

BUMPER STICKER

Whereas advocates of same-sex marriage argue that as long as same-sex couples cannot be legally married, they will not be regarded as legitimate families by the larger society, opponents do not want to legitimize homosexuality as a socially acceptable lifestyle. Opponents of same-sex marriage who view homosexuality as unnatural, sick, and/or immoral do not want their children to learn that homosexuality is an accepted "normal" lifestyle. The most common argument against same-sex marriage is that it subverts the stability and integrity of the heterosexual family. However, Sullivan (1997) suggests that homosexuals are already part of heterosexual families:

> [Homosexuals] are sons and daughters, brothers and sisters, even mothers and fathers, of heterosexuals. The distinction between "families" and "homosexuals" is, to begin with, empirically false; and the stability of existing families is closely linked to how homosexuals are treated within them. (p. 147)

Child Custody and Visitation Several respected national organizations—including the Child Welfare League of America, the American Psychological Association, the American Psychiatric Association, and the National Association of Social Workers—have taken the position that a parent's sexual orientation is irrelevant in determining child custody (Landis 1999). In a review of research on family relationships of lesbians and gay men, Patterson (2001) concluded that "the greater majority of children with lesbian or gay parents grow up to identify themselves as heterosexual" and that "concerns about possible difficulties in personal development among children of lesbian and gay parents have not been sustained by the results of research" (p. 279). Patterson (2001) notes that the "home environments provided by lesbian and gay parents are just as likely as those provided by heterosexual parents to enable psychosocial growth among family members" (p. 283). Nevertheless, some court judges are biased against lesbian and gay parents in custody and visitation disputes. For example, in 1999, the Mississippi Supreme Court denied custody of a teenage boy to his gay father and instead awarded custody to his heterosexual mother who remarried into a home "wracked with domestic violence and excessive drinking" ("Custody and Visitation" 2000, 1).

Adoption and Foster Care   In some states and counties, gay and lesbian people can adopt; in other states and counties, they cannot. Whether gay and lesbian individuals can adopt within a state may depend upon what county they live in, what judge they get, and whether they are seeking to adopt as an individual or as a couple. Gay and lesbian people have adopted in at least 22 states and the District of Columbia; however, Florida and Mississippi forbid adoption by gay and lesbian people and Utah forbids adoption by any unmarried couple, which includes all same-sex couples ("Adoption" 2000). In Arkansas, lesbians, gay men, and their partners are prohibited from serving as foster parents (ACLU Fact Sheet 1999). Most adoptions by gay people have been by individual gay men or lesbians who adopt the biological children of their partners. Such second-parent adoptions ensure that the children can enjoy the benefits of having two legal parents, especially if one parent dies or becomes incapacitated (ACLU Fact Sheet 1999). In early 1998, New Jersey became the first state to set a policy allowing same-sex (and unmarried heterosexual) couples to jointly adopt children under the same qualification standards as married couples.

Sodomy Laws

Sodomy refers to oral and anal sexual acts. Other terms for sodomy include "crimes against nature," "unnatural intercourse," "buggery," "sexual misconduct," and "lewd and lascivious acts." In 1986, the U.S. Supreme Court ruled in Bowers v. Hardwick that the Constitution allows states to criminalize sodomy, even between consenting adults. Sodomy laws were once on the books in all 50 U.S. states, but they have been repealed or struck down by courts in more than 30 states. Seventeen states still ban oral and anal sex between consenting adults. Penalties for engaging in sodomy range from a $200 fine to 20 years imprisonment.

In four states, sodomy laws target only same-sex acts (Elliot 2001). These states include Kansas, Missouri, Oklahoma, and Texas. A 1999 Gallup poll of U.S. adults found that 43 percent think that homosexual relations between consenting adults should not be legal (Gallup Organization 2000). In states that criminalize both same- and opposite-sex sodomy, sodomy laws are usually not used against heterosexuals but are primarily used against gay men and lesbians. For example, some courts have taken children away from lesbian and gay parents on the basis that these parents violate sodomy laws and are, therefore, lawbreakers.

Hate Crimes against Sexual Orientation Minorities

On October 6, 1998, Matthew Shepard, a 21-year-old student at the University of Wyoming, was abducted and brutally beaten. He was found tied to a wooden ranch fence by two motorcyclists who had initially thought that he was a scarecrow. His skull had been smashed, and his head and face had been slashed. The only apparent reason for the attack: Matthew Shepard was gay. On October 12, Matthew died of his injuries. Media coverage of his brutal attack and subse-

All of the research to date has reached the same unequivocal conclusion about gay parenting: the children of lesbian and gay parents grow up as successfully as the children of heterosexual parents.

OVERVIEW OF LESBIAN AND GAY PARENTING, ADOPTION AND FOSTER CARE
American Civil Liberties Union Fact Sheet

quent death focused nationwide attention on hate crimes against sexual orientation minorities.

Other incidents of brutal hate crimes against gays include the murder of Billy Jack Gaither in Alabama in February 1999. Two men who claimed to be angry over a sexual advance made by Gaither plotted his murder, beat Gaither to death with an ax handle and then burned his body on a pyre of old tires. In July 1999 at Fort Campbell, Kentucky, Private First Class Barry Winchell was beaten to death by another soldier with a baseball bat because Winchell was perceived to be homosexual.

In eighteenth-century America, where laws against homosexuality often included the death penalty, violence against gays and lesbians was widespread and included beatings, burnings, various kinds of torture, and execution (Button et al. 1997). Although such treatment of sexual orientation minorities is no longer legally condoned, gays, lesbians, and bisexuals continue to be victimized by hate crimes. Sexual orientation hate crimes are crimes against individuals or their property that are based on bias against the victim because of his or her perceived sexual orientation. Such crimes include verbal threats and intimidation, vandalism, sexual assault and rape, physical assault, and murder.

According to the FBI Uniform Crime Reports, in 1999 1,317 incidents of sexual orientation hate crimes were reported. The National Coalition of Anti-violence Programs (2000) reported 1,960 incidents of antigay hate crimes in 1999. The percentage of hate crimes based on sexual orientation increased from 8.9 percent in 1991 to 16.7 percent in 1999 (Human Rights Campaign 2000c). In 1999, 29 anti-LGBT murders were reported to the National Coalition of Anti-Violence Programs (2000).

Antigay Hate in Schools

America's schools are not safe places for gay, lesbian, and bisexual youth. More than two-thirds (69 percent) of gay and lesbian students have been verbally, physically, or sexually harassed at school (Chase 2000). A survey of more than 3,000 Massachusetts high school students found that students who reported having engaged in same-sex relations were more than three times as likely to report not going to school because they felt unsafe and more than twice as likely to report having been threatened or injured with a weapon at school (Faulkner & Cranston 1998). These students were also significantly more likely to report that their property was deliberately damaged or stolen at school.

Students who are victims of antigay hate often feel that they have no one in the school to go to for help. In a survey of 496 LGBT youth ages 12 to 19 from 32 states, over one-third (37 percent) reported hearing homophobic remarks from faculty or school staff ("GLSEN's National School Climate Survey" 1999). Nearly half (2 out of 5) of the youth did not feel safe in school. Tragically, over one-fourth of gay youth drop out of school—usually to escape the harassment, violence, and alienation they endure there (Chase 2000).

Antigay hate is also common among college students. In a survey of 484 young adults at six community colleges in California, 10 percent reported physically assaulting or threatening people whom they believed to be homosexual and 24 percent reported calling homosexuals insulting names (Franklin 2000). The researcher concluded that overall, findings suggest that many young adults believe that antigay harassment and violence is socially acceptable.

Next we highlight some of the strategies for reducing and responding to prejudice and discrimination toward sexual orientation minorities. These strategies include efforts to reduce employment discrimination, provide family benefits to same-sex couples, recognize parental rights of LGBT parents and co-parents, respond to antigay hate crimes, and implement educational programs to reduce prejudice against sexual orientation minorities and to provide programs and services for gay and lesbian students.

Strategies for Action: Reducing Antigay Prejudice and Discrimination

Many of the efforts to change policies and attitudes regarding sexual orientation minorities have been spearheaded by organizations such as the Human Rights Campaign (HRC); the National Gay and Lesbian Task Force (NGLTF); Gay and Lesbian Alliance against Defamation (GLAAD); the International Lesbian and Gay Association (ILGA); Lambda Legal Defense and Education Fund; the National Center for Lesbian Rights; and the Gay, Lesbian, and Straight Education Network (GLSEN). But the effort to reduce antigay prejudice and discrimination is not just a "gay agenda;" many heterosexuals and mainstream organizations (such as the National Education Association) have worked on this agenda as well.

Many of the advancements in gay rights have been the result of political action and legislation. Barney Frank (1997), an openly gay U.S. representative, emphasizes the importance of political participation in influencing social outcomes. He notes that demonstrative and cultural expressions of gay activism, such as "gay pride" celebrations, marches, demonstrations, or other cultural activities promoting gay rights are important in organizing gay activists. However, he points out:

> Too many people have seen the cultural activity as a substitute for democratic political participation. In too many cases over the past decades we have left the political arena to our most dedicated opponents [of gay rights], whose letter writing, phone calling, and lobbying have easily triumphed over our marching, demonstrating, and dancing. The most important lesson . . . for people who want to make America a fairer place is that politics—conventional, boring, but essential politics—will ultimately have a major impact on the extent to which we can rid our lives of prejudice. (Frank 1997, xi)

Reducing Employment Discrimination against Sexual Orientation Minorities

The majority of Americans agree that no one should be denied employment, fired, passed over for promotion, or otherwise discriminated against at work

Courtesy of Tammy Baldwin

Representative Tammy Baldwin (D–Wisconsin) is the first out lesbian and the first openly gay nonincumbent elected to Congress.

Effects of Sexual Orientation Antidiscrimination Legislation

In *Private Lives, Public Conflicts: Battles over Gay Rights in American Communities*, researchers James Button, Barbara Rienzo, and Kenneth Wald (1997) present survey data of all U.S. cities and counties with laws or policies prohibiting discrimination on the basis of sexual orientation. When the survey began in mid-1993, there were 126 such cities and counties.

Methods and Sample

To examine the effects of gay rights legislation in these communities, the researchers mailed questionnaires to either the director of the enforcement agency that was responsible for the antidiscrimination policy or ordinance or the local official who was perceived to be most knowledgeable about the legislation and its effect. Surveys were also sent to public school officials. For comparison purposes, the researchers also surveyed a random sample of 125 U.S. communities without antigay discrimination legislation. In addition, the researchers interviewed a variety of individuals who were knowledgeable about the passage and impact of gay rights legislation, including elected city officials and administrators, gay and lesbian activists, religious and other leaders of opposition groups, members of the business community, minority group leaders, and school personnel.

Among other topics, this research investigated the various effects of antidiscrimination legislation.

Findings and Conclusions

The most frequently cited (by 27 percent of respondents) positive effect of antidiscrimination legislation was that such legislation reduced discrimination against lesbians and gays in public employment and other institutions covered by the law.

Antidiscrimination legislation also made lesbians and gay men feel safer, more secure, comfortable, and generally more accepted in the community. As a result, lesbians and gay men were more likely to "come out of the closet"—to reveal their sexual orientation to others (or to at least stop hiding it). Legal measures ban-

solely because of his or her sexual orientation. Yet, as of this writing, federal law only protects individuals from discrimination based on race, religion, national origin, sex, age, and disability. A bill called the **Employment NonDiscrimination Act (ENDA)** would make it illegal to discriminate based on sexual orientation. The ENDA would not apply to religious organizations, businesses with fewer than 15 employees, or the military. ENDA was introduced in Congress in 1994. In 1996, the bill came within one vote of passing the Senate. It was reintroduced in Congress in 1999.

With the absence of federal legislation prohibiting antigay discrimination, some state and local governments as well as private employers have taken measures to prohibit employment discrimination based on sexual orientation. The scope of these measures varies, from prohibiting discrimination in public employment only to comprehensive protection against discrimination in public and private employment, education, housing, public accommodations, credit, and union practices.

State and Local Bans on Antigay Employment Discrimination In 1974, Minneapolis became the first municipality to ban antigay job discrimination. By August 2000, 116 U.S. cities and counties prohibited employment discrimination based on sexual orientation (Human Rights Campaign 2000b).

In 1982, Wisconsin became the first state to prohibit antigay employment discrimination. In April 2001, Maryland became the 12th state to ban antigay job discrimination. Other states that prohibit anti-gay employment discrimination include California, Connecticut, Hawaii, Massachusetts, Minnesota, Nevada, New Hampshire, New Jersey, Rhode Island, Vermont, and Wisconsin. Eight states ban antigay discrimination in their public work force by executive order of the governor or civil service regulations. In addition, more than 200 local

ning discrimination based on sexual orientation also gave gays and lesbians a sense of legitimacy that helps them overcome "internalized homophobia"—a sense of personal failure and self-hatred among lesbians and gay men resulting from social rejection and stigmatization.

The second most frequently cited (by 22 percent of respondents) positive effect of antidiscrimination legislation was the recognition that discrimination based on sexual orientation is legally wrong and not permissible. One survey respondent noted, "The law sends a message that no kind or type of harassment or hiring discrimination against gays will be tolerated" (p. 118). Respondents in this study indicated that through public hearings and publicity surrounding proposed anti-bias legislation, gay and lesbian groups were able to describe how they have been discriminated against by employers, landlords, the police, and others. "With many Americans expressing ignorance and confusion about the degree to which gays are stigmatized and faced with discrimination, these public messages provided an invaluable means by which to educate the public" (p. 130).

Based on their research, Button, Rienzo, and Wald concluded that antidiscrimination legislation has had only modest effects on police prejudice and discrimination against lesbians and gay men. "Indications of progress in law enforcement . . . seem fragile and incomplete . . . Police raids on gay gatherings still occur, often with violence and abuse" (p. 124). According to a Cincinnati city official, "There's still real prejudice here against gays, and intolerance among police on the street . . . . Gay jokes are still common. The police don't treat gays well generally" (p. 124).

The majority (51 percent) of survey respondents in this study reported no negative effects of antidiscrimination legislation. However, "a modest number of surveyed public officials mentioned that the debate over and passage of legislation protecting gays resulted in divisions in the community, greater controversy or tension, or the increased mobilization of those opposed to gay rights" (p. 120).

Source: James W. Button, Barbara A. Rienzo, and Kenneth D. Wald. 1997. *Private Lives, Public Conflicts: Battles over Gay Rights in American Communities*. Washington, D.C.: CQ Press.

governments or governmental agencies provide some form of protection against antigay discrimination in their public work force. In 1998, President Clinton signed an executive order banning sexual orientation discrimination among federal employees ("1998 in Review" 1999).

Nondiscrimination Policies in the Workplace In 1975, AT&T became the first employer to add sexual orientation to its nondiscrimination policy. By August 2000, more than 1,000 private sector employers had nondiscrimination policies that included sexual orientation. In addition, at least 308 colleges and universities prohibit employment discrimination based on sexual orientation (Human Rights Campaign 2000b). Since 1996, the majority of Fortune 500 companies have included sexual orientation in their nondiscrimination policies. According to the Human Rights Campaign (2000b), the closer a company is to the top of the Fortune 500 list, the more likely it is to prohibit antigay discrimination, suggesting that "the most successful companies in America are those that embrace diversity and work toward providing an inclusive work environment for lesbian and gay employees" (p. 8).

A list of employers that prohibit sexual orientation discrimination is available at the Human Rights Campaign WorkNet website at www.hrc.org/worknet. Whereas gay rights advocates promote doing business with companies that are "gay-friendly," groups that oppose gay rights advocate boycotting such businesses. For example, the website www.godhatesfags.com, urges its visitors to avoid doing business with any company that is listed on the Human Rights Campaign WorkNet website. This chapter's *Social Problems Research Up Close* feature looks at a landmark study of the various social effects of sexual orientation antidiscrimination legislation.

Providing Support to Lesbian and Gay Couples, Parents, and Co-Parents

In recent years, the gay rights movement has witnessed significant progress in the provision of benefits to lesbian and gay couples, parents, and co-parents.

Benefits to Same-Sex Couples   In spring of 2000, the Vermont Congress passed a "civil union" bill, allowing same-sex couples from any state—where each person is at least age 18, competent, and not closely related by blood—to apply to a Vermont town clerk for a civil union license, have that license "certified" by a judge, justice of the peace, or clergy, and then receive a civil union certificate ("A Historic Victory" 2000). A **civil union** is a legal status parallel to civil marriage for all purposes under Vermont state law. Civil union status entitles same-sex couples to all 300 or more benefits available under state law to married couples. Slightly more than half of Vermonters are in favor of the civil union law—52 percent support the law and 46 percent oppose it ("Post-Election Analysis" 2000).

If Vermonters with civil union status travel out-of-state, their rights may not be recognized. For example, a Vermont lesbian couple may have an auto accident while on vacation in Maine. One partner could be barred from visiting her lover in the hospital because the hospital does not recognize her as a family member despite their civil union. In a backlash to Vermont's civil union law, Nebraska has proposed the first anti-marriage legislation in the country that would specifically ban civil unions ("A Historic Victory" 2000).

Some states, counties, cities, and workplaces allow unmarried couples, including gay couples, to register as domestic partners. The benefits of **domestic partnerships** vary from place to place but may include coverage under a partner's health and pension plan, rights of inheritance and community property, and tax benefits. In 1991 the Lotus Development Corporation became the first major American firm to extend domestic partnership benefits to gay and lesbian employees. By August 2000, the Human Rights Campaign (2000b) identified 3,572 private companies, colleges and universities and state and local governments that offer domestic partner health insurance to their employees—an increase of 25 percent from the previous year. In 2000, the Big Three automakers—Daimler-Chrysler, General Motors, and Ford, as well as the United Auto Workers—announced their landmark decision to provide domestic partner benefits for their more than 400,000 employees. "This marked the first time that virtually an entire sector of American commerce, along with its leading union, decided collectively to provide domestic partner benefits" (Human Rights Campaign 2000b, 9). In eight states (Delaware, Massachusetts, Vermont, New York, Oregon, California, Connecticut, and Washington), lesbian and gay state employees and their domestic partners are eligible for some benefits, ranging from bereavement leave only to medical, dental, life insurance, and long-term care (Ettelbrick 2000). However, these benefits represent a fraction of those granted to married couples.

Gay and Lesbian Parental Rights   A substantial number of same-sex couples, especially lesbian couples, have children under age 18 living in the home. In the United States, an estimated 28 percent of lesbians and 14 percent of gay men have children in the household (Black et al. 2000). A number of recent policies and rulings reflect the increasing provision of parental rights to gay and lesbian parents and co-parents.

- In an unprecedented ruling, the California Board of Equalization granted head-of-household tax status to a nonbiological lesbian co-parent ("NCLR

No matter how comfortable lesbians may feel— especially those of us who are accepted by family, friends and neighbors— we all experience discrimination that damages our families . . . I hope our victory will inspire other lesbians to demand full recognition for their families.

Helmi Hisserich

A lesbian nonbiological co-parent who won head-of-household tax status in California

Wins Equal Tax Benefits for Non-biological Lesbian Mother" 2000). This is the first ruling by any state tax board that provides equitable tax status for gay and lesbian families.

- In 2000, a New Jersey appeals court awarded visitation to a nonbiological lesbian co-parent, reasoning that the woman had a close "parent-type relationship" with the children ("Custody and Visitation" 2000).
- A Massachusetts court ruled that two women may be listed as "mother" on a birth certificate, when one of the women donated the egg for the child and the other carried the child (*LAWbriefs* 2000).
- A New York family court judge granted visitation to a lesbian co-parent, who lived with and helped raise the 2- and 4-year-old children her former partner had by artificial insemination before the couple separated. The judge stated that "these children have the right of any other children to continue a loving relationship with their parents" (*LAWbriefs* 2000, p. 4).

Antigay Hate Crimes Legislation

Hate crimes laws call for tougher sentencing when prosecutors can prove that the crime committed was a hate crime. In June 2000, the U.S. Senate made history by passing hate crimes legislation covering sexual orientation, gender, and disability (*LAWbriefs* 2000). Formerly known as the Hate Crimes Prevention Act, the renamed Local Law Enforcement Enhancement Act of 2000 makes hate crimes based on bias against sexual orientation, gender, and disability, a federal offense. As of January 2001, 21 states have hate crime laws that include sexual orientation; 16 states have hate crime laws that do not include sexual orientation (National Gay and Lesbian Task Force 2001b). Three states have hate crime laws based on prejudice and bias and do not list specific categories and six states do not have any hate crimes laws.

Educational Strategies: Policies and Programs in the Public Schools

A survey of youths' risk behavior conducted by the Massachusetts Department of Education in 1999 found that 30 percent of gay teens attempted suicide in the previous year, compared with 7 percent of their straight peers (reported in Platt 2001). Forty percent of gay youth report schoolwork being negatively affected by conflicts around sexual orientation and, as noted earlier, over one-quarter of gay youth drop out of school (Chase, 2000; *Homophobia 101*, 2000). These findings suggest that if schools are to promote the health and well-being of all students, they must address the needs of gay, lesbian, and bisexual youth and promote acceptance of sexual orientation diversity within the school setting.

One strategy for promoting tolerance for diversity among students involves establishing and enforcing a school policy prohibiting antigay behavior. In 1994, Massachusetts state legislators passed such a policy prohibiting discrimination against gay and lesbian students. This law permits students who have suffered antigay discrimination, and who were not protected by the school administration, to bring lawsuits against their schools. Other school districts in Ohio, Indiana, and Nebraska have school policies to prohibit harassment of students based on sexual orientation (*LAWbriefs* 2000). But the majority of schools do not have any policies prohibiting antigay harassment (Button et al. 1997). Schools that do not protect students against harassment may face legal challenges. In a federal

By ensuring that today's teachers teach the lessons of respect for all, we ensure that the next generation of Americans will live in a world without antigay prejudice.

BOB CHASE
President, National Education Association

lawsuit, a Kansas City high school student alleged that he was taunted and beaten because classmates thought he was gay, and that school officials did nothing to stop it. Attorney General Janet Reno filed a motion in this lawsuit, noting that school officials exhibited a "deliberate indifference" to the harassment the student suffered, and asked the court to order the school to develop a plan to create a "harassment-free" educational environment (*LAWbriefs* 2000, 6).

One resource for creating a "harassment-free" climate is the Gay, Lesbian and Straight Education Network (GLSEN)—the leading national organization fighting against harassment and discrimination in K-12 schools. GLSEN has conducted training for nearly 400 school staffs around the country and developed the faculty training program of the Massachusetts Department of Education's "Safe Schools for Gay and Lesbian Students" program—the first statewide effort aimed at ending homophobia in schools (*Homophobia 101* 2000). The National Education Association has also implemented national training programs to educate teachers in every state about the role they can and must play to stop antigay harassment in their schools (Chase 2000).

Some schools have established school-based support groups for LGBT students. Such groups can help students increase self-esteem, overcome their sense of isolation, provide information and resources, and provide a resource for parents. School counselors may be trained to work with gay, lesbian, and bisexual youth and their parents. Education about sexual orientation can be implemented in sex education or health education classes or in conflict resolution or diversity curricula. In-service training for teachers and other staff is important and may include examining the effects of antigay bias, dispelling myths about homosexuality, and brainstorming ways to create a more inclusive environment (Mathison 1998). Gay-straight alliances (GSAs) are school-sponsored clubs for gay teens and their straight peers. First established in a Los Angeles high school in 1984, most of the approximately 750 gay-straight alliances in U.S. schools were founded within two years following the widely publicized 1998 murder of Matthew Shepard (Platt 2001). However, most public schools offer little to no support and education regarding sexual orientation diversity. Most schools have no support groups or special counseling services for gay and lesbian youth (Button et al. 1997).

Gay rights opponents have spearheaded efforts to deny gay and lesbian students access to school-based programs and services. In Oregon, Measure 9 would have prevented Oregon public schools from providing instruction that would "encourage, promote or sanction homosexual behavior" ("Post-Election Analysis" 2000, 3). The measure would also have restricted policies that offer counseling support to gay and lesbian students. In the 2000 election, Oregon voters narrowly defeated Measure 9. In another case, the Salt Lake City School Board had banned all extracurricular student clubs, ranging from the Young Republicans to the Black Students Union, in order to block a gay/straight student alliance, or GSA, that East High students tried to form in 1995. To block the GSA, the school had to ban all school clubs because, under the Equal Access Act, schools accepting federal funds may not censor some school clubs if any are allowed on campus. In 2000, under the shadow of lawsuits and national publicity, the school board voted to lift the club bans (Lambda Legal Defense and Educational Fund 2000a).

Campus Policies Regarding Homosexuality

Student groups have been active in the gay liberation movement since the 1960s. Because of the activism of students and the faculty and administrators who support them, almost 300 U.S. colleges and universities have nondiscrimination, antiharassment or misconduct policies that include sexual orientation (Lambda Legal Defense and Education Fund 2000b).

D'Emilio (1990) suggests that colleges and universities have the ability and the responsibility to promote gay rights and social acceptance of homosexual people:

> For reasons that I cannot quite fathom, I still expect the academy to embrace higher standards of civility, decency, and justice than the society around it. Having been granted the extraordinary privilege of thinking critically as a way of life, we should be astute enough to recognize when a group of people is being systematically mistreated. We have the intelligence to devise solutions to problems that appear in our community. I expect us also to have the courage to lead rather than follow (p. 18).

In addition to including sexual orientation in discrimination policies, colleges and universities have also taken more proactive measures to support the lesbigay student population. Such measures have included offering gay and lesbian studies programs, social centers, and support groups; and sponsoring events and activities that celebrate diversity. "These programs serve as both a refuge for lesbians and gay men on campus, and common ground from which to launch educational projects that foster respect for difference" (Lambda Legal Defense and Education Fund 2000b, 10).

Strategies for reducing antigay prejudice and discrimination are influenced largely by politicians, religious leaders, courts, and educators who will continue to make decisions that either promote the well-being of sexual orientation minorities or hinder it. Ultimately, however, each individual must decide to embrace either an inclusive or exclusive ideology; collectively, those individual decisions will determine the future treatment of sexual orientation minorities.

In addition, lesbigay individuals must find their own strategies for living in a homophobic and biphobic society. Representative Tammy Baldwin (D–Wisconsin)—the first out lesbian and the first openly gay nonincumbant elected to Congress—offered the following advice in a speech entitled, "Never Doubt," which she delivered at the Millennium March on Washington in April 2000:

> If you dream of a world in which you can put your partner's picture on your desk, then put his picture on your desk . . . and you will live in such a world.
>
> If you dream of a world in which there are more openly gay elected officials, then run for office . . . and you will live in such a world.
>
> And if you dream of a world in which you can take your partner to the office party, even if your office is the U.S. House of Representatives, then take her to the party. I do, and now I live in such a world.
>
> Remember, there are two things that keep us oppressed—them and us. We are half of the equation.

Understanding *Sexual Orientation*

As both functionalists and conflict theorists note, alternatives to a heterosexual lifestyle are threatening, for they challenge traditional definitions of family, childrearing, and gender roles. The result is economic, social, and legal discrimination by the majority. Gay, lesbian, and bisexual individuals are also victimized by hate crimes. In some countries, homosexuality is formally sanctioned with penalties ranging from fines to imprisonment and even death.

In the past, homosexuality was thought to be a consequence of some psychological disturbance. More recently, some evidence suggests that homosexuality, like handedness, may have a biological component. The debate between biological and social explanations is commonly referred to as the "nature versus

No matter how visible or successful individual gay men and lesbians become, no matter how encouraging is their social progress in America today, they remain second-class citizens without full protection of the law.

BYRNE FONE

nurture" debate. Research indicates that both forces affect sexual orientation, although debate over which is dominant continues. Sociologists are interested in society's response to sexual orientation diversity and how that response affects the quality of life of society's members. Because individuals' views toward homosexuality are related to their beliefs about what "causes" it, the question of the origins of sexual orientation diversity has sociological significance.

Prejudice and discrimination toward sexual orientation minorities are rooted in various aspects of culture, such as religious views, rigid gender roles, and negative myths and stereotypes. Hate crimes against homosexuals and bisexuals and lack of protection against job discrimination are examples of discrimination that even those who disapprove of homosexuality do not endorse.

Attitudes toward lesbians and gays have become more accepting in recent years. One explanation for these changing attitudes is that personal contact with openly gay individuals has increased in recent years as more gay and lesbian Americans are coming out to their family and friends. *Time* magazine reported in 1998 that 41 percent of Americans reported that they had a family member or close friend who is gay or lesbian, a 9 percent increase over a similar survey just 4 years before (reported in Wilcox & Wolpert 2000). "This is likely to increase support for gay and lesbian equality because contacts with openly gay individuals reduces negative stereotypes and ignorance" (p. 414).

Another explanation for changing attitudes toward homosexuality is the positive depiction of gays and lesbians in the popular media. In 1992 "Roseanne" and "Melrose Place" had gay and lesbian characters, in 1998 Ellen DeGeneres came out on her sitcom "Ellen," and by 2000 there were gay and lesbian characters in many television shows, including the popular "Dawson's Creek," "Will & Grace" and "Buffy the Vampire Slayer."

> We still live in a country in which there is greater outrage over a gay, lesbian, or bisexual schoolteacher's being hired than there is over that teacher's being fired.
>
> ALAN YANG
> *Columbia University*

But, as one scholar expressed, "the new confidence and social visibility of homosexuals in American life have by no means conquered homophobia. Indeed it stands as the last acceptable prejudice" (Fone 2000, 411). Although the gay rights movement has made significant gains in the last few decades, it has also suffered losses and defeat because of gay rights opposition groups and politicians. Many strategies for promoting gay rights and tolerance have been successful, and the American public is becoming increasingly supportive of gay rights. But, as Yang (1999) points out, as the antigay minority diminishes in size, "it often becomes more dedicated and impassioned" (p. ii).

> Our task in the coming years is to get the heterosexual Americans who support our cause to feel as passionately outraged by the injustices we face and to be as strongly motivated to act in support of our rights as our adversaries are in their opposition to our rights." (p. iii)

Critical Thinking

1 Research findings suggest that gay men generally earn less than other men with equal skills, whereas lesbian women generally earn more than other women (Black et al. 2000). Why do you think this is so?

2 Butler (2000) found that although homophobia appears to be greater among African Americans compared to whites, African-American men were more likely than white men to report having engaged in same-sex sexual activity. What might explain the higher likelihood of same-sex sexual activity among African American men compared to white men?

3 How is the homosexual population similar to and different from other minority groups?

4 Do you think that social acceptance of homosexuality leads to the creation of laws that protect lesbians and gays? Or does the enactment of laws that protect lesbians and gays help to create more social acceptance of gays?

5 A Virginia court ruling denied Sharon Bottoms (a lesbian) custody of her son and awarded custody to Sharon's mother. Some people support this court decision on the grounds that they believe that lesbian mothers are more likely to influence their children to be homosexual. Do you see any contradictions in the court's ruling in the Sharon Bottoms case?

Key Terms

biphobia	ex-gay ministries	master status
bisexuality	heterosexism	reparative therapy
civil union	heterosexuality	sexual orientation
conversion therapy	homophobia	sodomy laws
Defense of Marriage Act	homosexuality	transgendered individuals
domestic partnerships	lesbigay population	
Employment Nondiscrimination Act (ENDA)	LGBT	

Media Resources

The Wadsworth Sociology Resource Center: Virtual Society

http://sociology.wadsworth.com/

See the companion Web site for this book to access general sociology resources and text-specific features that can further your understanding of this chapter. The site contains Internet links, Internet exercises, online practice quizzes, information on InfoTrac College Edition, and many more valuable materials designed to enrich your learning experience in social problems.

InfoTrac College Edition

You can access InfoTrac College Edition either from the Wadsworth Sociology Resource Center at **http://sociology.wadsworth.com** or directly from your web browser at **http://www.infotrac-college.com/wadsworth/**. InfoTrac College Edition is an online university library that includes over 700 popular and scholarly journals in which you can find articles related to the topics in this chapter such as information on religion and homosexuality, homophobia, domestic partnerships, and antidiscrimination legislation.

Interactions CD-ROM for *Understanding Social Problems*, Third Edition

Go the "Interactions" CD-ROM to access additional interactive learning tools, such as in-depth review materials, corresponding practice quizzes, and other engaging resources and activities to help you study the concepts in this chapter.

Problems of Inequality and Power

The story of Harrison Bergeron is set in a futuristic society where absolute equality is rigidly enforced (Vonnegut 1968). If people can run faster, they must wear weights in their shoes; if they are brighter, disruptive transistors are implanted in their brains; if they have better vision, blinders are placed over their eyes. The point of the story is that equality, although a cultural ideal, in reality is not always the optimal way to live. The quality of life becomes so unbearable for Harrison Bergeron that he decides that it is better to commit one courageous act of grace and beauty and be killed than to live in a society where everyone is equal. Differences between people in and of themselves are not what is meant by inequality as a social problem, for few would want to live in a society where everyone was the same. Rather, problems of inequality concern inequities in the quality of life—between the haves and the have-nots (Chapter 10). Social inequality affects the opportunity to work and prosper (Chapter 11), to attend quality schools (Chapter 12), and to live in a healthy and safe environment (Chapter 13).

To a large extent, differences between the rich and the poor are a consequence of changes in the nature

of work and the resulting increase in unemployment for some people, in some jobs, in some parts of the world. A shift to a global economy and the accompanying deconcentration and deindustrialization of urban areas has resulted in high inner city unemployment, displaced workers, deteriorating neighborhoods, and eroding tax bases. Those who can get out do so, as evidenced by patterns of suburbanization; those who remain are involuntarily segregated.

As a consequence of these urban processes, city schools, which are heavily dependent on municipal funds, attract fewer and fewer good teachers, facili-

tics erode, the quality of education declines, and the number of poor minority students increases. Debates begin over the relevance of traditional curricula and the need for multiculturalism. Students graduate without knowing how to read, work simple math problems, or resolve interpersonal conflicts. The chapters in Section 3—"The Haves and the Have-Nots," "Work and Unemployment," "Problems in Education," and "Cities in Crisis"—are highly interrelated and speak to the need for examining both the cultural and the structural underpinnings of the social problems described.

10

The Haves and the Have-Nots

Is It True?

1. The United States has the lowest poverty rate of all industrialized nations.

2. Nearly half of the world's population live on less than $2 per day.

3. In 1999, a U.S. worker with a full-time minimum-wage job could keep a family of three above the poverty line.

4. Families receiving welfare tend to be large, with an average of 4 children.

5. The average cost of child care for a 4-year-old in an urban area is $10,000 or more per year, more than the average annual cost of public college tuition in all but one state.

Answers to "Is It True?": 1 = F; 2 = T; 3 = F; 4 = F; 5 = T

If all of the afflictions of the world were assembled on one side of the scale and poverty on the other, poverty would outweigh them all.

RABBA, MISHPATIM 31:14

From the outside, Jamal's building looks like an ordinary house that has seen better days. . . . But once you walk through the front door, all resemblance to a real home disappears. . . . The building has been broken up into separate living quarters, a rooming house with whole families squeezed into spaces that would not even qualify as bedrooms in most homes . . . Six families take turns cooking their meals in the only kitchen . . .

The plumbing breaks down without warning . . . Windows . . . are cracked and broken, pieced together by duct tape that barely blocks the steady, freezing draft blowing through on a winter evening. Jamal is of the opinion that for the princely sum of $300 per month, he ought to be able to get more heat. . . .

Jamal feels the walls closing in on him, on his relations with [his common law wife] Kathy, because they are so cramped . . . It does not help his temper, which has its explosive side, to be stuck in such a dump . . .

At the age of 22, he does a full shift whenever Burger Barn [a pseudonym for a national chain of fast food restaurants] . . . will give him the hours . . . .When their daughter was born . . . the young couple had nothing to live on besides Jamal's part-time wages . . . They wedged the crib into their single room in a Brooklyn tenement and struggled to manage the piles of Pampers and a squalling child in this tiny, claustrophobic space. But Tammy developed colic and became difficult to handle . . . To his eternal regret, Jamal lost his temper one day and lashed out at the helpless child . . . He insisted that he had accidentally pushed the baby's crib and the little one fell out. Social Services didn't buy this though, and they removed Tammy from her home, charging Jamal with abuse and Kathy with neglect . . . .

They were absolutely determined to get their baby back. Every week they visited her . . . supervised closely by the foster mother who had temporary custody of Tammy. They attended parenting classes, trying to learn how to take better care of the little girl . . . .The biggest problem, though, was that they couldn't meet the court's conditions for the return of Tammy. Somehow on his Burger Barn salary, Jamal was supposed to provide an apartment with a separate bedroom for Tammy. Kathy was forbidden to work, since the law's position was that she had neglected the baby's care. The cheapest place they had seen that met these conditions, deep in the heart of the South Bronx, a place known nationwide for its mean streets, was still $600 a month . . . .Landlords were asking for a security deposit as well, a reservoir of cash that was completely beyond the family's means. In the very best month Jamal had ever had, he earned only $680 before taxes—and that meant working full-time, something he could never count on. (Newman 1999, 3–9)

The substandard housing, stress, economic insecurity, and family problems experienced by Jamal and Kathy are problems common among the more than 32 million people in the United States who are officially poor (Dalaker & Proctor 2000). Despite its enormous wealth, the United States is characterized by persistent economic inequalities that divide the population into haves and have-nots. This chapter examines the extent of poverty globally and in the United States, focusing on the consequences of poverty for individuals, families, and societies. Theories of poverty and economic inequality are presented and strategies for rectifying economic inequality and poverty are considered.

The Global Context: Poverty and Economic Inequality around the World

Who are the poor? Are rates of world poverty increasing, decreasing, or remaining stable? The answers depend on how we define and measure poverty.

Defining and Measuring Poverty

Poverty has traditionally been defined as the lack of resources necessary for material well-being—most importantly food and water, but also housing, land, and health care. This lack of resources that leads to hunger and physical deprivation is known as **absolute poverty**. **Relative poverty** refers to a deficiency in material and economic resources compared with some other population. Consider, for example, Jamal and Kathy, whom we introduced in the opening of this chapter. Although they are poor compared with the majority of Americans, they have resources and a level of material well being that millions of people living in absolute poverty can only dream of.

Various measures of poverty are used by governments, researchers, and organizations. Next, we describe international and U.S. measures of poverty.

International Measures of Poverty The World Bank sets a "poverty threshold" of $1 per day to compare poverty in most of the developing world; $2 per day in Latin America, $4 per day in Eastern Europe and the Commonwealth of Independent States (CIS); and $14.40 per day in industrial countries (which corresponds to the income poverty line in the United States). An estimated 2.8 billion people, nearly half of the world's population, survive on less than $2 per day and a fifth of the world's population (1.2 billion people), live on less than $1 per day (Flavin 2001).

Another poverty measure used by the World Health Organization (WHO) is based on a household's ability to meet the minimum calorie requirements of its members. According to this poverty measure, a household is considered poor if it cannot meet 80 percent of the minimum calorie requirements (established by WHO), even when using 80 percent of its income to buy food.

In industrial countries, national poverty lines are sometimes based on the median household income of a country's population. According to this relative poverty measure, members of a household are considered poor if their household income is less than 50 percent of the median household income in that country.

Human poverty is more than income poverty—it is the denial of choices and opportunities for living a tolerable life.

HUMAN DEVELOPMENT REPORT 1997

Recent poverty research concludes that poverty is multidimensional and includes such dimensions as food insecurity; poor housing; unemployment; psychological distress; powerlessness; hopelessness; lack of access to health care, education, and transportation; and vulnerability (Narayan 2000). To capture the multidimensional nature of poverty, the United Nations Development Programme (1997) developed a composite measure of poverty: the **Human Poverty Index** (HPI). Rather than measure poverty by income, three measures of deprivation are combined to yield the Human Poverty Index: (1) deprivation of a long, healthy life, (2) deprivation of knowledge, and (3) deprivation in de-

Table 10.1 *Measures of Human Poverty in Developing and Industrialized Countries*

	Longevity	Knowledge	Decent Standard of Living
For developing countries	Probability at birth of not surviving to age 40	Adult illiteracy	A composite measure based on: 1. Percentage of people without access to safe water 2. Percentage of people without access to health services 3. Percentage of children under five who are underweight
For industrialized countries	Probability at birth of not surviving to age 60	Adult functional illiteracy rate	Percentage of people living below the income poverty line, which is set at 50% of median disposable income

Source: Adapted from the United Nations Development Programme. 2000. *Human Development Report 2000*. New York: Oxford University Press.

cent living standards. As shown in Table 10.1, the Human Poverty Index for developing countries (**HPI-1**) is measured differently than for industrialized countries (**HPI-2**). Among the 18 industrialized countries for which the HPI-2 was calculated, Norway has the lowest level of human poverty (7.3 percent), followed by Sweden (7.6 percent), and the Netherlands (8.2 percent) (United Nations Development Programme 2000). Those with the highest rates of human poverty are the United States (15.8 percent), Ireland (15.0 percent), and the United Kingdom (14.6 percent). The Human Poverty Index is a useful complement to income measures of poverty and "will serve as a strong reminder that eradicating poverty will always require more than increasing the income of the poorest" (United Nations Development Programme 1997, 19).

Measures of poverty tell us how many, or what percentage of people are living in poverty in a given year. Another way to assess poverty is to note the degree to which those who are poor stay in poverty from year to year. This can be done by calculating the average annual poverty exit rate—the share of the poor in 1 year that have left poverty by the following year. One study followed the same families over a 5-year period in six countries—Canada, Germany, Netherlands, Sweden, United Kingdom, and United States (reported in Mishel, Bernstein, & Schmitt 2001). The results indicated that the poor in the United States are less likely than the poor in other countries to leave poverty from one year to the next. According to this research, 28.6 percent of the poor in the United States escape poverty each year, compared with 29.1 percent in the United Kingdom, 36 percent in Sweden, 37 percent in Germany, 42 percent in Canada, and 44 percent in the Netherlands. The poor in the United States are also more likely to fall back into poverty once they make it out.

U.S. Measures of Poverty   In 1964 the Social Security Administration devised a poverty index based on a 1955 Agriculture Department survey that estimated the cost of an economy food plan for a family of four. Because families with three or more members spent one-third of their income on food at the time, the poverty line was set at three times the minimum cost of an adequate diet. Poverty thresholds differ by the number of adults and children in a family and, for some family types, by the age of the family head of household (see Table 10.2). Poverty thresholds are adjusted each year for inflation. Anyone liv-

■ **Table 10.2** *Poverty Thresholds: 1999 (householder under 65 years old)*

One adult	$8,667
Two adults	$11,214
One adult, one child	$11,483
Two adults, one child	$13,410
Two adults, two children	$16,895

Source: Dalaker, Joseph and Bernadette D. Proctor. 2000. *Poverty in the United States: 1999.* Current Population Reports P60-210. U.S. Census Bureau. Washington, D.C.: U.S. Government Printing Office. www.census.gov/hhes/poverty/threshold/thresh99.html

ing in a household with income (before tax) below the official poverty line is considered "poor." Individuals living in households that are above the poverty line, but not very much above it, are classified as "near poor" and those living below 50 percent of the poverty line live in "deep poverty."

The U.S. poverty line has been criticized on several grounds. First, the poverty line is based on the assumption that low-income families spend one-third of their household income on food. That was true in 1955, but because other living costs (e.g., housing, medical care, and child care) have risen more rapidly than food costs, low-income families today spend closer to one-fifth (rather than one-third) of their income on food (Pressman 1998). So current poverty lines should be based on multiplying food costs by five rather than three. This would raise the official poverty line by two-thirds, making the poverty level consistent with public opinion regarding what a family needs to escape poverty.

Another shortcoming of the official poverty line is that it is based solely on money income and does not take into consideration noncash benefits received by many low-income persons, such as food stamps, Medicaid, and public housing. Family assets, such as savings and property, are also excluded in official poverty calculations. The poverty index also fails to account for tax burdens that affect the amount of disposable income available to meet basic needs. The U.S. poverty line also disregards regional differences in the cost of living and, because poverty rates are based on surveys of households, the homeless—the most destitute of the poor—are not counted among the poor.

Another important shortcoming of the official poverty lines is that "they do not reflect overall income growth and thus fail to capture changing standards of what is an acceptable poverty threshold" (Mishel, Bernstein and Schmitt 2001, 297). As overall standards of living rise, the economic "distance" between those above and below the poverty line expands. A relative measure of poverty is better suited to capture this distance. The relative poverty threshold of 50 percent of median family income yields a higher poverty rate than the poverty rate based on the official poverty line. For example, whereas in 1998 12.7 percent of the U.S. population were poor according to the official poverty measure, 22.3 percent of the population were in families with incomes below one-half of the median family income (Mishel et al. 2001).

The Extent of Global Poverty and Economic Inequality

The *Global Poverty Report* (2000) finds that globally, the proportion of people living on less than $1 per day fell from 29 percent in 1987 to 26 percent in 1998. Social indicators have also improved over the last three decades. In developing countries, life expectancy rose from 55 years in 1970 to 65 years in 1998 and infant mortality rates have fallen from 107 per 1,000 live births to 59. However, sub-Saharan Africa, Central Asia, and Eastern Europe have not shared in this progress (*Global Poverty Report* 2000). South Asia has the greatest number of people affected by poverty and sub-Saharan Africa has the highest proportion of people in poverty. Indeed, in every region of the world except Africa, the share of the population that is hungry is diminishing (Brown 2001). Recent estimates of world hunger and malnourishment vary from 830 million (Wren 2001) to 1.1 billion people worldwide (Flavin 2001). In some African countries, such as Kenya, Zambia, and Zimbabwe, as much as 40 percent of the population is malnourished (Flavin 2001).

The 1997 *Report on the World Social Situation* notes that "infant mortality has fallen almost steadily in all regions and life expectancy has risen all over the globe. Educational attainment is rising, health care and living conditions are improving in most countries and the quantity, quality and range of goods and services available to a large majority of the world's population is increasing" (United Nations 1997, 80). But the report goes on to say that "not everyone has shared in this prosperity. Economic growth has been slow or non-existent in many of the world's poorest countries . . . The plight of the poor stands in stark contrast to the rising standards of living enjoyed by those favoured by growing abundance" (p. 80).

Global economic inequality has reached unprecedented levels. Consider that in 1999, the combined wealth of the world's 200 richest people was $1 trillion. In the same year, the combined incomes of the 582 million people living in the 43 least developed countries was $146 billion (United Nations Development Programme 2000). And global economic inequality widening the gap between the haves and the have-nots is likely to increase, if past trends continue. In 2000, the average income of the richest 20 countries was 37 times that of the poorest 20 countries—a gap that has doubled in the past 40 years (World Bank 2001). As the global gap between the rich and the poor increases, wealthy populations increase their consumption. As we discuss in chapter 14, overconsumption by the rich is a significant environmental problem. The wealthiest 25 percent of the world population consumes 85 percent of the world's resources and produces 90 percent of its wastes (Cracker & Linden 1998).

> Global Inequalities in income increased in the 20th century by orders of magnitude out of proportion to anything experienced before.
>
> HUMAN DEVELOPMENT REPORT 2000

Sociological Theories of Poverty and Economic Inequality

The three main theoretical perspectives in sociology—structural-functionalism, conflict theory, and symbolic interactionism—offer insights into the nature, causes, and consequences of poverty and economic inequality. Before reading further, you may want to take the "Attitudes Toward Economic Opportunity in the United States" survey in this chapter's *Self and Society* feature.

> In a nation as smart, inventive, and rich as America, the continuation of widespread poverty is a choice, not a necessity.
>
> MICHAEL KATZ
> *University of Pennsylvania*

Structural-Functionalist Perspective

According to the structural-functionalist perspective, poverty and economic inequality serve a number of positive functions for society. Decades ago, Davis and Moore (1945) argued that because the various occupational roles in society re-

Attitudes toward Economic Opportunity in the United States

In a 1998 Gallup poll of 5,001 U.S. adults, respondents were asked questions to assess their attitudes toward economic opportunity in the United States. After responding to the questions below, compare your answers with the results from a national sample.

1. Using a one-to-five scale, where "1" means not at all important, and "5" means extremely important, indicate how important each of the following is as a reason for a person's success.

Item	Ranking (1 = not at all important; 5 = extremely important) 1 2 3 4 5
a. Hard work and initiative	_____
b. Member of a particular race/ethnic group	_____
c. Getting right education and training	_____
d. Dishonesty and willingness to take whatever one can get	_____
e. Parents and family	_____
f. Willingness to take risks	_____
g. Gender (whether one is male or female)	_____
h. Connections/knowing the right people	_____
i. Money inherited from family	_____
j. Ability or talent one is born with	_____
k. Good luck/in right place at right time	_____
l. Physical appearance/good looks	_____

2. For the following two questions, indicate your answer from the choices provided:

a. Why are some people poor?

Answer choices: _____ lack of effort

_____ circumstances beyond their control

_____ both   _____ don't know

b. Why are some people rich?

Answer choices: _____ strong effort

_____ circumstances beyond their control

_____ both   _____ don't know

quire different levels of ability, expertise, and knowledge, an unequal economic reward system helps to assure that the person who performs a particular role is the most qualified. As people acquire certain levels of expertise (e.g., B.A., M.A., Ph.D., M.D.), they are progressively rewarded. Such a system, argued Davis and Moore, motivates people to achieve by offering higher rewards for higher achievements. If physicians were not offered high salaries, for example, who would want to endure the arduous years of medical training and long, stressful hours at a hospital?

3. Complete the following sentence with one of the two choices provided:

   a. The economic system in the United States:

   _____ is basically fair, since all Americans have an equal opportunity to succeed.

   _____ is basically unfair, since all Americans do not have an equal opportunity to succeed.

HOW DO YOUR ANSWERS COMPARE WITH A NATIONAL SAMPLE OF U.S. ADULTS?

1. This figure reveals the percentages of U.S. adults who rated the items in question #1 as important for success.

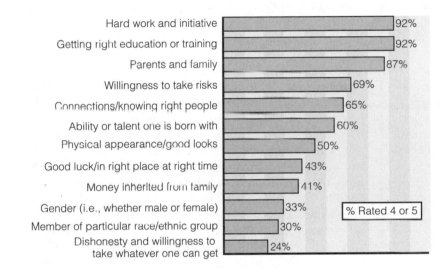

Hard work and initiative	92%
Getting right education or training	92%
Parents and family	87%
Willingness to take risks	69%
Connections/knowing right people	65%
Ability or talent one is born with	60%
Physical appearance/good looks	50%
Good luck/in right place at right time	43%
Money inherited from family	41%
Gender (i.e., whether male or female)	33%
Member of particular race/ethnic group	30%
Dishonesty and willingness to take whatever one can get	24%

% Rated 4 or 5

2a. In explaining why some people are poor, 43 percent of respondents indicated "lack of effort," 41 percent indicated "circumstances beyond their control," and 16 percent indicated "both" or "don't know."

2b. In explaining why some people are rich, 53 percent of respondents indicated "strong effort," 12 percent indicated "circumstances beyond their control," and 15 percent indicated "both" or "don't know."

3. Sixty-eight percent of respondents indicated that they believe the nation's economic system is basically fair; 29 percent believe it is basically unfair; and 3 percent had no opinion.

Source: Adapted from Gallup News Service Social Audit. 1998. (April 23 to May 31).
http://www.gallup.com/poll/socialaudits/have_havenot.asp (Used by permission).

The structural-functionalist view of poverty suggests that a certain amount of poverty has positive functions for society. Although poor people are often viewed as a burden to society, having a pool of low-paid, impoverished workers ensures that someone will be willing to do dirty, dangerous, and difficult work that others refuse to do. Poverty also provides employment for those who work in the "poverty industry" (e.g., welfare workers) and supplies a market for inferior goods such as older, dilapidated homes and automobiles (Gans 1972).

The structural-functionalist view of poverty and economic inequality has received a great deal of criticism from contemporary sociologists, who point out that many important occupational roles such as child care workers are poorly paid (the average salary of a child care worker is less than $15,000 per year) (Children's Defense Fund 2000), whereas many individuals in nonessential roles (e.g., professional sports stars and entertainers) earn astronomical sums of money. Functionalism also ignores the role of inheritance in the distribution of wealth.

Conflict Perspective

Although I have made a fortune in the financial markets, I now fear that the untrammeled intensification of laissez-faire capitalism and the spread of market values into all areas of life is endangering our open and democratic society. The main enemy . . . is no longer the communist threat but the capitalist threat.

GEORGE SOROS
Billionaire financier

Conflict theorists regard economic inequality as resulting from the domination of the **bourgeoisie** (owners of the means of production) over the **proletariat** (workers). The bourgeoisie accumulate wealth as they profit from the labor of the proletariat, who earn wages far below the earnings of the bourgeoisie. The U.S. educational institution furthers the ideals of capitalism by perpetuating the belief in equal opportunity, the "American Dream," and the value of the work ethic. The proletariat, dependent on the capitalist system, continue to be exploited by the wealthy and accept the belief that poverty is a consequence of personal failure rather than a flawed economic structure.

Conflict theorists note how laws and policies benefit the wealthy and contribute to the gap between the haves and the have-nots. Laws and policies that favor the rich—sometimes referred to as **wealthfare** or **corporate welfare**—include low-interest government loans to failing businesses, special subsidies and tax breaks to corporations, and other laws and policies that benefit corporations and the wealthy. A study of 250 large companies found that 41 companies paid no federal income tax in at least one year from 1996 to 1998 (McIntyre & Nguyen 2000). In those tax-free years, the 41 companies reported $25.8 billion in profits. But instead of paying $9 billion in federal income tax at the 35 percent rate, these companies received $3.2 billion in rebate checks from the U.S. Treasury. In 1998, 24 corporations—including General Motors, Pfizer, PepsiCo, Goodyear, Texaco, and Chevron—received tax rebates totaling $1.3 billion. The study found that 71 of the 250 companies paid taxes at less than half the official 35 percent corporate rate during 1996 to 1998. Companies use a variety of means to lower their federal income tax, including the growing use of stock options, which are an expense for tax purposes but do not count against profits reported to shareholders. Microsoft and Cisco Systems paid no federal income taxes in 1999 because stock options exercised by employees canceled profits for tax purposes (McIntyre & Nguyen 2000). The authors note that corporate tax avoidance is achieved "with significant help from Congress." Wealthy corporations use financial political contributions to influence politicians to enact corporate welfare policies that benefit the wealthy.

Corporate welfare is provided by government, but it is taxpayers and communities who pay the price. Consider the case of Seaboard Corporation, an agribusiness corporate giant that received at least $150 million in economic incentives from federal, state, and local governments between 1990 and 1997 to build and staff poultry- and hog-processing plants in the United States, support its operations in foreign countries, and sell its products (Barlett & Steele 1998). Taxpayers picked up the tab not just for the corporate welfare, but also for the costs of new classrooms and teachers (for schooling the children of Seaboard's employees, many of whom are immigrants), homelessness (because Seaboard's low-paid employees are unable to afford housing), and dwindling property values resulting from smells of hog waste and rotting hog carcasses in areas surrounding Seaboard's hog plants. Meanwhile, wealthy investors in Seaboard have earned millions in increased stock values.

Conflict theorists also note that throughout the world, "free-market" economic reform policies have been hailed as a solution to poverty. Yet, while such economic reform has benefited many wealthy corporations and investors, it has also resulted in increasing levels of global poverty. As companies relocate to countries with abundant supplies of cheap labor, wages decline. Lower wages lead to decreased consumer spending, which leads to more industries closing plants, going bankrupt, and/or laying off workers (downsizing). This results in higher unemployment rates and a surplus of workers, enabling employers to lower wages even more. Chossudovsky (1998) suggests that "this new international economic order feeds on human poverty and cheap labor" (p. 299).

Symbolic Interactionist Perspective

Symbolic interactionism focuses on how meanings, labels, and definitions affect and are affected by social life. This view calls attention to ways in which wealth and poverty are defined and the consequences of being labeled as "poor." Individuals who are viewed as poor—especially those receiving public assistance (i.e., welfare)—are often stigmatized as lazy; irresponsible; and lacking in abilities, motivation, and moral values. Wealthy individuals, on the other hand, tend to be viewed as capable, motivated, hard working, and deserving of their wealth.

The symbolic interaction perspective also focuses on the meanings of being poor. A qualitative study of over 40,000 poor women and men in 50 countries around the world explored the meanings of poverty from the perspective of those who live in poverty (Narayan 2000). Among the study's findings is that the experience of poverty involves psychological dimensions such as powerlessness, voicelessness, dependency, shame, and humiliation.

Meanings and definitions of wealth and poverty vary across societies and across time. For example, the Dinka are the largest ethnic group in the sub-Saharan African country of Sudan. By global standards, the Dinka are among the poorest of the poor, being among the least modernized peoples of the world. In the Dinka culture, wealth is measured in large part according to how many cattle a person owns. But, to the Dinka, cattle have a social, moral, and spiritual value as well as an economic value. In Dinka culture, a man pays an average "bridewealth" of 50 cows to the family of his bride. Thus, men use cattle to obtain a wife to beget children, especially sons, to ensure continuity of their ancestral lineage and, according to Dinka religious beliefs, their linkage with God. Although modernized populations might label the Dinka as poor, the Dinka view themselves as wealthy. As one Dinka elder explained, "It is for cattle that we are admired, we, the Dinka . . . All over the world, people look to us because of cattle . . . because of our great wealth; and our wealth is cattle" (Deng 1998, 107). Deng (1998) notes that many African peoples who are poor by U.S. standards resist being labeled as poor.

Definitions of poverty also vary within societies. For example, in Ghana men associate poverty with a lack of material assets, whereas for women poverty is defined as food insecurity (Narayan 2000).

The symbolic interactionist perspective emphasizes that norms, values, and beliefs are learned through social interaction and that social interaction influences the development of one's self-concept. Lewis (1966) argued that, over time, the

In today's economy a woman is considered lazy when she's at home taking care of her children. And to me that's not laziness . . . Let some of these men that work in the government, let some of them stay home and do that. They'll find that a woman is not lazy when she's taking care of her family.

DENISE TURNER
Welfare recipient

poor develop norms, values, beliefs, and self-concepts that contribute to their own plight. According to Lewis, the **culture of poverty** is characterized by female-centered households, an emphasis on gratification in the present rather than in the future, and a relative lack of participation in society's major institutions. "The people in the culture of poverty have a strong feeling of marginality, of helplessness, of dependency, of not belonging . . . Along with this feeling of powerlessness is a widespread feeling of inferiority, of personal unworthiness" (Lewis 1998, 7). Early sexual activity, unmarried parenthood, joblessness, reliance on public assistance, illegitimate income-producing activities (e.g., selling drugs), and substance use are common among the **underclass**—people living in persistent poverty. The culture of poverty view emphasizes that the behaviors, values and attitudes exhibited by the chronically poor are transmitted from one generation to the next, perpetuating the cycle of poverty. Critics of the culture of poverty approach point out that behaviors, values, and attitudes of the underclass emerge from the constraints and blocked opportunities that have resulted largely from the disappearance of work as jobs have moved out of inner-city areas to the suburbs (Van Kempen 1997; Wilson 1996) (see also chapters 7 and 13).

> Where jobs are scarce . . . and where there is a disruptive or degraded school life purporting to prepare youngsters for eventual participation in the workforce, many people eventually lose their feeling of connectedness to work in the formal economy; they no longer expect work to be a regular, and regulating, force in their lives . . . These circumstances also increase the likelihood that the residents will rely on illegitimate sources of income, thereby further weakening their attachment to the legitimate labor market. (Wilson 1996, 52–53)

> Many of today's problems in the inner-city ghetto neighborhoods—crime, family dissolution, welfare, low levels of social organization, and so on—are fundamentally a consequence of the disappearance of work.
>
> **William Julius Wilson**
> *Sociologist*

Wealth, Economic Inequality, and Poverty in the United States

The United States is a nation of tremendous economic variation ranging from the very rich to the very poor. Signs of this disparity are visible everywhere, from opulent mansions perched high above the ocean in California to shantytowns in the rural South where people live with no running water or electricity.

Wealth in the United States

> The accumulation of material goods is at an all-time high, but so is the number of people who feel an emptiness in their lives.
>
> **Al Gore**
> *Former U.S. Vice President*

Wealth refers to the total assets of an individual or household, minus liabilities (mortgages, loans, and debts). Wealth includes the value of a home, investment real estate, the value of cars, unincorporated business, life insurance (cash value), stocks/bonds/mutual funds/trusts, checking and savings accounts, individual retirement accounts (IRAs), and valuable collectibles. In the United States, wealthy households tend to have much of their wealth in stocks and bonds, whereas the less well-to-do typically hold most of their wealth in housing equity (Mishel et al. 2001).

Economic Inequality in the United States

The 1990s was a decade of U.S. economic growth: interest rates were down, unemployment low, and stock market averages reached record levels before declining at the end of 1999. At the close of the twentieth century, the United States had experienced the longest period of peacetime economic expansion in history.

But contrary to the adage that "a rising tide lifts all boats," economic prosperity has not been equally distributed in the United States. Economic inequality—the gap between the haves and the have-nots—has increased considerably.

From 1950 to 1978, all U.S. social classes enjoyed increases in economic prosperity. Family income for the bottom fifth of the U.S. population increased substantially more than for the top fifth of the population (a 138 percent increase for the former versus a 99 percent increase for the latter) (Briggs 1998). However, between 1979 and the end of the 1990s, inflation-adjusted income of the top 20 percent of the population grew by 26 percent while for the poorest it decreased by 9 percent. The average CEO pay, which increased 535 percent in the 1990s, is 475 times the pay of the average worker (Anderson, Cavanagh, Collins, Hartman, & Yeskel 2000). If the minimum wage had risen as fast as CEO pay in the 1990s, it would be $23.13 per hour instead $5.15.

The distribution of wealth is much more unequal than the distribution of wages or income. In 1998, the one percent of households with the highest incomes received 16.6 percent of all income. In the same year, the wealthiest one percent of households owned 38.1 percent of all wealth (Mishel et al. 2001).

Patterns of Poverty in the United States

Poverty is not as widespread or severe in the United States as it is in many less developed countries. Nevertheless, poverty represents a significant social problem in the United States. In 1999, the U.S. poverty rate of 11.8 percent was the lowest rate since 1979 (Dalaker & Proctor 2000), but that was no consolation to the more than 32 million Americans living in poverty in that year. Poverty rates vary considerably among the states, from 7.2 percent in Maryland to 20.5 percent in New Mexico. The average dollar amount needed to raise a poor family out of poverty in 1999 was $6,687 (Dalaker & Proctor 2000).

Poverty rates vary according to age, education, sex, family structure, race/ethnicity, and labor force participation. As discussed in this chapter's *Social Problems Research Up Close* feature, media portrayals of the poor do not accurately reflect the demographic characteristics of the poor.

Age and Poverty Children are more likely than adults to live in poverty (see Table 10.3). The 1999 U.S. poverty rate for people under age 18 was 16.9 percent—the lowest child poverty rate since 1979 (Dalaker & Proctor 2000). The poverty rate for young children is at least one-third higher and usually two to three times as high in the United States as in any other Western industrialized nation (Levitan, Mangum, & Mangum 1998).

Since the 1970s the poverty rate among the elderly has experienced a downward trend, largely as a result of more Social Security benefits and the growth

> We talk about the American Dream, and want to tell the world about the American Dream, but what is that dream, in most cases, but the dream of material things? I sometimes think that the United States for this reason, is the greatest failure the world has ever seen.
>
> EUGENE O'NEILL
> *Playwright*

Table 10.3 *U.S. Poverty Rates, by Age, 1999*

Age	Poverty Rate
Under 18 years	16.9
18 to 64	10.0
65 years and over	9.7

Source: Dalaker, Joseph and Bernadette D. Proctor. 2000. *Poverty in the United States: 1999.* Current Population Reports P60-210. U.S. Census Bureau. Washington, D.C.: U.S. Government Printing Office. www.census.gov/hhes/poverty/threshold/thresh99.html

Media Portrayals of the Poor

In the 1990s, intense political activity surrounding welfare reform placed poverty and welfare high on the nation's agenda. Throughout this period, the media focused significant attention on poverty and welfare reform issues. Researchers Clawson and Trice (2000) examined photographs of the poor found in newsmagazines during this period to determine whether the media perpetuate inaccurate and stereotypical images of the poor.

Sample and Methods

The sample consisted of every story on the topics of poverty, welfare, and the poor that appeared between January 1, 1993, and December 31, 1998 in five newsmagazines (*Business Week, Newsweek, New York Times Magazine, Time,* and *U.S. News & World Report*). A total of 74 stories were included in the sample, with a total of 149 photographs of 357 poor people.

In analyzing the photographs, researchers noted the race/ethnicity (white, black, Hispanic, Asian American, or undeterminable), sex (male or female), age (young: under 18; middle-aged: 18-64; or old: 65 and over); residence (urban or rural), and employment status (working/job training or not working). The researchers also analyzed whether each poor individual was portrayed in stereotypical ways, such as pregnant, engaging in criminal behavior, taking or selling drugs, drinking alcohol, smoking cigarettes, or wearing expensive clothing or jewelry. After coding all the photographs according to the aforementioned variables, the researchers compared the portrayal of poverty in their sample of photographs to poverty statistics reported by the U.S. Census Bureau or the U.S. House of Representatives Committee on Ways and Means.

Findings and Conclusions

Clawson and Trice found that the newsmagazine photographs overestimated the percentage of the poor who are black. U.S. Census data (from 1996) showed that African Americans made up 27 percent of the poor, but in the magazine portrayals, they made up nearly half (49 percent) of the poor. Whites, who according to Census data made up 45 percent of the poor, were depicted in the magazine portrayals as 33 percent of the poor. There were no magazine portrayals of Asian Americans in poverty, and Hispanics were underrepresented by 5 percent. The researchers suggest that "this underrepresentation of poor Hispanics and Asian Americans may be part of a larger phenomenon in which these groups are ignored by the media in general" (pp. 56-57).

The elderly were also underrepresented in the magazine portrayals of the poor. According to Census data, the elderly made up 9 percent of the true poor, yet they were only

of private pensions (see also Chapter 6). In 1970, the poverty rate among U.S. elderly was 24.6 percent; this rate fell to 15.7 in 1980 and in 1999 reached a record low of 9.7 percent (Dalaker & Proctor 2000; Levitan et al. 1998).

Education and Poverty Education is one of the best insurance policies to protect against an individual living in poverty. In general, the higher a person's level of educational attainment, the less likely that person is to be poor (see also Chapter 12). Among adults over age 25, those without a high school diploma are the most vulnerable to poverty (see Table 10.4).

Sex and Poverty Women are more likely than men to live below the poverty line—a phenomenon referred to as the **feminization of poverty.** The 1999 poverty rate of U.S. females was 13.2 percent, compared to 10.3 percent for males (U.S. Census Bureau 2000). As discussed in Chapter 8, women are less likely than men to pursue advanced educational degrees and tend to be concentrated in low-paying jobs, such as service and clerical jobs. However, even with the same level of education and the same occupational role, women still earn significantly less than men. Women who are minorities and/or who are single mothers are at increased risk of being poor.

4 percent of the magazine poor. Magazine portrayals of the poor exaggerated the percentage of the poor who are women, depicting 76 percent of the poor as women, compared to the Census figure of 62 percent.

Magazine depictions implied that poverty is primarily an urban problem. Ninety-six percent of the poor were shown in urban settings, compared to Census data showing that 77 percent of the poor resided in metropolitan areas. The authors note that "the urban underclass is often linked with various pathologies and antisocial behavior. Thus, this emphasis on the urban poor does not promote a positive image of those in poverty" (p. 60). And, the media portrayals of the poor in this study created the impression that most poor people do not work: less than one-third (30 percent) of poor adults were shown working or participating in job training programs. In reality, half of the poor worked in full-or part-time jobs, according to Census Bureau data.

Finally, the researchers analyzed the extent to which the news-magazines portrayed the poor as having stereotypical traits. They found that media portrayals did *not* portray poor mothers has having large numbers of children. And, the researchers noted that the media did not overly emphasize other stereotypical characteristics associated with the poor. Of the 357 people in the sample of photographs, only three were shown engaging in criminal behavior, and another three were shown with drugs. No alcoholics were presented, and only one person was smoking a cigarette. "However, of those seven stereotypical portrayals, only the person smoking was white—the others were either black or Hispanic" (p. 61). Only one poor woman was pregnant, so the media were not suggesting that poor women have babies to obtain larger welfare checks. However, the one pregnant woman shown was Hispanic. The researchers also noted whether the portrayals of the poor supported the image of the "welfare queen" stereotype (welfare recipients who do not really need assistance and who spend their welfare checks on luxuries). Of the 39 individuals who were shown with flashy jewelry or fancy clothes, "blacks and Hispanics were somewhat more likely to be portrayed this way than whites" (p. 61).

Clawson and Trice found that, overall, the portrayals of poor people in the five newsmagazines they analyzed did not reflect the reality of poverty; instead, they provided an inaccurate and stereotypical picture of poverty. The portrayals of poverty are important because they affect public opinion, which in turn, affects public policy. "Thus, if attitudes on poverty-related issues are driven by inaccurate and stereotypical portrayals of the poor, then the policies favored by the public (and political elites) may not adequately address the true problems of poverty" (p. 61).

Source: Clawson, Rosalee A. and Rakuya Trice. 2000. "Poverty As We Know It: Media Portrayals of the Poor." *Public Opinion Quarterly* 64:53–64.

Family Structure and Poverty Poverty is much more prevalent among female-headed, single-parent households than among other types of family structures (see Figure 10.1). The relationship between family structure and poverty helps to explain why women and children have higher poverty rates than men (see also Chapter 5).

Table 10.4 *Poverty Rates of U.S. Adults Aged 25 and Older, by Educational Attainment, 1998*

Level of Education	Percent Living in Poverty
No high school diploma	23.4%
High school diploma (no college)	10.1%
Some college (no bachelor's degree)	6.5%
Bachelor's degree or more	3.0%

Source: U.S. Census Bureau. 1999. "Annual Demographic Survey, March Supplement. Table 7." http://ferret.bls.census.gov/macro/031999/pov/new7_000.htm

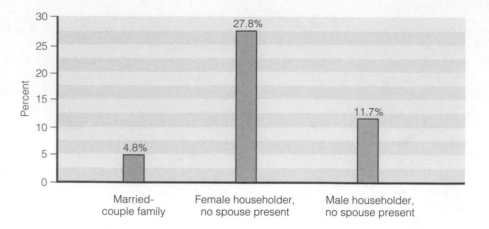

Figure 10.1 *U.S. Poverty Rates, by Family Structure: 1999*

Source: Dalaker, Joseph and Bernadette D. Proctor. 2000. *Poverty in the United States: 1999.* Current Population Reports P60-210. U.S. Census Bureau. Washington, D.C.: U.S. Government Printing Office.

In other countries, poverty rates of female-headed families are lower than those in the United States. For example, poverty rates of female-headed households are less than 10 percent in Belgium, France, Great Britain, Ireland, Luxembourg, Netherlands, Norway, and Poland (Pressman 1998). Unlike the United States, these countries offer a variety of supports for single mothers, such as income supplements, tax breaks, universal child care, national health care, and higher wages for female-dominated occupations.

Race/Ethnicity and Poverty Nearly half (46 percent) of the poor in the United States are non-Hispanic whites (Dalaker & Proctor 2000). However, as displayed in Figure 10.2, poverty rates are higher among blacks, Hispanics, and Native American/Alaska Natives than among non-Hispanic whites.

As discussed in Chapter 7, past and present discrimination has contributed to the persistence of poverty among minorities. Other contributing factors include the loss of manufacturing jobs from the inner city, the movement of whites and middle-class blacks out of the inner city, and the resulting concentration of poverty in predominantly minority inner-city neighborhoods (Massey 1991; Wilson 1987; 1996). Finally, blacks and Hispanics are more likely to live in fe-

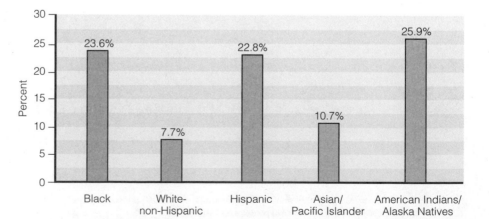

Figure 10.2 *U.S. Poverty Rates, by Race and Ethnicity: 1999*

Source: Dalaker, Joseph and Bernadette D. Proctor. 2000. *Poverty in the United States: 1999.* Current Population Reports P60-210. U.S. Census Bureau. Washington, D.C.: U.S. Government Printing Office.

male-headed households with no spouse present—a family structure that is associated with high rates of poverty.

Labor Force Participation and Poverty A common image of the poor is that they are jobless and unable or unwilling to work. Although the poor in the United States are primarily children and adults who are not in the labor force, many U.S. poor are classified as **working poor**. The working poor are individuals who spend at least 27 weeks per year in the labor force (working or looking for work), but whose income falls below the official poverty level. In 1999, 43 percent of all U.S. poor (ages 16 and over) worked; 12 percent worked year-round full time (U.S. Census Bureau 2000).

> In our society, it is murder, psychologically, to deprive a man of a job or an income. You are in effect saying to that man that he has no right to exist.
>
> MARTIN LUTHER KING JR.
> *Civil rights activist*

Consequences of Poverty and Economic Inequality

Poverty is associated with health problems, problems in education, problems in families and parenting, and housing problems. These various problems are interrelated and contribute to the perpetuation of poverty across generations, feeding a cycle of intergenerational poverty. In addition, poverty and economic inequality breed social conflict and war.

Health Problems and Poverty

In Chapter 2, we noted that poverty has been identified as the world's leading health problem. Persistent poverty is associated with higher rates of infant mortality and childhood deaths and lower life expectancies among adults. Poverty often causes chronic malnutrition, which can result in permanent brain damage, learning disabilities, and mental retardation in infants and children (Hill 1998). Poor children and adults also receive inadequate and inferior health care, which exacerbates their health problems (see also Chapter 2).

Economic inequality also affects psychological and physical health. Streeten (1998) cited research that suggests that "perceptions of inequality translate into psychological feelings of lack of security, lower self-esteem, envy and unhappiness, which, either directly or through their effects on life-styles, cause illness" (p. 5). Poor and middle income adults who live in states with the greatest gap between the rich and the poor are much more likely to rate their own health as poor or fair than people who live in states where income is more equitably distributed (Kennedy, Kawachi, Glass, & Prothrow-Stith 1998).

Educational Problems and Poverty

Research indicates that children living in poverty are more likely to suffer academically than are children who are not poor. "Overall, poor children receive lower grades, receive lower scores on standardized tests, are less likely to finish high school, and are less likely to attend or graduate from college than are non-poor youth" (Seccombe 2001, 323). The various health problems associated with childhood poverty contribute to poor academic performance. The poor often attend schools that are characterized by lower-quality facilities, overcrowded classrooms, and a higher teacher turnover rate (see also Chapter 12). Children living in poor inner-city ghettos "have to contend with public schools plagued by unimaginative curricula, overcrowded classrooms, inadequate plant and facilities, and only a small proportion of teachers who have confidence in their students and expect them to learn" (Wilson 1996, xv). Because poor parents have less schooling, on average, than do nonpoor parents, they may be less able to encourage and help their children succeed in school. However, research suggests that family income is a stronger predictor of ability and achievement outcomes than are measures of parental schooling or family structure (Duncan & Brooks-Gunn 1997). Poor parents have fewer resources to provide educational experiences (e.g., travel), private tutoring, books, and computers for their children. Not surprisingly, parental wealth strongly influences college enrollment (Conley 2001).

Poverty also presents obstacles to educational advancement among poor adults. Women and men who want to further their education in order to escape poverty may have to work while attending school or may be unable to attend school because of unaffordable child care, transportation, and/or tuition/fees/books.

Family Stress and Parenting Problems Associated with Poverty

In some cases, family problems contribute to poverty. For example, domestic violence causes some women to flee from their homes and live in poverty without the economic support of their husbands. Poverty also contributes to family problems. The stresses associated with poverty contribute to substance abuse, domestic violence, child abuse and neglect, divorce, and questionable parenting practices. For example, poor parents unable to afford child care expenses are more likely to leave children home without adult supervision. Poor parents are more likely than other parents to use harsh physical disciplinary techniques, and they are less likely to be nurturing and supportive of their children (Mayer 1997a; Seccombe 2001).

Another family problem associated with poverty is teenage pregnancy. Poor adolescent girls are more likely to have babies as teenagers or become young single mothers. Early childbearing is associated with numerous problems, such as increased risk of premature or low-birthweight babies, dropping out of school, and lower future earning potential as a result of lack of academic achievement. Luker (1996) notes that "the high rate of early childbearing is a measure of how bleak life is for young people who are living in poor communities and who have no obvious arenas for success" (p. 189). For poor teenage women who have been excluded from the American dream and disillusioned with education, "childbearing . . . is one of the few ways . . . such women feel they can make a change in their lives . . . " (p. 182).

> Having a baby is a lottery ticket for many teenagers: it brings with it at least the dream of something better, and if the dream fails, not much is lost . . . . In a few cases it leads to marriage or a stable relationship; in many others it motivates a woman to push herself for her baby's sake; and in still other cases it enhances the woman's self-esteem, since it enables her to do something productive, something nurturing and socially responsible . . . To the extent that babies can be ill or impaired, mothers can be unhelpful or unavailable, and boyfriends can be unreliable or punitive, childbearing can be just another risk gone wrong in a life that is filled with failures and losses. (Luker 1996, 182)

Housing Problems and Homelessness

Housing for the poor in the United States is largely inadequate. Many poor families and individuals live in housing units that lack central heating and air conditioning, sewer or septic systems, electric outlets in one or more rooms, and that have no telephone. Housing units of the poor are also more likely to have holes in the floor, a leaky roof, and open cracks in the walls or ceiling. In addition, poor individuals are more likely than the nonpoor to live in high-crime neighborhoods (Mayer 1997b).

The lack of affordable housing has produced a housing crisis that increasingly affects the middle class, as well as the poor. In the United States, 13.7 million families (14 percent) who have "critical housing needs" spend half or more of their income on housing or live in substandard housing, and 3 million of them are employed families (Stegman, Quercia, & McCarthy 2000). In the late 1990s, rents rose about twice as much as the consumer price index; and rent increases now exceed inflation everywhere across the country.

Even substandard housing would be a blessing to the 730,000 men, women, and children in the United States who are homeless on a given day—a number that reaches as high as 2 million during a year ("America's Poorest People Have No Place to Go" 2000). Homelessness has become a growing problem in the United States. Research on the duration of homelessness has identified three categories of homeless people (National Law Center on Homelessness and Poverty 2000):

- *Transitionally homeless* have a single episode of homelessness lasting an average of 58 days,
- *Episodically homeless* have four to five episodes of homelessness lasting a total of 265 days,
- *Chronically homeless* have an average of two episodes, lasting a total of 650 days.

The homeless population is diverse (see Table 10.5). Some homeless individuals have been forced out of their houses or apartments by rising rents or the inability to pay the mortgage. The homeless population also includes individu-

> Unfortunately, there are a lot more $6-an-hour jobs than $6-an-hour apartments.
>
> ANDREW CUOMO
> *Secretary of Housing and Urban Development*

© Elena Rooraid/PhotoEdit

Some cities and communities have laws or ordinances that prohibit the homeless from sleeping on public benches and soliciting money.

Life on the Streets: New York's Homeless

While a sociology graduate student at Columbia University, Gwendolyn Dordick undertook a study of homeless people living in New York City. She spent 15 months with four groups of homeless people: inhabitants of a large bus terminal, residents of a shantytown, occupants of a large public shelter, and clients of a small, church-run private shelter. In this *Human Side* feature, we present some of Dordick's observations, as well as excerpts from her conversations with homeless individuals she encountered at the Station and the Shanty.

The Station

The Station, located in Manhattan's West Side, encompasses bus terminals and depots, ticketing windows, shops, and fast-food restaurants. Scattered among the commuters and visitors are homeless women and men, and young adolescent boys and girls who have either run away or were kicked out by their guardians. One homeless man described living on the edge:

> Living on the streets makes you do a lot of things that you wouldn't normally do . . . Comin' into this environment I've done a lot of things I said I wouldn't do . . . There was some people that came along in a van and just threw sandwiches on the street and I picked them up and ate them . . . The guilt almost killed me . . . but my stomach said, "Hey, listen, you better eat this food." (pp. 5–6)

Homeless individuals often rely on one another, offering each other companionship, friendship, and protection. Although they have little to share, they often share what they have with their fellow homeless friends.

These people, when I got down here, these people reached out to me because they knew, they already knew what it was like. They're not afraid to help their fellow man. As soon as I got down here I met Ron and the fellows and they didn't push me away. I mean I didn't know where to go, I didn't know where to eat, I didn't know where to sleep. They just invited me right in. And ever since then at least I've been healthy, and I've been clean since I've met them (p. 13).

Homeless individuals are often treated harshly by police. One homeless man lamented:

> You may have an invalid laying down here. He's got problems and the [Station] cops will come up and kick him. Like he's an animal with no rights. (p. 11)

Pregnancy is common among homeless women who do not have access to or cannot afford contraception. Pregnancy can have disastrous consequences for homeless women, who fear having a baby and having to care for a baby. According to one man in the Station:

> It's one thing being homeless, but pregnant and homeless? Some women have their babies right out here; others get rid of them . . . Some of them abort theirself by sticking hangers up their vaginas. I've seen that myself. This young girl didn't want a baby and she stuck a hanger up her vagina. She had to go to the hospital. . . . (p. 25)

The Shanty

A barricaded makeshift community of 20 or so residents sits on a formerly vacant lot visible from the nearby streets and a bridge that crosses the East River. The Shanty consists of 15 makeshift dwellings, or "huts" as they are called by the residents, which are made of a vari-

Table 10.5 *Characteristics of the Homeless in the United States*

25% to 40% work

37% are families with children

25% are children

25 to 30% are mentally disabled

30% are veterans

40% are drug or alcohol dependent

Source: "Homelessness and Poverty in America." 2000. National Law Center on Homelessness and Poverty. http://www.nlchp.org/h&pusa.htm

ety of discarded materials such as pieces of wood and boards, cardboard, mattresses, fabric, and plastic tarps. The materials are fastened with nails, twine, or fabric. One of the residents has tapped into a source of electricity by running a wire from a lamppost into several huts, providing electricity for light and heat. The researcher notes that as in the case of the residents of the Station, welfare plays a minimal role in the lives of residents of the Shanty. "So difficult is negotiating the system that most forgo their entitlements" (p. 58). One resident explains:

> I don't get welfare. I just can't . . . do it. I hate those people in there. They make you . . . sit and ask you questions that don't make any sense . . . You're homeless but you have to have an address. What kind of shit is that? Give me a break. They want you to get so . . . upset that you do get up and walk out. (p. 58)

Although drug use is common among residents of the Shanty, using drugs in public is a violation of norms of "etiquette." One resident explained that using drugs in the presence of a nonuser is disrespectful:

> For me to just take my works out and shoot, I would feel uncomfortable in front of you. It's not right . . . very disrespectful. God forbid, I could be an influence. I could cause you to do it. (pp. 71–72)

Although survival among the homeless requires hustling, buying, trading, and selling, not everything is for sale. Some belongings have sentimental value that outweighs their economic value. One resident of the Shanty treasures a small gold key:

> There's a golden rule about gifts. You treasure them. You don't give them away, you don't sell them. I have right here a little key, a skeleton key. A little kid handed it to me four years ago. And every time somebody see that and say, "what is that?" I say it's a key to the world. I wouldn't give it to anyone if it was given to me. I treasure it. (p. 78)

Friendships and love relationships among the homeless suffer from the stresses of drug addictions and impoverished conditions. Nevertheless, Dordick explains that the homeless "survive through their personal relationships" (p. 193). Relationships are critical to securing the material resources needed to survive and to creating—to the extent that it is possible—a safe and secure environment. One resident of the Shanty, Richie, conveys the importance of a love relationship:

> . . . Regardless of what people might think and say, most of us that might have a woman it's all we want. We don't look for anyone else. We really don't . . . I'm happy just to take care of my woman . . . I happen to love my girl . . . Really, I know it sounds corny, but that's the truth . . . . As a matter of fact, we're gonna be married soon.

A Final Note

Virtually all the homeless individuals Dordick encountered expressed the desire to escape homelessness and be self-reliant—and they want to be understood. In the words of one homeless man at the Station:

> You never see me sleep in the street. I worked 32 years of my life. Went to prison in '85 . . . . I was brought up with a certain degree of independence . . . . And now all I need is two dollars to go and sit in a movie all night long. My pride is too good to beg. I don't want you to help me, Miss. I want you to understand me (p. 5).

als who have been released from mental hospitals as a result of the movement to deinstitutionalize individuals with psychiatric disorders. This chapter's *The Human Side* feature presents glimpses of what it is like to be homeless.

Intergenerational Poverty

As we have seen, problems associated with poverty, such as health and educational problems, create a cycle of poverty from one generation to the next. Poverty that is transmitted from one generation to the next is called **intergenerational poverty**. In a study of intergenerational poverty using a national longitudinal survey of families, researchers found considerable mobility out of childhood poverty: three-quarters of white poor children and over half of black poor children escaped poverty in early adulthood (Corcoran & Adams 1997). However, both white and black children in poor families were still much more likely to be poor in early adulthood than were children raised in nonpoor families.

Intergenerational poverty creates a persistently poor and socially disadvantaged population sometimes referred to as the underclass. The term **underclass** usually refers to impoverished individuals, often those who live in economically distressed neighborhoods (ghettos, slums, or barrios) with low educational attainment, chronic unemployment or underemployment, criminal involvement, unstable family structures, and welfare dependency. Although the underclass is stereotyped as being composed of minorities living in inner city or ghetto communities, the underclass is a heterogeneous population that includes poor whites living in urban and nonurban communities (Alex-Assensoh 1995).

Mead (1992) argues that intergenerational poverty may be caused by welfare dependency. According to Mead, when poor adults rely on welfare, the stigma of welfare fades, and welfare recipients develop poor work ethics that are passed on to their children. William Julius Wilson attributes intergenerational poverty and the underclass to a variety of social factors, including the decline in well-paid jobs and their movement out of urban areas, the resultant decline in the availability of marriageable males able to support a family, declining marriage rates and an increase in out-of-wedlock births, the migration of the middle-class to the suburbs, and the impact of deteriorating neighborhoods on children and youth (Wilson 1987; 1996).

War and Social Conflict

Poverty is often the root cause of conflict and war within and between nations, as "the desperation of the poor is never quiet for long" (Speth 1998, 281). Not only does poverty breed conflict and war, but war also contributes to poverty. For example, war contributes to homelessness as individuals and families are forced to flee from their homes. Military and weapons spending associated with war also divert resources away from economic development and social spending on health and education.

In the United States, the widening gap between the rich and poor may lead to class warfare (hooks 2000). Briggs (1998) asks how long the United States can maintain social order "when increasing numbers of persons are left out of the banquet while a few are allowed to gorge?" (p. 474). Although Karl Marx predicted that the have-nots would revolt against the haves, Briggs does not foresee a revival of Marxism; "the means of surveillance and the methods of suppression by the governments of industrialized states are far too great to offer any prospect of success for such endeavors" (p. 476). Instead, Briggs predicts that American capitalism and its resulting economic inequalities will lead to social anarchy—a state of political disorder and weakening of political authority.

Strategies for Action: Antipoverty Programs, Policies, and Proposals

In the United States, federal, state, and local governments have devoted considerable attention and resources to antipoverty programs for the last 50 years. Here we describe some of these programs and proposals, discuss international responses to poverty, and note the role of charity and the nonprofit sector in poverty alleviation.

Government Public Assistance and Welfare Programs in the United States

Many public assistance programs stipulate that households are not eligible for benefits unless their income and/or assets fall below a specified guideline. Programs that have eligibility requirements based on income are called **means-tested programs**. Government public assistance programs designed to help the poor include cash support, food programs, housing assistance, medical care, educational assistance, job training programs, child care, child support enforcement, and the earned income tax credit (EITC).

In 1996, President Clinton signed into law the **Personal Responsibility and Work Opportunity Reconciliation Act (PRWOR)** with the promise of "ending welfare as we know it." As you read further, you will note that the 1996 welfare reform legislation has affected numerous public assistance programs, primarily in the form of cutbacks and eligibility restrictions.

Cash Support Publicly funded cash support programs include Supplemental Security Income (SSI) and Temporary Assistance to Needy Families (TANF).

Supplemental Security Income Federal SSI, administered by the Social Security Administration, provides a minimum income to poor people who are aged 65 or older, blind, or disabled. Under the 1996 welfare reforms, the definition of disability has been sharply restricted and the eligibility standards tightened.

Temporary Assistance to Needy Families Before 1996, a cash assistance program called **Aid to Families with Dependent Children (AFDC)** provided single parents (primarily women) and their children with a minimum monthly income. The 1996 welfare reform legislation ended AFDC and replaced it with a program called **Temporary Assistance to Needy Families (TANF)**. In 1999, 98 percent of TANF families received cash and cash-equivalent assistance, in the monthly average amount of $357 (U.S. Department of Health and Human Services 2000). Under the new welfare legislation, after 2 consecutive years of receiving cash aid, welfare recipients are required to work at least 20 hours per week or to participate in a state-approved work program (exceptions may be made). A lifetime limit of 5 years is set for families receiving benefits, and able-bodied recipients aged 18 to 50 and without dependents have a 2-year lifetime limit. To qualify for TANF benefits, unwed mothers under the age of 18 are required to live in an adult-supervised environment (e.g., with their parents) and to receive education and job training. Legal immigrants (with few exceptions) are not eligible to receive benefits. Some states have implemented "behavior modification" measures, linking welfare receipt to a set of expected behaviors. Such state programs include (Albelda & Tilly 1997):

- Learnfare, which suspends aid if a child misses a certain number of days of school or gets a failing grade;
- The Family Cap, which freezes benefits at their current level when a poor mother receiving cash aid has another child (rather than raising benefits as was standard in the AFDC program);
- Incentives for implanting the contraceptive Norplant;
- Bridefare (or Wedfare), which gives monetary benefits for marrying the father of a child;
- Shotfare, in which a family loses benefits if immunization records are incomplete for any child.

People on welfare are just like you and me. They have the same basic hopes and fears. They want a job that brings self-worth and validation. They want to support their families and contribute to their communities. They want pride and dignity, just like those of us who have been lucky enough never to need public assistance.

ALEXIS M. HERMAN
Secretary, U.S. Department of Labor

A SMALL REASON TO FIND OUT IF YOU QUALIFY FOR FOOD STAMPS.

U.S. Department of Agriculture

USDA

CALL 1-800-221-5689

This poster is a public service message designed to encourage economically distressed families to apply for food stamps.

Food Benefits Food benefits include food stamps; school lunch and breakfast programs; the Special Supplemental Food Program for Women, Infants, and Children (WIC); and nutrition programs for the elderly. The largest food assistance program is the food stamp program, which issues monthly benefits through coupons or Electronic Benefits Transfer (EBT), using a plastic card similar to a credit card and a personal identification number (PIN). In 2000 the average gross monthly income per food stamp household was $584, and the average household benefit was $173 (U.S. Department of Agriculture 2000). The WIC program provides low-income pregnant and postpartum women, infants, and children up to 5 years of age with vouchers to purchase food items such as milk, cheese, fruit juice, and cereal, which are deemed important for pregnant and nursing mothers and their young children. Welfare reform legislation of 1996 resulted in cutbacks in federal food assistance programs, especially the food stamp program.

Housing Assistance Housing costs represent a major burden for the poor. The median cost of housing for poor households is 60 percent of household income; for nonpoor households the median cost of housing is only 20 percent of household income (Levitan et al. 1998). In 1999, more than 4 million households received some form of federal public or assisted housing (HUD 2000a). Among families with children living in public and assisted housing, nearly half earned wages as the primary source of income (HUD 2000a). Federal housing programs include public housing, Section 8 housing, and other private project-based housing.

The **public housing** program, initiated in 1937, provides federally **subsidized housing** owned and operated by local public housing authorities (PHAs). Public housing has been plagued by problems. To save costs and avoid public opposition, high-rise public housing units were built in inner-city projects. The concentration of poor families in deteriorating neighborhoods led to increases in crime, drugs, vandalism, and violence. One survey found that one in five residents living in public housing reported feeling unsafe in their neighborhood (HUD 2000b). The 2.6 million residents of public housing are more than twice as likely to suffer from firearm-related crimes than other U.S. residents (HUD 2000b). The Hope VI Urban Demonstration Program was established in 1992 to transform the nation's most distressed public housing projects by rebuilding the physical structure of public housing developments, expanding opportunities of its residents, and building a sense of community among residents. Between fiscal years 1993 and 2000, Hope VI funds were used to demolish nearly 97,000 severely distressed public housing units and will produce over 61,000 revitalized developments (Hope VI Fact Sheet 2000).

Rather than build new housing units for low-income families, Section 8 housing relies on existing housing. With **Section 8 housing**, federal rent sub-

Metropolitan Gardens, Alabama's largest public housing project, consists of 55 dilapidated apartment buildings with 910 housing units—some of which are 60 years old. Using funds from HOPE VI, a U.S. Housing and Urban Development program, this housing project will be replaced with about 600 townhouses for rent and purchase. Only 260 or so homes will be offered as low-income public housing, meaning many current residents will be displaced.

sidies are provided either to tenants (in the form of certificates and vouchers) or to private landlords. Other private project-based housing includes privately owned housing units that do not receive rent subsidies but receive other Federal subsidies such as interest rate reductions. Unlike public housing that confines low-income families to high-poverty neighborhoods, Section 8 and other private project-based housing attempts to disperse low-income families. However, because of opposition by residents in middle-class neighborhoods, most Section 8 housing units remain in low-income areas. In many communities, low-income families may remain on a waiting list for Section 8 housing for as long as 5 years.

Medical Care Medical care assistance programs include Indian Health Services, maternal and child health services, and Medicaid, which provides medical services and hospital care for the poor through reimbursing physicians and hospitals. However, many low-income individuals and families do not qualify for Medicaid and either cannot afford health insurance or cannot pay the deductible and co-payments under their insurance plan. In the earlier AFDC welfare program, all recipients were automatically entitled to Medicaid. Under the TANF program, states decide who is eligible for Medicaid; eligibility for cash assistance does not automatically convey eligibility for Medicaid. A provision of the 1996 welfare reform legislation guarantees at least 1 year of transitional Medicaid when leaving welfare for work.

Educational Assistance Educational assistance includes Head Start and Early Head Start programs and college assistance programs (see also Chapter 12). Head Start and Early Head Start programs provide educational services for disadvantaged infants, toddlers and preschool-age children and their parents.

Evaluations of Head Start programs indicate that they improve school performance and employability in adulthood; "compared with children in similar circumstances who do not receive early education, Head Start enrollees are more likely to graduate from high school and to find work" (Levitan et al. 1998, 163). However, because Head Start programs are not adequately funded, only one in five poor children are enrolled (Levitan et al. 1998).

To alleviate economic barriers for low-income persons wanting to attend college, the federal government offers grants, loans, and work opportunities. The

© AP/Wide World Photos

Pell grant program aids students from low-income families. The guaranteed student loan program enables college students and their families to obtain low-interest loans with deferred interest payments. The federal college-work-study program provides jobs for students with "demonstrated need." The federal government pays 70 to 80 percent of student wages.

Job Training Programs Various employment and job training programs are available to help individuals out of poverty (see also Chapter 11). These include summer youth employment programs, Job Corps, and training for disadvantaged adults and youth. These programs fall under the Job Training and Partnership Act (JTPA), a federally funded program passed in 1982 and amended in 1992. A primary shortcoming of job training programs has been that "they spread too little money among too many trainees, with the result that few are in training long enough for it to make a sufficient impact on their posttraining wages" (Levitan et al. 1998, 29).

Child Care Assistance In the United States, lack of affordable quality child care is a major obstacle to employment for single parents and a tremendous burden on dual-income families and employed single parents. The average cost of child care for a 4-year-old in a child care center ranges between $4,000 and $6,000 per year. The average cost of child care for a 4-year-old in an urban area is $10,000 or more per year, more than the average annual cost of public college tuition in all but one state (Children's Defense Fund 2000). In many cases, low-income families have placed their children in low-cost, often lower quality, and unstimulating care and nearly 7 million children are left home alone each week.

> If we are serious about rewarding work and helping people stay off welfare and keep their jobs, then we must make quality child care affordable now.
>
> MARIAN WRIGHT EDELMAN
> *President of the Children's Defense Fund*

Some public and private sector programs and policies provide limited assistance with child care. The Dependent Care Assistance Plan provisions of the 1981 Economic Recovery Tax Act permits individuals to exclude the value of employer-provided child care services from their gross income. However, few employers provide on-site child care or subsidies for child care. At the same time, Congress increased the amount of the child care tax credit and modified the federal tax code to allow taxpayers to shelter pretax dollars for child care in "flexible spending plans." The Family Support Act of 1988 offered additional funding for child care services for the poor (in conjunction with mandatory work requirements) and the Child Care and Development Block Grant, which became law in 1990, targeted child care funds to low-income groups. The Personal Responsibility and Work Opportunity Reconciliation Act of 1996 appropriated $16 billion over 5 years for child care, yet this amount is insufficient (Michel 1998). Of the nearly 15 million children eligible for child care assistance, only 12 percent receive any help (Children's Defense Fund 2000). According to Sonya Michel (1998), "the reluctance to make adequate provision for childcare is . . . symptomatic of a deeper aversion on the part of many legislators and public officials to helping poor and low-income women become truly economically independent, a status which is, in turn, essential to their ability to form autonomous households" (pp. 47–48).

> We must face the fact that families with extremely low wages do not earn enough to raise their children out of poverty. Without help like child care, transportation, training, and wage supplements, families are one crisis away from joblessness or hunger.
>
> DEBORAH WEINSTEIN
> *Children's Defense Fund, Family Income Division Director*

Child Support Enforcement The Personal Responsibility and Work Opportunity Act of 1996 requires states to set up child support enforcement programs. The new law establishes a Federal Case Registry and National Directory of New Hires to track delinquent parents across state lines, expands and streamlines procedures for direct withholding of child support from wages, and streamlines procedures for establishing paternity. Individuals who fail to cooperate with paternity establishment will have their monthly cash assistance reduced by at least 25 percent. The law allows for tough penalties for failure to pay child support, enabling states to seize assets and to revoke driving and professional licenses for parents who fail to pay child support. However, because many absent fathers earn low wages, collecting child support offers minimal economic relief to many single mothers.

Earned Income Tax Credit The federal **earned income tax credit (EITC),** created in 1975, is a refundable tax credit based upon a working family's income and number of children. The EITC is designed to offset Social Security and Medicare payroll taxes on working poor families and to strengthen work incentives. Increases in the EITC were enacted in 1986, 1990, and 1993. In 2000, families with more than one child and earning less than $31,152 could receive a credit of up to $3,888 (Friedman 2000).

The federal EITC lifts more children out of poverty than any other program (Johnson & Lazere 1998). The EITC is most effective in lifting Hispanic children out of poverty because poor Hispanic children are more likely than other poor children to live in families where a family member works full time. This is significant because poor families with a full-time worker receive the largest EITC benefits.

As of April 2000, 12 states and one locality (Montgomery County, Maryland) have enacted EITCs to offset state and local tax burdens on poor families (Friedman 2000). State sales and local taxes, for example, are **regressive taxes**, meaning that they absorb a much higher proportion of the incomes of lower-income households than of higher-income households. State and local EITCs help to offset the disproportionate burden that regressive taxes place on the poor.

Welfare in the United States: Myths and Realities

Public attitudes toward welfare assistance and welfare recipients are generally negative. Rather than view poverty as the problem, many Americans view welfare as the problem. What are some of the common myths about welfare that perpetuate negative images of welfare and welfare recipients?

MYTH 1 People receiving welfare are lazy and have no work ethic.

Reality First of all, single parents on welfare already do work—they do the work of parenting. Albelda and Tilly (1997) emphasize that "raising children is work. It requires time, skills, and commitment. While we as a society don't place a monetary value on it, it is work that is invaluable—and indeed, essential to the survival of our society" (p. 111). Second, many adults receiving public assistance are either employed or in the labor force looking for work. In 1999, more than one-quarter of adult welfare recipients were employed (earning an average of $598 per month) (U.S. Department of Health and Human Services 2000). Finally, a significant segment of welfare recipients are unable to work because of mental or physical disabilities or other medical conditions. "Among parents who are current TANF recipients who are not working, generally at least one-fifth but possibly as high as one-half in some states have health problems that they believe prevent them from working" (Sweeney 2000). These health problems include mental illness (e.g., clinical depression, post-traumatic stress disorder), physical impairments, and substance abuse problems.

MYTH 2 Welfare benefits are granted to many people who are not really eligible to receive them.

Reality Although some people obtain welfare benefits through fraudulent means, it is much more common for people who are eligible to receive welfare to not receive benefits. For example, the Department of Agriculture reported that at least 12 million people are not receiving food stamps even though they are eligible (Becker 2001). A main reason for not receiving benefits is lack of information; people don't know they are eligible. Many people who are eligible for public assistance do not apply for it because they do not want to be stigmatized as lazy people who just want a "free ride" at the taxpayers' expense—their sense of personal pride prevents them from receiving public assistance. Others want to avoid the administrative and transportation hassles involved in obtaining it.

MYTH 3 Most welfare parents are teenagers.

Reality In 1999, the average age of adult welfare recipients was 31.8 years. Only 6.2 percent of adult recipients were teenagers; 20 percent were 39 years and older (U.S. Department of Health and Human Services 2000). In addition, more than one-fourth (29 percent) of TANF families had no adult recipients—only children were eligible to receive assistance.

MYTH 4 Most welfare mothers have large families with many children.

Reality Mothers receiving welfare have no more children, on average, than mothers in the general population. In FY 1999, the average number of persons in TANF families was 2.8. The TANF families averaged 2 recipient children; two in five had only one child and one in 10 families had more than three children (U.S. Department of Health and Human Services 2000).

The economic, social, and psychological situation in which women on welfare find themselves is simply not conducive to desiring more children. Such women would appear to be motivated by cost-benefit considerations, but it is the costs that outweigh the benefits, not the reverse. Becoming pregnant and having a child are perceived as making the situation worse, not better.

MARK RANK
Sociologist

MYTH 5 Unmarried women have children so they can receive benefits. And if single mothers already receive benefits, they have additional children in order to receive increased benefits.

Reality Research consistently shows that receiving welfare does not significantly increase out-of-wedlock births (Albelda & Tilly 1997). In states that had the lowest AFDC cash benefits to single mothers, the teenage birthrates were among the highest (reported in Albelda & Tilly 1997).

MYTH 6 Public assistance to the poor creates an enormous burden on taxpayers.

Reality Public assistance to the poor is a small part of government spending. In the 2001 federal budget, 7 percent was allocated to Medicaid (health insurance for the poor) and only 6 percent was allocated to other means-tested forms of public assistance (Office of Management and Budget 2001) (see Figure 10.3). Meanwhile 16 percent of the 2001 federal budget was allocated for national defense and 23 percent was earmarked for providing Social Security benefits. Although Social Security benefits are paid to disabled workers and their children,

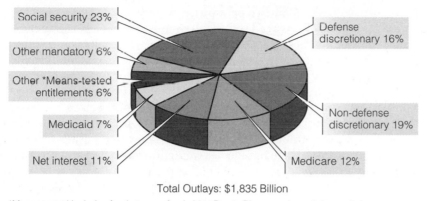

Total Outlays: $1,835 Billion

*Means tested includes food stamps, food aid to Puerto Rico, supplemental security income, child nutrition, earned income tax credit, and veterans' pensions

Figure 10.3 *2001 Federal Budget.*

Source: "A Citizen's Guide to the Federal Budget." 2001. Office of Management and Budget. Washington, D.C.: U.S. Government Printing Office.

they are also paid to retired workers and dependents and survivors, even those who are affluent. In addition, the U.S. government spends about $60 billion each year on tax subsidies to U.S. businesses, significantly more than it spends on welfare benefits (Stensel & Moore 1997). Despite the public perception that welfare benefits are too generous, cash and other forms of assistance to the poor do not meet the basic needs for many individuals and families who receive such benefits (see the *Social Problems Research Up Close* feature in Chapter 11).

Minimum Wage Increase and "Living Wage" Laws

An estimated 10.3 million workers (8.7 percent of the work force) make between $5.15 (the minimum wage) and $6.15 an hour. Of these workers, 71 percent are adults and 60 percent are women (Economic Policy Institute 2000). A full-time worker earning the $5.15 an hour would earn $10,712 per year, well below the 1999 federal poverty line of $13,410 for a family of three. Further, the purchasing power of the minimum wage has declined because increases in the minimum wage have not kept up with inflation. The result is that the minimum wage, when adjusted for inflation, is worth 21 percent less today than it was in 1979 (Mishel et al. 2001).

Clearly, raising the minimum wage would benefit low-wage workers and reduce poverty. Some states have established a minimum wage higher than the federal minimum wage. In January 2000, states with higher minimum wages included Oregon and Washington ($6.50); Massachusetts ($6.00); Delaware, Connecticut, and the District of Columbia ($6.15); Vermont and California ($5.75); and Alaska and Rhode Island ($5.65) (Economic Policy Institute 2000).

Those opposed to increasing the minimum wage argue that such an increase would result in higher unemployment, as businesses would reduce wage costs by hiring fewer employees. However, research has failed to find any systematic, significant job loss associated with minimum wage increases (Economic Policy Institute 2000).

We continue to believe that it is unacceptable that anyone who works full-time should have to rely on charity or the taxpayers to make ends meet. It's time for a real living wage.

BARRY HERMANSON
Co-Chair of the San Francisco Living Wage Coalition

In August 2000, San Francisco became the 51st locality in the nation to pass a living wage law. **Living wage laws** require state or municipal contractors, recipients of public subsidies or tax breaks, or, in some cases, all businesses to pay employees wages significantly above the federal minimum, enabling families to live above the poverty line. A wage earner must earn at least $8 an hour, working a 40-hour week in order to support a family of four above the poverty line (Kraut, Klinger, & Collins 2000). But over one quarter of U.S. jobs pay less than $8 an hour. Living wage laws are not only good for individuals and families; they are also good for business. Research findings show that businesses that pay their employees a living wage have lower worker turnover and absenteeism, reduced training costs, higher morale and higher productivity, and a stronger consumer market (Kraut et al. 2000). Over 50 business owners have signed a Living Wage Covenant, pledging to pay their employees over $8 an hour, as well as publicly advocate higher wages for all low-income workers.

Charity, Nonprofit Organizations, and Nongovernmental Organizations

Various types of aid to the poor are provided through individual and corporate donations to charities and nonprofit organizations. In 1998, 70 percent of U.S. households contributed to charity; the average contribution per household was $1,075 (Independent Sector 1999). Most households donated food or clothing and purchased goods or services from charitable organizations. Only 1.2 percent of contributors reported giving over the Internet.

Charity involves giving not only money and valuable goods, but time and effort in the form of volunteering. A national survey found that over half (56 percent) of Americans volunteered an average 3.5 hours per week in 1998 (Independent Sector 1999). The National Student Campaign against Hunger and Homelessness (NSCAHH) is the largest student volunteer network fighting hunger and homelessness in the United States, with more than 600 participating campuses in 45 states.

Nongovernmental organizations (NGOs) address many issues related to human rights, social justice, and environmental concerns. The number of international NGOs grew from fewer than 400 in 1900 to 26,000 in the year 2000—more than four times as many as existed just ten years earlier (Knickerbocker 2000; Paul 2000). At the Millennium Forum meeting in 2000, representatives from over 1,000 NGOs called for a UN Global Poverty Eradication Fund to ensure that poor people have access to credit (Deen 2000). The NGOs declared that poverty is the most widespread violation of human rights and called upon the United Nations and governments around the world to make poverty alleviation a priority.

All too many of those who live in affluent America ignore those who exist in poor America; in doing so, the affluent American will eventually have to face themselves with the question . . . : How responsible am I for the well-being of my fellows? To ignore evil is to become an accomplice to it.

MARTIN LUTHER KING JR.
Civil rights activist

International Responses to Poverty

Alleviating worldwide poverty continues to be a major concern of both developing and developed countries. Approaches to poverty reduction include promoting economic growth and investing in "human capital." Conflict resolution and the promotion of peace are also important for reducing poverty worldwide.

Promoting Economic Growth Economic growth, over the long term, generally reduces poverty (United Nations 1997). An expanding economy creates new employment opportunities and increased goods and services. In 1998, 150 million of the world's workers were unemployed (United Nations Development

Programme 2000). As employment prospects improve, individuals are able to buy more goods and services. The increased demand for goods and services, in turn, stimulates economic growth. As emphasized in Chapter 14, economic development requires controlling population growth and protecting the environment and natural resources, which are often destroyed and depleted in the process of economic growth.

However, economic growth does not always reduce poverty; in some cases it increases it. For example, growth resulting from technological progress may reduce demand for unskilled workers. Growth does not help poverty reduction when public spending is diverted away from meeting the needs of the poor and instead is used to pay international debt, finance military operations, and support corporations that do not pay workers fair wages. The World Bank loans about $30 billion a year to developing nations to pay primarily for roads, bridges, and industrialized agriculture that mostly benefit corporations. "Relatively little attention or money has been given to developing basic social services, building schools and clinics, and building decent public sanitation and clean water systems in some of the world's poorest countries" (Mann 2000, 2). Thus, "economic growth, though essential for poverty reduction, is not enough. Growth must be pro-poor, expanding the opportunities and life choices of poor people" (*Human Development Report* 1997, 72–73). Because three-fourths of poor people in most developing countries depend on agriculture for their livelihoods, economic growth to reduce poverty must include raising the productivity of small-scale agriculture. Not only does improving the productivity of small-scale agriculture create employment, it also reduces food prices. The poor benefit the most because about 70 percent of their income is spent on food (*Human Development Report* 1997). This chapter's *Focus on Technology* feature examines agricultural biotechnology as a strategy for alleviating global hunger.

Investing in Human Capital   Promoting economic development in a society requires having a productive work force. Yet, in many poor countries, large segments of the population are illiterate and without job skills, and/or are malnourished and in poor health. Thus a key feature of poverty reduction strategies involves investing in human capital. The term **human capital** refers to the skills, knowledge, and capabilities of the individual. Investments in human capital involve programs and policies that enhance the individual's health, skills, knowledge, and capabilities. Such programs and policies include those that provide adequate nutrition, sanitation, housing, health care (including reproductive health care and family planning), and educational and job training. Nobel Laureate Gary Becker has concluded that "the case is overwhelming that investments in human capital are one of the most effective ways to raise the poor to decent levels of income and health" (reported in Hill 1998, 279).

Poor health is both a consequence and a cause of poverty; improving the health status of a population is a significant step toward breaking the cycle of poverty. Increasing the educational levels of a population better prepares individuals for paid employment and for participation in political affairs that affect poverty and other economic and political issues. Improving the educational level and overall status of women in developing countries is also associated with lower birth rates, which in turn fosters economic development.

One way to help poor countries invest in human capital and reduce poverty is to provide debt relief. If African countries were relieved of their national debts, they would have funds that would save the lives of millions of children and provide basic education to millions of girls and women. Providing debt re-

> Trying to eradicate hunger while population continues to grow rapidly is like trying to walk up a down escalator.
>
> LESTER R. BROWN
> *World Watch Institute*

Global Hunger: Is Agricultural Biotechnology the Solution?

Biotechnology is any technique that uses living organisms or substances from those organisms to make or modify a product or develop microorganisms for specific uses (see also Chapter 15). **Agricultural biotechnology** involves the application of biotechnology to agricultural crops and livestock; however, our discussion here focuses on crops. Various terms refer to products that have been created or modified through agricultural biotechnology, including **genetically modified organisms (GMOs)**, genetically-improved organisms (GIOs), and genetically engineered foods (GE foods).

In 1999, over 70 genetically modified varieties of crops were registered for commercial cultivation worldwide (Persley 2000). Such crops include cotton, potato, pumpkin, corn, soybean, tobacco, papaya, squash, tomato, and canola (rapeseed). In the United States, at least 40 genetically modified products have completed all the federal regulatory requirements and may be sold commercially (United States Department of Agriculture 2001). An estimated 30,000 products in American supermarkets today —from ice cream to cantaloupes to corn flakes—contain genetically engineered ingredients (Environmental Defense 2000). Global areas planted with GM crops grew from 1.7 million hectares in 1996 to nearly 40 million in 1999; 72 percent of this area was in the United States, followed by Argentina (17 percent) and Canada (10 percent) (Serageldin 2000).

Scientists, academics, environmentalists, public health officials, policy makers, corporations, farmers, and citizens throughout the world are deeply divided over the use of agricultural biotechnology. Not surprisingly, supporters of agricultural technology emphasize its potential benefits, while critics focus on the potential risks. A number of ethical and other issues are also at the center of the controversy concerning biotechnology,

Health and Environmental Benefits

- *Alleviation of Poverty, Hunger, and Malnutrition*. Worldwide, 70 percent of poor and food insecure people live in rural areas, and most of these poor depend on agriculture for their livelihood (Pinstrup-Andersen & Cohen 2000). Any technology that improves agricultural productivity can potentially alleviate the poverty and hunger among the rural poor. Agricultural biotechnology can enable farmers to produce more food with higher nutritional value. Some GMOs are

designed to have a higher yield and earlier maturation. Others are engineered to be more durable during harvest or transportation. Some crops have improved tolerance to drought and poor soil conditions. Other crops are designed to resist herbicides (chemicals used to kill weeds), insects, and diseases.

Genes that increase vitamin A production and iron have been incorporated experimentally in rice. This could enhance the diets of the 180 million children who suffer from vitamin A deficiency—a deficiency that causes 2 million deaths annually and 14 million cases of eye damage. Iron deficiency, which affects 1 billion people in the developing world, leads to anemia—a condition that can diminish learning capacity and contributes to illness and death (Persley 2000).

- *Reduction in Pesticide Use*. Because some GMOs are designed to repel insects, they could replace chemical (pesticide) control and reduce the excessive use of pesticides that poisons field workers and contaminates land, water, and animals.

- *Less Deforestation*. As farmers have access to less land on which to plant crops, they are forced to clear forest area. As discussed in Chapter 14, deforestation contributes to environmental problems. GMOs enable farmers to reap higher yields from crops, and to plant crops on soil that otherwise would not be usable in farming. Thus, GMOs can potentially reduce deforestation.

Health and Environmental Risks

"Although no clear cases of harmful effects on human health have been documented from new genetically improved food, that does not mean that risks do not exist" (Persley 2000, 12). One potential health risk is food allergens in GM foods. Other health concerns are related to possible toxicity, carcinogenicity, food intolerance, antibiotic resistance buildup, and decreased nutritional value.

Another potential environmental risk is the spread of traits from GM plants to other plants, the effects of which are unknown. Additionally, insect populations can potentially build up resistance to GM plants with insect-repelling traits. And a potential threat to biodiversity is posed by the widespread uniformity of GM crops.

Ethical Issues and Other Concerns

- *"Playing God."* To some, the use of agricultural biotechnology is offensive because it involves "tinkering with the natural order" and "playing God."

Leisinger (2000) answers that criticism by saying "If God created humans as intelligent creatures, it should be compatible with God's intentions that they use their intelligence to improve living conditions" (p. 175).

- *Intellectual Property Rights.* Corporations that develop biotechnologies may obtain intellectual property rights, such as patents for their biotechnological inventions. A patent is a monopoly granted to the owner of an invention for a limited period of up to 20 years. Critics suggest that such rights result in increased prices, as other companies cannot offer the same technology at a competitive price. In response to this concern, Richer (2000) points out that intellectual property rights enable companies to recoup their investment in developing new technologies (the average cost of developing a GM plant is about $150 million).

  Others are concerned that intellectual property rights give corporations ownership of life forms—an idea that is intuitively unappealing. Yet, we already accept ownership rights of plants and animals. Serageldin (2000) asks how owning a "building block of life" is different from owning the actual living thing itself?

  Finally, critics are concerned about the "terminator gene"—the first patented technology aimed at protection of intellectual property rights regarding biotechnology. The terminator gene is designed to genetically switch off a plant's ability to germinate a second time. The terminator gene prevents farmers from planting seeds that are harvested from GMO crops and forces them to buy a fresh supply of seeds each year (if they want GMO seeds). Seeds with the terminator gene are not appropriate for small farmers in developing countries because the existing production process may not keep fertile and infertile seeds apart. If farmers accidentally planted infertile seeds, their losses could be devastating (Pinstrup-Andersen & Cohen 2000). In response to protests against terminator technology, Monsanto announced it would not market it (Shah 2001). But other approaches to property rights protection are under development. GE seeds are being developed which can only be activated through chemical treatment. Otherwise, the seed maintains its normal characteristics (without genetic modifications), but is still fertile. The farmer would have the choice to plant the seed as is, or to activate the genetically modified traits by applying the chemical (that is also made and sold by the same corporation that makes the seeds).

- *Corporate Greed.* Corporations that develop agricultural biotechnology do so in hopes of earning profit. However, according to Pinstrup-Andersen and Cohen (2000), corporations are not the only economic beneficiaries of GMOs. They cite a study of the distribution of the economic benefits of herbicide-tolerant soybean seed in the U.S. in 1997: the patent-holding company Monsanto received 22 percent of the economic benefits, seed companies gained 9 percent, consumers gained 21 percent, and farmers worldwide obtained 48 percent.

- *Neglect of the Needs of Developing Countries.* The development of GMOs to meet specific needs of developing countries is not likely to be profitable; consequently, relatively little biotechnology research has focused on the needs of poor farmers and consumers in developing countries. Those concerned with this neglect of the needs of developing countries call for stronger public sector involvement in the research and development of GMOs (Pinstrup-Andersen & Cohen 2000). Globally, 80 percent of biotechnology research is done by the private sector (Persley 2000).

- *Increased Poverty and Economic Inequality.* A criticism of agricultural biotechnology is that it is too expensive and inaccessible to small farmers. Persley (2000) suggests that unless countries have policies to ensure that small farmers have access to GM crop seeds and markets, agricultural biotechnology could lead to *increased* inequality of income and wealth if large farmers reap most of the benefits.

- *Scientific Apartheid.* Biotechnological knowledge and research is skewed to the potential markets of the affluent, excluding the concerns of the poor. This contributes to **scientific apartheid**—the growing gap between the industrial and developing countries in the rapidly evolving knowledge frontier (Serageldin 2000). To address this concern, poor farmers and other low-income populations in developing countries must have a voice in the development of and policies concerning agricultural biotechnology.

- *Insufficient Safeguards and Regulatory Mechanisms.* In 2000, Taco Bell taco shells, made by Kraft Foods, were recalled after traces of a genetically engineered variety of corn that had not been approved for human consumption were found in the taco shells (Union of Concerned Scientists 2001). No one—from farmers to grain dealers to Kraft—could explain how it got mixed into corn meant for taco shells. Further, the traces of unapproved corn were not found by the United States Department of Agriculture's Food

Continued

Safety and Inspection Service, nor by the Department of Health and Human Service's Food and Drug Administration. Rather, the traces were discovered by Genetically Engineered Food Alert—a coalition of biotech skeptics and foes. This widely publicized incident raised disturbing questions about the regulatory oversight of GE foods. Indeed, there is widespread agreement that the pursuit of agricultural biotechnology requires regulatory systems to govern food safety, assess risks, monitor compliance, and enforce regulations. But such safeguards are nonexistent in some countries and, as the Taco Bell incident suggests, even when regulatory systems are in place, they are not foolproof. However, some progress toward improving biotechnology safeguards was made in February 2000 when the landmark Biosafety Protocol was signed in Montreal by 130 nations. (Signing the protocol indicates general support for the Protocol and the intention to become legally bound by it. However, the Protocol does not become legally binding until a country ratifies the treaty by submitting a letter of acceptance to the United Nations.) The Biosafety Protocol includes the requirement that producers of a GMO must demonstrate it is safe before it is widely used. Prevailing U.S. policy, in contrast, required critics to prove that GMOs were potentially dangerous—"a deploy-now-ask-questions-later approach which places industry aspirations above public interest and safety" (Halweil 2000). The

Biosafety Protocol also allows countries to ban the import of GM crops based on suspected health, ecological, or social risks.

Concluding Remarks

Many citizens have clearly taken a stand either for or against agricultural biotechnology. However, many more are uncertain and struggle to make sense out of the competing claims of the benefits and safety versus the potential hazards of GMOs, and the complex ethical and sociopolitical implications of using these technologies.

As we strive to make sense of these issues, is it helpful to consider that many technologies with known risks have broad social acceptance? Consider the automobile—a technology that contributes to pollution and global warming and kills about 50,000 people each year and maims another 500,000 in the United States alone (Serageldin 2000). Yet, few individuals would agree to ban the automobile.

Is it helpful to consider the role of the media in shaping public views and opinions on biotechnology? Some have charged the media with biased reporting on biotechnology that fuels anti-biotech sentiments. For example, according to Leisinger (2000), when the Federal Institute of Technology in Zurich informed the world in 1999 of the possibility to genetically modify rice to contain vitamin A and iron—a major health benefit to the 250 million poor malnourished who subsist on rice—the media had little reac-

---

Whether there is proved to be life on Mars, and whether you may conduct your affairs electronically without leaving your armchair, the new century is not going to be a new century at all in terms of progress of humanity if we take along with us acceptance of the shameful shackles of the past. The shackles of poverty are not just a metaphor.

NADINE GORDIMER
Speaking at the United Nations on the occasion of International Day for the Eradication of Poverty

lief to the 20 worst-affected countries would cost between $5.5 billion and $7.7 billion—less than the cost of one Stealth bomber and roughly the cost of building the Euro-Disney theme park in France (*Human Development Report* 1997).

Understanding *The Haves and the Have-Nots*

As we have seen in this chapter, economic prosperity has not been evenly distributed; the rich have become richer, while the poor have become poorer. Meanwhile, the United States has implemented welfare reform measures that essentially weaken the safety net for the impoverished segment of the population—largely children. Welfare reform legislation of 1996 has achieved its goal of reducing welfare rolls across the country. From 1996 to 1999, the welfare caseload has been cut by nearly half (U.S. Department of Health and Human Services 2000). Advocates of welfare reform argue that transitions from welfare to work benefit children by creating positive role models in their working mothers, promoting maternal self-esteem, and fostering career advancement and higher family earnings. Critics of welfare reform argue that reforms increase stress on partion. Four months later when news broke that larvae of the monarch butterfly were damaged in a GM crop experiment, the media picked up on the story and focused on the potential harm of biotechnology to biodiversity.

Lester Brown (2001) of the World Watch Institute suggests that "perhaps the largest question hanging over the future of biotechnology is the lack of knowledge about the possible environmental and human health effects of using genetically modified crops on a large scale over the long term" (p. 52). This lack of knowledge calls for more research to answer questions about the potential risks of GM crops. But efforts to conduct such research are impeded by anti-biotechnology activists who have destroyed test sites and research offices. "Open debate about the issues involved is essential, but physical attacks on research and testing efforts contribute little to the free exchange of ideas or the formulation of policies that will advance food security" (Pinstrup-Andersen & Cohen 2000, p. 168).

Finally, even supporters of agricultural biotechnology remind us that such technology is not a "silver bullet" that will end poverty and hunger. Rather, "It is critical that biotechnology be viewed as one part of a comprehensive sustainable poverty alleviation strategy, not a technological 'quick-fix' for world hunger and poverty" (Persley 2000, p. 16).

Sources:

Brown, Lester R. 2001. "Eradicating Hunger: A Growing Challenge." In *State of the World 2001*, eds. Lester R. Brown, Christopher Flavin, and Hilary French, pp. 43–62. New York: W.W. Norton & Co.

Environmental Defense. 2000. "Annual Report 2000." http://www.environmentaldefense.org/pub/AnnualReport/2000/AR00.pdf

Halweil, Brian. 2000 (May/June). "Politically Modified Foods." *World Watch Magazine*. http://www.worldwatch.org/mag/2000/00-3a.html

Leisinger, Klaus M. 2000. "Ethical Challenges of Agricultural Biotechnology for Developing Countries." In *The Use of Agricultural Biotechnology to Feed the Poor*, eds. G. J. Persley and M. M. Lantin, pp. 173–180. Washington D. C.: Consultive Group on International Research. The World Bank.

Persley, G. J. 2000. "Agricultural Biotechnology and the Poor: Promethean Science." In *The Use of Agricultural Biotechnology to Feed the Poor*, eds. G. J. Persley and M. M. Lantin, pp. 3–21. Washington D.C.: Consultive Group on International Research. The World Bank.

Pinstrup-Andersen, Per and Marc J. Cohen. 2000. "Modern Biotechnology for Food and Agriculture: Risks and Opportunities for the Poor." In *The Use of Agricultural Biotechnology to Feed the Poor*, eds. G. J. Persley and M. M. Lantin, pp. 159–168. Washington D. C.: Consultive Group on International Research. The World Bank.

Richer, David L. 2000. "Intellectual Property Protection: Who Needs It?" In *The Use of Agricultural Biotechnology to Feed the Poor*, eds. G. J. Persley and M. M. Lantin, pp. 159–168. Washington D.C.: Consultive Group on International Research. The World Bank.

Serageldin, Ismail. 2000. "The Challenge of Poverty in the 21st Century: The Role of Science." In *The Use of Agricultural Biotechnology to Feed the Poor*, eds. G. J. Persley and M. M. Lantin, pp. 25–32. Washington D.C.: Consultive Group on International Research. The World Bank.

Shah, Anup. 2001 (January 1). "Terminator Technology." *Genetically Engineered Food*. http://www.globalissues.org/EnvIssues/GEFood/Terminator.asp?Print=True

Union of Concerned Scientists. 2001. http://www.ucsusa.org/agriculture/0biotechnology.html

United States Department of Agriculture. 2001. "Agricultural Biotechnology." http://www.aphis.usda.gov/biotechnology/faqs.html

ents, force young children into inadequate childcare, reduce parents' abilities to monitor the behavior of their adolescents, and deepen the poverty of many families. Although the long-term effects of welfare reform are not yet known, one study of the impact of welfare reform on children concluded that reforms will help some children and hurt others (Duncan & Chase-Lansdale 2001). As we discuss in the next chapter (Work and Unemployment), many of the jobs available to those leaving welfare for work are low-paying, have little security, and offer few or no benefits. Without decent wages, and without adequate assistance in child care, housing, health care, and transportation, many families who leave welfare for work find their situation becomes worse, not better (Children's Defense Fund and the National Coalition for the Homeless 1998).

A common belief among U.S. adults is that the rich are deserving and the poor are failures. Blaming poverty on individual rather than structural and cultural factors implies not only that poor individuals are responsible for their plight, but also that they are responsible for improving their condition. If we hold individuals accountable for their poverty, we fail to make society accountable for making investments in human capital that are necessary to alleviate poverty. Such human capital investments include providing health care, adequate food and housing, education, child care, and job training. Economist Lewis Hill (1998) believes that "the fundamental cause of perpetual poverty is the failure of the American people to invest adequately in the human capital represented by impoverished children" (p. 299). Blaming the poor for their plight also fails to recognize that there are not enough jobs for those who want to work and that many jobs fail to pay wages that enable families to escape poverty. And lastly, blaming the poor for their condition diverts attention away from the recognition that the wealthy—individuals and corporations—receive far more benefits in the form of wealthfare or corporate welfare, without the stigma of welfare.

Ending or reducing poverty begins with the recognition that doing so is a worthy ideal and an attainable goal. Imagine a world where everyone had comfortable shelter, plentiful food, adequate medical care, and education. If this imaginary world were achieved, and absolute poverty were effectively eliminated, what would the effects be on such social problems as crime, drug abuse, family problems (e.g., domestic violence, child abuse, divorce, and unwed parenthood), health problems, prejudice and racism, and international conflict? But it would be too costly to eliminate poverty—or would it? According to one source, the cost of eradicating poverty worldwide would be only about 1 percent of global income—and no more than 2 to 3 percent of national income in all but the poorest countries (*Human Development Report* 1997). Certainly the costs of allowing poverty to continue are much greater than that.

Critical Thinking

1 How could you counter the argument that although the United States has a higher poverty rate than many other industrialized countries, the opportunity to rise above poverty is greater in the U.S.?

2 Does a decline in the official U.S. poverty rate necessarily mean that fewer people are experiencing economic hardship? Or is it possible for a decline in the poverty rate to be accompanied by an increase in the numbers of individuals experiencing economic hardship?

3 Should someone receiving welfare benefits be entitled to spend some of his or her money on "nonessentials" such as cosmetics, eating out, lottery tickets, and cable TV? Why or why not?

4 Parenti (1998) points out that reports of income inequality based on U.S. census data are misleading because they do not take into account the super rich—the top 1 percent of income earners. For years, the Census Bureau never interviewed anyone who had an income higher than $300,000. The reportable upper limit of $300,000 was the top figure allowed by the bureau's computer program. In 1994, the bureau raised the upper limit to $1 million. But this figure still excludes the richest 1 percent—the hundreds of billionaires and thousands of multimillionaires who make many times more than $1 million a year. "The super rich simply have been computerized out of the Census Bureau's picture" (Parenti 1998, 36). How does the exclusion of the super rich from census data distort reports of economic inequality? Who benefits from this distortion?

5 The poor in the United States have low rates of voting and thus have minimal influence on elected government officials and the policies they advocate. What strategies might be effective in increasing voter participation among the poor?

6 Oscar Lewis (1998) noted that "some see the poor as virtuous, upright, serene, independent, honest, secure, kind, simple and happy, while others see them as evil, mean, violent, sordid and criminal" (p. 9). Which view of the poor do you tend to hold? How have various social influences, such as parents, peers, media, social class, and education, shaped your views toward the poor?

Key Terms

absolute poverty	HPI-1	proletariat
agricultural biotechnology	HPI-2	public housing
Aid to Families with Dependent Children (AFDC)	human capital	regressive taxes
	Human Poverty Index (HPI)	relative poverty
bourgeoisie	intergenerational poverty	scientific apartheid
corporate welfare	living wage laws	Section 8 housing
culture of poverty	means-tested programs	subsidized housing
earned income tax credit (EITC)	Personal Responsibility and Work Opportunity Reconciliation Act (PRWOR)	Temporary Assistance to Needy Families (TANF)
feminization of poverty		underclass
genetically modified organisms (GMOs)	poverty	wealth
		wealthfare
		working poor

Media Resources

The Wadsworth Sociology Resource Center: Virtual Society

http://sociology.wadsworth.com/

See the companion Web site for this book to access general sociology resources and text-specific features that can further your understanding of this chapter. The site contains Internet links, Internet exercises, online practice quizzes, information in InfoTrac College Edition, and many more valuable materials designed to enrich your learning experience in social problems.

InfoTrac College Edition

You can access InfoTrac College Edition either from the Wadsworth Sociology Resource Center at **http://sociology.wadsworth.com** or directly from your web browser at **http://www.infotrac-college.com/wadsworth/**. InfoTrac College Edition is an online university library that includes over 700 popular and scholarly journals in which you can find articles related to the topics in this chapter such as global poverty, economic inequality, welfare reform, corporate welfare, and homelessness.

Interactions CD-ROM

Go to the Interactions CD-ROM for *Understanding Social Problems*, Third Edition to access additional interactive learning tools, such as in-depth review materials, corresponding practice quizzes, and other engaging resources and activities to help you study the concepts in this chapter.

11

Work and Unemployment

Is It True?

1. Child labor and sweatshops are found in poor countries but are virtually nonexistent in the United States.

2. Federal regulations mandate that employers allow employees to use the bathroom when they need to.

3. The federal Fair Labor Standards Act allows youths as young as 12 years old to work in agriculture with no limit on the number of hours they can work on a school day.

4. Over the last few decades, U.S. labor union membership has increased steadily.

5. In 1999, at least 140 trade unionists around the world were assassinated, disappeared, or committed suicide after they were threatened as a result of their labor advocacy.

Answers to "Is It True?": 1 = F; 2 = T; 3 = T; 4 = F; 5 = T

All that serves labor serves the nation. All that harms is treason...If a man tells you he loves America, yet hates labor, he is a liar...There is no America without labor, and to fleece one is to rob the other.

ABRAHAM LINCOLN

In February 2000, 24-year-old Molly McGrath, a women's studies major at the University of Wisconsin, organized a takeover of Baskin Hall to protest sweatshop labor. Police arrested McGrath and 53 other students. At one point, school security guards showered pepper spray on the students, who shot back with a fire extinguisher. McGrath commented, "I think a big part of why they (arrested us) was to deplete our resources . . . Most of us spent the rest of the Spring semester dealing with legal repercussions." Those repercussions included a $150 court fine and academic probation (Harris 2001).

Molly McGrath, and thousands of other students who have joined the antisweatshop movement reflect the growing concern among young adults over workers' rights. In this chapter, we examine problems of work and unemployment, including child and sweatshop labor, health and safety hazards in the workplace, job dissatisfaction and alienation, work/family concerns, and declining labor strength and representation. We begin by looking at the global economy.

Before reading further, you may want to complete the "Attitudes toward Corporations" survey in the *Self and Society* feature of this chapter. It might be interesting to retake this survey after reading this chapter and see how your attitudes may have changed.

The Global Context: The Economy in the Twenty-first Century

In 1999, 11 of the 15 European Union nations began making the transition from their national currency to a new common currency—the euro—which will lock them together financially. The euro will replace German marks, French francs, and Italian lire. The adoption of the euro reflects the increasing globalization of economic institutions. The **economic institution** refers to the structure and means by which a society produces, distributes, and consumes goods and services.

In recent decades, innovations in communication and information technology have spawned the emergence of a **global economy**—an interconnected network of economic activity that transcends national borders and spans the world. The globalization of economic activity means that increasingly our jobs, the products and services we buy, and our nation's political policies and agendas influence and are influenced by economic activities occurring around the world. After summarizing the two main economic systems in the world—capitalism and socialism—we look at the emergence of corporate multinationalism. Then we describe how industrialization and postindustrialization have changed the nature of work.

319

Attitudes toward Corporations

PART ONE

How good a job do you think corporations are doing these days? Using letter grades like those in school, give corporations an A, B, C, D, or F in:

Letter Grade

1. Paying their employees good wages   _____

2. Being loyal to employees   _____

3. Making profits   _____

4. Keeping jobs in America   _____

PART TWO

Here are some things some large corporations are doing that some people think are serious problems, whereas others think they are not serious problems. For each of the following practices, indicate whether you think this is a serious problem or not.

	Serious Problem	Not a Serious Problem	Don't Know
5. Not providing health care and pensions to employees	_____	_____	_____
6. Not paying employees enough so that they and their families can keep up with the cost of living	_____	_____	_____
7. Paying CEOs 200 times what their employees make	_____	_____	_____
8. Laying off large numbers of workers even when they are profitable	_____	_____	_____

PART THREE

Which of the following statements comes closer to your view? (check one)

9A. A major problem with the economy today is government waste and inefficiency. Excessive government spending and high taxes burden middle class families and slow economic growth. Our government debt drives up interest rates, making it much harder for businesses to invest and create jobs.

OR

9B. A major problem with the economy today is politicians catering to the interests of powerful corporations and wealthy campaign contributors at the expense of working families. That is why politicians are not doing anything to stop large corporations from laying off large numbers of employees, denying health benefits, moving jobs overseas, and raiding pension funds.

9A. _____   9B. _____

10A. Wasteful and inefficient government is preventing the middle class from getting ahead and doing better. Excessive government spending and high taxes burden working families and slow economic growth. The budget deficit drives up interest rates and taxes, hurts consumers and business, and reduces job-creating investments. Red tape and excessive regulation are hurting business.

OR

Socialism and Capitalism

Socialism is an economic system in which the means of producing goods and services are collectively owned. In a socialist economy, the government controls income-producing property. Theoretically, goods and services are equitably distributed according to the needs of the citizens. Socialist economic systems emphasize collective well-being, rather than individualistic pursuit of profit.

10B. Corporate greed is preventing the middle class from getting ahead and doing better. In the past when people did their jobs well they could earn a decent wage and provide a better life for their children. Now, corporate America is squeezing their employees—cutting wages, downsizing jobs, and eliminating pensions and health benefits. Companies say they can't afford to treat employees better, but many have growing profits, record stock prices, and huge salaries for their executives.

10A. _____   10B. _____

11A. Large corporations are laying people off, cutting benefits, and moving jobs overseas mainly because they have gotten greedy and are squeezing employees to maximize profits.

OR

11B. Large corporations are laying people off, cutting benefits, and moving jobs overseas mainly because they have to in order to stay in business and provide jobs.

11A. _____   11B. _____

RESULTS OF A NATIONAL SAMPLE

You may want to compare your answers to this survey with responses from a national sample of U.S. adults.

PART ONE RESPONSES

Percentages do not total 100 because some individuals responded "Don't know."

	A	B	C	D	F
1.	11%	26%	36%	12%	7%
2.	10%	16%	28%	23%	19%
3.	52%	26%	9%	3%	2%
4.	12%	16%	31%	20%	18%

PART TWO RESPONSES

	Serious	Not Serious	Don't Know
5.	82%	15%	3%
6.	76%	19%	5%
7.	79%	14%	7%
8.	81%	14%	5%

PART THREE RESPONSES

9A. 33%;   9B. 40%; (21% answered "Both," and 6% answered "Don't know")

10A. 28%;   10B. 46%; (22% answered "Both," and 4% answered "Don't know")

11A. 70%;   11B. 22%; (7% answered "Don't know")

Source: Adapted from "Corporate Irresponsibility: There Ought to be Laws." 1996. EDK Poll, Washington, D.C.: Preamble Center for Public Policy. http://www.preamble.org/polledk.html (December 12, 1998). Used by permission.

Under **capitalism** private individuals or groups invest capital (money, technology, machines) to produce goods and services to sell for a profit in a competitive market. Whereas socialism emphasizes social equality, capitalism emphasizes individual freedom. Capitalism is characterized by economic motivation through profit, the determination of prices and wages primarily through supply and demand, and the absence of government intervention in

> Capitalism is the extraordinary belief that the nastiest of men, for the nastiest of reasons, will somehow work for the benefit of us all.
>
> JOHN MAYNARD KEYNES
> *Economist*

the economy. More people are working today in a capitalist economy than ever before in history (Went 2000). Critics of capitalism argue that it creates too many social evils, including alienated workers, poor working conditions, near-poverty wages, unemployment, a polluted and depleted environment, and world conflict over resources.

Both capitalism and socialism claim that they result in economic well being for society and its members. In reality, capitalist and socialist countries have been unable to fulfill their promises. Although the overall standard of living is higher in capitalist countries, so is economic inequality. Some theorists have suggested that capitalist countries will adopt elements of socialism, and socialist countries will adopt elements of capitalism. This idea, known as the **convergence hypothesis,** is reflected in the economies of Germany, France, and Sweden, which are sometimes called "integrated economies" because they have elements of both capitalism and socialism.

Corporate Multinationalism

Corporate multinationalism is the practice of corporations to have their home base in one country and branches, or affiliates, in other countries. Corporate multinationalism allows businesses to avoid import tariffs and costs associated with transporting goods to another country. Access to raw materials, cheap foreign labor, and the avoidance of government regulations also drive corporate multinationalism. "By moving production plants abroad, business managers may be able to work foreign employees for long hours under dangerous conditions at low pay, pollute the environment with impunity, and pretty much have their way with local communities. Then the business may be able to ship its goods back to its home country at lower costs and bigger profits" (Caston 1998, 274–75).

Although multinationalization provides jobs for U.S. managers, secures profits for U.S. investors, and helps the United States compete in the global economy, it also has "far-reaching and detrimental consequences" (Epstein, Graham, & Nembhard 1993, 206). First, it contributes to the trade deficit in that more goods are produced and exported from outside the United States than from within. Thus, the United States imports more than it exports. Second, multinationalism contributes to the budget deficit. The United States does not get tax income from U.S. corporations abroad, yet multinationals pressure the government to protect their foreign interests; as a result, military spending increases. Third, multinationalism contributes to U.S. unemployment by letting workers in other countries perform labor that could be performed by U.S. employees. Finally, corporate multinationalization must take its share of the blame for an array of other social problems such as poverty resulting from fewer jobs, urban decline resulting from factories moving away, and racial and ethnic tensions resulting from competition for jobs.

Industrialization, Postindustrialization, and the Changing Nature of Work

The nature of work has been shaped by the Industrial Revolution, the period between the mid–eighteenth century and the early nineteenth century when the factory system was introduced in England. **Industrialization** dramatically altered the nature of work: machines replaced hand tools; steam, gasoline, and electric power replaced human or animal power. Industrialization also led to the development of the assembly line and an increased division of labor as goods began to be mass-produced. The development of factories contributed to the emergence of large cities where the earlier informal social interactions dominated by primary relationships were replaced by formal interactions centered around secondary groups. Instead of the family-centered economy characteristic of an agricultural society, people began to work outside the home for wages.

Postindustrialization refers to the shift from an industrial economy dominated by manufacturing jobs to an economy dominated by service-oriented, information-intensive occupations. Postindustrialization is characterized by a highly educated workforce, automated and computerized production methods, increased government involvement in economic issues, and a higher standard of living (Bell 1973). Like industrialization before it, postindustrialization has transformed the nature of work.

The three fundamental work sectors (primary, secondary, and tertiary) reflect the major economic transformations in society—the Industrial Revolution and the Postindustrial Revolution. The **primary work sector** involves the production of raw materials and food goods. In developing countries with little industrialization, about 60 percent of the labor force works in agricultural activities; in the United States less than 2 percent of the workforce is in farming (*Report on the World Social Situation* 1997; *Statistical Abstract* 2000). The **secondary work sector** involves the production of manufactured goods from raw materials (e.g., paper from wood). The **tertiary work sector** includes professional, managerial, technical-support, and service jobs. The transition to a postindustrialized society is marked by a decrease in manufacturing jobs and an increase in service and information-technology jobs in the tertiary work sector. For example, in 2000, 234,000 manufacturing jobs in the United States were lost, while service employment continued to expand (Economic Policy Institute 2001).

The postindustrial economy of the United States requires many workers who are highly skilled in technology. The U.S. Department of Commerce projects that by 2006, nearly half of the U.S. workforce will be employed by industries that are either producers or intensive users of information technology products and services (Bowles 2000). But many U.S. workers, particularly women and minorities, are not educated and skilled enough for many of these positions (Koch 1998). U.S. employers claim that a shortage of skilled high-tech workers is hurting business, and the employers have lobbied Congress to allow them to admit more foreign workers. But critics believe that employers want more foreign workers because they are cheaper (Koch 1998). In developing countries, many individuals with the highest level of skill and education leave the country in search of work abroad, leading to the phenomenon known as the **brain drain.** Although U.S. employers benefit as they pay lower wages to foreign workers, U.S. workers are displaced and developing countries lose valuable labor.

Sociological Theories of Work and the Economy

Numerous theories in economics, political science, and history address the nature of work and the economy. In sociology, structural-functionalism, conflict theory, and symbolic interactionism serve as theoretical lenses through which we may better understand work and economic issues and activities.

Structural-Functionalist Perspective

According to the structural-functionalist perspective, the economic institution is one of the most important of all social institutions. It provides the basic necessities common to all human societies, including food, clothing, and shelter. By providing for the basic survival needs of members of society, the economic institution contributes to social stability. After the basic survival needs of a society are met, surplus materials and wealth may be allocated to other social uses, such as maintaining military protection from enemies, supporting political and religious leaders, providing formal education, supporting an expanding population, and providing entertainment and recreational activities. Societal development is dependent on an economic surplus in a society (Lenski & Lenski 1987).

Although the economic institution is functional for society, elements of it may be dysfunctional. For example, before industrialization, agrarian societies had a low degree of division of labor in which few work roles were available to members of society. Limited work roles meant that society's members shared similar roles and thus developed similar norms and values (Durkheim [1893] 1966). In contrast, industrial societies are characterized by many work roles, or a high degree of division of labor, and cohesion is based not on the similarity of people and their roles but on their interdependence. People in industrial societies need the skills and services that others provide. The lack of common norms and values in industrialized societies may result in *anomie*—a state of normlessness—which is linked to a variety of social problems including crime, drug addiction, and violence (see Chapters 3 and 4).

Conflict Perspective

The country is governed for the richest, for the corporations, the bankers, the land speculators, and for the exploiters of labor.

HELEN KELLER, 1911
Social activist

According to Karl Marx, the ruling class controls the economic system for its own benefit and exploits and oppresses the working masses. Whereas structural-functionalism views the economic institution as benefiting society as a whole, conflict theory holds that capitalism benefits an elite class that controls not only the economy but other aspects of society as well—the media, politics and law, education, and religion.

As an indication of the ties between business and government, consider that both President George W. Bush and Vice President Dick Cheney come from the oil industry. Also, the Bush cabinet includes as many or more corporate executives than any previous administration, and Bush's transition teams for the Department of Energy, Department of Health and Human Services, and the Department of Labor are almost entirely made up of people affiliated with or working for corporate interests (Mokhiber & Weissman 2001).

Corporate interests also find their way into politics through large political contributions. During the 2000 national election cycle, the Democratic and Republican parties raised a record $463.1 million in soft money contributions—nearly double the amount raised in the 1996 election. And soft money contributions to 2000 Congressional campaigns were triple that of 1996 (Common Cause 2001). **Soft money** is money that flows through a loophole to provide political parties, candidates, and contributors a means to evade federal limits on political contributions. Critics of this system of campaign financing argue that corporations and interest groups purchase political influence through financial contributions. "The high cost of political campaigns forces many candidates to rely upon funds from special interests, who then expect (and usually get) special treatment in the law"

(*Granny D Home Page* 2000). The special treatment can be in the form of business taxes, environmental loopholes, subsidies, or lower standards of consumer and worker protection, for example. Although the top soft money donors in the 2000 elections included two labor unions, most large contributors were corporations, including Philip Morris, AT&T, Amway, American Financial Group Insurance Company, and Microsoft (Common Cause 2000).

A survey of business leaders' views on political fundraising found that the main reasons U.S. corporations make political contributions is fear of retribution and to buy access to lawmakers ("Big Business for Reform" 2000). Although 75 percent say political donations give them an advantage in shaping legislation, nearly three-quarters (74 percent) say business leaders are pressured to make large political donations. Half of the executives said their colleagues "fear adverse consequences for themselves or their industry if they turn down requests" for contributions.

Penalties for violating health and safety laws in the workplace provide an example of legal policy that favors corporate interests. Suppose a corporation is guilty of a serious violation of health and safety laws, where "serious violation" is defined as one that poses a substantial probability of death or serious physical harm to workers. What penalty do you think that corporation should pay for such a violation? According to a report by the AFL-CIO (2000), serious violations of workplace health and safety laws carry an average penalty of only $776.00. As discussed in Chapter 4, penalties for corporate crimes tend to be much less severe than those applied to individuals who violate the law.

Corporate power is also reflected in the policies of the International Monetary Fund (IMF) and the World Bank, which pressure developing countries to open their economies to foreign corporations, promoting export production at the expense of local consumption, encouraging the exploitation of labor as a means of attracting foreign investment, and hastening the degradation of natural resources as countries sell their forests and minerals to earn money to pay back loans. Ambrose (1998) asserts that "for some time now, the IMF has been the chief architect of the global economy, using debt leverage to force governments around the world to give big corporations and billionaires everything they want—low taxes, cheap labor, loose regulations—so they will locate in their countries" (p. 5). Treaties such as the North American Free Trade Agreement (NAFTA), and the General Agreement on Tariffs and Trade (GATT) also benefit corporations at the expense of workers by providing U.S. corporations greater access to foreign markets. "These laws increasingly allow corporations to go anywhere and do anything they like, and prohibit workers and the governments that supposedly represent them from doing much about it" (Danaher 1998, 1). Meanwhile, both foreign and U.S. workers suffer negative effects of trade agreements. Consider the effects of NAFTA, which required Mexico to allow free entry and exit of investment, and lifted trade barriers for Mexican exports to the United States, making production there for export to the United States more profitable. In the 7 years after NAFTA went into effect, Mexican manufacturing wages dropped from $2.10 an hour to $1.90 an hour and workers' efforts to organize and demand higher wages have been systematically repressed (Anderson 2001). As of September 2000, at least 260,000 U.S. workers had lost their jobs because their employer moved production to Mexico or Canada or was hurt by import competition from those countries (Anderson 2001).

Our principal motivations in world affairs have been largely economic.

RICHARD J. CASTON
Sociologist

Symbolic Interactionist Perspective

No race can prosper til it learns there is as much dignity in tilling a field as in writing a poem.

BOOKER T. WASHINGTON
Address to the Atlanta Exposition, September 18, 1895

According to symbolic interactionism, the work role is a central part of a person's identity. When making a new social acquaintance, one of the first questions we usually ask is, "What do you do?" The answer largely defines for us who that person is. For example, identifying a person as a truck driver provides a different social meaning than identifying someone as a physician. In addition, the title of one's work status—maintenance supervisor or university professor— also gives meaning and self-worth to the individual. An individual's job is one of his or her most important statuses; for many, it comprises a "master status," that is, the most significant status in a person's social identity.

As symbolic interactionists note, definitions and meanings influence behavior. Meanings and definitions of child labor (discussed later) contribute to its perpetuation. In some countries, children learn to regard working as a necessary and important responsibility and rite of passage, rather than an abuse of human rights. Some children look forward to becoming bonded to a master "in the same way that American children look forward to a first communion or getting a driver's license" (Silvers 1996, 83).

Symbolic interactionism emphasizes that attitudes and behavior are influenced by interaction with others. The applications of symbolic interactionism in the workplace are numerous—employers and managers are concerned with using interpersonal interaction techniques that achieve the attitudes and behaviors they want from their employees; union organizers are concerned with using interpersonal interaction techniques that persuade workers to unionize; and job-training programs are concerned with using interpersonal interaction techniques that are effective in motivating participants.

Problems of Work and Unemployment

Next, we examine unemployment and other problems associated with work. Problems of workplace discrimination based on gender, race and ethnicity, and sexual orientation are addressed in other chapters. Minimum wage and living wage issues are discussed in Chapter 10. Here we discuss problems concerning sweatshop and child labor, health and safety hazards in the workplace, job dissatisfaction and alienation, work/family concerns, unemployment and underemployment, and labor unions and the struggle for workers' rights.

Sweatshop Labor

When a man tells you that he got rich through hard work, ask him whose.

DON MARQUIS
Journalist

A U.S. Department of Labor investigation of the Daewoosa Samoa garment factory in American Samoa—a factory that produces men's sportswear for J.C. Penney—found that garment factory workers lived and worked under conditions of poor sanitation, malnutrition, electrical hazards, fire hazards, machinery hazards, illegally low wages, sexual harassment and invasion of privacy, workplace violence and corporal punishment, and overcrowded barracks where two workers were forced to share each bed (National Labor Committee 2001). Female workers reported that the company owner routinely enters their barracks to watch them shower and dress. Workers reported incidents in which security guards slapped and kicked workers. The food provided to the workers at the Daewoosa Samoa garment factory consisted of a watery broth of rice and cabbage. The factory owner, Mr. Kil Soo Lee, ignores court orders to improve worker conditions at his factory and when investigators show up to inspect the worksite, Mr. Lee "happens to have $60,000 in a paper bag to flaunt in front of investigators" (U.S. Department of Labor, quoted in National Labor Committee 2001).

The workers at the Daewoosa Samoa garment factory are among the millions of people worldwide who work in **sweatshops**—work environments that are characterized by less-than-minimum wage pay, excessively long hours of work (often without overtime pay), unsafe or inhumane working conditions, abusive treatment of workers by employers, and/or the lack of worker organizations aimed at negotiating better work conditions. Sweatshop labor conditions occur in a wide variety of industries, including garment production, manufacturing, mining, and agriculture. The dangerous conditions of sweatshops result in high rates of illness, injury, and death. The International Labor Organization estimates 1.1 million workers worldwide die on the job or from occupational disease each year (Multinational Monitor 2000). In one tragic example of death resulting from sweatshop conditions, at least 53 workers, including 10 children, were burned to death in a fire at a Sagar Chowdury garment factory in Bangladesh (Hargis 2001). The fire, caused by an electrical short circuit, engulfed the entire factory with 900 workers who were *locked inside*. Local residents and firefighters broke open the locked gates of the building and rescued survivors.

Sweatshop Labor in the United States   Sweatshop conditions in overseas garment and footwear industries have been widely publicized. However, many Americans do not realize the extent to which sweatshops exist in the United States. The Department of Labor estimates that over half of the country's 22,000 sewing shops violate minimum wage and overtime laws and 75 percent violate safety and health laws ("The Garment Industry" 2001). The majority of garment workers in the United States are immigrant women who typically work 60 to 80 hours a week, often earning less than minimum wage, with no overtime, and many face verbal and physical abuse.

Migrant farm workers, who produce more than 85 percent of the fruits and vegetables grown in the United States, also work under sweatshop conditions. About 300,000 farm workers suffer acute pesticide poisoning each year (Human Rights Watch 2000a). Many live in substandard and crowded housing provided by their employer. They lack access to safe drinking water, bathing and sanitary toilet facilities, and risk injuries from using sharp and heavy farm equipment. Working 12-hour days under hazardous conditions, the average annual pay in 1999 for migrant farm workers was $7,500 (Human Rights Watch 2000a).

Child Labor: A Global Problem

Child labor involves children performing work that is hazardous; that interferes with a child's education; or that harms a child's health or physical, mental, spiritual, or moral development (U.S. Department of Labor 1995). Even though virtually every country in the world has laws that limit or prohibit the extent to which children can be employed, child labor persists throughout the world. An estimated 250 million children between 5 and 14 work for a living (Human Rights Watch 2001). To grasp the scale of child labor, imagine a country as populous as the United States, in which the entire population consists of child laborers.

> As long as the existence of hazardous jobs is tolerated, the most economically and socially disadvantaged workers will continue to be at the greatest risk.
>
> DANA LOOMIS AND DAVID RICHARDSON
> *University of North Carolina at Chapel Hill*

Child laborers work in factories, workshops, construction sites, mines, quarries, and fields, on deep-sea fishing boats, at home, and on the street. They make bricks, shoes, soccer balls, fireworks and matches, furniture, toys, rugs, and clothing. They work in manufacturing of brass, leather goods, and glass. They tend livestock and pick crops. In Egypt, over 1 million children ages 7 to 12 work each year in cotton pest management. They endure routine beatings by their foremen, as well as exposure to heat and pesticides ("Underage and Unprotected" 2001). Children typically earned the equivalent of about $1 per day and worked from 7:00 AM to 6:00 PM daily, with one midday break, 7 days a week. Supervising foremen routinely beat children with wooden switches whenever they perceived a child to be slowing down or overlooking leaves.

Children as young as 5 or 6 also work in domestic service. In one Latin American country, an estimated 22 percent of all working children are employed as servants (International Labour Organization 2000). This form of child labor is difficult to monitor, because of the hidden nature of the practice.

Child labor is most prevalent in Africa, Asia, and Central and South America. India has the largest child labor force in the world, with between 20 and 80 million working children (Parker 1998). This chapter's *The Human Side* feature depicts child labor in Pakistan.

Bonded labor—an extreme form of child labor—refers to the repayment of a debt through labor. Typically, an employer loans money to parents, who then give the employer their children as laborers to repay the debt. Sometimes the child is taken far away from the family to work; other times the child works in the same village and continues to live at home. The children are unable to work off the debt because of high interest rates; low wages; and wage deductions for meals, lodging, and mistakes made at work (U.S. Department of Labor 1995). Bonded labor is like slavery; a bonded worker is not free to leave the workplace. About 10 to 20 million children in the world are forced to work as bonded laborers (Parker 1998). Bonded labor is most common in India, Nepal, Bangladesh, and Pakistan.

Child Labor in the United States Illegal and oppressive employment of children also occurs in the United States in restaurants, grocery stores, meatpacking industries, sweatshops in urban garment districts, and in agriculture. Conservative estimates suggest that 301,000 children and youth work in violation of federal or state labor laws at some point during the year (Kruse & Mahony 2000).

Over 300,000 U.S. child workers labor on commercial farms, frequently under dangerous and grueling conditions (Human Rights Watch 2001). Child farm workers in the United States often work 12-hour days, sometimes beginning at 3:00 or 4:00 AM. Many are exposed to dangerous pesticides that cause cancer and brain damage, with short-term symptoms including rashes, headaches, dizziness, nausea, and vomiting. They often work in 100°F temperatures without adequate access to drinking water, and are sometimes forced to work without access to toilets or handwashing facilities (Human Rights Watch 2001). Agriculture is also one of the most dangerous occupations, and children (as well as adults) sustain high rates of injury from work with knives, other sharp tools, and heavy equipment.

Long hours of work interfere with the education of children working in the fields, making them miss school and leaving them too exhausted to study or stay awake in class. Only 55 percent of farm worker children in the United States finished high school (Human Rights Watch 2001).

Child Labor in Pakistan

Like most other countries, Pakistan has laws prohibiting child labor and indentured servitude. However, these laws are largely ignored, and about 11 million children aged 4 to 14 work under brutal and squalid conditions. Children make up about a quarter of the unskilled work force in Pakistan and can be found in virtually every factory, field, and workshop. They earn on average a third of the adult wage. The following excerpts from an *Atlantic Monthly* report describe child labor in Pakistan (Silvers 1996).

Soon after I arrived in Pakistan, I arranged a trip to a town whose major factories were rumored to enslave very young children. I found myself hoping during the journey there that the children I saw working in fields, on the roads, at the marketplaces, would prepare me for the worst. They did not. No amount of preparation could have lessened the shock and revulsion I felt on entering a sporting-goods factory in the town of Sialkot . . . where scores of children, most of them aged 5 to 10, produce soccer balls by hand for forty rupees, or about $1.20, a day. The children work 80 hours a week in near-total darkness and total silence. According to the foreman, the darkness is both an economy and a precautionary measure; child-rights activists have difficulty taking photographs and gathering evidence of wrongdoing if the lighting is poor. The silence is to ensure product quality: "If the children speak, they are not giving their complete attention to the product and are liable to make errors." The children are permitted one 30-minute meal break each day; they are punished if they take longer. They are also punished if they fall asleep, if their workbenches are sloppy, if they waste material or miscut a pattern, if they complain of mistreatment to their parents or speak to strangers outside the factory . . . . Punishments are doled out in a storage closet at the rear of the factory . . . . Children are hung upside down by their knees, starved, caned, or lashed . . . . The punishment room is a standard feature of a Pakistani factory, as common as a lunchroom at a Detroit assembly plant.

The town's other factories are no better, and many are worse. Here are brick kilns where 5-year-olds work hip-deep in slurry pits, where adolescent girls stoke furnaces in 160 degree heat. Here are tanneries where nursing mothers mix vats of chemical dye, textile mills where 8-year-olds tend looms and breathe air thick with cotton dust . . . .

A carpet workshop . . . was . . . about the size of a subway car, and about as appealing. The long, narrow room contained a dozen upright looms. On each rough-hewn workbench between the looms squatted a carpet weaver. The room was dark and airless . . . A thermometer read 105 degrees, and the mud walls were hot to the touch . . . .

Of the twelve weavers, five were 11 to 14, and four were under 10. The two youngest were brothers named Akbar and Ashraf, aged 8 and 9. They had been bonded to the carpet master at age 5, and now worked 6 days a week at the shop. Their workday started at 6:00 AM and ended at 8:00 PM, except, they said, when the master was behind on his quotas and forced them to work around the clock. They were small, thin, malnourished, their spines curved from lack of exercise and from squatting before the loom. Their hands were covered with calluses and scars, their fingers gnarled from repetitive work. Their breathing was labored, suggestive of tuberculosis. Collectively these ailments, which pathologists call captive-child syndrome, kill half of Pakistan's working children by age 12 . . . .

A hand-knotted carpet is made by tying short lengths of fine colored thread to a lattice of heavier white threads. The process is labor-intensive and tedious: a single four-by-six-foot carpet contains well over a million knots and takes an experienced weaver 4 to 6 months to complete . . . . Each carpet Akbar completed would retail in the United States for about $2,000—more than the boy would earn in ten years. Akbar revealed that, "the master screams at us all the time, and sometimes he beats us . . . . We're slapped often. Once or twice he lashed us with a cane. I was beaten 10 days ago after I made many errors of color in a carpet. He struck me with his fist quite hard on the face . . . . I was fined one thousand rupees and made to correct the errors by working two days straight." The fine was added to Akbar's debt, and would extend his "apprenticeship" by several months . . . Akbar declared that "staying here longer fills me with dread. I know I must learn a trade. But my parents are so far away, and all my friends are in school. My brother and I would like to be with our family. We'd like to play with our friends. This is not the way children should live."

Source: Jonathon Silvers. 1996 (February). "Child Labor in Pakistan." *The Atlantic Monthly.* Reprinted by permission.

Juvenile farm workers are less protected under U.S. law than are juveniles working in safer occupations. Under the federal **Fair Labor Standards Act (FLSA):** (1) children working on U.S. farms may be employed at age 12, whereas the minimum age for employment in other occupations is 14; (2) the number of hours child farm workers may work on school days is not limited, whereas in other occupations, children under age 16 are limited to 3 hours of work per day when school is in session; (3) FLSA does not require overtime pay for agricultural workers as it does for other occupations; and (4) juveniles working in agriculture may engage in hazardous work at age 16, whereas for other occupations, the minimum age for hazardous work is 18 (Human Rights Watch 2000a). Because 85 percent of migrant and seasonal farm workers nationwide are racial and ethnic minorities, the FLSA's bias against farm worker children is a form of institutional discrimination.

Child Prostitution and Trafficking   One of the worst forms of child labor is child prostitution and child trafficking. Although it is impossible to identify how prevalent child prostitution is, research estimates suggest that the problem is widespread. For example, surveys have identified 2930 child prostitutes in Athens, 3000 in Montreal, and, according to the U.S. Department of Health and Human Services, up to 300,000 in the United States (Dorman 2001). In poor countries, the sexual services of children are often sold by their families in an attempt to get money. Some children are kidnapped or lured by traffickers with promises of employment, only to end up in a brothel. The U.S. Immigration and Naturalization Service identified 250 brothels in 26 American cities where forced prostitutes, including children, are taken. Americans also engage in child-sex tourism abroad, particularly in Southeast Asia. Of 240 identified cases in which legal action was taken against foreigners for sexually abusing children in this region, researchers found that about one-quarter of the violators were from the United States (Dorman 2001).

© Jim Bourg/Liaison Agency

Iqbal Masih, who worked in a carpet factory in Pakistan when he was just 4 years old, escaped the factory when he was 10. With the help of the Bonded Labor Liberation Front (BLLF), Iqbal entered school and worked with the BLLF to enforce labor laws, free bonded children, and educate the public about child labor. In 1995, at age 12, Iqbal was shot and killed in a rural village where he was visiting relatives. Although the killer was never found, some suspect that Iqbal's murder was arranged by the carpet manufacturers, who were losing child workers as a result of Iqbal's campaigning and activism.

Causes of Child Labor   Poverty, economic exploitation, social values, and lack of access to education are factors contributing to the persistence of child labor. One mother in Bangladesh whose 12-year-old daughter works up to 14 hours a day in a garment sweatshop explained, "Children shouldn't have to work . . . But if she didn't, we'd go hungry" (Parker 1998, 47). The economic advantages to industries that profit from child labor also perpetuate the practice. In the United States, youths working illegally in hazardous jobs earn an average of $1.38 less per hour than legal young adults in the same occupations, resulting in employer cost savings of about $136 million per year (Kruse & Mahony 2000).

Traditional social values have also contributed to child labor. In the words of one employer in Pakistan who uses child labor, "Child labor is a tradition the West cannot understand and must not attempt to change" (Silvers 1996, 86). Finally, child labor results from failure to provide education to all children. The education system in Pakistan, for example, can only accommodate about one-third of the country's school-age children, leaving the remainder to join the child labor pool (Silvers 1996).

Consequences of Child Labor   Child laborers are at risk for a wide variety of health problems such as injuries, stunted growth, and many diseases. Child carpet weavers develop gnarled fingers from the repetitive work, and their spines are curved from sitting at looms all day. Young brickworkers breathe in dust from the dry bricks and sand, causing scarring of the lungs and early death. Child farm workers are exposed to harmful pesticides. In rural areas, more child workers in agriculture die from pesticide poisoning than from all of the most common childhood diseases put together (UNICEF 2000). Child prostitutes are often physically abused by their pimps and customers, are at high risk for acquiring HIV and other sexually transmitted infections, and suffer the emotional scars of their exploitation. Child laborers are fed inadequate diets and must endure harsh punishment from their employers. One girl who was forced into prostitution in Bangkok said, "One time I refused to sleep with a man and they slapped me, hit me with a cane and bashed my head against the wall. One of my friends tried to run away but unfortunately she was caught and very badly beaten" (Parker 1998, 42).

Child labor also increases poverty by depressing already low wages. Parker (1998) explains:

> For every child who works, there may be an adult who cannot find a job. Children are usually paid less than adult workers—sometimes only one-third of what adults earn. As a result, adult workers' wages stay low or go down. When parents cannot find jobs, they are more likely to send their children to work. They have more children in the hope of increasing their income. Each generation of poor, uneducated child workers becomes the next genre of poor parents who must send their kids to work. Then the cycle of poverty and illiteracy continues. (Parker 1998, 48)

Health and Safety Hazards in the U.S. Workplace

Accidents at work and hazardous working conditions contribute to illnesses, injuries, and deaths. Globally, an estimated 1.1 million workers die on the job or from occupational disease each year (Multinational Monitor 2000). Some occupational health and safety hazards are attributed to willful disregard of information and guidelines concerning worker safety. A PBS documentary called *Trade Secrets* revealed that for decades the chemical industry knew that vinyl chloride (used in plastics manufacturing) was harmful to workers exposed to this chemical (PBS 2001). Bernie Skaggs, who worked at a BF Goodrich plant in Kentucky, described the effects of workplace exposure to vinyl chloride:

> My hands began to get sore, and they began to swell some. My fingers got so sore on the ends, I couldn't button a shirt, couldn't dial a phone. And I had thick skin

like it was burned all over the back of my hand, back of my fingers, all the way up under my arm, almost to my armpit. And after enough time, I got thick places on my face right under my eyes. . . . (PBS 2001)

Although BF Goodrich sent a memo to Union Carbide, Imperial Chemical Industries, and The Monsanto Company indicating that vinyl chloride could cause the bones in the hands of their workers to dissolve, Bernie Skaggs was not told that workplace exposure to vinyl chloride caused his disabling health problem.

In 1969, ten years after Bernie Skaggs first complained to the company doctor about the pain in his hands, members of the chemical industry's trade association met to discuss a report from a group of medical researchers they had hired. The report advised companies to improve workplace ventilation to reduce employees' exposure to vinyl chloride by ninety percent—from 500 parts per million to 50 parts per million. But the industry's Occupational Health Committee rejected the recommendation. After scientists in the 1970s found cancer in laboratory rats exposed to vinyl chloride, representatives from European and chemical companies (including Conoco, BF Goodrich, Dow, Shell, Ethyl, and Union Carbide) signed a secrecy agreement promising to withhold this information from the public and from their workers. Dan Ross, who worked at a vinyl-producing Conoco plant in Louisiana for 23 years, died in 1990 at age 43 of brain cancer. Before he died, his wife, Elaine, found a report from the plant indicating that Dan had been exposed to more dangerous fumes than government standards allowed, leading her and her dying husband to sue Conoco. In the last words he was able to speak, Dan told his wife, "Mama, they killed me" (PBS 2001).

Workplace Illnesses and Injuries   Between 1994 and 1999, the rate of workplace nonfatal injuries and illnesses declined from 8.4 to 6.3 per 100 full-time workers (Bureau of Labor Statistics 2000a). This was the lowest rate since the Department of Labor began reporting this information in the 1970s. Of the 5.7 million nonfatal workplace illnesses and injuries reported in 1999, most (5.3 million) were injuries. The highest rate of nonfatal illnesses and injuries were in manufacturing, followed by construction, transportation and public utilities, and agriculture, forestry, and fishing. Some conditions (e.g., long-term latent illnesses caused by exposure to carcinogens) often are difficult to relate to the workplace and are not adequately recognized and reported.

The most common type of workplace illness, comprising 66 percent of total illness cases, were disorders associated with repeated trauma, such as carpal tunnel syndrome (a wrist disorder that can cause numbness, tingling, and severe pain), tendonitis (inflammation of the tendons), and noise-induced hearing loss. Such disorders—referred to by a number of terms, including **cumulative trauma disorders, repetitive strain disorders,** and **repeated trauma disorders**—are muscle, tendon, vascular, and nerve injuries that result from repeated or sustained actions or exertions of different body parts. Jobs that are associated with high rates of upper-body cumulative trauma disorders include computer programming, manufacturing, meatpacking, poultry processing, and clerical/office work (National Safety Council 1997). Cumulative trauma disorders are classified as illness, not as injury, because they are not sudden, instantaneous traumatic events.

It's difficult to think of an area of my life that has not been affected . . . . I have changed everything from going from a pump hair spray to an aerosol, to a new bra I could hook, to a new car I didn't have to shift.

ALLI ROBERTSON
Developed a cumulative trauma disorder while programming software at a Utah company

Job Stress and Chronic Fatigue Another work-related health problem is job stress and chronic fatigue. Prolonged job stress, also known as **job burnout,** can cause physical problems, such as high blood pressure, ulcers, and headaches, as well as psychological problems. In a survey of 1,298 employees (Bond, Galinsky, & Swanberg 1997), nearly one-quarter of employees felt nervous or stressed often or very often in the 3 months prior to the survey; 13 percent had difficulty coping with the demands of everyday life often or very often. The survey also found that substantial numbers of employees felt burned-out by their jobs. In the 3 months prior to the survey, 26 percent felt emotionally drained by their work often or very often, and 36 percent felt used up at the end of the workday often or very often.

A study of workers in Finland, Germany, Poland, the United Kingdom, and the United States found that one in ten U.S. workers suffers from depression, anxiety, stress, or burnout each year ("S.O.S. Stress at Work" 2000). Researchers concluded that in the United States, new workplace technologies and methods of work organization aimed at rising productivity requirements is causing more cases of depression and work-induced stress.

Workplace Fatalities Although many workplaces are safer today than in generations past, fatal occupational injuries and illnesses still occur in troubling numbers. In 1999, 6,023 on-the-job deaths occurred in the United States (Bureau of Labor Statistics 2000b). The leading cause of workplace fatalities in 1999 was highway accidents, followed by falls and homicides (see Figure 11.1). Occupations with large numbers of fatal injuries included truck driving, construction, and farming. Other industries with high rates of workplace fatalities include forestry, fishing, public utilities, and mining.

Dissatisfaction and Alienation

An advertisement for Army recruits claims that joining the Army enables one to "be all that you can be." Indeed, if you read the classified ad section of any newspaper, you are likely to find job advertisements that entice applicants with

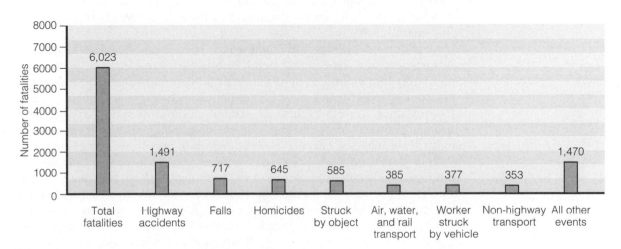

■ **Figure 11.1** *Causes of U.S. Workplace Fatalities: 1999*

Source: Bureau of Labor Statistics. 2000. "National Census of Fatal Occupational Injuries, 1999." United States Department of Labor, Washington, D.C.

claims such as "discover a rewarding and challenging career . . ." and "we offer opportunities for advancement and travel . . . ." Unfortunately, most jobs do not allow workers to "be all that they can be." In reality, most employers want you to "be all you can be for them" with limited concern for your career satisfaction.

Millions of U.S. workers are dissatisfied with their work. In one study of employees in firms with 25 or more workers, one-third of workers reported being dissatisfied with their jobs, although two-thirds say they "usually look forward" to going to work (Freeman & Rogers 1999). Dissatisfaction was highly correlated with a sense of nonparticipation in workplace decisions.

Factors that contribute to job satisfaction include income, prestige, a feeling of accomplishment, autonomy, a sense of being challenged by the job, opportunities to be creative, congenial coworkers, the feeling that one is making a contribution, fair rewards (pay and benefits), promotion opportunities, and job security (Bavendam 2000; Gordon 1996). These factors often overlap—for example, high-paying jobs tend to have more prestige, be more autonomous, provide more benefits, and permit greater creativity. Yet many jobs lack these qualities, leaving workers dissatisfied. And even during recent periods of low unemployment rates, 23 percent of employed adults are worried that they may lose their jobs in the next few years (Simmons 2001). Thirty percent of employees feel that their company does not have a strong sense of loyalty to them (Bond, Galinsky, & Swanberg 1997). Workers are also dissatisfied with declining wages. The decline in wages in the 1980s and 1990s occurred in spite of the increased economic productivity achieved in the same time period. Economist Lester Thurow (1996) commented that "never before have a majority of American workers suffered real wage reductions while the real per capita GDP was advancing" (p. 24).

One form of job dissatisfaction is a feeling of **alienation.** Work in industrialized societies is characterized by a high degree of division of labor and specialization of work roles. As a result, workers' tasks are repetitive and monotonous and often involve little or no creativity. Limited to specific tasks by their work roles, workers are unable to express and utilize their full potential—intellectual, emotional, and physical. According to Marx, when workers are merely cogs in a machine, they become estranged from their work, the product they create, other human beings, and themselves. Marx called this estrangement "alienation."

Alienation usually has four components: powerlessness, meaninglessness, normlessness, and self-estrangement. Powerlessness results from working in an environment in which one has little or no control over the decisions that affect one's work. Meaninglessness results when workers do not find fulfillment in their work. Workers may experience normlessness if workplace norms are unclear or conflicting. For example, many companies that have family leave policies informally discourage workers from using them. Or workplaces that officially promote nondiscrimination in reality practice discrimination. Alienation also involves a feeling of self-estrangement, which stems from the workers' inability to realize their full human potential in their work roles and lack of connections to others. In general, traditional women's work is more alienating than men's work (Ross & Wright 1998). "Homemaking exposes women to routine, unfulfilling, isolated work; and part-time employment exposes them to routine, unfulfilling work, with little decision-making autonomy" (p. 343).

Clearly the most unfortunate people are those who must do the same thing over and over again, every minute, or perhaps twenty to the minute. They deserve the shortest hours and the highest pay.

JOHN KENNETH GALBRAITH
American economist

Without work, all life goes rotten, but when work is soulless, life stifles and dies.

ALBERT CAMUS
Philosopher

Work/Family Concerns

Spouses, parents, and adult children caring for elderly parents increasingly struggle to balance their work and family responsibilities. When Hochschild (1997) asked a sample of employed parents, "Overall, how well do you feel you can balance the demands of your work and family?" only 9 percent said "very well" (pp. 199–200).

In 1999, both parents worked in nearly two-thirds (64 percent) of married couple families with children under 18, whereas the father, but not the mother, worked in 29 percent of these families (Bureau of Labor Statistics 2000c). More than three-quarters (79 percent) of unmarried mothers and 70 percent of married mothers were employed in 1999. Between 1979 and 1998, middle-income families have added 619 hours of paid work (more than 3 months) per year (Mishel, Bernstein, & Schmitt 2001). This increase in hours worked by family members is the result, in part, of the increase in the number of women in the workforce. In addition, individuals are spending more time on the job. Average annual hours worked is higher in the United States than in any of the 15 industrialized OECD countries. And although nearly every other OECD country reduced its average hours worked per year between 1979 and 1998, average hours worked by U.S. employees *increased* by 61 hours (Mishel et al. 2001). More hours in the workplace means fewer hours available to take care of household and family responsibilities.

Increasingly, the work/family balancing act also involves caring for elderly family members. The Families and Work Institute estimates that about 42 percent of workers will provide some form of elder care by 2002 (reported in U.S. Department of Labor 1999). In the *Strategies for Action* section later in this chapter, we discuss policies and programs that address the work/family concerns of Americans today.

Unemployment and Underemployment

The International Labour Organization (2001) reports that one-third of the world's workforce is unemployed or underemployed. Poor countries tend to suffer high rates of unemployment. For example, recent unemployment rates in South Africa and Lesotho (a southern African country) were 23 percent and 42 percent, respectively (International Labour Organization 2001).

Measures of **unemployment** in the United States consider an individual to be unemployed if he or she is currently without employment, is actively seeking employment, and is available for employment. Unemployment figures do not include discouraged workers, who have given up on finding a job and are no longer looking for employment. **Underemployment** is a broader term which includes unemployed workers as well as (1) those working part-time but who wish to work full-time; and (2) those who want to work but have been discouraged from searching as a result of lack of success ("discouraged" workers).

Compared with other industrialized countries, the United States has a low rate of unemployment (see Table 11.1). However, other industrialized countries have higher wages than the United States and provide more social supports, such as universal health care and unemployment insurance, for their citizens. And although the 1999 U.S. unemployment rate of 4.2 percent is considered

> How can you ask your people to unselfishly support the needs of the corporation if the corporation will not support the needs of its people?
>
> ROGER MEADE
> *Scitor Corporation*

■ **Table 11.1** *Unemployment Rates in Nine Countries: 2000*

Country	Unemployment Rate
United States	4.0%
Canada	5.8%
Australia	6.6%
Japan	4.8%
France	9.7%
Germany	8.3%
Italy	10.7%
Sweden	5.8%
United Kingdom	5.5%

Source: Bureau of Labor Statistics. 2001. "Unemployment Rates in Nine Countries, 1990–2001." Washington, D.C.: U.S. Department of Labor.

low, that still translates into more than 136 million unemployed adults. Further, the rate of underemployment in 1999 was 7.5 percent (Mishel et al. 2001).

Many of those underemployed are **contingent workers** (also called "disposable workers")—involuntary part-time workers, temporary employees, and workers who do not perceive themselves as having an explicit or implicit contract for ongoing employment. In 1999, 17.1 percent of U.S. workers were part-time (Mishel et al. 2001). Of these, 85 percent voluntarily chose part-time work, whereas 15 percent (3.3 million workers) were working part-time jobs because they could not find full-time employment. The underemployed also include workers in the temporary help industry. "Temping" has grown from 0.5 percent of total U.S. employment in 1982 to 2.3 percent in 1999 (Mishel et al. 2001). Although the overall percentage of "temp" employees is low, it has increased more than four-fold in the last two decades.

Types and Causes of Unemployment Unemployment can be either discriminatory or structural. **Discriminatory unemployment** involves high rates of unemployment among particular social groups, such as racial and ethnic minorities and women (see Chapters 7 and 8). For example, the unemployment rate for black workers in 1999 was 8.0 percent, compared with 3.7 percent for whites (Mishel et al. 2001). **Structural unemployment** exists when not enough jobs are available for those who want them. Structural unemployment is the result of social factors rather than personal inadequacies of the unemployed.

For example, some unemployment results from **corporate downsizing**—the corporate practice of discharging large numbers of employees. Simply put, the term *downsizing* is a euphemism for mass firing of employees (Caston 1998). Another cause of U.S. unemployment is **job exportation,** the relocation of jobs to other countries where products can be produced more cheaply. **Automation,** or the replacement of human labor with machinery and equipment also contributes to unemployment. Structural unemployment has also resulted from the shift in the U.S. economy away from manufacturing and toward highly skilled jobs, such as those in information technology. The Educational Testing Service projects that fewer than 10 percent of new jobs

Table 11.2 *Unemployment Rates of U.S. Adults 25 Years and Over, by Education: January 2001*

Level of Educational Attainment	Unemployment Rate
Less than high school diploma	6.8%
High school graduate	3.8%
College graduate	1.6%

Source: "Employment Situation Summary." 2001 (February 2). Bureau of Labor Statistics. http://stats.bls.gov/news.release/empsit.nr0htm

created through 2006 will be in the minimal skill range and that just over 20 percent will be in the basic skill range (Coalition on Human Needs 2000). As shown in Table 11.2, higher unemployment rates are associated with lower levels of education.

Labor Unions and the Struggle for Workers' Rights

Labor unions originally developed to protect workers and represent them at negotiations between management and labor. Labor unions have played an important role in fighting for fair wages and benefits, healthy and safe work environments, and other forms of worker advocacy. In 1997, the average wage of union workers was $17.60, compared to $14.29 for non-union workers, and unionized workers received insurance and pension benefits worth more than double those of non-union employees (Mishel et al. 2001). Labor unions are also influential in achieving better working conditions. For example, the United Food and Commercial Workers (UFCW), the country's largest union representing poultry processing workers, was instrumental in the formation of an Occupational Safety and Health Administration (OSHA) rule that established a federal workplace "potty" policy governing when employees can use the bathroom while on the job. According to UFCW international president Doug H. Dority, "For years workers in food processing industries have had to suffer the indignity of being denied the right to go to the bathroom when needed, just to maintain ever-increasing assembly-line speeds" ("New OSHA Policy Relieves Employees" 1998, 8). Dority claims that "poultry processors often have no other choice than to relieve themselves where they stand on the assembly line because their floor boss will not let them leave their workstation" (p. 8). The new OSHA rule mandates that employers must make toilet facilities available so that employees can use them when they need to. The employer may not impose unreasonable restrictions on employee use of the facilities to ensure that employees need not wait an unreasonably long time to use the bathroom.

Labor unions also help to win better wages, benefits (e.g., health insurance), and full-time jobs. In the *Justice for Janitors 2000* campaign, approximately 100,000 janitors who are members of the Service Employees International Union (SEIU) successfully sought to raise standards for the people who clean, sweep, and mop about 4,000 commercial real estate properties in the United States (Wright 2001). The janitors won increased wages, expanded health care benefits, and the restoration of full-time jobs. The wage raises won will help janitors lift their families out of poverty, which was a major goal of the national campaign. The campaign included strikes, rallies, and/or protests in cities across the country.

The Labor Movement: the folks who brought you the weekend.

FROM A BUMPER STICKER, 1995

I consider it important, indeed urgently necessary, for intellectual workers to get together, both to protect their own economic status and, also generally speaking, to secure their influence in the political field.

ALBERT EINSTEIN
Commenting on why he joined the American Federation of Teachers, AFL-CIO

Despite the successes of labor unions' efforts to improve wages, benefits, and working conditions, the strength and membership of unions in the United States have declined over the last several decades (Western 1995). **Union density**—the percentage of workers who belong to unions—grew in the 1930s and peaked in the 1940s and 1950s, when 35 percent of U.S. workers were unionized. In the 1960s and 1970s, U.S. corporations mounted an offensive attack on labor unions, "aiming to tame them or maim them" (Gordon 1996, 207). Corporations hired management consultants to help them develop and implement antiunion campaigns. They threatened unions with decertification, fired union leaders and organizers, and threatened to relocate their plants unless the unions and their members "behaved."

One management consultant firm . . . was unusually blunt in broadcasting its methods. A late-1970s blurb promoting its manual promised: "We will show you how to screw your employees (before they screw you)—how to keep them smiling on low pay—how to maneuver them into low-pay jobs they are afraid to walk away from" (Gordon 1996, 208).

In 2000, the percentage of American workers belonging to unions had fallen to 13.5 percent, its lowest point in six decades (Greenhouse 2001). Reasons for the decline in union representation include the loss of manufacturing jobs, which tend to have higher rates of unionization than other industries (Greenhouse 2001). Job growth has been in high technology and financial services where unions have little presence. In addition, globalization has led to layoffs and plants closing at many unionized worksites, as companies move to other countries to find cheaper labor.

Although the 1935 **National Labor Relations Act** guarantees the right to unionize and to strike, workers risk their jobs by doing so. In the 1990s, more than 20,000 U.S. workers each year were fired or discriminated against because of their union-related activities (Human Rights Watch 2001). For example, when workers at the Smithfield Packing Company's slaughterhouse in North Carolina—the world's largest pork processing plant—held unionizing campaigns, 11 workers were illegally fired and other workers were threatened and improperly interrogated about their union activities (Sack 2001). The company warned of layoffs and possible plant closing if the unionizing campaign succeeded and one pro-union employee had been assaulted for his organizing efforts.

One study found that in more than half of all union-organizing drives, employers threatened to close the plant (Mokhiber & Weissman 1998). Where union organizing drives are successful, employers carry out the threat and close the plant, in whole or in part, 15 percent of the time. In today's climate of expanding trade agreements and skyrocketing levels of corporate migration, plant-closing threats continue to be among the most powerful anti-union strategies. A study of survey data collected from lead organizers in 407 National Labor Relations Board union certification elections in 1998 and 1999 (representing 5 percent of the 6,207 NLRB union certification elections in 1998–99) found that half of all employers in 1998–1999 made threats to close all or part of the facility if the union was to win the certification election campaigns (Bronfenbrenner 2000). In addition, most employers in the 1998–1999 survey sample aggressively opposed the union's organizing efforts through a combination of threats, discharges, promises of improvements, unscheduled changes in wages and benefits, bribes, and surveillance.

The Taiwanese-owned Chentex factory in Nicaragua, employs nearly 2,000 workers who sew 20,000 to 25,000 pairs of jeans a day for the U.S. military, Kohl's, J.C. Penney, Kmart, and Wal-Mart, along with various other labels including Gloria Vanderbilt and Bugle Boy. The workers earn just 18 cents for each $24-pair of jeans they sew. When Chentex refused to negotiate a wage increase with the worker's union in 2000, workers called a 1-hour work stoppage. Chentex management responded by firing nine union leaders, hiring thugs to terrorize the workers, putting up barbed wire and surveillance cameras, bringing in 20 armed National Police, and threatening to shut the Chentex factory down and move productions elsewhere.

Labor Union Struggles around the World   International norms established by the United Nations and the International Labor Organization declare the rights of workers to organize, negotiate with management, and strike (Human Rights Watch 2000b). In European countries, labor unions are generally strong (Mishel et al. 2001). However, in many less developed countries and countries undergoing economic transition, workers and labor unions struggle to have a voice in matters of wages and working conditions. A survey of 113 countries by the International Confederation of Free Trade Unions (ICFTU) found that both corporations and governments repressed union efforts. Among the survey's findings are: ("Global Labor Repression" 2000):

- At least 140 trade unionists around the world in 1999 were assassinated, disappeared, or committed suicide after they were threatened as a result of their labor advocacy. Colombia was found to be the most dangerous country for union activists. In 1999, 76 trade unionists in Colombia were assassinated or reported missing.
- Nearly 3,000 people were arrested, more then 1,500 were injured, beaten, or tortured, and at least 5,800 were harassed because of their trade union activities. Another 700 trade unionists received death threats.
- About 12,000 workers were unfairly dismissed or refused reinstatement, sometimes with the complicity of the government, because they were active members of a trade union. At least 140 strikes or demonstrations were repressed by governments, sometimes with the support of the employers using strikebreakers, and 80 of the 113 countries surveyed restrict the right to strike.

Strategies for Action: Responses to Workers' Concerns

Government, private business, human rights organizations, labor organizations, college student activists, and consumers play important roles in responding to the concerns of workers. Next we look at responses to sweatshop and child labor, health and safety concerns, work-family policies and programs, workforce development programs, efforts to strengthen labor, and challenges to corporate power and economic globalization.

Responses to Sweatshop and Child Labor

The International Programme on the Elimination of Child Labour has been working to remove child laborers from oppressive work conditions and provide them with education and provide their parents with jobs or income. Since the program began in 1992, it has grown from 6 participating countries to more than 200 participating countries in 2000 (Human Rights Watch 2001).

In 1999, 174 nations adopted the new Convention on the Worst Forms of Child Labor, and by September 2000, 37 countries had ratified the convention (Human Rights Watch 2001). This represents a global consensus to end the most severe forms of child labor and requires nations to take immediate measures to abolish child slavery, trafficking, debt bondage, child prostitution and pornography, and forced labor. The U.S. Congress adopted a landmark trade bill, which for the first time makes U.S. trade benefits dependent on a country's progress in eliminating the worst forms of child labor. The Trade and Development Act of 2000 denies U.S. trade preferences to countries in sub-Saharan Africa, the Caribbean, and Central America that have not implemented their commitments to eliminate the worst forms of child labor (Human Rights Watch 2001). In 1997, U.S. law was enacted prohibiting the U.S. Customs Service from allowing the importation of any product that is made by "forced or indentured child labor" (International Labor Rights Fund 1997).

Human rights organizations such as the International Labour Organization, UNICEF, and the Child Labor Coalition, are active in the campaign against child labor. Another organization, the Bonded Labor Liberation Front (BLLF) has led the fight against bonded and child labor in Pakistan, freeing 30,000 adults and children from brick kilns, carpet factories, and farms, and placing 11,000 children in its own primary school system (Silvers 1996). However, employers in Pakistan have threatened workers with violence if they talk with "the abolitionists" or possess "illegal communist propaganda" (Silvers 1996). Human rights activists campaigning against child labor have also been victims of threats and violence.

The United Nations Children's Fund recommends that national and international corporations adopt codes of conduct guaranteeing that neither they nor their subcontractors will employ children in conditions that violate their rights (UNICEF 1997). Some industries, including rug and clothing manufacturers, use labels and logos to indicate that their products are not made by child laborers. A recently formed Fair Labor Association (FLA) involves six leading apparel and footwear companies who voluntarily participate in a monitoring system to inspect their overseas factories and require them to meet minimum labor standards, such as not requiring workers to work more than 60 hours a week. However, critics point out a number of problems with the Fair Labor Association, including (1) standards are too low (allows below-poverty wages and excessive overtime); (2) requires only 10 percent of companies' factories to be monitored yearly; (3) companies can influence which factories are inspected and who does the inspection; and (4) FLA does not uphold workers' right to organize (Benjamin 1998). Critics suggest that companies use their participation in FLA as a marketing tool. Once "certified" by the FLA, companies can sew a label into their products saying they were made under fair working conditions.

Pressure from college students and other opponents of child and sweatshop labor and consumer boycotts of products made by child and sweatshop labor have resulted in some improvements in factories that make goods for companies such as Nike and Gap, which have cut back on child labor, use less dangerous chemicals, and require fewer employees to work 80-hour weeks (Greenhouse 2000). At many factories, supervisors have stopped hitting employees, have improved ventilation, and have stopped requiring workers to obtain permission to use the toilet. But improvements are not widespread and oppressive forms of labor continue throughout the world. According to the National Labor Committee, two areas where "progress seems to grind to a halt" are efforts to form unions and efforts to achieve wage increases (Greenhouse 2000).

Changes in the legal protections for U.S. juvenile farm workers are also needed, and the limited laws that apply to such workers must be better enforced. The U.S. Department of Labor cited only 104 cases of child labor violations in agriculture in 1998, even though an estimated one million child labor violations occur in U.S. agriculture every year (Human Rights Watch 2001). A recent report noted that although agriculture is one of the most dangerous jobs, less than 3 percent of all OSHA inspections were devoted to agriculture (Human Rights Watch 2000a). And penalties were typically too weak to discourage employers from using illegal child labor.

As discussed earlier, juvenile farm workers do not have the same legal protections under the Fair Labor Standards Act as do other young workers. Representative Tom Lantos has introduced the Young American Workers' Bill of Rights every year for 12 years. This bill, which has not passed into legislation, would amend the Fair Labor Standards Act to forbid employment of migrant workers age 13 and under, and would restrict the hours juvenile farm workers could work on school days (Human Rights Watch 2000a). In 2000, Senator Tom Harkin introduced The Children's Act for Responsible Employment which would amend the FLSA to protect all children equally under U.S. labor laws. The bill was pending before the Senate Committee on Health, Education, Labor, and Pensions at the time of this writing.

Efforts to eliminate child labor cannot succeed unless the impoverished conditions that contribute to its practice are alleviated. A living minimum wage must be established for adult workers so they won't need to send their children to work. One labor rights advocate said, "We will not end child labor merely by attacking it in the export sector in poor nations. If children's parents in all countries around the world are not earning a living wage . . . , children will be driven into working in dangerous informal sector jobs. Labor rights for adults are essential if we truly want to eliminate child labor" (Global March against Child Labor 1998).

© AP/Wide World Photos

Members of the United Students Against Sweatshops protest the use of sweatshops at a Nike store in New York. Nike supports the Fair Labor Association in the fight against sweatshop labor, but critics believe the organization is not effective in ensuring safe and humane working conditions.

The appropriate societal response to exploitative child labor is straightforward: eliminate it and meet the economic needs of individuals forced to work under such conditions in other ways.

LETITIA DAVIS
Director, Occupational Health Surveillance Massachusetts Department of Public Health

Responses to Worker Health and Safety Concerns

Over the last few decades, health and safety conditions in the U.S. workplace have improved as a result of media attention, demands by unions for change, more white-collar jobs, and regulations by the Occupational Safety and Health Administration (OSHA). Through OSHA, the government develops, monitors, and enforces health and safety regulations in the workplace. Since OSHA was created three decades ago, workplace fatalities have dropped by 75 percent (*Multinational Monitor* 2000). But much work remains to be done to improve worker safety and health. Inadequate funding leaves OSHA unable to do its job effectively. In FY 1999, OSHA's 2,145 federal and state inspectors were responsible for monitoring and enforcing laws at more than 7 million workplaces (AFL-CIO 2000). Consequently, in 1999 OSHA inspected less than 2 percent of U.S. workplaces. Because "the task of monitoring and enforcement simply cannot be effectively carried out by a government administrative agency," Kenworthy (1995) suggests that the United States follow the example of many other industrialized countries: turn over the bulk of responsibility for health and safety monitoring to the workforce (p. 114). Worker health and safety committees are a standard feature of companies in many other industrialized countries and are mandatory in most of Europe. These committees are authorized to inspect workplaces and cite employers for violations of health and safety regulations.

In developing countries, governments fear that strict enforcement of workplace regulations will discourage foreign investment (*Multinational Monitor* 2000). Investment in workplace safety in developing countries, whether by domestic firms or foreign multinationals, is far below that in the rich countries. Unless global standards of worker safety are implemented and enforced in *all* countries, millions of workers throughout the world will continue to suffer under hazardous work conditions. Low unionization rates, as well as workers' fears of losing their jobs—or their lives—if they demand health and safety protections leave most workers powerless to improve their working conditions.

Business and industry often fight against efforts to improve safety and health conditions in the workplace. For example, in November 1999, after a 10-year struggle between labor and business, the Occupational Safety and Health Administration issued ergonomic standards requiring employers to implement ergonomic programs in jobs where musculoskeletal disorders occur. **Ergonomics** refers to the designing or redesigning of the workplace to prevent and reduce cumulative trauma disorders. According to OSHA, the new ergonomic standards would prevent 4.6 million workers over the next 10 years from experiencing painful, potentially debilitating work-related musculoskeletal disorders (U.S. Department of Labor 2001). But business and industry representatives pressured Congress and President George W. Bush to repeal the ergonomic standard.

Behavioral-Based Safety Programs   A controversial health and safety strategy used by business management is behavioral-based safety programs. Instead of examining how work processes and conditions compromise health and safety on the job, **behavioral-based safety programs** direct attention to workers themselves as the problem. Behavior-based safety programs claim that 80 to 96 percent of job injuries and illnesses are caused by workers' own carelessness and unsafe acts (Frederick & Lessin 2000). These programs focus on teaching employees and managers to identify, "discipline," and change unsafe worker behaviors that cause accidents, and to encourage a work culture that recognizes and rewards safe behaviors.

Critics contend that behavior-based safety programs divert attention away from the employer's failure to provide safe working conditions. Consider the following example (Frederick & Lessin 2000):

> In a Midwest tire manufacturer with a behavior-based safety program, the official accident report written up after a worker slipped and fell on ice in the parking lot stated, "Worker's eyes not on path," as the cause of the injury. The report did not mention the need to have ice and snow removed from the parking lot. It did not mention that the sidewalk had not been cleared of snow and ice for several weeks, even though workers were required to use the sidewalk periodically.

Critics also say that the real goal of behavior-based safety programs is to discourage workers from reporting illness and injuries. Workers whose employers have implemented behavior-based safety programs describe an atmosphere of fear in the workplace, such that workers are reluctant to report injuries and illnesses for fear of being labeled an "unsafe worker." At one factory that had implemented a behavioral safety program, when a union representative asked workers during shift meetings to raise their hands if they were afraid to report injuries, about half of 150 workers raised their hands (Frederick & Lessin 2000). Worried that some workers feared even raising their hand in response to the question, the union representative asked a subsequent group to write "yes" on a piece of paper if they were afraid to report injuries. Seventy percent indicated they were afraid to report injuries. Asked why they would not report injuries, workers said, "we know that we will face an inquisition," "we would be humiliated," and "we might be blamed for the injury."

Work/Family Policies and Programs

The influx of women into the workforce has been accompanied by an increase in government and company policies designed to help women and men balance their work and family roles. In 1993, President Clinton signed into law the **Family and Medical Leave Act** (FMLA), which requires all companies with 50 or more employees to provide eligible workers (who work at least 25 hours a week and have been working for at least a year) with up to 12 weeks of job-protected, unpaid leave so they can care for a seriously ill child, spouse, or parent; stay home to care for their newborn, newly adopted, or newly placed child; or take time off when they are seriously ill. Yet, nearly half of the workforce is not covered by the FMLA (Lovell & Rahmanou 2000). In addition, many workers do not take advantage of the FMLA because they cannot afford to take leave without pay. A 2000 survey found that the most commonly noted reason for not taking leave was being unable to afford it; 88 percent of those who needed time off but did not take it said they would have taken leave if they could have received pay during their absence (Cantor et al. 2001). And more than half of the workers who did take family or medical leave were concerned about having enough money to pay bills. Another reason for not taking family or medical leave was fear of losing one's job. Nearly one-third of all workers who needed leave but did not take it cited worry about losing their job as a reason for not taking leave (Cantor et al. 2001). Another problem concerns awareness; among employees in covered workplaces, only 38 percent correctly reported that the FMLA applied to them and about one-half said they did not know if it did (Cantor et al. 2001).

In a national survey of 500 parents, three-fourths favored legislation requiring companies (with 25 or more employees) to offer up to twelve weeks of *paid* job-protected family leave (Hewlett & West 1998). Because of the inadequate

At one Midwest plant, all workers who did not report an injury during the course of a year were invited to dinner. At the dinner, one of the workers' names was pulled from a hat. That worker was given a check for $10,000.

JAMES FREDERICK
AND NANCY LESSIN
United Steelworkers of America
and AFL-CIO (respectively)

In the long run, no work–family balance will ever fully take hold if the social conditions that might make it possible— men who are willing to share parenting and housework, communities that value work in the home as highly as work on the job, and policymakers and elected officials who are prepared to demand family-friendly reform— remain out of reach.

ARLIE HOCHSCHILD
Sociologist

Our leaders talk as though they value families, but act as though families were a last priority.

SYLVIA HEWLETT AND
CORNELL WEST
Family advocates

provisions of the federal Family and Medical Leave Act provision, at least 21 states have considered initiatives for some type of paid family leave (Lovell & Rahmanou 2000). The Department of Labor issued new regulations in 2000 that permit states to use unemployment insurance funds for income replacement when new parents take leave to care for their children.

Aside from government-mandated work/family policies, some corporations and employers have "family-friendly" work policies and programs. Only 2 percent of U.S. workers have paid family leave provided by their employers (Lovell & Rahmanou 2000). Other more common forms of employer-provided work/family assistance include child care assistance (e.g., plans that allow employees to pay for child care with pre-tax dollars), assistance with elderly parent care, and flexible work options.

Offering employees more flexibility in their work hours helps parents balance their work and family demands. Flexible work arrangements, which benefit child-free workers as well as employed parents, include flextime, job sharing, a compressed workweek, and teleworking. **Flextime** allows the employee to begin and end the workday at different times as long as 40 hours per week are maintained. For example, workers may begin at 6 AM and leave at 2 PM instead of the traditional 9 AM to 5 PM. With **job sharing,** two workers share the responsibility of one job. A **compressed workweek** allows employees to condense their work into fewer days (e.g., four 10-hour days each week). **Telework** allows employees to work part- or full-time at home or at a satellite office (see this chapter's *Focus on Technology* feature). A study of U.S. companies found that the more women and minorities a company has in managerial positions, the more likely that company is to offer flexible work options (Galinsky & Bond 1998).

Fran Rodgers, president of Work/Family Directions, explains the need for work/family policies:

> For over 20 years we at Work/Family Directions have asked employees in all industries what it would take for them to contribute more at work. Every study found the same thing: They need aid with their dependent care, more flexibility and control over the hours and conditions of work, and a corporate culture in which they are not punished because they have families. These are fundamental needs of our society and of every worker. (Galinsky, Riesbeck, Rodgers, & Wohl 1993, 51)

Workforce Development and Job-Creation Programs

The International Labour Organization (2001) estimates that over 500 million new jobs are needed by 2010 to accommodate new entrants to the workforce and to reduce current unemployment levels by half. Developing a workforce and creating jobs involves far-reaching efforts, including those designed to improve health and health care, alleviate poverty and malnutrition, develop infrastructures, and provide universal education.

In the United States, workforce development programs have provided a variety of services, including assessment to evaluate skills and needs, career counseling, job search assistance, basic education, occupational training (classroom and on-the-job), public employment, job placement, and stipends or other support services for child care and transportation assistance (Levitan, Mangum, & Mangum 1998). Workforce development programs primarily assist youths, the handicapped, welfare recipients, displaced workers, the elderly, farm workers, Native Americans, and veterans. Numerous studies have looked at the effectiveness of workforce development programs. In general, "evaluations indicate

Telework: The New Workplace of the 21st Century

The ever-widening use of modern technology in the workplace, such as computers, the Internet, e-mail, fax machines, copiers, mobile phones, and personal digital assistants, makes it possible for many workers to perform their jobs at a variety of locations. The term **telework** (also known as "telecommuting") refers to flexible and alternative work arrangements that involve use of information technology. There are four types of telework (Pratt 2000): (1) homebased telework; (2) satellite offices where all employees telework for one employer; (3) telework centers, which are occupied by employees from more than one organization; and (4) mobile workers. Most (89 percent) teleworkers are home-based (Bowles 2000). Some people telework full-time, but a larger number telework 1 or 2 days a week.

The number of companies offering telecommuting increased from 19.5 percent in 1996 to 28 percent in 1999. There are an estimated 13 to 19 million teleworkers in the United States today and a predicted 137 million worldwide by 2003. The largest fraction of teleworkers are professionals (37 percent), followed by clerical and sales (14 percent). Technical, service, and managerial jobs each make up 12 percent of the teleworkforce.

Telework holds potential benefits for employers, workers, and the environment. After presenting some of these benefits, we discuss concerns related to telework.

Benefits of Telework for Employers

Attracts and Helps Retain Employees   Companies regard telework and other flexible work arrangements as important in recruiting and maintaining good employees. Telework can also lower turnover, and thus save companies expenses associated with hiring and training replacement employees. A 1997 AT&T survey of telecommuters showed that 36 percent of employees would quit or find another work-at-home job if their employer decided they could not work at home (cited in Lovelace 2000). However, Bowles (2000) reports that companies are beginning to express dissatisfaction with telework "because they believe that it causes resentment among office-bound colleagues and weakens corporate loyalty" (p. 2).

Reduces Costs   AT&T saved about $550 million from 1991 to 1998 by eliminating offices that teleworkers don't need, consolidating others, and reducing related overhead costs (Lovelace 2000).

Increases Worker Productivity   Several studies of managers and employees at large companies conclude that telework increases worker productivity (Lovelace 2000).

Benefits of Telework for Employees

Increases Job and Life Satisfaction   Studies have shown that employee satisfaction among teleworkers is higher than for their non-teleworking counterparts (Lovelace 2000). Much of the job satisfaction among telecommuters is related to the job flexibility that enables them to balance work and family demands.

Helps Balance Work/Family Demands   Telework can provide flexibility to working parents and adults caring for aging parents, thus reducing role conflict and strengthening family life. One father of three children described his being home when his children came back from school as being "the most significant impact" of his telecommuting (Riley, Mandavilli & Heino 2000, p. 5). He also took time during the day to take his children to school, to the doctor's office, and to run errands. However, one national study of children whose parents work at home found that older children (grades 7 to 12) were more likely to agree that "my father does not have the energy to do things with me because of his job" and "my father has not been in a good mood with me because of his job" than the children of fathers who work in an office (Galinsky & Kim 2000). The effects of telework on parent/child relationships seems to depend then on how each parent interacts with his or her children.

Expands Work Opportunities for Americans Outside the Economic Mainstream   Telework may expand job opportunities for rural job seekers who lack local employment opportunities, and for low-income urban job seekers who lack access to suburban jobs (Kukreja & Neely 2000). Telework can also bring work opportunities to individuals with disabilities. Some of the technologies that have been developed for individuals with serious disabilities include "Eye Gaze" (a communication system that allows people to operate a computer with their eyes); "Magic Wand Keyboard" (for people with limited or no hand movement); and "Switched Adapted Mouse and Trackball" (that allows clicking the mouse with other parts of the body) (Bowles 2000).

Continued

Avoids the Commute For many teleworkers, the primary motivation for working from home is to reduce or eliminate the long and stressful commute that so many Americans now endure. In one AT&T unit, the average teleworker gained nearly 5 weeks per year by eliminating a 50-minute daily commute (Lovelace 2000).

Environmental Benefits of Telework

Telework can reduce pollution by reducing the need for transportation to the workplace, thus reducing the pollution associated with vehicle emissions. The National Environmental Policy Institute says that "telecommuting presents a non-coercive way for corporations to help the nation achieve environmental goals and improve quality of life" (quoted in Lovelace 2000, p. 3).

Concerns about Telework

Blurred Boundaries between Home and Work People who work at home may find themselves on call around the clock, responding to e-mail, pagers, faxes, and voice mail. Without clear boundaries between home and work, teleworkers may feel that they are unable to escape the work environment and mindset (Pratt 2000). Questions about overtime pay may arise when work spills over into personal time. Having a separate office within the home and a routine work schedule may help create the psychological boundary between work and family/leisure. But for some teleworkers, learning to "log off" is a challenge.

Zoning Regulations Teleworkers who work at home full time must contend with zoning regulations that may prohibit residents from having an "office" in their home.

Losing Benefits as a Contract Employee Some employers attempt to convert the teleworker into a contract worker. This type of worker lacks job protections and benefits (Bowles 2000).

Social Isolation Does telework lead to social isolation for those who live and work at home? Evidence suggests

that teleworkers are able to maintain personal relationships with co-workers and are included in office networks. However, for rural and disabled individuals, telework may contribute to social isolation.

Exclusion of the Disenfranchised Lower socioeconomic groups are less likely than more affluent populations to have access to and skills in the Internet and other modern forms of information technology. Bowles (2000) suggests that "the eventual success of telework programs in the future must . . . account for the masses of people left behind . . . All must be included in the new economy; it is not a luxury, but a must" (p. 9).

Sources:

Bowles, Diane O. 2000. "Growth in Telework." Paper presented at the symposium *Telework and the New Workplace of the 21st Century,* Xavier University, New Orleans, October 16, 2000. U.S. Department of Labor. http://www.dol.gov/dol/asp/public/telework/htm

Galinsky, Ellen and Stacy S. Kim. 2000. "Navigating Work and Parenting by Working at Home: Perspectives of Workers and Children Whose Parents Work at Home." Paper presented at the symposium *Telework and the New Workplace of the 21st Century,* Xavier University, New Orleans, October 16, 2000. U.S. Department of Labor. http://www.dol.gov/dol/asp/public/telework/htm

Kukreja, Anil and George M. Neely, Sr. 2000. "Strategies for Preventing the Digital Divide." Paper presented at the symposium *Telework and the New Workplace of the 21st Century,* Xavier University, New Orleans, October 16, 2000. U.S. Department of Labor. http://www.dol.gov/dol/asp/public/telework/htm

Lovelace, Glenn. 2000. "The Nuts and Bolts of Telework." Paper presented at the symposium *Telework and the New Workplace of the 21st Century,* Xavier University, New Orleans, October 16, 2000. U.S. Department of Labor. http://www.dol.gov/dol/asp/public/telework/htm

Pratt, Joanne H. 2000. "Telework and Society—Implications for Corporate and Societal Cultures." Paper presented at the symposium *Telework and the New Workplace of the 21st Century,* Xavier University, New Orleans, October 16, 2000. U.S. Department of Labor. http://www.dol.gov/dol/asp/public/telework/htm

Riley, Patricia, Anu Mandavilli, and Rebecca Heino. 2000. "Observing the Impact of Communication and Information Technology on 'Net-Work.'" Paper presented at the symposium *Telework and the New Workplace of the 21st Century,* Xavier University, New Orleans, October 16, 2000. U.S. Department of Labor. http://www.dol.gov/dol/asp/public/telework/htm

that employment and training programs enhance the earnings and employment of participants, although the effects vary by service population, are often modest because of brief training durations and the inherent difficulty of alleviating long-term deficiencies, and are not always cost effective" (Levitan, Mangum, & Mangum 1998, 199).

Efforts to prepare high school students for work include the establishment of technical and vocational high schools and high school programs and school-to-work programs. School-to-work programs involve partnerships between business, labor, government, education, and community organizations that help prepare high school students for jobs (Leonard 1996). Although school-to-work programs vary, in general, they allow high school students to explore different careers, and they provide job skill training and work-based learning experiences, often with pay (Bassi & Ludwig 2000).

One strategy for creating jobs involves local, state, and federal government providing benefits to corporations in the form of subsidies, tax breaks, real estate, and low-interest loans to corporations with the hope that this "corporate welfare" will result in new jobs (see also Chapter 10). However, most recent job creation in the United States is with small- and medium-sized companies. Although Fortune 500 companies are the biggest beneficiaries of corporate welfare, they have eliminated more jobs than they have created in the past decade (Barlett & Steele 1998). And many of the jobs that are created are part-time or temporary jobs.

In examining efforts to create jobs we must consider where the jobs are being created. The U.S. economy is described as a **split labor market** (or dual economy), because it is made up of two labor markets. The **primary labor market** refers to jobs that are stable, economically rewarding, and come with benefits. The most educated and trained individuals (e.g., corporate attorney, teacher, or accountant) usually occupy these jobs, most often white males. The **secondary labor market** refers to jobs that involve low pay, no security, few benefits, and little chance for advancement. Domestic servants, clerks, and food servers are examples of these jobs. Women and racial and ethnic minorities are disproportionately represented in the secondary labor market. These workers often have no union to protect them and are more likely to be dissatisfied with their job than workers in the primary labor market.

With the new limits on welfare (see Chapter 10), more adults with low levels of education and job training are going to be entering the workforce in the coming years. The cuts in welfare benefits exacerbate the need to provide not only workforce development programs, but also jobs that pay a living wage. As shown in this chapter's *Social Problems Research Up Close* feature, single mothers who work in low-wage jobs often have more hardships than those who are dependent on welfare.

U.S. Bureau of Labor

Maximizing employment also requires matching job seekers with hiring employers. The U.S. Bureau of Labor's website America's Job Bank enables job-seekers to search for employment opportunities by occupation and region.

No business which depends for existence on paying less than living wages to its workers has any right to continue in this country. By living wages I mean more than a bare subsistence level—I mean the wages of decent living.

F.D. ROOSEVELT

Efforts to Strengthen Labor

Although efforts to strengthen labor are viewed as problematic to corporations and employers, such efforts have potential to remedy many of the problems facing workers. In an effort to strengthen their power, some labor unions have merged with one another. Labor union mergers result in higher membership

Making Ends Meet: Survival Strategies among Low-Income and Welfare Single Mothers

As welfare recipients reach the time limit established by welfare legislation of 1996 for receiving welfare benefits, they are forced into the workforce. But as individuals leave welfare for work, they often find themselves in low-paying jobs, often with no or few benefits. How do individuals in low-income jobs compare with those dependent on welfare in terms of their well-being? And how do both low-wage earners and welfare recipients survive on income that does not meet their basic needs? Researchers Kathryn Edin and Laura Lein (1997) conducted research to answer these questions.

Sample and Methods

The sample consisted of 379 African-American, white, and Mexican-American single mothers from four cities (Chicago, San Antonio, Boston, and Charleston, South Carolina), who either received welfare cash assistance (Aid to Families with Dependent Children, or AFDC) (N = 214) or nonrecipients who

held low-wage jobs earning $5 to $7 an hour between 1988 and 1992 (N = 165). Edin and Lein used a "snowball sampling" technique in which each mother who was interviewed was asked to refer researchers to one or two friends who might also participate in interviews. Nearly 90 percent of the mothers contacted agreed to be interviewed.

Researchers collected data through conducting multiple semi-structured in-depth interviews with women in the sample. Interview topics included the mothers' income and job experience, types and amount of welfare benefits they received, spending behavior, housing situation, use of medical care and child care, and hardships the women and their children experienced because of lack of financial resources.

Interviewing the mothers more than once was an important research strategy in gathering accurate information. Mothers who were unclear about their expenditures in the first interview could keep careful track of

what they spent between interviews and give a more precise accounting of their spending in a later interview. Also, some mothers who insisted they received no child support later revealed that the child's father "helped out" every week by providing cash. "Most mothers only termed absent fathers' cash contributions as 'child support' if it was collected by the state" (p. 13).

Findings and Conclusions

Low-wage–earning single mothers had a higher monthly reported income than welfare-reliant mothers. However, the expenses of wage-earning mothers were also higher. This is because employed mothers usually have to pay for child care, transportation to work, and additional clothing to wear to work. If newly employed mothers have a federal housing subsidy, every extra $100 in cash income raises their rent by $30 (Jencks 1997). And employed mothers are usually not eligible for Medicaid, which means that they have more out-of-pocket medical expenses and often go uninsured.

A woman's place is in her union.

COALITION OF LABOR UNION WOMEN (CLUW)

numbers, thereby increasing the unions' financial resources, which are needed for successful recruiting and to withstand long strikes.

The increasing numbers of women in labor unions, which nearly doubled from 20 to 39 percent between 1960 and 1998, has helped to strengthen labor unions' advocacy for women ("Labor's 'Female Friendly' Agenda" 1998). For example, a number of unions have been successful in bargaining for expanded family leave benefits, subsidized child care, elder care, and pay equity.

Because workers must fight for labor protections within a globalized economic system, their unions must cross national boundaries to build international cooperation and solidarity. Otherwise, employers can play working and poor people in different countries against each other. An example of international union cooperation occurred in 1994 when Ford factory workers in Cuatitlan, Mexico went on strike to protest layoffs and poor working conditions. Members of the United Auto Workers in a U.S. Ford factory sent money to support the action. "Their reasoning was: if the Mexicans win, that's good for us as

The monthly expenses of both groups of women exceeded their reported monthly income, forcing women to use various strategies to make ends meet. Cash welfare and food stamps covered only three-fifths of welfare-reliant mothers' expenses. The main job of low-wage earning mothers covered only 63 percent of their expenses. Edin and Lein found that women relied on three basic strategies to make ends meet: work in the formal, informal, or underground economy; cash assistance from absent fathers, boyfriends, relatives, and friends; and cash assistance or help from agencies, community groups, or charities in paying overdue bills. Welfare recipients had to keep their income-generating activities hidden from their welfare caseworkers and other government officials. Otherwise, their welfare checks would be reduced by nearly the same amount as their earnings. Many of the wage-earning mothers also concealed income generated "on the side" in order to maintain food stamps, housing subsidies, or other benefits that would have been reduced or eliminated if they had reported this additional income.

Most of the single mothers in the study described experiencing serious material hardship during the previous 12 months. Material hardships included not having enough food and clothes, not receiving needed medical care, not having health insurance, having the utilities or phone cut off, not having a phone, and being evicted and/or homeless. An important finding was that wage-reliant mothers experienced more hardship than welfare-reliant mothers. In addition to the increased financial pressures of child care costs, transportation, health care, and work clothing, employed mothers worried about not providing adequate supervision of their children and struggled with balancing work and parenting responsibilities, especially when their children were sick. Nevertheless, almost all of the mothers said they would rather work than rely on welfare. They believed that work provided important psychological benefits and increased self-esteem, avoided the stigma of welfare, and enabled them to be good role models for their children.

Harvard University scholar Christopher Jencks (1997) comments on the implications of Edin and Lein's (1997) research:

As the new time limits on welfare receipt begin to take effect, more and more single mothers will have to take jobs. Most of these newly employed mothers will have more income than they had on welfare, so their official poverty rate will fall. But they will also have more expenses than they had on welfare, and they will get fewer noncash benefits. Edin and Lein's findings dramatize the likely result. Between 1988 and 1992, mothers who held low-wage jobs reported substantially more income than those who collected welfare, but they also reported more hardship. If this pattern persists in the years ahead, time limits will probably bring both a decline in the official poverty rate and an increase in material hardship. (p. x)

Source: Based on Kathryn Edin and Laura Lein. 1997. *Making Ends Meet*. New York: Russell Sage Foundation. Used with permission.

well as them because Ford won't be so quick to threaten to move production to Mexico" (Went 2000, p. 126). Another example of internationalizing labor efforts is the Industrial Workers of the World— an international labor union for all workers, rather than for a particular trade.

The National Labor Relations Board (NLBR) and the Courts play an important role in upholding workers' rights to unionize and sanctioning employers who violate these rights. In 1998, the National Labor Relations Board issued nearly 24,000 reinstatement and "back-pay" orders or other remedial orders to workers wrongfully fired or demoted for participating in union-related activities (Human Rights Watch 2001). The National Labor Relations Board and the courts have held that employer threats to close the plant if the union succeeds in organizing can be unlawful under certain circumstances (Bronfenbrenner 2000). For example, Guardian Industries Corp. v. NLRB held that it was unlawful for a supervisor to say to an employee, "If we got a union in there, we'd be in the unemployment line." However, under the employer free speech provi-

I haven't seen as much raw anger as I see in the workplace today. One thing I've heard repeatedly around the country from unorganized workers is the following: "I never thought about joining a union, but for the first time I'm now thinking about it, because I need somebody to protect me."

ROBERT REICH
Former U.S. Secretary of Labor

For once we have a judge who says that the right of workers to organize in unions of their own choosing should be given as much, if not more, weight as the freedom of corporations to use the threat of capital mobility to hold down wages and thwart organizing drives.

KATE BRONFENBRENNER
Cornell University, referring to Judge Moreno's injunction against Quadrtech Corporation

sions of the Taft-Hartley Act, the courts have permitted the employer to predict a plant closing in situations where it is based on an objective assessment of the economic consequences of unionization. When an employer is found guilty of making an unlawful threat of plant closure, the typical remedy is a cease and desist order coupled with the posting of a notice promising not to make such statements in the future.

In December 2000, a court action set a precedent that could help workers' efforts to unionize for better wages and working conditions. One day after workers at Quadrtech Corporation, a Los Angeles-based jewelry manufacturer, voted to join the International Union of Electronic Workers, Quadrtech owner Vladimir Reil announced he would lay off the majority of workers and move his company to Tijuana. With the support of the union, the workers appealed to the National Labor Relations Board (NLRB) to seek an injunction against the move. U.S. District Court Judge Carlos Moreno ruled in favor of an injunction to block the move and forced the return of two truckloads of equipment that had already been shipped across the border. This injunction is believed to be the first ruling of its kind in the United States ("Stay Put, Judge Tells Co." 2000).

Ninety-year-old Doris Haddock, known as "Granny D," walked from Los Angeles to Washington DC—a 3200 mile, 14-month trip—to protest the system of political campaign financing that allows special interest groups to purchase political influence through large financial contributions. Here she is shown with Senators John McCain and Russ Feingold, who worked to pass the McCain-Feingold campaign finance reform bill.

Challenges to Corporate Power and Globalization

In the United States, advocates for campaign finance reform have challenged the power that corporations have in influencing laws and policies. Efforts to reform the system of political campaign financing that allows corporations to purchase political influence were rewarded when the Senate passed the McCain-Feingold bill in April 2001. As of this writing, this bill which bans soft-money political contributions awaits a vote in the House. According to a survey of senior executives at the nation's largest businesses, more than half support a soft-money ban ("Big Business for Reform" 2000). Such a ban would protect businesses against the pressure that businesses feel from political parties and candidates to make large political contributions. It would also help the public image of big business. Nearly three-quarters (71 percent) of executives in the survey said that stories about large political contributions are hurting the images of U.S. companies.

Challenges to corporate globalization have also taken root in the United States and throughout the world. Antiglobalization activists have targeted the World Trade Organization (WTO), the International Monetary Fund (IMF), and World Bank as forces that advance corporate-led globalization at the expense of social goals like justice, community, national sovereignty, cultural diversity, ecological sustainability, and workers' rights.

In 1999, 50,000 street protesters and Third World delegates demonstrated in opposition to the policies of the World Trade Organization that promoted corporate-led globalization. The brutal assaults on largely peaceful demonstrators by Seattle police dressed in their Darth Vader-like uniforms in full view of television cameras has made the Seattle WTO protest the "grand symbol of the crisis of globalization" (Bello 2001).

Another confrontation between pro-globalization and anti-globalization forces occurred at the 2000 meeting of the International Monetary Fund (IMF) and the World Bank in Washington, D.C. About 30,000 protesters descended on America's capital and found a large section of the northwest part of the city walled off by some 10,000 police. For four days, the protesters tried, unsuccessfully, to break through the police barrier to reach the IMF-World Bank complex at 19th and H streets, resulting in hundreds of arrests. Media attention to the protest contributed to the growing worldwide awareness of the forces of corporate globalization and its social, environmental, and economic effects.

Understanding *Work and Unemployment*

On December 10, 1948, the General Assembly of the United Nations adopted and proclaimed the Universal Declaration of Human Rights. Among the articles of that declaration are the following:

> Article 23. Everyone has the right to work, to free choice of employment, to just and favourable conditions of work and to protection against unemployment.
>
> Everyone, without any discrimination, has the right to equal pay for equal work.
>
> Everyone who works has the right to just and favourable remuneration ensuring for himself and his family an existence worthy of human dignity, and supplemented, if necessary, by other means of social protection.
>
> Everyone has the right to form and to join trade unions for the protection of his interests.
>
> Article 24. Everyone has the right to rest and leisure, including reasonable limitation of working hours and periodic holidays with pay.

The 1999 Seattle protest against the World Trade Organization brought national and international media attention to the growing concerns over globalization and corporate power.

More than half a century later, workers around the world are still fighting for these basic rights as proclaimed in the Universal Declaration of Human Rights.

To understand the social problems associated with work and unemployment, we must first recognize the power and influence of governments and corporations on the workplace. We must also be aware of the role that technological developments and postindustrialization have on what we produce, how we produce it, where we produce it, and who does the producing. In regard to what we produce, the United States is moving away from producing manufactured goods to producing services. In regard to production methods, the labor-intensive blue-collar assembly line is declining in importance, and information-intensive white-collar occupations are increasing. Because of increasing corporate multinationalization, U.S. jobs are being exported to foreign countries where labor and raw materials are cheap, and regulations are lax. In developing countries, investment in workplace safety is far below that in the rich nations.

Decisions made by U.S. corporations about what and where to invest influence the quantity and quality of jobs available in the United States. As conflict

If there ever is another war in this country, it will be between capital and labor.

FRANK JAMES
JESSE'S BROTHER

What the public wants is called "politically unrealistic." Translated into English, that means power and privilege are opposed to it.

NOAM CHOMSKY

I see in the near future a crisis approaching that unnerves me and causes me to tremble for the safety of my country . . . . corporations have been enthroned and an era of corruption in high places will follow, and the money power of the country will endeavor to prolong its reign by working upon the prejudices of the people until all wealth is aggregated in a few hands and the Republic is destroyed.

ABRAHAM LINCOLN

theorists argue, such investment decisions are motivated by profit, which is part of a capitalist system. Profit is also a driving factor in deciding how and when technological devices will be used to replace workers and increase productivity. But if goods and services are produced too efficiently, workers are laid off and high unemployment results. When people have no money to buy products, sales slump, recession ensues, and social welfare programs are needed to support the unemployed. When the government increases spending to pay for its social programs, it expands the deficit and increases the national debt. Deficit spending and a large national debt make it difficult to recover from the recession, and the cycle continues.

What can be done to break the cycle? Those adhering to the classic view of capitalism argue for limited government intervention on the premise that business will regulate itself via an "invisible hand" or "market forces." For example, if corporations produce a desired product at a low price, people will buy it, which means workers will be hired to produce the product, and so on.

Ironically, those who support limited government intervention also sometimes advocate that the government intervene to bail out failed banks and lend money to troubled businesses. Such government help benefits the powerful segments of our society. Yet, when economic policies hurt less powerful groups, such as minorities, there has been a collective hesitance to support or provide social welfare programs. It is also ironic that such bail-out programs, which contradict the ideals of capitalism, are needed because of capitalism. For example, the profit motive leads to multinationalization, which leads to unemployment, which leads to the need for government programs. The answers are as complex as the problems. The various forces transforming the American economy are interrelated—technology, globalization, capital flight through multinationalization, and the movement toward a service economy (Eitzen & Zinn 1990). For the individual worker, the concepts of work, job, and career have changed forever.

Critical Thinking

1 Union membership is higher among black employees than among white employees. For example, in 1997, 15.9 percent of whites belonged to a union, compared to 20.5 percent of blacks (Mishel et al. 2001). What might explain the higher rate of unionization among black workers?

2 U.S. industries dominated by female workers have a higher "win rate" in organizing elections compared with industries dominated by men. Female-dominated industries have an average win rate of 60 percent, compared to 38 percent for industries employing more males ("Labor's 'Female Friendly' Agenda" 1998). Why do you think this is so?

3 In the late 1990s, a glut of media attention has been given to the positive economic indicators in the United States: low inflation rates, low unemployment rates, and surging stock market values. How might a conflict theorist view the media's portrayal of the U.S. economy?

Key Terms

alienation	behavioral-based safety programs	brain drain
anomie		capitalism
automation	bonded labor	child labor

compressed workweek

contingent workers

convergence hypothesis

corporate downsizing

corporate multinationalism

cumulative trauma
disorders

discriminatory
unemployment

economic institution

ergonomics

Family and Medical
Leave Act

Fair Labor Standards Act
(FLSA)

flextime

global economy

industrialization

job burnout

job exportation

job sharing

labor unions

National Labor Relations
Act

National Labor Relations
Board

postindustrialization

primary labor market

repeated trauma disorders

repetitive strain disorders

secondary labor market

socialism

soft money

split labor market

structural unemployment

sweatshop

telecommuting

telework

tertiary labor market

underemployment

unemployment

union density

work sectors (primary,
secondary, tertiary)

Media Resources

The Wadsworth Sociology Resource Center: Virtual Society

http://sociology.wadsworth.com/

See the companion Web site for this book to access general sociology resources and text-specific features that can further your understanding of this chapter. The site contains Internet links, Internet exercises, online practice quizzes, information on InfoTrac College Edition, and many more valuable materials designed to enrich your learning experience in social problems.

InfoTrac College Edition

You can access InfoTrac College Edition either from the Wadsworth Sociology Resource Center at **http://sociology.wadsworth.com** or directly from your web browser at **http://www.infotrac-college.com/wadsworth/**. InfoTrac College Edition is an online university library that includes over 700 popular and scholarly journals in which you can find articles related to the topics in this chapter such as labor unions, global economic issues, sweatshop and child labor, worker health and safety issues, and work/family concerns.

Interactions CD-ROM

Go to the Interactions CD-ROM for *Understanding Social Problems*, Third Edition to access additional interactive learning tools, such as in-depth review materials, corresponding practice quizzes, and other engaging resources and activities to help you study the concepts in this chapter.

12

Problems in Education

Is It True?

1. Court-ordered busing and a national emphasis on the equality of education have largely eliminated segregation in the public schools.

2. When a national sample of American adults were asked about education in America, over half replied that they were dissatisfied with the education their children were receiving.

3. More than 1 out of every 10 Americans aged 16 to 24 has not completed high school.

4. Less than 40 percent of all 3- to 5-year-olds are enrolled in either nursery school or kindergarten.

5. American schoolchildren consistently outperform their international counterparts in mathematics and science.

Answers to "Is It True?": 1 = F; 2 = T; 3 = T; 4 = F; 5 = F

The quality of our public schools directly affects us all—as parents, as students, and as citizens. Yet too many children in American are segregated by low expectations, illiteracy, and self-doubt. In a constantly changing world that is demanding increasingly complex skills from its workforce, children are literally being left behind.

GEORGE W. BUSH
U.S. President

Sean, Dan, and Lance were eating lunch together in the school cafeteria when they decided to go outside. As they were walking up a hill, they saw two figures with guns in the distance. Must be Annihilation, Sean thought, a paintball game that seniors played. Odd. Those guns looked real—not like the plastic models he had seen before. "Pop, Pop, Pop." The guns were suddenly turned toward the school and before Sean knew it, he was the only one of his three friends standing. Sean turned to look for the paintball that had just grazed his neck. They must be frozen he thought—he was bleeding—but as he turned he was shot three times in the abdomen. He began to run. Why am I running from paintballs he thought, as he headed toward school. Then it hit—the bullet that really hurt—in the back, striking the spine, exiting through the hip.

Sean survived the attack at Columbine High School in Littleton, Colorado, as did Lance. But Dan Rohrbough did not, leaving others the grim task of trying to make some sense out of his death and the deaths of 12 others. Sean tries not to think about it, preferring to pretend it never happened, wanting to put it and his wheelchair behind him. But, embarrassingly, people keep staring at him. What are they looking at, he wonders. Then he thought, maybe that's how Dylan and Eric felt. (Pollock 2000)

Violence is just one of the many issues that must be addressed in today's schools (see this chapter's *Social Problems Research Up Close* feature). Students continue to graduate from high school being unable to read, work simple math problems, or write grammatically correct sentences. Graduates discover that they are ill prepared for corporations who demand literate, articulate, informed employees. Teachers leave the profession because of uncontrollable discipline problems, inadequate pay, and overcrowded classrooms. Students and teachers alike are "dumbing down," lowering their standards, expectations, and role performances to fit increasingly undemanding and unresponsive systems of learning.

And yet it is education that is often claimed as a panacea—the cure-all for poverty and prejudice, drugs and violence, war and hatred, and the like. Can one institution, riddled with problems, be a solution for other social problems? In this chapter we focus on this question and what is being called one of the major sociopolitical issues of the next century—the educational crisis (Associated Press 1998). We begin with a look at education around the world.

Guns, Kids, and Schools

Reducing school violence is consistently listed as one of the top education priorities in public opinion polls, and is a focus of President Bush's educational reform package entitled "No Child Left Behind." In an effort to curb violence, schools have established "zero-tolerance policies." For example, nationwide, over 90 percent of schools have zero-tolerance for students carrying firearms or other weapons to school (Executive Summary 2001). Sociologist Pamela Roundtree (2000) addresses the issue of violence in schools by asking adolescents why they carry weapons.

Sample and Methods

The respondents in this study were sixth through twelfth grade students who had participated in a state-sponsored research project, the Kentucky Youth Survey. Because weapon carrying is likely to vary by region, data from three distinct areas of the state were collected: (1) Urban County (wealthier, north-central part of state), (2) Western County (rural, tobacco growing area), and (3) Eastern County (poorer, high unemployment, mining region). Race and sex distributions varied between county samples with whites and females the majority of each of the three samples.

The dependent variable is possession of weapons at school, measured by whether or not a student reported carrying a weapon to school in the 30 days prior to the survey. In addition to the standard demographics variables of sex, race, age, and family socioeconomic background, variables thought to be predictive of carrying weapons were measured. The independent variables include: (1) fear of crime indicators (prior victimization, fear of victimization), (2) criminal involvement indicators (previous arrest, drug involvement), (3) pro-weapon socialization indicators (weapon ownership or use by respondent, weapon ownership by parent, weapon carrying by peers), and (4) social isolation indicators (disattachment from school, church). Each of these categories of indicators was predicted to be directly related to the likelihood of carrying a weapon to school, that is, as fear of crime, criminal involvement, pro-weapon socialization, and social isolation increase, the likelihood of carrying weapons increases.

The Global Context: Cross-Cultural Variations in Education

Looking only at the American educational system might lead one to conclude that most societies have developed some method of formal instruction for their members. After all, the United States has more than 91,000 schools, 4 million teachers and faculty, and 68 million students (Digest of Educational Statistics 2001). In reality, many societies have no formal mechanism for educating the masses. As a result, worldwide, over 120 million children do not attend school—42 million in sub-Saharan Africa alone—and over 880 million adults are illiterate. In 2000, the World Education Forum met in Dakar, Senegal where over 1,000 leaders from 145 countries recommitted their energies to improving basic education in developing countries (U.S. Newswire 2000a).

On the other end of the continuum are societies that emphasize the importance of formal education. Public expenditures on education in developed countries are 25 times higher than in less-developed countries (Population Reference Bureau 1999). In Japan, students attend school on Saturday, and the school calendar is 40 days longer than in the United States. Spain, China, Korea, Israel, Switzerland, Scotland, and Canada also have more mandatory school days than the United States (Youth Indicators 1996).

Some countries also empower professionals to organize and operate their school systems. Japan, for example, hires professionals to develop and implement a national curriculum and to administer nationwide financing for its schools. In contrast, school systems in the United States are often run at the lo-

Education is an indispensable tool for the improvement of the quality of life.

UNITED NATIONS
Conference on Population and Development

Findings and Conclusion

Carrying a weapon to school is a relatively rare event, with 5 percent or less of students reporting that they had carried a weapon to school in the previous 30 days. Possession was slightly lower in Urban than in the Eastern or Western County, Urban County being the most industrialized and the wealthiest of the three.

In general, age, race, and sex were unrelated to the likelihood of a student taking a weapon to school. Only in Eastern County did sex significantly predict carrying a weapon with males 700 percent more likely than females to possess a weapon in school. Surprisingly, prior victimization and fear of crime were unrelated to weapon carrying. However, drug dealing was predictive of weapon carrying in both Eastern and Western Counties, and student drug use as an indicator of criminal involvement was significantly related to weapon possession in each of the three counties.

Pro-weapon socialization had an even stronger effect than criminal involvement. Of the three measures of this variable—weapon ownership or use by respondent, ownership by parent, or carrying by peers—carrying by peers had the strongest relationship with carrying a weapon. With each "best friend" the respondent reported as having carried a gun to school, the likelihood of the respondent carrying a gun to school increased by between 75 and 100 percent. Social isolation variables were not related to carrying a weapon in any consistent way.

Unlike many studies on adult weapon carrying, the results of this study indicate that peer-based socialization has a much larger impact on the probability of carrying a weapon than fear of crime variables. However, consistent with research on adults, carrying a weapon to school was significantly related to criminal involvement and, specifically, to drug involvement.

Reference: Executive Summary. 2001. "Indicators of School Crime and Safety." http://www.nces.ed.gov/pubs2001/crime2000

Source: Pamela Wilcox Rountree. 2000. "Weapons at School: Are the Predictors Generalizable across Context?" *Sociological Spectrum* 20:291–324.

cal level by school boards and PTAs composed of lay people. In effect, local communities raise their own funds and develop their own policies for operating the school system in their area; the result is a lack of uniformity from district to district. Thus, parents who move from one state to another often find the quality of education available to their children differs radically. In societies with

Because teaching is a highly respected vocation in Japan, students treat their teachers with respect and obedience. Teaching is less well regarded in the United States, and the consequences are felt by our teachers every day.

professionally operated and institutionally coordinated schools such as Japan, teaching is a prestigious and respected profession. Consequently, Japanese students are attentive and obedient to their teachers. In the United States, the phrase "Those that can, do; those that can't, teach" reflects the lack of esteem for the teaching profession. The insolence and defiance among students in American classrooms are evidence of the disrespect that many American students have for their teachers.

Sociological Theories of Education

The three major sociological perspectives—structural-functionalism, conflict theory, and symbolic interactionism—are important in explaining different aspects of American education.

Structural-Functionalist Perspective

According to structural-functionalism, the educational institution serves important tasks for society, including instruction, socialization, the provision of custodial care, and the sorting of individuals into various statuses. Many social problems, such as unemployment, crime and delinquency, and poverty, may be linked to the failure of the educational institution to fulfill these basic functions (see Chapters 4, 10, and 11). Structural-functionalists also examine the reciprocal influences of the educational institution and other social institutions, including the family, political institution, and economic institution.

Instruction   A major function of education is to teach students the knowledge and skills that are necessary for future occupational roles, self-development, and social functioning. Although some parents teach their children basic knowledge and skills at home, most parents rely on schools to teach their children to read, spell, write, tell time, count money, and use computers. As discussed later, many U.S. students display a low level of academic achievement. The failure of schools to instruct students in basic knowledge and skills both causes and results from many other social problems.

> The beautiful thing about learning is that nobody can take it away from you.
>
> **B. B. KING,**
> *Musician*

Socialization   The socialization function of education involves teaching students to respect authority—behavior that is essential for social organization (Merton 1968). Students learn to respond to authority by asking permission to leave the classroom, sitting quietly at their desks, and raising their hands before asking a question. Students who do not learn to respect and obey teachers may later disrespect and disobey employers, police officers, and judges.

The educational institution also socializes youth into the dominant culture. Schools attempt to instill and maintain the norms, values, traditions, and symbols of the culture in a variety of ways, such as celebrating holidays (Martin Luther King Jr. Day, Thanksgiving); requiring students to speak and write in standard English; displaying the American flag; and discouraging violence, drug use, and cheating.

As the number and size of racial and ethnic minority groups have increased, American schools are faced with a dilemma: Should public schools promote only one common culture, or should they emphasize the cultural diversity reflected in the American population? Some evidence suggests that most Americans believe that schools should do both—they should promote one common culture and emphasize diverse cultural traditions (Elam, Rose & Gallup 1994).

Multicultural education, education that includes all racial and ethnic groups in the school curriculum, promotes awareness and appreciation for cultural diversity. In a recent national survey, 68 percent of respondents agreed with the statement that at least one cultural and ethnic diversity course should be required for college graduation (Conciatore 2000).

Sorting Individuals into Statuses Schools sort individuals into statuses by providing credentials for individuals who achieve various levels of education, at various schools, within the system. These credentials sort people into different statuses—for example, "high school graduate," "Harvard alumna," and "English major." Further, schools sort individuals into professional statuses by awarding degrees in such fields as medicine, nursing, and law. The significance of such statuses lies in their association with occupational prestige and income—the higher one's education, the higher one's income. Further, unemployment rates are tied to educational status (see Figure 12.1).

Custodial Care The educational system also serves the function of providing custodial care (Merton 1968), which is particularly valuable to single-parent and dual-earner families, and the likely reason for the increase in enrollments of 3- to 5-year-olds. In 1970, 37 percent of 3- to 5-year-olds were enrolled in formal classes. Today, over 60 percent of this age group are enrolled in either nursery school or kindergarten (*Statistical Abstract* 2000).

The school system provides supervision and care for children and adolescents until they are 16—12 years of school totaling almost 13,000 hours per pupil! Yet, some school districts are increasing class hours. For example, the *Knowledge is Power* program in Houston, Texas requires students to attend school several weeks in the summer, on alternate Saturdays, and from 7:25 AM to 5:00 PM during the regular school year. Working parents, the hope that increased supervision will reduce delinquency rates, and higher educational standards that require longer hours of study are some of the motivations behind the "more time" movement (Wilgoren 2001).

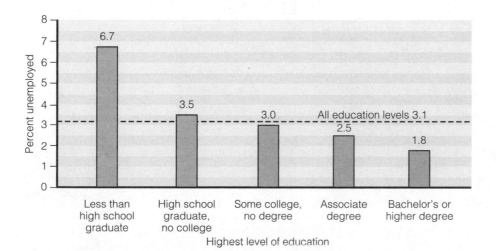

■ **Figure 12.1**
Unemployment Rate of Persons 25 Years and Over, by Highest Level of Education, 1999.

Source: Digest of Education Statistics. Figure 24. U.S. Department of Education. NCES 2001-034.

Conflict Perspective

Conflict theorists emphasize that the educational institution solidifies the class positions of groups and allows the elite to control the masses. Although the official goal of education in society is to provide a universal mechanism for achievement, in reality educational opportunities and the quality of education are not equally distributed.

Conflict theorists point out that the socialization function of education is really indoctrination into a capitalist ideology. In essence, students are socialized to value the interests of the state and to function to sustain it. Such indoctrination begins in kindergarten. Rosabeth Moss Kanter (1972) coined the term "the organization child" to refer to the child in nursery school who is most comfortable with supervision, guidance, and adult control. Teachers cultivate the organization child by providing daily routines and rewarding those who conform. In essence, teachers train future bureaucrats to be obedient to authority.

Further, to conflict theorists, education serves as a mechanism for **cultural imperialism**, or the indoctrination into the dominant culture of a society. When cultural imperialism exists, the norms, values, traditions, and languages of minorities are systematically ignored. A Mexican-American student recalls his feelings about being required to speak English (Rodriquez 1990, 203):

> When I became a student, I was literally "remade"; neither I nor my teachers considered anything I had known before as relevant. I had to forget most of what my culture had provided, because to remember it was a disadvantage. The past and its cultural values became detachable, like a piece of clothing grown heavy on a warm day and finally put away.

Conflict theorists are also quick to note that learning is increasingly a commercial enterprise as necessary financial support for equipment, laboratories, and technological upgrades are funded by corporations anxious to bombard students with advertising and other pro-capitalist messages. For example, a free-to-schools Chevron-produced video on the greenhouse effect is accompanied by a student discussion guide that asks the question, "What, if anything, can be done

Per-pupil expenditure, averaging about $7,000 a year, varies dramatically between school districts and states.

© James D. Wilson/Woodfin Camp & Associates

© Dan Habib/Impact Visuals

to safeguard the public against errors in judgement by policymakers, who may be reacting to political pressures?" (quoted in Consumer Reports 2000).

Finally, the conflict perspective focuses on what Kozol (1991) calls the "savage inequalities" in education that perpetuate racial disparities. Kozol documents gross inequities in the quality of education in poorer districts, largely composed of minorities, compared with districts that serve predominantly white middle-class and upper-middle-class families. Kozol reveals that schools in poor districts tend to receive less funding and have inadequate facilities, books, materials, equipment, and personnel. For example, nationally, the richest school districts spend 56 percent more per pupil than the poorest school districts (CDF 2000).

Symbolic Interactionist Perspective

Whereas structural-functionalism and conflict theory focus on macro-level issues such as institutional influences and power relations, symbolic interactionism examines education from a micro perspective. This perspective is concerned with individual and small group issues, such as teacher-student interactions, the students' self-esteem, and the self-fulfilling prophecy.

Teacher-Student Interactions   Symbolic interactionists have examined the ways in which students and teachers view and relate to each other. For example, children from economically advantaged homes may be more likely to bring to the classroom social and verbal skills that elicit approval from teachers. From the teachers' point of view, middle-class children are easy and fun to teach: they grasp the material quickly, do their homework, and are more likely to "value" the educational process. Children from economically disadvantaged homes often bring fewer social and verbal skills to those same middle-class teachers, who may, inadvertently, hold up social mirrors of disapproval. Teacher disapproval contributes to the lower self-esteem among disadvantaged youth.

Self-Fulfilling Prophecy   The **self-fulfilling prophecy** occurs when people act in a manner consistent with the expectations of others. For example, a teacher who defines a student as a slow learner may be less likely to call on that student or to encourage the student to pursue difficult subjects. As a consequence of the teacher's behavior, the student is more likely to perform at a lower level.

A study by Rosenthal and Jacobson (1968) provides empirical evidence of the self-fulfilling prophecy in the public school system. Five elementary school students in a San Francisco school were selected at random and identified for their teachers as "spurters." Such a label implied that they had superior intelligence and academic ability. In reality, they were no different from the other students in their classes. At the end of the school year, however, these five students scored higher on their intelligence quotient (IQ) tests and made higher grades than their classmates who were not labeled as spurters. In addition, the teachers rated the spurters as more curious, interesting, and happy and more likely to succeed than the non-spurters. Because the teachers expected the spurters to do well, they treated the students in a way that encouraged better school performance. This chapter's *The Human Side* feature in this chapter illustrates how negative labeling affects a student's self-concept and performance.

If we are to continue to have an essentially free and classless society in this country, we must proceed from the premise that there are no educational privileges...

JAMES BRYANT CONNANT
Former Harvard President and educational reformer

They Think I'm Dumb

The following excerpt poignantly recounts a conversation between author-ethnographer Thomas Cottle and Ollie, an 11-year-old boy, labeled as "dumb."

You know what, Tom?" he said, looking down at his ice cream as though it suddenly had lost its flavor, "nobody, not even you or my Dad, can fix things now. The only thing that matters in my life is school, and there they think I'm dumb and always will be. I'm starting to think they're right. . . ."

"Even if I look around and know that I'm the smartest in my group, all that means is that I'm the smartest of the dumbest, so I haven't gotten anywhere at all, have I? I'm right where I always was. Every word those teachers tell me, even the ones I like most, I can hear in their voice that what they're really saying is, 'alright you dumb kids, I'll make it easy as I can, and if you don't get it then, then you'll never get it.' That's what I hear every day, man. From every one of them. Even the other kids talk to me that way too. . . ."

"I'll tell you something else," he was saying, unaware of the ice cream that was melting on his hand. "I used to think, man, that even if I wasn't so smart, that I could talk in any class in that school, if I did my studying, I mean, and have everybody in that class, all the kids and the teacher too, think I was alright. Maybe better than alright. . . ."

"Then they told me, like on a Friday, that today would be my last day in that class. That I should go to it today, you know, but that on Monday I had to switch to this other one. They just gave me a different room number, but I knew what they were doing. Like they were giving me one more day with the brains, and then I had to go to be with the dummies, where I was suppose to be. . . ."

"So I went with the brains one more day, on that Friday like I said, in the afternoon. But the teacher didn't know I was moving, so she acted like I belonged there. Wasn't her fault. All the time I was just sitting there thinking this is the last day for me. This is the last time I'm ever going to learn anything, you know what I mean? Real learning."

He had not looked up at me even once since leaving the ice cream store....."From now on," he was saying, "I knew I had to go back where they made me believe I belonged . . . then the teacher called on me, and this is how I know just how not smart I am. She called on me, like she always did, like she'd call on anybody, and she asked me a question. I knew the answer, 'cause I'd read it the night before in my book. . . . So I began to speak, and suddenly I couldn't say nothing. Nothing, man. Not a word. Like my mind died in there. And everybody was looking at me, you know, like I was crazy or something. My heart was beating real fast, I knew the answer, man. And she was just waiting, and I couldn't say nothing. And you know what I did? I cried. I sat there and cried, man, 'cause I couldn't say nothing. That's how I know how smart I am. That's when I really learned at that school, how smart I was. I mean, how smart I thought I was. I had no business being there. Nobody smart's sitting in no class crying. That's the day I found out for real. That's the day that made me know for sure."

Ollie's voice had become so quiet and hoarse that I had to lean down to hear him. We were walking in silence, I was almost afraid to look at him. At last he turned toward me, and for the first time I saw the tears pouring from his eyes. His cheeks were bathed in them. Then he reached over and handed me his ice cream cone.

"I can't eat it now, man," he whispered. "I'll pay you back for it when I get some money."

Source: Excerpted from Thomas Cottle. 1976. *Barred from School*. Washington, D.C.: The New Republic Book Company, Inc., pp.138–40. Used by permission.

Who Succeeds? The Inequality of Educational Attainment

As noted earlier, conflict theory focuses on inequalities in the educational system. Educational inequality is based on social class and family background, race and ethnicity, and gender. Each of these factors influences who succeeds in school.

Social Class and Family Background

One of the best predictors of educational success and attainment is socioeconomic status. Children whose families are in middle and upper socioeconomic brackets are more likely to perform better in school and to complete more years of education than children from lower socioeconomic-class families. For example, Muller and Schiller (2000) report that students from higher socioeconomic backgrounds are more likely to enroll in advanced mathematics course credits, and to graduate from high school—two indicators of future educational and occupational success.

Families with low incomes have fewer resources to commit to educational purposes. Low-income families have less money to buy books, computers, tutors, and lessons in activities such as dance and music and are less likely to take their children to museums and zoos. Parents in low-income brackets are also less likely to expect their children to go to college, and their behavior may lead to a self-fulfilling prophecy. Disproportionately, children from low-income families do not go to college (Levinson 2000).

Disadvantaged parents are also less involved in their children's education. Seventy percent of children of college-educated mothers are read to daily compared with 38 percent of children whose mothers have not completed high school (NCES 2000a). Although working-class parents may value the education of their children, in contrast to middle- and upper-class parents, they are intimidated by their child's schools and teachers, don't attend teacher conferences, and have less access to educated people to consult about their child's education (Lareau 1989).

Because low-income parents are often themselves low academic achievers, their children are exposed to parents who have limited language and academic skills. Children learn the limited language skills of their parents, which restricts their ability to do well academically. Low-income parents may be unable to help their children with their math, science, and English homework because they often do not have the academic skills to do the assignments. Interestingly, Call, Grabowski, Mortimer, Nash, and Lee (1997) report that even among impoverished youths, parental education is one of the best predictors of a child's academic success.

Children from poor families also have more health problems and nutritional deficiencies. In 1965 Project **Head Start** began to help preschool children from disadvantaged families. It provided an integrated program of health care, parental involvement, education, and social services. Today, over 800,000 3- to 5-year-olds are enrolled in Head Start (Fact Sheet 2000). Early Head Start, a recently developed program for infants and toddlers from low-income families, has also met with great success. Evaluation of the program found that "after a year or more of program services, when compared with a randomly assigned control group, 2-year-old Early Head Start children performed significantly better on a range of measures of cognitive, language, and socio-emotional development" (Summary Report 2001).

Lack of adequate funding for Head Start is a problem as is equality of educational funding in general. Children who live in lower socioeconomic conditions receive fewer public educational resources. Schools that serve low socioeconomic districts are largely overcrowded and understaffed, lacking adequate building space and learning materials.

The U.S. tradition of decentralized funding means that local schools depend upon local taxes, usually property taxes; about 45 percent of school funding comes from local sources. The amount of money available in each district varies

Educational reform measures alone can have only modest success in raising the educational achievements of children from low-income families. The problems of poverty must be attacked directly.

RICHARD J. MURNANE
Harvard University

The family plays a key role in their children's success in school.

ALBERT BANDURA
Social scientist

by the socioeconomic status, or SES, of the district. This system of depending on local communities for financing has several consequences:

- Low-SES school districts are poorer because less valuable housing means lower property values; in the inner city, houses are older and more dilapidated; less desirable neighborhoods are hurt by "white flight," with the result that the tax base for local schools is lower in deprived areas.
- Low-SES school districts are less likely to have businesses or retail outlets where revenues are generated; such businesses have closed or moved away.
- Because of their proximity to the downtown area, low-SES school districts are more likely to include hospitals, museums, and art galleries, all of which are tax-free facilities. These properties do not generate revenues.
- Low-SES neighborhoods are often in need of the greatest share of city services; fire and police protection, sanitation, and public housing consume the bulk of the available revenues. Precious little is left for education in these districts.
- In low-SES districts, a disproportionate amount of the money has to be spent on maintaining the school facilities, which are old and in need of repair, so less is available for the children themselves.

Although the state provides additional funding to supplement local taxes, it is not enough to lift schools in poorer districts to a level that even approximates the funding available to schools in wealthier districts.

Race and Ethnicity

Socioeconomic status interacts with race and ethnicity. Because race and ethnicity are so closely tied to socioeconomic status, it appears that race or ethnicity alone can determine school success. Although race and ethnicity also have independent effects on educational achievement (Bankston & Caldas 1997; Jencks & Phillips 1998), their relationship is largely a result of the association between race/ethnicity and socioeconomic status. As Table 12.1 indicates, edu-

Table 12.1 *Educational Attainment by Race and Ethnicity, 1975 and 1999*

	1975	1999
Four Years of High School or More		
White	64.5	84.3
Black	42.5	77.0
Hispanic	37.9	56.1
Asian	NA	84.7
Total	62.5	83.4
Four Years of College or More		
White	14.5	25.9
Black	6.4	15.4
Hispanic	NA	10.9
Asian	NA	42.4
Total	13.9	25.2

NA = Not Available

Source: *Statistical Abstract of the United States*, 2000, 120th edition. Table 249.

The debate over bilingual education is likely to grow. By 2040, less than half of all school-age children will be non-Hispanic whites.

cational attainment has increased over time, and varies by race and ethnicity. In general, the high school graduation gap between racial and ethnic groups is narrowing; the college graduation gap, however, is getting wider—whites and Asians on one side Hispanics and African Americans on the other (Rand Institute 1999).

One reason why some minority students have academic difficulty is that they did not learn English as their native language. For example, in 1999, 70 percent of U.S.-born Hispanics graduated from high school compared with 44 percent of foreign-born Hispanics living in the U.S. (Armas 2000). Between 2000 and 2020 the number of white and black children is expected to remain fairly constant; however, the number of Hispanic and Asian children is projected to increase by 60 and 64 percent, respectively (U.S. Department of Education 2000).

To help American children who do not speak English as their native language, some educators advocate **bilingual education**—teaching children in both English and their non-English native language. Advocates claim that bilingual education results in better academic performance of minority students, enriches all students by exposing them to different languages and cultures, and enhances the self-esteem of minority students. Critics argue that bilingual education limits minority students and places them at a disadvantage when they compete outside the classroom, reduces the English skills of minorities, costs money, and leads to hostility with other minorities who are also competing for scarce resources.

Another factor that hurts minority students academically is the fact that many tests used to assess academic achievement and ability are biased against minorities. Questions on standardized tests often require students to have knowledge that is specific to the white middle-class majority culture, and students for whom English is not their native language are seriously disadvantaged.

In addition to being hindered by speaking a different language and being from a different cultural background, minority students in white school systems are also disadvantaged by overt racism and discrimination. Much of the educa-

Reducing the black-white test score gap would probably do more to promote racial equality than any other strategy that commands broad political support.

CHRISTOPHER JENCKS
Social scientist

tional inequality experienced by poor children is a result of the fact that a high percentage of them are also nonwhite or Hispanic. Discrimination against minority students takes the form of unequal funding, as discussed earlier, and school segregation.

In 1954, the U.S. Supreme Court ruled in *Brown v. Board of Education* that segregated education was unconstitutional because it was inherently unequal. Despite this ruling, many schools today are racially segregated. In 1966, a landmark study entitled Equality of Educational Opportunity (Coleman et al. 1966) revealed the extent of segregation in U.S. schools. In this study of 570,000 students and 60,000 teachers in 4,000 schools, the researchers found that almost 80 percent of all schools attended by whites contained 10 percent or fewer blacks, and that whites outperformed minorities (excluding Asian Americans) on academic tests. Coleman and his colleagues emphasized that the only way of achieving quality education for all racial groups was to desegregate the schools. This recommendation, known as the **integration hypothesis**, advocated busing to achieve racial balance.

> Give me children from a two-parent Asian-American family. They will outscore almost everybody because they are taught at home that school is special, that teachers must be respected.
>
> WILLIAM BENNETT
> *Former Secretary of Education*

In spite of the Coleman report, court-ordered busing, and an emphasis on the equality of education, public schools in the 1990s and into the twenty-first century remain largely segregated. Research documents the harmful effects of such continued practices. After examining the reading and mathematics achievement levels of a nationally representative sample of high school students, Roscigno (1998, 1051) concludes that "school racial composition matters...in the direction one would expect, even with class composition and other familial and educational attributes accounted for. Attending a black segregated school continues to have a negative influence on achievement." Nonetheless, for financial as well as political reasons, busing has essentially been abandoned (Orfield & Eaton 1997).

Gender

Worldwide, women receive less education than men. Two-thirds of the more than 920 million illiterate people in the world are women (United Nations Population Fund 1999).

Historically, U.S. schools have discriminated against women. Prior to the 1830s, U.S. colleges accepted only male students. In 1833, Oberlin College in Ohio became the first college to admit women. Even so, in 1833, female students at Oberlin were required to wash male students' clothes, clean their rooms, and serve their meals and were forbidden to speak at public assemblies (Fletcher 1943; Flexner 1972).

In the 1960s, the women's movement sought to end sexism in education. Title IX of the Education Amendments of 1972 states that no person shall be discriminated against on the basis of sex in any educational program receiving federal funds. These guidelines were designed to end sexism in the hiring and promoting of teachers and administrators. Title IX also sought to end sex discrimination in granting admission to college and awarding financial aid. Finally, the guidelines called for an increase in opportunities for female athletes by making more funds available to their programs. The push toward equality in education for women has had a positive effect on the aspirations of women. In 1970, 8.2 percent of women had completed 4 years of college or more; in 2000, 24 percent had completed 4 years of college or more (U.S. Bureau of the Census 2000a; U.S. Bureau of the Census 2000b).

Traditional gender roles account for many of the differences in educational achievement and attainment between women and men. As noted in Chapter 8, schools, teachers, and educational materials reinforce traditional gender roles in several ways. Some evidence suggests, for example, that teachers provide less attention and encouragement to girls than to boys, and that textbooks tend to stereotype females and males in traditional roles (Evans & Davies 2000).

Studies of academic performance indicate that females tend to lag behind males in math and science. One explanation is that women experience workplace discrimination in these areas, which restricts their occupational and salary opportunities. The perception of restricted opportunities, in turn, negatively affects academic motivation and performance among girls and women (Baker & Jones 1993).

Most of the research on gender inequality in the schools focuses on how female students are disadvantaged in the educational system. But what about male students? For example, schools fail to provide boys with adequate numbers of male teachers to serve as positive role models. To remedy this, some school systems actively recruit male teachers, especially in the elementary grades, where female teachers are in the majority.

The problems that boys bring to school may indeed require schools to devote more resources and attention to them. More than 70 percent of students with learning disabilities such as dyslexia are male, as are 75 percent of students identified as having serious emotional problems. Boys are also more likely than girls to have speech impairments, to be labeled as mentally retarded, to exhibit discipline problems, to drop out of school, and to feel alienated from the learning process (this chapter's *Self and Society* feature assesses student alienation) (Bushweller 1995; Goldberg 1999; Sommers 2000). As discussed in Chapter 8, the argument that girls have been educationally shortchanged has recently come under attack as some academicians charge that it is boys, not girls, who have been left behind (Sommers 2000).

Problems in the American Educational System

A recent poll indicates that 61 percent of Americans are either "somewhat dissatisfied" or "completely dissatisfied" with the quality of education in the United States, and dissatisfaction is growing (Gallup 2001). Consistent with these results is President Bush's emphasis on educational reform and the problems his administration hope to address—low academic achievement, high dropout rates, questionable teacher training, and school violence. These and other problems contribute to the widespread concern over the quality of education in America.

Low Levels of Academic Achievement

The most recent national data available indicate that more than 40 million adults in this country cannot "read basic signs or maps, complete simple forms, or carry out many of the tasks required of an adult" in today's society, that is, they are **functionally illiterate** (Literacy 2000, 1). Functionally illiterate adults are dis-

Student Alienation Scale

Indicate your agreement to each statement by selecting one of the responses provided:

1. It is hard to know what is right and wrong because the world is changing so fast.
   _____ Strongly agree   _____ Agree   _____ Disagree   _____ Strongly disagree

2. I am pretty sure my life will work out the way I want it to.
   _____ Strongly agree   _____ Agree   _____ Disagree   _____ Strongly disagree

3. I like the rules of my school because I know what to expect.
   _____ Strongly agree   _____ Agree   _____ Disagree   _____ Strongly disagree

4. School is important in building social relationships.
   _____ Strongly agree   _____ Agree   _____ Disagree   _____ Strongly disagree

5. School will get me a good job.
   _____ Strongly agree   _____ Agree   _____ Disagree   _____ Strongly disagree

6. It is all right to break the law as long as you do not get caught.
   _____ Strongly agree   _____ Agree   _____ Disagree   _____ Strongly disagree

7. I go to ball games and other sports activities at school.
   _____ Always   _____ Most of the time   _____ Some of the time   _____ Never

8. School is teaching me what I want to learn.
   _____ Strongly agree   _____ Agree   _____ Disagree   _____ Strongly disagree

9. I go to school parties, dances, and other school activities.
   _____ Always   _____ Most of the time   _____ Some of the time   _____ Never

10. A student has the right to cheat if it will keep him or her from failing.
    _____ Strongly agree   _____ Agree   _____ Disagree   _____ Strongly disagree

11. I feel like I do not have anyone to reach out to.
    _____ Always   _____ Most of the time   _____ Some of the time   _____ Never

12. I feel that I am wasting my time in school.
    _____ Always   _____ Most of the time   _____ Some of the time   _____ Never

13. I do not know anyone that I can confide in.
    _____ Strongly agree   _____ Agree   _____ Disagree   _____ Strongly disagree

14. It is important to act and dress for the occasion.
    _____ Always   _____ Most of the time   _____ Some of the time   _____ Never

---

> The more you know, the more you know you don't know.
>
> **UNKNOWN**

proportionately poor, over the age of 55, uneducated, and members of racial or ethnic minority groups (NAAL 2000).

Among children, illiteracy tends to be highest among students who attend the poorest schools, yet apathy and ignorance may be found among students at more affluent schools as well. For example, an ABC News Special entitled "Burning Questions: America's Kids: Why They Flunk" began with the following interview with students from middle-class high schools:

Interviewer: Do you know who's running for president?
First Student: Who, run? Ooh. I don't watch the news.
Interviewer: Do you know when the Vietnam War was?
Second Student: Don't even ask me that. I don't know.
Interviewer: Which side won the Civil War?
Third Student: I have no idea.
Interviewer: Do you know when the American Civil War was?
Fourth Student: 1970.

15. It is no use to vote because one vote does not count very much.

_____ Strongly agree   _____ Agree   _____ Disagree   _____ Strongly disagree

16. When I am unhappy, there are people I can turn to for support.

_____ Always   _____ Most of the time   _____ Some of the time   _____ Never

17. School is helping me get ready for what I want to do after college.

_____ Strongly agree   _____ Agree   _____ Disagree   _____ Strongly disagree

18. When I am troubled, I keep things to myself.

_____ Always   _____ Most of the time   _____ Some of the time   _____ Never

19. I am not interested in adjusting to American society.

_____ Strongly agree   _____ Agree   _____ Disagree   _____ Strongly disagree

20. I feel close to my family.

_____ Always   _____ Most of the time   _____ Some of the time   _____ Never

21. Everything is relative and there just aren't any rules to live by.

_____ Strongly agree   _____ Agree   _____ Disagree   _____ Strongly disagree

22. The problems of life are sometimes too big for me.

_____ Always   _____ Most of the time   _____ Some of the time   _____ Never

23. I have lots of friends.

_____ Strongly agree   _____ Agree   _____ Disagree   _____ Strongly disagree

24. I belong to different social groups.

_____ Strongly agree   _____ Agree   _____ Disagree   _____ Strongly disagree

INTERPRETATION

This scale measures four aspects of alienation: powerlessness, or the sense that high goals (e.g., straight A's) are unattainable; meaninglessness, or lack of connectedness between the present (e.g., school) and the future (e.g., job); normlessness, or the feeling that socially disapproved behavior (e.g., cheating) is necessary to achieve goals (e.g., high grades); and social estrangement, or lack of connectedness to others (e.g., being a "loner"). For items 1, 6, 10, 11, 12, 13, 15, 18, 19, 21, and 22, the response indicating the greatest degree of alienation is "strongly agree" or "always." For all other items, the response indicating the greatest degree of alienation is "strongly disagree" or "never."

Source: Rosalind Y. Mau. 1992. "The Validity and Devolution of a Concept: Student Alienation." Adolescence 27:107, pp. 739–40. Used by permission of Libra Publishers, Inc., 3089 Clairemont Drive, Suite 383, San Diego, California 92117.

However, results from the National Assessment of Educational Progress (NAEP), a nationwide testing effort, have improved over time. For example, mathematics proficiency of 9-, 13-, and 17-year-olds increased between 1973 and 1994 and has remained fairly stable since. Reading scores, although varying by age, are also generally improved (Digest of Educational Statistics 2001). The most recent test results also indicate that: (1) parental education and student performance are directly related, (2) private school students outperform public school students , (3) use of scientific equipment is linked to achievement, and (4) reading more and watching television less are associated with higher performance (NCES 2000b; Zernike 2001).

Although some international comparisons are improving, American students are still outperformed by their foreign counterparts on many tests. The Third International Mathematics and Science Study (TIMSS) assesses the math and science performance of students worldwide. TIMSS indicates that U.S. fourth graders scored above the 26-nation average in both mathematics and science; eighth graders scored above the international average in science and below it in mathematics; and twelfth graders scored below the international average in both math and science (Digest of Educational Statistics 2001). It must be noted, however, that comparisons between U.S. students and students in other countries may be unfair—in many other countries students are channeled into specialty areas (e.g., math, science). Such specializations may account for achievement differences (Pinkerton 1998).

School Dropouts

> Boys who are morally neglected have unpleasant ways of getting themselves noticed. All children need clear, unequivocal rules.
>
> CRISTINA SOMMERS
> *Philosopher, author*

Approximately 11 percent (4 million) of 16- to 24-year-olds are dropouts, that is, they are not presently enrolled in and did not graduate from high school (NCES 2000b). As Figure 12.2 indicates, dropout rates have decreased significantly since the 1970s and, with the exception of Hispanics, racial and ethnic differences have narrowed over time.

Why do students drop out of high school? An analysis of data from the National Educational Longitudinal Study lends some insight. Teachman, Paasch, and Carver (1997) found that staying in school between tenth and twelfth grade is a function of several variables. Children who attend Catholic schools, are from intact, two-parent families, and come from families with greater income and educational levels have lower odds of dropping out of school. The authors conclude that social capital, that is, resources that are a consequence of specific social relationships, is predictive of school continuation.

The economic and social consequences of dropping out of school are significant. Dropouts are more likely than those who complete high school to be unemployed (see Figure 12.1) and to earn less when they are employed (Digest of Educational Statistics 2001). Individuals who do not complete high school are also more likely to engage in criminal activity, have poorer health and lower rates of political participation, and require more government services such as welfare and health care assistance (Rumberger 1987; Natriello 1995).

Violence in the Schools

The number of high school students carrying guns, knives, or other weapons has decreased since 1992, from 12 to 9 percent and, despite the horrific images of the killings at Columbine High School, fewer than 1 percent of school-aged children's deaths take place on or around school property. However, the number of "multiple-victim events" has increased in recent years (U.S. Department

Figure 12.2 *Student Dropout Rate, Ages 16 to 24 by Race/Ethnicity*

Source: Kwon Kaufman, J.Y. Klein, and S. Chapman, C.D. 2000. *Dropout Rates in the United States: 1999*. Washington, D.C.: National Center for Education Statistics.

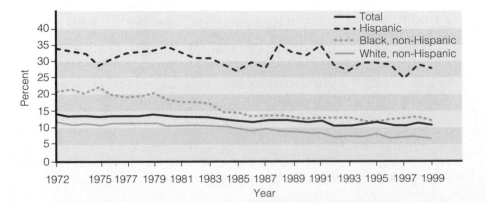

of Justice 2000; CDC 2000), and one-third of high school students report hearing a classmate threaten to kill someone (ABCNews 2001).

In response to violence, many schools throughout the country have police officers patrolling the halls, require students to pass through metal detectors before entering school, and conduct random locker searches. Video cameras set up in classrooms, cafeterias, halls, and buses purportedly deter some student violence. More than 2,000 schools nationwide conduct peer mediation and conflict resolution programs to help youth resolve conflict in nonviolent ways. One successful antiviolence program, Peace Builders of Phoenix, Arizona, enlists the aid of parents, trained community volunteers, local media, and marketing campaigns to promote a "pro-social way of life" among elementary school students (Embry, Flannery, Alexander-Vazsonvi, Powell, & Atha 1996).

Inadequate School Facilities and Personnel

In June of 2000, The National Center for Educational Statistics released the most recent report on the conditions of U.S. schools. The estimated cost of renovations and repairs is over $125 billion, with one in five schools reporting "less than adequate conditions for life safety features, roofs and electric power" (U.S. Department of Education 2000, 2). Further, over one third of U.S. schools rely on portable classrooms to accommodate the expanding student population. By 2010, the number of children attending school will have increased by over 42 million students, leading to fears of higher student-to-teacher ratios and overcrowded classrooms (U.S. Department of Education 2000).

In *Savage Inequalities*, Jonathan Kozol (1991) provides vivid and shocking descriptions of some of the consequences of overcrowding in our nation's poorer schools. His descriptions reveal the gross inadequacies in school facilities and personnel that plague our school system. Inadequate school facilities not only impede learning, but also contribute to behavioral problems, low self-esteem, and violence among students. Consider the following description:

> At Irvington High School...the gym is used by up to seven classes at a time. To shoot one basketball, according to the coach, a student waits for 20 minutes. There are no working lockers. Children lack opportunities to bathe. They fight over items left in lockers they can't lock. They fight for their eight minutes on the floor. Again, the scarcity of things that other children take for granted in America—showers, lockers, space and time to exercise—creates the overheated mood that also causes trouble in the streets. The students perspire. They grow dirty and impatient. They dislike who they are and what they have become. (Kozol 1991, 159)

School districts with inadequate funding often have difficulty attracting qualified school personnel. Unfortunately, these poorer districts are in dire need of talented teachers who can meet the needs of children from diverse backgrounds and of varying abilities. The number of minority teachers who can serve as role models, have similar life experiences, and have similar language and cultural backgrounds is far too few for the number of minority students (NCES 2000c). Black and Hispanic male children may perceive that white female teachers are less understanding of what they experience than minority male teachers.

Undisciplined and violent students and inadequate teaching materials and school facilities contribute to low teacher morale. Low morale interferes with teaching effectiveness and drives some teachers out of the profession. The relatively low pay of teachers also contributes to morale problems and the questionable quality of instruction.

Instead of building schools for the 1950s, let's build schools for 2050. Schools designed with the community and for the community. Schools that reflect a dedication to excellence and innovation. . . .

RICHARD RILEY
Former U.S. Secretary of Education

The actual success of a teacher depends in large measure on the capacity to state the subject matter of instruction in terms of the experience of the student.

GEORGE HERBERT MEAD
Sociologist

Deficient Teachers

In a National Press Club speech, former Department of Education Secretary Richard Riley spoke of the need for "a talented, dedicated, well-prepared teacher in every classroom." Given that over the next decade over 2 million additional teachers will be needed, and nearly 50 percent of today's teachers are over the age of 45, the challenge is not an insignificant one (ED Initiatives 1998; Congressional Budget Office 2000).

Individuals who enter and remain in the teaching profession are not necessarily competent and effective teachers. There is evidence that those who go into teaching, on the average, have lower college entrance exam scores than the average college student (Finsterbusch 1999). In response to such concerns, many states have implemented mandatory competency testing, which requires prospective teachers to pass the National Teacher's Examination or some other test of knowledge on the subject they intend to teach. In 1997, over 50 percent of prospective teachers in Massachusetts failed the state's competency exam, some unable to spell such basic words as "burned" and "abolished" (Leo 1998; Silber 1998). In response to such scandalous results, federal monies are now tied to "state 'report cards' that aggregate the test scores of would-be teachers by teacher preparation institution and thus produce a passing or failing 'grade' for each institution and state" (Cochran-Smith 2000, 163).

> Education is the key to prosperity and the wisest investment we can make in our children's and our nation's future.
>
> **RICHARD RILEY**
> *Former U.S. Secretary of Education*

Additionally, more than half of the states have adopted **alternative certification programs**, whereby college graduates with degrees in fields other than education may become certified if they have "life experience" in industry, the military, or other relevant jobs. *Teach for America*, originally conceived by a Princeton University student in an honors thesis, is an alternative teacher education program aimed at recruiting liberal arts graduates into teaching positions in economically deprived and socially disadvantaged schools. After completing an 8-week training program, recruits are placed as full-time teachers in rural and inner city schools. Critics argue that these programs may place unprepared personnel in schools.

Strategies for Action: Trends and Innovations in American Education

Americans rank improving education as one of their top priorities (Gallup 2001). Recent attempts to improve schools include raising graduation requirements, barring students from participating in extracurricular activities if they are failing academic subjects, lengthening the school year, and prohibiting dropouts from obtaining driver's licenses. Further, there is a nationwide movement to eliminate **social promotion**, the passing of students from grade to grade even if they are failing. However, educational reformers are calling for changes that go beyond get-tough policies that maintain the status quo.

> You don't fatten cattle by weighing them more often, and you don't improve student learning by giving more low quality tests.
>
> **DR. MONTY NEILL**
> *Testing researcher*

National Educational Policy

During the Clinton administration, Congress passed Goals 2000: The Educate America Act. To achieve the academic objectives in Goals 2000, the United States moved toward a system of national standards that would have defined what students should know and be able to do at each grade level. The administration's testing plans, however, suffered a serious blow in October of 1998 when Congress passed an amendment prohibiting "any new national tests that are not specifically and explicitly authorized by Congress" (Legislative Action 1998).

Calling for bipartisan support, President Bush's 2001 education plan, "No Child Left Behind," advocates trimming federal involvement in education. In his first proposed reform, President Bush argues that the "program for every problem" approach has failed to heal America's ailing schools and, furthermore, has cost taxpayers billions of dollars (see Fig. 12.3). The recommended legislation is designed, among other things, to increase student achievement, enhance teacher quality, improve English proficiency of non–native-speaking students, empower parents through school choice, make schools a safer place to learn, and hold teachers and schools accountable through annual academic assessments.

Character Education

Seventy percent of over 8,000 students surveyed by the Institute of Ethics reported cheating on a test at least once in the previous year, and almost half report cheating more than once. Moreover, in the last year, 92 percent reported lying to their parents, 78 percent reported lying to a teacher, and 68 percent reported hitting someone when they were mad. One in four said they would lie to get a job (Durham 2000). Although the data were no worse than in previous years, to many educators and the general public, they signify the need for **character education.**

Most schools' curricula neglect this side of education—the moral and interpersonal aspects of developing as an individual and as a member of society. President Bush's proposed educational reform includes support for character education through "grants to states and districts to train teachers in methods of incorporating character-building lessons and activities into the classroom" (2001, 6). Proponents of character education argue

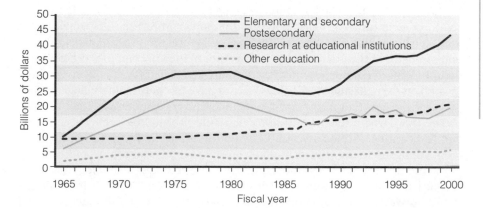

Dr. Roderick R. Page is the seventh U.S. Secretary of Education. He will oversee 4,700 employees and a budget of approximately $35 billion.

Our society does not need to make its children first in the world in mathematics and science. It needs to care for its children—to reduce violence, to respect honest work of every kind, to reward excellence at every level, to ensure a place for every child . . .

NEL NODDINGS
Educational reformer

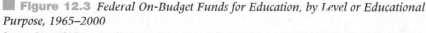

© Reuters/Kevin Lamarque/Hulton Archive

Figure 12.3 *Federal On-Budget Funds for Education, by Level or Educational Purpose, 1965–2000*
Source: Digest Of Education Statistics. Figure 21. U.S. Department of Education. NCES 2001-034.

that students should "be engaged in a general education that guides them in caring for self, intimate others, global others, plants, animals, the environment, objects and instruments, and ideas "(Noddings 1995, 369). For example, service learning programs are increasingly popular at universities and colleges nationwide. Service learning programs are community-based initiatives in which students volunteer in the community and receive academic credit for doing so. Studies on student outcomes have linked service learning to enhanced civic responsibility and moral reasoning, increased tolerance of diversity and promotion of racial equality, and a strong commitment to volunteerism (Zlotkowski 1996; Waterman 1997; Aberle-Grasse 2000).

How does the public feel about character education? In a national poll, the majority of public school parents approved of teaching courses on values and ethical behavior in the public schools. Although public schools are constitutionally prohibited from teaching any particular religion, about two-thirds of public school parents favor nondevotional instruction about various religions (Elam, Rose, & Gallup 1994). Character education also occurs to some extent in schools that have peer mediation and conflict resolution programs. Such programs teach the value of nonviolence, collaboration, and helping others, as well as skills in interpersonal communication and conflict resolution.

Computer Technology in Education

Computers in the classroom allow students to access large amounts of information (see this chapter's *Focus on Technology* feature). The proliferation of computers both in school and at home may mean that teachers will become facilitators and coaches rather than sole providers of information. Not only do computers enable students to access enormous amounts of information including that from the World Wide Web, but they also allow students to progress at their own pace. However, computer technology is not equally accessible by all students despite survey results which indicate that teenagers define computer courses as the single most important courses they can take (*State of Our Nation's Youth* 2001). Students in poorer school districts are less likely to have access to computers either in school or at home (see Chapter 15).

Interestingly, the conclusion of one of the largest studies of school computers was that students who use computers often scored lower on math tests than their low-use counterparts. The Educational Testing Service's study of 14,000 fourth and eighth graders concluded that how the computers were used—repetitive math drills versus real-life simulation—was responsible for the test variations. Also of note, students in classrooms where teachers are trained in computer use do better than students in classrooms where teachers are less skilled (Weiner 2000). Minority and low-income students are less likely to have teachers highly skilled in computer technology than their middle- and upper-income counterparts.

The U.S. Department of Education has issued a 2000 technology plan entitled "e-Learning: Putting a World-Class Education at the Fingertips of All Children." Contrary to the 1996 plan that concentrated on establishing a technology infrastructure, the new plan focuses on computer literacy, technology as a pedagogical tool, and the evaluation of existing strategies. The shift in emphasis from hardware to program implementation was expedited by the success of the 1996 plan. In 1994, 35 percent of schools and 3 percent of classrooms had Internet access; in 1999, 95 percent of schools and 63 percent of classrooms had Internet access (Stellin 2001).

Distance Learning and the New Education

Imagine never having an 8 o'clock class or walking into the lecture room late. Imagine no room and board bills, or having to eat your roommate's cooking. Imagine going to class when you want, even 3 o'clock in the morning. Imagine not worrying about parking! The future of higher education? Maybe. It's possible that the World Wide Web and other information technologies have so revolutionized education that the above scenarios are a *fait accompli*.

What is distance learning? **Distance learning** separates, by time or place, the teacher from the student. They are, however, linked by some communication technology: video-conferencing, satellite, computer, audiotape or videotape, real-time chat room, closed-circuit television, electronic mail, or the like. Distance learning has become so popular that today it is possible to earn a bachelor's degree, graduate degree and/or professional degree via the net.

The benefits of distance learning are clear. It provides a less expensive, accessible, and often more convenient way to complete a college degree. There are even pedagogical benefits. Research suggests that "students of all ages learn better when they are actively engaged in a process, whether that process comes in the form of a sophisticated multimedia package or a low-tech classroom debate on current events" (Carvin 1997). Distance education also benefits those who have historically been disadvantaged in the classroom. A review of research on gender differences suggests that females outperform males in distance learning environments (Koch 1998).

But all that glitters is not gold. Evidence suggests that students feel more estranged from their distance learning instructors than from teachers in conventional classrooms. The results of a study by Freitas, Meyers, and Avtgis (1998) indicate that college students in distance learning classes perceived significantly lower levels of teacher nonverbal immediacy (e.g., eye contact) when compared with students in conventional classes.

Additionally problematic is the proliferation of "virtual degrees." A Miami elementary school teacher enrolled in an online university to complete a master's degree in special education. After paying $800 of a total $2,000 bill, she was sent a book to summarize as part of her degree requirements. Shortly after returning her summary she was sent not only a master's degree, but a Ph.D. and transcripts of courses she had never taken with a recorded 3.9 grade point average (Koeppel 1998)! Although not a new problem, (e.g., correspon-

dence schools), there is the concern that the Internet lends itself to such fraudulent practices.

Further, teachers, particularly in higher education, are concerned about the quality of distance education. Several regulatory and advisory bodies, including the Congressional Web-based Education Commission, are presently establishing quality standards and guidelines (Carnevale 2001). And although a committee of the American Association of University Professors acknowledged that distance learning may be a "valuable pedagogical tool," it also questioned whether "academic quality, academic freedom, intellectual property rights and instructor's workloads and compensation" will be compromised (Arenson 1998, A14).

Nonetheless, distance education continues to grow, in part, because it is a moneymaker—a multibillion dollar industry. Even one-time critic William Bennett, former U.S. Secretary of Education, has recently thrown his hat into the corporate ring opening *K12*, a company that markets elementary and secondary school courses to home schoolers and to parents who want their children to have academic help outside of the classroom (Wildavsky 2001). Additionally, many commercial sites now offer "educational" courses and, alternatively, educational sites increasingly carry advertising banners, consumer discounts, product photographs, and the like (Guernsey 2000).

Will distance learning solve all the problems facing education today? The answer is clearly no. Although not the technological fix some are looking for, distance education, from digital libraries to "virtual" charter schools, does provide a provocative and financially lucrative alternative to traditional education providers.

Sources:

Laren Arenson.1998. "More Colleges Plunging into the Unchartered Waters of On-Line Courses." *New York Times*, November 2, A14.

William M. Bulkeley.1998. "Education: Kaplan Plans a Law School via the Web." *Wall Street Journal*, September 16, B1.

Dan Carnevale. 2001. "Commission Says Federal Rules on Distance Education Must be Updated." *The Chronicle of Higher Education*, January 5, A46.

Andy Carvin. 1997. EdWeb: Exploring Technology and School Reform. http://edweb.gsn.org

Frances Anne Freitas, Scott Meyers, and Theodore Avtgis. 1998. "Student Perceptions of Instructor Immediacy in Conventional and Distributed Learning Classrooms." *Communication Education* 47(4):366–72.

Lisa Guernsey. 2000. "Education: Web's New Come-On. *New York Times*, March 16, C1, C7.

David Koeppel. 1998. "Easy Degrees Proliferate on the Web." *New York Times*, August 2, 17.

James V. Koch. 1998. "How Women Actually Perform in Distance Education." *The Chronicle of Higher Education* 45:A60; Joe Rudich. 1998. "Internet Learning." *Link-Up* 15:23–25.

School Choice

Traditionally, children have gone to school in the district where they live. School vouchers, charter schools, home schooling, and private schools provide parents with alternative school choices for their children. **School vouchers** are tax credits that are transferred to the public or private school the parent selects for their child. President Bush's education plan calls for federally funded vouchers for students attending failing schools that don't improve test scores for 3 consecutive years. The $1500 vouchers could be used as tuition at a private school, or for out-of-class tutoring (Walsh & Wildavsky 2001).

Proponents of the voucher system argue that it reduces segregation and increases the quality of schools because they must compete for students to survive. Opponents argue that vouchers increase segregation because white parents use the vouchers to send their children to private schools with few minorities. Research by Saporito and Lareau (1999) supports this contention:

> We find compelling evidence that race is a very powerful force in guiding family choices . . . After eliminating "black" schools from consideration, white families choose "white" schools, many of which have inferior safety records, test scores, and higher rates of poverty. In contrast, African American families do not show a similar sensitivity to race; instead they tend to select schools with lower poverty rates. (p. 418)

Vouchers, opponents argue, are also unfair to economically disadvantaged students who are not able to attend private schools because of the high tuition. Further, they argue, the use of vouchers for religious schools violates the constitutional guarantee of separation of church and state. As one indication of the volatility of this issue, when a sample of registered voters was asked whether they were opposed or in favor of a policy that would permit school vouchers to be used for private or religious school attendance, 49 percent were in favor, 47 percent were opposed, and 3 percent were unsure (*Washington Post* 2001).

Vouchers can also be used for charter schools. **Charter schools** originate in contracts, or charters, which articulate a plan of instruction that must be approved by local or state authorities. Although charter schools can be funded by foundations, universities, private benefactors, and entrepreneurs, many are supported by tax dollars (Toch 1998; Mollison 2001). The Department of Education estimates that more than 2,000 charter schools are operating in the United States today, up from just two a decade ago. Charter schools, like school vouchers, were designed to expand schooling options and increase the quality of education through competition. Recent evidence, however, suggests that charter schools have little positive impact on student achievement and, in some cases, may result in lower academic performance (*New York Times* 2000; U.S. Newswire 2000b).

Some parents are choosing not to send their children to school at all but to teach them at home. More than 1 million students in the United States are home-schooled, their numbers increasing by 15 percent annually (Winters 2000). For some parents, **home schooling** is part of a fundamentalist movement to protect children from perceived non-Christian values in the public schools. Other parents are concerned about the quality of their children's education and their safety. How does being schooled at home instead of attending public school affect children? Some evidence suggests that home-schooled chil-

Every adult American has the right to vote, the right to decide where to work, where to live. It's time parents were free to choose the schools that their children attend. This approach will create the competitive climate that stimulates excellence. . . .

GEORGE BUSH
Former U.S. President

dren perform as well or better than their institutionally schooled counterparts (Webb 1989; Winters 2000).

Another choice parents may make is to send their children to a private school. The primary reason parents send their children to private schools is for religious instruction. The second most common reason is the belief that private schools are superior to public schools in terms of academic achievement. Research suggests, however, that when controlling for parents' education and income, few differences in private and pubic school educational outcomes may exist (Shanker 1996; Ascher, Fruchter, & Berne 1997; Cohen 1998). Parents also choose private schools for their children in order to have greater control over school policy, to avoid busing, or to obtain a specific course of instruction such as dance or music.

Understanding *Problems in Education*

What can we conclude about the educational crisis in the United States? Any criticism of education must take into account that just over a century ago the United States had no systematic public education system at all. Many American children did not receive even a primary school education. Instead, they worked in factories and on farms to help support their families. Whatever education they received came from the family or the religious institution. In the mid-1800s, educational reformer Horace Mann advocated at least 5 years of mandatory education for all U.S. children. Mann believed that mass education would function as the "balanced wheel of social machinery" to equalize social differences among members of an immigrant nation. His efforts resulted in the first compulsory education law in 1852, which required 12 weeks of attendance each year. By World War I, every state mandated primary school education, and by World War II, secondary education was compulsory as well.

Public schools are supposed to provide all U.S. children with the academic and social foundations necessary to participate in society in a productive and meaningful way. But, as conflict theorists note, for many children the educational institution perpetuates an endless downward cycle of failure, alienation, and hopelessness.

Breaking the cycle requires providing adequate funding for teachers, school buildings, equipment, and educational materials, a task that has become increasingly difficult as student enrollments swell and diversify, a consequence of the "baby-boom echo" and changing immigration patterns. Between 1989 and 2009, elementary school enrollment will have increased 12 percent, high school enrollment 19 percent, and full-time college student enrollment 11 percent (U.S. Department of Education 2000).

The public is, however, supportive of government spending on education. In a national survey of persons 18 and over, 66 percent responded that the government spends too little on public education (Harris Poll 1998). Legislation such as the 1998 amendments to the Higher Education Act which allocated $300 billion to teacher preparation and recruitment, reduced interest on college student loans, and increased the maximum in Pell grants, is essential. But, even with financial support, as functionalists argue, education alone cannot bear the burden of improving our schools.

> In order for all children to enter elementary school prepared to learn, we will have to eliminate child poverty in this country and provide every child who needs them not only with adequate health and social services but also with an early childhood education program well beyond anything envisioned by Head Start.
>
> EVANS CLINCHY
> *Educational reformer*

Jobs must be provided for those who successfully complete their education. Students with little hope of school success will continue to experience low motivation as long as job prospects are bleak and earnings in available jobs are low. Ray and Mickelson (1993) explain:

> . . . School reforms of any kind are unlikely to succeed if non-college bound students cannot anticipate opportunity structures that reward diligent efforts in school. Employers are not apt to find highly disciplined and motivated young employees for jobs that are unstable and low paying. (pp. 14–15)

Finally, "if we are to improve the skills and attitudes of future generations of workers, we must also focus attention and resources on the quality of the lives children lead outside the school" (Murnane 1994, 290). We must provide support to families so children grow up in healthy, safe, and nurturing environments. Children are the future of our nation and of the world. Whatever resources we provide to improve the lives and education of children are sure to be wise investments in our collective future.

Critical Thinking

1 Clearly, home schooling has both advantages and disadvantages. After making a list of each, would you want your child to be home-schooled? Why or why not?

2 As discussed in Chapter 6, the proportion of elderly in the United States is increasing dramatically as we move into the twenty-first century. Because the elderly are unlikely to have children in public schools, how will the allocation of necessary school funds be affected by this demographic trend?

3 Student violence is one of the most pressing problems in U.S. public schools. One response is defensive, that is, police patrolling halls, metal detectors, etc. Other than such defensive tactics, what violence prevention techniques should be instituted?

4 Students who drop out of school are often blamed for their lack of motivation. How may a teenager's dropping out of high school be explained as a failure of the educational system rather than as a "motivation" problem?

Key Terms

alternative certification programs	cultural imperialism	integration hypothesis
bilingual education	distance learning	multicultural education
character education	functionally illiterate	school vouchers
charter schools	Head Start	self-fulfilling prophecy
	home schooling	social promotion

Media Resources

 The Wadsworth Sociology Resource Center: Virtual Society

http://sociology.wadsworth.com/

See the companion Web site for this book to access general sociology resources and text-specific features that can further your understanding of this chapter. The site contains Internet links, Internet exercises, online practice quizzes, information on InfoTrac College Edition, and many more valuable materials designed to enrich your learning experience in social problems.

 InfoTrac College Edition

You can access InfoTrac College Edition either from the Wadsworth Sociology Resource Center at **http://sociology.wadsworth.com** or directly from your web browser at **http://www.infotrac-college.com/wadsworth/**. InfoTrac College Edition is an online university library that includes over 700 popular and scholarly journals in which you can find articles related to the topics in this chapter such as school vouchers, Title IX, distance learning, and school violence.

Interactions CD-ROM

Go to the Interactions CD-ROM for *Understanding Social Problems*, Third Edition to access additional interactive learning tools, such as in-depth review materials, corresponding practice quizzes, and other engaging resources and activities to help you study the concepts in this chapter.

13

Cities in Crisis

Is It True?

1. Most of the world's largest cities are in industrialized countries.

2. Most poor adults living in inner cities are unemployed.

3. In the United States, rates of HIV/AIDS are higher in metropolitan areas than in rural areas.

4. The automobile industry (U.S. and worldwide) spends more money on advertising than any other industry.

5. More people die as a result of car accidents than of urban violent crime.

Answers to "Is It True?": 1 = F; 2 = F; 3 = T; 4 = T; 5 = T

I ncreasing urbanization has the potential for improving human life or increasing human misery. The cities can provide opportunities or frustrate their attainment; promote health or cause disease; empower people to realize their needs and desires or impose on them a simple struggle for basic survival.

—United Nations' 1996 State of the World Population Report

In June 2000, the enjoyment of a warm afternoon in Central Park turned into a day of horror for more than 50 women who were attacked in the park after the Puerto Rican Parade. Up to 60 men participated in the attack, harassing, stripping, molesting, and in some cases, robbing the victims. Anne Peyton Bryant, who was attacked while skating, said she approached police several times for help but they did nothing. "I never before felt in my entire life that I couldn't protect myself until then," said Bryant, a kickboxing instructor. "I felt confused. I felt terrified. I felt traumatized" (Associated Press 2000). A teenaged British tourist said that about 15 men surrounded her, ripped off her clothes, and fondled her before stealing $200. "They were trying to pull my top up, and at the same time guys were trying to pull my shorts down," the teenager said. "I was touched everywhere, and it was really humiliating. There must have been 14 guys touching me, just trying to take my clothes off" (Associated Press 2000). Another victim of the attack described her experience: "Before I knew it, I was surrounded by what seemed to be 20 guys, all pouring water on me...They were coming at me from all directions, and they were grabbing my butt, groping my butt, and I was screaming and I was trying to get through, trying to get away...I was scared. I felt like I had no control...I didn't know what would end up happening; that was the scariest part of all" (CNN 2000). One of the perpetrators, 18-year-old Lonnie Hopson, described, "Everyone was throwing water on the ladies, including me. Like 50 people on one person, like wild hyenas, grabbed...whatever they could grab" (Claffey 2000).

The attack in Central Park shocked the nation, in large part because of the number of both victims and perpetrators involved. As many as 10 onlookers videotaped some of the attacks, but few in the crowd of onlookers offered any type of assistance to the victims. A stunned nation asked how such an attack could have occurred. Could a similar event have occurred in a small town or rural setting? What elements of city life contribute to urban crime?

Cities may be viewed as representing both the best and the worst of social conditions. Many U.S. citizens are intrigued by cities and their diverse populations; their history; their architectural wonders; and their assorted offerings of cultural events, entertainment, sports, museums, conventions, universities, government centers, tourist attractions, restaurants, and innumerable consumer products and services. Many argue, however, that the problems of urban living outweigh the advantages. In this chapter, we look at the historical and social forces that have created our cities, the modern social forces that are shaping our cities today, the social problems that affect and are affected by cities in the United States and throughout the world, and strategies for revitalizing cities and improving the quality of life among urban dwellers.

The Global Context: A World View of Urbanization

As early as 5000 B.C., cities of 7,000 to 20,000 people existed along the Nile, Tigris-Euphrates, and Indus River valleys. But it was not until the Industrial Revolution in the nineteenth century that **urbanization**, the transformation of a society from a rural to an urban one, spread rapidly. According to the 1990 U.S. census definition, an **urban population** consists of persons living in cities or towns of 2,500 or more inhabitants. An **urbanized area** refers to one or more places and the adjacent densely populated surrounding territory that together have a minimum population of 50,000. Urbanization has accelerated in the past 40 years and is expected to continue into the twenty-first century. The share of the global population living in urban areas has increased from one-tenth in 1900, to one-third in 1960, to half in 2000 and by 2030, an estimated 61 percent of the world's projected 8.1 billion people will live in cities (Sheehan 2001a; United Nations Population Fund 1999). Africa and Asia remain the least urbanized of the developing regions, with less than 38 percent of the population living in urban areas. In Latin America, the Caribbean, Europe, North America, and Japan, between 75 and 79 percent of the population live in urban areas (United Nations Population Fund 1999).

The number of **megacities**—cities with 10 million residents or more is also increasing. Most megacities are in less developed countries. In 1999, there were 17 megacities, 13 in less-developed countries. It is projected that 256 megacities will exist by 2015 (United Nations Population Fund 1999). Nearly all of the urban population growth is occurring in developing countries. Between 1970 and 2020, 93 percent of urban population growth will have occurred in the developing world (United Nations Population Fund 1996). Table 13.1 presents the 10 most populated cities in the world.

Increasing urbanization results about equally from births in urban areas and from migration of people from rural areas to the cities (United Nations Population Fund 1999). Rural dwellers migrate to urban areas to flee from war or natural disasters or to find employment. As foreign corporate-controlled commer-

▌ **Table 13.1** *Ten Most Populated Cities in the World*

City	Projected Population in 2000
1. Tokyo-Yokohama, Japan	29,971,000
2. Mexico City, Mexico	27,872,000
3. Sao Paulo, Brazil	25,354,000
4. Seoul, South Korea	21,976,000
5. Bombay, India	15,357,000
6. New York City, USA	14,648,000
7. Osaka-Kobe-Kyoto, Japan	14,287,000
8. Tehran, Iran	14,251,000
9. Rio de Janeiro, Brazil	14,169,000
10. Calcutta, India	14,088,000

Source: *Statistical Abstract of the United States: 1994*, 114th ed. U.S. Bureau of the Census. Washington, D.C.: U.S. Government Printing Office, Table 1355.

cial agriculture displaces traditional subsistence farming in poor rural areas, peasant farmers flock to the city looking for employment. Some rural dwellers migrate to urban areas in search of a better job—one that has higher wages and better working conditions. Governments have also stimulated urban growth by spending more to improve urban infrastructures and services, while neglecting the needs of rural areas (Clark 1998). Lastly, urban populations in developing countries have increased because of high fertility rates (see Chapter 14).

History of Urbanization in the United States

Urbanization of the United States began as early as the 1700s, when most major industries were located in the most populated areas, including New York City, Philadelphia, and Boston. Unskilled laborers, seeking manufacturing jobs, moved into urban areas as industrialization accelerated in the nineteenth century. The "pull" of the city was not the only reason for urbanization, however. Technological advances were making it possible for fewer farmers to work the same amount of land. Thus "push" factors were also involved—making a living as a farmer became more and more difficult as technology, even then, replaced workers.

Urban populations continued to multiply as a large influx of European immigrants in the late 1800s and early 1900s settled in U.S. cities. This was followed by a major migration of southern rural blacks to northern urban areas. People were lured to the cities by the promise of employment and better wages and such urban amenities as museums, libraries, and entertainment. Immigrants are also attracted to cities with large ethnic communities that provide a familiar cultural environment in which to live and work. Today, a large percentage of immigrants into inner cities are Asian or Hispanic, and they settle disproportionately in certain cities, including Miami, San Francisco, Los Angeles, and Chicago.

Figure 13.1 depicts the tremendous growth of the U.S. urban population over the last 200 years. Between 1800 and 1990, the percentage of the U.S. population that is urban grew from 6.1 percent to 75.2 percent.

Suburbanization

In the late nineteenth century, railroad and trolley lines enabled people to live outside the city and still commute into the city to work. As more and more people moved to the **suburbs**—urban areas surrounding central cities—America underwent **suburbanization**. As city residents left the city to live in the suburbs, cities lost population and experienced **deconcentration**, or the redistribution of the population from cities to suburbs and surrounding areas.

■ **Figure 13.1**
Percentage of U.S. Population That Is Urban, 1800–1990

An urban population consists of persons living in cities or towns of 2,500 or more residents.

Sources: *Statistical Abstract of the United States: 1993*, 113th ed.; *Statistical Abstract of the United States: 2000*, 120th ed. U.S. Bureau of the Census, Washington, D.C.: U.S. Government Printing Office.

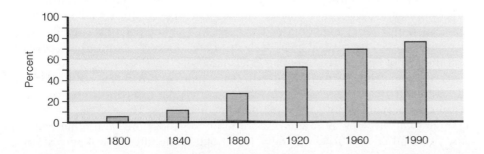

Many factors have contributed to suburbanization and deconcentration. Following World War II, many U.S. city dwellers moved to the suburbs out of concern for the declining quality of city life and the desire to own a home on a spacious lot. Suburbanization was also spurred by racial and ethnic prejudice, as the white majority moved away from cities that, because of immigration, were becoming increasingly diverse. Mass movement into suburbia was encouraged by the federal road building program, which promoted automobile use, and by federal grants that have allowed the extensive construction of sewers. Because of the perceived threat of nuclear attack during the Cold War, federal, state, and local governments decentralized their offices and facilities, and defense contractors located new plants outside city centers (Kelbaugh 1997). In the 1950s, Veterans Administration (VA) and Federal Housing Administration (FHA) loans made housing more affordable, enabling many city dwellers to move to the suburbs. Suburb dwellers who worked in the central city could commute to work or work in a satellite branch in suburbia that was connected to the main downtown office. The new technologies of fax machines, home computers, cellular phones, pagers, and e-mail have also facilitated the movement to the suburbs. As increasing numbers of people moved to the suburbs, so did businesses and jobs. Without a strong economic base, city services and the quality of city public schools declined, which furthered the exodus from the city.

U.S. Metropolitan Growth and Urban Sprawl

You can't keep spreading out. The cost to make roads and sewers gets to the point where it doesn't work.

MIKE BURTON
Executive director
Portland's metro government

Simply defined, a **metropolitan area** is a densely populated core area, together with adjacent communities. The largest city in each metropolitan area is designated as a **central city**. Another term for metropolitan area is **metropolis**, from the Greek meaning "mother city."

Current standards require that each newly qualifying metropolitan area must include at least one city or urban area with at least 50,000 residents and a total metropolitan population of at least 100,000 (75,000 in New England). The most populated U.S. metropolitan area encompasses several cities in New York, New Jersey, Connecticut, and Pennsylvania.

Metropolises have grown rapidly in the United States. As new areas reach the minimum required city or urbanized area population, and as adjacent towns, cities, and counties satisfy the requirements for inclusion in metropolitan areas, both the number and size of metropolitan areas have grown. The number of metropolitan areas in the United States increased from 243 in 1970 to 280 in 2000 (*Statistical Abstract 2000;* U.S. Census Bureau 2001). One U.S. state—New Jersey—is entirely occupied by metropolitan areas as designated by the U.S. census *(Statistical Abstract 2000).*

The growth of metropolitan areas is often referred to as **urban sprawl**—the ever-increasing outward growth of urban areas. Urban sprawl results in the loss of green, open spaces, the displacement and endangerment of wildlife, traffic congestion and noise, and pollution liabilities. In a Gallup poll of U.S. adults, 73 percent said they worried "a great deal" or "a fair amount" about urban sprawl and the loss of open spaces (Dunlap & Saad 2001).

Those who enjoy the conveniences and amenities of urban life, yet find large metropolitan areas undesirable, may choose to live in a **micropolitan area**—a small city located beyond congested metropolitan areas. These areas are large enough to attract jobs, restaurants, community organizations, and other benefits, yet small enough to elude traffic jams, high crime rates, and high costs of housing and other living expenses associated with larger cities. One in 20 Americans lives in a micropolitan area (Heubusch 1998).

Sociological Theories of Urbanization

The three main sociological theories—structural-functionalism, conflict theory, and symbolic interactionism—focus on different aspects of urbanization.

Structural-Functionalist Perspective

Structural-functionalists view the development of urban areas as functional for societal development. Although cities initially functioned as centers of production and distribution, today they are centers of finance, administration, education, and information.

Metropolitan areas play a vital role in the nation's economy. In 1999, U.S. metropolitan economies accounted for 85 percent of Gross Domestic Product; 84 percent of national employment; and 88 percent of total labor income (Coles 2001). And two of the fastest growing segments of the U.S. economy, high-tech and business services, are heavily concentrated in metropolitan areas. In 1999, more than 90 percent of new business services jobs and high-tech jobs were in metropolitan areas (Coles 2001).

The expansion of urban areas, although functional, also leads to increased rates of anomie, or normlessness, as the bonds between individuals and social groups become weak (see Chapters 1 and 4). Whereas in rural areas social cohesion is based on shared values and beliefs, in urban areas social cohesion is a consequence of specialization and interdependence spurred by increased social diversity. Urbanization thus leads to higher rates of deviant behavior, including crime, drug addiction, and alcoholism. To functionalists, these increased rates are an indication of the social disorganization of the area.

Historically, another major dysfunction of cities is that, by concentrating large numbers of people in small areas, they have facilitated the rapid spread of infectious disease. Overcrowding, poverty, and environmental destruction are also viewed as dysfunctions associated with urbanization.

The structural-functionalist concern with latent functions (unintended consequences) is also applicable to understanding urban life. For example, the economic development in many urban areas has unintentionally created a housing crisis. "The economic growth that is pushing up employment and homeownership in most of the Nation's cities is also driving increases in rents more than one-and-a-half times faster than inflation—and creating staggering jumps in home prices as well" (U.S. Department of Housing and Urban Development 2000, vii). In a survey of the mayors of 25 U.S. cities, nearly every one cited the lack of affordable housing as the primary cause of homelessness (U.S. Conference of Mayors 2000a).

Conflict Perspective

Conflict theorists emphasize the role of power, wealth, and the profit motive in the development and operations of urban areas. According to the conflict perspective, capitalism requires that the production and distribution of goods and services be centrally located, thus, at least initially, leading to urbanization. Today, global capitalism and corporate multinationalism, in search of new markets, cheap labor, and raw materials, have largely spurred urbanization of the developing world. Capitalism also contributes to migration from rural areas into cities as peasant farmers who have traditionally produced goods for local consumption are being displaced by commercial agriculture that is geared to producing fruits, flowers, and vegetables for export to the developed world. Displaced from their traditional occupations, peasant farmers are flocking to cities to find employment (Clark 1998).

The conflict perspective also focuses on how individuals and groups with wealth and power influence decisions that affect urban populations. For example, according to citizens' groups working to stop urban sprawl in Central and Eastern Europe, city officials may be bribed to approve a new shopping mall or other development project (Sheehan 2001b). And in the United States, "bribery" has taken the form of financial contributions to political parties and candidates. In the 1998 congressional election, industries with a stake in transportation and land use decisions contributed $128 million to political parties and candidates. During the 2000 presidential election campaign, a construction industry lobbyist in favor of increased highway building told the *Wall Street Journal* that he looked forward to meeting with the President if the candidate he supported with a large contribution won: "I'm assuming, since we've supported him throughout his gubernatorial career, that we'd have an opportunity to go visit if we wanted to" (quoted in Sheehan 2001b, 120). (Increased highway building encourages urban sprawl and diverts funds away from improvements in public transportation that are needed in urban areas.)

Symbolic Interactionist Perspective

The symbolic interaction perspective emphasizes how meanings, labels, and definitions affect behavior. One application of this perspective is found in efforts to change the public's negative definitions of cities. Cities are often viewed as dangerous and crime-ridden, decaying, dirty places. These negative definitions of cities discourage people from living and vacationing in cities and deter businesses from locating there. Negative definitions of cities may also contribute to a self-fulfilling prophecy. Detroit mayor Dennis Archer (1998) observed that "negative perceptions of the City of Detroit and low expectations for our future convinced many observers that Detroit was beyond hope. This made issues that many large cities face—unemployment—crime—flight of the middle class—dissatisfaction with public schools—much harder to handle in Detroit" (p. 341). Efforts to redefine cities in positive terms are reflected in advertising campaigns sponsored by cities' convention and visitors bureaus. For example, an advertising slogan for Detroit indicates, "It's a Great Time in Detroit."

The symbolic interactionist perspective also focuses on how social forces influence individuals' self-concepts, values, and behaviors. People tend to identify with the place where they live, and this is especially true of urban dwellers. The distinctive cultures and lifestyles of cities shape the self-concepts, values, and behaviors of their residents. Although each city is unique, city dwellers tend to share a basic culture and lifestyle, known as **urbanism**. Individualistic and cosmopolitan norms, values, and styles of behavior characterize this urban culture and lifestyle. In contrast to the slow-paced and traditional nature of rural life, life in the city is fast-paced and always on the "cutting edge" in terms of the latest technology, entertainment, fashions, food trends, and other forms of cultural change.

Symbolic interactionists emphasize the effects of urbanization on interaction patterns and social relationships. The classical and modern views represent different observations of how urban living affects social relationships.

Classical Theoretical View   Cities have the reputation of being cold and impersonal. As early as 1902, George Simmel observed that urban living involved an overemphasis on punctuality, individuality, and a detached attitude toward interpersonal relationships (Wolff 1978). It is not difficult to find evidence to support Simmel's observations: witness New Yorkers pushing each other to get onto the subway during rush hour, hear motorists curse each other in Los Angeles traffic jams, watch Chicago residents ignore the homeless man asleep on the sidewalk.

Louis Wirth (1938), a second-generation student of Simmel, argued that urban life is disruptive for both families and friendships. He believed that because of the heterogeneity, density, and size of urban populations, interactions become segmented and transitory, resulting in weakened social bonds. Those bonds that do develop are superficial and detract from the closeness of primary relationships. Wirth held that as social solidarity weakens, people exhibit loneliness, depression, stress, and antisocial behavior.

Modern Theoretical View   In contrast to Wirth's pessimistic view of urban areas, Herbert Gans ([1962] 1984) argued that cities do not interfere with the development and maintenance of functional and positive interpersonal relationships. Among other communities, he studied an Italian urban neighborhood in Boston called the West End and found such neighborhoods to be community oriented and marked by close interpersonal ties. Rather than finding the social disorganization described by Wirth, Gans observed that kinship and ethnicity helped bind people together. Intimate small groups with strong social bonds characterized these enclaves. Thus Gans saw the city as a patchwork quilt of different neighborhoods or **urban villages**, each of which helped individuals deal with the pressures of urban living.

A Theoretical Synthesis   Fisher (1982) interviewed more than 1,000 respondents in various urban areas and found evidence for both the classical and the modern theoretical views of urbanism. From the classical perspective, he found that heterogeneity (the diversity among urban residents) does make community integration and consensus difficult—community cohesion is less. Ties that do exist are less often kin-related than in nonurban areas and are more often based on work relationships, memberships in voluntary and professional organizations, and proximity to neighbors.

Fisher also found, however, that the diversity of urban populations facilitates the development of subcultures that have a sense of community ties. For example, large urban areas include such diverse groups as gays, ethnic and racial minorities, and artists. These individuals find each other and develop their own unique subcultures.

Other research has also found support for a synthesis of the classical and modern views of urban life. Tittle (1989) found that the larger the size of the community, the weaker a respondent's social bonds and the higher his or her anonymity, tolerance, alienation, and reported incidence of deviant behavior. He also found that racial and ethnic groups created their own sense of community in urban neighborhoods.

> A community must simultaneously nurture both a respect for group values and a tolerance for individuality . . .
>
> DOUGLAS KELBAUGH
> *University of Washington*

Cities and Social Problems

Given that urban areas comprise high concentrations of people, it is not surprising that they also experience high concentrations of various social problems. Urban areas tend to be plagued by poverty and unemployment, poor and unaffordable housing, and inadequate schools. Other urban problems include HIV/AIDS, addiction, and crime; and transportation and traffic problems. Another social problem related to urban and suburban sprawl is the loss of natural habitat and the displacement of wildlife. Whereas many urban problems in the United States are exacerbated by the loss of inner-city residents to the suburbs, in developing countries, urban problems are compounded by rapid urban population growth.

Urban Poverty and Unemployment

The number of urban poor is growing rapidly worldwide. Over one-quarter of the developing world's urban population lives below the official poverty line. In some of the world's poorest developing countries, half of the urban population lives in conditions of extreme deprivation (United Nations Population Fund 1996; World Health Organization and United Nations Joint Programme on HIV/AIDS 1998).

> We simply can't allow our cities to become nothing but massive shelters for the poor.
>
> KURT L. SCHMOKE
> *Mayor of Baltimore*

In the United States, poverty is concentrated in central cities. The U.S. poverty rate in 1999 was 16.4 percent inside central cities, compared with the overall U.S. poverty rate of 11.8 percent (Dalaker & Proctor 2000). The 1999 poverty rate in rural areas was 14.3 percent and 8.3 percent in metropolitan areas outside of central cities.

Urban poverty is reflected in the high numbers of homeless in urban areas. In a survey of 25 U.S. cities, the average demand for emergency shelter increased by 15 percent from 1999 to 2000—the largest increase in 10 years (U.S. Conference of Mayors 2000a).

High poverty rates in cities are related to unemployment, the lack of jobs, and the decrease in high-paying jobs. Many central cities face unemployment rates that are much higher than the national average: one out of six has an unemployment rate 50 percent or more above the national rate (U.S. Department of Housing and Urban Development 1999). However, as emphasized in this chapter's *Social Problems Research Up Close* feature, most inner-city poor adults are employed, but their jobs do not pay well enough to lift them out of poverty.

> Our culture's value system is built around being "productive," which is defined as having a job. Even if you're poor, when you're employed you have a sense of being valuable in society.
>
> LUIS RODRIGUEZ
> *Author and founder of the organization* Youth Struggling for Survival

In the United States and other industrialized countries, urban unemployment and poverty are partly the results of **deindustrialization**, or the loss and/or relocation of manufacturing industries. Since the 1970s, many urban factories have closed or relocated, forcing blue-collar workers into unemployment. However, even when manufacturing jobs are in urban areas, manual workers are now expected to do more than just manual labor. They may be required, for example, to gather statistical quality-control data, develop innovative solutions, and reorganize assembly lines. "The average high-school dropout, and many an inner-city grad, cannot compete against more advantaged job-seekers when it comes to the written tests manufacturers now routinely employ to find people who can fit into the new, high-tech factory" (Newman 1999, xiii). When prospects of finding decent employment are low, resulting feelings of frustration and worthlessness can lead to drug use, crime, and violence. Rodriguez (2000) explains, "When your job disappears because of a new technology or moving factories to Central America, it's easy to internalize that feeling of worthlessness, instead of connecting your personal experience to larger economic and social issues" (p. 6).

Urban Housing Problems

Urban housing problems include lack of affordable housing, substandard housing, and housing segregation. Many cities are experiencing a housing crisis. The Center for Budget Policy Priorities reported that in the past 25 years the number of low-income renters increased by 70 percent, while the number of low-cost rental units has dropped ("Crisis in Low-Income Rental Housing" 1998). Housing that is available and affordable is often substandard, characterized by outdated plumbing and wiring, overcrowding, rat infestations, toxic lead paint, and fire hazards (see also Chapter 10).

Low-income housing tends to be concentrated in inner-city areas of concentrated poverty. But jobs are increasingly moving to the suburbs, and central-city residents often lack transportation to get to these jobs. The more affluent suburbs restrict development of affordable housing to keep out "undesirables" and maintain their high property values. Suburban zoning regulations that require large lot sizes, minimum room sizes, and single-family dwellings serve as barriers to low-income development in suburban areas (Orfield 1997).

Concentrated areas of poverty and poor housing in urban neighborhoods are called **slums**. Slums that are occupied primarily by African Americans are known as **ghettos,** and those occupied primarily by Latinos are called **barrios**. U.S. minorities, who are disproportionately represented among the poor, tend to be segregated in concentrated areas of low-income housing (see also Chapter 7). About 30 percent of U.S. whites live in cities but they typically live in urban neighborhoods that are 72 percent white (Schmitt 2001). More than 60 percent of blacks live in cities and the typical black urban resident lives in a neighborhood that is 76 percent minorities. Some of the most segregated U.S. cities include New York, Newark, Milwaukee, Chicago, Cleveland, Cincinnati, St. Louis, and Miami (Schmitt 2001).

> Once upon a time, not so long ago, poorly educated job-seekers might have found refuge in high-wage work in the auto factories or the steel mills. Those days are history now.
>
> KATHERINE S. NEWMAN
> *Harvard University*

Inadequate Schools

When jobs and middle-class residents left the city for the suburbs in mass exodus, local revenues decreased, and the quality of city schools, particularly in inner-city neighborhoods, declined. With some exceptions, "schools in poor inner-city neighborhoods...are dangerous, overcrowded, and ineffective. Moreover, children in those schools, seeing the unemployment and poor earnings of the adults in their neighborhood, may conclude that education has little value" (Jargowsky 1997, 110). Inner-city schools face a chronic shortage of qualified teachers, and often hire underprepared and inexperienced teachers, many of whom are hired to teach subjects outside their areas of preparation (Wilson 1996). The poor quality of inner-city schools contributes further to middle-class residents leaving the city, and deters businesses and potential residents from moving to those neighborhoods. As a result, property values continue to decline, the tax base decreases, schools further erode, and the cycle repeats itself.

> Cities have been ignored because they are blacker, browner, poorer, and more female than the rest of the nation.
>
> JULIANNE MALVEAUX
> *Writer/Scholar*

No Shame in My Game: The Working Poor in the Inner City

Researcher Katherine Newman (1999) conducted research on the working poor in the inner city. Unlike many social scientists who believe that research should be "objective" with no social or political agendas, Newman is unapologetic about her research agenda. In an interview with the Russell Sage Foundation (undated), she states,

> I really wanted to reshape the way Americans think about poverty. They have heard so much about . . . people who don't want to work, who are deliberately irresponsible. . . . But the vast majority of poor people do work. They just don't earn enough to pull themselves out of poverty. I wanted to look at those working poor people as a way of recasting the whole image of the inner city.

Newman presents new data that counters the common misconception that residents of impoverished inner cities are typically unemployed. She notes, for example, that in central Harlem, 69 percent of the families have at least one worker. A brief overview of the sample, methods, and findings of her study follows.

Sample and Methods

Newman used a qualitative design in which she and her research team interviewed and followed the lives of 200 African-American, Dominican, and Puerto Rican residents of Harlem and Washington Heights over an 18-month period. Newman also surveyed more than 100 business owners and managers in the Harlem and Washington Heights areas using written questionnaire surveys and interviews.

Newman chose a qualitative study to get a "deeper understanding of the daily lives and real values of inner city workers." She explains,

> Most of the information we have on labor markets and the workforce naturally comes from economists or sociologists who work with large data sets. That research is crucial, especially for explaining the big picture. But it doesn't help us understand how ordinary people in poor communities view their lives, their options, or how they put the resources together to survive, to raise their kids, to balance going to school and keeping a job. (Russell Sage Foundation, undated)

Originally, Newman's study included only respondents who were employed. During the middle of the project, her colleagues criticized her research design as being "too limiting," in that by studying only working adults, they "were going to end up finding out only about the 'diamonds in the rough,' the hardworking people who might or might not be representative of central Harlem" (Newman 1999, xvii). Consequently, Newman added a phase to her research that involved studying residents who were looking for work, largely without success. She found respondents by tracking down applicants for fast food jobs throughout New York City.

In interviews with respondents, Newman and her research team focused on finding answers to a number of questions, including: Where do residents of Harlem look for employment? How are they greeted by employers when they come knocking? How do their peers react when they "stoop to seeking 'hamburger-flipping' jobs that have become the butt of countless parodies and sar-

The Urban "Three-Headed Monster": Drug Addiction, AIDS, and Crime

In a speech delivered at the Conference on the Successful City of the 21st Century, Baltimore's mayor Kurt Schmoke (1998), used the term "three-headed monster" to refer to three problems that plague cities: drug addiction, HIV/AIDS, and crime.

Drug Addiction High rates of drug use and drug addiction in urban areas are related to joblessness, poverty, and hopelessness that characterize many inner-city neighborhoods. With little hope of finding legitimate means to escape the joblessness and poverty, many turn to selling drugs, "while others have become so demoralized that they are ready candidates for alcohol and drug addiction"

castic asides?" (p. xiv). How do they cope with the stigma of a low status job in the fast-food industry?

Findings and Conclusions

Some of the findings of Newman's research include the following:

- *Why Do Residents Seek Low-Wage Jobs?* Newman found that for many of the young job seekers in her study, a job provided an alternative to crime, violence, and drugs. A job at a fast-food chain represents a "safe haven" from hanging out on the streets where one is vulnerable to trouble and danger. Young job seekers also wanted the independence and pride of paying their own way and freeing their families from the obligation to take care of their financial needs.

  Some fast-food workers, especially young single mothers, wanted to avoid being on welfare and could not find a better paying job. And for many working mothers, "working is an insurance policy against men who may not be around for the long haul" (p. 70).

- *How Do Residents Find Jobs?* The most important asset job seekers have in finding work is the network of friends and acquaintances who are already working somewhere and who can provide a personal connection to an employer. One resident explained, "You might not know anything [about the job demands]. But if you know that right person, they can pull you in and hook you up with a job, a title, and you're in there" (p. 81).

- *How Do Workers in Fast Food Restaurants Handle the Stigma of a Low-Status Job?* Many of the respondents in Newman's study did not buy into the idea that their job was something to be embarrassed about. They have a sense of dignity and pride that over rides the stigma of having a "McJob." One respondent said, "I'm not ashamed because I have a job...and I'm proud of myself that I decided to get up and do something at an early age. So as I look at it, I'm not on welfare. I'm doing something"(p. 98). Another fast-food worker had a more mixed response:

  > Regardless of what kind of work you do, you still can be respected. Ain't saying I'm ashamed of my job, but I wouldn't walk down the street wearing the uniform. . . . I ain't gonna lie and say I'm not ashamed, period. But I'm

  proud that I'm working. (p. 99–100)

In conclusion, Newman suggests that the daily concerns of the inner-city working poor mirror those of their suburban counterparts. "They worry about job security and the quality of their children's schools. They put in a full day at tiring, boring jobs" and rush home to get dinner on the table and supervise the homework routine of their children (p. 229). But, says Newman, the working poor in inner cities face a more problematic environment where crime, drugs, and other social ills exacerbate the difficulties of survival and achievement. Nevertheless, Newman suggests that we recognize that one of the greatest assets of the working poor is the commitment they have to the work ethic. "These are not people whose values need reengineering. They work hard at jobs the rest of us would not want because they believe in the dignity of their work" (p. xv).

Sources:

Newman, Katherine S. 1999. *No Shame in My Game: The Working Poor in the Inner City*. New York: Alfred A. Knopf and the Russell Sage Foundation.

Russell Sage Foundation. Undated. "An Interview with Katherine S. Newman." http://www.russellsage.org/about/newman_interview.html

(Anderson 1992, 81). A 28-year-old mother from a public housing project in Chicago described selling drugs to feed her family: "Me myself I have sold marijuana, I'm not a drug pusher, but I'm just tryin' to make ends—I'm tryin' to keep bread on the table—I have two babies" (quoted in Wilson 1996, 58). Another resident from the west side of Chicago sold drugs to augment his income from a part-time job:

> Four years I been out here trying to find a steady job. Going back and forth all these temporary jobs and this 'n' that. Then you know you gotta give money at home, you know you gotta buy clothes . . . food in the house too, . . . Well, . . . I have been selling drugs lately on the side after I get off work and, ah, it has been going all right. . . . you can make more money dealing drugs than your job . . . . (quoted in Wilson 1996, 58–59)

Urban Crime   Although the 1999 U.S. crime rate was the lowest it had been since 1973, the rates of violent crime and property crime continue to be higher in cities and metropolitan areas than in rural areas (see Figure 13.2). High rates of urban crime are linked to high rates of unemployment and poverty, widespread drug trafficking and addiction, single-parent households where children have little supervision, inadequate schools that fail to channel students into productive roles, and lack of positive role models. Orfield (1997) explains:

> In neighborhoods lacking successful middle-class role models, gang leaders, drug dealers, and other antisocial figures are often the only local residents with money and status. Tightly knit gangs replace nonexistent family structures. These factors interact with anger, frustration, isolation, boredom, and hopelessness, and create a synergism of disproportionate levels of crime, violence, and other antisocial behavior. (p. 19)

Rodriguez (2000) describes how joblessness and economic marginalization that plague urban areas can lead to involvement in gang-related violence and other criminal behavior:

> Many Mexicans moved [to Los Angeles] to work in the factories and sweatshops, but in the seventies and eighties, all the big shops, steel mills, auto plants, and meat-packing plants began shutting down. Whole communities built on these industries suffered, and the kids in these communities grew up not only unable to go to college, but unable even to get a job. They started feeling expendable. And when you see yourself as expendable—when you've been told often enough that you *are* expendable—it's very easy to "go crazy." (p. 6)

HIV/AIDS   Rates of HIV and AIDS are higher in urban areas—especially inner cities— than in rural areas because individuals who live in cities have higher rates of risk behavior (see Figure 13.3). For example, as noted earlier, rates of drug addiction are higher among city dwellers than the rest of the population. Drug use impairs judgment and interferes with safer sex practices, and some addicts resort to the high-risk behavior of prostitution to pay for their drugs. Most importantly, drug use that involves needle injection accounts for a significant percentage of HIV transmission and the disease that often follows: AIDS. In 1999, about one-quarter of U.S. men living with AIDS and 41 percent of U.S. women living with AIDS had contracted HIV from injecting drugs (Centers for Disease Control and Prevention 2000, Table 25).

■ **Figure 13.2** *Crime Rates by Area: 1999*

Source: Federal Bureau of Investigation. 2000. *Uniform Crime Reports: Crime in the United States—1999.*

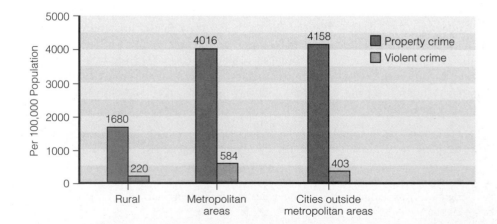

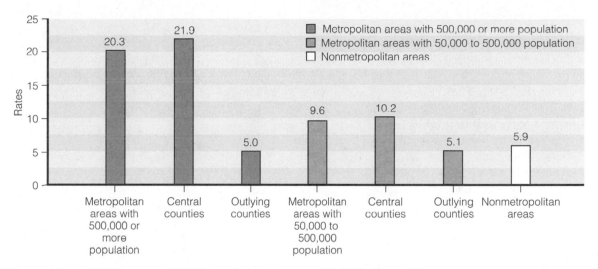

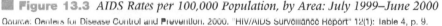

Figure 13.3 *AIDS Rates per 100,000 Population, by Area: July 1999–June 2000*

Source: Centers for Disease Control and Prevention. 2000. "HIV/AIDS Surveillance Report" 12(1): Table 4, p. 9.

Transportation and Traffic Problems

Urban areas are often plagued with transportation and traffic problems. In a survey of 2000 U.S. households, the majority (79 percent) cited heavier traffic as the most negative aspect of urban growth (National Association of Home Builders 1999).

In a telephone survey of randomly selected registered voters, the majority (89 percent) of respondents agreed that traffic congestion has worsened nationwide (U.S. Conference of Mayors 2001a). When asked if traffic had gotten better, worse, or stayed the same in their areas over the past 5 years, 79 percent said traffic had gotten worse and one out of two respondents said that traffic was currently "much worse" than it was 5 years ago.

Many public roads in urban areas are afflicted with what some call **autosclerosis**—defined as "clogged vehicular arteries that slow rush hour traffic to a crawl or a stop, even when there are no accidents or construction crews ahead" ("The Bridge to the 21st Century Leads to Gridlock in and around Decaying Cities" 1997, 1). According to Lundberg (in Jensen 2001), "the average vehicle speed for crosstown traffic in New York City is less than six miles per hour—slower than it was in the days of horse-drawn buggies" (p. 6). And "traffic jams in Atlanta have been so entangled that babies have been born in traffic standstills, and some desperate drivers have had to leave their cars to relieve themselves behind roadside bushes" (Shevis 1999, 2). In some foreign cities such as Sao Paulo, Brazil; Bangkok, Thailand; and Cairo, Egypt, traffic jams are even worse than those in the United States. According to one report, "it sometimes takes so long to reach the Bangkok airport from downtown—from 3 to 6 traffic-paralyzed hours—that roadside entrepreneurs sell minitoilet kits for use by desperate riders in traffic-jammed cars" ("The Bridge to the 21st Century..." 1997, 1). Traffic congestion creates stress on drivers, which sometimes leads to aggressive driving and violent reactions to other drivers—a phenomenon known as **road rage** (see this chapter's *Self and Society* feature).

Natives who beat drums to drive off evil spirits are objects of scorn to smart Americans who blow horns to break up traffic jams.

MARY ELLEN KELLY
QUOTED IN *THE SUN*

Road Rage Survey

Increased congestion of urban highways has led to a phenomenon called road rage. Please answer the questions below to assess your level of road rage and then compare your answers to a national sample of respondents.

1. Which of the following do you ever do?

   Category I (check all that apply)

   Going 5 to 10 mph over the speed limit ☐

   Making rolling stops ☐

   Making illegal turns ☐

   Lane hopping without signaling ☐

   Following very close as a habit ☐

   Going through red lights ☐

   Denying right of way (failure to yield) ☐

   Swearing, cursing, name calling ☐

   Combined, how regularly do you do the things in Category I (circle one)?

       Never 1 2 3 4 5 6 7 8 9 10 Quite regularly

   Category II (check all that apply)

   Going 15 to 25 mph over the speed limit ☐

   Yelling at another driver ☐

   Honking in protest ☐

   Revving the engine ☐

   Making an insulting gesture ☐

   Tailgating dangerously ☐

   Shining the headlights to retaliate ☐

   Cruising in the passing lane (forcing others to pass on the right) ☐

   Combined, how regularly do you do the things in Category II (circle one)?

       Never 1 2 3 4 5 6 7 8 9 10 Quite regularly

   Category III (check all that apply)

   Braking suddenly to punish a tailgater ☐

   Deliberately cutting someone off to retaliate ☐

   Using the car to block the way ☐

   Using the car as a weapon to threaten someone ☐

   Chasing another car in pursuit ☐

   Getting into a physical fight with another driver ☐

   Other things ☐

   Combined, how regularly do you do the things in Category III (circle one)?

       Never 1 2 3 4 5 6 7 8 9 10 Quite regularly

2. Which of the following emotions do you experience when driving?

   Anger or rage behind the wheel

   Never 1 2 3 4 5 6 7 8 9 10 Regularly

   Enjoying fantasies of violence

   Never 1 2 3 4 5 6 7 8 9 10 Regularly

   Experiencing fear for self or family in the car

   Never 1 2 3 4 5 6 7 8 9 10 Regularly

   Feeling compassion for another driver

   Never 1 2 3 4 5 6 7 8 9 10 Regularly

   Feeling competitive with other drivers

   Never 1 2 3 4 5 6 7 8 9 10 Regularly

Feeling impatient and the constant urge to rush
Never   1   2   3   4   5   6   7   8   9   10   Regularly
Wanting to drive dangerously
Never   1   2   3   4   5   6   7   8   9   10   Regularly
Feeling levelheaded and calm
Never   1   2   3   4   5   6   7   8   9   10   Regularly

Below are the results of a national survey of 761 U.S. drivers. Numbers indicate the percentage of people who have engaged in the behavior ±3 percent.

Category I	Percentage
Going 5 to 10 mph over the speed limit	92
Making rolling stops	50
Making illegal turns	15
Lane hopping without signaling	25
Following very close as a habit	14
Going through red lights	10
Denying right of way (failure to yield)	8
Swearing, cursing, name calling	58
Mean regularity rating	6.0

Category II	Percentage
Going 15 to 25 mph over the speed limit	39
Yelling at another driver	32
Honking in protest	35
Revving the engine	11
Making an insulting gesture	24
Tailgating dangerously	12
Shining the headlights to retaliate	18
Cruising in the passing lane (forcing others to pass on the right)	11
Mean regularity rating	3.9

Category III	Percentage
Braking suddenly to punish a tailgater	32
Deliberately cutting someone off to retaliate	15
Using the car to block the way	18
Using the car as a weapon to threaten someone	2
Chasing another car in pursuit	8
Getting into a physical fight with another driver	2
Other things	8
Mean regularity rating	2.7

Emotions	Mean Regularity Rating
Anger or rage behind the wheel	4.5
Enjoying fantasies of violence	2.4
Experiencing fear for self or family in the car	4.0
Feeling compassion for another driver	4.9
Feeling competitive with other drivers	4.9
Feeling impatient and the constant urge to rush	5.7
Wanting to drive dangerously	2.7
Feeling levelheaded and calm	6.7

Source: James, Leon. 1998. "World Road Rage Survey" as appeared at http://www.drdriving.org. Reprinted by permission.

Everybody says that living in the inner city is dangerous, but the truth is that, if you take car crashes into account, the suburbs are statistically far more dangerous places to live.

JAN LUNDBERG
Director of the Alliance for a Paving Moratorium

In the United States, private automobiles, often carrying only one person, are the dominant mode of travel. Cars and light trucks are the largest single source of air pollution (Union of Concerned Scientists 1999). Air pollution and traffic congestion have been major forces driving residents and businesses away from densely populated urban areas (Warren 1998). Health problems associated with congested traffic include stress, respiratory problems, and death. Indeed, far more people are killed and injured in automobile accidents than by violent crime. Therefore, it has been argued that despite higher crime rates in the inner city, the suburbs are the more dangerous place to live, because suburbanites "drive three times as much, and twice as fast, as urban dwellers" (Durning 1996, 24).

Dependence on cars has been encouraged by a number of factors: unwillingness to tax gasoline commensurate with its cost to society, free and tax-free parking typically provided by corporations to their employees outside of the city, the glamorization of automobiles—largely perpetuated by the automobile industry—and federal subsidies that favor the building of highways over investments in public transit. The conversion of the United States to an automobile-based system of transportation was heavily influenced by industries that profit from automobile use. In the 1930s, National City Lines, a company backed by the three major automakers, major oil companies, tire manufacturers, and the trucking and construction industries, succeeded in systematically buying and closing down more than 100 electric trolley lines in 45 U.S. cities. Although National City Lines was convicted of this conspiracy in 1949, the dismantling of the rail transit system had already been accomplished (Warren 1998).

Sprawl and the Displacement and Endangerment of Wildlife

The spread of urban and suburban areas increasingly is replacing natural habitats with pavement, buildings, and human communities. The loss of open green space, trees, and plant life not only affects the quality of life for humans, but for the animals whose homes are turned into parking lots, shopping centers, office buildings, and housing developments. The majority of U.S. adults (81 percent) in a Gallup poll reported that they worried "a great deal" or "a fair amount" about the loss of natural habitat for wildlife (Dunlap & Saad 2001).

Evidence of wildlife displacement resulting from sprawl is found across the nation. Coyotes, normally found only in the West and in Appalachia, are now being sighted in every state (Shevis 1999). Other displaced species include bear, Canada geese, and deer. "With no place to go, they bound into suburban backyards in search of food and water and across highways, frequently injuring themselves and causing harm to drivers...Deer cause an estimated half-million vehicle accidents a year, killing 100 people and injuring thousands more" (Shevis 1999, 2–3). About 1 million animals are killed on U.S. roads every day (Lundberg, in Jensen 2001).

In 1999, the U.S. Fish and Wildlife Service (USFWS) reported that 357 animal species and 568 plants were listed as "endangered" and another 123 animal species and 135 plant species were listed as "threatened," for a total of 1,183 plant and animal species in need of protection. In the 1990s, the USFWS has added an average of 85 plant and animal species to the list each year. According to the USFWS, habitat loss resulting from urban and suburban sprawl is the number-one reason why wildlife species are becoming increasingly endangered (Shevis 1999).

In various parts of the Western United States, the survival of bears, mountain lions, coyotes and black-tailed deer is the most pressing problem; in the Eastern and Midwestern parts of the country the most visible problems are with white-tailed deer and Canada geese. In every case, their enemy is man, relentlessly taking over their territory with backhoes and earthmovers (Shevis 1999, 3).

Strategies for Action: Saving Our Cities

Numerous federal, state, and private policies and programs have been implemented to attempt to revitalize the cities and address the various urban problems discussed in this chapter. Here we summarize strategies to alleviate poverty and stimulate economic development in inner cities, tame the urban "three-headed monster" (AIDS, addiction, and crime), improve transportation and alleviate traffic congestion, and curb urban growth in developing countries.

Urban Economic Development and Revitalization

A number of strategies have been proposed and implemented to restore prosperity to U.S. cities and well-being to their residents, businesses, and workers, including strategies to attract new businesses, create jobs, and repopulate cities. The economic development and revitalization of cities also involves improving affordable housing options (discussed in Chapter 10), and alleviating urban problems related to HIV/AIDS, addiction, crime, and traffic and transportation—topics we address later in this chapter.

> It does not make sense to ignore the fact that so many of our citizens are born and grow up in neighborhoods that may well limit their ability to become productive citizens.
>
> PAUL A. JARGOWSKY
> *Economist*

Empowerment Zone/Enterprise Community Program   The federal **Empowerment Zone/Enterprise Community Initiative**, or EZ/EC program, provides tax incentives, grants, and loans to businesses to create jobs for residents living within various designated zones or communities, many of which are in urban areas. Federal money provided to Empowerment Zones and Enterprise Communities is also used to train and educate youth and families and to improve child care, health care, and transportation. The program provides grant funding so communities can design local solutions that empower residents to participate in the revitalization of their neighborhoods.

When the EZ/EC Initiative began in 1994, 105 distressed communities were designated as Enterprise Zones and Enterprise Communities. In 1999, the initiative expanded to include 40 additional areas, and in 2000, Congress approved legislation adding nine new Empowerment Zones (U.S. Department of Housing and Urban Development 2001).

Some cities have attributed significant overall economic gains and improvements to the EZ/EC program. Detroit mayor Dennis Archer (1998) claimed that,

> Since winning our Empowerment Zone designation in December 1994...one hundred million dollars in federal money has been provided to 49 Detroit community organizations for construction projects and social programs. Motivated by tax breaks for employment within the Empowerment Zone, businesses have invested at least $3.9 billion in the Zone since early 1995. At least 2,750 new jobs—with 400 more on the way—have been created within the 18.3-square mile area during that period. (pp. 341–42)

Infrastructure Improvements   Urban revitalization often involves making improvements in the **infrastructure**—the underlying foundation that enables a city to function. Infrastructure includes such things as water and sewer lines, phone lines, electricity cables, sidewalks, streets, curbs, lighting, and storm drainage systems. Improving infrastructure may help attract business to an area. Infrastructure improvements in urban areas also increase property values and renews residents' sense of pride in their neighborhood (Cowherd 2001).

Brownfield Redevelopment   **Brownfields** are abandoned or undeveloped sites that are located on contaminated land. A survey of mayors in 231 U.S. cities found that 210 cities estimated that they had more than 21,000 brownfield sites, ranging in size from a quarter of an acre to a single site that measures 1,300 acres (U.S. Conference of Mayors 2000b). The majority of mayors in this survey reported that through the redevelopment of brownfields, their cities could significantly increase jobs, population, and tax revenues. In addition, "the positive impacts of brownfields redevelopment can extend beyond the locality where the brownfields are located by providing an alternative to sprawl and thereby preserving greenfields" (Greenberg, Miller, Lowrie, & Mayer 2001, 21).

But cleaning contaminated sites and redeveloping them is costly. In 1993, the Environmental Protection Agency (EPA) initiated a brownfields pilot program to help local and regional governments clean up brownfield sites and turn them into houses, recreational facilities, and businesses. EPA has awarded more than 300 brownfields pilot grants to local governments; however, thousands of places are without federal or state support for brownfields redevelopment (Greenberg et al. 2001). A survey of mayors in 231 U.S. cities found that the most commonly cited obstacle to redeveloping brownfields is lack of funding (U.S. Conference of Mayors 2000b).

Gentrification and Incumbent Upgrading   **Gentrification** is a type of neighborhood revitalization in which middle- and upper-income persons buy and rehabilitate older homes in a depressed neighborhood. They may live there or sell or rent to others. The city provides tax incentives for investing in old housing with the goal of attracting wealthier residents back into these neighborhoods and increasing the tax base. However, low-income residents are often forced into substandard housing as less and less affordable housing is available. In effect, gentrification often displaces the poor and the elderly (Johnson 1997).

An alternative to gentrification is **incumbent upgrading**, in which aid programs help residents of depressed neighborhoods buy or improve their homes and stay in the community. Both gentrification and incumbent upgrading improve decaying neighborhoods, which attracts residents as well as businesses.

This old mill in Groton, Massachusetts, designated as a brownfields site renovation project, will be torn down and replaced by a senior assisted living center. This project is one of about 175 brownfields projects underway in Massachusetts.

Improving Education   Any serious effort to revitalize cities must include making improvements in the educational system so that young people have access to training that will enable them to compete in the job market. In a survey of 110 U.S. cities, officials in four of five survey cities reported a shortage of highly skilled workers; over half said that this was affecting their ability to attract new businesses (U.S. Conference of Mayors 2000c). Although the majority of the officials indicated that the education and training institutions and organizations in their cities had the potential to develop the skills needed by employers, about half of the officials believed that more public and private resources were needed to develop the skills of the local workforce.

Another study on business location decision making found that businesses look to locate in areas that have quality education and a high-quality labor force as opposed to availability of land, buildings, and transportation systems (Cohen 2000). In choosing a location, businesses look at the quality of elementary and secondary schools, as well as community colleges and universities. They want to know that the local workforce is well educated. Further, in order to attract highly skilled workers from other regions of the country, they must advertise that a good school system is in place for their employees' children.

Community-Based Urban Renewal Efforts   Some residents of deteriorating urban areas have begun grassroots programs to improve living conditions in their neighborhoods. Community-based development programs involve small-scale developers and volunteers working with a small professional staff who are concerned with improving the community. For example, 26,000 volunteers spent a Saturday cleaning Detroit streets, parks, and playgrounds during the fourth annual spring Clean Sweep campaign (Archer 1998). In Detroit's annual Paint the Town event, individuals, community organizations, and corporate volunteers fix and paint the homes of the poor and elderly.

Taming the Urban "Three-Headed Monster": Strategies to Reduce HIV/AIDS, Addiction, and Crime

Although strategies to reduce HIV/AIDS, addiction, and crime are discussed in other chapters in this text (see Chapters 2, 3, and 4), here we highlight how some U.S. cities have responded to these problems.

Cities Respond to HIV/AIDS and Drug Addiction   As noted earlier, HIV is often transmitted through sharing infected needles used to inject drugs. In Baltimore, 85 percent of new HIV infections are attributed to drug addicts using infected needles (Schmoke 1998). Some cities have succeeded in reducing HIV transmission by setting up needle-exchange programs whereby drug users can exchange used (and perhaps infected) needles for free clean ones. The largest needle-exchange program operated by a local government is in Baltimore. An evaluation of the program found that enrolled addicts lowered their risk of contracting HIV by almost 40 percent (Schmoke 1998). Other studies have found that HIV infection rates are over three times lower in drug injectors who participate in needle-exchange programs than in those who do not (World Health Organization and United Nations Joint Programme on HIV/AIDS 1998). Needle-exchange programs represent a public health approach to drug addiction, as they not only reduce the spread of HIV and AIDS, but also serve as a stepping-stone to drug treatment.

America and its cities are at a crossroads. If they choose to make their separate ways into the future, decline and decay almost surely lie ahead for both. But if they stride together as partners into the twenty-first century, the nation and its great urban areas will compete and brilliantly succeed in the new global economy.

FINAL REPORT OF THE
82ND AMERICAN ASSEMBLY

Strategies and Techniques Used by Inner-City Residents to Avoid Crime-Victimization

University of Pennsylvania sociologist Elijah Anderson spent 14 years studying the culture, experiences, and interactions of the residents living in the two adjacent inner-city communities in Philadelphia that he refers to by the pseudonym "Village-Northton." His research, described in *Streetwise: Race, Class, and Change in an Urban Community* (1992), illuminates many aspects of social life in urban America, including the causes and effects of unemployment, drug use, crime, out-of-wedlock teenage pregnancies and single-parent families, and race and class relations. Anderson's research also examines the norms, values, beliefs, and patterns of interpersonal interaction that characterize inner-city life. In this *Human Side* feature, we describe one aspect of his findings: how residents of crime-ridden inner-city neighborhoods attempt to avoid being victims of crime.

Residents of Village-Northton described a number of defensive strategies and techniques used in attempts to avoid criminal victimization, including the following:

- *Conceal valuable assets.* Residents made efforts to conceal anything of value in their home. After getting a new refrigerator, one woman "just cut the box up and put out a little each week; she'd put it in a big Hefty bag and just put it right out. She didn't want her neighbors to know she'd gotten something, because that meant either she had money or there was something in there to steal." (pp. 78–79)

- *Don't tempt thieves.* Whenever one resident went to church, *"she would put her pocketbook over her shoulder, and then put her coat on over the pocketbook. She'd never carry her pocketbook hanging out. That was her strategy. And no jewelry. Even if she was gonna get a ride. Sometimes when she'd be picked up and taken to church, she wouldn't wear the jewelry out of the house. She'd wait till she got in the car, and then on the way to where she was going put it on. On the way back, she'd take it off. It's a hell of a way to live, but that's how she avoided problems."* (pp. 78–79)

- *Dress down.* Anderson observed that dress is an important consideration when walking the Village-Northton streets. Women wear unisex jackets, blue jeans, and sneakers—clothing that negates stereotypical 'female frailty' and symbolizes aggressiveness. Women wear 'sexy' dresses only when they are in a group, accompanied by a man, or traveling by car. Men also wear "nonshowy" clothing, such as blue jeans or a sweat suit. More expensive clothing is reserved for daytime work hours or traveling by car.

- *Get a dog.* Anderson observed that "the company of a dog allows residents, particularly whites, to feel more secure on the street and gives them more power in anonymous black-white interactions"

I'd rather sleep in the middle of nowhere than in any city on earth.

STEVE MCQUEEN
Actor

Cities Respond to Crime   Individuals respond to living in crime-ridden areas in a number of ways, from moving out of the neighborhood, to joining in the criminal activity, to developing strategies and techniques to avoid criminal victimization (see this chapter's *The Human Side* feature).

Collective responses to urban crime include strategies and policies implemented by city planners; local, state, and federal officials; and other organizations. For example, city planners attempt to reduce urban crime by designing housing, public spaces, and roads in such a way as to prevent crime. A national program, Crime Prevention through Environmental Design, emphasizes the security effects of walkways, building entries, lighting, parking areas, and landscaping (Johnson 1997). "Proper placement of lights, vegetation, and access doors will reduce a building's vulnerability to burglary and vandalism" (p. 153). Regarding traffic, some cities have implemented traffic diverters at intersections that force drivers to make a right-hand turn, making alleys into cul-de-sacs to discourage easy escape from crime scenes.

A survey of 490 U.S. cities found that 337 cities have established nighttime curfews over the last 20 years to deter crime and violence among youth, and 35 cities are considering one (National League of Cities 2000a). Most curfews

(pp. 223-224). Even when residents own dogs for the purpose of having a pet rather than a protector, dogs discourage potential attackers. Anderson explains that, "many working-class blacks are easily intimidated by strange dogs, either off or on the leash" (p. 222). One young black man said: *"I tell you, when I see a strange dog, I am very careful...But white people have a whole different attitude. Some of them want to go up and pet the dog."* (p. 222)

- *Make friends with neighbors.* Among residents of inner-city ghetto neighborhoods, concern about the drug culture breeds fear of criminal victimization. Residents know that "a person on drugs is dangerous and out of control and is thus capable of robbing his own family and friends or turning on them violently" (p. 78). Because of the widespread use of drugs, particularly highly addictive crack cocaine, residents feel vulnerable to criminal victimization within their own neighborhoods. "The perpetrator of a crime might be a nephew on drugs, one's best friend's son, or simply a young man 'down the street'" (p. 78). Residents try to protect themselves from crimes committed by other neighborhood residents by "'making friends' with their neighbors in an effort to 'be known,' to ingratiate themselves with potential criminals" (p. 78). One resident described how a local woman would try to make friends with the neighborhood boys:

She'd bake cakes for some of the young boys, if they had a birthday. And when she'd pass by and see a bunch of guys on the corner and they'd be coughing and sniffing [implied cocaine habit] or whatever, she'd have these Hall's cough drops. She'd reach her hand in her pocket and say, 'Here, son. Take this for your cough.' She got to be known as 'the lady with the cough drops.' And she'd just walk up and hug him [a local street corner youth] and say, 'How's your cold, son?' She knew all the time that the cough and sniffles were from drugs." (p. 79)

- *Act like you know them.* One man described his mother attempting to prevent criminal victimization by acting like she knew the suspected potential perpetrator. *"She'd say things like, 'How's your sister?' or "How's Bea?' or whatever. Out of ten young guys, if you can just connect with one, then the whole group will be cool . . . Sometimes she would act like the aunt or mother of a boy whom she was suspicious of. On one occasion, she was standing at a bus stop and a youth approached her and she said, "Son! Where's your coat? You're gon' catch your death of cold out here . . . What would your mother say? . . . What's your name? Take better care of yourself!"* (p. 80)

Source: Anderson, Elijah. 1992. *Streetwise: Race, Class, and Change in an Urban Community.* Chicago: University of Chicago Press. Used by permission.

expect children under 18 to be off the street by 11:00 P.M. on weekdays and midnight on weekends. The majority (96 percent) of mayors of cities with curfews views their laws as effective in combating youth crime and violence. To deter truancy, 68 cities also have daytime youth curfews that require youth to be off the street during school hours (National League of Cities 2000a).

Other cities have taken a "get tough on crime" approach as state governments have lengthened sentences and built more prisons. However, Orfield (1997) notes that "these very large expenditures and their small effects strongly suggest that it is time to seek solutions to crime by reducing the concentration of its breeding ground, the poverty core" (p. 24). Orfield argues that providing poor inner-city residents with access to opportunity is likely to reduce crime. Opportunities for poor city residents increase by integrating low-income residents among the middle-class through mixed-income housing, improving city schools, expanding job training programs, and developing jobs. "Young people, whether or not they're working, have to be given a sense of who they are and the opportunity to do something with themselves" (Rodriguez 2000, 6).

Improving Transportation and Alleviating Traffic Congestion

A number of strategies to improve transportation and alleviate traffic congestion have been implemented and proposed. Some states have established high-occupancy vehicle (HOV) lanes designed to reduce traffic congestion (and pollution-causing car emissions) by inducing more commuters to carpool. Cars carrying two or more persons are permitted to drive in less crowded HOV lanes. However, efforts to encourage carpooling, such as HOV lanes and park-and-ride lots, generally have not been successful (Nelessen 1997).

The **New Urbanists**, a growing group of planners, architects, developers, and traffic engineers support neighborhood designs that create a strong sense of community by incorporating features of traditional small towns. New Urbanists advocate designing **mixed-use neighborhoods** that combine residential and commercial elements along with public and private facilities such as schools, recreation centers, and places of worship. The idea of mixed-use neighborhoods is to provide suburban residents with convenient access to jobs, stores and service providers, schools, and other facilities, thus reducing driving distances. In many cases, mixed-use neighborhoods are designed to allow residents to walk or bike to various destinations, minimizing the need to drive. New Urbanists envision pedestrian communities in which it is possible to walk to schools, parks, recreation, jobs, and transit stops.

Some cities have turned to high-tech solutions to ease traffic congestion. Thanks to the Integrated Traveler Information Sharing System, commuters in some California counties may use their car radios and cell phones to access traffic advisory information regarding accidents, alternative routes, and lane closures ("Technology Smooths the Ride for Santa Ana Commuters" 1998). However, a more effective way to ease traffic congestion may be to return to a simpler technology: the bicycle (see this chapter's *Focus on Technology* feature).

Another strategy for reducing traffic congestion involves increasing the use of public transit, such as buses, trains, and subways. A resurgence in public transportation may be underway: between 1995 and 2000, ridership on the nation's public transportation systems grew by over 20 percent (American Public Transportation Association 2001). The United States Conference of Mayors (2001b) urged President Bush and Congress to make passenger rail service a top priority and a solution to the growing crisis of traffic and air congestion. The 1998 Transportation Equity Act for the 21st Century (TEA21) provided $169.5 billion for highway improvements and $42 billion for mass transit over 6 years. The findings from a national survey suggest that the majority of Americans (80 percent) support building more rail systems serving cities, suburbs, and entire regions to give them the option of not driving their cars. Only 16 percent were opposed (U.S. Conference of Mayors 2001a).

> Adding highway capacity to solve traffic congestion is like buying larger pants to deal with your weight problem.
>
> MICHAEL REPLOGLE
> *Transportation Specialist*
> *Environmental Defense*

Opposition to building more roads and highways is growing. Transportation planners increasingly recognize that building more roads does not necessarily ease traffic problems. The U.S. public agrees: in a poll of randomly selected registered voters the majority (66 percent) said they do not think that traffic congestion will be eased if more roads are built (U.S. Conference of Mayors 2001a). And more importantly, concern is growing over the social and environmental problems related to road-building and motor vehicle use. According to Jan Lundberg, Director of a grass-roots group called the *Alliance for a Paving Moratorium*, nearly half of all urban space is paved; more land is devoted to cars than

A Return to Simpler Technology: Bicycles as a Solution to Urban Problems

The application of technology to solving modern social problems often involves relatively current, state-of-the-art technologies. However, sometimes an effective solution to a social problem can be found in an older, simpler technology.

In cities around the world, bicycles are emerging as a solution to some of today's urban problems. "For safer streets, less congestion, and cleaner air, the bicycle is poised to become an integral part of urban transportation systems in the 21st century" (Gardner 1999, 23). The use of bicycles can also result in improved health and lower health care costs—benefits associated with cleaner air, reduced noise, and more exercise. Studies in the United Kingdom have found that the health benefits of cycling far outweigh the risks of bicycle accidents (Sheehan 2001). The exercise involved in cycling decreases the risk of heart disease and diabetes. Increased bicycle use could also help alleviate the problem of noise pollution and its associated adverse health effects. Sheehan (2001) explains:

> Noise is perceived by many urban residents as one of the greatest problems associated with road traffic. It contributes to stress disturbances, cardiovascular disease, and hearing loss. The problem is particularly acute in Japan, where 30 percent of people experience noise levels greater than 65...decibels—a level considered unacceptable in many countries—and in Europe, where the figure is 17 percent. (p. 111)

Increased bicycle use would also result in fewer people being injured and killed by cars. Nearly a million people are killed on the world's roads each year, most of whom are pedestrians. And, replacing motor vehicle trips with bike trips could improve health by reducing vehicle emissions and improving air quality. Sheehan (2001) cites a report by the World Health Organization that finds:

> In some parts of the world, vehicular air pollution actually kills more people than traffic accidents do. In Austria, France, and Switzerland, in 1996 the premature mortality brought about by particulate emissions from vehicles stood at about twice the number from traffic accidents. (p. 110)

Although bicycles could contribute to improving urban life by reducing traffic congestion, reducing pollution-causing vehicle emissions , reducing noise pollution, and improving public health, their use in urban areas is not widespread. Indeed, between 1995 and 1998, bicycle production worldwide fell by 25 percent (Sheehan 2001). A number of factors discourage widespread bicycle use, including unsafe roads and a lack of safe bike parking. Ironically, air pollution, largely caused by vehicle emissions, discourages the use of bicycles, thus hindering a mode of transportation that would help to alleviate the air pollution.

However, bicycles have become a major mode of transportation in some cities throughout the world. In several major cities in Germany, Denmark, and the Netherlands, bicycles now account for 20 to 30 percent of all trips and in some Asian cities, more than half of all trips are by bicycle (Gardner 1999). In contrast, bicycles are used for fewer than 1 percent of all trips in the

© AP/Wide World Photos

Portland, Oregon incorporates many features of a model American city, with ample bicycle lanes, plenty of public transportation, and a limit on urban sprawl.

Continued

403

United States and Canada, despite the fact that in the United States, 40 percent of travel involves trips that are two miles or shorter and more than a quarter of all trips are under a mile (Gardner 1999).

Encouraging the use of bicycles requires changes in urban design and policy to make cycling a more safe, viable, and/or necessary option. Some European cities, such as Munich, Vienna, and Copenhagen, have commercial centers that restrict vehicle traffic to ambulances, delivery trucks, and cars owned by local residents. About 20 car-free communities are in various stages of development in Germany (Sheehan 2001). Programs in Lima, Peru help low-income residents buy bicycles and a program in Copenhagen, Denmark provides bikes for public use (Gardner 1999). Organic food growers participating in a local project in California called *Pedal Power Produce* use bike carts to bring organic foods to market (Jensen 2001). The United Kingdom has built an 8,000-kilometer National Cycle Network that will pass within 4 kilometers of half of the country's population (Brown 2000). It is hoped that the accessibility of safe cycling paths will induce people to shift from cars to bicycles on short trips.

To encourage bike use for longer trips, cities can establish convenient connections between cycling and public transit. "Bicycles and transit can complement each other when people are able to carry their bikes aboard buses or trains, or park them at stations" (Shee-han 2001, 17). Urban planning based on "mixed use" designs can also facilitate bike use by combining housing, public facilities (such as schools, parks, and libraries), and commercial sites (such as grocery stores and banks) in the same neighborhood.

Even if cities are designed to promote safe bicycling as a mode of transportation, people in the United States and other industrialized countries are not likely to trade their car keys for a bicycle helmet. People throughout the developed world have acquired a love for the automobile and its images of freedom, power, adventure, and sexiness. These images are perpetuated by the automobile industry, which spends more money on advertising than any other industry both in the United States and worldwide (Sheehan 2001). Until our love for cleaner air and better health outweighs our love of the automobile, most Americans will continue to leave their bicycles (if they own them) in the closet or garage.

Sources:

Brown, Lester. 2000. "Overview: The Acceleration of Change." In *Vital Signs: The Environmental Trends that Are Shaping Our Future*, ed. Linda Starke, pp. 17–29. New York: W.W. Norton & Co.

Gardner, Gary. 1999. "Cities Turning to Bicycles to Cut Costs, Pollution, and Crime." *Public Management* 81(1):23.

Jensen, Derrick. 2001 (February). "Road to Ruin: An Interview with Jan Lundberg." *The Sun*, 302(4–13).

Sheehan, Molly O'Meara. 2001. "Making Better Transportation Choices." In *State of the World 2001*, ed. Linda Starke, pp. 103–122. New York: W.W. Norton & Company.

to housing, and every year, nearly 100,000 people are displaced by highway construction (Jensen 2001). The *Alliance for a Paving Moratorium* advocates halting road-building. "In the Alliance's view, a paving moratorium would limit the spread of population, redirect investment from suburbs to inner cities, and free up funding for mass transportation and maintenance of existing roads" (Jensen 2001, 6).

> The goal of smart growth is not no growth or even slow growth. Rather, the goal is sensible growth that balances our need for jobs and economic development with our desire to save our natural environment.
>
> PARRIS GLENDENING
> *Governor of Maryland*

Responding to Urban Sprawl: Growth Boundaries and Smart Growth

Some cities have tried to manage urban sprawl by establishing growth boundaries. Since 1973, Oregon has required each city to draw a growth boundary based on its estimate of economic development and community needs in the next 20 years (Geddes 1997). A number of cities in California have also enacted urban growth boundaries (Froehlich 1998). Rather than simply put a limit on urban growth, another approach to managing urban sprawl is to pursue what is called "smart growth." Rather than take a hard stand either for or against urban growth, advocates of **smart growth** encourage development that serves the economic, environmental, and social needs of communities. A smart growth ur-

Previously a landfill, this area in Palos Verdes, California, is now the site of the South Coast Botanic Garden.

© PhotoEdit

ban development plan entails the following principles (Froehlich 1998; Smart Growth Network 1999):

- Mixed-use land use, whereby residences, jobs, schools, grocery stores, etc. are located within close proximity of each other; ample sidewalks for a walkable community
- Compact building design
- Housing and transportation choices
- Distinctive and attractive community design
- The preservation of open space, farmland, natural beauty, and critical environmental areas
- The redevelopment of existing communities, rather than letting them decay and building new communities around them
- Regional planning and collaboration between business, private residents, community groups, and policy makers on development/redevelopment issues

However, most local zoning codes prohibit smart growth design by mandating large housing setbacks, wide streets, and separation of residential and commercial areas (Pelley 1999).

Regionalism

Addressing the various social problems facing urban areas may best be achieved through **regionalism**—a form of collaboration among central cities and suburbs that encourages local governments to share common responsibility for common problems. Central cities, declining inner suburbs, and developing suburbs are often in conflict over the distribution of government-funded resources, zoning and land use plans, transportation and transit reform, and development plans. Rather than compete with each other, regional government provides a mechanism for achieving the interests of an entire region. A metropolitan-wide government would handle the inequities and concerns of both suburban and urban areas. As might be expected, suburban officials resist regionalization because they believe it will hurt their neighborhoods economically by draining off money for the cities. But some regions have had success with regionalism. Minnesota state representative Myron Orfield (1997) successfully formed a political coalition between legislators from Minneapolis-St. Paul, their declining blue-collar suburbs, and more affluent developing suburbs.

Strategies for Reducing Urban Growth in Developing Countries

In developing countries, limiting population growth is essential for alleviating social problems associated with rapidly growing urban populations. Strategies for lowering the birthrate and reducing population growth are discussed in detail in Chapter 14.

Another strategy for minimizing urban growth in less developed countries involves redistributing the population from urban to rural areas. Such redistribution strategies include the following (United Nations 1994): (1) promote agricultural development in rural areas, (2) provide incentives to industries and businesses to relocate from urban to rural areas, (3) provide incentives to encourage new businesses and industries to develop in rural areas, (4) develop the infrastructure of rural areas, including transportation and communication systems, clean water supplies, sanitary waste disposal systems, and social services. Of course, these strategies require economic and material resources, which are in short supply in less-developed countries.

Understanding *Cities in Crisis*

What can we conclude from our analysis of urban life? First, attention to urban problems and issues is increasingly important, as the United States and the rest of the world are becoming increasingly and rapidly urbanized. Second, the social forces affecting urbanization in industrialized countries are different from those in developing countries. Countries such as the United States have experienced urban decline as a result of deindustrialization, deconcentration, and the shift to a service economy in which jobs that pay well and come with full benefits are scarce. At the same time, developing countries have experienced rapid urban growth as a result of industrialization, fueled in part by a global economy in which corporate multinationals locate industry in developing countries to gain access to cheap labor, raw materials, and new markets.

We will neglect our cities to our peril, for in neglecting them we neglect the nation.

JOHN F. KENNEDY
Former U.S. President

Efforts to repopulate U.S. cities have been somewhat successful: eight of the ten largest U.S. cities gained population in the 1990s (Perry & Mackun 2001). Only Philadelphia and Detroit declined in size. As discussed in the next chapter, cities in developing countries continue to grapple with the problem of limiting population growth. The population of urban areas in developing countries is expected to double in the next 30 years ("Urbanization Trend Seen as Accelerating" 2001). Aside from population concerns, problems affecting urban residents include high rates of HIV/AIDS, drug use and addiction, and crime (also referred to as the "three-headed monster"), poverty and unemployment, inadequate schools, inferior or unaffordable housing, pollution, and traffic and transportation problems.

People living in rural areas are also affected by urban processes. In developing countries, peasant farmers are being displaced by the growth of large-scale commercial agriculture geared toward producing products for export rather than for local consumption. Displaced rural dwellers migrate to cities looking for work, but often find that urban conditions are not as promising as they had hoped. In the United States and other developed countries, rural and suburban dwellers who leave or avoid cities find increasingly that the city is coming after them as urban sprawl replaces green open places with concrete. Also, as central cities deteriorate, the surrounding suburbs tend to do so as well.

Urban problems are multifaceted and interrelated. For example, urban joblessness and poverty lead to underfunded and inadequate schools, which results in a poorly trained workforce, leading to the loss of more jobs (or discouraging new jobs from locating in the city), which leads to more poverty, which causes high rates of crime, drug addiction, and HIV/AIDS (and other health problems). All of these problems are confounded by racial and ethnic prejudice and discrimination, family stress and disruption, and increasing economic inequalities in society. Each problem facing cities seems to reinforce other problems.

Although the number and scope of problems facing cities is daunting, the progress some U.S. cities have made is encouraging. Many formerly decaying cities are bustling with new waterfront development, sports arenas, and beautification projects. And there is optimism for the future. The fourteenth annual State of the Cities survey of municipal officials in cities with populations above 10,000 found improvement in most indicators of municipal conditions. Among the 30 indicators, 23 (or 77 percent) were ranked by more officials as improved rather than worsened (National League of Cities 1998). The majority of municipal officials in this survey reported improved overall economic conditions and less unemployment during the past year, improved recreation services, and increased range of city services. Nearly 9 out of 10 are optimistic about the direction in which their city appears to be heading and rated local services as good or very good in relation to community needs. And a more recent survey of city officials found that nearly 3 of 4 (73 percent) described their financial situation as better than a year ago (National League of Cities 2000b).

However, given the growing inequality between the haves and the havenots, the lack of affordable housing, and the projected increase in poverty and hardship caused by welfare reform and the lack of well-paying, low-skill jobs (see also Chapters 10 and 11), cities may see their poor populations increase as well. Donald J. Borut, Executive Director of the National League of Cities, summed up the fourteenth annual State of the Cities survey as follows:

> This report describes a remarkable record of sustained progress and continuing recovery in cities and towns all across America, but it's not a one-dimensional picture. There are shadows that linger in each area of gleaming success. There are needs that remain even as each measure of progress is recorded (National League of Cities 1998, 2).

One of the shadows lingering over cities throughout the world is cast by environmental problems. Global environmental problems, discussed in the next chapter, have many urban roots. "Cities generate some three-quarters of the carbon dioxide that is released from fossil fuel burning worldwide; a similar share of industrial timber is used in cities" (Sheehan 2000, 104). Cities consume the largest share of the world's natural resources and contribute most of the pollution and waste that compromise the health of our planet. As we ponder ways to improve cities, we must include the larger environment in the equation.

The battle to achieve a sustainable balance between Earth's resource base and its human energy will be largely won or lost in the world's cities.

MOLLY O. SHEEHAN
Worldwatch Institute

Critical Thinking

1 Today, one of the most significant forces in the United States and throughout the developed world is the aging of the population. As the population ages, what issues must cities confront in order to accommodate the needs of their elderly residents?

2 How might the growing trend of telework affect cities?

3 E-commerce is a growing trend. One part of e-commerce is the increased tendency for goods and services to be purchased over the Internet. How might this trend affect urban life?

4 What could be done in your community to encourage the use of bicycles as an alternative to motor vehicles?

Key Terms

autosclerosis	incumbent upgrading	slum
barrio	infrastructure	smart growth
brownfields	megacities	suburbanization
central city	metropolis	suburbs
deconcentration	metropolitan area	urbanism
deindustrialization	micropolitan area	urbanization
Empowerment Zone/Enterprise Community Initiative	mixed-use neighborhoods	urban population
	New Urbanists	urbanized area
gentrification	regionalism	urban sprawl
ghetto	road rage	urban villages

Media Resources

 The Wadsworth Sociology Resource Center: Virtual Society

http://sociology.wadsworth.com/

See the companion Web site for this book to access general sociology resources and text-specific features that can further your understanding of this chapter. The site contains Internet links, Internet exercises, online practice quizzes, information on InfoTrac College Edition, and many more valuable materials designed to enrich your learning experience in social problems.

InfoTrac College Edition

You can access InfoTrac College Edition either from the Wadsworth Sociology Resource Center at **http://sociology.wadsworth.com** or directly from your web browser at **http://www.infotrac-college.com/wadsworth/**. InfoTrac College Edition is an online university library that includes over 700 popular and scholarly journals in which you can find articles related to the topics in this chapter such as smart growth, New Urbanism, sustainable communities, and road rage.

Interactions CD-ROM

Go to the Interactions CD-ROM for *Understanding Social Problems*, Third Edition to access additional interactive learning tools, such as in-depth review materials, corresponding practice quizzes, and other engaging resources and activities to help you study the concepts in this chapter.

Section 4

Problems of Modernization

Section 4 focuses on problems of modernization—the cultural and structural changes that occur as a consequence of society changing from traditional to modern. Both Durkheim and Marx were concerned with the impact of modernization. Each theorized that as societies moved from "mechanical" to "organic solidarity" (in Durkheimian terms) or from "production for use" to "production for exchange" (in Marxian terms), fundamental changes in social organization would lead to increased social problems. Although modernization has contributed to many of the social problems we have already discussed, it is more directly related to the four problems we examine in this section—population, the environment, technology, and global conflict.

One of the difficulties in understanding social problems involves sorting out the numerous social forces that contribute to social problems. Every social problem is related, in some way, to many other social problems. Nowhere is this more apparent than in the final three chapters. For example, even though scientific and technological advances are designed and implemented to enhance the quality and conditions of social life, they contribute to other social problems. Science and technology have extended life through various medical advances and have successfully lowered the infant mortality rate in many developing nations. However, these two "successes" (fewer babies dying and an increased life expectancy), when coupled with a relatively high fertility rate, lead to expanding populations. Many nations struggle to feed, clothe, house, and provide safe drinking water and medical care for their increased numbers. Fur-

ther, in responding to problems created by overpopulation, science and technology have contributed to the growing environmental crisis. For example, many developing countries use hazardous pesticides to increase food production for their growing populations; overuse land, which leads to desertification; and, out of economic necessity, agree to deforestation by foreign investors.

Developed countries also contribute to the environmental crisis. Indeed, modernization itself, independent of population problems, exacerbates environmental concerns as the fragile ecosystem is overburdened with the by-products of affluent societies and scientific and technological triumphs: air pollution from the burning of fossil fuels, groundwater contamination from chemical runoff, nuclear

waste disposal, and destruction of the ozone layer by chlorofluorocarbons. Thus, population and the environment (Chapter 14), as well as science and technology (Chapter 15) are inextricably related.

While population patterns and the resultant increased scarcity of resources provide a motivation for global conflict (Chapter 16), science and technology provide more efficient means of worldwide destruction. Conversely, global conflict has devastating effects on the environment (e.g., nuclear winter) and has inspired much scientific and technological research and development such as laser-based and nuclear technologies. Thus, each of the chapter topics to follow is both an independent and a dependent variable in a complex web of cause and effect.

14

Population and Environmental Problems

Is it True?

1. Most of the world's population growth is occurring in the least developed countries.

2. Globally, the 1990s was the warmest decade on record, providing evidence of global warming.

3. Computer monitors contain an average of 5 to 7 pounds of lead, most of which ends up in landfills.

4. Most Americans rate themselves as having very little knowledge about environmental issues.

5. Most estimates suggest that at least 1,000 species of life are lost per year.

Answers to "Is It True?": 1 = T; 2 = T; 3 = T; 4 = F; 5 = T

The beginning of a new millennium finds the planet Earth poised between two conflicting trends. A wasteful and invasive consumer society, coupled with continued population growth, is threatening to destroy the resources on which human life is based. At the same time, society is locked in a struggle against time to reverse these trends and introduce sustainable practices that will ensure the welfare of future generations.

UNITED NATIONS ENVIRONMENT PROGRAMME

The hit movie "Erin Brockovich" was based on a true story of a single mother with three children who worked as a file clerk for the law firm of Masry and Vititoe in southern California. Not long after being hired, she found some medical records in a file on a real estate case. After getting permission from her boss, she began an investigation to find out why medical records were in a real estate case file. Through her investigation, Erin discovered that hundreds of people who lived in and around Hinkley, California in the 1960s, '70s, and '80s had been devastated by exposure to chromium 6—a toxic substance that had leaked into the groundwater from the nearby Pacific Gas and Electric Company's compressor station. Erin's discovery resulted in the largest direct action lawsuit of its kind. The Pacific Gas and Electric Company was ordered to make the largest legal settlement in U.S. history, paying about $333 million in damages to more than 600 Hinkley residents. No longer a file clerk, Erin is now director of environmental research at Masry and Vititoe.

The movie "Erin Brockovich," told the story of how hundreds of people suffered from environmental pollution. The popularity of the movie stems not only from the quality of the acting and filmmaking, but also from the movie's subject matter: Americans are concerned about environmental issues. In this chapter, we discuss population growth and environmental problems. As Hunter (2001) explains:

> Global population size is inherently connected to land, air, and water environments because each and every individual uses environmental resources and contributes to environmental pollution. While the scale of resource use and the level of wastes produced vary across individuals and across cultural contexts, the fact remains that land, water, and air are necessary for human survival. (p. 12)

After discussing population growth in the world and in the United States, we view population and environmental problems through the lens of structural-functionalism, conflict theory, and symbolic interactionism. We also explore how population growth contributes to a variety of social problems and examine strategies for limiting population growth. The second half of the chapter focuses on environmental problems, examining their social causes and exploring strategies that attempt to reduce or alleviate environmental problems.

The Global Context: A World View of Population Growth

For thousands of years, the world's population grew at a relatively slow rate. During 99 percent of human history, the size of hunting and gathering societies was restricted by disease and limited food supplies. Around 8000 BC, the development of agriculture and the domestication of animals led to increased food supplies and population growth, but even then harsh living conditions and disease still put limits on the rate of growth. This pattern continued until the mid- eighteenth century when the Industrial Revolution improved the standard of living for much of the world's population. The improvements included better food, cleaner drinking water, and improved housing, as well as advances in medical technology such as antibiotics and vaccinations against infectious diseases; all contributed to rapid increases in population (see Figure 14.1).

World Population Growth

Population may be the key to all the issues that will shape the future: economic growth; environmental security; and the health and well-being of countries, communities, and families.

NAFIS SADIK
Executive Director,
UN Population Fund

In the year A.D. 1, the world's population was about 250 million. It took until 1830 for the world population to reach one billion. Since then the population has increased exponentially. The **doubling time**—or time it takes for a population to double in size from any base year—decreases as the population grows. Although the population in A.D. 1 took 1,650 years to double, the second doubling took only 200 years, and the third 80 years. In 2000, the doubling time for the world's population was 51 years (Population Reference Bureau 2000a). When world population reached six billion on October 12 1999, "the five billionth baby had not even reached adolescence yet, having been born in 1986" (Population Institute 2000, 8). It took less time—only 12 years—to add this last billion to the world's population than any previous billion, despite an annual population growth rate of only 1.3 percent, which is the lowest growth rate in a half-century (Halweil 2000). Figure 14.2 illustrates world population growth from 1950 projected until 2050.

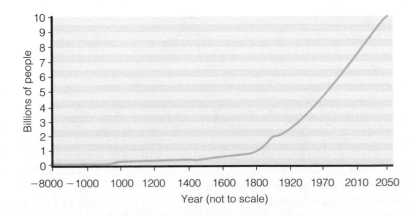

■ **Figure 14.1** *The World's Population is Exploding in Size*

Source: John R. Weeks. 2001. *Population: An Introduction to Concepts and Issues*, 7ᵗʰ ed., p. 10. Belmont, CA: Wadsworth Publishing Co

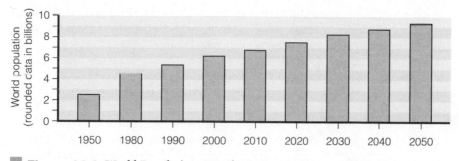

Figure 14.2 *World Population Growth 1950 to 2050*

SOURCES: Brown, Lester R., Gary Gardner, and Brian Halweil. 1998. *Beyond Malthus: Sixteen Dimensions of the Population Problem.* World Watch Paper 143. Washington, D.C.: World Watch Institute; *Statistical Abstract of the United States: 1998,* 118th ed. U.S. Bureau of the Census. Washington, D.C.: U.S Government Printing Office.

As much as 97 percent of annual world population growth is occurring in poor developing countries that are already overburdened with economic, environmental, and public health problems (Population Institute 1999). Consider that Africa's share of global population is expected to rise to 20 percent by 2050, as compared to only 9 percent in 1960 (Hunter 2001). Populations of 65 countries are expected to double within 30 years (Population Institute 2000). As shown in Table 14.1, the population doubling time in less developed countries is much shorter than the population doubling time in more developed countries.

Many Americans lack knowledge about world population size. A national survey of U.S. adults found that only 14 percent correctly identified the world's population as being in the 5 billion to 6 billion range (DaVanzo, Adamson, Belden, & Patterson 2000). The largest percentage of people (38 percent) said they did not know the world population size, a quarter underestimated it, and a quarter overestimated it.

Falling Fertility Rates and the "Birth Dearth"

While many countries, especially in the developing world, are experiencing continued population growth, other countries are declining in population. According to the United Nations Population Division (2001), 39 countries, including Japan, Germany, Italy, Hungary, and Russia, are expected to decrease in population size over the next 50 years. In these countries the fertility rate—the average number of children born to each woman—has fallen below 2.1, the replacement level required to maintain the population. Declining birthrates in some countries are thought to result from high unemployment and a high cost of living. Spain has the lowest birthrate in the world: each woman has an aver-

Table 14.1 Population Doubling Time*

Area	Years to doubling
World	51 years
More developed countries	809 years
Less developed countries	42 years
Less developed countries (excluding China)	36 years

* Based on rates of population growth in 2000
Source: Population Reference Bureau. 2000. "2000 World Population Data Sheet."

age of 1.15 children in her lifetime. Falling population levels are also a concern in France, Germany, Greece, Italy, Russia, and Japan. According to Steve Mosher, president of the Population Research Institute, "humanity's long-term problem is not too many children being born but too few...Over time, the demographic collapse will extinguish entire cultures" (Cooper 1998, 612).

United States Population Growth

During the colonial era (c. 1650), the U.S. population included about 50,000 colonists and 750,000 Native Americans. By 1859, disease and warfare had reduced the American Indian population to 250,000, but the European population had increased to 23 million. The U.S. population continued to increase into the 1900s. From the mid-1940s to the late 1960s the U.S. birthrate increased significantly. This period of high birthrates, commonly referred to as the baby boom, peaked in 1957 when the crude birthrate (number of live births per 1,000 people) reached 25.3. Although the birthrate dropped to 15.3 in 1986, the U.S. population continues to increase. U.S. total population was more than 280,000,000 in 2000 and is expected to reach more than 400,000,000 by 2050 (Perry & Mackun 2001; *Statistical Abstract* 2000). The population growth of 32.7 million people between 1990 and 2000 represents the largest census-to-census increase in U.S. history (Perry & Mackun 2001).

Sociological Theories of Population and Environmental Problems

The three main sociological perspectives—structural-functionalism, conflict theory, and symbolic interactionism—may be applied to the study of population and environmental problems.

Structural-Functionalist Perspective

> The challenge facing us today is to realize that we are a part of the entire earth community. We already understand that we are individuals—we've got that down pat here in America—but we're also part of the whole.
>
> BRIAN SWIMME
> *California Institute of Integral Studies*

Structural-functionalism emphasizes the interdependence between human beings and the natural environment. From this perspective, human actions, social patterns, and cultural values affect the environment and in turn, the environment affects social life. For example, population growth affects the environment, as more people utilize natural resources and contribute to pollution. However, the environmental impact of population growth varies tremendously according to a society's patterns of economic production and consumption (Hunter 2001).

Structural-functionalism focuses on how changes in one aspect of the social system affect other aspects of society. For example, the **demographic transition theory** of population describes how industrialization has affected population growth. According to this theory, in traditional agricultural societies, high fertility rates are necessary to offset high mortality and to ensure continued survival of the population. As a society becomes industrialized and urbanized, improved sanitation, health, and education lead to a decline in mortality. The increased survival rate of infants and children, along with the declining economic value of children, leads to a decline in fertility rates.

Other changes in social structure and culture that affect population include the shift in values toward individualism and self-fulfillment. The availability and cultural acceptability of postnatal forms of family size limitation also affect fertility rates (Mason 1997). In some countries, traditional values permit parents to control family size postnatally by "returning" children at birth (i.e., killing them), selling them to families in need of a child, sending them into bonded labor or prostitution, or marrying them off in early childhood.

The structural-functional perspective is concerned with latent functions—consequences of social actions that are unintended, and not widely recognized. For example, the more than 840,000 dams worldwide provide water to irrigate farmlands and currently supply 19 percent of the world's electricity ("A Prescription for Reducing the Damage Caused by Dams" 2001). Yet dam building has had unintended negative consequences for the environment, including the loss of wetlands and wildlife habitat, the emission of methane (a gas that contributes to global warming) from rotting vegetation trapped in reservoirs, and the altering of river flows downstream killing plant and animal life. Dams have also displaced millions of people from their homes. As philosopher Kathleen Moore points out, "sometimes in maximizing the benefits in one place, you create a greater harm somewhere else...While it might sometimes seem that small acts of cruelty or destruction are justified because they create a greater good, we need to be aware of the hidden systematic costs" (Jensen 2001, 11). Being mindful of latent functions means paying attention to the unintended and often hidden environmental consequences of human activities.

Conflict Perspective

The conflict perspective focuses on how wealth, power, or the lack thereof affect population and environmental problems. In 1798, Thomas Malthus predicted that population would grow faster than the food supply and that masses of people were destined to be poor and hungry. According to Malthusian theory, food shortages would lead to war, disease, and starvation that would eventually slow population growth. However, conflict theorists argue that food shortages result primarily from inequitable distribution of power and resources (Livernash & Rodenburg 1998).

Conflict theorists also note that population growth results from pervasive poverty and the subordinate position of women in many less developed countries. Poor countries have high infant and child mortality rates. Hence, women in many poor countries feel compelled to have many children to increase the chances that some will survive into adulthood. The subordinate position of women prevents many women from limiting their fertility. For example, in 14 countries around the world, a woman must get her husband's consent before she can receive any contraceptive services (United Nations Population Fund 1997). Thus, according to conflict theorists, population problems result from continued inequality within and between nations.

The conflict perspective also emphasizes how wealth, power, and the pursuit of profit underlie many environmental problems. Wealth is related to consumption patterns that cause environmental problems. Wealthy nations have higher per capita consumption of petroleum, wood, metals, cement, and other commodities that deplete the earth's resources, emit pollutants, and generate large volumes of waste. Americans, for example, make up 4.5 percent of the world's population, but consume 20 percent of the world's resources (Population Institute 2000). The capitalistic pursuit of profit encourages making money from industry regardless of the damage done to the environment. Further, to maximize sales, manufacturers design products intended to become obsolete. As

> Do not live with a vocation that is harmful to humans and nature. Do not invest in companies that deprive others of their chance to live.
>
> THICH NHAT HANH
> *Buddhist monk and teacher*

a result of this planned obsolescence, consumers continually throw away used products and purchase replacements. Industry profits at the expense of the environment, which must sustain the constant production and absorb ever-increasing amounts of waste. Industries also use their power and wealth to resist environmental policies that would hurt the industries' profits. For example, to fight the attack on fossil-fuel use (a major polluter and contributor to global warming), petroleum, automobile, coal, and other industries that profit from fossil-fuel consumption created and fund the Global Climate Coalition, an industry-front group that proclaims global warming a myth and characterizes hard evidence of global climate change as "junk science" (Jensen 1999). Industries also take their opposition to environmental policies to the courts. In 1997, the Environmental Protection Agency (EPA) set air quality rules limiting pollution levels of smog and fine, sooty particles. A coalition of industry groups including the American Trucking Associations and the United States Chamber of Commerce legally challenged the EPA's authority to make these rules. In 2001, the Supreme Court ruled unanimously in favor of the Environmental Protection Agency, in what has been called one of the court's most important environmental rulings in years (Greenhouse 2001). The court upheld the right of the EPA to set national air quality standards based on public health and safety considerations, rather than on consideration of cost to polluting industries.

A final example of how wealthy and powerful industries pursue profit at the expense of environmental and public health is the "cancer establishment," which includes organizations such as the American Cancer Society, the National Cancer Institute, and corporations that (1) profit from medical and pharmaceutical products and services designed to detect and treat cancer and/or (2) have a vested interest in keeping public attention on detecting and treating cancer, rather than on preventing it. Not surprisingly, these industries play major roles in providing funding for the American Cancer Society and the National Cancer Institute. Rather than focus on preventing cancer by examining the role that environmental pollutants and hazardous substances play in cancer, industries—who are major sources of pollutants and hazardous substances—focus attention on detection and treatment. Barbara Brenner, director of Breast Cancer Action (a grassroots organization in San Francisco) criticizes the cancer establishment's promotion of National Breast Cancer Awareness Month:

> Because of Breast Cancer Awareness Month, everybody thinks wearing a pink ribbon and racing for the cure will solve the problem. The soft sell approach, combined with the exclusive focus on early detection, keeps people from understanding "what is really going on.". . . While women who detect breast cancer early are better off than those who detect it late, early detection is certainly no guarantee. On the other hand, the focus on detection comes at the expense of any critical questioning about [cancer] causation . . . (quoted in Mokhiber & Weissman 2000, 11)

Symbolic Interactionist Perspective

The symbolic interaction perspective focuses on how meanings, labels, and definitions learned through interaction affect population and environmental problems. For example, many societies are characterized by **pronatalism**—a cultural value that promotes having children. Throughout history, many religions have worshiped fertility and recognized it as being necessary for the continuation of the human race. In many countries, religions prohibit or discourage birth control, contraceptives, and abortion. Women in pronatalistic societies learn through interaction with others that deliberate control of fertility is defined as deviant and socially unacceptable. Once some women learn new definitions of fertility control, they become role models and influence the attitudes and behaviors of others in their personal networks (Bongaarts & Watkins 1996).

Meanings, labels, and definitions learned through interaction and the media also affect environmental problems. Whether or not an individual recycles, carpools, or joins an environmental activist group is influenced by the meanings and definitions of these behaviors that the individual learns through interaction with others.

Some businesses and industries have been criticized for attempting to increase profits and improve their public image through a strategy called **greenwashing**. The term greenwashing refers to the way in which environmentally damaging companies portray their corporate image and products as being "environmentally friendly" or socially responsible. As described by Peter Dykstra of Greenpeace, "greenwashing corporations depict 5 percent of environmental virtue to mask 95 percent of environmental vice" (Hager & Burton 2000). Switzer (1997) explains that greenwashing is commonly used by public relations firms that specialize in damage control for clients whose reputations and profits have been hurt by poor environmental practices. Philip Morris, the infamous cigarette and food producer donated $60 million to charity in 1999, but spent another $108 million in advertising to tell the world about their generosity ("Corporate Spotlight" 2001). DuPont, the biggest private generator of toxic waste in the United States, attempted to project a "green" image by producing a TV ad showing seals clapping, whales and dolphins jumping, and flamingos flying. In an Earth Day event on the National Mall in Washington, D.C., the National Association of Manufacturers (NAM) highlighted renewable and energy-efficient technologies. Yet members of NAM have spent millions of dollars lobbying against use of these very same technologies (Karliner 1998). A logging company facing opposition from environmentalists in New Zealand described their activities as "sustainable harvesting of indigenous production forests"—a

Symbolic interactionists focus on how meanings influence our behavior. For more than two years, Julia Butterfly Hill, 24, lived 18 stories high in a redwood tree near Elinor, California, to protest the cutting of old-growth redwoods. What do you think the meaning of trees is for Julia?

© Jeff Greenberg/ PhotoEdit

phrase that sounds more environmentally friendly than "logging of old growth forests" (Hager & Burton 2000).

Although greenwashing involves manipulation of public perception to maximize profits, many corporations make genuine and legitimate efforts to improve their operations, packaging, or overall sense of corporate responsibility toward the environment (Switzer 1997). For example, in 1990, McDonald's announced it was phasing out foam packaging and switching to a new, paper- based packaging that is partially degradable. Later in this chapter we discuss ways in which industries are participating in alleviating environmental problems.

Social Problems Related to Population Growth

Some of the most urgent social problems today are related to population growth. They include poor maternal and infant health, shortages of food and water, environmental degradation, overcrowded cities, and conflict within and between countries. When asked to rate how serious the problem of rapid population growth is on a scale of 1 to 10 (10 = very serious), only 20 percent of respondents rated rapid population growth as very serious (a rating of 10); the average rating was 6.5 (DaVanzo et al. 2000).

Poor Maternal and Infant Health

As noted in Chapter 2, maternal deaths (deaths related to pregnancy and childbirth) are the leading cause of mortality for reproductive-age women in the developing world. Having several children at short intervals increases the chances of premature birth, infectious disease, and death for the mother or the baby. Childbearing at young ages has been associated with anemia and hemorrhage, obstructed and prolonged labor, infection, and higher rates of infant mortality (Zabin & Kiragu 1998). In developing countries, one in four children are born unwanted, increasing the risk of neglect and abuse. In addition, the more children a woman has, the fewer parental resources (parental income and time, and maternal nutrition) and social resources (health care and education) are available to each child (Catley-Carlson & Outlaw 1998). The adverse health effects of high fertility on women and children are, in themselves, compelling reasons for providing women with family planning services. "Reproductive health and choice are often the key to a woman's ability to stay alive, to protect the health of her children and to provide for herself and her family" (Catley-Carlson & Outlaw 1998, 241).

Increased Global Food Requirements

As populations expand, more food must be grown to feed them. This presents a challenge, especially for countries that have not been able to meet the food needs of their current populations. In 1950, 500 million people (20 percent of the world's population) were considered malnourished; in the late 1990s, more than 3 billion people (one-half of the world's population) suffered from malnutrition (Pimentel et al. 1998).

Countries with large populations, few resources, and limited land are particularly vulnerable to food shortages. An estimated 420 million people live today in countries that have less than 0.7 hectare of cultivated land per person—less

The greatest challenge of the coming century is the maintenance of growth in global food production to match or exceed the projected doubling (at least) of the human population.

PAUL EHRLICH, ANNE H. EHRLICH , AND GRETCHEN C. DAILY
Demographers and biologist

than one-seventh the size of an American football field—the minimum parcel capable of supplying a vegetarian diet for one person without costly chemicals and fertilizers (Engelman, Cincotta, Dye, Gardner-Outlaw, & Wisnewski 2000).

As global food requirements increase with population growth, so do demands on the environment. Agricultural activities contribute to the destruction of forests and the species that inhabit them; nearly one-third of temperate, tropical, and subtropical forests have been converted to agriculture (Wood, Sebastian, & Scherr 2000). Expanding agricultural activities also depletes water supplies, as agriculture consumes 70 percent of the fresh water used by humans (primarily for irrigation). Pesticides and fertilizers used in agriculture contaminate soil and water. Although genetically modified crops are extolled as providing solutions to some of the environmental problems associated with feeding an expanding population, these crops pose potential environmental problems as well (see *Focus on Technology* feature in Chapter 10).

Water Shortages and Depletion of Other Natural Resources

Global water use has tripled since 1950. Americans tend to take fresh water availability for granted, as they turn on a faucet to water lawns and clean their cars, take long showers, and leave the water running as they brush their teeth. In the United States, daily per capita water consumption is currently about 185 gallons for domestic use (drinking, cooking, and washing) (Hunter 2001). But the authors of *State of the World* 1998 suggest that "one of the most underrated issues facing the world as it enters the third millennium is spreading water scarcity" (Brown, Flavin, & French 1998, 5). About 40 percent of the world population faces water shortages at some time during the year (Zwingle 1998) and 33 countries are expected to have chronic water shortages by 2025 ("Water Wars Forecast if Solutions Not Found" 1999). The number of people living in countries facing severe or chronic water shortages is projected to increase more than four-fold over the next 25 years, from an estimated 505 million people in 2000 to between 2.4 and 3.2 billion people by 2025 (Engelman et al. 2000).

Water shortages are exacerbating international conflict. Jordan, Israel, and Syria compete for the waters of the Jordan River basin. Jordan's King Hussein declared that water was the only issue that could lead him to declare war on Israel (Mitchell 1998). Speaking of the water shortage, General Federico Mayor, director of UNESCO, warned, "As it becomes increasingly rare, it becomes coveted, capable of unleashing conflicts. More than petrol or land, it is over water that the most bitter conflicts of the near future may be fought" ("Water Wars Forecast if Solutions Not Found" 1999).

Population growth also contributes to the depletion of other natural resources such as forests, oil, gas, coal, and certain minerals. Later in this chapter, we discuss environmental problems associated with the use of these natural resources.

Urban Crowding and High Population Density

Population growth contributes to urban crowding and high population density. India has one-third the land area of the United States, but more than three times the population. Imagine tripling the U.S. population. Then imagine that this tripled population all lived in the eastern third of the United States. That will give you an idea of how crowded countries like India are.

Without economic and material resources to provide for basic living needs, urban populations in developing countries often live in severe poverty. Urban

One defining characteristic of a civilized society is a sense of responsibility to the next generation. If we do not assume that responsibility, environmental deterioration leading to economic decline and social disintegration could threaten the survival of civilization as we know it.

LESTER R. BROWN
AND JENNIFER MITCHELL
World Watch Institute

poverty in turn produces environmental problems such as unsanitary disposal of waste. In Nigerian urban ghettos, for example, "mounds of refuse (including human wastes) that litter everywhere—gutters, schools, roads, market places and town squares—have been accepted as part of the way of life" (Nzeako, quoted in Agbese 1995). The World Health Organization estimates that half the people in the world do not have access to a decent toilet. Unsanitary disposal of human waste contaminates water supplies. Half the people in the developing world suffer from diseases caused by poor sanitation. Diarrhea caused by many of these diseases is the leading killer of children today (Gardner 1998).

Densely populated urban areas facilitate the spread of disease among people. Infectious diseases cause more than one-third of all deaths worldwide. Crowded conditions in urban areas provide the ideal environment for the culture and spread of diseases such as cholera and tuberculosis (Pimentel et al. 1998).

Strategies for Action: Slowing Population Growth

In some countries with below-replacement-level birthrates, government policies have attempted to encourage rather than discourage childbearing. For example, Italy, Germany, and France have implemented generous child subsidies, in the form of tax credits for every child born, extended maternal leave with full pay, guaranteed employment upon returning to work, and free child care (Cooper 1998). In Japan, a country with one of the world's lowest birth rates, (1.4 children per woman), a Japanese toy company is offering its employees $10,000 for every baby born after the second child ("Japanese Toy Firm Offers Employees Fertility Incentives" 2000).

Most strategies related to population involve attempts to slow population growth by reducing fertility levels. Strategies for slowing population growth include providing access to birth control methods, improving the status of women, increasing economic development and improving health status, and imposing governmental regulations and policies. However, even if every country in the world achieved replacement-level fertility rates (an average of 2.1 births per woman), populations would continue to grow for several decades because of **population momentum**—continued population growth as a result of past high fertility rates which have resulted in large numbers of young women who are currently entering their childbearing years.

Provide Access to Family Planning Services

An estimated 120 million women worldwide do not have access to safe family planning services because such services do not exist or cultural and religious barriers prevent their use (Halweil 2000). Another 350 million women—nearly a third of all women of reproductive age in developing countries—have incomplete or sporadic access to safe family planning services (Halweil 2000). In some countries, methods of birth control are provided only to married women. Family planning personnel often refuse, or are forbidden by law or policy, to make referrals for contraceptive and abortion services for unmarried women. Finally, many women throughout the world do not have access to legal, safe abortion. Without access to contraceptives, many women who experience unwanted

The question that society must answer is this: Shall family limitation be achieved through birth control or abortion? Shall normal, safe, effective contraceptives be employed, or shall we continue to force women to the abnormal, often dangerous surgical operation? Contraceptives or Abortion—which shall it be?

MARGARET SANGER
Birth control advocate

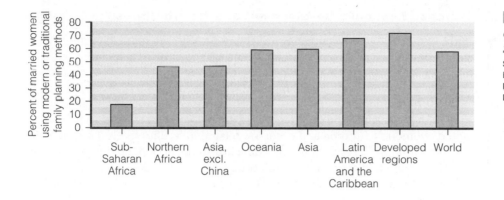

■ **Figure 14.3**
*Contraceptive Use in
Selected Regions*
Source: Population Reference
Bureau. 1999. "1999 World
Population Data Sheet." Washington
D.C.: Population Reference Bureau.

pregnancy resort to abortion—even under illegal and unsafe conditions (see also Chapter 2).

Significant gains have been made in increasing contraceptive use in less developed countries. Today, more than half of married women in less developed countries use some form of contraceptive, compared with 10 percent in 1960 (Population Reference Bureau 2000b). But use of contraceptives is still very low in some countries, particularly those in sub-Saharan Africa (see Figure 14.3). In Yemen, only 10 percent of married women use modern contraceptive methods, and only 8 percent do so in Uganda (Hunter 2001).

Cuts in U.S. assistance to international family planning programs are largely the result of opposition to abortion practices in some countries. One of George W. Bush's first official acts as president was to impose a global "gag rule," which prohibits health-care providers who receive federal funds for family planning assistance from providing abortions or referrals (even when they do so with their own non-federal funds). The gag order also bars recipients of federal family planning funds from engaging in political speech on abortion laws ("News & Views" 2001). However, a bipartisan coalition of family planning supporters in Congress introduced a bill (The Global Democracy Promotion Act of 2000) that would repeal President Bush's gag rule on family planning providers in developing countries.

Improve the Status of Women

Throughout the developing world, the status of women is primarily restricted to that of wife and mother. Women in developing countries traditionally have not been encouraged to seek education or employment, but rather to marry early and have children. In some countries, a woman must obtain the consent of her husband before she can receive contraceptive services. In a study in Zambia, one man interviewed said: "I cannot allow my wife to become a whore. Women who use contraceptives cannot be trusted" (quoted in Population Reference Bureau 2000b, 3).

Improving the status of women is vital to curbing population growth. Education plays a key role in improving the status of women and reducing fertility rates. Educated women are more likely to delay their first pregnancies, to use safe and effective contraception, and to limit and space their children (Population Reference Bureau 2000c; Population Institute 2000). Providing employment opportunities for women is also important to slow population growth; high levels of female labor force participation and higher wages for women are associated with smaller family size (Population Reference Bureau 2000c).

■ If we are to make genuine progress in economic and social development, if we are to make progress in achieving population goals, women increasingly must have greater freedom of choice in determining their roles in society.

Jay Rockefeller
U.S. Senator

Increase Economic Development and Health Status

The greatest single obstacle to the economic and social advancement of the majority of the peoples of the underdeveloped world is rapid population growth.

ROBERT MCNAMARA
Former Secretary of Defense

Although fertility reduction may be achieved without industrialization, economic development may play an important role in slowing population growth. Families in poor countries often rely on having many children to provide enough labor and income to support the family. Economic development decreases the economic value of children. Economic development is also associated with more education for women and greater gender equality. As previously noted, women's education and status are related to fertility levels. Economic development tends to result in improved health status of populations. Reductions in infant and child mortality are important for fertility decline, as couples no longer need to have many pregnancies to ensure that some children survive into adulthood. Finally, the more developed a country is, the more likely women will be exposed to meanings and values that promote fertility control through interaction in educational settings and through media and information technologies (Bongaarts & Watkins 1996).

Impose Government Regulations and Policies

Some countries have imposed strict governmental regulations on reproduction. In the 1970s, India established mass sterilization camps and made the salaries of public servants contingent on their recruiting a quota of citizens who would accept sterilization. The Indian state of Maharashtra enacted compulsory sterilization legislation and in a 6-month period sterilized millions of people, many of them against their will (Boland, Rao, & Zeidenstein 1994). This policy has since been rescinded.

Governmental population regulations in China are less extreme, but still controversial. In 1979, China developed a one-child family policy whereby parents get special privileges for limiting their family size. Specifically, parents who pledge to have only one child receive an allowance for health care for 5 years that effectively increases their annual income by 10 percent. Parents who have only one child also have priority access to hospitals and receive a pension when they reach old age. In addition, their only child is given priority enrollment in nursery school and is exempted from school tuition. However, China has begun to phase out its one-child-per-family policy, largely because Chinese officials conclude that there will not be enough adult children to care for aging parents unless the one-child policy is ended ("China Phasing Out One-Child Policy" 2000). To begin the phase-out, the Chinese government now offers exemptions to adults who have no siblings. When two of these adults with no siblings marry, they are allowed to have two children and prevent a second generation in which one couple must be responsible for four elderly parents with no siblings to help.

International Assistance for Slowing Population Growth

Industrialized countries provided $2.1 billion in international assistance for reproductive health and population programs in 2000—far less than the $5.7 billion suggested by The International Conference on Population and Development (United Nations Population Fund 2000). The majority of U.S. adults favor U.S. economic aid for family planning programs in developing countries, but only half favor U.S. economic aid to provide voluntary, safe abortion as part of reproductive health care in developing countries that request it (DaVanzo et al.

2000). Two days after his inauguration, President George W. Bush imposed a ban known as the "global gag rule" that denies overseas family planning family organizations U.S. funding if they perform abortion, counsel clients regarding abortion, or lobby foreign governments on abortion policy ("News & Views" 2001). Senator Barbara Boxer (D-California) says the global gag rule will mean "organizations that are reducing unsafe abortions and providing contraceptives will be forced either to limit their services or to simply close their doors to women across the world" ("Senators Seek Gag Rule Overturn" 2001, 1). As this book goes to press, a bipartisan group of U.S. Senators is leading a move to overturn the global gag rule.

Environmental Problems

An expanding population is one of many factors that contribute to environmental problems such as land, water, and air pollution and depletion of natural resources. Before reading further, you may want to take the Environmental Knowledge Quiz in this chapter's *Self and Society* feature.

Air Pollution

Transportation vehicles, fuel combustion, industrial processes (such as the burning of coal and wood), and solid waste disposal have contributed to the growing levels of air pollutants, including carbon monoxide, sulfur dioxide, nitrogen dioxides, and lead. Air pollution levels are highest in areas with both heavy industry and traffic congestion, such as Los Angeles, New Delhi, Jakarta, Bangkok, Tehran, Beijing, and Mexico City. In the mid-1990s, breathing the air in Mexico City was like smoking two packs of cigarettes a day (Weiner 2001).

Indoor Air Pollution   When we hear the phrase "air pollution," we typically think of smokestacks and vehicle exhausts pouring gray streams of chemical matter into the air. But much air pollution is invisible to the eye and exists

Many women and children in poor countries are exposed to harmful indoor pollution every day from cooking and heating.

Environmental Knowledge Survey

DIRECTIONS: After answering each of the following items, check your answers for items 2 through 11 using the answer key provided. For items 2 through 11, calculate the total number of items you answered correctly. Compare your score to the results of a national survey.

1. In general, how much do you feel you yourself know about environmental issues and problems—would you say you know a lot, a fair amount, only a little, or practically nothing?
   a. A lot
   b. A fair amount
   c. Only a little
   d. Practically nothing
   e. Don't know

2. How is most of the electricity in the U.S. generated?
   a. By burning oil, coal, and wood
   b. With nuclear power
   c. Through solar energy
   d. At hydroelectric power plants
   e. Don't know

3. What is the most common cause of pollution of streams, rivers, and oceans?
   a. Dumping of garbage by cities
   b. Surface water running off yards, city streets, paved lots, and farm fields
   c. Trash washed into the ocean from beaches
   d. Waste dumped by factories
   e. Don't know

4. What do you think is the main cause of global climate change, that is, the warming of the planet Earth?
   a. A recent increase in oxygen in the atmosphere
   b. Sunlight radiating more strongly through a hole in the upper ozone layer
   c. More carbon emissions from autos, homes, and industry
   d. Increased activity from volcanoes worldwide
   e. Don't know

5. To the best of your knowledge, what percentage of the world's water is fresh and available for use?
   a. 1%
   b. 5%
   c. 10%
   d. 33%
   e. Don't know

6. The current worldwide reduction in the number of ocean fish is PRIMARILY the result of which of the following?
   a. Pollution in coastal waters worldwide
   b. Increased harvesting by fishing vessels
   c. Changes in ocean temperature
   d. Loss of fishing shoals and other deep sea habitats
   e. Don't know

7. What is the leading cause of childhood death worldwide?
   a. Malnutrition and starvation
   b. Asthma from dust in the air
   c. Auto and home accidents
   d. Germs in the water
   e. Don't know

8. What is the most common reason that an animal species becomes extinct?
   a. Pesticides are killing them.
   b. Their habitats are being destroyed by humans.
   c. There is too much hunting.
   d. Climate changes are affecting them.
   e. Don't know.

where we least expect it—in our homes, schools, workplaces, and public buildings. Common household, personal, and commercial products contribute to indoor pollution. Some of the most common indoor pollutants include carpeting (which emits nearly 100 different chemical gases), mattresses (which may emit formaldehyde and aldehydes), drain cleaners, oven cleaners, spot removers,

9. Thousands of waste disposal areas—dumps and landfills—in the United States contain toxic waste. The greatest threat posed by these waste disposal areas is:
   a. Chemical air pollution
   b. Contact with farm animals and household pets
   c. Contamination of water supplies
   d. Human consumption through contaminated food
   e. Don't know

10. Many communities are concerned about running out of room in their community trash dumps and landfills. What is the greatest source of landfill material?
    a. Disposable diapers
    b. Lawn and garden clippings, trimmings, and leaves
    c. Paper products including newspapers, cardboard, and packaging
    d. Glass and plastic bottles and aluminum and steel cans
    e. Don't know

11. Some scientists have expressed concern that chemicals and certain minerals accumulate in the human body at dangerous levels. These chemicals and minerals enter the body primarily through
    a. Breathing air
    b. Living near toxic waste dumps
    c. Household cleaning products
    d. Drinking water
    e. Don't know

ANSWER KEY FOR ITEMS 2 THROUGH 11: 2a, 3b, 4c, 5a, 6b, 7d, 8b, 9c, 10c, 11d

COMPARISON DATA FROM A NATIONAL SAMPLE

Item	# Percent of National Sample Who Answered Correctly
2	70%
3	52
4	45
5	31
6	28
7	25
8	24
9	23
10	13
11	7

AVERAGE NUMBER OF CORRECT RESPONSES: 3.2

Source: Based on *1999 NEETF/Roper Report Card.* 1999. The National Environmental Education & Training Foundation & Roper Starch Worldwide. Used by permission.

shoe polish, dry-cleaned clothes, paints, varnishes, furniture polish, potpourri, mothballs, fabric softener, and caulking compounds. Air fresheners, deodorizers, and disinfectants emit the pesticide paradichlorobenzene. Potentially harmful organic solvents are present in numerous office supplies, including glue, correction fluid, printing ink, carbonless paper, and felt-tip markers.

In poor countries fumes from cooking and heating are major contributors to indoor air pollution. About half of the world's population and up to 90 percent of rural households in developing countries rely on wood, dung, and crop residues for fuel (Bruce, Perez-Padilla, & Albalak 2000). These fuels are typically burnt indoors in open fires or poorly functioning stoves, producing hazardous emissions such as soot particles, carbon monoxide, nitrous oxides, sulphur oxides, and formaldehyde. An estimated 2 million deaths in developing countries each year result from exposure to indoor air pollution (Bruce et al. 2000).

Destruction of the Ozone Layer   The use of human-made chlorofluorocarbons (CFCs), which are used in refrigerators, cleaning computer chips, hospital sterilization, solvents, dry cleaning, and aerosols, has damaged the ozone layer of the earth's atmosphere. The depletion of the ozone layer allows hazardous levels of ultraviolet rays to reach the earth's surface. Ultraviolet light has been linked to increases in skin cancer and cataracts, declining food crops, rising sea levels, and global warming. The largest ozone hole is over Antarctica and spans 11.5 million square miles—more than three times the size of the United States (Stiefel 2000). Previously, this ozone hole exposed only ocean and barren land in Antarctica. But now it exposes Puntas Arenas, Chile, a populous city near the southern tip of South America.

Acid Rain   Air pollutants, such as sulfur dioxide and nitrogen oxide, mix with precipitation to form **acid rain**. Polluted rain, snow, and fog contaminate crops, forests, lakes, and rivers. Due to the effects of acid rain, 140 Minnesota lakes are totally devoid of fish (Miller 2000). Because pollutants in the air are carried by winds, industrial pollution in the midwest falls back to Earth as acid rain on southeast Canada and the northeast New England states. Acid rain is not just a problem in North America; it decimates plant and animal species around the globe.

Global Warming

A trend of increasing air temperature provides evidence of global warming. Globally, the 1990s was the warmest decade on record. Average global air temperature rose by 0.6 degrees C over the 20th century and, between 1990 and 2100, is expected to rise another 1.4 to 5.8 degrees C (Intergovernmental Panel on Climate Change 2001a). Over the 20th century, the average annual U.S. temperature has risen 1 degree F, and is expected to rise by about 5 to 9 degrees F over the next 100 years (National Assessment Synthesis Team 2000).

Causes of Global Warming   The main cause of global warming is the accumulation of various gases that collect in the atmosphere and act like the glass in a greenhouse, holding heat from the sun close to the earth and preventing it from rising back into space. **Greenhouse gases** include carbon dioxide, methane, and nitrous oxide. Since 1750, atmospheric concentrations of carbon dioxide have increased by 31 percent; methane by 151 percent, and nitrous oxide by 17 percent (Intergovernmental Panel on Climate Change 2001a). Three-fourths of human- activity-related emission of carbon dioxide during the last 20 years is the result of fossil fuel burning; the rest is predominately caused by deforestation. Trees and other plant life utilize carbon dioxide and release oxygen into the air. As forests are cut down, fewer trees are available to absorb the carbon dioxide. With less than 5 percent of the world's population, the United States contributes more than 20 percent of all fossil fuel burning emissions (Engelman et al. 2000).

Effects of Global Warming   Numerous effects of global warming have been observed and are anticipated in the future (Intergovernmental Panel on Climate Change 2001b; National Assessment Synthesis Team 2000). As temperature increases, some areas will experience heavier rain, while other regions will get drier. Whereas some regions will experience increased water availability and crop yields, other regions, particularly tropical and subtropical regions, are expected to experience decreased water availability and a reduction of crop yields. Global warming results in shifts of plant and animal habitats and the increased risk of extinction of some species. Regions that experience increased rainfall as a result of increasing temperatures may face increases in waterborne diseases and diseases transmitted by insects. Over time, climate change may produce opposite effects on the same resource. For example, in the short-term, forest productivity is likely to increase in response to higher levels of carbon dioxide in the air, but over the long term, forest productivity is likely to decrease because of drought, fire, insects, and disease.

Global warming also threatens to melt glaciers and permafrost, resulting in a rise in sea level. Global average sea levels rose 0.1 to 0.2 meters during the twentieth century, and are expected to rise by 0.09 to 0.88 meters from 1990 to 2100. As seas levels rise, some island countries, as well as some barrier islands off the U.S. coast are likely to disappear and low-lying coastal areas will become increasingly vulnerable to storm surges and flooding. Although U.S. coastal counties (excluding Alaska) constitute only 11 percent of U.S. land area, they are home to 53 percent of the population (Hunter 2001).

In urban areas, flooding can be a problem where storm drains and waste management systems are inadequate. Increased flooding associated with global warming is expected to result in increases in drownings, and diarrheal and respiratory diseases. An increase in the number of people exposed to insect and water-related diseases, such as malaria and cholera, are also expected. Even if greenhouse gases are stabilized, global air temperature and sea level are expected to continue to rise for hundreds of years. Many Americans are unaware of the seriousness of global warming. In a 2000 Gallup poll, fewer than half (40 percent) of Americans said they worry "a great deal" about global warming, whereas more than half worry "a great deal" about water pollution, air pollution, and soil contamination (Carlson 2001).

Land Pollution

About 30 percent of the world's surface is land, which provides soil to grow the food we eat. Increasingly, humans are polluting the land with nuclear waste, solid waste, and pesticides. In 1999, 1,274 hazardous waste sites were on the National Priority List (also called Superfund sites) (*Statistical Abstract 2000*). States with the most Superfund sites include New Jersey (114 sites), Pennsylvania (98 sites), California (97 sites), New York (86 sites), and Michigan (70 sites).

Nuclear Waste   The global community generates more than a million tons of hazardous waste each day (Brown 1995). Decades of nuclear weapons production has polluted over 2.3 million acres of land in at least 24 states; cleanup will

We have found the sources of hazardous waste and they are us.

EVERYBODY'S PROBLEM: HAZARDOUS WASTE
U.S. EPA booklet, 1980

extend well into the twenty-first century at a cost of over $200 billion (Black 1999). Radioactive waste from nuclear power plants and major weapons production sites is associated with cancer and genetic defects. Radioactive plutonium, used in both nuclear power and weapons production, has a half-life (the time it takes for its radioactivity to be reduced by half) of 24,000 years (Mead 1998). The disposal of nuclear waste, is particularly problematic. The U.S. Nuclear Regulatory Commission licenses nuclear reactors for 40 years. When nuclear reactors reach the end of their 40-year licenses, the radioactive waste must be stored somewhere. About 28,000 metric tons of nuclear waste are already stored around the United States. This figure is expected to grow to 48,000 metric tons by 2003 and 87,000 metric tons by 2030 (not including waste from military nuclear operations) (Stover 1995). The United States has also promised to accept nuclear waste from about 40 nations with nuclear research reactors. The Nuclear Regulatory Commission has recently given a 20-year operating extension to the Calvert Cliffs nuclear power plant in Maryland, and is considering license renewal extensions for six other nuclear plants ("Nukes Rebuked" 2000).

The cost of storing nuclear waste and cleaning other hazardous waste sites is immense. Cleaning up hazardous waste sites in the United States would require an estimated $750 billion, roughly three-fourths of the U.S. federal budget in 1990 (Brown 1995). But not cleaning these sites is also costly. Brown (1995) remarks that "one way or another, society will pay—either in clean up bills or in rising health care costs" (p. 416). Under the Nuclear Waste Policy Act (passed in 1982), beginning in 1998 the Department of Energy has handled the storage of nuclear waste from commercial nuclear power plants. Scientists have proposed several options for disposing of nuclear waste, including burying it in rock formations deep below the earth's surface or under Antarctic ice, injecting it into the ocean floor, or hurling it into outer space. Each of these options is risky and costly. Recognizing the environmental hazards of nuclear power plants and their waste, Germany has become the first country to order all of its 19 nuclear power plants shut down by 2020 ("Nukes Rebuked" 2000). Nevertheless, at the end of 1999, 431 nuclear reactors were operating around the world (Lenssen 2000).

The high-tech electronics industry uses vast amounts of dangerous chemicals and significantly depletes natural resources to fuel its global expansion and rapidly changing product lines. There are few other products for which the sum of the environmental impacts of raw material extraction, industrial refining and production, use and disposal is so extensive.

SILICON VALLEY TOXICS COALITION

Solid Waste   In 1960, each U.S. citizen generated 2.7 pounds of garbage on average every day. By 1998, this figure had increased to 4.5 pounds (*Statistical Abstract 1995; Statistical Abstract 2000*). In the United States, we discarded 220 million tons of solid waste in 1998. Some of this waste is converted into energy by incinerators; more than half is taken to landfills. The availability of landfill space is limited, however. Some states have passed laws that limit the amount of solid waste that can be disposed of; instead, they require that bottles and cans be returned for a deposit or that lawn clippings be used in a community composting program. This chapter's *Focus on Technology* feature examines the hazards of disposing computers in landfills (and other environmental and health hazards of computers).

Pesticides   Pesticides are used worldwide in the growing of crops and gardens; outdoor mosquito control; the care of lawns, parks, and golf courses; and indoor pest control. Pesticides contaminate food, water, and air and can be absorbed through the skin, swallowed, or inhaled. Many common pesticides are considered potential carcinogens and neurotoxins. Recognizing the health risks associated with exposure to the pesticide Dursban, the Environmental Protection Agency issued a ban on the popular pesticide in 2000 (Kaplan & Morris 2000). Even when a pesticide is found to be hazardous and is banned in the

Environmental and Public Health Hazards Associated with Computers

Computers play an important role in improving and protecting the environment in several ways. First, computers are an important part of environmental education, as they provide improved public access to information about environmental issues. Second, through videoconferencing and e-mail, computers help reduce the use of paper. Finally, computers have enabled hundreds of thousands of individuals to participate in online activism, which has become a vital part of the environmental movement. Yet, the production and disposal of computers pose significant hazards to public and environmental health.

The Hazards of Computer Manufacturing, Disposal, and Recycling

Computer equipment is composed of more than 1,000 materials, many of which are highly toxic. Toxic components in computers include lead and cadmium in computer circuit boards, lead oxide and barium in computer monitors' cathode ray tubes, mercury in switches and flat screens, brominated flame retardants on printed circuit boards, cables and plastic casing, polychlorinated biphenyls (PCBs) present in older capacitors and transformers, and polyvinyl chloride (PVC)-coated copper cables and plastic computer casings that release highly toxic dioxins and furans when burned (Silicon Valley Toxics Coalition 2001a).

The average computer now has a lifespan of about 2 years. Hardware and software companies constantly generate new programs that demand more speed, memory and power, and it is usually cheaper and more convenient to buy a new machine than it is to upgrade the old. Experts estimate that by the year 2004, the United States will have over 315 million obsolete computers. Estimates suggest that over three-quarters of all computers ever bought in the United States are currently stored in people's attics, basements, office closets, and pantries (Silicon Valley Toxics Coalition 2001b). Many of these computers are destined for landfills, incinerators, or hazardous waste exports. Only about 14 percent of unwanted computers are recycled or donated.

The toxic materials used in computer manufacturing pose significant hazards for workers in the production of computers. In addition, the toxic **e-waste** (waste from electronic equipment) makes the disposal and recycling of computers a human and environmental hazard.

Hazards of Computer Production. The production of semiconductors, printed circuit boards, disk drives, and monitors uses particularly hazardous chemicals, and workers in chip manufacturing are reporting cancer clusters and birth defects (Silicon Valley Toxics Coalition 2001b).

© A. Ramey/ PhotoEdit

These obsolete computers from a Manhattan office building will eventually end up in a landfill. Highly toxic materials in these computers, such as lead, mercury, and cadmium, leach out of landfills and contaminate the soil and groundwater. In 2000, Massachusetts became the first state to ban dumping of computers in landfills.

Continued

Hazards of Disposing E-waste. Most discarded computers end up in landfills, and when we consider that all landfills leak and eventually allow a certain amount of chemical and metal leaching, the mountains of e-waste destined for landfills are particularly disturbing. In many city landfills, cathode ray tubes from computers and TVs are the largest source of lead (Motavalli 2001). Computer monitors (14 and 15 inch size) contain an average of 5 to 8 pounds of lead (Fisher 2000). Lead can cause damage to the central and peripheral nervous systems, endocrine system, circulatory system, and kidneys. Its serious negative effects on children's brain development has been well documented. Lead accumulates in the environment and has high acute and chronic toxic effects on plants, animals, and microorganisms. The main concern in regard to the presence of lead in landfills is the potential for the lead to leach and contaminate drinking water supplies. Between 1997 and 2004, obsolete computers in the United States will contain up to about 1.2 billion pounds of lead (Silicon Valley Toxics Coalition 2001b). Other toxic materials that leach into the soil and groundwater from landfilled computers include mercury and cadmium.

Some computer waste is disposed of through incineration (burning). Incineration of scrap computers generates extremely toxic dioxins which are released in air emissions. Incineration of computers also results in high concentrations of toxic heavy metals in the flue gas residues, fly ash, and filter cake.

Hazards of Recycling Computers. The presence of toxic polybrominated flame retardants in computers makes recycling dangerous and difficult. High concentrations of these substances have been found in the blood of workers in recycling plants. In addition, a Swedish study found that when computers, fax machines or other electronic equipment are recycled, dust containing toxic flame retardants is spread in the air (Silicon Valley Toxics Coalition 2001b).

The Exportation of Hazardous E-Waste. The majority of the world's hazardous waste is generated by industrialized countries. Exporting this waste to less developed countries has been one way in which the industrialized world has cut costs of waste disposal. An estimated 1 million of the 1.7 million monitors recycled in 1997 were shipped abroad for processing (Fisher 2000). The export of scrap is profitable because the labor costs are cheap and regulations are lax compared to U.S. law. One pilot program that collected electronic scrap in San Jose, California estimated that it was 10 times cheaper to ship computer

monitors to China than it was to recycle them in the United States (Silicon Valley Toxics Coalition 2001b).

In 1989 the world community established the Basel Convention on the Transboundary Movement of Hazardous Waste for Final Disposal to stop industrialized nations from dumping their waste on less developed countries. The United States, however, has declined to sign the Convention.

Over 60 countries have agreed to ban exports of hazardous waste to poor countries. But this ban did not stop the transport of waste that countries claimed was being exported for recycling purposes. China and 77 non-OECD countries supported a ban on the shipping of waste for recycling. As a result, the **Basel Ban** was adopted, promising to end the export of hazardous waste from rich OECD countries to poor non-OECD countries—even when the waste was exported for recycling. The United States has declined to participate in the ban and has lobbied Asian governments to establish bilateral trade agreements to allow continued exporting of hazardous waste after the Basel Ban came into effect on January 1, 1998. The amount of computer scrap exported from the United States is expected to grow as product obsolescence increases (Silicon Valley Toxics Coalition 2001b).

Solutions to the Problem of Toxic E-Waste

U.S. Government Solutions. Effective April 1, 2000, the state of Massachusetts became the first state to ban dumping of video monitors (computer screens and televisions) (Massachusetts Department of Environmental Protection 2001). Washington State is also expected to pass legislation banning video monitors from its landfills (Motavalli 2001). The city of Seattle has contracts with Boeing that require Boeing to take back computer packaging for reuse or recycling and requires on-line manuals and preinstalled programs (Northwest Product Stewardship Council 2000).

The European Union's WEEE Directive. The European Union (comprising 15 European nations) has proposed legislation on Waste from Electrical and Electronic Equipment (the WEEE Directive) that would make producers responsible for taking back their old products, phasing out the use of certain toxic materials in computer manufacturing, and would encourage cleaner product design and less waste generation. This strategy, known as **Extended Producer Responsibility**, is based on the principle that producers bear a degree of responsibility for all the environmental impacts of their products. This includes impacts arising from the choice of materials and from the manufacturing process as well as the use and disposal of products. By making manufacturers financially responsible for waste management of their products, electronics producers will have a financial incentive to design their products with less hazardous and more recyclable materials. Various European countries (the Netherlands, Denmark, Sweden, Austria, Belgium, Italy, Finland, and Germany) have already drafted legislation requiring Extended Producer Responsibility. The WEEE Directive would harmonize all these countries' initiatives to allow industry to operate uniformly throughout Europe. The WEEE Directive calls for the following (Silicon Valley Toxics Coalition 2001b):

- The use of mercury, cadmium, hexavalent chromium, and two classes of brominated flame retardants in electronic and electrical goods must be phased out by the year 2004.
- Producers of electronics must bear full financial responsibility for setting up collection, recycling, and disposal systems.
- 70 percent of collected computers and monitors must be recycled. "Recycling" does not include incineration, so companies won't be able to meet recycling goals by burning the waste.
- For disposal, incineration with energy recovery is allowed for the 10 percent to 30 percent of waste remaining. However, components containing certain toxic substances (such as lead, mercury, and cadmium) must be removed from any equipment that is destined for landfill, incineration, or recovery.
- Member states shall encourage producers to integrate an increasing quantity of recycled material in new products. Originally the EU stipulated that by 2004 new equipment must contain at least 5 percent of recycled plastic content but this provision was recently dropped because of intense industry lobbying.
- Producers must design equipment that includes labels for recyclers that identify plastic types and location of all dangerous substances.
- Producers can undertake the treatment operation in another country, but this should not lead to shipments of e-waste to non-European Union (EU) countries where no or lower treatment standards than in the EU exist. Accordingly, producers shall deliver e-waste only to those establishments that comply with the treatment and recycling requirements set out in the proposal.

The U.S. Trade Representative, the U.S. Mission in Brussels, and U.S. trade associations such as the American Electronics Association (AEA) and the Electronics Industry Alliance (EIA) have expressed strong opposition to the WEEE Directive. The AEA and EIA have taken legal action against the WEEE Directive, asserting that the directive would violate World Trade Organization international trade rules. A coalition of public advocacy groups, organized by the International Campaign for Responsible Technology based in Silicon Valley, California petitioned the European Union not to cave in to U.S. opposition to the WEEE Directive.

Industry Initiatives. A number of industries have taken steps to alleviate the environmental hazards of computers. Some computer manufacturers such as Dell and Gateway lease out their products, thereby ensuring that they get them back to further upgrade and lease out again. In 1998 IBM introduced the first computer that uses 100 percent recycled resin in all major plastic parts. Hewlett-Packard Company has developed a safe cleaning method for chips using carbon dioxide cleaning as a substitute for hazardous solvents. Toshiba is working on a modular upgradable and customizable computer to reduce product obsolescence. Pressures to eliminate halogenated flame retardants and design products for recycling have led to the use of metal shielding in computer housings and a range of lead-free solders is now available.

As governments and industries continue to develop strategies for reducing environmental hazards associated with computer manufacturing and disposal, consumers can also make a difference by choosing manufacturers who practice "Product Stewardship" by making products that are less toxic, conserve natural resources, and reduce waste. *A Guide to Environmentally Preferable Computer Purchasing* is available from the Northwest Product Stewardship Council at http://www.govlink.org/nwpsc/.

Sources:

Fisher, Jim. 2000 (September 18). "Poison PCs." http://www.salon.com/tech/feature/2000/09/18/toxic_pc/

Massachusetts Department of Environmental Protection. 2001. "TV and Computer Reuse and Recycling." www.state.ma.us/dep/recycle/crt/crthome.htm

Motavalli, Jim. 2001 (March). "Is There an Afterlife For Your Computer? Grappling with America's Techno-Trash Dilemma." *Environmental Defense* XXXII(2): 6.

Northwest Product Stewardship Council (Computer Subcommittee). 2000 (October). *A Guide to Environmentally Preferable Computer Purchasing.* http://www.govlink.org/nwpsc/

Silicon Valley Toxics Coalition. 2001a. "Why Focus on Computers?" http://www.svtc.org/cleancc/focus/htm

Silicon Valley Toxics Coalition. 2001b. *Just Say No to E-Waste: Background Document on Hazards and Waste from Computers.* http://www.svtc.org

United States, other countries may continue to use it. Although DDT is banned in the United States, the pesticide is used in many other countries from which we import food. About one-third of the foods purchased by U.S. consumers have detectable levels of pesticides; from 1 to 3 percent of the foods have pesticide levels above the legal level (Pimentel & Greiner 1997).

Water Pollution

According to the Environmental Protection Agency, 40 percent of all U.S. waters are not fishable or swimmable (Smith, Coequyt, & Wiles 2000). Our water is being polluted by a number of harmful substances, including pesticides, industrial waste, acid rain, and oil spills. In 1998, more than 8,000 oil spills occurred in and around U.S. waters alone, totaling more than one and one-half million gallons of spilled oil (*Statistical Abstract 2000*). However, worldwide, more disease and death are caused by water contaminated from feces, not chemicals (Kemps 1998).

About 1.2 billion people lack access to clean water. In developing nations, as much as 95 percent of untreated sewage is dumped directly into rivers, lakes, and seas that are also used for drinking and bathing (Pimentel et al. 1998). Approximately 35,000 river miles in the United States are polluted by runoff from livestock operations ("New Rules for Feedlots" 1998).

Depletion of Natural Resources

Overpopulation contributes to the depletion of natural resources, such as coal, oil, and forests. The demand for new land, fuel, and raw materials has resulted in **deforestation**—the conversion of forest land to non-forest land (Intergovernmental Panel on Climate Change 2000b). Global forest cover has been reduced by at least 20 percent and as much as 50 percent since preagricultural times (World Resources Institute 2000). The major causes of deforestation are the expansion of agricultural land, commercial logging, and road building.

As we explained earlier, deforestation contributes to global warming. Deforestation also displaces people and wild species from their lands. About half of the world's approximately 14 million species live in tropical forests (World Resources Institute 1998) Finally, soil erosion caused by deforestation may lead to flooding.

Deforestation contributes to **desertification**—the degradation of semiarid land, which results in the expansion of desert land that is unusable for agriculture. Overgrazing by cattle and other herd animals also contributes to desertification. The problem of desertification is most severe in Africa (Reese 2001). As more land turns into desert land, populations can no longer sustain a livelihood on the land, and so they migrate to urban areas or other countries, contributing to social and political instability.

Environmental Illness

Exposure to pollution, toxic substances, and other environmental hazards is associated with numerous illnesses and even death. Worldwide, pesticide poisoning results in about 1 million deaths and chronic illnesses each year (Pimentel et al. 1998). In 1996, the National Environmental Protection Agency of China reported 3 million deaths in cities during the preceding two years from chronic bronchitis resulting from urban air pollution (Brown 1998a). The destruction of the ozone layer increases ultraviolet B (UVB) radiation, and the incidence of skin cancer has quadrupled over the past 20 years, resulting in nearly 10,000 deaths per year (Pimentel et al. 1998). Destruction of the ozone layer also results in increased exposure of the eyes to radiation in sunlight, which increases the risk of cataracts (Bergman 1998).

Current estimates suggest that in the United States, 1 in 2 men and 1 in 3 women will develop cancer at some point in his or her lifetime and most cancer researchers believe that many cancer cases are associated with the environment in which we live and work (U.S. Department of Health and Human Services 2001). The *Ninth Report on Carcinogens* (U.S. Department of Health and Human Services 2001) classifies 218 substances as "known to be human carcinogens" or "reasonably anticipated to be human carcinogens."

Many of the chemicals we are exposed to in our daily lives can cause not only cancer, but other health problems such as infertility and impaired intellectual development in children (McGinn 2000). Persistent organic pollutants (POPs), which accumulate in the food chain and persist in the environment, taking centuries to degrade, contribute to increasing rates of birth defects, fertility problems, greater susceptibility to disease, diminished intelligence, and certain cancers (Fisher 1999). Chemicals in the environment are being investigated as causes of a number of childhood developmental and learning problems (Kaplan & Morris 2000).

Substances found in common household, personal, and commercial products can result in a variety of temporary acute symptoms such as drowsiness, disorientation, headache, dizziness, nausea, fatigue, shortness of breath, cramps, diarrhea, and irritation of the eyes, nose, throat, and lungs. Long-term exposure can affect the nervous system, reproductive system, liver, kidneys, heart, and blood ("The Delicate Balance" 1994). Fragrances, which are found in many consumer products, may produce sensory irritation, pulmonary irritation, decreases in expiratory airflow velocity, and possible neurotoxic effects (Fisher 1998). Fragrance products can cause skin sensitivity, rashes, headache, sneezing, watery eyes, sinus problems, nausea, wheezing, shortness of breath, inability to concentrate, dizziness, sore throat, cough, hyperactivity, fatigue, and drowsiness (DesJardins 1997). More and more businesses are voluntarily limiting fragrances in the workplace or banning them altogether to accommodate employees who experience ill effects from them.

Sick building syndrome (SBS) is a situation in which occupants of a building experience symptoms that seem linked to time spent in a building, but no specific illness or cause can be identified (Lindauer 1999). Building occupants complain of symptoms such as headache; eye, nose, and throat irritation; a dry cough; dry or itchy skin; dizziness and nausea; difficulty in concentrating; fatigue; and sensitivity to odors. Although specific causes of SBS remain unknown, contributing factors may include polluted outdoor air that enters a building through poorly located air intake vents, windows, and other openings and indoor air pollution from sources inside the building, such as adhesives, upholstery, carpeting, copy machines, manufactured wood products, cleaning agents, and pesticides.

Sufferers of a controversial condition known as **multiple chemical sensitivity** (MCS) say that after one or more acute or traumatic exposures to a chemical or group of chemicals, they began to experience adverse effects from low levels of chemical exposure that do not produce symptoms in the general public. Some individuals with MCS build houses made from materials that do not contain chemicals that are typically found in building materials. Sufferers of MCS often avoid public places and/or wear a protective breathing filter to avoid inhaling the many chemical substances in the environment.

When scientists tested journalist Bill Moyers' blood as part of a documentary on the chemical industry, they found traces of 84 of the 150 chemicals they had tested (PBS 2001). Sixty years ago, when Moyers was 6 years old, only one of these chemicals—lead—would have been present in his blood. Of the 80,000 chemicals registered with the Environmental Protection Agency for commercial use, only a small fraction have been adequately tested for their health effects on humans. Of the 15,000 chemicals that are produced in major quantities, less than half (43 percent) have been properly tested for their health effects on humans (PBS 2001). With more than 1,000 new chemicals being introduced to the global market each year (McGinn 2000), the need for testing health effects of chemicals in our environment is clear. This chapter's *Social Problems Research Up Close* feature describes the first National Report on Human Exposure to Environmental Chemicals.

Environmental Injustice

Although environmental pollution and degradation and depletion of natural resources affect us all, some groups are more affected than others. **Environmental injustice**, also referred to as **environmental racism**, refers to the tendency for socially and politically marginalized groups to bear the brunt of environmental ills. In the United States, polluting industries, industrial and waste facilities, and transportation arteries (that generate vehicle emissions pollution) are often located in minority communities (Bullard 2000; Bullard & Johnson 1997). U.S. communities that are predominantly African American, Hispanic, or Native American are disproportionately affected by industrial toxins, contaminated air and drinking water, and the location of hazardous waste treatment and storage facilities. One study found, for example, that hog industries—and the associated environmental and health risks associated with hog waste—in eastern North Carolina tend to be located in communities with high black populations, low voter registration, and low incomes (Edwards & Ladd 2000). Another study of 370 communities in Massachusetts found that compared with communities with higher median household incomes and lower percentages of people of color, communities with lower household incomes and higher percentages of people of

This sugar cane factory in central Mexico has been closed twice because of violations of pollution laws, but has reopened again. The factory emits toxic fumes, which leads many local residents to wear gauze filters over their mouths and noses.

Courtesy of Caroline Schacht

National Report on Human Exposure to Environmental Chemicals

The National Report on Human Exposure to Environmental Chemicals (Centers for Disease Control and Prevention 2001) is a landmark publication that provides measures of exposure for 27 environmental chemicals, 24 of which have never been measured before in a nationally representative sample of the U.S. population. Prior research has assessed the U.S. population's exposure to lead, cadmium, and cotinine—a metabolite of nicotine that tracks exposure to environmental tobacco smoke (ETS) among nonsmokers. The current Report provides new data about these chemicals, as well as first-time human exposure data on 24 additional chemicals.

Methods and Sample

The first National Report on Human Exposure to Environmental Chemicals is based on 1999 data from the National Center for Health Statistics National Health and Nutrition Examination Survey (NHANES). The NHANES is a series of surveys designed to collect data on the health and nutritional status of the U.S. population. NHANES collects information about a wide range of topics, from the prevalence of infectious diseases to risk factors for cardiovascular disease. Beginning in 1999, NHANES became a continuous and annual survey.

The 1999 NHANES selected a representative sample of the noninstitutionalized, civilian U.S. population, with an oversampling of African Americans, Mexican Americans, adolescents (aged 12–19 years), older Americans (aged 60 years or older), and pregnant women to produce more reliable estimates for these groups. The 1999 NHANES was conducted in 12 counties across the United States. From these locations, 5,325 people were selected to participate in the survey. Of these, 3,812 (71 percent) participated in the survey. The sample did *not* specifically target people who were believed to have high or unusual exposures to environmental chemicals.

Assessing research participants' exposure to environmental chemicals involved measuring the chemicals or their metabolites (breakdown products) in the participants' blood or urine—a method known as **biomonitoring**. Blood was obtained by venipuncture for participants aged 1 year and older, and urine specimens were collected for people aged 6 years and older.

The 27 chemicals measured in the research participants' blood and urine are grouped into four categories:

1. Metals (antimony, barium, beryllium, cadmium, cesium, cobalt, lead, mercury, molybdenum, platinum, thallium, tungsten, and uranium).
2. Cotinine (tracks human exposure to tobacco and tobacco smoke).
3. Organophosphate pesticide metabolites (track exposure to 28 pesticides).
4. Phthalate metabolites (track chemicals commonly used in such consumer products as soap, shampoo, hair spray, nail polish, and flexible plastics).

The researchers conducting this study had several objectives, including the following:

- To determine whether selected environmental chemicals are getting into the bodies of Americans and the levels of these chemicals in blood and urine.
- For chemicals that have a known toxicity level, to measure the prevalence of people in the U.S. population who have elevated levels of the environmental chemical.
- To assess the effectiveness of public health efforts to reduce the exposure of Americans to specific environmental chemicals (e.g., pesticides).
- To track, over time, trends in the levels of exposure of the U.S. population to environmental chemicals.
- To determine whether levels of three chemicals (cadmium, cotinine, and lead) vary by sex, age, and race/ethnicity. Future Reports will contain information about sex, age, and race/ethnicity for the remaining 24 chemicals and will provide additional information about such variables as education level, income, and urban or rural residence.

Findings and Conclusions

Selected findings of the first National Report on Human Exposure to Environmental Chemicals include the following:

- *Establishment of reference ranges.* The levels of chemicals found in the blood and urine of the research participants provide reference ranges (also known as background exposure levels) for 24 of the environmental chemicals

Continued

tested. Reference ranges are extremely helpful to physicians and health researchers because levels above the reference range indicate high exposure to a particular chemical. For example, if a physician was concerned about a patient's potential exposure to mercury and measured the level of mercury in the patient's urine, the results could be compared with the population reference range in the 1999 Report. A mercury level much higher than those found in the Report would indicate that there may have been an unusual exposure to mercury worthy of further investigation.

- *Reduced exposure of the U.S. population to environmental tobacco smoke.* Cotinine is a metabolite of nicotine that tracks exposure to environmental tobacco smoke (ETS) among nonsmokers; higher continine levels reflect more exposure to ETS. Prior research from 1988 through 1991 documented the median level of cotinine among nonsmokers in the United States. Results from the 1999 Report showed that the median continine level among people aged 3 years and older has decreased by more than 75 percent. These findings represent a dramatic reduction in exposure to the general U.S. population to environmental tobacco smoke since 1988–1991. However, since more than half of American youth are still exposed, ETS remains a major public health concern.
- *Decline in blood lead levels among children since 1991–1994.* Since 1976, CDC has measured blood lead levels as part of the NHANES surveys. Results in the

current Report show that the average blood lead level for children aged 1–5 years is lower than it was for the period 1991–1994. The researchers attribute this decrease to public health efforts to decrease the exposure of children to lead. Nevertheless, children at high risk for lead exposure (e.g., those living in homes containing lead-based paint or lead-contaminated dust) remain a major public health concern.

- *Differences by sex/age/race/ethnicity.* The data show no differences in cadmium levels by sex or race/ethnicity. The results of cotinine analyses indicated that children and adolescents aged 1–19 years had higher levels than adults aged 20 years and older. In addition, non-Hispanic blacks had higher continine levels than Mexican Americans and those of other racial or ethnic groups. For lead, the Report found that males had higher levels than females, children aged 1–5 years had higher levels than children aged 6–11 years, and adults older than 60 years had the highest levels of all adults.
- *Evidence that chemicals in consumer products are absorbed by humans.* This study found that seven phthalates—compounds found in such products as soap, shampoo, hair spray, many types of nail polish, and flexible plastics—are absorbed by humans. Surprisingly, human exposure levels to the two phthalates that are produced in greatest quantity were relatively low, while exposure levels to other phthalates produced in much lower quantities were high. Hence, future research should focus on the

health effects of phthalates that are least common in the environment, because their levels are much higher in the general population.

As the presence of an environmental chemical in a person's blood or urine does not necessarily mean that the chemical will cause disease, more studies are need to determine which levels of these chemicals in people result in disease. Research has already documented adverse health effects of lead and environmental tobacco smoke. However, for many environmental chemicals, few studies are available, and more research is needed to assess health risks associated with different blood or urine levels of these chemicals. Future studies must also consider duration of exposure. In the future, the National Report on Human Exposure to Environmental Chemicals will expand to include 100 environmental chemicals. CDC will update the Report with new data at least once per year. Data sources for future reports include NHANES and other studies that CDC conducts or participates in which examine human exposure to environmental chemicals. Future releases of the Report will include data from CDC studies of groups of people in special-exposure situations (e.g., pesticide applicators, people living near hazardous waste sites, people working in lead smelters).

Source: Centers for Disease Control and Prevention, National Center for Environmental Health. 2001. *National Report on Human Exposure to Environmental Chemicals.* Atlanta, GA: Centers for Disease Control and Prevention.

color contained significantly higher levels of chemical emissions from industrial facilities and had more hazardous waste sites (Leutwyler 2001). Not only are minority communities more likely to be polluted than others, but environmental regulations are also less likely to be enforced in these areas (Stephens 1998).

Margie Eugene Richard, a resident of the predominantly African American community of Norco, Louisiana, and president of Concerned Citizens of Norco, describes what it is like living next to Norco's Shell Oil refinery and chemical plant:

> Norco is situated between Shell Oil Refinery on the east and Shell Chemical plant on the west. The entire town of Norco is only half the size of the oil refineries. Nearly everyone in the community suffers from health problems caused by industry pollution. The air is contaminated with bad odors from carcinogens, and benzene, toluene, sulfuric acid, ammonia, xylene and propylene—run-off and dumping of toxic substances also pollute land and water. . . . My youngest daughter and her son suffer from severe asthma; my mother has breathing problems and must use a breathing machine daily. Many of the residents suffer from sore muscles, cardiovascular diseases, liver, blood and kidney toxicant. Many die prematurely from poor health caused by pollution from toxic chemicals. . . . Daily, we smell foul odors, hear loud noises, and see blazing flares and black smoke that emanates from those foul flares. . . . Norco and many other communities of color across our nation suffer the same ills. (quoted in Bullard 2000, 157)

Environmental injustice affects marginalized populations around the world, including minority groups, indigenous peoples, and other vulnerable and impoverished communities such as peasants and nomadic tribes (Renner 1996). These groups are often powerless to fight against government and corporate powers that sustain environmentally damaging industries. For example, in the early 1970s, a huge copper mine began operations on the South Pacific island of Bougainville. While profits of the copper mine benefited the central government and foreign investors, the lives of the island's inhabitants were being destroyed. Farming and traditional hunting and gathering suffered as mine pollutants covered vast areas of land, destroying local crops of cocoa and bananas, contaminating rivers and their fish. Indigenous groups in Nigeria, such as the Urhobo, Isoko, Kalabare, and Ogoni, are facing environmental threats caused by oil production operations run by multinational corporations. Oil spills, natural gas fires, and leaks from toxic waste pits have polluted the soil, water, and air and compromised the health of various local tribes. "Formerly lush agricultural land is now covered by oil slicks, and much vegetation and wildlife has been destroyed. Many Ogoni suffer from respiratory diseases and cancer, and birth defects are frequent" (Renner 1996, 57). The environmental injustices experienced by Bougainville and the Ogoni reflect only the tip of the iceberg. Renner (1996) warns that "minority populations and indigenous peoples around the globe are facing massive degradation of their environments that threatens to irreversibly alter, indeed destroy, their ways of life and cultures" (p. 59).

Threats to Biodiversity

Biodiversity refers to the great variety of life forms on Earth. The diverse forms of life provide food, fibers, and many other products and "natural services." For example, about 25 percent of drugs prescribed in the United States include chemical compounds derived from wild species (Tuxill 1998). Insects, birds, and bats provide pollination services that enable us to feed ourselves. Frogs, fish, and

■ **Table 14.2** *Threatened and Endangered Species* U.S. and Worldwide: 2000*

	Mammals	Birds	Reptiles	Amphibians	Fishes
Item					
Total Listings	339	274	115	27	123
Endangered species, total	314	253	79	18	79
United States	63	77	14	10	68
Threatened species, total	25	21	36	9	44
United States	9	15	22	8	44

* An endangered species is one in danger of becoming extinct throughout all or a significant part of its natural range. A threatened species is one likely to become endangered in the foreseeable future.
†— represents zero.
Source: *Statistical Abstract of the United States 2000*, 120th edition. 2000. Table 404. Washington D.C.: U.S. Bureau of the Census.

> People everywhere must come to understand that they depend completely on biodiversity—for food, much medicine, and many other products; for global sustainability and stability; and for spiritual renewal. Acting on this realization will put human beings in their proper place, as only one among millions of kinds of organisms, with responsibility for all our fellow humans and for the future of the planet as a whole.
>
> PETER RAVEN
> *Missouri Botanical Garden*

birds provide natural pest control. Various aquatic organisms filter and cleanse our water and plants and microorganisms enrich and renew our soil. In recent decades, we have witnessed mass extinction rates of diverse life forms. Unlike the extinction of the dinosaurs millions of years ago, humans are the primary cause of disappearing species today. Air, water, and land pollution; deforestation; disruption of native habitats; and overexploitation of species for their meat, hides, horns, or medicinal or entertainment value threaten biodiversity and the delicate balance of nature. The primary cause of species decline is human-induced habitat destruction (Hunter 2001). Most estimates suggest that at least 1,000 species of life are lost per year (Tuxill 1998). Biologists expect that at least half of the roughly 10 million species alive today will become extinct during the next few centuries from habitat losses alone; more could disappear as a result of other causes, including pollution, over-harvesting, and global warming (Cincotta & Engelman 2000). Table 14.2 lists the numbers of species worldwide and in the United States that are endangered or threatened.

Environmental Problems and Disappearing Livelihoods

Environmental problems threaten the ability of the earth to sustain its growing population. Agriculture, forestry, and fishing provide 50 percent of all jobs worldwide and 70 percent of jobs in sub-Saharan Africa, East Asia, and the Pacific (World Resources Institute 2000). As croplands become scarce or degraded, as forests shrink, and as marine life dwindles, millions of people who make their living from these natural resources must find alternative livelihoods. In Canada's maritime provinces, collapse of the cod fishing industry in the 1990s from overfishing left 30,000 fishers dependent on government welfare payments and decimated the economies of hundreds of communities (World Resources Institute 2000).

Global estimates suggest that as many as 25 million people may be **environmental refugees**—individuals who have migrated because they can no longer secure a livelihood because of deforestation, desertification, soil erosion, and

Snails	Clams	Crustaceans	Insects	Arachnids	Plants
32	71	21	42	6	705
21	63	18	34	6	566
20	61	18	30	6	565
11	8	3	8	—	139
11	8	3	8	—	139

other environmental problems. As individuals lose their source of income, so do nations. In one-quarter of the world's nations, crops, timber, and fish contribute more to the nation's economy than industrial goods (World Resources Institute 2000).

Social Causes of Environmental Problems

In addition to population growth, various other structural and cultural factors have also played a role in environmental problems.

Industrialization and Economic Development

Many of the environmental problems confronting the world have been caused by industrialization and economic development. Industrialized countries, for example, consume more energy and natural resources and contribute more pollution to the environment than poor countries. A child born in the United States will produce 10 times the pollution of a child born in Bangladesh (Hunter 2001). The relationship between level of economic development and environmental pollution is not necessarily a linear one. For example, industrial emissions are minimal in regions with low levels of economic development, yet high in the middle-development range as developing countries move through the early stages of industrialization. At more advanced industrial stages, some environmental pressures ease because heavy polluting manufacturing industries decline and "cleaner" service industries increase and because rising incomes are associated with a greater demand for environmental quality and cleaner technologies (Hunter 2001). However, a positive linear correlation has been demonstrated between per capita income and national carbon dioxide emissions (Hunter 2001). Industrialization and economic development have been the primary cause of global environmental problems such as the ozone hole and global warming (Koenig 1995). In less developed countries, environmental problems are largely

The faster populations grow, the more environmental problems are aggravated.

WERNER FORNOS
President, Population Institute

Americans are the world's hyper-consumers and must begin to consume differently. We must demand and purchase products designed for durability, repair, and reuse. And we must seek to have more fun and be genuinely content with less stuff.

BETSY TAYLOR
Executive Director of Center for a New American Dream

the result of poverty and the priority of economic survival over environmental concerns. "Poverty can induce people to behave as if they are myopic, making use of local resources without consideration of global effects or the implications for future generations" (Hunter 2001, 32). Vajpeyi (1995) explains:

> Policymakers in the Third World are often in conflict with the ever-increasing demands to satisfy basic human needs—clean air, water, adequate food, shelter, education—and to safeguard the environmental quality. Given the scarce economic and technical resources at their disposal, most of these policymakers have ignored long-range environmental concerns and opted for short- range economic and political gains. (p. 24)

Measures of economic development and the "health" of economies are based primarily on gross domestic product (GDP) and consumer spending. News reports claiming that increases in GDP and consumer spending mean the U.S. economy is "strong" overlook the social and environmental costs of the production and consumption of goods and services. Until definitions and measurements of "economic development" and "economic health" reflect these costs, the pursuit of economic development will continue to contribute to environmental problems.

Cultural Values and Attitudes

What good is a house if you don't have a decent planet to put it on?

HENRY D. THOREAU
Author

Cultural values and attitudes that contribute to environmental problems include individualism, and materialism. Individualism, which is a characteristic of U.S. culture, puts individual interests over collective welfare. When a Gallup poll asked U.S. adults if they would be willing to have the United States take steps to reduce global warming if costs for gasoline or electricity went up a great deal, nearly half (48 percent) said no (Gallup & Saad 1997). Even though recycling is good for our collective environment, many individuals do not recycle because of the personal inconvenience involved in washing and sorting recyclable items. Similarly, individuals often indulge in countless behaviors that provide enjoyment and convenience at the expense of the environment: long showers, use of dishwashing machines, recreational boating, meat eating, and use of air conditioning, to name just a few.

In America we have two dominant religions: Christianity and accumulation.

BRIAN SWIMME
California Institute of Integral Studies

Finally, the influence of materialism, or the emphasis on worldly possessions, also encourages individuals to continually purchase new items and throw away old ones. The media bombard us daily with advertisements that tell us life will be better if we purchase a particular product. Materialism contributes to pollution and environmental degradation by supporting industry and contributing to garbage and waste.

The cultural value of militarism also contributes to environmental degradation. This issue is discussed in detail in Chapter 16.

Strategies for Action: Responding to Environmental Problems

Lowering fertility, as discussed earlier, helps reduce population pressure on the environment. Environmental activism, environmental education, the development of alternative sources of energy, modifications in consumer behavior, government regulations, sustainable economic development, and international cooperation and assistance can also alleviate environmental problems.

Environmental Activism

In a national survey of first-year college students, only 17.5 percent said that "becoming involved in programs to clean up the environment" was an essential or very important objective (American Council on Education & University of California 2000). A Gallup poll found that the percentage of Americans who consider themselves "environmentalists" dropped from three quarters in 1989 to half in 1999 (DeCourcy/Public Agenda 2001). Nevertheless, significant numbers of individuals who are concerned about environmental problems join (or form) environmental activist groups, which exert pressure on government and private industry to initiate or intensify actions related to environmental protection. For example, lawsuits filed in the 1970s by the Natural Resources Defense Council (NRDC) resulted in the phase-out of leaded gasoline. Lobbying by the same organization led to a 1993 White House order directing that the federal government use recycled paper. Environmentalist groups also design and implement their own projects, and disseminate information to the public about environmental issues.

In the United States, environmentalist groups date back to 1892 with the establishment of the Sierra Club, followed by the Audubon Society in 1905. Other environmental groups include the National Wildlife Federation, World Wildlife Fund, Environmental Defense Fund, Friends of the Earth, Union of Concerned Scientists, Greenpeace, Environmental Action, Natural Resources Defense Council, World Watch Institute, National Recycling Coalition, World Resources Institute, and Rainforest Alliance, to name a few.

Online Activism The Internet and e-mail provide important tools for environmental activism. For example, the organization Environmental Defense sends Member Action Alerts by e-mail to more than 650,000 activists, informing them when Congress and other decision makers threaten the health of the environment ("Online Activism Lives Up to Its Promise" 2001). These members can then send e-mails and faxes to Congress, the President, and business leaders. "Our online activists pressed EPA for strict controls on diesel trucks and buses and helped persuade former President Clinton to place one third of our national forests off limits to road building, saving them from oil exploration and logging" (p. 8).

The Role of Industry in the Environmental Movement Industries are major contributors to environmental problems and often fight against environmental efforts that threaten their profits. However, some industries are joining the environmental movement for a variety of reasons, including pressure from consumers and environmental groups, the desire to improve their public image, and genuine concern for the environment. For example, in 1997, British Petroleum (BP) publicly acknowledged that global warming is a threat to human and environmental health and began working with the environmental organization Environmental Defense to develop ways to reduce its greenhouse gas emissions ("Global Corporations Join Us To Reduce Greenhouse Gas Emissions" 2000). This lead to the development of the Partnership for Climate Action—a collaboration between Environmental Defense and seven of the world's largest corporations (including companies in the petroleum, aluminum, chemical, and electric power sectors) to take action against global warming. Participating companies have committed to report their emissions publicly, and each has set a firm target for emissions reductions. Environmental Defense is working with each company to

> Protection of the environment is a goal that virtually everyone believes is worthwhile and would enhance the quality of life. The disagreement comes in deciding how to implement policies to achieve the goal.
>
> RANDALL G. HOLCOMBE
> *Florida State University*

> That's not the way to save the planet!
>
> BENJAMIN LEWIS
> *Four-year-old who yelled at a man who threw a soda can out of his car window*

devise ways to improve energy efficiency, use renewable energy sources and improve manufacturing processes.

BP, along with several other industries, have withdrawn their membership to the Global Climate Coalition—an industry group that denies global warming and the role that industries play in the emission of greenhouse gases. Other industries that have quit the Global Climate Coalition include Ford, Shell Oil, Dow Chemical, General Motors, and Daimler-Chrysler (Associated Press 2000).

Ecoterrorism   An extreme form of environmental activism is **ecoterrorism**, defined as any crime intended to protect wildlife or the environment that is violent, puts human life at risk or results in damages of $10,000 or more (Denson 2000). A group known as the Earth Liberation Front set fire to the nation's busiest ski resort at Vail, Colorado, resulting in $12 million in damages. The reason: to express opposition to the resort's proposed expansion into the forest habitat of the Canada lynx. This act was the most destructive act of ecoterrorism in U.S. history. Since 1980, 100 major ecoterrorist crimes involving nearly $43 million in damages have occurred in the western United States (Denson 2000). Documented crimes of ecoterrorism include arson, bombings, and mailing letters rigged with razors to primate researchers nationwide. In addition to the Earth Liberation Front, other groups known to engage in ecoterrorism include the Animal Liberation Front and a group calling itself the Justice Department. Although critics of ecoterrorism cite the damage done by such tactics, members and supporters of ecoterrorist groups claim that the real terrorists are corporations that plunder the Earth.

Environmental Education

One goal of environmental organizations and activists is to educate the public about environmental issues and the seriousness of environmental problems. In a national survey, U.S. adults were asked to rate the seriousness of "threats to the global environment" on a scale of 1 to 10 (10 = "very serious"). Only 25 percent of the respondents rated "threats to global environment" as "very serious" (a rating of 10); the average rating was 7.2 (DaVanzo et al. 2000). Some Americans may underestimate the seriousness of environmental concerns because they lack accurate information about environmental issues. In a national survey of U.S. adults, most (69 percent) rated themselves as having either "a lot" (10 percent) or "a fair amount" (59 percent) of knowledge about environmental issues and problems (National Environmental Education & Training Foundation & Roper Starch Worldwide 1999). Yet, on seven of ten questions asked about emerging environmental issues, more Americans gave incorrect answers than correct ones. The average was 3.2 correct answers out of 10 questions (the ten questions are found in this chapter's *Self and Society* feature). Even those who rated themselves as having "a lot" of environmental knowledge, as well as those with a college degree, only scored four out of ten items correctly. The authors of the study concluded:

> Increased knowledge is the key to changing attitudes and behaviors on issues critical to our environmental future. If Americans can answer an average of only 3 of 10 simple knowledge questions, there is a clear need to provide environmental information in a form that the American public can easily digest and act upon.
> (National Environmental Education & Training Foundation & Roper Starch Worldwide 1999, 41)

A main source of information about environmental issues for most Americans is the media. However, because media are owned by corporations and wealthy individuals with corporate ties, unbiased information about environmental impacts of corporate activities may not readily be found in mainstream media channels. Indeed, the public must consider the source in interpreting information about environmental issues. Propaganda by corporations sometimes comes packaged as "environmental education." Hager and Burton (2000) explain: "Production of materials for schools is a growth area for public relations companies around the world. Corporate interests realise the value of getting their spin into the classrooms of future consumers and voters" (p. 107).

Technological Innovations in Sources and Uses of Energy

In the 1990s, the world's primary sources of energy were oil (34 percent), coal (24 percent), gas (19 percent), renewables or biomass (e.g., fuel wood, crops, animal wastes) (18 percent), and nuclear (5 percent) (Fridleifsson 2000). Cleaner alternative sources of energy include solar power, geothermal power, ocean thermal power, tidal power, and energy from fuel cells. The fastest developing alternative source of energy is wind power, which could supply 10 percent of the world's electricity by 2020 (Flavin 2000). A study by the U.S. Department of Energy concluded that North Dakota, South Dakota, and Texas had enough harnessable wind energy to meet all U.S. electricity needs (Brown 1998b).

In addition to developing cleaner forms of energy, efforts are being made to develop products that use energy more efficiently. In 1992 the U.S. Environmental Protection Agency introduced the ENERGY STAR label for products that meet EPA energy efficiency standards. Orginially applied to personal computers and monitors, the ENERGY STAR program has expanded to include new homes, and a variety of electronics, home appliances, and office equipment.

Several car manufacturers have devloped eco-friendly cars powered by nonpolluting electric motors or "hybrid" gasoline-electric motors (Buss 2001). And cars with hydrogen-powered fuel cells are expected to be available within the next decade.

The development of cleaner sources of energy and more efficient uses of energy does not appear to be a priority of the Bush administration. In the first George W. Bush budget, government support of wind and solar energy development was cut in half, and research on more fuel-efficient cars was cut by 27 percent ("President Bush's Tax and Budget Bumbles Label Him Commander in Cheap" 2001). And in response to the recent energy crisis in California, Vice President Dick Cheney (who is also the head of the White House task force on energy policy), outlined a plan that emphasized more oil drilling, more power plants, and more nuclear power, with less emphasis on conservation and alternatives to fossil fuel. His policy would require building one new electricity-generating plant a week for 20 years (Alternative Energy Institute 2001).

Modifications in Consumer Behavior

Increasingly, consumers are making choices in their behavior and purchases that reflect concern for the environment. For example, consumers are increasingly buying energy-efficient compact fluorescent lamps (CFLs) (Scholand 2000). CFLs last about 10 times longer than the more commonly used incandescent light bulb and they use 75 percent less electricity. The 275 million CFLs used in North America in 2000 avoided 3.5 million tons of carbon emissions and 69,000 tons of sulfur emissions during the year (Scholand 2000). Recycling has also increased. Between 1975 and 1997, the share of paper used that is recycled globally increased from 38 percent to more than 43 percent and is expected to reach 45 percent by 2010 (Abramovitz 2000). A national survey found that 64 percent of Americans say they frequently recycle newspapers, cans, and glass (National Environmental Education & Training Foundation & Roper Starch Worldwide 1999). This same survey found that 83 percent of Americans frequently turn off lights and electrical appliances when not in use and more than half say they frequently try to cut down on the amount of trash their household creates and conserve water in their homes and yards. Sales of organic foods (that are grown without the use of pesticides) are increasing by 15 to 20 percent a year, primarily because of consumers' concern for their own health, rather than for the health of the planet (Lipke 2001).

> As the environment continues to become a core American value, consumers will respond increasingly to environmentally unfriendly products or processes. If consumers make a statement through their purchases, corporations and politicians are sure to follow.
>
> EUGENE LINDEN
> *Science Writer,* **Time** *Magazine*

From credit card and long-distance telephone companies that donate a percentage of their profits to environmental causes, to socially responsible investment services, consumers are increasingly voting with their dollars to support environmentally responsible practices. However, many Americans continue to disregard the environmental impact of their consumer behavior. Between 1975 and 1999, light trucks (sport utility vehicles or SUVs, minivans, and pickup trucks) increased from 20 percent to 46 percent of new car sales in the United States (Renner 2000). Light trucks not only contribute to rising fuel use, but they are also far more polluting than regular cars. This chapter's *The Human Side* feature offers examples of how consumers can resist materialistic urges and reduce waste and overconsumption.

Government Regulations and Funding

Worldwide, governments spend about $700 billion (U.S.) per year subsidizing environmentally unsound practices in the use of water, agriculture, energy, and transportation (World Resources Institute 2000). Yet, through regulations and the provision of funds, governments also play a vital role in protecting and restoring the environment. Governments have responded to concerns related to air and water pollution, ozone depletion, and global warming by imposing regulations affecting the production of pollutants. Governments have also been involved in the preservation of deserts, forests, reefs, wetlands, and endangered plant and animal species.

For example, to reduce air pollution caused by automobiles, the U.S. government requires cars to have catalytic converters. In 1990, Congress amended the Clean Air Act to provide marketable pollution rights for sulfur dioxide emissions by electric companies. The Environmental Protection Agency (EPA) issued more than 5 million marketable permits, each allowing the emission of one ton of sulfur dioxide annually, to 110 companies, which must have permits for all of their sulfur dioxide emissions or face large fines. Companies that want to emit more sulfur dioxide can buy permits from other companies, whereas those that reduce their own pollution can profit from selling their permits (Holcombe 1995). More recently, the EPA issued new rules that tighten emissions standards for diesel trucks and buses and requires oil companies to remove 97 percent of the sulfur in diesel fuel (sulfur clogs catalytic converters and other pollution-control equipment) ("EPA Cracks Down on Diesel Trucks" 2000). The EPA has also issued new rules (which will be phased in between 2004 and 2009) that re-

Affluenza

A 1997 PBS television program, "Affluenza," "explores the high social and environmental costs of materialism and overconsumption." In a 1998 sequel, "Escape from Affluenza," people working to reduce waste and over-consumption are profiled. Below is a list of some of the ways you can "Escape from Affluenza" (KCTS 1998).

Shopping

- Biggest shopping trap—it was on sale. If it wasn't something you identified as a need, you didn't SAVE money, you SPENT money. Learn this mantra: It's not a bargain if I don't need it.
- Mail-order shopping is a great way to save time, fuel, and money. But avoid the companies that make up the cost of glossy catalogs by charging twice as much. About 94 percent of these catalogs go unused into the waste stream: don't be one of the 6 percent who pay more to make this practice profitable. . . !
- The average American family of four spends an esti-mated 10 percent of its annual income on clothing. Using principles you already know (only buying what you need, buying secondhand items that are as good as new, for instance), you can easily cut your clothing expenditures in half. . . .

Waste and Clutter

- A poorly insulated house can easily waste 30 percent to 50 percent of the energy poured into it . . . Simple changes around the house: run the dishwasher half as often as you do now, and save 50 percent in water, energy, and time. Wash only full loads of cloth-ing, and because 90 percent of the cost of washing clothes is to heat the water, avoid using hot water.
- Half of the average family's household energy goes to heating and/or cooling. Put a dent in this expense by keeping the thermostat set at 65° F in the daytime, and 55° or 60° F at night.
- Planting a large tree to shade your home can save you an estimated $73 a year in air-conditioning bills. If every household did this, the country would save more than $4 billion in energy costs.
- About half of the waste stream in the U.S. is dis-carded packaging. Be very aware of packaging ex-cess, and when all else is equal, choose the least-packaged. Help raise awareness of this issue by being a packaging activist: ask your grocers to phase out pre-wrapped fruits and vegetables.

Home and Hearth

- When home-hunting, pick the smallest amount of space in which you are comfortable. This will limit the amount of stuff you can accumulate, and take far less of your time and resources to furnish, clean, maintain, insure, and pay for.
- Eighty percent of the dirt in your home is brought in on shoes. Save time and cleaning expenses by start-ing a no-shoes policy.
- Start a neighborhood swap of seldom-used tools. Why should a street of 10 houses have 10 lawnmow-ers, 10 paint-sprayers, and 10 band saws?

Food

- Eat a local diet as much as possible. This creates jobs in the region, reduces transportation costs and energy consumption, ensures higher nutritional value and encourages local small-scale agriculture that protects land from development. Ask your grocers to mark foods local. Hint: If it's out of season (e.g., strawber-ries in December in Seattle), you know it's not local.
- To dramatically reduce your ecological footprint, save about 50 percent in food costs and maximize your prospects for a longer, healthier life, become a vege-tarian or reduce your meat consumption.

Health and Happiness

- Prevention is the cheapest and best health insurance. Get an annual checkup (even frugal fanatics agree this is worth the $$). Take care of your body by eat-ing and sleeping right, and stop smoking.
- Stop equating the amount of fun and pleasure you get with the amount of money you spend to get it. Sit down and make a list of 25 things you like to do that cost little or no money, and keep it where you can see it every day.
- Instead of spending your time and energy buying and taking care of stuff, volunteer to feed the hungry, care for the suffering, visit the lonely, or mentor a child.
- Be happy with what you have. If you make a habit of thinking in terms of what you have, rather than what you don't, you may well find that you've got enough.

Source: From "100 Ways to Escape from Affluenza" as appears on www.pbs.org/kcts/affluenza/escape by Vivia Boe. Reprinted by permis-sion. Films are available from BULLFROG FILMS, 800-543-FROG or www.bullfrogfilms.com.

quire light trucks and SUVs to meet the same nitrogen oxide emissions standards as regular cars (Renner 2000).

City and state governments also play a role in environmental issues. In a landmark city ordinance, San Francisco banned the use of pesticides by all city departments, with the exception of schools and public housing. The city has mandated that integrated pest management methods, which utilize least-toxic alternatives, be used in public parks, golf courses, and office buildings ("San Francisco Bans Pesticides" 1997). As mentioned in this chapter's *Focus on Technology* feature, the state of Massachusetts has banned the dumping of video monitors (from computers and TVs) in landfills.

Some environmentalists advocate using taxes to discourage polluting practices. If government heavily taxed destructive activities such as carbon emissions, sulfur emission, the generation of toxic waste, the use of pesticides, and the use of virgin raw material, these activities would decrease (Brown & Mitchell 1998). As of early 2000, nine countries had raised taxes on environmental harm and used the revenue to pay for cuts in income taxes (Roodman 2000).

In his first months of presidency, George W. Bush has weakened rather than strengthened governmental protection of the environment. Bush has withdrawn regulations approved by the Clinton administration that set stricter limits on arsenic levels in drinking water. Bush also broke his campaign promise to curb carbon dioxide emissions—the main cause of global warming—and has threatened to pull the United States out of the Kyoto Protocol (an international agreement to reduce greenhouse gases). Referring to these actions, Joshua Karliner (2001) suggests that:

> Unless he dramatically reverses course, it's going to be "G" for global, "W" for warming, Bush from now on. (Karliner 2001)

Sustainable Economic Development

It will not be possible for the community of nations to achieve any of its major goals—not peace, not environmental protection, not human rights or democratization, not fertility reduction, not social integration—except in the context of sustainable development.

1994 HUMAN DEVELOPMENT REPORT

Achieving global cooperation on environmental issues is difficult, in part, because developed countries (primarily in the Northern Hemisphere) have different economic agendas from developing countries (primarily in the Southern Hemisphere). The northern agenda emphasizes preserving wealth and affluent lifestyles whereas the southern agenda focuses on overcoming mass poverty and achieving a higher quality of life (Koenig 1995). Southern countries are concerned that northern industrialized countries—having already achieved economic wealth—will impose international environmental policies that restrict the economic growth of developing countries just as they are beginning to industrialize. Global strategies to preserve the environment must address both wasteful lifestyles in some nations and the need to overcome overpopulation and widespread poverty in others.

Development involves more than economic growth, it involves sustainability—the long-term environmental, social, and economic health of societies. **Sustainable development** involves meeting the needs of the present world without endangering the ability of future generations to meet their own needs. "The aim here is for those alive today to meet their own needs without making it impossible for future generations to meet theirs...This in turn calls for an economic structure within which we consume only as much as the natural environment can produce, and make only as much waste as it can absorb" (McMichael, Smith, & Corvalan 2000, 1067).

International Cooperation and Assistance

Global environmental concerns such as global warming and climate change require international governmental cooperation. Since the first U.N. Stockholm Conference on the Human Environment in 1972, about 240 international environmental agreements have been reached (French 2000). For example, the 1992 Earth Summit in Rio de Janeiro brought together heads of state, delegates from more than 170 nations, nongovernmental organizations, representatives of indigenous peoples, and the media to discuss an international agenda for both economic development and the environment. The 1992 Earth Summit resulted in the Rio Declaration—"a nonbinding statement of broad principles to guide environmental policy, vaguely committing its signatories not to damage the environment of other nations by activities within their borders and to acknowledge environmental protection as an integral part of development" (Koenig 1995, 15).

The 1987 Montreal Protocol forged an agreement made by 70 nations to curb the production of CFCs, (which contribute to ozone depletion and global warming) (Koenig 1995). Under the Montreal Protocol, CFC emissions have dropped nearly 90 percent (French 2000). In 1997, delegates from 160 nations met in Kyoto, Japan, and forged the **Kyoto Protocol**—the first international agreement to place legally binding limits on greenhouse gas emissions from developed countries. However, as of this writing, none of the major industrial countries has ratified the agreement. More recently, representatives from 122 countries drafted a legally binding agreement to phase out a group of dangerous chemicals known as persistent organic pollutants (POPs) (Cray 2001). However, "even as the number of treaties climbs, the condition of the biosphere continues to deteriorate . . . The main reason that many environmental treaties have not yet turned around the environmental trends they were designed to address is because the governments that created them permitted only vague commitments and lax enforcement" (French 2000, 135). In addition, some countries do not have the technical or economic resources necessary for implementing the requirements of environmental treaties.

One way developed countries can help less developed countries address environmental concerns is through economic aid. In a national survey of U.S. adults, "protecting the global environment" was the second most commonly cited top priority for U.S. economic international assistance ("improving children's health" was the most commonly cited top priority) (DaVanzo et al. 2000).

Because industrialized countries have more economic and technological resources, they bear primary responsibility for leading the nations of the world toward environmental cooperation. Jan (1995) emphasizes the importance of international environmental cooperation and the role of developed countries in this endeavor:

Advanced countries must be willing to sacrifice their own economic well-being to help improve the environment of the poor, developing countries. Failing to do this will lead to irreparable damage to our global environment. Environmental protection is no longer the affair of any one country. It has become an urgent global issue. Environmental pollution recognizes no national boundaries. No country, especially a poor country, can solve this problem alone. (pp. 82–83)

Understanding *Population and Environmental Problems*

Man is here for only a limited time, and he borrows the natural resources of water, land and air from his children who carry on his cultural heritage to the end of time....One must hand over the stewardship of his natural resources to the future generations in the same condition, if not as close to the one that existed when his generation was entrusted to be the caretaker.

DELANO SALUSKIN
Yakima Indian Nation

Because population increases exponentially, the size of the world's population has grown and will continue to grow at a staggering rate. Given the problems associated with population growth—deteriorating socioeconomic conditions, depletion of natural resources, and urban crowding—most governments recognize the value of controlling population size. Efforts to control population must go beyond providing safe, effective, and affordable methods of birth control. Slowing population growth necessitates interventions that change the cultural and structural bases for high fertility rates. These interventions include increasing economic development and improving the status of women, which includes raising their levels of education, their economic position, and their (and their children's) health.

Globally, the average number of children born to each woman has fallen from 5 in 1960 to less than 3 (Engelman et al. 2000), as more women today are using methods of birth control. High mortality rates resulting from HIV/AIDS and war have also contributed to the slowing of population growth in many countries. Although birthrates have decreased in most countries throughout the world, population continues to grow as a result of population momentum. In addition, many African and Middle Eastern nations continue to have high birth rates: 6.9 in Uganda, 7.0 in Somalia, 7.5 in Niger, 6.4 in Saudi Arabia, 6.5 in Yemen, and 7.1 in Oman (Hunter 2001). These factors combined—population momentum and continued high fertility rates in some countries—contribute to escalating world population. Even in the 29 African nations that have been hardest hit by AIDS, overall population of these 29 countries is projected to increase by more than half between 1995 and 2015 (Hunter 2001). An estimated 65 countries are expected to double their populations in 30 years or less, and 14 countries will triple or nearly triple their populations by 2050. "The impact of population growth in these fast-growing countries will more than offset the gains of low fertility in 80 other countries" (Population Institute 1998, 3).

Rapid and dramatic population growth, along with expanding world industrialization and patterns of consumption, has contributed to environmental problems. The greater numbers of people make increased demands on natural resources and generate excessive levels of pollutants. Environmental problems result not only from large populations, but also the ways in which these populations live and work. As conflict theorists argue, individuals, private enterprises, and governments have tended to choose economic and political interests over environmental concerns. However, growing evidence of the irreversible effects of global warming and loss of biodiversity, and increased concerns about the health effects of toxic waste and other forms of pollution are making it difficult for even profit-seeking industries to turn their backs on environmental problems.

Many Americans believe in a "technological fix" for the environment—that science and technology will solve environmental problems. Paradoxically, the same environmental problems that have been caused by technological progress may be solved by technological innovations designed to clean up pollution and preserve natural resources. Although technological innovation is important in resolving environmental concerns, other social changes are needed as well. Addressing the values that guide choices, the economic contexts in which the choices are made, and the governmental policies that encourage various choices is critical to resolving environmental and population problems.

Global cooperation is also vital to resolving environmental concerns, but is difficult to achieve because rich and poor countries have different economic development agendas: developing poor countries struggle to survive and provide for the basic needs of their citizens; developed wealthy countries struggle to maintain their wealth and relatively high standard of living. Can both agendas be achieved without further pollution and destruction of the environment? Is sustainable economic development an attainable goal? The answer must be yes. But sustainable development will not occur on its own or as the inevitable outcome of current trends. If it is to happen, we must, collectively, make it happen. The motivation for finding and implementing sustainable economies may come from understanding the consequences if we don't. "We must work together to get onto a path of ecological sustainable social and economic development. The level of health attained by the world's population will be the ultimate criterion of how well we succeed" (McMichael et al. 2000, 1067).

Critical Thinking

1 One writer observed that "on a certain November Day an obscure woman in Iowa gives birth to seven babies; we marvel and rejoice. On the same day an obscure woman in Nigeria gives birth to her seventh child in a row; we are distressed and appalled" (Zwingle 1998, 38). Why might reactions to these two events be so different?

2 In many developing countries that have strict laws against abortion, the use of child labor is common. Do you think there may a connection between laws prohibiting abortion and child labor in developing countries? How could you find out if there is a connection?

3 Do you think governments should regulate reproductive behavior to control population growth? If so, how?

4 Babies eat more, drink more, and breathe more, proportionally, than adults—which means babies are more susceptible to environmental toxins than are adults (Dionis 1999). Yet, federal environmental standards are largely set at levels designed to protect adults. What social forces do you think discourage the government from setting environmental standards based on what is safe for infants and young children?

5 In Chapters 7 and 9, we discussed hate crimes, noting that hate crime laws impose harsher penalties on the perpetrator of a crime if the motive for that crime was hate or bias. Should motives be considered in imposing penalties on persons who are convicted of acts of ecoterrorism? For example, should a person who sets fire to a business to protest that business' environmentally destructive activities receive a lighter penalty than a person who sets fire to a business for some other reason?

Key Terms

acid rain	environmental injustice	Kyoto Protocol
Basel Ban	environmental racism	Malthusian theory
biodiversity	environmental refugees	multiple chemical
biomonitoring	e-waste	sensitivity
deforestation	Extended Producer	planned obsolescence
demographic transition	Responsibility	population momentum
theory	fertility rate	pronatalism
desertification	global warming	replacement level
doubling time	greenhouse gases	sick building syndrome
ecoterrorism	greenwashing	sustainable development

Media Resources

The Wadsworth Sociology Resource Center: Virtual Society

http://sociology.wadsworth.com

See the companion Web site for this book to access general sociology resources and text-specific features that can further your understanding of this chapter. The site contains Internet links, Internet exercises, online practice quizzes, information on InfoTrac College Edition, and many more valuable materials designed to enrich your learning experience in social problems.

InfoTrac College Edition

You can access InfoTrac College Edition either from the Wadsworth Sociology Resource Center at **http://sociology.wadsworth.com** or directly from your web browser at **http://www.infotrac-college.com/wadsworth/**. InfoTrac College Edition is an online university library that includes over 700 popular and scholarly journals in which you can find articles related to the topics in this chapter such as population growth, population policies, environmental education, global warming, pollution, environmental injustice, and the environmental movement.

Interactions CD-ROM

Go to the "Interactions" CD-ROM for *Understanding Social Problems*, Third Edition to access additional interactive learning tools, such as in-depth review materials, corresponding practice quizzes, and other engaging resources and activities to help you study the concepts in this chapter.

Science and Technology

Is It True?

1. Non-English speakers constitute the fastest growing group of Internet users.

2. Partial birth abortions are illegal in all U.S. states.

3. Industry's expenditure on research and development exceeds that of academic institutions combined.

4. The clock was invented by capitalistic industrialists as a means of controlling the time workers spent on the job.

5. Over 50 percent of American homes have a personal computer.

Answers to "Is It True?": 1 = T; 2 = F; 3. = T; 4 = F; 5 = T

> M ost of the consequences of technology that are causing concern at the present time— pollution of the environment, potential damage to the ecology of the planet, occupational and social dislocations, threats to the privacy and political significance of the individual, social and psychological malaise . . . are with us in large measure because it has not been anybody's explicit business to foresee and anticipate them.

> EMMANUEL MESTHENE,
> *Former Director of the Harvard Program on Technology and Society*

Mitch Maddox was a 26-year-old computer systems manager working and living in Dallas, Texas when, on January 1, 2000, he moved into an empty house with little more than a laptop and the clothes on his back. After legally changing his name to DotComGuy, the former Mr. Maddox pledged to "live off the Internet for a year and never leave the apartment" (quoted in Brand 2000, 1). Financed by advertisers, including United Parcel Services, his former employer, Mr. DCG then went about furnishing his new apartment— shopping online, of course. Throughout his year of confinement DotComGuy relied completely on the Internet for all of life's basic needs: groceries (Peapod.com, a grocery delivery service), entertainment (online games, CDs, videos), companionship (chat rooms and dating services), recreation (online golf lessons), and of course pet supplies for his beagle, DotComDog.

Mr. DCG "made internet and marketing history" with the "first 24-hour-366 day uninterrupted multi-camera production" including a live Net concert performed by 70s rock group Kansas (DotComGuy, Inc. 2001; Witt 2000). DotComGuy, however, doesn't want to be known as an entertainer; he wants to "show the ease of e-commerce [and] how it can simplify your life" (Stenger 2000, 1). He also wants to provide a service to the online community—a cyber Consumer Reports if you will. As a very, very regular online shopper, Mr. DCG visits hundreds of sites daily and ranks them from 1 to 5 based on their "usability, reliability, value, selection, and customer support" (Witt 2000). What's next for DotComGuy? He'll continue to help surfers find the right product at the right price. A comic strip may even be on the horizon, or should we say a DotComStrip.

Living in an apartment for a year with your only means of communication being a computer, a monitor, and a modem is a little futuristic. But, like DotComGuy, virtual reality, cloning, and teleportation are no longer just the stuff of popular sci-fi movies such as *The Cell*, *Star Trek*, and *Matrix*. Virtual reality is now used to train workers in occupations as diverse as medicine, engineering, and professional football. The ability to genetically replicate embryos has sparked worldwide debate over the ethics of reproduction, and California Institute of Technology scientists have beamed a ray of light from one location to another. Just as the telephone, the automobile, the television, and countless other technological innovations have forever altered social life, so will more recent technologies (see this chapter's *Focus on Technology* feature).

Science and technology go hand in hand. **Science** is the process of discovering, explaining, and predicting natural or social phenomena. A scientific approach to understanding AIDS, for example, might include investigating the molecular structure of the virus, the means by which it is transmitted, and pub-

New Technological Inventions

Here are some of the finalists from *Discovery Magazine's* annual awards for technological innovation.

- **Micro Disks.** Size does matter. The size of a disk determines storage, i.e., the capacity for memory—one of the most important characteristics in selecting a personal appliance, whether it be a digital camera or a hand-held computer. Researchers have recently created a Microdrive, a drive as powerful as those in full size desktop computers but made for a disk the size of a quarter. Such an innovation is likely to lead to the further miniaturization of electronic appliances.

- **Tendon-activated Hand.** Prosthetics are generally clumsy and not very responsive to fine-tuned actions. However, a new "tendon-activated hand" connects sensors in the artificial hand to remnant tendons providing flexible and natural movement in the prosthetic. One recent accident victim has even returned to playing the piano!

- **Lego Mindstorms.** Researchers at the Massachusetts Institute of Technology have developed computer controlled building blocks. Described as half robot and half Lego, the custom designed software helps children write computer programs by piecing together instructions on a computer screen which are transmitted to a robot. The robot then carries out the instructions—whether it be dealing a hand of gin rummy or playing a game of hide-and-seek. Besides being fun, children learn the elementary principles of programming.

- **Talking Lights.** Imagine you are visually impaired and in search of the "Main Street" exit of a building. As you walk through a door the badge you are wearing says, "This is the Main Street exit." Designed by electrical engineers at MIT, talking lights transform ordinary fluorescent light bulbs into "global positioning satellites" that can be used to guide people around malls, senior centers, hospitals, and the like. The lights simply emit information to a decoder that then translates it to verbal messages. The unit is wireless, inexpensive, and consumes no more energy than a regular light bulb.

- **Recodable Lock.** One of the biggest concerns of the 21st century is computer security—Internet privacy, virus-free computer environments, secure e-mail, etc. Although software security continues to be a multibillion dollar industry, two researchers may have discovered a way to secure a computer from the inside out. This new microscopic lock is almost impossible to crack and may do to electronic snoopers what software programs have failed to do—put a lock on cyberspace.

- **Microsystems *Jini*.** All of us have spent hours pouring over the manuals of our latest high tech purchases trying to figure out the operating instructions of each new gadget. *Jini,* however, is a new software package that once installed allows your computer to talk to other "intelligent" appliances installed in to the *Jini* system. From the *Jini* web page simply click on an icon to warm your coffee, retrieve pictures from a digital camera, or toast a Pop Tart.

- **Toyota Prius.** For years, car manufacturers have tried to find the perfect balance of economy, efficiency, and comfort. Toyota's Prius, as one of the finalist in the transportation category, may be the perfect mix. The new automobile combines a conventional gasoline engine with a nickel-metal hydride battery resulting in an attractive, roomy, economy size care that gets 66 city miles per gallon. It is presently on sale in the United States and Japan.

- **Virtual Touch.** Using a computer is a multisensory experience—bright lights, moving images, and real life audio but, until recently, no tactile stimuli. Researchers at MIT have now created what they call a haptic interface, that is, a joystick, which gives the computer user a sense of touch. The advantages of such an innovation are unlimited allowing doctors to actually feel tissue during cybersurgery or engineers to experience the goodness of fit of parts just designed.

- **Breaking Through.** That all too familiar sound of a jackhammer may be a thing of the past. In trying to design a quiet way to break up concrete, researchers have developed RAPTOR, a lightweight "gun" that fires penny nails at speeds of 5,000 feet per second. When the nails are fired into the concrete in consecutive lines, stress fractures occur and even the thickest of concrete begins to crumble. Because RAPTOR can be fitted with a silencer, the firing of nails not only breaks up the concrete it does so quietly and with much less effort on the part of the operator.

Sources:

Joseph D'Aguese. 2000. "The 11th Annual Discover Awards." *Discover* 21(7): July. http://www.discover.com/jul_00featawards.htm

"1999 Emerging Technology Finalists." http://www.discover.com/awards/awards_emerging.shtml

lic attitudes about AIDS. **Technology**, as "a form of human cultural activity that applies the principles of science and mechanics to the solution of problems," is intended to accomplish a specific task—in this case, the development of an AIDS vaccine.

Societies differ in their level of technological sophistication and development. In agricultural societies, which emphasize the production of raw materials, **mechanization**, or the use of tools to accomplish tasks previously done by hand, dominates. As societies move toward industrialization and become more concerned with the mass production of goods, automation prevails. **Automation** involves the use of self-operating machines, as in an automated factory where autonomous robots assemble automobiles. Finally, as a society moves toward post-industrialization, it emphasizes service and information professions (Bell 1973). At this stage, technology shifts toward **cybernation**, whereby machines control machines—making production decisions, programming robots, and monitoring assembly performance.

What are the effects of science and technology on humans and their social world? How do science and technology help to remedy social problems and how do they contribute to social problems? Is technology, as Postman (1992) suggests, both a friend and a foe to humankind? This chapter addresses each of these questions.

The Global Context: The Technological Revolution

If religion was formerly the opiate of the masses, then surely technology is the opiate of the educated public . . .

JOHN MCDERMOTT
State University of New York

Less than 50 years ago, traveling across state lines was an arduous task, a long-distance phone call was a memorable event, and mail carriers brought belated news of friends and relatives from far away. Today, travelers journey between continents in a matter of hours, and for many, e-mail, faxes, video-conferencing, and electronic fund transfers have replaced conventional means of communication.

The world is a much smaller place than it used to be and will become even smaller as the technological revolution continues. The Internet is projected to have over 750 million users in over 100 countries by the year 2003 (Global Reach 2000). Americans constitute the largest share of Internet users (111 million), followed by the Japanese (18 million), English, Canadians, and Germans (14 million each). English-speakers comprise the largest language group online (48 percent), but non-English speakers constitute the fastest growing group on the Internet. For example, Brazil (6.9 million), China (6.3 million), and South Korea (5.7 million) were recently added to the list of top ten online countries edging out several European nations. Although 87 percent of all Internet users live in industrialized countries, there is some movement toward the Internet becoming a truly global medium as Africans and Latin Americans increasingly "get online" (Sampat 2000; World Employment Report 2001).

The movement toward globalization of technology is, of course, not limited to the use and expansion of the Internet. The world robot market, and America's share of it, continues to expand (IFR 1997); Latin America's computer industry is dominated by Texas-based manufacturer Compaq; Microsoft's Internet platform and support products are sold overseas; "globe-trotting scientists" collect skin and blood samples from remote islanders for genetic research (Shand 1998); and a global treaty regulating trade of genetically altered products between over a hundred nations has been signed (Pollack 2000).

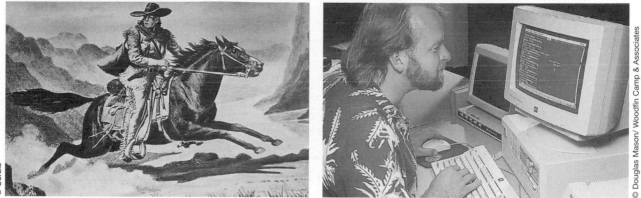

The world was made a smaller place in the mid- to late-1800s by the Pony Express. And it gets smaller all the time, particularly now that much of the world is connected to the Internet.

To achieve such scientific and technological innovations, sometimes called research and development (R&D), countries need material and economic resources. Research entails the pursuit of knowledge; development refers to the production of materials, systems, processes, or devices directed toward the solution of a practical problem. In 2000, the United States spent over $250 billion dollars on R&D. As in most other countries, U.S. funding sources were primarily from four sectors: private industry (69 percent), the federal government (27 percent), colleges and universities (3 percent), and other nonprofit organizations such as research institutes (1 percent) (NSF 2000).

Scientific discoveries and technological developments also require the support of a country's citizens and political leaders. For example, although abortion has been technically possible for years, millions of the world's citizen's live in countries where abortion is either prohibited or permitted only when the life of the mother is in danger. Thus the degree to which science and technology are considered good or bad, desirable or undesirable, is to a large extent socially constructed.

Postmodernism and the Technological Fix

Many Americans believe that social problems can be resolved through a **technological fix** (Weinberg 1966) rather than through social engineering. For example, a social engineer might approach the problem of water shortages by persuading people to change their lifestyle: use less water, take shorter showers, and wear clothes more than once before washing. A technologist would avoid the challenge of changing people's habits and motivations and instead concentrate on the development of new technologies that would increase the water supply. Social problems may be tackled through both social engineering and a technological fix. In recent years, for example, social engineering efforts to reduce drunk driving have included imposing stiffer penalties for drunk driving and disseminating public service announcements such as "Friends Don't Let Friends Drive Drunk." An example of a technological fix for the same problem is the development of car air bags, which reduce injuries and deaths resulting from car accidents.

Not all individuals, however, agree that science and technology are good for society. **Postmodernism,** an emerging world view, holds that rational thinking

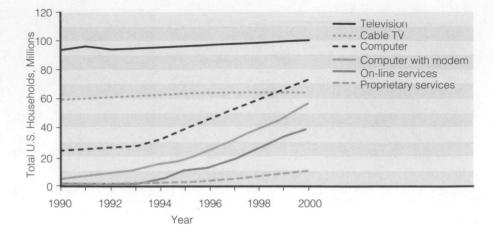

Figure 15.1
Technology in U.S. Households, 1990–2000
Source: FAS. 1998a. From Cyber-Strategy Project, http://www.fas.org/house.gif. Reprinted by permission.

I've got gigabytes. I've got megabytes. I'm voice-mailed. I'm e-mailed. I surf the Net. I'm on the Web. I am a Cyber-Man. So how come I feel so out of touch?

VOLKSWAGEN TELEVISION COMMERCIAL

and the scientific perspective have fallen short in providing the "truths" they were once presumed to hold. During the industrial era, science, rationality, and technological innovations were thought to pave the way to a better, safer, and more humane world. Today, postmodernists question the validity of the scientific enterprise, often pointing to the unforeseen and unwanted consequences of resulting technologies. Automobiles, for example, began to be mass-produced in the 1930s in response to consumer demands. But the proliferation of automobiles also led to increased air pollution and the deterioration of cities as suburbs developed, and today, traffic fatalities are the number one cause of accident-related deaths. Examine Figure 15.1 and consider the negative consequences of each of these modern-day technologies.

Sociological Theories of Science and Technology

Each of the three major sociological frameworks helps us to better understand the nature of science and technology in society.

Structural-Functionalist Perspective

Functionalists view science and technology as emerging in response to societal needs—that "[science] was born indicates that society needed it" (Durkheim [1925] 1973). As societies become more complex and heterogeneous, finding a common and agreed-upon knowledge base becomes more difficult. Science fulfills the need for an assumed objective measure of "truth" and provides a basis for making intelligent and rational decisions. In this regard, science and the resulting technologies are functional for society.

If society changes too rapidly as a result of science and technology, however, problems may emerge. When the material part of culture (i.e., its physical elements) changes at a faster rate than the nonmaterial (i.e., its beliefs and values) a **cultural lag** may develop (Ogburn 1957). For example, the typewriter, the conveyor belt, and the computer expanded opportunities for women to work outside the home. With the potential for economic independence, women were able to remain single or to leave unsatisfactory relationships and/or establish careers. But although new technologies have created new opportunities for women, beliefs about women's roles, expectations of female behavior, and values concerning equality, marriage, and divorce have lagged behind.

Robert Merton (1973), a functionalist and founder of the sub-discipline sociology of science, also argued that scientific discoveries or technological innovations may be dysfunctional for society and create instability in the social system. For example, the development of time-saving machines increases production, but also displaces workers and contributes to higher rates of employee alienation. Defective technology can have disastrous effects on society. In 1994, a defective Pentium chip was discovered to exist in over 2 million computers in aerospace, medical, scientific, and financial institutions, as well as schools and government agencies. Replacing the defective chip was a massive undertaking, but was necessary to avoid thousands of inaccurate computations and organizational catastrophe.

Conflict Perspective

Conflict theorists, in general, emphasize that science and technology benefit a select few. For some conflict theorists, technological advances occur primarily as a response to capitalist needs for increased efficiency and productivity and thus are motivated by profit. As McDermott (1993) notes, most decisions to increase technology are made by "the immediate practitioners of technology, their managerial cronies, and for the profits accruing to their corporations" (p. 93). In the United States, private industry spends more money on research and development than the federal government does. The Dalkon Shield and silicone breast implants are examples of technological advances that promised millions of dollars in profits for their developers. However, the rush to market took precedence over thorough testing of the products' safety. Subsequent lawsuits filed by consumers who argued that both products had compromised the physical well being of women resulted in large damage awards for the plaintiffs.

Science and technology also further the interests of dominant groups to the detriment of others. The need for scientific research on AIDS was evident in the early 1980s, but the required large-scale funding was not made available as long as the virus was thought to be specific to homosexuals and intravenous drug users. Only when the virus became a threat to mainstream Americans were millions of dollars made available for AIDS research. Hence, conflict theorists argue that granting agencies act as gatekeepers to scientific discoveries and technological innovations. These agencies are influenced by powerful interest groups and the marketability of the product rather than by the needs of society.

Finally, conflict theorists as well as feminists argue that technology is an extension of the patriarchal nature of society that promotes the interest of men and ignores the needs and interests of women. For example, washing machines, although time-saving devices, disrupted the communal telling of stories and the resulting friendships among women who gathered together to do their chores. Bush (1993) observes that in a " . . . society characterized by a sex-role division of labor, any tool or technique . . . will have dramatically different effects on men than on women (p. 204).

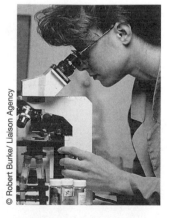

© Robert Burke/ Liaison Agency

Motivated by profit, private industry spends more money on research and development than the federal government does.

The Social Construction of the Hacking Community

Cyberstalking, pornography on the Internet, identity theft—crimes almost unheard of before the computer revolution and the enormous growth of the Internet. One such "high tech" crime, computer hacking, ranges from childish pranks to deadly viruses that shut down corporations. Below Jordan and Taylor (1998) enter the world of hackers, analyzing the nature of this illegal activity, hackers' motivations, and the social construction of the "hacking community."

Sample and Methods

Jordan and Taylor (1998) researched computer hackers and the hacking community through 80 semi-structured interviews, 200 questionnaires, and an examination of existing data on the topic. As is often the case in crime, illicit drug use, and other similarly difficult research areas, a random sample of

hackers was not possible. **Snowball sampling** is often the preferred method in these cases; that is, one respondent refers the researcher to another respondent, who then refers the researcher to another respondent, and so forth. Through their analysis the authors lend insight into this increasingly costly social problem and the symbolic interactionist notion of "social construction"—in this case, of an online community.

Findings and Conclusions

Computer hacking, or "unauthorized computer intrusion," is an increasingly serious problem, particularly in a society dominated by information technologies. Unlawful entry into computer networks or databases can be achieved by several means including (1) guessing someone's password, (2) tricking a computer about the identity of another

computer (called "IP spoofing"), or (3) "social engineering," a slang term referring to getting important access information by stealing documents, looking over someone's shoulder, going through their garbage, etc.

Hacking carries with it certain norms and values for, according to Jordan and Taylor, the hacking community can be thought of as a culture within a culture. The two researchers identify six elements of this socially constructed community:

- *Technology.* The core of the hacking community is the technology that allows it to occur. As one professor interviewed stated, the young today have " . . . lived with computers virtually from the cradle, and therefore have no trace of fear, not even a trace of reverence."
- *Secrecy.* The hacking community must, on the one hand, commit secret acts because their "hacks"

Symbolic Interactionist Perspective

New technologies alter the structure of our interests: the things we think about. They alter the character of our symbols: the things we think with. And they alter the nature of community: the arena in which thoughts develop.

NEIL POSTMAN
New York University

Knowledge is relative. It changes over time, over circumstances, and between societies. We no longer believe that the world is flat or that the earth is the center of the universe, but such beliefs once determined behavior as individuals responded to what they thought to be true. The scientific process is a social process in that "truths"—socially constructed truths—result from the interactions between scientists, researchers, and the lay public.

Kuhn (1973) argues that the process of scientific discovery begins with assumptions about a particular phenomenon (e.g., the world is flat). Because unanswered questions always remain about a topic (e.g., why don't the oceans drain?), science works to fill these gaps. When new information suggests that the initial assumptions were incorrect (e.g., the world is not flat), a new set of assumptions or framework emerges to replace the old one (e.g., the world is round). It then becomes the dominant belief or paradigm.

Symbolic interactionists emphasize the importance of this process and the impact social forces have on it. Conrad (1997), for example, describes the media's contribution in framing societal beliefs that alcoholism, homosexuality, and racial inequality are genetically determined. Technological innovations

are illegal. On the other hand, much of the motivation for hacking requires publicity in order to achieve the notoriety often sought. Further, hacking is often a group activity that bonds members together. As one hacker stated, hacking ". . . can give you a real kick some time. But it can give you a lot more satisfaction and recognition if you share your experiences with others. . . ."

- *Anonymity.* Whereas secrecy refers to the hacking act, anonymity refers to the importance of the hacker's identity remaining anonymous. Thus, for example, hackers and hacking groups take on names such as Legion of Doom, the Inner Circle I, Mercury, and Kaos, Inc.

- *Membership Fluidity.* Membership is fluid rather than static, often characterized by high turnover rates, in part, as a response to law enforcement pressures. Unlike more structured organiza-

tions, there are no formal rules or regulations.

- *Male Dominance.* Hacking is defined as a male activity and, consequently, there are few female hackers. Jordan and Taylor also note, after recounting an incidence of sexual harassment, that " . . . the collective identity hackers share and construct . . . is in part misogynist" (p. 768).

- *Motivation.* Contributing to the articulation of the hacking communities' boundaries are the agreed-upon definitions of acceptable hacking motivations, including: (1) addiction to computers, (2) curiosity, (3) excitement, (4) power, (5) acceptance and recognition, and (6) community service through the identification of security risks.

Finally, Jordan and Taylor (1998, 770) note that hackers also maintain group boundaries by distinguishing between their community and other social groups, including

"an antagonistic bond to the computer security industry (CSI)." Ironically, hackers admit a desire to be hired by the CSI, which would not only legitimize their activities but give them a steady income as well.

The authors conclude that the general fear of computers and of those who understand them underlies the common although inaccurate portrayal of hackers as pathological, obsessed computer "geeks." When journalist Jon Littman asked hacker Kevin Mitnick if he was being demonized because of increased dependence on and fear of information technologies, Mitnick replied, "Yeah . . . That's why they're instilling fear of the unknown. That's why they're scared of me. Not because of what I've done, but because I have the capability to wreak havoc" (Jordan & Taylor 1998, 776).

Source: Tim Jordan and Paul Taylor. 1998. "A Sociology of Hackers." *The Sociological Review* (November): 757–78.

are also affected by social forces, and their success is, in part, dependent upon the social meaning assigned to any particular product. If a product is defined as impractical, cumbersome, inefficient, or immoral, it is unlikely to gain public acceptance. Such is the case with RU486, an oral contraceptive that is widely used in France, Great Britain, and China, but whose availability, although legal in the United States, is opposed by a majority of Americans (Gallup 2000b; Gottlieb 2000).

Not only are technological innovations subject to social meaning, but who becomes involved in what aspects of science and technology is also socially defined. Men, for example, outnumber women three to one in earning computer science degrees. They also score higher on measures of computer aptitude and report higher computer use than women (Lewin 1998; Papadakis 2000; AAUW 2000). Societal definitions of men as being rational, mathematical, and scientifically minded and having greater mechanical aptitude than women are, in part, responsible for these differences. This chapter's *Social Problems Research Up Close* feature highlights one of the consequences of the masculinization of technology, as well as the ways in which computer hacker identities and communities are socially constructed.

Technology and the Transformation of Society

A number of modern technologies are considerably more sophisticated than technological innovations of the past. Nevertheless, older technologies have influenced the social world as profoundly as the most mind-boggling modern inventions. Postman (1992) describes how the clock—a relatively simple innovation that is taken for granted in today's world—profoundly influenced not only the workplace but the larger economic institution:

> The clock had its origin in the Benedictine monasteries of the twelfth and thirteenth centuries. The impetus behind the invention was to provide a more or less precise regularity to the routines of the monasteries, which required, among other things, seven periods of devotion during the course of the day. The bells of the monastery were to be rung to signal the canonical hours; the mechanical clock was the technology that could provide precision to these rituals of devotion . . . . What the monks did not foresee was that the clock is a means not merely of keeping track of the hours but also of synchronizing and controlling the actions of men. And thus, by the middle of the fourteenth century, the clock had moved outside the walls of the monastery, and brought a new and precise regularity to the life of the workman and the merchant . . . . In short, without the clock, capitalism would have been quite impossible. The paradox . . . [is] that the clock was invented by men who wanted to devote themselves more rigorously to God; it ended as the technology of greatest use to men who wished to devote themselves to the accumulation of money. (pp. 14–15)

Technology has far-reaching effects not only on the economy but on every aspect of social life. The following sections discuss societal transformations resulting from various modern technologies, including workplace technology, computers, the information highway, and science and biotechnology.

Technology in the Workplace

All workplaces, from doctors' offices to factories and from supermarkets to real estate corporations, have felt the impact of technology. The Office of Technology Assessment of the U.S. Congress estimates that over 7 million U.S. workers are under some type of computer surveillance, which lessens the need for supervisors and makes control by employers easier. Further, technology can make workers more accountable by gathering information about their performance. Through such time-saving devices as personal digital assistants and battery-powered store shelf labels, technology can enhance workers' efficiency. Technology is also changing the location of work. Over 3 million employees **telework** that is, complete all or part of their work away from the workplace. Fifty-one percent of corporations now allow at least some of their employees to work from home (Carey 1998).

Information technologies are also changing the nature of work. Lilly Pharmaceutical employees communicate via their own "intranet" on which all work-related notices are posted. Federal Express not only created a FedEx network for their 30,000 employees, but allowed customers to enter their package-tracking database, saving the Memphis-based company $2 million a year. It is estimated that one-fifth of all corporations are now using such

The time is fast approaching when anything an entire company can do, its every single employee can do. From his [her] desk. Anywhere on earth.

GLOBAL INTERNET PROJECT

telecommunication devices, leading to the potential of a paperless workplace (GIP 1998).

Robotic technology, sometimes called computer-aided manufacturing (CAM), has also revolutionized work, particularly in heavy industry such as automobile manufacturing. An employer's decision to use robotics depends on direct (e.g., initial investment) and indirect (e.g., unemployment compensation) costs, the feasibility and availability of robots performing the desired tasks, and the increased rate of productivity. Use of robotics may also depend on whether labor unions resist the replacement of workers by machines.

© Adam Lubroth/STONE

Automation means that machines can now perform the labor originally provided by human workers, such as the robots that perform tasks on automobile assembly lines.

The Computer Revolution

Early computers were much larger than the small machines we have today and were thought to have only esoteric uses among members of the scientific and military communities. In 1951, only about half a dozen computers existed (Ceruzzi 1993). The development of the silicon chip and sophisticated microelectronic technology allowed tens of thousands of components to be imprinted on a single chip smaller than a dime. The silicon chip led to the development of laptop computers, mini-television sets, cellular phones, electronic keyboards, and singing birthday cards. The silicon chip also made computers affordable. Although the first personal computer (PC) was developed only 20 years ago, today over 50 percent of American homes have one; 47 percent of adults use a computer in one or more places—64 million at work, 56 million at home and 11 million at schools (see Table 15.1) (Papadakis 2000; U.S. Department of Commerce 1999).

Not surprisingly, computer education has mushroomed in the last two decades (see Chapter 12). In 1971, 2,388 U.S. college students earned a bachelor's degree in computer and information sciences; by 1997 that number had increased to 24,768 (*Statistical Abstract 2000*). Universities are moving toward mandatory laptop policies for their students, and college and university spending on hardware and software needs are at an all-time high.

Computers are also big business, and the United States is one of the most successful producers of computer technology in the world, boasting several of the top selling desktops—Hewlett-Packard, Compaq, and Gateway. Retail sales of computers exceeds $25 billion with an average home computer cost of $1,100 (*Statistical Abstract 2000*; Gibson 2001). Americans spend over $400 million on educational software alone, and by the year 2006 spending on home computers is predicted to grow tenfold as consumers increasingly define PCs as a necessity rather than a luxury (Tanaka 2000; Klein 1998).

Computer software is also big business and in some cases too big. In 2000, a federal judge found that Microsoft Corporation was in violation of antitrust laws, which prohibit unreasonable restraint of trade. At issue were Microsoft's Windows operating system and the vast array of Windows-based applications (e.g., spreadsheets, word processors, tax software) i.e., applications that *only*

I think there is a world market for maybe five computers.

THOMAS WATSON
Chairman of IBM, speaking in 1943

■ **Table 15.1** *Public's Use of Home Computers, Work Computers, and the Internet*

Variable	1995	1997	1999
Percentage of public with			
Access to a home computer	37	43	54
Access to a computer at work	39	38	42
Subscription to online service at home	7	18	32
Average time spent per year			
On home computer for home computer users in hours	278	302	283
On work computer for work computer users in hours	818	971	957
Average time spent online at home per year in hours			
For the general public	6	29	86
For home computer users	15	67	159
For Internet users	80	161	296

Source: National Science Foundation. *Science and Engineering Indicators 2000*. Chapter 8, Table 8-5. http://www.nsf.gov

work with Windows. The court held that the 70,000 programs written exclusively for Windows made "competing against Microsoft impractical" (Markoff 2000). In an agreement reached between the U.S. Department of Justice and Microsoft Corporation, Microsoft will be divided into two companies—an operating-systems company and an applications company.

The Information Highway

Information technology, or **IT** for short, refers to any technology that carries information. Most information technologies were developed within a 100-year span: photography and telegraphy (1830s), rotary power printing (1840s), the typewriter (1860s), transatlantic cable (1866), telephone (1876), motion pictures (1894), wireless telegraphy (1895), magnetic tape recording (1899), radio (1906), and television (1923) (Beniger 1993). The concept of an "information society" dates back to the 1950s when an economist identified a work sector he called "the production and distribution of knowledge." In 1958, 31 percent of the labor force were employed in this sector—today more than 50 percent are. When this figure is combined with those in service occupations, more than 78 percent of the labor force are involved in the information society.

The development of a national information infrastructure was outlined in the Communications Act of 1994. An information infrastructure performs three functions (Kahin 1993). First, it carries information, just as a transportation system carries passengers. Second, it collects data in digital form that can be understood and used by people. Finally, it permits people to communicate with one another by sharing, monitoring, and exchanging information based on common standards and networks. In short, an information infrastructure facilitates telecommunications, knowledge, and community integration.

The **Internet** is an international information infrastructure—a network of networks—available through universities, research institutes, government agencies, and businesses. Today about half of all Americans use the Internet. U.S. users are more likely to be male (52 percent), white (85 percent), and between the ages of 12 and 29 (CyberAtlas 2000a; Headcount 1998). However, the fastest growing group of U.S. Internet users is 55 and older and the number of online seniors is projected to triple by 2004 (CyberAtlas 2000b).

Despite dramatic growth, there is some evidence that Internet use is slowing down. Researchers from the Pew Internet and American Life Project found that of current non-users, 57 percent said they had no plans of getting online. The most common reasons given included expense, difficulty of use, and that "[T]he Internet is a dangerous thing." According to the survey results, non-users "are less networked in their social lives, less trusting, and more concerned about their privacy being breached" than their user counterparts (Meeks 2000, 3).

Even **e-commerce**, or the buying and selling of goods and services over the Internet, has lost ground as companies like Amazon and Yahoo struggle to stay afloat. Other Internet businesses, for example e-Bay, have grown considerably. In 2000, e-Bay reported having 22.5 million registered users and revenues of over $400 million (Cohen 2001). Forty-eight percent of Internet users, one-fourth of the U.S. adult population, report making online purchases (Gallup Poll 2000a). For example, in 1999, 52 million Americans went online to make travel arrangements and reservations. Interestingly, online consumerism may actually help the environment. A team of energy experts report that, given present trends, by 2007 "e-commerce could prevent the annual release of 35 million tons of greenhouse gases by reducing the need for up to 3 billion square feet of energy-consuming office buildings and malls in the United States" (Sampat 2000, 94).

> The question pornography poses in cyberspace is the same one it poses everywhere else: whether anything will be done about it.
>
> CATHARINE MACKINNON
> *Law professor*

Science and Biotechnology

While recent computer innovations and the establishment of an information highway have led to significant cultural and structural changes, science and its resulting biotechnologies have led to not only dramatic changes, but also hotly contested issues (see Figure 15.2). Here we will look at some of the issues raised by developments in genetics and reproductive technology.

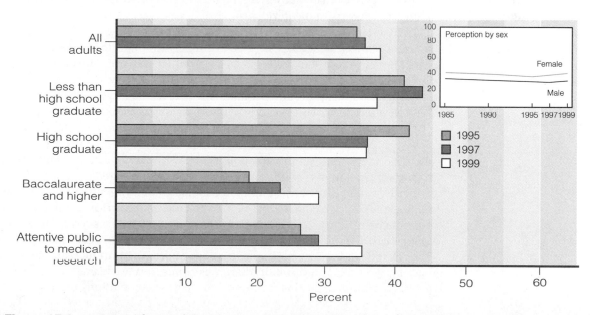

Figure 15.2 *Percentage of U.S. Adults Who View the Harmful Effects of Genetic Engineering as Outweighing the Benefits*

Source: National Science Foundation. *Science and Engineering Indicators 2000*. Chapter 8, Figure 8-13. http://www.nsf.gov

Genetics   Molecular biology has led to a greater understanding of the genetic material found in all cells—DNA (deoxyribonucleic acid)—and with it the ability for **genetic screening**. Currently, researchers are trying to complete genetic maps that will link DNA to particular traits. Already, specific strands of DNA have been identified as carrying such physical traits as eye color and height, as well as such diseases as breast cancer, cystic fibrosis, prostate cancer, and Alzheimer's.

The Human Genome Project, a 10-year effort to decode human DNA, was completed in 2000. The decoding entailed "identifying and placing in order the 3.1 billion-unit long sequence that make up human DNA" (Potter 2001,1). Conclusion of the project will revolutionize medicine. The hope is that if a defective or missing gene can be identified, it may be possible to get a healthy duplicate and transplant it to the affected cell. This is known as **gene therapy.** Alternatively, viruses have their own genes that can be targeted for removal. Experiments are now underway to accomplish these biotechnological feats.

Genetic engineering is the ability to manipulate the genes of an organism in such a way that the natural outcome is altered. Genetic engineering is accomplished by splicing the DNA from one organism into the genes of another. Often, however, unwanted consequences ensue. For example, through genetic engineering some plants are now self-insecticiding, that is, the plant itself produces an insect-repelling substance. Ironically, the plant's continual production of the insecticide, in contrast to only sporadic application by farmers, is leading to insecticide-resistant pests (Ehrenfeld 1998). The debate over genetically engineered crops is ongoing as advocates note the prospects for expanded food production and proponents question health and environmental consequences (see Chapter 10).

Reproductive Technologies   The evolution of "reproductive science" has been furthered by scientific developments in biology, medicine, and agriculture. At the same time, however, its development has been hindered by the stigma associated with sexuality and reproduction, its link with unpopular social movements (e.g., contraception), and the feeling that such innovations challenge the natural order (Clarke 1990). Nevertheless, new reproductive technologies have been and continue to be developed.

> It is now a matter of a handful of years before biologists will be able to irreversibly change the evolutionary wisdom of billions of years with the creation of new plants, new animals, and new forms of human and post human beings.
>
> TED HOWARD, JEREMY RIFKIN
> *Biotechnology critics*

In **in-vitro fertilization** (IVF), an egg and a sperm are united in an artificial setting such as a laboratory dish or test tube. Although the first successful attempt at IVF occurred in 1944, it was not until 1978 that the first test-tube baby, Louise Brown, was born. Today, more than 300 fertility clinics in the United States provide this procedure, resulting in over 10,000 live births. Criticisms of IVF are often based on traditional definitions of the family and the legal complications created when a child can have as many as five potential parental ties—egg donor, sperm donor, surrogate mother, and the two people who raise the child (depending on the situation, IVF may not involve donors and/or a surrogate). Litigation over who are the "real" parents has already occurred.

Perhaps more than any other biotechnology, abortion epitomizes the potentially explosive consequences of new technologies. **Abortion** is the removal of an embryo or fetus from a woman's uterus before it can survive on its own. Since the U.S. Supreme Court's ruling in *Roe v. Wade* in 1973, abortion has been legal in the United States. However, recent Supreme Court decisions have limited the *Roe v. Wade* decision. In *Planned Parenthood of Southeastern Pennsylvania v. Casey*, the Court ruled that a state may restrict the conditions under which an

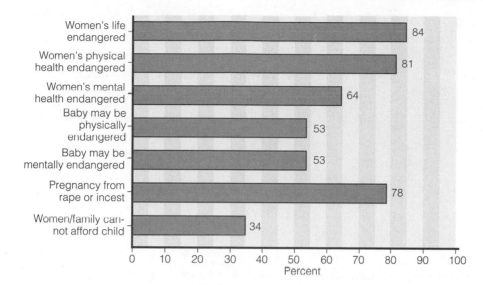

■ **Figure 15. 3** *Support for Legal Abortions under Specific Circumstances, 2000*
Source: *Gallup Poll. 2000.* "Abortion Issues." http://gallup.com/poll/indicators/indabortion.asp

abortion is granted, such as by requiring a 24-hour waiting period or parental consent for minors.

Most recent debates concern intact dilation and extraction (D&X) abortions. Opponents refer to such abortions as **partial birth abortions** because the limbs and the torso are typically delivered before the fetus has expired. D&X abortions are performed because the fetus has a serious defect, the woman's health is jeopardized by the pregnancy, or both. Polls indicate that public support for partial birth abortions has declined over the years (Gallup Poll 2000b). The U.S. Supreme Court is presently hearing arguments on whether or not states can legally ban the procedure.

Abortion is a complex issue for everyone, but especially for women, whose lives are most affected by pregnancy and childbearing. Women who have abortions are disproportionately poor, unmarried minorities who say they intend to have children in the future. Abortion is also a complex issue for societies, which must respond to the pressures of conflicting attitudes toward abortion and the reality of high rates of unintended and unwanted pregnancy. Figure 15.3 illustrates the percentage of Americans who support legal abortions under various circumstances.

Attitudes toward abortion tend to be polarized between two opposing groups of abortion activists—pro-choice and pro-life. Advocates of the pro-choice movement hold that freedom of choice is a central human value, that procreation choices must be free of government interference, and that because the woman must bear the burden of moral choices, she should have the right to make such decisions. Alternatively, pro-lifers hold that the unborn fetus has a right to live and be protected, that abortion is immoral, and that alternative means of resolving an unwanted pregnancy should be found. Assess your attitudes toward abortion in this chapter's *Self and Society* feature.

In July of 1996, scientist Ian Wilmut of Scotland successfully cloned an adult sheep named Dolly. To date, cattle, goats, mice, and pigs have also been cloned. This technological breakthrough has caused worldwide concern about the possibility of human cloning. One argument in favor of developing human cloning

Abortion Attitude Scale

This is not a test. There are no wrong or right answers to any of the statements, so just answer as honestly as you can. The statements ask you to tell how you feel about legal abortion (the voluntary removal of a human fetus from the mother during the first three months of pregnancy by a qualified medical person). Tell how you feel about each statement by circling one of the choices beside each sentence. Respond to each statement and circle only one response.

Strongly Agree 5	Agree 4	Slightly Agree 3	Slightly Disagree 2	Disagree 1	Strongly Disagree 0

	Strongly Agree	Agree	Slightly Agree	Slightly Disagree	Disagree	Strongly Disagree
1. The Supreme Court should strike down legal abortions in the United States.	5	4	3	2	1	0
2. Abortion is a good way of solving an unwanted pregnancy.	5	4	3	2	1	0
3. A mother should feel obligated to bear a child she has conceived.	5	4	3	2	1	0
4. Abortion is wrong no matter what the circumstances are.	5	4	3	2	1	0
5. A fetus is not a person until it can live outside its mother's body.	5	4	3	2	1	0
6. The decision to have an abortion should be the pregnant mother's.	5	4	3	2	1	0
7. Every conceived child has the right to be born.	5	4	3	2	1	0
8. A pregnant female not wanting to have a child should be encouraged to have an abortion.	5	4	3	2	1	0
9. Abortion should be considered killing a person.	5	4	3	2	1	0
10. People should not look down on those who choose to have abortions.	5	4	3	2	1	0
11. Abortion should be an available alternative for unmarried, pregnant teenagers.	5	4	3	2	1	0
12. Persons should not have the power over the life or death of a fetus.	5	4	3	2	1	0
13. Unwanted children should not be brought into the world.	5	4	3	2	1	0
14. A fetus should be considered a person at the moment of conception.	5	4	3	2	1	0

SCORING AND INTERPRETATION

As its name indicates, this scale was developed to measure attitudes toward abortion. It was developed by Sloan (1983) for use with high school and college students. To compute your score, first reverse the point scale for Items 1, 3, 4, 7, 9, 12, and 14. Total the point responses for all items. Sloan provided the following categories for interpreting the results:

    70–56 Strong proabortion
    55–44 Moderate proabortion
    43–27 Unsure
    26–16 Moderate prolife
    15–0   Strong prolife

RELIABILITY AND VALIDITY

The Abortion Attitude Scale was administered to high school and college students, Right to Life group members, and abortion service personnel. Sloan (1983) reported a high total test estimate of reliability (0.92). Construct validity was supported in that Right to Life members' mean scores were 16.2; abortion service personnel mean scores were 55.6; and other groups' scores fell between these values.

Source: "Abortion Attitude Scale" by L.A. Sloan. Reprinted with permission from the *Journal of Health Education* Vol. 14. No. 3, May/June 1983. The *Journal of Health Education* is a publication of the American Allegiance for Health, Physical Education, Recreation and Dance. 1900 Association Drive, Reston, Virginia 20191.

technology is its medical value; it may potentially allow everyone to have "their own reserve of therapeutic cells that would increase their chance of being cured of various diseases, such as cancer, degenerative disorders and viral or inflammatory diseases" (Kahn 1997, 54). Human cloning could also provide an alternative reproductive route for couples who are infertile and for those in which one partner is at risk for transmitting a genetic disease.

Arguments against cloning are largely based on moral and ethical considerations. Critics of human cloning suggest that, whether used for medical therapeutic purposes or as a means of reproduction, human cloning is a threat to human dignity. For example, cloned humans would be deprived of their individuality and, as Kahn (1997, 119) points out, "creating human life for the sole purpose of preparing therapeutic material would clearly not be for the dignity of the life created." Nonetheless, British physicians recently recommended that therapeutic cloning be permitted. **Therapeutic cloning** entails using stem cells from human embryos. Stem cells can produce any type of cell in the human body and, thus, can be used for "grow-your-own organ transplants and tissue for use in treating medical conditions ranging from paralysis to diabetes" (Reuters 2000,1). Because the use of stem cells entails the destruction of human embryos, many conservatives are opposed to the practice and are hoping that President Bush's anti-abortion stance will make him sympathetic to their cause.

Despite what appears to be a universal race to the future and the indisputable benefits of such scientific discoveries as the workings of DNA and the technology of IVF, some people are concerned about the duality of science and technology. Science and the resulting technological innovations are often life assisting and life giving; they are also potentially destructive and life threatening. The same scientific knowledge that led to the discovery of nuclear fission, for example, led to the development of both nuclear power plants and the potential for nuclear destruction. Thus, we now turn our attention to the problems associated with science and technology.

Societal Consequences of Science and Technology

Scientific discoveries and technological innovations have implications for all social actors and social groups. As such, they also have consequences for society as a whole.

Alienation, Deskilling, and Upskilling

As technology continues to play an important role in the workplace, workers may feel there is no creativity in what they do—they feel alienated (see Chapter 11). The movement from mechanization, to automation, to cybernation increasingly removes individuals from the production process, often relegating them to flipping a switch, staring at a computer monitor, or entering data at a keyboard. For example, many low-paid employees, often women, sit at computer terminals for hours entering data and keeping records for thousands of businesses, corporations, and agencies. The work that takes place in these "electronic sweatshops" is monotonous, solitary, and provides little autonomy.

Men [and women] have become the tools of their tools.

HENRY DAVID THOREAU
Author/social activist

Not only are these activities routine, boring, and meaningless, they promote **deskilling**, that is, "labor requires less thought than before and gives them [workers] fewer decisions to make" (Perrolle 1990, 338). Deskilling stifles development of alternative skills and limits opportunities for advancement and creativity as old skill sets become obsolete. To conflict theorists, deskilling also provides the basis for increased inequality for "[T]hroughout the world, those who control the means of producing information products are also able to determine the social organization of the 'mental labor' which produces them" (Perrolle 1990, 337).

Technology in some work environments, however, may lead to **upskilling**. Unlike deskilling, upskilling reduces alienation as employees find their work more rather than less meaningful, and have greater decision-making powers as information becomes decentralized. Futurists argue that upskilling in the workplace could lead to a "horizontal" work environment in which "employees do not so much what they are told to do, but what their expansive knowledge of the entire enterprise suggests to them needs doing" (GIP 1998).

Social Relationships and Social Interaction

> Online life is rich and rewarding, but it's no substitute for face-to-face interaction.
>
> JACQUES LESLIE
> *Journalist/writer*

Technology affects social relationships and the nature of social interaction. The development of telephones has led to fewer visits with friends and relatives; with the coming of VCRs and cable television, the number of places where social life occurs (e.g., movie theaters) has declined. Even the nature of dating has changed as computer networks facilitate cyberdates and "private" chat rooms. As technology increases, social relationships and human interaction are transformed.

Technology also makes it easier for individuals to live in a cocoon—to be self-sufficient in terms of finances (e.g., Quicken), entertainment (e.g., pay-per-view movies), work (e.g., telework), recreation (e.g., virtual reality), shopping (e.g., e-Bay), communication (e.g., Internet), and many other aspects of social life. Ironically, although technology can bring people together, it can also isolate them from each other, leading the Amish to ban any technology that "is seen as a threat to the cohesion of the community or might contaminate the culture's values" (Hafner 1999, 1). For example, some technological innovations replace social roles—an answering machine may replace a secretary, a computer-operated vending machine may replace a waitperson, an automatic teller machine may replace a banker, and closed circuit television, a teacher (Johnson 1988; Winner 1993; Schwartz 2001). These technologies may improve efficiency, but they also reduce necessary human contact. Some researchers believe that the recent reduction in the number of online shoppers reflects the need for social interaction in retail activities (Schwartz 2001).

Loss of Privacy and Security

When Massachusetts Institute of Technology (M.I.T.) professor of media technology Nicholas Negroponte was asked "What do you fear most in thinking about the digital future?," his response was "privacy and security."

> When I send you a message in the future, I want you to be sure it is from me; that when that message goes from me to you, nobody is listening in; and when it lives on your desk, nobody is snooping later. (Negroponte 1995, 88)

Schools, employers, and the government are increasingly using technology to monitor individuals' performance and behavior. Up from 27 percent in 1999,

38.2 percent of employers in 2000 reported monitoring employees' e-mail (Guernsey 2000). Today, through cybernation, machines monitor behavior by counting a telephone operator's minutes online, videotaping a citizen walking down a city street, or tracking the whereabouts of a student or faculty member on campus (Quick 1998; White 2000).

Employers and schools may subject individuals to drug testing technology (see Chapter 3) and annually nearly 500,000 Americans are victimized by identity theft (see Chapter 4) (Arnold 2001). Through computers, individuals can obtain access to phone bills, tax returns, medical reports, credit histories, bank account balances, and driving records. Some companies sell such personal information. Unauthorized disclosure is potentially devastating. If a person's medical records indicate that he or she is HIV-positive, for example, that person could be in danger of losing his or her job or health benefits. DNA testing of hair, blood, or skin samples can be used to deny individuals insurance, employment, or medical care. In response to the possibility of such consequences, Brin (1998), author of *The Transparent Society*, argues that because it is impossible to prevent such intrusions, "reciprocal transparency," or complete openness, should prevail. If organizations can collect the information, then citizens should have access to it and to its uses.

Technology has created threats not only to the privacy of individuals, but also to the security of entire nations. Computers and modems can be used (or misused) in terrorism and warfare to cripple the infrastructure of a society and tamper with military information and communication operations.

© David Sams/ Stock Boston

The development of new technologies has produced new forms of work and new demands for highly skilled workers in certain segments of the labor market.

> People aren't aware that mouse clicks can be traced, packaged, and sold.
>
> **LARRY IRVING**
> *U.S. Commerce Department*

Unemployment

Some technologies replace human workers—robots replace factory workers, word processors displace secretaries and typists, and computer-assisted diagnostics reduce the need for automobile mechanics. It is estimated, for example, that 52,000 branch banks will be replaced by automatic teller machines in the next decade. The cost of a teller transaction is $1.07; an electronic transaction, 7¢ (Fix 1994).

Technology is likely to cause worldwide increases in unemployment, according to activist Jeremy Rifkin, whose book, *The End of Work* (1996), predicts a global reduction in service-sector employees. Unlike previous decades, according to Rifkin, when uprooted workers moved from farms to factories to offices, today's technologically displaced workers have no place to go. It should be noted that all-time-low unemployment rates seem to contradict Rifkin's dire forecasts.

There is little doubt, however, that technology changes the nature of work and the types of jobs available. For example, fewer semiskilled workers are needed because many of these jobs have been replaced by machines. The jobs

> Human skills are subject to obsolescence at a rate perhaps unprecedented in American history.
>
> **ALAN GREENSPAN**
> *Chairman, Federal Reserve*

that remain, often white-collar jobs, require more education and technological skills. Technology thereby contributes to the split labor market as the pay gulf between skilled and unskilled workers continues to grow. For example, Addison, Fox, and Ruhm (2000) report that employees who use computers at work are at a lower risk of losing their jobs than non-computer users.

The Digital Divide

One of the most significant social problems associated with science and technology is the increased division between the classes. As Welter (1997, 2) notes,

> . . . it is a fundamental truth that people who ultimately gain access to, and who can manipulate, the prevalent technology are enfranchised and flourish. Those individuals (or cultures) that are denied access to the new technologies, or can not master and pass them on to the largest number of their offspring, suffer and perish.

The fear that technology will produce a "virtual elite" (Levy 1997) is not uncommon. Several theorists hypothesize that as technology displaces workers, most notably the unskilled and uneducated, certain classes of people will be irreparably disadvantaged—the poor, minorities, and women (Hayes 1998; Welter 1997). There is even concern that biotechnologies will lead to a "genetic stratification," whereby genetic screening, gene therapy, and other types of genetic enhancements are available only to the rich (Mehlman &Botkin 1998).

The wealthier the family, for example, the more likely the family is to have a computer. Of American families with an income of $75,000 a year or more, 74 percent have at least one PC (Lohr 1999). However, across all income groups only about 45 percent of U.S. homes have a computer. Further, 46.6 percent of white Americans own a computer compared to 23.2 percent of African Americans (Papadakis 2000).

Racial disparities also exist in Internet access—21 percent of white children have access to the Internet at school compared with 12 percent of Hispanic children and 15 percent of black children (Zehr 2000). Inner-city neighborhoods, disproportionately populated by racial and ethnic minorities, are simply less likely to be "wired," that is, to have the telecommunications hardware necessary for schools to access online services. In fact, cable and telephone companies are less likely to lay fiber optics in these areas—a practice called "information apartheid" or "electronic redlining." Students who live in such neighborhoods are technologically disadvantaged and may never catch up to their middle-class counterparts (Welter 1997).

The cost of equalizing such differences is enormous, but the cost of not equalizing them may be even greater. Employees who are technologically skilled have higher incomes than those who are not—up to 15 percent higher (Hancock 1995; World Employment Report 2001). Further, technological disparities exacerbate the structural inequities perpetuated by the split labor force and the existence of primary and secondary labor markets.

Mental and Physical Health

Some new technologies have unknown risks. Biotechnology, for example, has promised and, to some extent, has delivered everything from life-saving drugs to hardier pest-free tomatoes. Biotechnologies have also, however, created **technology-induced diseases** such as those experienced by Chellis

Glendinning (1990). Glendinning, after using the "Pill" and, later, the Dalkon Shield IUD, became seriously ill.

> Despite my efforts to get help, medical professionals did not seem to know the root of my condition lay in immune dysfunction caused by ingesting artificial hormones and worsened by chronic inflammation. In all, my life was disrupted by illness for twenty years, including six years spent in bed . . . . For most of the years of illness, I lived in isolation with my problem. Doctors and manufacturers of birth control technologies never acknowledged it or its sources.

Other technologies that pose a clear risk to a large number of people include nuclear power plants, DDT, automobiles, X rays, food coloring, and breast implants.

The production of new technologies may also place manufacturing employees in jeopardy. For example, the electronics industry uses thousands of hazardous chemicals, including freon, acetone, and sulfuric and nitric acids:

> [Semiconductor] workers are expected to dip wafer-thin silicon that has been painted with photoresist into acid baths. The wafers are then heated in gas-filled ovens, where the gas chemically reacts with the photosensitive chemicals. After drying, the chips are then bonded to ceramic frames, wires are attached to contacts and the chip is encapsulated with epoxy. These integrated circuits are then soldered onto boards, and the whole device is cleaned with solvents. Gases and silicon lead to respiratory and lung diseases, acids to burning and blood vessel damage, and solvents to liver damage. (Hosmer 1986, 2)

Finally, technological innovations are, for many, a cause of anguish and stress particularly when the technological changes are far-reaching (Hormats 2001). As many as 10 percent of Internet users are "addicted" to being online and, alternatively, nearly 60 percent of workers report being "technophobes," that is, fearful of technology (Papadakis 2000; Boles & Sunoo 1998). Says Dr. Michelle Weil, a clinical psychologist, the key to dealing with technophobia is to decide "which tools make sense in a person's life and will give them more control, and, ultimately, more enjoyment" (Kelly 1997, 4). This chapter's *The Human Side* feature deals directly with this issue.

■ The history of the "rocky road of progress" is made up of apparent technological wonders that turned out to be techno-threats or outright disasters.

ABIGAIL TRAFFORD AND
ANDREA GABOR
Journalists

The Challenge to Traditional Values and Beliefs

Technological innovations and scientific discoveries often challenge traditionally held values and beliefs, in part because they enable people to achieve goals that were previously unobtainable. Before recent advances in reproductive technology, for example, women could not conceive and give birth after menopause. Technology that allows postmenopausal women to give birth challenges societal beliefs about childbearing and the role of older women. Macklin (1991) notes that the techniques of egg retrieval, in vitro fertilization, and gamete intrafallopian transfer (GIFT) make it possible for two different women to each make a biological contribution to the creation of a new life. Such technology requires society to reexamine its beliefs about what a family is and what a mother is. Should family be defined by custom, law, or the intentions of the parties involved?

Medical technologies that sustain life lead us to rethink the issue of when life should end. The increasing use of computers throughout society challenges the traditional value of privacy. New weapons systems challenge the traditional idea of war as something that can be survived and even won. And cloning chal-

■ Technological systems are both socially constructed and society shaping.

THOMAS HUGHES
Social scientist

Data Smog Removal Techniques

The following excerpt from David Shenk's *Data Smog: Surviving the Information Glut* (1997), outlines techniques for simplifying your life and reducing the stress associated with a "high-tech" lifestyle.

Most of us have excess information in our lives, distracting us, pulling us away from our priorities and from a much-desired tranquility. If we stop just for a moment to look (and listen) around us, we will begin to notice a series of data streams that we'd be better off without, including some distractions we pay handsomely for.

- *Turn the television off.* There is no quicker way to regain control of the pace of your life, the peace of your home, and the content of your thinking than to turn off the appliance that supplies for all-too-many of us the ambiance of our lives . . .
- *Leave the pager and cell phone behind.* It is thrilling to be in touch with the world at all times, but it's also draining and interfering. Are wireless communicators instruments of liberation, freeing people to be more mobile with their lives—or are they more like electronic leashes, keeping people more plugged into their work and their info-glutted lives than is necessary and healthy?
- *Limit your e-mail.* If we're spending too much time each day reading and answering e-mail that has virtually no value, we must take steps to control it. Ask people (nicely) not to forward trivia indiscriminately.

"Unsubscribe" to the news groups that you're no longer really interested in. Tell spammers that you have no interest in their product, and ask them to remove you from their customer list.

- *Say no to dataveillance.* With some determination and a small amount of effort, one can also greatly reduce the amount of junk mail and unsolicited sales phone calls. It involves writing just a few letters, requesting to have your name put on "do-not-disturb" lists, which some 75 percent of direct marketers honor . . .
- *Resist upgrade mania.* Remember: Upgrades are designed to be sales tools, not to give customers what they've been clamoring for.
- *Cleanse your system with "data-fasts."* . . . Take some time to examine your daily intake and consider whether or not your info diet needs some fine-tuning. Take some data naps in the afternoon, during which you stay away from electronic information for a prescribed period. You could also consider limiting yourself to no more than a certain number of hours on the Internet each week, or at least balancing the amount of time spent online with an equal amount of time reading books . . . [P]eriodic fasts of a week or month have a remarkably rejuvenating effect. One sure way to gauge the value of something, after all, is to go without it for a while.

Source: Shenk, David. 1997. *Data Smog: Surviving the Information Glut.* San Francisco: HarperEdge, pp. 184–189. Reprinted by permission.

lenges our traditional notions of family, parenthood, and individuality. Toffler (1970) coined the term **future shock** to describe the confusion resulting from rapid scientific and technological changes that unravel our traditional values and beliefs.

Strategies for Action: Controlling Science and Technology

As technology increases, so does the need for social responsibility. Nuclear power, genetic engineering, cloning, and computer surveillance all increase the need for social responsibility: " . . . technological change has the effect of enhancing the importance of public decision making in society, because technology is continually creating new possibilities for social action as well as new problems that have to be dealt with" (Mesthene 1993, 85). In the following section, various aspects of the public debate are addressed, including science, ethics and the law, the role of corporate America, and government policy.

Science, Ethics, and the Law

Science and its resulting technologies alter the culture of society through the challenging of traditional values. Public debate and ethical controversies, however, have led to structural alterations in society as the legal system responds to calls for action. For example, in 2000 Michigan and Massachusetts passed laws prohibiting discrimination in insurance and employment decisions based on an individual's genetic characteristics. Massachusetts also banned genetic discrimination in financial services as well as housing (Lewin 2001). Further, the California legislature recently passed a bill that requires physicians to inform women, in writing, of the risks of egg donation, and "directed state health departments to consider . . . whether to set lifetime limits on the number of egg donations an individual may make and on the compensation an egg donor may receive" (Dresser 2000, 26).

Are such regulations necessary? In a society characterized by rapid technological and thus social change—a society where custody of frozen embryos is part of the divorce agreement—many would say yes. Cloning, for example, is one of the most hotly debated technologies in recent years. Bioethicists and the public vehemently debate the various costs and benefits of this scientific technique. Despite such controversy, however, the chairman of the National Bioethics Advisory Commission has said that human cloning will be "very difficult to stop" (McFarling 1998). Such comments have fueled state legislative action. As of January 1, 1999, California became the first state to outlaw human cloning (Eibert 1998).

Should the choices that we make, as a society, be dependent upon what we can do, or what we should do? Whereas scientists and the agencies and corporations who fund them often determine the former, who should determine the latter? Although such decisions are likely to have a strong legal component, that is, they must be consistent with the rule of law and the constitutional right of scientific inquiry (Eibert 1998; White 2000), legality or the lack thereof often fails to answer the question, what should be done? *Roe v. Wade* (1973) did little to squash the public debate over abortion and, more specifically, the question of when life begins. Thus it is likely that the issues surrounding the most controversial of technologies will continue into the twenty-first century and with no easy answers.

> Prohibiting scientific and medical activities would also raise troubling enforcement issues . . . Would they [FBI] raid research laboratories and universities? Seize and read the private medical records of infertility patients? Burst into operating rooms with their guns drawn? Grill new mothers about how their babies were conceived?
>
> **Mark Eibert**
> *Attorney*

Technology and Corporate America

As philosopher Jean-Francois Lyotard notes, knowledge is increasingly produced to be sold (Powers 1998). The development of genetically altered crops, the commodification of women as egg donors, and the harvesting of regenerated organ tissues are all examples of potentially market-driven technologies. Like the corporate pursuit of computer technology, profit-motivated biotechnology creates several concerns.

First and foremost is the concern that only the rich will have access to such life-saving technologies as genetic screening and cloned organs. Such fears are justified. Several "companies with enigmatic names such as Progenitor, Millennium Pharmaceuticals, and Darwin Molecular have been pinpointing and patenting human life with the help of $4.5 billion in investments from pharmaceuticals companies" (Shand 1998, 46). Millennium Pharmaceutical holds the patent on the melanoma gene and the obesity gene; Darwin Molecular controls the premature aging gene, and Progenitor the gene for schizophrenia.

These patents result in **gene monopolies**, which could lead to astronomical patient costs for genetic screening and treatment. One company's corporate literature candidly states that its patent of the breast cancer gene will limit competition and lead to huge profits (Shand 1998, 47). The biotechnology industry argues that such patents are the only way to recoup research costs which, in turn, lead to further innovations. To date, about 1,000 genes have been patented and thousands of others are in the process of being claimed (Regalado 2000).

The commercialization of technology causes several other concerns, including the tendency for discoveries to remain closely guarded secrets rather than collaborative efforts, and issues of quality control (Rabino 1998; Lemonick & Thompson 1999). Further, industry involvement has made government control more difficult as researchers depend less and less on federal funding. Over 65 percent of research and development in the United States is supported by private industry using their own company funds (NIST 2000).

Finally, although there is little doubt that profit acts as a catalyst for some scientific discoveries, other less commercially profitable but equally important projects may be ignored. As biologist Isaac Rabino states, "[I]magine if early chemists had thrown their energies into developing profitable household products before the periodic table was discovered . . . ." (Rabino 1998, 112).

Runaway Science and Government Policy

Science and technology raise many public policy issues. Policy decisions, for example, address concerns about the safety of nuclear power plants, the privacy of electronic mail, the hazards of chemical warfare, and the legality of surrogacy. In creating science and technology, have we created a monster that has begun to control us rather than the reverse? What controls, if any, should be placed on science and technology? And are such controls consistent with existing law? Consider the use of Napster software to download music files (the question of intellectual property rights and copyright infringement); a proposed federal regulation that would require schools to use blocking software on computers or lose federal funding (free speech issues); and a recent court decision ordering investigation of *Carnivore*, a FBI surveillance program that can search every message that passes through an Internet service provider (fourth amendment privacy issues) (White 2000; Kaplan 2000; Schaefer 2001).

As president, I will prohibit genetic discrimination, criminalize identity theft, and guarantee the privacy of medical and sensitive financial records.

GEORGE W. BUSH
U.S. President

The government, often through regulatory agencies and departments, prohibits the use of some technologies (e.g., assisted-suicide devices) and requires others (e.g., seat belts). For example, the Food and Drug Administration, the Environmental Protection Agency, and the Agriculture Department recently investigated the use of genetically altered corn—corn that had only been approved for animal feed—in the making of Taco Bell taco shells (Brasher 2000). Concern with genetically altered crops has led to an additional $1.7 million in federal funds for research and a tightening of federal guidelines (Petersen 2000). Recently, a 130-nation treaty was adopted which permits countries to "bar imports of genetically altered seeds, microbes, animals and crops that they deem a threat to their environment" (Pollack 2000,1)

Finally, the federal government has instituted several initiatives dealing with technology-related crime. For example, the U.S. Senate Judiciary Committee has approved the *Internet Integrity and Critical Infrastructure Protection Act* which clarifies the federal role in prosecuting computer hackers and establishes a National Cyber

Crime Technical Support Center (Johnson 2000). As a 1999 U.S. Attorney General report notes, "many of the attributes of this [Internet] technology—low cost, ease of use, and anonymous nature, among others—make it an attractive medium for fraudulent scams, child sexual exploitation, and . . . cyberstalking" (U.S. Department of Justice 1999,1). More than 50 bills concerning the protection of Internet users were introduced in Congress in 2000 (Schaefer 2001).

Understanding *Science and Technology*

What are we to understand about science and technology from this chapter? As functionalists argue, science and technology evolve as a social process and are a natural part of the evolution of society. As society's needs change, scientific discoveries and technological innovations emerge to meet these needs, thereby serving the functions of the whole. Consistent with conflict theory, however, science and technology also meet the needs of select groups and are characterized by political components. As Winner (1993) notes, the structure of science and technology conveys political messages including "[P]ower is centralized," "[T]here are barriers between social classes," "[T]he world is hierarchically structured," and "[T]he good things are distributed unequally" (Winner 1993, 288).

The scientific discoveries and technological innovations that are embraced by society as truth itself are socially determined. Research indicates that science and the resulting technologies have both negative and positive consequences—a **technological dualism.** Technology saves lives and time and money; it also leads to death, unemployment, alienation, and estrangement. Weighing the costs and benefits of technology poses ethical dilemmas as does science itself. Ethics, however, "is not only concerned with individual choices and acts. It is also and, perhaps, above all concerned with the cultural shifts and trends of which acts are but the symptoms" (McCormick 1994, 16).

Thus, society makes a choice by the very direction it follows. Such choices should be made on the basis of guiding principles that are both fair and just (Winner 1993; Goodman 1993; Eibert 1998; Buchanan, Brock, Daniels, & Wikler 2000):

1. Science and technology should be prudent. Adequate testing, safeguards, and impact studies are essential. Impact assessment should include an evaluation of the social, political, environmental, and economic factors.
2. No technology should be developed unless all groups, and particularly those who will be most affected by the technology, have at least some representation "at a very early stage in defining what that technology will be" (Winner 1993, 291). Traditionally, the structure of the scientific process and the development of technologies has been centralized (that is, decisions have been in the hands of a few scientists and engineers); decentralization of the process would increase representation.
3. Means should not exist without ends. Each new innovation should be directed toward fulfilling a societal need rather than the more typical pattern in which a technology is developed first (e.g., high-definition television) and then a market is created (e.g., "you'll never watch a regular TV again!"). Indeed, from the space program to research on artificial intelligence, the vested interests of scientists and engineers, whose discoveries and innovations build careers, should be tempered by the demands of society.

It would be nice if the scientist and the engineer were cognizant of, and deeply concerned about, the potential risks, secondary consequences, and other costs of their creations.

ALLAN MAZUR
Sociologist

What the twenty-first century will hold, as the technological transformation continues, may be beyond the imagination of most of society's members. Technology empowers; it increases efficiency and productivity, extends life, controls the environment, and expands individual capabilities. But, as Steven Levy (1995) notes, there is a question as to whether society can accommodate such empowerment (p. 26).

As we proceed into the first computational millennium, one of the great concerns of civilization will be the attempt to reorder society, culture, and government in a manner that exploits the digital bonanza, yet prevents it from running roughshod over the checks and balances so delicately constructed in those simpler pre-computer years.

Critical Thinking

1 Use of the Internet by neo-Nazi and white supremacist groups has recently increased. Despite such increases, the U.S. Supreme Court has strengthened First Amendment protections of Internet material (Whine 1997). Should such groups have the right to disseminate information about their organizations and recruit members through the Internet?

2 In 1996, President Clinton signed a bill that requires TV manufacturers to equip future television sets with the "V-chip"—a technological device designed to prevent children from watching programs their parents find objectionable. President Bush supports Clinton's initiative. Hollywood executives opposed the "V-chip" on the grounds that it is intrusive and violates constitutional guarantees of freedom of expression. How might a conflict theorist explain opposition to the "V-chip"?

3 What currently existing technologies have had more negative than positive consequences for individuals and for society?

4 Some research suggests that productivity actually declines with the use of computers (Rosenberg 1998). Assuming this "paradox of productivity" is accurate, what do you think causes the reduction in efficiency?

Key Terms

abortion	genetic engineering	snowball sampling
automation	genetic screening	technological dualism
cultural lag	Internet	technological fix
cybernation	in-vitro fertilization	technology
deskilling	IT	technology-induced diseases
e-commerce	mechanization	
future shock	partial birth abortion	telework
gene monopolies	postmodernism	therapeutic cloning
gene therapy	science	upskilling

Media Resources

 The Wadsworth Sociology Resource Center: Virtual Society

http://sociology.wadsworth.com/

See the companion Web site for this book to access general sociology resources and text-specific features that can further your understanding of this chapter. The site contains Internet links, Internet exercises, online practice quizzes, information on InfoTrac College Edition, and many more valuable materials designed to enrich your learning experience in social problems.

 InfoTrac College Edition

You can access InfoTrac College Edition either from the Wadsworth Sociology Resource Center at **http://sociology.wadsworth.com** or directly from your web browser at **http://www.infotrac-college.com/wadsworth/**. InfoTrac College Edition is an online university library that includes over 700 popular and scholarly journals in which you can find articles related to the topics in this chapter such as information on cloning, genetic engineering, the Microsoft law suit, and the Internet.

Interactions CD-ROM

Go to the "Interactions" CD-ROM for *Understanding Social Problems*, Third Edition to access additional interactive learning tools, such as in-depth review materials, corresponding practice quizzes, and other engaging resources and activities to help you study the concepts in this chapter.

16

Conflict around the World

Is It True?

1. Military spending has actually declined by almost one-third in the last decade (SIPRI 2000).

2. The resolution of conflict between nations also tends to result in the resolution of conflict within nations.

3. Although the military causes a great deal of environmental damage during wartime, during times of peace military forces are geared to help clean up the environment.

4. The 2002 U.S. budget spends more money on national defense than on education, the justice system, the environment, and transportation combined.

5. There has never been an act of biological terrorism in the United States.

Answers to "Is It True?": 1 = T; 2 = F; 3 = F; 4 = T; 5 = F

Every gun that is made, every warship launched, every rocket fired, signifies in the final sense a theft from those who hunger and are not fed, those who are cold and not clothed. The world in arms is not spending money alone. It is spending the sweat of its laborers, the genius of its scientists, and the hopes of its children.

DWIGHT D. EISENHOWER
Former U.S. President/Military Leader

He was just a "regular guy"—at least that's what neighbors thought. At 56, Bill Hanssen seemed to have everything. He and his wife, Bonnie, lived in a Virginia suburb in a $300,000 split level house. She taught part-time at the local Catholic high school and he worked for the government for the last 27 years. Every Sunday morning Bill and Bonnie Hanssen and their six children got into the family minivan and went to church (Gullo 2001). Sunday, February 18, 2001, was no exception. However, on Sunday afternoon Bill Hanssen drove to a nearby park and allegedly left a package to be picked up by Russian operatives. Bill Hanssen was a spy.

The FBI veteran was arrested and charged with "espionage and conspiracy to commit espionage" (Margasak 2001). It's believed that Hanssen had decided on a career in counterintelligence at the age of 14 after reading the memoirs of a famous British double agent.

According to FBI sources, his 15-year career as a spy was a lucrative one, netting him over $1.4 million in cash and diamonds (Margasak 2001). Information reveals that Hanssen's spy activities included passing along ". . . to Soviet and later Russian agents 6,000 pages of documents—a virtual catalogue of top secret and secret programs" (Margasak 2001,1). He is also accused of identifying three Russian double agents to the KGB resulting in the executions of two of them. Other victims include Bill Hanssen's family. When approached by the media about her son's activities, Vivian Hanssen responded, ". . . I just love him, that's all" (quoted in AP 2001). If convicted, Bill Hanssen could receive the death penalty.

Despite what has been called the end of the Cold War, U.S.–Russian relations remain characterized by traitorous acts, double agents, and counterintelligence. It is hard to deny that the history of the world is a history of conflict. Never in recorded history has there been a time when conflict didn't exist. The most violent form of conflict—**war**—refers to organized armed violence aimed at a social group in pursuit of an objective. Wars have existed throughout human history and continue in the contemporary world (see Figure 16.1).

War is one of the great paradoxes of human history. It both protects and annihilates. It creates and defends nations, but also destroys them. Whether war is just or unjust, defensive or offensive, it involves the most horrendous atrocities known to humankind. This chapter focuses on the causes and consequences of global conflict and war. Along with population and environmental problems, war and global conflict are among the most serious of all social problems in their threat to the human race and life on earth.

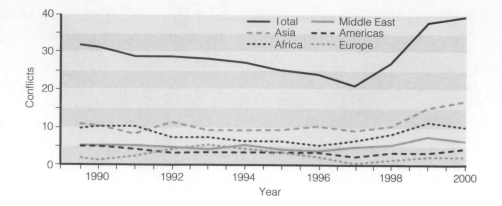

■ Figure 16.1 *Global Conflict, 1989–2000*

Source: *SIPRI Yearbook 2000;* Upsala University Conflict Project; Center for Defense Information.

The Global Context: Conflict in a Changing World

As societies have evolved and changed throughout history, the nature of war has also changed. Before industrialization and the sophisticated technology that resulted, war occurred primarily between neighboring groups on a relatively small scale. In the modern world, war can be waged between nations that are separated by thousands of miles, as well as between neighboring nations. In the following sections, we examine how war has changed our social world and how our changing social world has affected the nature of war in the industrial and post-industrial information age.

War and the Development of Civilization

The significance of wars is not just that they lead to major changes during the period of hostilities and immediately after. They produced transformations which have turned out to be of enduring significance.

ANTHONY GIDDENS
Sociologist

The very act that now threatens modern civilization—war—is largely responsible for creating the advanced civilization in which we live. Before large political states existed, people lived in small groups and villages. War broke the barriers of autonomy between local groups and permitted small villages to be incorporated into larger political units known as "chiefdoms." Centuries of warfare between chiefdoms culminated in the development of the state. The **state** is "an apparatus of power, a set of institutions—the central government, the armed forces, the regulatory and police agencies—whose most important functions involve the use of force, the control of territory and the maintenance of internal order" (Porter 1994, 5–6). The creation of the state in turn led to other profound social and cultural changes:

> And once the state emerged, the gates were flung open to enormous cultural advances, advances undreamed of during—and impossible under—a regimen of small autonomous villages . . . . Only in large political units, far removed in structure from the small autonomous communities from which they sprang, was it possible for great advances to be made in the arts and sciences, in economics and technology, and indeed in every field of culture central to the great industrial civilizations of the world (Carneiro 1994, 14–15).

Thus war, in a sense, gave rise to the state. Interestingly, the development of the state reduced the amount of lethal conflict (i.e., death through war, execution, homicide, or rebellion) in a society by providing alternative means of dispute resolution (Cooney 1997).

Industrialization may decrease a society's propensity to war, but at the same time, it increases the potential destructiveness of war because with industrialization, warfare technology becomes more sophisticated and lethal.

© T. Hartwell/ Sygma

The Influence of Industrialization on War

Industrialization and technology could not have developed in the small social groups that existed before military action consolidated them into larger states. Thus war contributed indirectly to the industrialization and technological sophistication that characterize the modern world.

Industrialization, in turn, has had two major influences on war. Cohen (1986) calculated the number of wars fought per decade in industrial and pre-industrial nations and concluded that "as societies become more industrialized, their proneness to warfare decreases" (p. 265). Cohen summarized the evidence for this conclusion: the pre-industrial nations had an overall mean of 10.6 wars per decade, whereas the industrial nations averaged 2.7 wars per decade. Perhaps industrialized nations have more to lose, so they avoid war and the risk of defeat. Indeed, Gentry (1998) notes that U.S. leaders "are increasingly reluctant to use [military forces] in roles that could lead to casualties in combat." The result—a "national aversion to danger" (pp. 179–180).

Although industrialization may decrease a society's propensity to war, it also increases the potential destruction of war. With industrialization, military technology became more sophisticated and more lethal. Rifles and cannons replaced the clubs, arrows, and swords used in more primitive warfare and in turn were replaced by tanks, bombers, and nuclear warheads. This chapter's *Focus on Technology* feature looks at how information technology is transforming warfare capabilities.

The Economics of Military Spending

The increasing sophistication of military technology has commanded a large share of resources totaling, worldwide, $780 billion in 1999. Although recent increases in military expenditures by the United States (36 percent) and Russia (24 percent) have contributed to the 2.6 percent worldwide increase between 1998 and 1999, military spending has declined by almost one-third in the last

War in the 21st Century

In the post-industrial information age, computer technology has revolutionized the nature of warfare and future warfare capabilities. Today, a "whole range of new technologies are offered for the next generations of weapons and military operations" (BICC 1998, 3) including the use of high-performance sensors, information processors, directed-energy technologies, and precision-guided munitions (O'Prey 1995; BICC 1998). With the increasing proliferation and power of computer technology, military strategists and political leaders are exploring the horizons of "information warfare," or **infowar**. Essentially, infowar utilizes technology to attack or manipulate the military and civilian infrastructure and information systems of an enemy. For example, infowar capabilities include the following (Waller 1995):

- Breaking into the communications equipment of the enemy army and disseminating incorrect information to enemy military leaders.
- Inserting computer viruses into the computer systems that control the phone system, electrical power, banking and stock exchanges, and air-traffic control of an enemy country.
- Jamming signals on the enemy's government television station and inserting a contrived TV program that depicts enemy leaders making unpopular statements that will alienate their people.
- Incapacitating the enemy's military computer systems, such as the one that operates the weapons system.

The U.S. Army, Navy, and Air Force are setting up infowar offices. In 1995, the first 16 infowar officers graduated from the National Defense University in Washington after being specially trained in everything from defending against computer attacks to using virtual reality computer technology to plan battle maneuvers.

Further, Air Force "battlelabs" have been established across the United States (Sietzen 2000). Because "the ability of American forces to deny access to space by any enemy of the United States or its allies" is paramount, battlelab personnel are conducting research on space control technologies. Today, the concept of space control includes controlling "everyday communications moving through space: voice, e-mail, paging signals, computer data and weather projections, . . . reconnaissance images of enemy forces and basic military communications between forces, fleets and communication centers" (Sietzen 2000, 2).

Two concerns exist, however, with the use of infowar technologies. Although the United States leads the pack, other nations are catching up and in some cases threatening to overcome our position as the infowar leader (Bernstein & Libicki 1998). For example, France now leads in digitizing three-dimensional battles, a process in which information technologies are used to "acquire, exchange and employ timely information throughout the battlespace." (Cook 1996, 602). Clearly, if other nations develop information technologies that rival or even surpass American capabilities, U.S. military preparedness could be jeopardized.

Another problem is that infowar is relatively inexpensive and readily available. With a computer, a modem, and some rudimentary knowledge of computers, anyone could initiate an information attack. In 2000, in the wake of renewed Middle East tensions, the web site of Hezbollah, an anti-Zionist organization, was infiltrated by pro-Israeli hackers. When accessing the page visitors were welcomed to the site by an Israeli flag and the national anthem of Israel. Pro-Arab groups counterattacked, flooding Israeli government sites with "a barrage of hundreds of thousands—possibly millions—of hostile electronic signals" (Hockstader 2000, A01).

The United States is also vulnerable to such attacks. As a matter of fact, the U.S. government's Joint Security Commission described U.S. vulnerability to infowar as "the major security challenge of this decade and possibly the next century" (quoted in Waller 1995, 40). In 1999 more than 22,000 cyber-attacks occurred on Defense Department networks, at least some of which are likely to have been initiated by foreign enemies (Reuters 2000).

Sources:

Alvin Bernstein and Martin Libicki. 1998. "High-Tech: The Future of War." *Commentary* 105:28–37.

BICC (Bonn International Center for Conversion). 1998. "Chapter Six." *Conversion Survey, 1998.* Bonn, Germany: BICC.

Nick Cook. 1996. "Battlespace 2000." *Interavia Business and Technology* 51:43–46.

Lee Hockstader. 2000. "Pings and E-Arrows Fly in Mideast Cyber-War." *Washington Post.* October 27, A01.

Kevin P. O'Prey. 1995. *The Arms Export Challenge: Cooperative Approaches to Export Management and Defense Conversion.* Washington, D.C.: The Brookings Institute.

Reuters. 2000. "Pentagon Still Under Attack from Hackers." http:www//nytimes.com/library/tech.

Frank Sietzen Jr. 2000. " 'Battlelabs' Beef Up Space Defense." http://www.msnbc.com/ews/4/2426; Douglas Waller. 1995. "Onward Cyber Soldiers." *Time,* August 21, 38–44.

decade (SIPRI 2000). U.S. military spending, approximately $300 billion in 2001 (Crock 2001), includes expenditures for salaries of military personnel, research and development, weapons, veterans' benefits, and other defense-related expenses.

The U.S. government not only spends money on its own military, but also sells military equipment to other countries either directly or by helping U.S. companies sell weapons abroad. Although the purchasing countries may use these weapons to defend themselves from hostile attack, foreign military sales may pose a threat to the United States by arming potential antagonists. For example, the United States, which provides more than half of the world's arms exports, supplied weapons to Iraq to use against Iran. These same weapons were then used against Americans in the Gulf War. A recent survey indicates that most Americans favor cutting back selling military aid to other countries (CCFR 2000).

The **Cold War,** the state of political tension and military rivalry that existed between the United States and the former Soviet Union, provided justification for large expenditures on military preparedness. The end of the Cold War, along with the rising national debt, has resulted in cutbacks in the U.S. military budget in 7 of the last 10 years (*Statistical Abstract 2000*).

The economic impact of these defense cutbacks, and cutbacks around the world, has been mixed. On the negative side, defense cutbacks result in job losses and the closure of military bases, facilities, factories, and plants in defense industries. Certain industries and the countries, regions, and states where they are located suffer disproportionately as a result of defense cutbacks. Worldwide, 8.3 million jobs were lost in the defense industry between 1987 and 1996—47 percent of the 1987 total (BICC 1998).

On the positive side, defense cutbacks provide a **peace dividend** in that resources previously spent on military purposes can be used for private or public investment or consumption. For example, the peace dividend could be used for new manufacturing plants and machinery (private investment) or for public education, transportation, housing, health, and the environment. The peace dividend could also be used to lower the national debt and/or taxes, and is one of the main reasons the federal budget is balanced (Perry & Shalikashvili 2000). Achieving a peace dividend requires the reallocation of resources from military forces and defense industries to other sectors of the economy—a process referred to as **economic conversion**. For example, military bases could be converted to civil airports, prisons, housing developments, or shopping centers, and workers in the defense industry could be employed in the civil sector.

Sociological Theories of Conflict and War

Sociological perspectives can help us understand various aspects of war. In this section, we describe how structural-functionalism, conflict theory, and symbolic interactionism may be applied to the study of conflict and war.

Structural-Functionalist Perspective

Structural-functionalism focuses on the functions war serves and suggests that war would not exist unless it had positive outcomes for society. It has already been noted that war has served to consolidate small autonomous social groups

Inter-societal conflicts have furthered the development of social structures.

HERBERT SPENCER
Sociologist

into larger political states. An estimated 600,000 autonomous political units existed in the world around the year 1000 B.C. Today, that number has dwindled to less than 200 (Carneiro 1994).

Another major function of war is that it produces social cohesion and unity among societal members by giving them a "common cause" and a common enemy. Unless a war is extremely unpopular, military conflict promotes economic and political cooperation. Internal domestic conflicts between political parties, minority groups, and special interest groups dissolve as they unite to fight the common enemy. During World War II, U.S. citizens worked together as a nation to defeat Germany and Japan.

In the short term, war also increases employment and stimulates the economy. The increased production needed to fight World War II helped pull the United States out of the Great Depression. The investments in the manufacturing sector during World War II also had a long-term impact on the U.S. economy. Hooks and Bloomquist (1992) studied the effect of the war on the U.S. economy between 1947 and 1972 and concluded that the U.S. government "directed, and in large measure, paid for a 65 percent expansion of the total investment in plant and equipment" (p. 304).

Another function of war is the inspiration of scientific and technological developments that are useful to civilians. Research on laser-based defense systems led to laser surgery, for example, and research in nuclear fission and fusion facilitated the development of nuclear power. The airline industry owes much of its technology to the development of air power by the Department of Defense, and the Internet was created by the Pentagon for military purposes. Other **dual-use technologies**, a term referring to defense-funded innovations with commercial and civilian applications, include SLICE, a high speed, twin hull water vessel originally made for the Office of Naval Research. SLICE has a variety of commercial applications "including its use as a tour or sport fishing

As structural functionalists argue, a major function of war is that it produces unity among societal members. They have a common cause and common identity. They feel a sense of cohesion and work together to defeat the enemy.

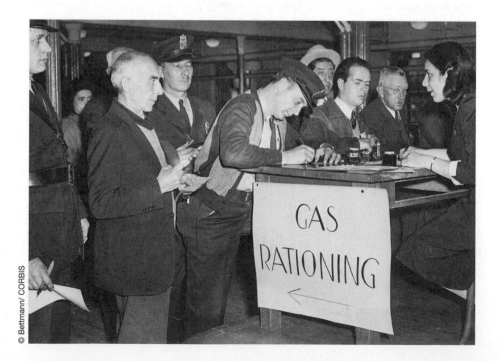

© Bettmann/ CORBIS

boat, oceanographic research vessel, oil spill response ship, and high-speed ferry" (Hawaii 2000).

Finally, war serves to encourage social reform. After a major war, members of society have a sense of shared sacrifice and a desire to heal wounds and rebuild normal patterns of life. They put political pressure on the state to care for war victims, improve social and political conditions, and reward those who have sacrificed lives, family members, and property in battle. As Porter (1994) explains, "Since . . . the lower economic strata usually contribute more of their blood in battle than the wealthier classes, war often gives impetus to social welfare reforms (p. 19).

Conflict Perspective

Conflict theorists emphasize that the roots of war often stem from antagonisms that emerge whenever two or more ethnic groups (e.g., Bosnians and Serbs), countries (United States and Vietnam), or regions within countries (the U.S. North and South) struggle for control of resources or have different political, economic, or religious ideologies. Further, conflict theory suggests that war benefits the corporate, military, and political elites. Corporate elites benefit because war often results in the victor taking control of the raw materials of the losing nations, thereby creating a bigger supply of raw materials for its own industries. Indeed, many corporations profit from defense spending. Under the Pentagon's bid and proposal program, for example, corporations can charge the cost of preparing proposals for new weapons as overhead on their Defense Department contracts. Also, Pentagon contracts often guarantee a profit to the developing corporations. Even if the project's cost exceeds initial estimates, called a cost overrun, the corporation still receives the agreed-upon profit. In the late 1950s, President Dwight D. Eisenhower referred to this close association between the military and the defense industry as the **military-industrial complex**.

The military elite benefit because war and the preparations for it provide prestige and employment for military officials. For example, Military Professional Resources, Inc. (MPRI), in Virginia, boasts that it can "perform any task or accomplish any mission requiring defense related expertise . . ." and lists such capabilities as war gaming, anti-terrorism/force protection, and peacekeeping (MPRI 2001). The company employs over 8,000 retired defense and law enforcement professionals, has clients around the world, and represents a movement toward the privatization of the military.

War also benefits the political elite by giving government officials more power. Porter (1994) observed that "throughout modern history, war has been the lever by which . . . governments have imposed increasingly larger tax burdens on increasingly broader segments of society, thus enabling ever-higher levels of spending to be sustained, even in peacetime" (p. 14). Political leaders who lead their country to a military victory also benefit from the prestige and hero status conferred on them.

> As war becomes more sophisticated, it continuously increases governmental authority and decreases the power of the people.
>
> J. C. L. Simonde de Sismondi, *Swiss economist*

Symbolic Interactionist Perspective

The symbolic interactionist perspective focuses on how meanings and definitions influence attitudes and behaviors regarding conflict and war. The development of attitudes and behaviors that support war begins in childhood. American children learn to glorify and celebrate the Revolutionary War, which created our nation. Movies romanticize war, children play war games with toy weapons, and various video and computer games glorify heroes conquering villains. Indeed, from 1938 to 1942 a series of "Horrors of Wars" cards were manufactured and distributed in the United States and collected by millions of American youth much like baseball cards (Nelson 1999).

Symbolic interactionism helps to explain how military recruits and civilians develop a mindset for war by defining war and its consequences as acceptable and necessary. The word "war" has achieved a positive connotation through its use in various phrases—the "war on drugs," the "war on poverty," and the "war on crime." Positive labels and favorable definitions of military personnel facilitate military recruitment and public support of armed forces. Military personnel wear uniforms that command public respect and earn badges and medals that convey their status as "heroes."

> Why do we kill people who are killing people to show that killing people is wrong?
>
> HOLLY NEAR
> *Singer-songwriter*

Many government and military officials convince the masses that the way to ensure world peace is to be prepared for war. Most world governments preach peace through strength rather than strength through peace. Governments may use propaganda and appeals to patriotism to generate support for war efforts and motivate individuals to join armed forces.

To legitimize war, the act of killing in war is not regarded as "murder." Deaths that result from war are referred to as "casualties." Bombing military and civilian targets appears more acceptable when nuclear missiles are "peacekeepers" that are equipped with multiple "peace heads." Finally, killing the enemy is more acceptable when derogatory and dehumanizing labels such as Gook, Jap, Chink, and Kraut convey the attitude that the enemy is less than human.

Causes of War

The causes of war are numerous and complex. Most wars involve more than one cause. The immediate cause of a war may be a border dispute, for example, but religious tensions that have existed between the two countries for decades may also contribute to the war. The following section reviews various causes of war.

Conflict over Land and Other Natural Resources

Nations often go to war in an attempt to acquire or maintain control over natural resources, such as land, water, and oil. Disputed borders are one of the most common motives for war. Conflicts are most likely to arise when borders are physically easy to cross and are not clearly delineated by natural boundaries, such as major rivers, oceans, or mountain ranges.

Water is another valuable resource that has led to wars. At various times the empires of Egypt, Mesopotamia, India, and China all went to war over irrigation rights. In 1998, 5 years after Eritrea gained independence from Ethiopia, forces clashed over control of the port city Assab and with it, access to the Red Sea ("Conflict" 2000).

Not only do the oil-rich countries in the Middle East present a tempting target in themselves, but war in the region can threaten other nations that are dependent on Middle Eastern oil. Thus, when Iraq seized Kuwait and threatened the supply of oil from the Persian Gulf, the United States and many other nations reacted militarily in the Gulf War. In a document prepared for the Center for Strategic and International Studies, Starr and Stoll (1989) warn that soon:

> . . . water, not oil will be the dominant resource issue of the Middle East. According to World Watch Institute, "despite modern technology and feats of engineering, a secure water future for much of the world remains elusive." The prognosis for Egypt, Jordan, Israel, the West Bank, the Gaza Strip, Syria, and Iraq is especially alarming. If present consumption patterns continue, emerging water shortages, combined with a deterioration in water quality, will lead to more competition and conflict. (p. 1)

Conflict over Values and Ideologies

Many countries initiate war not over resources, but over beliefs. World War II was largely a war over differing political ideologies: democracy versus fascism. The Cold War involved the clashing of opposing economic ideologies: capitalism versus communism. Wars over differing religious beliefs have led to some of the worst episodes of bloodshed in history, in part, because some religions are partial to martyrdom—the idea that dying for one's beliefs leads to eternal salvation. The Shiites (one of the two main sects within Islam) in the Middle East represent a classic example of holy warriors who feel divine inspiration to kill the enemy.

Conflicts over values or ideologies are not easily resolved. The conflict between secularism and Islam has lasted for 14 centuries (Lewis 1990). Conflict over values and ideologies are less likely to end in compromise or negotiation because they are fueled by people's convictions. For example, when a representative sample of American Jews was asked, "In the framework of a permanent peace with the Palestinians, should Israel be willing to compromise on the status of Jerusalem as a united city under Israeli jurisdiction?," 57 percent of the respondents said "no" and 36 percent said "yes"; the remainder were unsure (American Jewish Committee 2000).

If ideological differences can contribute to war, do ideological similarities discourage war? The answer seems to be yes; in general, countries with similar ideologies are less likely to engage in war with each other than countries with differing ideological values (Dixon 1994). Democratic nations are particularly disinclined to wage war against one another (Doyle 1986).

Racial and Ethnic Hostilities

Ethnic groups vary in their cultural and religious beliefs, values, and traditions. Thus, conflicts between ethnic groups often stem from conflicting values and ideologies. Racial and ethnic hostilities are also fueled by competition over land and other scarce natural and economic resources. Gioseffi (1993) notes that "experts agree that the depleted world economy, wasted on war efforts, is in great measure the reason for renewed ethnic and religious strife. 'Haves' fight with 'have-nots' for the smaller piece of the pie that must go around" (p. xviii). Racial and ethnic hostilities are also perpetuated by the wealthy majority to divert attention away from their exploitations and to maintain their own position of power (see Chapter 7).

As described by Paul (1998), sociologist Daniel Chirot argues that the recent worldwide increase in ethnic hostilities is a consequence of "retribalization," that is, the tendency for groups, lost in a globalized culture, to seek solace in the

> The most important divide in this hard world is not over bloodlines or ways of worship. It's between those who pursue reason and those who dismiss it. It's between those who build and those who bomb.
>
> ELLEN GOODMAN
> *Columnist*

"extended family of an ethnic group (p.56). Chirot identifies five levels of ethnic conflict: (1) multiethnic societies without serious conflict (e.g., Switzerland), (2) multiethnic societies with controlled conflict (e.g., United States, Canada), (3) societies with ethnic conflict that has been resolved (e.g., South Africa), (4) societies with serious ethnic conflict leading to warfare (e.g., Sri Lanka), and (5) societies with genocidal ethnic conflict including "ethnic cleansing" (e.g., Kosovo).

Defense against Hostile Attacks

The threat or fear of being attacked may cause the leaders of a country to declare war on the nation that poses the threat. The threat may come from a foreign country or from a group within the country. After Germany invaded Poland in 1939, Britain and France declared war on Germany out of fear that they would be Germany's next victims. Germany attacked Russia in World War I, in part out of fear that Russia had entered the arms race and would use its weapons against Germany. Japan bombed Pearl Harbor hoping to avoid a later confrontation with the U.S. Pacific fleet, which posed a threat to the Japanese military. In 2001, U.S. and British forces conducted air strikes against Iraq in response to growing concerns for the safety of allied patrols in the no-fly zone.

Revolution

Revolutions involve citizens warring against their own government and often result in significant political, economic, and social change. A revolution may occur when a government is not responsive to the concerns and demands of its citizens and when strong leaders are willing to mount opposition to the government (Renner 2000a; Barkan & Snowdan 2001).

Those who make peaceful evolution impossible, make violent revolution inevitable.

JOHN F. KENNEDY
Former U.S. President

The birth of the United States resulted from colonists revolting against British control. Contemporary examples of civil war include Sri Lanka, where the Tamils, a separatist group living in the northern region of the country, have been at war with the Sri Lankan government for over 10 years. According to Mylvaganam (1998) the Liberation of Tamil Tigers (LTTE) has successfully fought off the better prepared Sri Lankan government army through the "use of paramilitary and guerrilla warfare, coupled with their expert knowledge of the terrain . . ." (p. 1). Additionally, civil wars have erupted in newly independent republics created by the collapse of communism in Eastern Europe, as well as in Rwanda, Liberia, Guatemala, Chile, and Uganda.

Nationalism

Lift your head. Be proud to be German.

NAZI WAR CRIMINAL
Judgment at Nuremberg

Some countries engage in war in an effort to maintain or restore their national pride. For example, Scheff (1994) argues that "Hitler's rise to power was laid by the treatment Germany received at the end of World War I at the hands of the victors" (p. 121). Excluded from the League of Nations, punished by the Treaty of Versailles, and ostracized by the world community, Germany turned to nationalism as a reaction to material and symbolic exclusion.

In the late 1970s, Iranian fundamentalist groups took hostages from the American Embassy in Iran. President Carter's attempt to use military forces to free the hostages was not successful. That failure intensified doubts about America's ability to effectively use military power to achieve its goals. The hostages in Iran were eventually released after President Reagan took office, but doubts about the strength and effectiveness of America's military still called into question America's status as a world power. Subsequently, U.S. military forces invaded the small island of Grenada because the government of Grenada was building an airfield large enough to accommodate major military armaments. U.S. officials feared that this airfield would be used by countries in hostile attacks on the United States. From one point of view, the large scale and "successful" attack on Grenada functioned to restore faith in the power and effectiveness of the American military.

Social Problems Associated with War and Militarism

Social problems associated with war and militarism include death and disability; rape, forced prostitution, and displacement of women and children; disruption of social-psychological comfort; diversion of economic resources; and destruction of the environment.

Death and Disability

In the last century, war has taken the lives of more than 100 million persons—more than the total number of deaths in all previous wars or massacres in human history combined (Porter 1994). Many American lives have been lost in wars, including over 53,000 in World War I, 292,000 in World War II, 34,000 in Korea, and 47,000 in Vietnam (*Statistical Abstract 2000*). Globalization and sophisticated weapons technology combined with increased population density has made it easier to kill large numbers of people in a short amount of time. When the atomic bomb was dropped on the Japanese cities of Hiroshima and Nagasaki during World War II, 250,000 civilians were killed.

War's impact extends far beyond those who are killed. Many of those who survive war incur disabling injuries as well as diseases. For example, more than a quarter of a million people worldwide have been disabled by landmines. In 1999, The Landmine Ban Treaty, requiring that governments destroy stockpiles within 4 years and clear landmine fields within 10 years, became international law. To date, 139 countries have signed the agreement, 54 countries remain (Landmine 2001). This chapter's *Self and Society* feature tests your knowledge of the subject.

War-related deaths and disabilities also deplete the labor force, create orphans and single-parent families, and burden taxpayers who must pay for the care of orphans and disabled war veterans.

Persons who participate in experiments for military research may also suffer physical harm. U.S. Representative Edward Markey of Massachusetts identified 31 experiments dating back to 1945 in which U.S. citizens were subjected to harm from participation in military experiments. Markey charged that many of the experiments used human subjects who were captive audiences or populations considered "expendable," such as the elderly, prisoners, and hospital patients. Eda Charlton of New York was injected with plutonium in 1945. She and

> The object of war is not to die for your country but to make the other bastard die for his.
>
> GEN. GEORGE PATTON

The Landmine Knowledge Quiz

Carefully read each of the following statements and select the correct answer. When finished, compare your answers to those provided and rank your performance using the scale below.

1. The most heavily mined regions of the world are in Southeast Asia and Central America. True or False?

2. Worldwide, more than 600 different types of landmines exist. True or False?

3. Each landmine costs: (A) $1000 (B) $20 (C) $3 (D) $200

4. Annually, more than 25,000 people are victims of landmine violence. True or False?

5. What are the *only* two countries in the Western hemisphere not to have signed the 1997 Landmine Ban Treaty? (A) U.S. and Cuba (B) Canada and Belize (C) Brazil and Mexico (D) U.S. and Mexico.

6. Only 16 countries in the world produce landmines, including the United States, Russia, and China. True or False?

7. Landmines from WWII are still being discovered in Europe. True or False?

8. In Cambodia, one out of every 1,236 people is an amputee from a landmine. True or False?

9. In developing countries, basic recovery and rehabilitation from a single landmine is approximately: (A) $1,000 (B) $10,000 (C) $5,000 (D)$60,000

10. Today, approximately 1 million active landmines lie in thousands of minefields around the world. True or False?

ANSWERS: 1. False (Africa and the Middle East are the most heavily mined); 2. True ; 3. C; 4. True; 5. A ; 6. True; 7. True; 8. False (one out of every 236); 9. B; 10. False (there are over 80 million active landmines).

Rank your Performance—Number Correct

9 or 10	Excellent
7 or 8	Good
5 or 6	Average
3 or 4	Fair
0, 1 or 2	Poor

Source: Adapted from information from The Landmine Site. 2000. "About Landmines." http://www.thelandminesite.com. Used by permission.

17 other patients did not learn of their poisoning until 30 years later. Her son, Fred Shultz, said of his deceased mother:

> I was over there fighting the Germans who were conducting these horrific medical experiments . . . at the same time my own country was conducting them on my own mother (Miller 1993, 17).

Rape, Forced Prostitution, and Displacement of Women and Children

Half a century ago, the Geneva Conventions prohibited rape and forced prostitution in war. Nevertheless, both continue to occur in modern conflicts.

Before and during World War II, the Japanese military forced an estimated 100,000 to 200,000 women and teenage girls into prostitution as military "comfort women" (Amnesty International 1995). These women were forced to have sex with dozens of soldiers every day in "comfort stations." Many of the women died as a result of untreated sexually transmitted disease, harsh punishment, or indiscriminate acts of torture.

The use of rape in conflict reflects the inequalities women face in their everyday lives in peacetime. . . . Women are raped because their bodies are seen as the legitimate spoils of war.

Amnesty International

In 1994, militia forces roamed through the town of Kibuye, Rwanda, burning houses and killing civilians. "Women found sheltering in the parish church were raped, then pieces of wood were thrust into their vaginas, and they were left to die slowly" (Amnesty International 1995, 17). More recently, rapes of Albanian women by Serbian and Yugoslavian paramilitary soldiers were, according to a Human Rights Watch report, "used deliberately as an instrument to terrorize the civilian population, extort money from families, and push people to flee their homes" (quoted in Lorch & Mendenhall 2000, 2). Feminist analysis of wartime rape emphasizes that the practice reflects not only a military strategy, but ethnic and gender dominance as well (Card 1997). In 2001, three Serbian soldiers accused of mass rape and forced prostitution were convicted by a United Nations tribunal and sentenced to a combined total of 60 years in prison (CNN 2001a).

War also forces women and children to flee to other countries or other regions of their homeland. For example, as a result of civil war, Colombia has an estimated 1.5 million refugees—a million of whom are children (Dudley 2000). Refugee women and female children are particularly vulnerable to sexual abuse and exploitation by locals, members of security forces, border guards, or other refugees. In refugee camps, women and children may also be subjected to sexual violence. Recently the definition of war crimes has been broadened to include incidents of rape and sexual violence.

Disruption of Social-Psychological Comfort

War and living under the threat of war interferes with social-psychological well-being as well as family functioning (see this chapter's *Social Problems Research Up Close* feature). In a study of 269 Israeli adolescents, Klingman and Goldstein (1994) found a significant level of anxiety and fear, particularly among younger females, in regard to the possibility of nuclear and chemical warfare. Similarly, Myers-Brown, Walker, and Myers-Walls (2000) report that Yugoslavian children suffer from depression, anxiety, and fear as a response to recent conflicts in that region.

Civilians who are victimized by war and military personnel who engage in combat may experience a form of psychological distress known as **post–traumatic stress disorder** (PTSD), a clinical term referring to a set of symptoms that may result from any traumatic experience, including crime victimization, rape, or war. Symptoms of PTSD include sleep disturbances, recurring nightmares, flashbacks, and poor concentration (Novac 1998). For example, Canadian Lt. General Romeo Dallaire, head of the U.N. peacekeeping mission in Rwanda, witnessed horrific acts of genocide. Four years after his return he continues to have images of "being in a valley at sunset, waist deep in bodies, covered in blood" (quoted in Rosenberg 2000, 14). PTSD is also associated with other personal problems, such as alcoholism, family violence, divorce, and suicide.

One study estimates that about 30 percent of male veterans of the Vietnam War have experienced PTSD, and about 15 percent continue to experience it (Hayman & Scaturo 1993). In another study of 215 Army National Guard and Army Reserve troops who served in the Gulf War and who did not seek mental health services upon return to the states, 16 to 24 percent exhibited symptoms of PTSD (Sutker, Uddo, Brailey, & Allain 1993). Finally, research on PTSD in children reveals that females are generally more symptomatic than males and that PTSD may disrupt the normal functioning and psychological development of children (Cauffman, Feldman, Waterman, & Steiner 1998; Pfefferbaum 1997).

> I have seen enough of one war never to wish to see another.
>
> **THOMAS JEFFERSON**
> *Former U.S. President*

Family Adjustment to Military Deployment

Research indicates that long and/or frequent absences from home by a spouse may create a variety of family problems including feelings of isolation, depression, and marital conflict. Despite recent trends toward downsizing the military, deployment of military personnel has increased, and as a result, separations of military families have increased. The present investigation by Rohall, Segal, and Segal (1999), examines the impact of military deployment on family functioning.

Sample and Methods.

A survey was administered to 518 enlisted personnel stationed at air bases in Dsuwon, Kunson, and Osan, South Korea. All respondents were from one of two battalions. Battalion A personnel were separated from their families for a longer period of time (19 months) than those from Battalion B (7 months), and had been deployed more frequently. Thus, the first independent variable is *length and frequency of separation* from family. Respondents were also questioned about their *morale* (e.g., "How satisfied are you with your current job?"), perceived *organizational support* for family and self (e.g., support from commanding officer, chaplain, etc.), and satisfaction with *resources to communicate* with family members. The dependent variable is the *soldier's assessment of family adjustment* to their absence. The authors' (1999) research question is thus, "Are soldiers' perceptions of family adjustment affected by: (1) time and frequency of separation from family, (2) morale, (3) perceptions of organizational support, and/or (4) satisfaction with resources to communicate with family members?

Findings and Conclusions

As hypothesized, soldiers in Battalion B were more likely to report lower levels of family functioning than soldiers in Battalion A. Thus, longer and more frequent separations are associated with poorer family adjustment

Diversion of Economic Resources

As discussed earlier, maintaining the military and engaging in warfare requires enormous financial capital and human support. In 1999, worldwide military expenditures approached $800 billion (SIPRI 2000). This amount exceeds the combined government research expenditures on developing new energy technologies, improving human health, raising agricultural productivity, and controlling pollution.

Money that is spent for military purposes could be allocated for social programs. The decision to spend $1.4 billion for one Trident submarine, equal in cost to a 5-year immunization program that would prevent nearly 1 million childhood deaths annually (Renner 1993b), is a political choice. Similarly, allocating $100 million for one B-2 bomber while our schools continue to deteriorate is also a political choice (Weida 1998). Nonetheless, although the majority of Americans advocate no change in expenditures, for the first time since 1978 more Americans favor increased rather than decreased military spending (CCFR 2000). As Figure 16.2 indicates, the 2002 U.S. budget projects spending more money on national defense than on education, the justice system, the environment, and transportation combined (Office of Management and Budget 2001).

Destruction of the Environment

Traditional definitions of and approaches to national security have assumed that political states or groups constitute the principal threat to national security and welfare. This assumption implies that national defense and security are best served by being prepared for war against other states. The environmental costs of military preparedness are often overlooked or minimized in such discussions. Annually, for example, the U.S. Navy Public Works Center in San Diego pro-

or, at least, the perception of poorer family adjustment. Morale was also positively correlated with family adjustment—the higher a soldier's reported morale the greater their perception of positive family functioning. Similarly, respondents' ratings of organizational support and satisfaction with resources to communicate with family members were directly associated with higher family functioning ratings.

Although the results of the study indicate that each of the four independent variables is related to family adjustment, the relative importance of each is unclear. Further analysis by Rohall, Segal and Segal (1999) indicates that morale is the best single predictor of perceived family ad-

justment, followed by organizational support. Surprisingly, length and frequency of separation as measured by unit, that is, Battalion A versus Battalion B, contributed only minimally to variation in family adjustment.

As the authors' note, the results of the present study may, in part, reflect what is called selection bias. For example, soldiers who believe that their families are not adjusting to military life, have low morale, feel little organizational support, and/or are dissatisfied with communication resources, may be more likely to drop out of the military. If that is the case, the present research, by only measuring *active* duty personnel, sampled those *most* satisfied with military life and thus

used a biased sample. In this case, however, the possibility of selection bias does not negate the significance of the research results. If concern over family adjustment leads to soldiers leaving the military, keeping the length and frequency of separation low, morale high, organizational support strong, and communication resources satisfactory, should reduce dropout rates. In a voluntary army when concerns over the quality of military personnel and recruitment shortfalls are high, insight into retention is invaluable.

Source: David E. Rohall, Mady Wechsler Segal, and David R. Segal. 1999. "Examining the Importance of Organizational Supports on Family Adjustment to Army Life in a Period of Increasing Separation." *Journal of Political and Military Sociology* 27: 49–65.

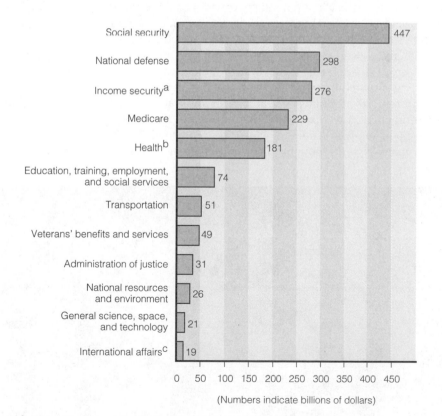

Figure 16.2 *Selected U.S. Federal Outlays, 2002 (projected)*

SOURCE: Office of Management and Budget. 2001. "A Citizen's Guide to the Federal Budget." Office of Management and Budget. Washington, D.C.

[a]Includes retirement and disability insurance, unemployment compensation, and housing and food assistance.
[b]Includes health research and health care services.
[c]Includes international development and humanitarian assistance.

duces more than 12,000 tons of paint cans. The depleted cans, enough to fill 12 semi-trucks, contain volatile chemicals making their disposal or any proposed recycling methods problematic (Lifsher 1999). A seemingly trivial matter—paint cans, in one year, at one facility. Imagine the environmental problems generated by the military, worldwide.

My God . . . what have we done?

ROBERT LEWIS
co-pilot of the **Enola Gay,** *after dropping the atomic bomb on Hiroshima*

Destruction of the Environment during War The environmental damage that occurs during war continues to devastate human populations long after war ceases. In Vietnam, 13 million tons of bombs left 25 million bomb craters. Nineteen million gallons of herbicides, including Agent Orange, were spread over the countryside. An estimated 80 percent of Vietnam's forests and swamplands were destroyed by bulldozing or bombing (Funke 1994).

The Gulf War also illustrates how war destroys the environment (Funke 1994; Renner 1993a;). In 6 weeks, 1,000 air missions were flown, dropping 6,000 bombs. In 1991, Iraqi troops set 650 oil wells on fire, releasing oil, which still covers the surface of the Kuwaiti desert and seeps into the ground, threatening to poison underground water supplies. The estimated 6–8 million barrels of oil that spilled into the Gulf waters are threatening marine life. This spill is by far the largest in history—25–30 times the size of the 1989 Exxon Valdez oil spill in Alaska.

The clouds of smoke that hung over the Gulf region for 8 months contained soot, sulfur dioxide, and nitrogen oxides—the major components of acid rain—and a variety of toxic and potentially carcinogenic chemicals and heavy metals. The U.S. Environmental Protection Agency estimates that in March 1991 about 10 times as much air pollution was being emitted in Kuwait as by all U.S. industrial and power-generating plants combined. Acid rain destroys forests and harms crops. It also activates several dangerous metals normally found in soil, including aluminum, cadmium, and mercury. Further, the over 900,000 depleted uranium munitions fired at Kuwait and Iraq are thought to have serious environmental as well as health consequences. This chapter's *The Human Side* feature poignantly describes the destructive course of Gulf War Syndrome—one of the many diseases thought to be associated with this radioactive material.

Oil smoke from the 650 burning oil wells left in the wake of the Gulf War contains soot, sulfur dioxide, and nitrogen oxides, the major components of acid rain, and a variety of toxic and potentially carcinogenic chemicals and heavy metals.

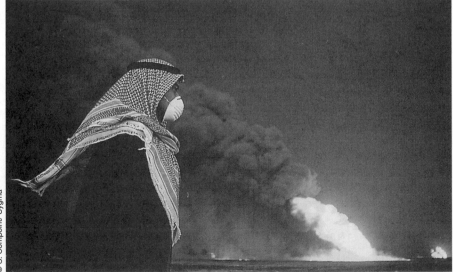

© S. Compoint/ Sygma

The Death of a Gulf War Veteran

The following letter to *American Legion Magazine* is from the parents of a Gulf War veteran.

Our son may be unique. We really don't know. We're desperate for information that is extremely hard to come by. We are very frustrated that what happened to us may have already happened to other families—or will—before something is done.

This is for Scott, our son, who said, "Go for it, Mom," when I told him I'd never quit trying to find out what made him so ill.

He was sent to the Persian Gulf with the 1133rd Transportation Co., National Guard, of Mason City, Iowa. He left for Saudi as a very healthy young man and he returned in much the same physical condition. Just tired, but not unusual considering where he'd been and what he'd seen.

. . . two years after he returned from home in the summer of 1993, he developed a rash on his torso. It would erupt, disappear and then come back again. It didn't always look the same. We thought it was a heat rash, something he ate, new laundry soap, etc.

In the fall of 1993 he developed sores in his mouth. He could eat very little and quickly lost about 40 pounds. He went from specialist to specialist, who sent tests many places to try and find a cause. He was put on steroids and depending on the dosage, it would get a little better and then flare up again. By winter he could only eat pureed foods and liquids. He could not use a straw as it hurt too much. Many doctors, many tests, many different medications. Nothing helped very much. And no concrete diagnosis.

In May 1994, he was examined by the VA hospital in Des Moines. They were not able to find the cause for the terrible sores in his mouth and the rash that now also affected his feet, hands and arms. He made many trips to Des Moines—a three-hour drive.

In early August 1994, the VA Hospital in Des Moines came up with a diagnosis of lupus. My heart just broke when I heard those words, but he was thrilled to know there was finally a name and treatment for his symptoms.

By mid-August he was hardly able to walk, his feet swollen and so extremely sore from the rash that seemed to get worse by the day. He went to the VA Hospital on Friday, August 19 . . . and was admitted. The rash had become blisters about the size of a 50-cent piece and were breaking and bleeding. He was running a fever . . .

He was transferred to University [hospital] in Iowa City and taken to surgery . . . [where] they removed all his skin and replaced it with what is called pig skin. Out of surgery, bandaged from head to toe, he was given a five percent chance of survival.

Our family gathered together to give him all the love and support we could. Ten days later the pig skin was removed. Infection had been found. It was too risky to take him back to surgery, so it was removed in a sterile room close to his room.

The next seven weeks are a blur. We had many ups—a good day for Scott—and many downs—a bad day for him. He had to endure burn baths every day. Water jets removed sloughed skin. He'd grow a tiny patch of skin only to lose it in a day or so later. He was fed through a tube. Many antibiotics were given in the hopes of warding off infection; morphine for the tremendous pain. We read to him—he'd correctly pronounce words we missed. His sense of humor never left. He worried about all of us.

Scott's life ended at 3 A.M. on October 15, 1994 . . .

We are convinced beyond anything we've ever felt as parents that Persian Gulf Syndrome killed our only son. We want someone to tell us the truth. We won't quit until we have believable answers to our question.

What do we tell his oldest son, now 11, who has nothing but pictures of his daddy. . . ?

What do we say to a three-year-old whose wish is to build a rocket so he can go get daddy and make everyone happy?

What do you say to a son who was born two weeks after his daddy died. . . ?

What do you say to a wife who longs only for her husband's love, strength and support?

If chemical and gases were used over there, tell us the truth. It kills innocent people and destroys innocent families. We are real people with real feelings and we deserve the truth. This was not easy to write. We did it in hopes it may help someone. Only then will Scott's death make sense to us. He would have wanted it that way. He was that special.

Ardie and Rollie Siefken
Plainfield, Iowa

Source: "The Sad Death of a Gulf War Veteran." *American Legion Magazine*, August 6, 1996. Copyright lifted for distribution. http://www.ascension-research.org/part-5.html

The ultimate environmental catastrophe facing the planet is a massive exchange thermonuclear war. Aside from the immediate human casualties, poisoned air, poisoned crops, and radioactive rain, many scientists agree that the dust storms and concentrations of particles would block vital sunlight and lower temperatures in the Northern Hemisphere, creating a **nuclear winter**. In the event of large-scale nuclear war, most living things on earth would die. The fear of nuclear war has greatly contributed to the military and arms buildup, which, ironically, also causes environmental destruction even in times of peace.

Destruction of the Environment in Peacetime   Even when military forces are not engaged in active warfare, military activities assault the environment. For example, modern military maneuvers require large amounts of land. The use of land for military purposes prevents other uses, such as agriculture, habitat protection, recreation, and housing. More importantly, military use of land often harms the land. In practicing military maneuvers, the armed forces demolish natural vegetation, disturb wildlife habitats, erode soil, silt up streams, and cause flooding. Military bombing and shooting ranges leave the land pockmarked with craters and contaminate the soil and groundwater with lead and other toxic residues.

Bombs exploded during peacetime leak radiation into the atmosphere and groundwater. From 1945 to 1990, 1,908 bombs were tested—that is, exploded—at more than 35 sites around the world. Although underground testing has reduced radiation, some still escapes into the atmosphere and is suspected of seeping into groundwater (Renner 1993a). Similarly, in Russia decommissioned submarines have been found to emit radioactive pollution (Editorial 2000).

Finally, although arms-control and disarmament treaties of the last decade have called for the disposal of huge stockpiles of weapons, no completely safe means of disposing of weapons and ammunition exist. Many activist groups have called for placing weapons in storage until safe disposal methods are found. Unfortunately, the longer weapons are stored, the more they deteriorate, increasing the likelihood of dangerous leakage.

Conflict in a Post–Cold War Era

In the discussion of the structural-functionalist view of global conflict and war, we noted that war functions as a catalyst for social cohesion. The corollary to the cohesive effect of war is that without a common enemy to fight, internal strife is likely to occur. Porter (1994) identifies a historical pattern in which "the end of an era of international rivalry and conflict has marked the beginning of internal conflict and disarray almost everywhere" (p. 300). According to Porter, the post–Cold War era is likely to be an era of political turmoil and divisiveness among social, racial, religious, and class groups. In support of his contention, 85 percent (92 of 108) of armed conflicts between 1989 and 1999 were internal, that is, within one country's borders (Renner 2000a).

Terrorism: A Growing Threat

Terrorism is the premeditated use, or threatened use, of violence by an individual or group to gain a political or social objective (INTERPOL 1998; Barkan & Snowden 2001). Terrorism may be used to publicize a cause, promote an ideology, achieve religious freedom, attain the release of a political prisoner, or rebel against a government. Terrorists use a variety of tactics, including assassinations, skyjackings, armed attacks, kidnaping and hostage taking, threats, and various forms of bombing. Acts of terrorism have increased in recent years with a total of 392 in 1999, a 43 percent increase from the previous year (USIS 2000).

Terrorism can be either domestic or transnational. Transnational terrorism occurs when a terrorist act in one country involves victims, targets, institutions, governments, or citizens of another country. The 1988 bombing of Pan-Am Flight 103 in Lockerbie, Scotland exemplifies transnational terrorism. The incident took the lives of 270 people and, after a 10-year investigation, resulted in the life sentence of a Libyan intelligence agent (CNN 2001b). Other examples of transnational terrorism include the 1993 bombing of the World Trade Center in New York, attacks of American embassies in Kenya and Tanzania, and the recent bombing of the naval ship the USS Cole moored in Aden Harbor, Yemen.

Domestic terrorism, sometimes called insurgent terrorism (Barkan & Snowden 2001), is exemplified by the 1995 truck bombing of a nine-story federal office building in Oklahoma City, resulting in 168 deaths and more than 200 injured. Gulf War veteran Timothy McVeigh, who, along with Terry Nichols, was convicted of the crime, is reported to have been a member of a paramilitary group that opposes the U.S. government. In 1997, McVeigh was sentenced to death for his actions.

A government can use both defensive and offensive strategies to fight terrorism. Defensive strategies include using metal detectors at airports and strengthening security at potential targets, such as embassies and military command posts. Offensive strategies include retaliatory raids such as the U.S. bombing of terrorist facilities in Afghanistan, group infiltration, and preemptive strikes. Unfortunately, efforts to stop one kind of terrorism may result in an increase in other types of terrorist acts. For example, after the use of metal detectors in airports increased, the incidence of skyjackings decreased, but the incidence of assassinations increased (Enders & Sandler 1993).

Combating terrorism is difficult, and recent trends will make it increasingly problematic (Zakaria 2000; Strobel, Kaplan, Newman, Whitelaw, & Grose,

> If you ask me what is the obstacle to the implementation [of the peace plan], it is terrorism. We face a unique kind of terrorism—the suicidal terror mission. There is no deterrent to a person who goes with high explosives in his car in his bag and explodes himself.
>
> YITZHAK RABIN
> *Former Prime Minister of Israel*

© Jeff Greenberg/PhotoEdit

The October, 2000 bombing of the USS Cole killed 17 sailors and injured 33. The terrorist act left a 30 by 40-foot hole in the hull of the 91,000-ton destroyer.

2001). First, data stored on computers can be easily accessed by hackers who illegally gain classified information. Interlopers obtained the fueling and docking schedules of the USS Cole. Second, the Internet permits groups with similar interests, once separated by geography, to share plans, fund-raising efforts, recruitment strategies, and other coordinated efforts. Worldwide, thousands of terrorists keep in touch through Hotmail.com Internet accounts. Finally, globalization contributes to terrorism by providing international markets where the tools of terrorism—explosives, guns, electronic equipment, and the like—can be purchased.

> War doesn't determine who's right, just who's left.
>
> GEORGE CARLIN
> *Comedian*

The possibility of terrorists using weapons of mass destruction is the most frightening scenario of all. Weapons of mass destruction **(WMD)** include the use of chemical, biological, and nuclear weapons. Anthrax for example, although usually associated with diseases in animals, is a highly deadly disease in humans and, although preventable by vaccine, has a "lethal lag time." In a hypothetical city of 100,000 people, delaying a vaccination program 1 day would result in 5,000 deaths; 6 days, 35,000 deaths (Garrett 2001).

Weapons of mass destruction have already been used. On at least eight occasions, Japanese terrorists dispersed aerosols of anthrax and botulism in Tokyo (Inglesby et al. 1999) and in 2000, a religious cult, hoping to disrupt elections in an Oregon county "contaminated local salad bars with salmonella, infecting hundreds . . ." (Garrett 2001, 76). Government officials have initiated a variety of laws, policies, and technological innovations designed to combat WMD and, in general, terrorism. Table 16.1 compares the opinions of a sample of U.S. adults with those of a sample of American "leaders" concerning terrorism and a number of other defense-related issues.

Table 16.1 *Opinions of Public and Leaders on Issues Related to National Defense*[a]

Item	Percent Holding an Opinion		
	Public	**Leader**	**Gap**[b]
1. It would be best for the country if we stayed out of world affairs rather than take an active part.	32	3	29
2. International terrorism is a critical threat.	86	61	24
3. Oppose an independent Palestinian state on the West Bank and the Gaza Strip.	42	19	24
4. Oppose troop use if North Korea invaded South Korea.	66	25	41
5. Oppose troop use if Iraq invaded Saudi Arabia.	48	20	28
6. Oppose troop use if Russia invaded Poland.	66	40	26
7. Military strength is more important than economic strength in determining a country's power.	30	8	22
8. Favor assassination of individual terrorist leaders in order to combat international terrorism.	61	35	26
9. In international crisis, do not take action alone without support of allies.	77	52	25
10. There will be more bloodshed and violence in the 21st century than in the 20th century.	57	24	33

[a]The public sample was composed of a random sample of 1,507 adults; the leaders sample was composed of 279 Americans in "senior positions with knowledge of international affairs" including business, government, media, and academic leaders (CCFR 2000, 3).

[b]The "gap" equals the public percent holding an opinion minus the leader percent holding an opinion.

Source: "American Public Opinion and U.S. Foreign Policy 1999." 2000. Chicago Council on Foreign Relations. Used by permission.

Guerrilla Warfare

Unlike terrorist activity, which targets civilians and may be committed by lone individuals, **guerrilla warfare** is committed by organized groups opposing a domestic or foreign government and its military forces. Guerrilla warfare often involves small groups who use elaborate camouflage and underground tunnels to hide until they are ready to execute a surprise attack. Since 1945, more than 120 armed guerrilla conflicts have occurred, resulting in the death of over 20 million people (Perdue 1993). Most of these conflicts occurred in developing countries. Fidel Castro's guerrillas in Cuba and the Vietcong in Vietnam are examples.

A more recent case of guerrilla warfare includes disturbances in the Congo (Renner 2000a; HRW 2001). The conflict involves troops from six countries, the Congolese Rally for Democracy (RCD) rebels, the Liberation of Congo (MLC) opposition group, and local warlords. In 3 years of conflict, more than 200,000 have died—twice the number of total Americans killed in Korea and Vietnam.

Two theories have been advanced to explain why guerrilla warfare occurs most often in less developed countries (Moaddel 1994). First, as societies change from a traditional to a modern industrialized society, they experience institutional instability, which leads to conflict as various groups compete for control and resources. A second theory holds that the conflicts arise from external international relations. Less developed countries are dependent on more developed countries, which exploit their labor forces and resources. The growing inequality between less developed and more developed countries creates conflict.

Strategies for Action: In Search of Global Peace

Various strategies and policies are aimed at creating and maintaining global peace. These include the redistribution of economic resources, the creation of a world government, peacekeeping activities of the United Nations, mediation and arbitration, and arms control.

Redistribution of Economic Resources

Inequality in economic resources contributes to conflict and war as the increasing disparity in wealth and resources between rich and poor nations fuels hostilities and resentment. Therefore, any measures that result in a more equal distribution of economic resources are likely to prevent conflict. John J. Shanahan (1995), retired U.S. Navy vice admiral and director of the Center for Defense Information, suggests that wealthy nations can help reduce social and economic roots of conflict by providing economic assistance to poorer countries. Nevertheless, U.S. military expenditures for national defense far outweigh U.S. economic assistance to foreign countries.

As we discussed in Chapter 14, strategies that reduce population growth are likely to result in higher levels of economic well being. Funke (1994) explains that "rapidly increasing populations in poorer countries will lead to environmental overload and resource depletion in the next century, which will most likely result in political upheaval and violence as well as mass starvation" (p. 326). Although achieving worldwide economic well being is important for

> Nuclear weapons are inherently dangerous, hugely expensive, militarily inefficient, and morally indefensible.
>
> **G. Lee Butler**
> *Former Commander-in-Chief of*
> *U.S. Strategic Air Command*

minimizing global conflict, it is important that economic development does not occur at the expense of the environment.

World Government

Some analysts have suggested that world peace might be attained through the establishment of a world government. The idea of a single world government is not new. In 1693, William Penn advocated a political union of all European monarchs, and in 1712, Jacques-Henri Bernardin de Saint-Pierre of France suggested an all-European Union with a "Senate of Peace." Proposals such as these have been made throughout history.

Although some commentators are pessimistic about the likelihood of a new world order, Lloyd (1998) identifies three global trends that, he contends, signify the "ghost of a world government yet to come" (p. 28). First is the increasing tendency for countries to engage in "ecological good behavior," indicating a concern for a global rather than national well being. Second, despite several economic crises worldwide, major world players such as the United States and Great Britain have supported rather than abandoned nations in need. And, finally, an International Criminal Court has been created with "powers to pursue, arraign, and condemn those found guilty of war and other crimes against humanity" (p. 28). On the basis of these three trends, Lloyd concludes, "[G]lobal justice, a wraith pursued by peace campaigners for over a century, suddenly seems achievable" (p. 28).

The United Nations

If you want peace, work for justice.

POPE PAUL VI

The United Nations (U.N.), whose charter begins, "We the people of the United Nations—Determined to save succeeding generations from the scourge of war . . ." has engaged in over 58 peacekeeping operations since 1948 (Renner 2000b). The U.N. Security Council can use force, when necessary, to restore international peace and security. Recently, the United Nations has been involved in overseeing multinational peacekeeping forces in East Timor, Sierra Leone, Kosovo, and the Congo.

In the last few years, the United Nations has come under heavy criticism. First, in recent missions, developing nations have supplied more than 75 percent of the troops while developed countries—United States, Japan, and Europe—have contributed 85 percent of the finances. As one U.N. official commented, ". . . you can't have a situation where some nations contribute blood and others only money . . ."(quoted in Vesely 2001, 8). Second, a recent review of U.N. peacekeeping operations noted several failed missions including an intervention in Somalia in which 44 American marines were killed (Lamont 2001).

Finally, the concept of the United Nations is that its members represent individual nations, not a region or the world. And because nations tend to act in their own best economic and security interests, U.N. actions performed in the name of world peace may be motivated by nations acting in their own interests.

Mediation and Arbitration

Mediation and arbitration are nonviolent strategies used to resolve conflicts and stop or prevent war. In mediation, a neutral third party intervenes and facilitates negotiation between representatives or leaders of conflicting groups. Mediators do not impose solutions, but rather help disputing parties generate options for resolving the conflict. Ideally, a mediated resolution to a conflict meets at least some of the concerns and interests of each party to the conflict. In other words, mediation attempts to find "win-win" solutions in which each side is satisfied with the solution. Although mediation is used to resolve conflict between individuals, it is also a valuable tool for resolving international conflicts. For example, in 2000, mediators were used to resolve the escalating violence between Israeli and Palestinian troops (Myre 2000).

Arbitration also involves a neutral third party who listens to evidence and arguments presented by conflicting groups. Unlike mediation, however, in arbitration the neutral third party arrives at a decision or outcome that the two conflicting parties agree to accept. As Sweet and Brunell (1998) note, **triad dispute resolution**, that is, resolution that involves two disputants and a negotiator—performs "profoundly political functions including the construction, consolidation, and maintenance of political regimes" (p. 64).

Arms Control

In the 1960s, the United States and the Soviet Union led the world in an arms race, each competing to build a more powerful military arsenal than its adversary. If either superpower were to initiate a full-scale war, the retaliatory powers of the other nation would result in the destruction of both nations. Thus, the principle of **mutually assured destruction** (MAD) that developed from nuclear weapons capabilities transformed war from a win-lose proposition to a lose-lose scenario. If both sides would lose in a war, the theory goes, neither side would initiate war.

Because of the end of the Cold War and the growing realization that current levels of weapons literally represented "overkill," governments have moved in the direction of arms control, which involves reducing or limiting defense spending, weapons production, and armed forces. Recent arms-control initiatives include SALT (Strategic Arms Limitation Treaty), START (Strategic Arms Reduction Treaty), NPT (Nuclear Nonproliferation Treaty), and the CTBT (Comprehensive Test Ban Treaty).

Strategic Arms Limitation Treaty   Under the 1972 SALT agreement (SALT I), the United States and the Soviet Union agreed to limit both their defensive weapons and their land-based and submarine-based offensive weapons. Also in 1972, Henry Kissinger drafted the Declaration of Principles, known as **detente**, which means "negotiation rather than confrontation." A further arms limitation agreement (SALT II) was reached in 1979, but was never ratified by Congress because of the Soviet invasion of Afghanistan. Subsequently, the arms race continued with the development of new technologies and an increase in the number of nuclear warheads.

Strategic Arms Reduction Treaty   Strategic arms talks resumed in 1982, but made relatively little progress for several years. During this period, President Reagan proposed the Strategic Defense Initiative, more commonly known as "Star Wars," which purportedly would be able to block missiles launched by another country against the United States (Brown 1994). Although some research was conducted on the system, Star Wars was never actually built. Today, research on the development of a national missile defense system continues with proposed funding for

Nonviolence appeals directly to the mysterious unity among all of us, which is the hidden glory of each of us.

MICHAEL NAGLER
Berkeley Peace and Conflict Studies Program

We are closer to nuclear war than we ever were in the last generation of the cold war.

DANIEL PATRICK MOYNIHAN
Former U.S. Senator

2001–2005 reaching $10.4 billion (Center for Defense Information 2000). U.S. use of a national defense system would be in violation of the 1972 Anti-Ballistic Missile Treaty signed with the former Soviet Union (MacIntyre 2001).

In 1991, the international situation changed dramatically. The communist regime in the Soviet Union had fallen, the Berlin Wall had been dismantled, and many Eastern European and Baltic countries were under self-rule. SALT was re-named START (Strategic Arms Reduction Treaty) and was signed in 1991. A second START agreement, signed in 1993 and ratified by the U.S. Senate, signaled the end of the Cold War. START II calls for the reduction of nuclear warheads to 3,500 by the year 2003, a significant reduction from present levels (Zimmerman 1997). To date, START II awaits ratification by the Russian Parliament, thereby delaying any official negotiations on START III (Arms Control Association 2000).

Nuclear Nonproliferation Treaty   The 1970 Nuclear Nonproliferation Treaty (NPT) was renewed in 2000, adopted by 187 countries. Only Cuba, India, Israel and Pakistan have not signed the agreement (Albright 2000). The treaty holds that countries without nuclear weapons will not to try to get them; in exchange, the nuclear-weapon countries (the United States, United Kingdom, France, China, and Russia) agree they will not provide nuclear weapons to countries that did not have them.

> The real danger is that we will be tempted more and more to resort to the use of military forces as we diminish our capacity to respond by other means. As the saying goes, if all you have is a hammer then every problem starts to look like a nail.
>
> VICE ADMIRAL JOHN
> J. SHANAHAN
> *U.S. Navy (retired)*

However, even if military superpowers honor agreements to limit arms, the availability of black market nuclear weapons and materials presents a threat to global security. For example, Kyl and Halperin (1997) note that U.S. security is threatened more by nuclear weapons falling into the hands of a terrorist group than by a nuclear attack from an established government such as Russia. The authors further conclude that if Russia were to launch missiles directed at the United States "the odds are overwhelming that it [would] be a Russian missile fired by accident or without authority" (p. 28).

Comprehensive Test Ban Treaty   On September 10, 1996, the UN General Assembly passed the Comprehensive Test Ban Treaty (CTBT) by a vote of 158 to 3. The treaty, a "prime disarmament goal for more than forty years" (UCS 1998, 1), would put an end to underground nuclear testing. President Clinton, the first world leader to sign the ban, submitted it to the U.S. Senate for ratification in 1997 but it was defeated when finally reaching a vote in 1999 (Shalikashvili 2001). Britain and France have already ratified the agreement. For the CTBT to be enforced, it must be ratified by the 44 members of the Conference on Disarmament. Both India and Pakistan, which tested nuclear devices in 1998, are members of the Conference.

Understanding *Conflict around the World*

As we come to the close of this chapter, how might we have an informed understanding of conflict around the world? Each of the three theoretical positions discussed in this chapter reflects the realities of war. As functionalists argue, war offers societal benefits—social cohesion, economic prosperity, scientific and technological developments, and social change. Further, as conflict theorists contend, wars often occur for economic reasons as corporate elites and political leaders benefit from the spoils of war—land and water resources and raw materials. The symbolic interactionist perspective emphasizes the role that meanings, labels, and definitions play in creating conflict and contributing to acts of war.

Ultimately, we are all members of one community—Earth—and have a vested interest in staying alive and protecting the resources of our environment for our own and future generations. But conflict between groups is a feature of social life and human existence that is not likely to disappear. What is at stake—human lives and the ability of our planet to sustain life—merits serious attention. Traditionally, nations have sought to protect themselves by maintaining large military forces and massive weapons systems. These strategies are associated with serious costs. In diverting resources away from other social concerns, militarism undermines a society's ability to improve the overall security and well being of its citizens. Conversely, defense-spending cutbacks can potentially free up resources for other social agendas, including lowering taxes, reducing the national debt, addressing environmental concerns, eradicating hunger and poverty, improving health care, upgrading educational services, and improving housing and transportation. Therein lies the promise of a "peace dividend."

Hopefully, future dialogue on the problem of global conflict and war will redefine national and international security to encompass social, economic, and environmental concerns. These other concerns play a vital role in the security of nations and the world. The World Commission on Environment and Development (1990) concluded:

> The deepening and widening environmental crisis presents a threat to national security—and even survival—that may be greater than well-armed, ill-disposed neighbors and unfriendly alliances . . . . The arms race in all parts of the world—pre-empts resources that might be used more productively to diminish the security threats created by environmental conflict and the resentments that are fueled by widespread poverty . . . . There are no military solutions to environmental insecurity.

National and global policies aimed at reducing poverty and ensuring the health of our planet and its present and future inhabitants are important aspects of world peace. But as long as we define national and global security in military terms, we will likely ignore the importance of nonmilitary social policies in achieving world peace. According to Funke (1994), changing the definition of national security "is the first step to changing policy" (p. 342).

> The first victim of war is not truth—it is conscience. Otherwise human beings could not bear the guilt of participating in a savage conflict in which armed force is not a last resort, but a hair trigger.
>
> WILLIAM PERDUE
> *Sociologist*

Critical Thinking

1 Certain actions constitute "war crimes." Such actions include the use of forbidden munitions such as biological weapons, purposeless destruction, killing civilians, poisoning of waterways, and violation of surrender terms. In addition to those listed above, what other actions should constitute "war crimes"?

2 Describe countries that have the highest probability of going to war in terms of their economic, social, and psychological makeup. Now, describe countries that are the least likely to go to war. Does history confirm your hypotheses?

3 Selecting each of the five major institutions in society, what part could each play in attaining global peace?

4 Make a list of famous war movies (e.g., Schindler's List). With specific movies in mind, list media sounds and images of war. Has the portrayal of war in movies changed over time? If so, how and why?

Key Terms

Cold War

détente

dual-use technologies

economic conversion

guerrilla warfare

infowar

military-industrial complex

mutually assured destruction (MAD)

nuclear winter

peace dividend

post-traumatic stress disorder

state

terrorism

triad dispute resolution

war

WMD

Media Resources

The Wadsworth Sociology Resource Center: Virtual Society

http://sociology.wadsworth.com/

See the companion Web site for this book to access general sociology resources and text-specific features that can further your understanding of this chapter. The site contains Internet links, Internet exercises, online practice quizzes, information on InfoTrac College Edition, and many more valuable materials designed to enrich your learning experience in social problems.

InfoTrac College Edition

You can access InfoTrac College Edition either from the Wadsworth Sociology Resource Center at **http://sociology.wadsworth.com** or directly from your web browser at **http://www.infotrac-college.com/wadsworth/**. InfoTrac College Edition is an online university library that includes over 700 popular and scholarly journals in which you can find articles related to the topics in this chapter such as the military, terrorism, the threat of nuclear war, and nuclear disarmament.

Interactions CD-ROM

Go to the "Interactions" CD-ROM for *Understanding Social Problems*, Third Edition to access additional interactive learning tools, such as in-depth review materials, corresponding practice quizzes, and other engaging resources and activities to help you study the concepts in this chapter.

Epilogue

Never doubt that a small group of thoughtful, committed citizens can change the world. Indeed, it's the only thing that ever has.

MARGARET MEAD
Anthropologist

Today, there is a crisis—a crisis of faith: faith in the ideals of equality and freedom, faith in political leadership, faith in the American dream, and, ultimately, faith in the inherent goodness of humankind and the power of one individual to make a difference. To some extent, faith is shaken by texts such as this one. Drug use is up; marriages down; political corruption is everywhere; bigotry's on the rise; the environment is killing us—if we don't kill it first. Social problems are everywhere, and what's worse, many solutions only seem to create more problems.

The transformation of American society in recent years has been dramatic. With the exception of the Industrial Revolution, no other period in human history has seen such rapid social change. The structure of society, forever altered by such macro-sociological processes as multi-nationalization, de-industrialization, and globalization, continues to be characterized by social inequities—in our schools, in our homes, in our cities, and in our salaries.

The culture of society has also undergone rapid change, leading many politicians and lay persons alike to call for a return to traditional values and beliefs and to emphasize the need for moral education. The implication is that somehow things were better in the "good old days," and if we could somehow return to those times, things would be better again. Some things were better—for some people.

Fifty years ago, there were fewer divorces and less crime. AIDS and crack cocaine were unheard of, and violence in schools was almost nonexistent. At the same time, however, in 1950, the infant mortality rate was over three times what it is today; racial and ethnic discrimination flourished in an atmosphere of bigotry and hate, and millions of Americans were routinely denied the right to vote because of the color of their skin; more than half of all Americans smoked cigarettes; and persons over the age of 25 had completed a median of 6.8 years of school.

The social problems of today are the cumulative result of structural and cultural alterations over time. Today's problems are not necessarily better or worse than those of generations ago—they are different and, perhaps, more diverse as

> A pessimist sees the difficulty in every opportunity; an optimist sees the opportunity in every difficulty.
>
> WINSTON CHURCHILL
> *British statesman*

507

The day will come when
nations will be judged
not by their military or
economic strength,
nor by the splendor of
their capital cities and
public buildings,
but by the well being of
their peoples;
by their levels of health,
nutrition and education;
by their opportunities to
earn a fair reward for
their labours;
by their ability to
participate in the
decisions that affect their
lives;
by the respect that is
shown for their civil and
political liberties;
by the provision that is
made for those who are
vulnerable and
disadvantaged;
and by the protection that
is afforded to the
growing minds and
bodies of their children.

UNICEF THE PROGRESS
OF NATIONS, 2000.

The only thing necessary
for the triumph of evil
is for good men [and
women] to do nothing.

EDMUND BURKE
English political writer/orator

a result of the increased complexity of social life. But, as surely as we brought the infant mortality rate down, prohibited racial discrimination in education, housing, and employment, increased educational levels, and reduced the number of smokers, we can continue to meet the challenges of today's social problems. But how does positive social change occur? How does one alter something as amorphous as society? The answer is really quite simple. All social change takes place because of the acts of individuals. Every law, every regulation and policy, every social movement and media exposé, and every court decision began with one person.

Sociologist Earl Babbie (1994) recounts how the behavior of one person—Rosa Parks—made a difference. Rosa Parks was a seamstress in Montgomery, Alabama, in the 1950s. Like almost everything else in the South in the 1950s, public transportation was racially segregated. On December 1, 1955, Rosa Parks was on her way home from work when the "white section" of the bus she was riding became full. The bus driver told black passengers in the first row of the black section to relinquish their seats to the standing white passengers. Rosa Parks refused.

She was arrested and put in jail, but her treatment so outraged the black community that a boycott of the bus system was organized by a new minister in town—Martin Luther King Jr. The Montgomery bus boycott was a success. Just 11 months later, in November of 1956, the U.S. Supreme Court ruled that racial segregation of public facilities was unconstitutional. Rosa Parks had begun a process that in time would echo her actions—the civil rights movement, the March on Washington, the 1963 Equal Pay Act, the 1964 Civil Rights Act, the 1965 Voting Rights Act, regulations against discrimination in housing, and affirmative action.

Was social change accomplished? In 1960, just 20 percent of blacks 25 years old and older had completed high school compared to 40 percent of whites. By 1999, 77 percent of blacks 25 years old and older had completed high school compared to 84 percent of whites (*Statistical Abstract* 2000). While many would point out that such changes and thousands like them have created other problems that need to be addressed, who among us would want to return to the "good old days" of the 1950s in Montgomery, Alabama?

Millions of individuals make a difference daily. Chuck Beattie and Bret Byfield, two social workers in Minneapolis, began Phoenix Group in 1991. Purchasing houses with a few grants and some private donations, they hired "street people" to renovate the houses and then let them move in. Today, Phoenix Group has more than 39 properties, 300 residents, and 11 businesses. Susan Brotchie, deserted by her husband and unable to get child support, began Advocates for Better Child Support, which has helped more than 9,000 people establish and collect child support payments. Pedro José Greer was an intern in Miami when he treated his first homeless patient. Appalled by the fatal incidence of tuberculosis, a curable disease, Dr. Greer opened a medical clinic in a homeless shelter. Today, he heads the largest provider of medical care for the poor in Florida—Camillus Health Concern—annually serving over 4,500 patients (Chinni et al. 1995, 34). After suffering chronic illness from exposure to air pollutants from a sewage plant and home renovation materials, Mary Lamielle founded the National Center for Environmental Health Strategies (NCEHS), a national nonprofit organization dedicated to the development of creative solutions to environmental health problems (Lamielle 1995). Through her work as director of the NCEHS, Lamielle has influenced policy development and research and has provided support and advocacy to sufferers of environmental pollution.

College students have also worked to bring about social change. College students have prompted the adoption of multi-cultural curriculums, helped the homeless, made their schools more environmentally accountable, and organized against-sweatshop labor (Loeb 1995; Alvarado 2000). College students also influenced universities to rid themselves of South African investments and played an important role in building the international movement that helped end apartheid.

While only a fraction of the readers of this text will occupy social roles that directly influence social policy, one need not be a politician or member of a social reform group to make a difference. We, the authors of this text, challenge you, the reader, to make individual decisions and take individual actions to make the world a more humane, just, and peaceful place for all. Where should we begin? Where Rosa Parks and others like her began—with a simple individual act of courage, commitment, and faith.

Some people see things that are and say, "Why?" He saw things that never were and said, "Why not?"

TED KENNEDY
Of his brother, Bobby Kennedy

Appendix A

Methods of Data Analysis

There are three levels of data analysis: description, correlation, and causation. Data analysis also involves assessing reliability and validity.

Description

Qualitative research involves verbal descriptions of social phenomena. Having a homeless and single pregnant teenager describe her situation is an example of qualitative research.

Quantitative research often involves numerical descriptions of social phenomena. Quantitative descriptive analysis may involve computing the following: (1) means (averages), (2) frequencies, (3) mode (the most frequently occurring observation in the data), (4) median (the middle point in the data; half of the data points are above, and half are below the median), and (5) range (the highest and lowest values in a set of data).

Correlation

Researchers are often interested in the relationship between variables. *Correlation* refers to a relationship among two or more variables. The following are examples of correlational research questions: What is the relationship between poverty and educational achievement? What is the relationship between race and crime victimization? What is the relationship between religious affiliation and divorce?

If there is a correlation or relationship between two variables, then a change in one variable is associated with a change in the other variable. When both variables change in the same direction, the correlation is positive. For example, in general, the more sexual partners a person has, the greater the risk of contracting a sexually transmissible disease. As variable A (number of sexual partners) increases, variable B (chance of contracting an STD) also increases. Similarly, as the number of sexual partners decreases, the chance of contracting an STD decreases. Notice that in both cases, the variables change in the same direction, suggesting a positive correlation (see Figure A.1).

When two variables change in opposite directions, the correlation is negative. For example, there is a negative correlation between condom use and contracting STDs. In other words, as condom use increases, the chance of contracting an STD decreases (see Figure A.2).

The relationship between two variables may also be curvilinear, which means that they vary in both the same and opposite directions. For example, suppose a researcher finds that after drinking one alcoholic beverage, research participants

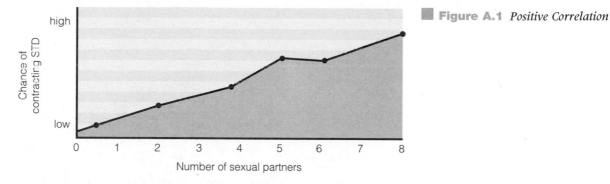

■ **Figure A.1** *Positive Correlation*

are more prone to violent behavior. After two drinks, violent behavior is even more likely, and this trend continues for three and four drinks. So far, the correlation between alcohol consumption and violent behavior is positive. After the research participants have five alcoholic drinks, however, they become less prone to violent behavior. After six and seven drinks, the likelihood of engaging in violent behavior decreases further. Now the correlation betweeen alcohol consumption and violent behavior is negative. Because the correlation changed from positive to negative, we say that the correlation is curvilinear (the correlation may also change from negative to positive) (see Figure A.3).

A fourth type of correlation is called a spurious correlation. Such a correlation exists when two variables appear to be related, but the apparent relationship occurs only because they are both related to a third variable. When the third variable is controlled through a statistical method in which the variable is held constant, the apparent relationship between the variables disappears. For example, blacks have a lower average life expectancy than whites do. Thus, race and life expectancy appear to be related. However, this apparent correlation exists because both race and life expectancy are related to socioeconomic status. Since blacks are more likely than whites to be impoverished, they are less likely to have adequate nutrition and medical care.

Causation

If the data analysis reveals that two variables are correlated, we know only that a change in one variable is associated with a change in another variable. We cannot assume, however, that a change in one variable *causes* a change in the other variable unless our data collection and analysis are specifically designed to assess causation. The research method that best assesses causality is the experimental method (discussed in Chapter 1).

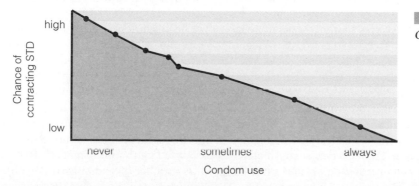

■ **Figure A.2** *Negative Correlation*

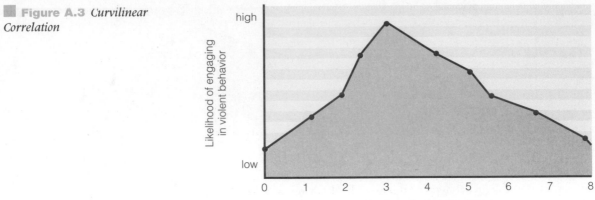

■ **Figure A.3** *Curvilinear Correlation*

To demonstrate causality, three conditions must be met. First, the data analysis must demonstrate that variable A is correlated with variable B. Second, the data analysis must demonstrate that the observed correlation is not spurious. Third, the analysis must demonstrate that the presumed cause (variable A) occurs or changes prior to the presumed effect (variable B). In other words, the cause must precede the effect.

It is extremely difficult to establish causality in social science research. Therefore, much social research is descriptive or correlative, rather than causative. Nevertheless, many people make the mistake of interpreting a correlation as a statement of causation. As you read correlative research findings, remember the following adage: "Correlation *does not* equal causation."

Reliability and Validity

Assessing reliability and validity is an important aspect of data analysis. *Reliability* refers to the consistency of the measuring instrument or technique; that is, the degree to which the way information is obtained produces the same results if repeated. Measures of reliability are made on scales and indexes (such as those in the *Self and Society* features in this text) and on information-gathering techniques, such as the survey methods described in Chapter 1.

Various statistical methods are used to determine reliability. A frequently used method is called the "test-retest method." The researcher gathers data on the same sample of people twice (usually one or two weeks apart) using a particular instrument or method and then correlates the results. To the degree that the results of the two tests are the same (or highly correlated), the instrument or method is considered reliable.

Measures that are perfectly reliable may be absolutely useless unless they also have a high validity. *Validity* refers to the extent to which an instrument or device measures what it intends to measure. For example, police officers administer "breathalyzer" tests to determine the level of alcohol in a person's system. The breathalyzer is a valid test for alcohol consumption.

Validity measures are important in research that uses scales or indices as measuring instruments. Validity measures are also important in assessing the accuracy of self-report data that are obtained in survey research. For example, survey research on high-risk sexual behaviors associated with the spread of HIV relies heavily on self-report data on such topics as number of sexual partners, types of sexual activities, and condom use. Yet, how valid are these data? Do survey respondents under-report the number of their sexual partners? Do people who say they use a condom every time they engage in intercourse really use a condom every time? Because of the difficulties in validating self-reports of number of sexual partners and condom use, we may not be able to answer these questions.

Ethical Guidelines in Social Problems Research

Social scientists are responsible for following ethical standards designed to protect the dignity and welfare of people who participate in research. These ethical guidelines include the following (Schutt 1999; Neuman 2000; American Sociological Association 2001):

1. *Freedom from coercion to participate.* Research participants have the right to decline to participate in a research study or to discontinue participation at any time during the study. For example, professors who are conducting research using college students should not require their students to participate in their research.

2. *Informed consent.* Researchers are required to inform potential participants of any aspect of the research that might influence a subject's willingness to participate. After informing potential participants about the nature of the research, researchers typically ask participants to sign a consent form indicating that the participants are informed about the research and agree to participate in it.

3. *Deception and debriefing.* Sometimes the researcher must disguise the purpose of the research in order to obtain valid data. Researchers may deceive participants as to the purpose or nature of a study only if there is no other way to study the problem. When deceit is used, participants should be informed of this deception (debriefed) as soon as possible. Participants should be given a complete and honest description of the study and why deception was necessary.

4. *Protection from harm.* Researchers must protect participants from any physical and psychological harm that might result from participating in a research study. This is both a moral and a legal obligation. It would not be ethical, for example, for a researcher studying drinking and driving behavior to observe an intoxicated individual leaving a bar, getting into the driver's seat of a car, and driving away.

   Researchers are also obligated to respect the privacy rights of research participants. If anonymity is promised, it should be kept. Anonymity is maintained in mail surveys by identifying questionnaires with a number coding system, rather than with the participants' names. When such anonymity is not possible, as is the case with face-to-face interviews, researchers should tell participants that the information they provide will be treated as confidential. Although interviews may be summarized and excerpts quoted in published material, the identity of the individual participants is not revealed. If a research participant experiences either physical or psychological harm as a result of participation in a research study, the researcher is ethically obligated to provide remediation for the harm.

5. *Reporting of research.* Ethical guidelines also govern the reporting of research results. Researchers must make research reports freely available to the public. In these reports, researchers should fully describe all evidence obtained in the study, regardless of whether the evidence supports the researcher's hypothesis. The raw data collected by the researcher should be made available to other researchers who might request it for purposes of analysis. Finally, published research reports should include a description of the sponsorship of the research study, its purpose, and all sources of financial support.

Glossary

abortion The intentional termination of a pregnancy.

absolute poverty The chronic absence of the basic necessities of life, including food, clean water, and housing.

acculturation Learning the culture of a group different from the one in which a person was originally raised.

achieved status A status assigned on the basis of some characteristic or behavior over which the individual has some control.

acid rain The mixture of precipitation with air pollutants, such as sulfur dioxide and nitrogen oxide.

acquaintance rape Rape that is committed by someone known by the victim.

activity theory A theory that emphasizes that the elderly disengage, in part, because they are structurally segregated and isolated with few opportunities to engage in active roles.

acute condition A health condition that can last no more than 3 months.

adaptive discrimination Discrimination that is based on the prejudice of others.

affirmative action A series of programs based on the 1964 Civil Rights Act that provide opportunities or other benefits to persons on the basis of their group membership (e.g., race, gender, ethnicity).

age grading The assignment of social roles to given chronological ages.

ageism The belief that age is associated with certain psychological, behavioral, and/or intellectual traits.

agricultural biotechnology The application of biotechnology to agricultural crops and livestock.

Aid to Families with Dependent Children (AFDC) Before 1996, a cash assistance program that provided single parents (primarily women) and their children with a minimum monthly income.

alienation The concept used by Karl Marx to describe the condition when workers feel powerlessness and meaninglessness as a result of performing repetitive, isolated work tasks. Alienation involves becoming estranged from one's work, the products one creates, other human beings, and/or one's self; it also refers to powerlessness and meaninglessness experienced by students in traditional, restrictive educational institutions.

alternative certification programs Programs that permit college graduates, without education degrees, to be certified to teach based on job and/or life experiences.

amalgamation The physical blending of different racial and/or ethnic groups, resulting in a new and distinct genetic and cultural population; results from the intermarriage of racial and ethnic groups over generations.

anomie A state of normlessness in which norms and values are weak or unclear; results from rapid social change and is linked to many social problems, including crime, drug addiction, and violence.

antimiscegenation laws Laws that prohibited interracial marriages.

ascribed status A status that society assigns to an individual on the basis of factors over which the individual has no control.

assimilation The process by which minority groups gradually adopt the cultural patterns of the dominant majority group.

asylees Immigrants who apply from within the United States for admission on the basis of persecution or fear of persecution for their political or religious beliefs.

authoritarian-personality theory A psychological theory of prejudice that suggests prejudice arises in people with a certain personality type. According to this theory, people with an authoritarian personality, who are highly conformist, intolerant, cynical, and preoccupied with power, are prone to being prejudiced.

automation A type of technology in which self-operated machines accomplish tasks formerly done by workers; develops as a society moves toward industrialization and becomes more concerned with the mass production of goods.

autosclerosis Clogged vehicular arteries that slow rush hour traffic to a crawl or a stop, even when there are no accidents or construction crews ahead.

aversive racism A subtle, often unintentional form of prejudice exhibited by many Americans who possess strong egalitarian values and who view themselves as non-prejudiced.

barrio A slum that is occupied primarily by Latinos.

Basel Ban A proposal that promises to end the export of hazardous waste from rich OECD countries to poor non-OECD countries even when the waste is exported for recycling.

behavioral-based safety programs A controversial health and safety strategy used by business management that attributes health and safety problems in the workplace to workers' behavior, rather than to work processes and conditions.

beliefs Definitions and explanations about what is assumed to be true.

bias-motivated crime See *hate crime.*

bilingual education Educational instruction provided in two languages—the student's native language and another language. In the United States, bilingual education involves teaching individuals in both English and their non-English native language.

biodiversity The variability of living organisms on earth.

biphobia Negative attitudes toward bisexuality and people who identify as bisexual.

biomonitoring A method of assessing a person's exposure to environmental chemicals or their metabolites (break down products) in blood or urine.

bisexuality A sexual orientation that involves cognitive, emotional, and sexual attraction to members of both sexes.

bonded labor The repayment of a debt through labor.

bourgeoisie The owners of the means of production.

Brady Bill A law that requires a 5-day waiting period for handgun purchases so that sellers can screen buyers for criminal records or mental instability.

brain drain The phenomenon whereby in developing countries, many individuals with the highest level of skill and education leave the country in search of work abroad.

brownfields Abandoned or undeveloped sites that are located on contaminated land.

burden of disease The number of deaths in a population combined with the impact of premature death and disability on that population.

capitalism An economic system in which private individuals or groups invest capital to produce goods and services, for a profit, in a competitive market.

central city The largest city in a metropolitan area.

character education Education that emphasizes the moral and interpersonal aspects of an individual.

charter schools Public schools founded by parents, teachers, and communities, and maintained by school tax dollars.

chemical dependency A condition in which drug use is compulsive, and users are unable to stop because of physical and/or psychological dependency.

child abuse The physical or mental injury, sexual abuse, negligent treatment, or maltreatment of a child under the age of 18 by a person who is responsible for the child's welfare.

child labor Children performing work that is hazardous, that interferes with a child's education, or that harms a child's health, or physical, mental, spiritual, or moral development.

chronic condition A long-term health problem, such as a disease or impairment.

civil union A legal status under Vermont state law that entitles same-sex couples who apply for and receive a civil union certificate nearly all the benefits available to married couples.

classic rape A rape committed by a stranger, with the use of a weapon, resulting in serious bodily injury.

clearance rate As used in the UCR, the percentage of cases in which an arrest and official charge have been made and the case turned over to the courts.

club drugs A general term used to refer to illicit, often synthetic drugs commonly used at nightclubs or all-night dances called "raves."

Cold War The state of political tension and military rivalry that existed between the United States and the former Soviet Union from the 1950s through the late 1980s.

colonialism When a racial and/or ethnic group from one society takes over and dominates the racial and/or ethnic group(s) of another society.

common couple violence Occasional acts of violence that result from

conflict that gets out of hand between person's in a relationship.

community policing A type of policing in which uniformed police officers patrol and are responsible for certain areas of the city as opposed to simply responding to crimes as they occur.

compressed workweek Workplace option in which employees work full-time, but in 4 rather than 5 days.

computer crime Any violation of the law in which a computer is the target or means of criminal activity.

conflict perspective A sociological perspective that views society as comprising different groups and interests competing for power and resources.

contingent workers (also called "disposable workers") Involuntary part-time workers, temporary employees, and workers who do not perceive themselves as having an explicit or implicit contract for ongoing employment.

control theory A theory that argues that a strong social bond between a person and society constrains some individuals from violating norms.

convergence hypothesis The argument that capitalist countries will adopt elements of socialism and socialist countries will adopt elements of capitalism, that is, they will converge.

conversion therapy See *reparative therapy.*

cooperative learning Learning in which a heterogeneous group of students, of varying abilities, help one another with either individual or group assignments.

corporal punishment The use of physical force with the intention of causing a child to experience pain, but not injury, for the purpose of correction or control of a child's behavior.

corporate downsizing The corporate practice of discharging large numbers of employees. Simply put, the term "downsizing" is a euphemism for mass firing of employees.

corporate multinationalism The practice of corporations to have their home base in one country and branches, or affiliates, in other countries.

corporate violence The production of unsafe products and the failure of corporations to provide a safe working environment for their employees.

corporate welfare Laws and policies that favor corporations, such as low-interest government loans to failing businesses and special subsidies and tax breaks to corporations.

covenant marriage A type of marriage offered in Louisiana that permits divorce only under condition of fault or after a marital separation of more than 2 years.

crack A crystallized illegal drug product made by boiling a mixture of baking soda, water, and cocaine.

crime An act or the omission of an act that is a violation of a federal, state, or local law and for which the state can apply sanctions.

cultural imperialism The indoctrination into the dominant culture of a society; when cultural imperialism exists, the norms, values, traditions, and languages of minorities are systematically ignored.

cultural lag A condition in which the material part of the culture changes at a faster rate than the nonmaterial part.

cultural sexism The ways in which the culture of society perpetuates the subordination of individuals based on their sex classification.

culture of poverty The set of norms, values, and beliefs and self-concepts that contribute to the persistence of poverty among the underclass.

cumulative trauma disorders The most common type of workplace illness in the United States; includes muscle, tendon, vascular, and nerve injuries that result from repeated or sustained actions or exertions of different body parts. Jobs that are associated with high rates of upper body cumulative stress disorders include computer programming, manufacturing, meat packing, poultry processing, and clerical/office work.

cybernation The use of machines that control other machines in the production process; characteristic of postindustrial societies that emphasize service and information professions.

date-rape drugs Drugs that are used to render victims incapable of resisting sexual assaults.

de facto segregation Segregation that is not required by law, but exists "in fact," often as a result of housing and socioeconomic patterns.

de jure segregation Segregation that is required by law.

deconcentration The redistribution of the population from cities to suburbs and surrounding areas.

decriminalization The removal of criminal penalties for a behavior, as in the decriminalization of drug use.

Defense of Marriage Act The Defense of Marriage Act states that marriage is a legal union between one man and one woman.

deforestation The destruction of the earth's rain forests.

deindustrialization The loss and/or relocation of manufacturing industries.

deinstitutionalization The removal of individuals with psychiatric disorders from mental hospitals and large residential institutions and into outpatient community mental health centers.

demographic transition theory A theory that attributes population growth patterns to changes in birthrates and death rates associated with the process of industrialization. In preindustrial societies, the population remains stable because, although the birthrate is high, the death rate is also high. As a society becomes industrialized, the birthrate remains high, but the death rate declines, causing rapid population growth. In societies with advanced industrialization, the birthrate declines, and this decline, in conjunction with the low death rate, slows population growth.

dependency ratio The number of societal members who are under 18 or 65 and over compared to the number of people who are between 18 and 64.

dependent variable The variable that the researcher wants to explain. See also *independent variable*.

deregulation The reduction of government control of, for example, certain drugs.

desertification The expansion of deserts and the loss of usable land due to the overuse of semiarid land on the desert margins for animal grazing and obtaining firewood.

deskilling The tendency for workers in a postindustrial society to make fewer decisions and for labor to require less thought.

detente The philosophy of "negotiation rather than confrontation" in reference to relations between the United States and the former Soviet Union; put forth by Henry Kissinger's Declaration of Principles in 1972.

deterrence The use of harm or the threat of harm to prevent unwanted behaviors.

devaluation hypothesis The hypothesis that argues that women are paid less because the work they perform is socially defined as less valuable than the work performed by men.

differential association A theory developed by Edwin Sutherland that holds that through interaction with others, individuals learn the values, attitudes, techniques, and motives for criminal behavior.

disability-adjusted life year (DALY) Years lost to premature death and years lived with illness or disability. More simply, 1 DALY equals 1 lost year of healthy life.

discrimination Differential treatment of individuals based on their group membership.

discriminatory unemployment High rates of unemployment among particular social groups such as racial and ethnic minorities and women.

disengagement theory A theory claiming that the elderly disengage from productive social roles in order to relinquish these roles to younger members of society. As this process continues, each new group moves up and replaces another, which, according to disengagement theory, benefits society and all of its members.

distance learning Learning in which, by time or place, the student is separated from the teacher.

diversity training Workplace training programs designed to increase employees' awareness of cultural differences in the workplace and how these differences may affect job performance.

divorce law reform Policies and proposals designed to change divorce law. Usually, divorce law reform measures attempt to make divorce more difficult to obtain.

divorce mediation A process in which divorcing couples meet with a neutral third party (mediator) who assists the individuals in resolving such issues as property division, child custody, child support, and spousal support in a way that minimizes conflict and encourages cooperation.

domestic partnership A status that grants legal entitlements such as health insurance benefits and inheritance rights to heterosexual cohabiting couples and homosexual couples.

double effect In reference to physician-assisted suicide (PAS), the use of medical interventions to relieve pain and suffering but that also may hasten death.

double jeopardy See *multiple jeopardy.*

doubling time The time it takes for a population to double in size from any base year.

drug Any substance other than food that alters the structure and functioning of a living organism when it enters the bloodstream.

drug abuse The violation of social standards of acceptable drug use, resulting in adverse physiological, psychological, and/or social consequences.

drug addiction. See *chemical dependency.*

dual-use technologies Defense-funded technological innovations with commercial and civilian use.

dumbing down The lowering of educational standards or expectations by students and/or teachers.

Earned Income Tax Credit (EITC) A refundable tax credit based on a working family's income and number of children. In addition to the federal EITC, many states also have a state EITC.

e-commerce The buying and selling of goods and services over the Internet.

economic conversion The reallocation of resources from military forces and defense industries to other sectors of the economy.

economic institution The structure and means by which a society produces, distributes, and consumes goods and services.

ecoterrorism Any violent crime intended to coerce, intimidate or change public policy on behalf of the natural world.

elder abuse The physical or psychological abuse, financial exploitation, or medical abuse or neglect of the elderly.

Employment Nondiscrimination Act The Employment Nondiscrimination Act (ENDA) has been promoted by gay rights activists in order to prohibit discrimination, preferential treatment, and quotas in the workplace on the basis of sexual orientation.

Empowerment Zone/Enterprise Community Initiative Also known as the "EZ/EC program"; a federal program that provides tax incentives, grants, and loans to businesses to create jobs for residents living within various designated zones or communities, many of which are in urban areas.

environmental injustice (see also *environmental racism*) The tendency for socially and politically marginalized groups to bear the brunt of environmental ills.

environmental racism The tendency for U.S. hazardous waste sites and polluting industries to be located in areas where the surrounding residential population is black, Native American, or Hispanic.

environmental refugees Individuals who have migrated because they can no longer secure a livelihood because of deforestation, desertification, soil erosion, and other environmental problems.

epidemiological transition The shift from a society characterized by low life expectancy and parasitic and infectious diseases to one characterized by high life expectancy and chronic and degenerative diseases.

epidemiologist A scientist who studies the social origins and distribution of health problems in a population and how patterns of illness and disease vary between and within societies.

epidemiology The study of the distribution of disease within a population.

ergonomics The designing or redesigning of the workplace to prevent and reduce cumulative trauma disorders.

ethnicity A shared cultural heritage and/or national origin.

e-waste Waste from electronic equipment.

extended producer responsibility An environmental strategy whereby producers are responsible for taking back their old products, phasing out the toxic materials in manufacturing, and designing cleaner products with less waste generation.

ex-gay ministries Programs sponsored by some religious organizations that claim to "cure" homosexuals and transform them into heterosexuals through prayer and other forms of therapy.

experiment A research method that involves manipulating the independent variable to determine how it affects the dependent variable.

expulsion When a dominant group forces a subordinate group to leave the country or to live only in designated areas of the country.

Fair Labor Standards Act (FLSA) Federal law that sets rules of employment.

familism A value system that encourages family members to put their family's well-being above their individual and personal needs.

family As defined by the U.S. Census Bureau, a group of two or more persons related by birth, marriage, or adoption who reside together. Some family scholars have redefined the family to include nonrelated persons who reside together and who are economically, emotionally, and sexually interdependent (e.g., cohab   heterosexual or homosexua'

Family and Medica'   (FMLA) A federa' that requires all more employe' workers (w' a week a' least a' job-r' can care for a seriously ill child, spouse or parent; stay home to care for their newborn, newly adopted, or newly placed child; or take time off when they are seriously ill.

family household As defined by the U.S. Census Bureau, a family household consists of two or more persons related by birth, marriage, or adoption who reside together.

family preservation programs In-home interventions for families who are at risk of having a child removed from the home because of abuse or neglect.

feminism The belief that women and men should have equal rights and responsibilities.

feminization of poverty The disproportionate distribution of poverty among women.

fertility rate The average number of births per woman.

field research A method of research that involves observing and studying social behavior in settings in which it naturally occurs; includes participant observation and nonparticipant observation.

flextime An option in work scheduling that allows employees to begin and end the workday at different times as long as they perform 40 hours of work per week.

folkway The customs and manners of society.

frustration-aggression theory Also known as the "scapegoating theory" of prejudice; a psychological theory of prejudice that suggests prejudice is a form of hostility resulting from frustration. According to this theory, minority groups serve as convenient targets of displaced aggression.

functionally illiterate The inability to carry out many of the tasks required of an adult in today's society, because of reading, writing, and/or quantitative deficiencies.

future shock The state of confusion resulting from rapid scientific and technological changes that challenge traditional values and beliefs.

gateway drug A drug (e.g., marijuana) that is believed to lead to the use of other drugs (such as cocaine and heroin).

gender The social definitions and expectations associated with being male or female.

gender tourism The recent tendency for definitions of masculinity and femininity to become less clear resulting in individual exploration of the gender continuum.

gene monopolies Exclusive control over a particular gene as a result of government patents.

gene therapy The transplantation of a healthy duplicate gene to replace a defective or missing gene.

genetic engineering The manipulation of an organism's genes in such a way that the natural outcome is altered.

genetic screening The use of genetic maps to detect predispositions to human traits or disease(s).

genetically modified organisms (GMOs) Also known as "genetically improved organisms (GIOs)" and "genetically engineered foods (GE foods)"; products that have been created or modified through agricultural biotechnology.

genocide The systematic annihilation of one racial and/or ethnic group by another.

gentrification The process by which private individuals and/or developers purchase and renovate housing in older neighborhoods.

gerontophobia Fear or dread of the elderly.

ghetto A slum section of a city occupied primarily by African Americans.

glass ceiling An invisible, socially created barrier that prevents some women and other minorities from being promoted into top corporate positions.

global economy An interconnected network of economic activity that transcends national borders.

global warming The increasing average global air temperature, caused mainly by the accumulation of various gases (see *greenhouse gases*) that collect in the atmosphere.

globalization The economic, political, and social interconnectedness among societies throughout the world.

greenhouse effect The collection of increasing amounts of chlorofluorocarbons (CFCs), carbon dioxide, methane, and other gases in the atmosphere, where they act like the glass in a greenhouse, holding heat from the sun close to the earth and preventing the heat from rising back into space.

greenwashing The corporate practice of displaying a sense of corporate responsibility for the environment. For example, many companies publicly emphasize the steps they have taken to help the environment. Another greenwashing strategy is to retool, repackage, or relabel a company's product.

guaranteed annual income A guarantee by the government that a poor person's income will not fall below a specified percentage of the country's median income.

guerrilla warfare Warfare in which organized groups oppose domestic or foreign governments and their military forces; often involves small groups of individuals who use camouflage and underground tunnels to hide until they are ready to execute a surprise attack.

harm reduction A recent public health position that advocates reducing the harmful consequences of drug use for the user as well as society as a whole.

hate crime Also known as a "bias-motivated crime"; an act of violence motivated by prejudice or bias.

Head Start A project begun in 1965 by the federal government to help preschool children from disadvantaged families.

health A state of complete physical, mental, and social well-being.

health expectancy Number of years an individual can expect to live in good health.

health maintenance organizations (HMOs) Health care organizations that provide complete medical services for a monthly fee.

heterosexism The belief that heterosexuality is the superior sexual orientation; results in prejudice and discrimination against homosexuals and bisexuals.

heterosexuality The predominance of cognitive, emotional, and sexual attraction to persons of the other sex.

home schooling The education of children at home instead of in a public or private school; often part of a fundamentalist movement to protect children from perceived non-Christian values in the public schools.

homophobia Negative attitudes toward homosexuality.

homosexuality The predominance of cognitive, emotional, and sexual attraction to persons of the same sex.

household All persons who share occupancy of a housing unit such as a house or an apartment.

HPI-1 The Human Poverty Index for developing countries.

HPI-2 The Human Poverty Index for industrialized countries.

human capital The skills, knowledge, and capabilities of the individual.

human capital hypothesis The hypothesis that female-male pay differences are a function of differences in women's and men's levels of education, skills, training, and work experience.

Human Poverty Index (HPI) A composite measure of poverty based on three measures of deprivation: (1) deprivation of a long, healthy life; (2) deprivation of knowledge; and (3) deprivation in decent living standards.

hypothesis A prediction or educated guess about how one variable is related to another variable.

in-vitro fertilization (IVF) The union of an egg and a sperm in an artificial setting such as a laboratory dish.

incapacitation A criminal justice philosophy that views the primary purpose of the criminal justice system as preventing criminal offenders from committing further crimes against the public by putting them in prison.

incidence The number of new cases of a specific health problem within a given population during a specified time period.

incumbent upgrading Aid programs that help residents of depressed neighborhoods buy or improve their homes and stay in the community.

independent variable The variable that is expected to explain change in the dependent variable.

index offenses Crimes identified by the FBI as the most serious, including personal crimes (homicide, rape, robbery, assault) and property crimes (burglary, larceny, car theft, arson).

individual discrimination Discriminatory acts by individuals.

individualism A value system that stresses the importance of individual happiness.

Industrial Revolution The period between the mid-eighteenth and the early nineteenth century when machines and factories became the primary means for producing goods. The Industrial Revolution led to profound social and economic changes.

infant mortality rate The number of deaths of infants under 1 year of age per 1,000 live births in a calendar year.

infantilizing elders The portrayal of the elderly in the media as childlike in terms of clothes, facial expression, temperament, and activities.

infotech An abbreviation for "information technology"; any technology that carries information.

infowar The utilization of technology to manipulate or attack an enemy's military and civilian infrastructure and information systems.

infrastructure The underlying foundation that enables a city to function. Infrastructure includes such things as water and sewer lines, phone lines, electricity cables, sidewalks, streets, curbs, lighting, and storm drainage systems.

institution An established and enduring pattern of social relationships. The five traditional social institutions are family, religion, politics, economics, and education. Institutions are the largest elements of social structure.

institutional discrimination Discrimination in which the normal operations and procedures of social institutions result in unequal treatment of minorities.

integration hypothesis A theory that states that the only way to achieve quality education for all racial and ethnic groups is to desegregate the schools.

intergenerational poverty Poverty that is transmitted from one generation to the next.

internalized homophobia A sense of personal failure and self-hatred among lesbians and gay men due to social rejection and stigmatization.

Internet An international information infrastructure available through many universities, research institutes, government agencies, and businesses; developed in the 1970s as a Defense Department experiment.

intimate partner violence Actual or threatened violent crimes committed against persons by their current or former spouses, boyfriends, or girlfriends.

intimate terrorism Almost entirely perpetrated by men, a from of violence that is motivated by a wish to control one's partner and involves the systematic use of not only violence, but economic subordination, threats, isolation, verbal and emotional abuse, and other control tactics. This form of violence is more likely to escalate over time and to involve serious injury.

Jim Crow laws Laws that separated blacks from whites by prohibiting blacks from using "white" buses, hotels, restaurants, and drinking fountains.

job burnout Prolonged job stress; can cause physical problems, such as high blood pressure, ulcers, and headaches as well as psychological problems.

job exportation The relocation of U.S. jobs to other countries where products can be produced more cheaply.

job sharing A work option in which two people, often husband and wife, share and are paid for one job.

Kyoto Protocol The first international agreement to place legally binding limits on greenhouse gas emissions from developed countries.

labeling theory A symbolic interactionist theory that is concerned with the effects of labeling on the definition of a social problem (e.g., a social condition or group is viewed as problematic if it is labeled as such) and with the effects of labeling on the self-concept and behavior of individuals (e.g., the label "juvenile delinquent" may contribute to the development of a self-concept and behavior consistent with the label).

labor unions Worker organizations that originally developed to protect workers and to represent them at negotiations between management and labor. Labor unions have played an important role in fighting for fair wages and benefits, healthy and safe work environments, and other forms of worker advocacy.

latent functions Consequences that are unintended and often hidden, or unrecognized; for example, a latent function of education is to provide schools that function as baby-sitters for employed parents.

law Norms that are formalized and backed by political authority.

legalization Making prohibited behavior legal; for example, legalizing marijuana or prostitution.

lesbigays A collective term sometimes used to refer to lesbians, gays, and bisexuals.

LGBT A term used to refer collectively to lesbians, gays, bisexuals, and transgendered individuals.

life chances A term used by Max Weber to describe the opportunity to obtain all that is valued in society, including happiness, health, income, and education.

life expectancy The average number of years that a person born in a given year can expect to live.

lifestyle A distinct subculture associated with a particular social class.

living wage laws Laws that require state or municipal contractors, recipients of public subsidies or tax breaks, or, in some cases, all businesses to pay employees wages significantly above the federal minimum, enabling families to live above the poverty line.

looking-glass self The idea that individuals develop their self-concept through social interaction.

macro sociology The study of large aspects of society, such as institutions and large social groups.

MADD Mothers Against Drunk Driving. A social action group committed to reducing drunk driving.

mailbox economy The tendency for a substantial portion of local economies to be dependent on pension and social security checks received in the mail by older residents.

Malthusian theory The theory proposed by Thomas Malthus in which he predicted that population would grow faster than the food supply and that masses of people were destined to be poor and hungry. According to Malthus, food shortages would lead to war, disease, and starvation that would eventually slow population growth.

managed care A type of medical insurance plan that controls costs through monitoring and controlling the decisions of health care providers.

manifest functions Consequences that are intended and commonly recognized; for example, a manifest function of education is to transmit knowledge and skills to youth.

marital assimilation Assimilation that occurs when different ethnic or racial groups become married or pair-bonded and produce children.

master status The status that is considered the most significant in a person's social identity.

maternal mortality Deaths that result from complications associated with pregnancy or childbirth.

means-tested programs Public assistance programs that have eligibility requirements based on income.

mechanization The use of tools to accomplish tasks previously done by workers; characteristic of agricultural societies that emphasize the production of raw materials.

Medicaid A public assistance program designed to provide health care for the poor.

medicalization The tendency to define negatively evaluated behaviors and/or conditions as medical problems in need of medical intervention.

Medicare A national public insurance program created by Title XVIII of the Social Security Act of 1965; originally designed to protect people 65 years of age and older from the rising costs of health care. In 1972, Medicare was extended to permanently disabled workers and their dependents and persons with end-stage renal disease.

Medigap The difference between Medicare benefits and the actual cost of medical care.

megacities Cities with 10 million residents or more.

melting pot The product of different groups coming together and contributing equally to a new, common culture.

mental disorder A behavioral or psychological syndrome or pattern that occurs in an individual, and that is associated with present distress or disability, or with a significantly increased risk of suffering, death, pain, disability, or loss of freedom.

mental health The successful performance of mental function, resulting in productive activities, fulfilling relationships with other people, and the ability to adapt to change and to cope with adversity.

mental illness A term used to refer collectively to all mental disorders.

metropolis From the Greek meaning "Mother City." See also *metropolitan area.*

metropolitan area A densely populated core area and any adjacent communities that have a high degree of social and economic integration with the core; a large city and its surrounding suburbs; also called a "metropolis."

micropolitan area A small city located beyond congested metropolitan areas.

micro-society school A simulation of the "real" or nonschool world where students design and run their own democratic, free-market society within the school.

micro sociology The study of the social psychological dynamics of individuals interacting in small groups.

military-industrial complex A term used by Dwight D. Eisenhower to connote the close association between the military and defense industries.

minority A category of people who are denied equal access to positions of power, prestige, and wealth because of their group membership.

mixed-use neighborhoods Communities that combine residential and commercial elements along with public and private facilities such as schools, recreation centers, and places of worship. The idea of mixed-use neighborhoods is to provide suburban residents with convenient access to jobs, stores and service providers, schools, and other facilities, thus reducing driving distances.

modern racism A subtle and complex form of racism in which individuals are not explicitly racist, but tend to hold negative views of racial minorities and blame minorities for their social disadvantages.

modernization theory A theory claiming that as society becomes more technologically advanced, the position of the elderly declines.

monogamy Marriage between two partners; the only legal form of marriage in the United States.

morbidity The amount of disease, impairment, and accidents in a population.

mores Norms that have moral basis.

mortality Death.

multicultural education Education that includes all racial and ethnic groups in the school curriculum and promotes awareness and appreciation for cultural diversity.

multiculturalism A philosophy that argues that the culture of a society should represent and embrace all racial and ethnic groups in that society.

multiple chemical sensitivity (MCS) A controversial health condition in which, after one or more acute or traumatic exposures to a chemical or group of chemicals, people experience adverse effects from low levels of chemical exposure that do not produce symptoms in the general public.

multiple jeopardy The disadvantages associated with being a member of two or more minority groups.

mutual violent control A rare pattern of abuse when two intimate terrorists battle for control. See also *intimate terrorism*.

mutually assured destruction (MAD) A perspective that argues that if both sides in a conflict were to lose in a war, neither would initiate war.

National Labor Relations Act An act passed in 1935 that guarantees the right to unionize and to strike.

naturalized citizen An immigrant who applied and met the requirements for U.S. citizenship.

needle exchange programs Programs designed to reduce transmission of HIV among injection drug users, their sex partners, and their children, by providing new, sterile syringes in exchange for used, contaminated syringes.

neglect A form of abuse involving the failure to provide adequate attention, supervision, nutrition, hygiene, health care, and a safe and clean living environment for a minor child or a dependent elderly individual.

New Urbanists A growing group of planners, architects, developers, and traffic engineers who support neighborhood designs that create a strong sense of community by incorporating features of traditional small towns.

no-fault divorce A divorce that is granted based on the claim that there are irreconcilable differences within a marriage (as opposed to one spouse being legally at fault for the marital breakup).

non-family household A household that consists of one person who lives alone, two or more people as roommates, and cohabiting heterosexual or homosexual couples involved in a committed relationship.

norms Socially defined rules of behavior, including folkways, mores, and laws.

nuclear winter The predicted result of a thermonuclear war whereby dust storms and concentrations of particles would block out vital sunlight, lower temperatures in the Northern Hemisphere, and lead to the death of most living things on earth.

objective element of social problems Awareness of social conditions through one's own life experience and through reports in the media.

occupational sex segregation The concentration of women in certain occupations and of men in other occupations.

one drop of blood rule A rule that specified that even one drop of Negroid blood defined a person as black and, therefore, eligible for slavery.

operational definition In research, a definition of a variable that specifies how that variable is to be measured (or was measured) in the research.

organized crime Criminal activity conducted by members of a hierarchically arranged structure devoted primarily to making money through illegal means.

overt discrimination Discrimination that occurs because of an individual's own prejudicial attitudes.

Parental Alienation Syndrome An emotional and psychological disturbance in which children engage in exaggerated and unjustified denigration and criticism of a parent.

parity The concept of equality between mental health care insurance coverage and other health care insurance.

partial birth abortion Also called an intact dilation and extraction (D&X) abortion, the procedure may entail delivering the limbs and the torso of the fetus before it has expired.

patriarchal Literally, rule by father; today, connotes rule by males.

patriarchal terrorism A form of abuse in which husbands control their wives by the systematic use of not only violence, but economic subordination, threats, isolation, and other control tactics.

patriarchy A tradition in which families are male-dominated.

peace dividend Resources that are diverted from military spending and channeled into private or public investment or consumption, used to reduce the deficit, and/or used to lower taxes.

perinatal transmission The transmission of a virus (such as HIV) from an infected mother to a fetus or newborn.

Personal Responsibility and Work Opportunity Reconciliation Act (PRWOR) The 1996 legislation that affected numerous public assistance programs, primarily in the form of cutbacks and eligibility restrictions. This law ended Aid to Families with Dependent Children (AFDC) and replaced it with Temporary Assistance to Needy Families (TANF).

pink-collar jobs Jobs that offer few benefits, often have low prestige, and are disproportionately held by women.

planned obsolescence The manufacturing of products that are intended to become inoperative or outdated in a fairly short period of time.

pluralism A state in which racial and/or ethnic groups maintain their distinctness, but respect each other and have equal access to social resources.

polyandry The concurrent marriage of one woman to two or more men.

polygamy A form of marriage in which one person may have two or more spouses.

polygyny The concurrent marriage of one man to two or more women.

population momentum Continued population growth that occurs even if a population achieves replacement-level fertility (2.1 births per woman) due to past high fertility rates which have resulted in large numbers of young women who are currently entering their childbearing years.

population transfer See *expulsion*.

postindustrialization The shift from an industrial economy dominated by manufacturing jobs to an economy dominated by service-oriented, information-intensive occupations.

postmodernism A world view that questions the validity of rational thinking and the scientific enterprise.

post-traumatic stress disorder A set of symptoms that may result from any traumatic experience, including crime victimization, war, natural disasters, or abuse.

poverty Lacking resources for an "adequate" standard of living (see also *absolute poverty* and *relative poverty*)

poverty gap The difference between the household income of the poor and the poverty line.

poverty line An annual dollar amount below which individuals or families are considered officially poor by the government.

preferred provider organizations (PPOs) Health care organizations in which employers who purchase group health insurance agree to send their employees to certain health care providers or hospitals in return for cost discounts.

prejudice An attitude or judgment, usually negative, about an entire category of people based on their group membership.

prevalence The total number of cases of a condition within a population that exist at a given time.

primary aging Biological changes associated with aging that are due to physiological variables such as cellular and molecular variation (e.g., gray hair).

primary assimilation The integration of different groups in personal, intimate associations such as friends, family, and spouses.

primary group Small groups characterized by intimate and informal interaction.

primary labor market See *split labor market*.

primary prevention strategies Family violence prevention strategies that target the general population.

primary work sector The set of work roles in which individuals are involved in the production of raw materials and food goods; develops when a society changes from a hunting and gathering society to an agricultural society.

proletariat Workers, often exploited by the bourgeoisie.

pronatalism A cultural value that promotes having children.

public assistance A general term referring to some form of support by the government to citizens who meet certain established criteria.

public housing An assistance program that provides federal subsidies for low-income housing units built, owned, and operated by local public housing authorities; also known as *subsidized housing*.

race A category of people who share distinct physical characteristics that are deemed socially significant.

racial profiling The law enforcement practice of targeting suspects based upon race.

racial steering A real estate practice whereby realtors discourage minorities from moving into certain areas by showing them homes only in minority neighborhoods.

racism The belief that certain groups of people are innately inferior to other groups of people based on their racial classification. Racism serves to justify discrimination against groups that are perceived as inferior.

redlining The practice whereby mortgage companies deny loans for the purchase of houses in minority neighborhoods, arguing that the financial risk is too great.

refugees Immigrants who apply from abroad for admission on the basis of persecution or fear of persecution for their political or religious beliefs.

regionalism A form of collaboration among central cities and suburbs that encourages local governments to share responsibility for common problems.

regressive taxes Taxes that absorb a much higher proportion of the incomes of lower-income households than of higher-income households.

rehabilitation A criminal justice philosophy that views the primary purpose of the criminal justice system as changing the criminal offender through such programs as education and job training, individual and group therapy, substance abuse counseling, and behavior modification.

relative poverty A deficiency in material and economic resources compared with some other population.

reparative therapy Various therapies that are aimed at changing homosexuals' sexual orientation.

repeated trauma disorders See *cumulative trauma disorders*.

repetitive strain disorders See *cumulative trauma disorders*.

replacement level The average number of births per woman (2.1) in a population below which the population begins to decline.

restorative justice A philosophy primarily concerned with reconciling conflict between the offender, the community, and the victim.

restrictive home covenants An illegal form of discrimination in which neighbors make a pact that they will not sell their homes to minority group members.

reverse discrimination The unfair treatment of members of the majority group (i.e., white males) that, according to some, results from affirmative action.

road rage Aggressive and violent driving behavior.

role A set of rights, obligations, and expectations associated with a status.

sample In survey research, the portion of the population selected to be questioned.

sanctions Social consequences for conforming to or violating norms. Types of sanctions include positive, negative, formal, and informal.

sandwich generation The generation that has the responsibility of simultaneously caring for their children and their aging parents.

scapegoating theory See *frustration-aggression theory*.

school vouchers Tax credits that are transferred to the public or private school of a parent's choice.

science The process of discovering, explaining, and predicting natural or social phenomena.

scientific apartheid The growing gap between the industrial and developing countries in the rapidly evolving knowledge frontier.

second shift The household work and childcare that employed parents (usually women) do when they return home from their jobs.

secondary aging Biological changes associated with aging that can be attributed to poor diet, lack of exercise, and increased stress.

secondary assimilation The integration of different groups in public areas and in social institutions, such as

neighborhoods, schools, the workplace, and in government.

secondary group A group characterized by impersonal and formal interaction.

secondary labor market See *split labor market*.

secondary prevention strategies Prevention strategies that target groups that are thought to be at high risk for family violence.

secondary work sector The set of work roles in which individuals are involved in the production of manufactured goods from raw materials; emerges when a society becomes industrialized.

Section 8 housing A federal low-income housing program in which federal rent subsidies are provided either to tenants (in the form of certificates and vouchers) or to private landlords.

segregation The physical and social separation of categories of individuals, such as racial or ethnic groups.

self-fulfilling prophecy A concept referring to the tendency for people to act in a manner consistent with the expectations of others.

senescence The biology of aging.

serial monogamy A succession of marriages in which a person has more than one spouse over a lifetime but is legally married to only one person at a time.

sex A person's biological classification as male or female.

sexism The belief that there are innate psychological, behavioral, and/or intellectual differences between females and males and that these differences connote the superiority of one group and the inferiority of another.

sexual aggression Sexual interaction that occurs against one's will through the use of physical force, threat of force, pressure, use of alcohol/drugs, or use of position of authority.

sexual harassment When an employer requires sexual favors in exchange for a promotion, salary increase, or any other employee benefit and/or the existence of a hostile environment that unreasonably interferes with job performance, as in the case of sexually explicit remarks or insults being made to an employee.

sexual orientation The identification of individuals as heterosexual, bisexual, or homosexual, based on their emotional and sexual attractions, relationships, self-identity, and lifestyle.

sick building syndrome (SBS) A situation in which occupants of a building experience symptoms that seem to be linked to time spent in a building, but no specific illness or cause can be identified.

single-payer system A single tax-financed public insurance program that replaces private insurance companies.

slavery A condition in which one social group treats another group as property to exploit for financial gain.

slums Concentrated areas of poor housing and squalor in heavily populated urban areas.

smart growth A strategy for managing urban sprawl that serves the economic, environmental, and social needs of communities.

social class Group of people who share a similar position or social status within the stratification system.

social group Two or more people who have a common identity, interact, and form a social relationship. Institutions are made up of social groups.

social problem A social condition that a segment of society views as harmful to members of society and in need of remedy.

social promotion The passing of students from grade to grade even if they are failing.

socialism An economic ideology that emphasizes public rather than private ownership. Theoretically, goods and services are equitably distributed according to the needs of the citizens.

socialized medicine National health insurance systems in other countries such as Canada, Great Britain, Sweden, Germany, and Italy.

sociological imagination A term coined by C. Wright Mills to refer to the ability to see the connections between our personal lives and the social world in which we live.

sodomy Oral and anal sexual acts. Other terms for sodomy include crimes against nature, unnatural intercourse, buggery, sexual misconduct, and lewd and lascivious acts.

soft money Money that flows through a loophole to provide political parties, candidates, and contributors a means to evade federal limits on political contributions.

split-labor market The existence of primary and secondary labor markets. A primary labor market refers to jobs that are stable and economically rewarding and have many benefits; a secondary labor market refers to jobs that offer little pay, no security, few benefits, and little chance for advancement.

state The organization of the central government and government agencies such as the armed forces, police force, and regulatory agencies.

status A position a person occupies within a social group.

stereotypes Oversimplified or exaggerated generalizations about a category of individuals. Stereotypes are either untrue or are gross distortions of reality.

stigma Refers to any personal characteristic associated with social disgrace, rejection, or discrediting.

strain theory A theory that argues that when legitimate means of acquiring culturally defined goals are limited by the structure of society, the resulting strain may lead to crime or other deviance.

structural-functionalism A sociological perspective that views society as a system of interconnected parts that work together in harmony to maintain a state of balance and social equilibrium for the whole; focuses on how each part of society influences and is influenced by other parts.

structural sexism The ways in which the organization of society, and specifically its institutions, subordinate individuals, and groups based on their sex classification.

structural unemployment Exists when there are not enough jobs available for those who want them; unemployment that results from structural variables such as government and business downsizing, job exportation, automation, a reduction in the number of

new and existing businesses, an increase in the number of people looking for jobs, and a recessionary economy where fewer goods are purchased and, therefore, fewer employees are needed.

subcultural theories A set of theories that argue that certain groups or subcultures in society have values and attitudes that are conducive to crime and violence.

subculture The distinctive lifestyles, values, and norms of discrete population segments within a society.

subjective element of social problems The belief that a particular social condition is harmful to society, or to a segment of society, and that it should and can be changed.

subsidized housing See *public housing*.

suburbanization The process in which city dwellers move to the suburbs due to concern about the declining quality of life in urban areas.

suburbs The urbanlike areas surrounding central cities.

survey research A method of research that involves eliciting information from respondents through questions; includes interviews (telephone or face-to-face) and written questionnaires.

sustainable development Societal development that meets the needs of current generations without threatening the future of subsequent generations.

sweatshops Work environments that are characterized by less than minimum wage pay, excessively long hours of work often without overtime pay, unsafe or inhumane working conditions, abusive treatment of workers by employers, and/or the lack of worker organizations aimed at negotiating better work conditions.

symbol Something that represents something else.

symbolic interactionism A sociological perspective that emphasizes that human behavior is influenced by definitions and meanings that are created and maintained through symbolic interaction with others.

technological dualism A term referring to the tendency for technology

to have both positive (e.g., time saving) and negative (e.g., unemployment) consequences.

technological fix The use of scientific principles and technology to solve social problems.

technology Activities that apply the principles of science and mechanics to the solution of specific problems.

technology-induced diseases Diseases that result from the use of technological devices, products, and/or chemicals.

telemedicine Using information and communication technologies to deliver a wide range of health care services, including diagnosis, treatment, prevention, health support and information, and education of health care workers.

telework A form of work that allows employees to work part- or full-time at home or at a satellite office.

Temporary Assistance to Needy Families (TANF) The welfare program that resulted from 1996 welfare reform legislation. Under the TANF program, which replaced Aid to Families with Dependent Children (AFDC), after 2 consecutive years of receiving aid, welfare recipients are required to work at least 20 hours per week or to participate in a state-approved work program (few exceptions are made). A lifetime limit of 5 years is set for families receiving benefits.

terrorism The premeditated use or threatened use of violence by an individual or group to gain a political objective.

tertiary labor market See *split labor market*.

tertiary prevention strategies Prevention strategies that target families who have experienced family violence.

therapeutic cloning Therapeutic cloning entails using stem cells from human embryos to produce body cells that can be used to grow needed organs or tissues.

therapeutic communities Organizations where approximately 35–100 individuals reside for up to 15 months to abstain from drugs, develop marketable skills, and receive counseling.

tracking An educational practice in which students are grouped together on the basis of similar levels of academic achievement and abilities.

traditional family Families in which the husband is the breadwinner and the wife is a homemaker.

transgendered individuals Persons who do not fit neatly into either the male or female category, or their behavior is not congruent with the rules and expectations for their sex in the society in which they live.

transnational crime Crime that, directly or indirectly, involves more than one country.

triad dispute resolution Dispute resolution that involves two disputants and a negotiator.

triangulation The use of multiple methods and approaches to study a social phenomenon.

triple jeopardy See *multiple jeopardy.*

underclass A persistently poor and socially disadvantaged group that disproportionately experiences joblessness, welfare dependency, involvement in criminal activity, single-parent families, and low educational attainment.

underemployment Employment in a job that is underpaid; is not commensurate with one's skills, experience, and/or education; and/or involves working fewer hours than desired.

under-5 mortality rate The rate of deaths among children under age 5.

unemployment Measures of U.S. unemployment consider an individual to be unemployed if he or she is currently without employment, is actively seeking employment, and is available for employment. Unemployment figures do not include discouraged workers, who have given up on finding a job and are no longer looking for employment.

union density The percentage of workers who belong to unions.

universal health care See *socialized medicine.*

upskilling The opposite of deskilling; upskilling reduces employee alienation and increases decision-making powers.

urban population Persons living in cities or towns of 2,500 or more inhabitants.

urban sprawl The ever-increasing outward growth of urban areas.

urban villages Neighborhood communities, within urban settings, that facilitate the formation of intimate and strong social bonds among community members.

urbanism The culture and lifestyle of city dwellers, often characterized by individualistic and cosmopolitan norms, values, and styles of behavior.

urbanization The transformation of a society from a rural to an urban one.

urbanized area One or more places and the adjacent densely populated surrounding territory that together have a minimum population of 50,000.

values Social agreements about what is considered good and bad, right and wrong, desirable and undesirable.

variable Any measurable event, characteristic, or property that varies or is subject to change.

victimless crimes Illegal activities, such as prostitution or drug use, that have no complaining party; also called "vice crimes."

violent resistance Acts of violence by a partner that are committed in self-defense. Violent resistance is almost exclusively perpetrated by women against a male partner.

virtual reality Computer-generated three-dimensional worlds that change in response to the movements of the head or hand of the individual; a simulated experience of people, places, sounds, and sights.

war Organized armed violence aimed at a social group in pursuit of an objective.

wealth The total assets of an individual or household, minus liabilities.

wealthfare Governmental policies and regulations that economically favor the wealthy.

white-collar crime Includes both occupational crime, where individuals commit crimes in the course of their employment, and corporate crime, where corporations violate the law in the interest of maximizing profit.

WMD Weapons of mass destruction including chemical, biological, and nuclear weapons.

working poor Individuals who spend at least 27 weeks a year in the labor force (working or looking for work), but whose income falls below the official poverty line.

work sectors The division of the labor force into distinct categories (primary, secondary, and tertiary) based on the types of goods/services produced.

References

Chapter 1

Blumer, Herbert. 1971. "Social Problems as Collective Behavior." *Social Problems* 8(3):298–306.

Caldas, Stephen and Carl L. Bankston III. 1999. "Black and White TV: Race, Television Viewing, and Academic Achievement." *Sociological Spectrum* 19:39–61.

Catania, Joseph A., David R. Gibson, Dale D. Chitwook, and Thomas J. Coates. 1990. "Methodological Problems in AIDS Behavioral Research: Influences on Measurement Error and Participation Bias in Studies of Sexual Behavior." *Psychological Bulletin* 108:339–62.

Coleman, John R. 1990. "Diary of a Homeless Man." In *Social Problems*, ed. James M. Henslin, pp. 160–69. Englewood Cliffs, NJ: Prentice Hall.

Dordick, Gwendolyn A. 1997. *Something Left to Lose: Personal Relations and Survival Among New York's Homeless.* Philadelphia: Temple University Press.

Dunn, Jennifer. 2000. "What Love Has to Do with It: The Cultural Construction of Emotion and Sorority Women's Responses to Forcible Interaction." *Social Problems* 46:440–459.

Eitzen, Stanley and Maxine Baca Zinn. 2000. *Social Problems.* Boston: Allyn and Bacon.

Gallup Poll. 2000. "Long Term Gallup Poll Trends." http://gallup.com/poll/releases/pr00062.asp

Hewlett, Sylvia Ann. 1992. *When the Bough Breaks: The Cost of Neglecting Our Children.* New York: Harper Perennial.

Hills, Stuart L. 1987. *Corporate Violence: Injury and Death for Profit*, ed. Stuart L. Mills. Lanham, MD: Rowman & Littlefield.

Jekielek, Susan M. 1998. "Parental Conflict, Marital Disruption and Children's Emotional Well-Being." *Social Forces* 76:905–35.

May, Richard. 2000. "Human Development Report." *Journal of American Planning Association* 66: 219.

Merton, Robert K. 1968. *Social Theory and Social Structure.* New York: Free Press.

Mills, C. Wright. 1959. *The Sociological Imagination.* London: Oxford University Press.

Miringoff, Marc and Marque-Luisa Miringoff. 1999. *The Social Health of the Nation: How America is Really Doing.* New York: Oxford University Press.

Mirowsky, John, Catherine E. Ross, and Marieke Van Willigen. 1996. "Instrumentalism in the Land of Opportunity: Socioeconomic Causes and Emotional Consequences." *Social Psychology Quarterly* 59:322–37.

Mouw, Ted and Yu Xie. 1999. "Bilingualism and Academic Achievement of Asian Immigrants." *American Sociological Review* 64:232–252.

Romer, D., R. Hornik, B. Stanton, M. Black, X. Li, I. Ricardo, and S. Feigelman. 1997. "Talking Computers: A Reliable and Private Method to Conduct Interviews on Sensitive Topics with Children." *The Journal of Sex Research* 34:3–9.

Skeen, Dick. 1991. *Different Sexual Worlds: Contemporary Case Studies of Sexuality.* Lexington, MA: Lexington Books.

Thomas, W. I. [1931] 1966. "The Relation of Research to the Social Process." In *W. I. Thomas on Social Organization and Social Personality,* ed. Morris Janowitz, pp. 289–305. Chicago: University of Chicago Press.

Troyer, Ronald J., and Gerald E. Markle. 1984. "Coffee Drinking: An Emerging Social Problem." *Social Problems* 31:403–16.

Ukers, William H. 1935. *All about Tea, vol. 1.* The Tea and Coffee Trade Journal Co.

Wilson, John. 1983. *Social Theory.* Englewood Cliffs, NJ: Prentice-Hall.

Chapter 2

Abeysinghe, Devinka. 1998. "First Lady Stresses Family Planning on World Health Day." *Popline* (March-April): 3–4.

Adetunji, Jacob. 2000. "Trends in Under-5 Mortality Rates and the HIV/AIDS Epidemic." *Bulletin of the World Health Organization* 78(10):1200–1206.

Anderson, John E., James W. Carey, and Samuel Taveras. 2000. "HIV Testing Among the General US Population and Persons at Increased Risk: Information from National Surveys, 1987–1996." *American Journal of Public Health* 90(7):1089–1095.

American Psychiatric Association. 1999a. "American Psychiatric Association Mental Health Parity—Its Time Has Come." *APA Online.* www.psycho.org/

————. 1999b. "Addressing the Mental Health Needs of America's Children." *APA Online.* www.psych.org

————. 2000. *Diagnostic and Statistical Manual of Mental Disorders,* 4th edition Text Revision DSM–TR. American Psychiatric Association. Washington D.C.

Antezana, Fernando S., Claire M. Chollat-Traquet, and Derik Yach. 1998. "Health for All in the 21st Century." *World Health Statistics Quarterly* 51:3–6.

Associated Press. 2000 (August 9). "Millions of Kids Could Have Insurance." MSNBC. www.msnbc.com/news

Brown, June Gibbs. 2000 (June). *Annual Report: State Medicaid Fraud Control Units*. The Department of Health and Human Services. http://www.hhs.gov/oig/oi/mcfu/index.htm

Brundtland, Gro Harlem. 2000. "Mental Health in the 21st Century." *Bulletin of the World Health Organization* 78(4):411.

Centers for Disease Control and Prevention. 1997. "Youth Risk Behavior Surveillance: National College Health Risk Behavior Survey—United States, 1995." *Surveillance Summaries* 46(SS-6):1–54.

Centers for Disease Control and Prevention. 1999. "Young People at Risk: HIV/AIDS Among America's Youth." www.cdc.gov/hiv/pubs/facts/youth.htm

Centers for Disease Control and Prevention. 2000a. "Top Health Stories of 1999, Based on Data From the National Center for Health Statistics, Centers for Disease Control and Prevention." www.cdc.gov/nchs/releases/00facts/hlstorie.htm

Centers for Disease Control and Prevention. 2000b. "HIV/AIDS Surveillance Report." In *Readings in the Sociology of AIDS*, eds. Anthony J. Lemelle, Jr.; Charlene Harrington; and Allen J. LeBlanc, pp. 33–40. Upper Saddle River, NJ: Prentice Hall.

Cherner, Linda L. 1995. *The Universal Healthcare Almanac*. Phoenix: Silver & Cherner.

Children's Defense Fund. 2000. "The State of America's Children Yearbook 2000." www.childrensdefense.org/keyfacts.htm

Cockerham, William C. 1998. *Medical Sociology*, 7th ed. Upper Saddle River, NJ: Prentice Hall.

Conrad, Peter, and Phil Brown. 1999. "Rationing Medical Care: A Sociological Reflection." In *Health, Illness, and Healing: Society, Social Context, and Self*, eds. Kathy Charmaz and Debora A. Paterniti, pp. 582–90. Los Angeles: Roxbury Publishing Co.

Diamond, Catherine and Susan Buskin. 2000. "Continued Risky Behavior in HIV-Infected Youth." *American Journal of Public Health* 90(1):115–118.

Everett, Sherry A., Rae L. Schnuth, and Joanne L. Tribble. 1998. "Tobacco and Alcohol Use in Top-Grossing American Films." *Journal of Community Health* 23:317–24.

Family Care International. 1999. "Safe Motherhood." www.safemotherhood.org/init_facts.htm

Feachum, Richard G. A. 2000. "Poverty and Inequality: A Proper Focus for the New Century." *The International Journal of Public Health* (Bulletin of the World Health Organization.) 78:1–2.

Feldman, Debra S., Dennis H. Novack, and Edward Gracely. 1998. "Effects of Managed Care on Physician Patient Relationships, Quality of Care, and the Ethical Practice of Medicine: A Physician Survey." *Archives of Internal Medicine* 158:1626–32.

Garfinkel, P. E. and Goldbloom, D. S. 2000. "Mental Health—Getting Beyond Stigma and Categories." *Bulletin of the World Health Organization* 78(4):503–505.

"Global Summary of the HIV/AIDS Epidemic, December 2000." 2001. "News." *Bulletin of the World Health Organization* 79(1):78

Goldstein, Michael S. 1999. "The Origins of the Health Movement." In *Health, Illness, and Healing: Society, Social Context, and Self*, eds. Kathy Charmaz and Debora A. Paterniti, pp. 31–41. Los Angeles: Roxbury Publishing Co.

Gottlieb, Scott. 2000. "Oral AIDS Vaccine to Be Tested in the Republic of Uganda." *Bulletin of the World Health Organization* 78(7):946–956.

Health Care Financing Administration. 1998. "The Clinton Administration's Comprehensive Strategy to Fight Health Care Fraud, Waste, and Abuse." http://www.hcfa.gov/facts/f980316.htm

_____. 2000. "The State Children's Health Insurance Program." www.hcfa.gov/facts/fs000224.htm

Hoffman, Earl D. Jr.; Barbara S. Klees, and Catherine A. Curtis. 2000 (July 1). "National Health Care Expenditures." Health Care Financing Administration. www.hcfa.gov/pubforms/actuary/ormedmed/DEFAULT2htm

Hyman, Steven E. 2000 (May 18). "Hearing on Mental Health Insurance Parity." National Institute of Mental Health. www.nimh.nih.gov/about/paritytestimony.cfm

Iglehart, John K. 1999. "The American Health Care System: Expenditures." *New England Journal of Medicine* 340:70–76.

Inciardi, James A., and Lana D. Harrison. 1997. "HIV, AIDS, and Drug Abuse in the International Sector." *Journal of Drug Issues* 27:1–8.

Insure.com 2000. "Average Health Insurance Cost in January and February 2000." http://www.insure.com/health/ceridian200.html

Johnson, Tracy L., and Elizabeth Fee. 1997. "Women's Health Research: An Introduction." In *Women's Health Research: A Medical and Policy Primer*, eds. Florence P. Haseltine and Beverly Greenberg Jacobson, pp. 3–26. Washington, D.C.: Health Press International.

Joint United Nations Programme on HIV/AIDS. 2000a. "Report on the Global HIV/AIDS Epidemic—June 2000." www.unaids.org/epidemic update/report/Epi_report_chap_glo_estim.htm

_____. 2000b. "AIDS and Population." www.unaids.org

_____. 2000c. "HIV/AIDS and Development." www.unaids.org

_____. 2000d. "AIDS: Men Make a Difference: World AIDS Campaign." www.unaids.org

_____. 2000e. "Innovative Approaches to HIV Prevention." www.unaids.org

Kaiser Commission on Medicaid and the Uninsured. 2000 (May). "The Uninsured and Their Access to Health Care." Washington, D.C.: The Kaiser Family Foundation

Kaiser Family Foundation. 2000. "Preliminary Findings from a New National Survey of Teens on HIV/AIDS, 2000." www.kff/org/content/2000/3066/

Kann, Laura, Steven A. Kinchen, Barbara I. Williams, James G. Ross, Richard Lowry, Jo Anne Grunbaum, Lloyd J. Kolbe, and State and Local YRBSS Coordinators. 2000. "Youth Risk Behavior Surveillance—United States, 1999." *Journal of School Health* 70(7):271–285.

Kessler, Ronald C., Katherine A. McGonagle, Shanyang Zhao, Christopher B. Nelson, Michael Hughes, Suzann Eshleman, Hans-Ulrich Wittchen, and Kenneth S.

Kendler. 1994. "Life-time and 12-Month Prevalence of DSM-III-R Psychiatric Disorders in the United States." *Archives of General Psychiatry* 51:8–19.

Lantz, Paula M., James S. House, James M. Lepkowski, David R. Williams, Richard P. Mero, and Jieming Chen. 1998. "Socioeconomic Factors, Health Behaviors, and Mortality: Results from a Nationally Representative Prospective Study of U.S. Adults." *Journal of the American Medical Association* 279:1703–8.

LaPorte, Ronald E. 1997. "Improving Public Health via the Information Superhighway." http://www.the-scientist.library.upenn.edu/yr1997/August/opin_97018.html

Lay, Carolyn. 2000. "Family Planning Access Is Seen as Key Determinant in Maternal Well Being." *Popline* 22:3

Lerer, Leonard B., Alan D. Lopez, Tord Kjellstrom, and Derek Yach. 1998. "Health for All: Analyzing Health Status and Determinants." *World Health Statistics Quarterly* 51:7–20.

Link, Bruce G. and Jo Phelan. 1998. "Social Conditions as Fundamental Causes of Disease." In *Readings in Medical Sociology*, eds. William C. Cockerham, Michael Glasser, and Linda S. Heuser, pp. 23–36. Upper Saddle River, NJ: Prentice Hall.

Mathews, T. J., Sally C. Curtin, and Marian F. MacDorman. 2000 (July 20). "Infant Mortality Statistics from the 1998 Period Linked Birth/Infant Death Data Set." *National Vital Statistics Reports* 48(12):1–28.

Miller, K., and A. Rosenfield. 1996. "Population and Women's Reproductive Health: An International Perspective." *Annual Review of Public Health* 17:359–82.

Mills, Robert J. 2000. "Health Insurance Coverage: 1999." *Current Population Reports* (September). U.S. Census Bureau. Washington D.C: U.S. Government Printing Office.

Mischel, Lawrence, Jared Bernstein, and John Schmitt. 2001. The State of Working America 2000/2001. Ithaca, NY: Cornell University Press.

Monardi, Fred, and Stanton A. Glantz. 1998. "Are Tobacco Industry Campaign Contributions Influencing State Legislative Behavior?" *American Journal of Public Health* 88:918–23.

Morse, Minna. 1998. "The Killer Mosquitoes." *Utne Reader* (May–June):14–15.

Murray, C. and A. Lopez, eds. 1996. *The Global Burden of Disease.* Boston: Harvard University Press.

National Alliance for the Mentally Ill. 2000a. "Parity in Insurance Coverage." www.nami.org/

National Alliance for the Mentally Ill. 2000b. "Mental Health Early Intervention, Treatment and Prevention Act of 2000." www.nami.org/

National Center for Health Statistics. 2000. *Health, United States, 2000 With Adolescent Health Chartbook.* Hyattsville, Md: U.S. Government Printing Office.

Ostrof, Paul. 1998. "Readers Write: My Chair." *The Sun* (August):33–40.

Parsons, Talcott. 1951. *The Social System.* New York: The Free Press.

Pate, Russell R., Michael Pratt, Steven N. Blair, William L. Haskell, Caroline A. Macera, Claude Bouchard, David Buchner, Walter Ettiger, Gregory W. Health, Abby C. King, Andrea Kriska, Arthur L. Leon, Bess H. Marcus, Jeremy Morris, Ralph S. Paffenbarger, Kevin Patrick, Michael L. Pollock, James M. Rippe, James Sallis, and Jack H. Wilmore. 1995. "Physical Activity and Public Health: A Recommendation from the Centers for Disease Control and Prevention and the American College of Sports Medicine." *Journal of the American Medical Association* 273(5):402–6.

Peeno, Linda, M.D. 2000. "Taking On the System." *Hope* (Spring)(22):18–21.

PNHP Data Update. 1997 (Dec.). *Physicians for a National Health Program Newsletter.* http://www.pnhp.org.Data/dataD97.html

_____. 1998. *Physicians for a National Health Program Newsletter.* http://www.pnhp.org/dataM98.html

PNHP Data Update. 2000 (Sept.). *Physicians for a National Health Program Newsletter.* http://www.pnhp.org/Press/2000/data_update0900.htm

"Poverty Threatens Crisis." 1998. *Popline* 20(May–June):3.

Rice, Amy L, Lisa Sacco, Adnan Hyder, and Robert E. Black. 2000.

"Malnutrition as an Underlying Cause of Childhood Deaths Associated with Infectious Diseases in Developing Countries." *Bulletin of the World Health Organization* 78(10):1207–1221.

Rustein, Shea O. 2000. "Factors Associated with Trends in Infant and Child Mortality in Developing Countries During the 1990s." *Bulletin of the World Health Organization* 78(10):1256–1270.

Robins, Douglas N. 1998. "Testimony of Douglas N. Robins, M.D." *Physicians for a National Health Program* (PNHP). http://www.pnhp.org/press/robins.html

Safe Motherhood Initiative. 1998. "Fact and Figures." http://www.safemotherhood.org/init_facts.htm

"Single Payer Fact Sheet." 1999. http://www.pnhp.org/fctsht.html

Stine, Gerald J. 1998. *Acquired Immune Deficiency Syndrome: Biological, Medical, Social, and Legal Issues.* Upper Saddle River, NJ: Prentice Hall.

Szasz, Thomas. 1970 (orig. 1961). *The Myth of Mental Illness: Foundations of a Theory of Personal Conduct.* New York: Harper & Row.

UNICEF. 2001. *The State of the World's Children, 2001.* New York: UNICEF. www.unicef.org/sowco1/pdf/

United Nations Population Fund. 2000. *The State of World Population Report 2000.* www.unfpa.org/SWP/2000/english/index

U.S. Department of Health and Human Services. 1998. "Needle Exchange Programs: Part of a Comprehensive HIV Prevention Strategy." www.hhs.gov/news/press/1998pres/980420.html

U.S. Department of Health and Human Services. 1999. *Mental Health: A Report of the Surgeon General: Executive Summary.* Rockville, MD: U.S. Government Printing Office.

Verbrugge, Lois M. 1999. "Pathways of Health and Death." In *Health, Illness, and Healing: Society, Social Context, and Self,* eds. Kathy Charmaz and Debora A. Paterniti, pp. 377–94. Los Angeles: Roxbury Publishing Co.

Visschedijk, Jan, and Silvere Simeant. 1998. "Targets for Health for All in the 21st Century." *World Health Statistics Quarterly* 51:56–67.

Ward, Darrell E. 1999. *The AmFAR AIDS Handbook*. New York: W.W. Norton & Company.

Weitz, Rose. 2001. *The Sociology of Health, Illness, and Health Care: A Critical Approach*. 2nd ed. Belmont, CA: Wadsworth Publishing Co.

WHO International Consortium in Psychiatric Epidemiology. 2000. "Cross-National Comparisons of the Prevalences and Correlates of Mental Disorders." *Bulletin of the World Health Organization* 78(4):413–426.

Williams, David R., and Chiquita Collins. 1999. "U.S. Socioeconomic and Racial Differences in Health: Patterns, and Explanations." In *Health, Illness, and Healing: Society, Social Context, and Self*, eds. Kathy Charmaz and Debora A. Paterniti, pp. 349–76. Los Angeles: Roxbury Publishing Co.

World Health Organization. 1946. "Constitution of the World Health Organization." New York: World Health Organization Interim Commission.

_____. 1997. "Fact Sheet No. 178: Reducing Mortality from Major Childhood Killer Diseases." www.cdc.gov/ogh/frames.htm

_____. 1998. "Fifty Facts from the World Health Report 1998." www.who.int/whr/1998/factse/htm

_____. 1999. *The World Health Report 1999*. www.who.int

_____. 2000. *The World Health Report 2000*. www.who.int

_____. 2001. "Prevalence Rates for FGM, Updated February 2001." http://www.who.int/frh-whd/FGM/prevalence_rates_for_fgm_htm

Chapter 3

AP (Associated Press). 2000. "MIT pays $4.75M in Drinking Death." *Philadelphia Daily News*, September 14, 11.

_____. 1999. "Alcoholism touches Millions." http://abcnews.go.com. December 30.

AAP (Australian Associated Press). 1998. "Genetics of Alcoholism." *Institute of Alcohol Studies Update*. London: IAS Publications.

Alcoholism and Drug Abuse Weekly. 2000. "Hawaii is Seventh State to permit Medical Marijuana. June 26, 8.

Alcohol Alert. 2000. "Mechanisms of Addiction." *National Institute on Alcohol Abuse and Alcoholism* 46 (April), 2.

ACS (American Cancer Society). 2000a. "Costs of Tobacco." *Statistics*. http://cancer.org/statistics

_____. 2000b. "Cigarette Exports." *Statistics*. http://cancer.org/statistics

American Heart Association. 2000. "Tobacco Industry's Targeting of Youth, Minorities, and Women." *AHA Advocacy Position Paper*. http://www.americanheart.org

Becker, H. S. 1966. *Outsiders: Studies in the Sociology of Deviance*. New York: Free Press.

CDC (Centers for Disease Control and Prevention). 2000. "Adolescent Health Chartbook." *Health, United States, 2000*. National Center for Health Statistics. U.S. Department of Health and Human Services. Washington, D.C.

Cloud, John. 2000. "The Lure of Ecstacy." *Time*. June 5, 63–72.

DEA (Drug Enforcement Administration). 2000. "An overview of Club Drugs." *Drug Intelligence Brief*, February. 1–10. Washington, D.C.: U.S. Department of Justice.

Dembo, Richard, Linda Williams, Jeffrey Fagan, and James Schmeidler. 1994. "Development and Assessment of a Classification of High Risk Youths." *Journal of Drug Issues* 24:25–53.

Duke, Steven, and Albert C. Gross. 1994. *America's Longest War: Rethinking Our Tragic Crusade against Drugs*. New York: G. P. Putnam & Sons.

Easley, Margaret, and Norman Epstein. 1991. "Coping with Stress in a Family with an Alcoholic Parent." *Family Relations* 40: 218–24.

Ebinger, Nick, 2000. "Congress Agrees on Controversial Aid Plan for Colombia." July 31. http://www.policy.com/news

Economist, The. 2000. "Country Boy Crack Heads." 354:43 (February).

Feagin, Joe R., and C. B. Feagin. 1994. *Social Problems*. Englewood Cliffs, NJ.: Prentice-Hall.

Final Report. 2000. "Methamphetamine Interagency Task Force Final Report." Federal Advisory Committee. Washington D.C.: Department of Justice.

Fletcher, Michael A. 2000. "War on Drugs Sends More Blacks to Prison than Whites." *Washington Post*. June 8, A10.

Francis, Craig. 2000. "Europe mellows out over Cannabis." CNN.com. October 9. http://www.cnn.com/2000/world/europe/10/09/drugs.law

Gallup Poll. 2000a. "What Would You Say Is the Most Urgent Health Care...." *Gallup Poll Archives*, September 20.

_____. 2000b. "A–Z: Illegal Drugs." *Gallup Poll Topics*. http://gallup.com/poll/indicators/mddrugs.html

Gentry, Cynthia. 1995. "Crime Control through Drug Control." In *Criminology*, 2d ed., ed. Joseph F. Sheley, pp. 477–93. Belmont, CA: Wadsworth.

Gusfield, Joseph. 1963. *Symbolic Crusade: Status Politics and the American Temperance Movement*. Urbana: University of Illinois Press.

Healthy People 2000 Review. 2000. Office of Disease Prevention and Health Promotion. U.S. Department of Health and Human Services. National Center for Health Statistics. Washington, D.C. http://odphp.dhhs.gov/pubs/hp2000

HHS (U.S. Department of Health and Human Services). 2000. "1999 National Household Survey on Drug Abuse." Substance Abuse and Mental Health Service Administration. Washington, D.C.: U.S. Government Printing Office.

Heroin Drug Conference. 1997. "Administrator's Message." U.S. Department of Justice: Drug Enforcement Administration. http://udsdoj.gov/dea/pubs/special/heroin.html

Hunt, Terence. 2000. "Clinton Signs Bill to Toughen Drunken Driving Standards." *News*. Voter.com. October 23. http://voter.excite.com/home/news

ISDD (Institute for Study of Drug Dependence-Drug Scope). 1999. "UK Trends and Updates." http://www.isdd.co.uk/trends/introduction1.html

International Narcotics Control Strategy Report. 2000. "Policy and Program Development." Bureau for International Narcotics and Law Enforcement Affairs. Washington, D.C.: U.S. Department of State.

Jarvik, M. 1990. "The Drug Dilemma: Manipulating the Demand." *Science* 250:387–92.

Klutt, Edward C. 2000. "Pathology of Drug Abuse." http://www.medlib-utah.edu/WebPath.

Leinwad, Donna. 2000. "20% Say They Used Drugs with Their Mom or Dad..." *USA Today*. August 24, 1A.

Leonard, K. E., and H. T. Blane. 1992. "Alcohol and Marital Aggression in a National Sample of Young Men." *Journal of Interpersonal Violence* 7:19–30.

MacCoun, Robert J. and Peter Reuter. 2001. "Does Europe Do It Better? Lessons from Holland, Britain and Switzerland." In *Solutions to Social Problems*, eds. D. Stanley Eitzen and Craig S. Leedham, pp. 260–264. Boston: Allyn and Bacon.

MADD (Mothers Against Drunk Driving). 2000. "New Statistics for 1999." *Statistics*. http://madd.org/stats

Mann, Judy. 2000a. "Make War on the War on Drugs." *Washington Post*. July 26, C13.

_____. 2000b. "Drug War's Failure Opens Door to New Tactic." *Washington Post*. August 16, C13.

Mayell, Hillary. 1999. "Tobacco on Course to Become World's Leading Cause of Death." *National Geographic News*. http://ngnews/news/1999/121499

McCaffrey, Barry. 1998. "Remarks by Barry McCaffrey, Director, Office of National Drug Control Policy, to the United Nations General Assembly: Special Session on Drugs." Office of National Drug Control Policy. http://www.whitehousedrugpolicy.gov/news/speeches

Moore, Martha T. 1997. "Binge Drinking Stalks Campuses." *USA Today*, October 1, A3.

Morgan, Patricia A. 1978. "The Legislation of Drug Law: Economic Crisis and Social Control." *Journal of Drug Issues* 8: 53–62.

MTF (Monitoring the Future). University of Michigan. 2000. "Infofax, Marijuana." National Institute on Drug Abuse. National Institute of Health. Washington, D.C. http://www.nida.nih.gov/InfoFax/marijuana.html

NIAAA (National Institute on Alcohol Abuse and Alcoholism). 2000. "Tenth Special Report on Alcohol and Health to the U.S. Congress." Washington, D.C.

NIDA (National Institute on Drug Abuse). 1999. "Principles of Effective Treatment." NIDA. National Institute of Health. Publication No. 99-4180. Washington, D.C.

_____. 2000a. "Researcher Announces Latest study on Drug Dependence and Abuse." News Releases. http://www.nida.nih.gov

_____. 2000b. "Cocaine and Alcohol Combined are More Damaging to Mental Ability Than Either Drug Alone." NIDA News Release, June 26, 1.

_____. 2000c."Club Drugs." Community Alert Bulletin. http://www.niga.nih.gov/ClubAlert/Clubdrugalert.html

_____.2000d. "Heroin Abuse and Addiction, "Research Report Series. NIH Publication No. 00-4165. http://www.nida.nih/gov/ResearchReports/Heroin/Heroin.html

_____. 2000e. "Dr. Drew Pinsky to Join National Institute on Drug Abuse in Launching...." News Release. August 29. http://nida.nih.gov/MedAdv

Nylander, Albert, Tuk-Ying Tung, and Xiaohe Xu. 1996. "The Effect of Religion on Adolescent Drug Use in America: An Assessment of Change." American Sociological Association Meetings. San Francisco, CA, August.

ONDCP (Office of National Drug Control Policy). 2000a. "The Link between Drugs and Crime." Chapter II. *The National Drug Control Strategy 2000 Annual Report*. http://whitehousedrugpolicy.gov

_____. 2000b. "The Consequence of Illegal Drug Use." Chapter II. *The National Drug Control Strategy 2000 Annual Report*. http://whitehousedrugpolicy.gov

_____. 1998. "Focus on the Drug Problem." http://www.whitehousedrugpolicy.gov

Reitox (European Information Network on Drugs and Drug Addiction). 2000. "International Comparisons." Drug Monitoring Center of Finland. http://www.stakes.fi/reitox/index.html

Rorabaugh, W. J. 1979. *The Alcoholic Republic: An American Tradition*. New York: Oxford University Press.

Rychtarik, Robert G., Gerald J. Connors, Kurt H Dermen, and Paul Stasiewicz. 2000. "Alcoholics Anonymous and the Use of Medications to Prevent Relapse." *Journal of Studies on Alcohol* 61:134–141.

Sheldon, Tony. 2000. "Cannabis Use among Dutch Youth." *British Medical Journal* 321:655.

State Legislatures. 2000. "All You Ever Wanted to Know about Drunk Drivers." *State Legislatures* 26:7.

Straus, Murry, and S. Sweet. 1992. "Verbal/Symbolic Aggression in Couples: Incidence Rates and Relationships to Personal Characteristics." *Journal of Marriage and the Family* 54:346–57.

Sullivan, Thomas, and Kenrick S. Thompson. 1994. *Social Problems*. New York: Macmillan.

Thompson, Don. 2000. "States Ballot Questions Focus on Drug Rehab Instead of Prison." *Excite News*. http://news.excite.com/news

Tubman, J. 1993. "Family Risk Factors, Parental Alcohol Use, and Problem Behaviors among School-Aged Children." *Family Relations* 42:81–86.

Van Dyck, C., and R. Byck. 1982. "Cocaine." *Scientific American* 246: 128–41.

Van Kammen, Welmoet B., and Rolf Loeber. 1994. "Are Fluctuations in Delinquent Activities Related to the Onset and Offset in Juvenile Illegal Drug Use and Drug Dealing?" *Journal of Drug Issues* 24:9–24.

Wechsler, Henry, Jae Eun Lee, Meichun Kuo, and Hang Lee. 2000. "College Binge Drinking in the 1990s: A Continuing Problem." *Journal of American College Health* 48: 199–210.

White, Helene Raskin, and Erich W. Labouvie. 1994. "Generality versus Specificity of Problem Behavior: Psychological and Functional Differences." *Journal of Drug Issues* 24:55–74.

Wilson, Catherine. 2000. "Tobacco Industry Told to pay $145B." *Excite News*. http://news.excite.com/news

Witters, Weldon, Peter Venturelli, and Glen Hanson. 1992. *Drugs and Society*, 3d ed. Boston: Jones & Bartlett.

Worden, Amy. 2000. "MADD Calls for Stricter Law at 20th Birthday." *Crime, Justice and Safety*. September 6. http://www.apbnews.com

Wysong, Earl, Richard Aniskiewicz, and David Wright. 1994. "Truth and Dare: Tracking Drug Education to Graduation and as Symbolic Politics." *Social Problems* 41:448–68.

Chapter 4

ABCNews. 2001. "U.S.–Russia Child Porn Bust." http://www.abcnews. go.com/sections/world/Daily/News/childpornbust_010326.htm

Albanese, Jay. 2000. *Criminal Justice.* Boston: Allyn and Bacon.

Anderson, David. 1999. "The Aggregate Burden of Crime. *Journal of Law and Economics* XLII:611–642.

Anderson, Elijah. 1994. "The Code of the Streets: Sociology of Urban Violence." *The Atlantic* 273(5):80–91.

Barkan, Steven. 1997. *Criminology: A Sociological Understanding.* Englewood Cliffs, NJ: Prentice-Hall.

Bartollas, Clemens. 2000. *Juvenile Delinquency,* 5ᵗʰ ed. Boston: Allyn and Bacon.

Becker, Howard S. 1963. *Outsiders: Studies in the Sociology of Deviance.* New York: Free Press.

BJS (Bureau of Justice Statistics). 2001. "Capital Punishment Statistics." http://www.ojp. usdoj.gov/bjs/cp.html

———. 2000a. *Sourcebook of Criminal Justice Statistics.* U.S. Department of Justice, Office of Justice Programs. Washington, D.C.

———. 2000b. "U.S. Correctional Population." http://www.ojp.usdoj. gov/bjs/pub/press/pp99pr.pr

CATW (Coalition Against Trafficking in Women). 1997. "Promoting Sex Work in the Netherlands." *Coalition Report* 4(1). http://www.uri.edu/ artsci/wms/hughes/catw

Chesney-Lind, Meda and Randall G. Shelden. 1998. *Girls, Delinquency and Juvenile Justice.* Belmont, CA: Wadsworth.

Conklin, John E. 1998. *Criminology,* 6th ed. Boston: Allyn and Bacon.

COPS. 1998. "About the Office of Community Oriented Policing Services (COPS)." http://communitypolicing. org/copspage.html

Dickerson, Debra. 2000. "Racial Profiling: Are We All Really Equal in the Eyes of the Law?" *Los Angeles Times.* July 16. http://www.latimes.com

DiIulio, John. 1999. "Federal Crime Policy: Time for a Moratorium." *Brookings Review* 17(1) (Winter):17.

Dixon, Travis L. and Daniel Linz. 2000. "Race and the Misrepresentation of Victimization on Local Television News." *Communication Research* 27 547–74.

Dorning, Michael. 2000. "U.S. Crime Continues to Fall." *Chicago Tribune.* August 28, A1.

Doyle, Roger. 2000. "The Roots of Homicide." *Scientific American.* October. http://www.sciam. com/2000

Economist, The. 2000. "Dead Man Walking Out." June 10, 21–23.

Elliot, D., and S. Ageton. 1980. "Reconciling Race and Class Differences in Self-Reported and Official Estimates of Delinquency." *American Sociological Review* 45:95–110.

Erikson, Kai T. 1966. *Wayward Puritans.* New York: John Wiley & Sons.

(FBI) Federal Bureau of Investigation. 2000. *Crime in the United States, 1999.* Uniform Crime Reports. Washington, D.C.: U.S. Government Printing Office

Felson, Marcus. 1998. *Crime and Everyday Life,* 2nd ed. Thousand Oaks, CA: Pine Forge Press.

Fields, Gary. 2000. "Victims of Identity Theft Often Unaware They've Been Stung." *USA Today.* March 15, 6A.

Finckenauer, James O. 2000. "Meeting the Challenge of Transnational Crime. *National Institute of Justice Journal.* July, 2–7.

Fletcher, Michael A. 2000. "War on drugs sends more Blacks to prison than Whites." *The Washington Post.* June 8, A10.

Gallup Poll. 2000a. "Most Important Problem." June 22–25. http://www.gallup.com/poll/indicators

———. 2000b. "Crime Tops List of Americans' Local Concerns." June 21. http://www.gallup.com/poll/releases

———. 1999."Racial Profiling is Seen as Widespread...." December 9. http://www.gallup.com/ poll/releases

Garey, M. 1985. "The Cost of Taking a Life: Dollars and Sense of the Death Penalty." *U.C. Davis Law Review* 18:1221–73.

Gest, Ted, and Dorian Friedman. 1994. "The New Crime Wave." *U.S. News and World Report,* August 29, 26–28.

Global Report on Crime and Justice. 1999. United Nations: Office of Drug Control and Crime Prevention. http://www.uncjin.org/ special/overview.html

Gullo, Karen. 2001. "More Behind Bars." http://www.abcnews.go.com/ sections/US/Daily/News/ prisonpopulation0326.htm

"Guns". 2000. Polling Report. http:// www.pollingreport.com/guns/html

Hagan, John, and Ruth Peterson. 1995. "Criminal Inequality in America: Patterns and Consequences." In *Crime and Inequality,* ed. John Hagan and Ruth Peterson, pp. 14–36. Stanford, Calif: Stanford University Press.

Heubusch, Kevin. 1997. "Teens on the Trigger." *American Demographics,* February. http://www. demographics.com

Hinds, Michael DeCourcy. 2000. "Gun-Weary Americans Applaud Controls." *American Demographics.* (April). http://www.demographics. com/publications/ad/0004_ad

Hirschi, Travis. 1969. *Causes of Delinquency.* Berkeley: University of California Press.

Hochstetler, Andrew, and Neal Shover. 1997. "Street Crime, Labor Surplus, and Criminal Punishment, 1980–1990." *Social Problems* 44(3):358–67.

Human Rights Watch. 2000. *Human Rights Watch World Report 2000. United States.* http://www.hrw.org/ wr2k/us.html

INTERPOL. 1998. "INTERPOL Warning: Nigerian Crime Syndicate's Letter Scheme Fraud Takes on New Dimension." *Press Releases.* http:// www.kenpubs.co.uk/INTERPOL. COM/English/pres/nig.html

Jacobs, David. 1988. "Corporate Economic Power and the State: A Longitudinal Assessment of Two Explanations." *American Journal of Sociology* 93:852–81.

Kong, Deborah and Jon Swartz. 2000 "Experts See Rash of Hack Attacks Coming..." *USA Today,* September 27, 1B.

Laub, John, Daniel S. Nagin, and Robert Sampson. 1998. "Trajectories of Change in Criminal Offending: Good Marriages and the Desistance Process." *American Sociological Review* 63 (April):225–38.

Lehrur, Eli. 1999. "Communities and Cops Join Forces." *Insight on the News* 15(3):16. (January 25).

Liebman, James S., Jeffery Fagan, and Valerie West. 2000. "A Broken System: Error Rates in Capital Cases, 1973-1995." http://justice.policy.net/jpreport

Lindberg, Kirsten, Joseph Petrenko, Jerry Gladden, and Wayne Johnson. 1997."The Changing Face of Organized Crime in America: Asian Organized Crime." *Crime and Justice International* 13(10). http://www.acsp.uic.edu/oicj/pvbs/cjintl/310/131005.shtml

Lipsey, M.W. and D.B. Wilson. 1998. "Effective Intervention for Serious Juvenile Offenders: A Synthesis of Research." In *Serious and Violent Offenders*, eds. R Loeber and David Farrington. Thousand Oaks, CA: Sage.

Lott, John R. Jr. 2000. *More Guns, Less Crime*. Chicago: University of Chicago Press.

Madriz, Esther. 2000. "Nothing Bad Happens to Good Girls." In *Social Problems of the Modern World*, ed. Frances Moulder, pp. 293–297. Belmont, CA: Wadsworth.

Meek, James Gordon. 2000. "Gun Injuries on Decline." http://abpnews.com/newscenter/breakingnews/2000/10/09/guns1009_01.html

Merton, Robert. 1957. "Social Structure and Anomie." In *Social Theory and Social Structure*. Glencoe, Ill.: Free Press.

Miller, Melissa. 1999. "Identity Theft Is a Growing Problem and Legislatures Are Responding." *Missouri Digital News*. http://mdn.org/1999/stories/theft.html

Moore, Elizabeth, and Michael Mills. 1990. "The Neglected Victims and Unexamined Costs of White Collar Crime." *Crime and Delinquency* 36: 408–18.

Murray, Mary E., Nancy Guerra, and Kirk Williams. 1997. "Violence Prevention for the Twenty-First Century." In *Enhancing Children's Awareness*, ed. Roger P. Weissberg, Thomas Gullota, Robert L. Hampton, Bruce Ryan, and Gerald Adams, pp. 105–128. Thousand Oaks, Calif: Sage Publications.

Myths and Facts about the Death Penalty. 1998. "Death Penalty: Focus on California." http://members.aol.com/Dpfocus/facts.htm

National Research Council. 1994. *Violence in Urban America: Mobilizing a Response*. Washington, D.C.: National Academy Press.

NNO (National Night Out). 2000, "Welcome to National Night Out." http://nationaltownwatch.org/nno/intro.htm

OJJDP (Office of Juvenile Justice and Delinquency Prevention). 2000. "Delinquency Cases in Juvenile Court, 1997." *OJJDP Fact Sheet*. March, #04.

PBS (Public Broadcasting System). 2000. "Police Divide." *Online NewsHour*. February 28. http://www.pbs.org/newshour

Pertossi, Mayra. 2000. "ANALYSIS—Argentine Crime Rate Soars." September 27. http://news.excite.com

Petersilia, Joan. 2000. "When Prisoners Return to the Community: Political, Economic and Social Consequences." *Sentencing and Corrections* 9:1–8.

Pew Research Center. 2000. "Respondents' Perception of Safety." The Pew Research Center for the People and the Press. May 12. http://www.people-press.org/april00rpt.htm

Pickler, Nedra. 2000. "Documents Point to Tire Problem." September 6. http://news.excite.com

Rosoff, Stephen, Henry Pontell, and Robert Tillman. 1998. *Profit Without Honor: White Collar Crime and the Looting of America*. Englewood Cliffs, NJ: Prentice-Hall.

Sanday, P. R. 1981. "The Socio-cultural Context of Rape: A Cross-Cultural Study." *Journal of Social Issues* 37:5–27.

Shabalin, Victor, J.J. Albini, and R.E. Rogers. 1995. "The New Stage of the Fight against Organized Crime in Russia." *IASOC: Criminal Organization* 10 (1):19–21.

Sherrill, Robert. 2000. "A Year in Corporate Crime." In *Social Problems in the Modern World*. ed. Frances Moulder, pp.302–308. Belmont, CA: Wadsworth.

Siegel, Larry. 2000. *Criminology*. Belmont, CA: Wadsworth.

Sileo, Chi Chi. 2000. "Crime Fighters Get Streetwise." In *Social Problems 00/01*, ed. Kurt Finsterbusch, pp.189–191. Guilford, CN: Dushkin/McGraw-Hill.

Steffensmeier, Darrell, and Emilie Allan. 1995. "Criminal Behavior: Gender and Age." In *Criminology: A Contemporary Handbook*, 2d ed., ed. Joseph F. Sheley, pp. 83–113. Belmont, CA: Wadsworth.

Sutherland, Edwin H. 1939. *Criminology*. Philadelphia: Lippincott.

United Nations. 2000. "UN Acts to Advance Restorative Justice." http://www.restorativejustice.org/conference/UN.RJ_UNbody.htm

United Nations. 1997. "Crime Goes Global." Document No. DPI/1518/SOC/CON/30M. New York: United Nations.

U.S. Department of Justice. 1993. "Highlights of 20 years of Surveying Crime Victims." Bureau of Justice Statistics. Washington, D.C.: U.S. Government Printing Office (NCJ-144525).

(VORP) Victim-Offender Reconciliation Program. 1998. "About Victim-Offender Mediation and Reconciliation." http://www.igc.org/vorp

Walker, Samuel, Cassia Spohn, and Miriam Delone. 1996. *The Color of Justice: Race, Ethnicity, and Crime in America*. Belmont, CA: Wadsworth.

Warner, Barbara, and Pamela Wilcox Rountree. 1997. "Local Social Ties in a Community and Crime Model." *Social Problems* 4(4):520–36.

Weed and Seed. 2000. "Operation Weed and Seed." Executive Office. http://www.ojp.usdoj.gov/eows/nutshell.htm

Williams, Linda. 1984. "The Classic Rape: When Do Victims Report?" *Social Problems* 31: 459–67.

Wolfgang, Marvin, Robert Figlio, and Thorstein Sellin. 1972. *Delinquency in a Birth Cohort*. Chicago: University of Chicago Press.

Worden, Amy. 2000a. "Community Policing Has Impact, Study Says." http://www.apbnews.com/newscenter/breakingnews/2000/09/07/cops0907.html

_____. 2000b. "More Whites than Blacks Evade Death Penalty." http://www.apbnews.com/newscenter/breakingnews/2000/07/24deathpleas0/24_01.html

_____. 2000c. "Crime Bills Pushed in Last Month of Congress." http://www.apbnews.com/newscenter/

breakingnews/2000/09/09/
legislation0908.01.html

Zimring, F.E., and G. Hawkins. 1997. *Crime Is Not the Problem: Lethal Violence in America.* New York: Oxford University Press.

Chapter 5

Amato, Paul R. 2001. "The Consequences of Divorce for Adults and Children." In *Understanding Families Into the New Millennium: A Decade in Review,* ed. Robert M. Milardo, pp. 488–506. Minneapolis: National Council on Family Relations.

American Council on Education and University of California. 2000. *The American Freshman: National Norms for Fall, 2000.* Los Angeles: Los Angeles Higher Education Research Institute.

Anderson, Kristin L. 1997. "Gender, Status, and Domestic Violence: An Integration of Feminist and Family Violence Approaches." *Journal of Marriage and the Family* 59:655–69.

Bachu, Amara. 1999. "Trends in Premarital Childbearing." *Current Population Reports* p. 23–197. Washington, D.C.: U.S. Bureau of the Census.

Beitchman, J. H., K. J. Zuker, J. E. Hood, G. A. daCosta, D. Akman, and E. Cassavia. 1992. "A Review of the Long-Term Effects of Child Sexual Abuse." *Child Abuse and Neglect* 16: 101–19.

Browning, Christopher R., and Edward O. Laumann. 1997. "Sexual Contact between Children and Adults: A Life Course Perspective." *American Sociological Review* 62:540–60.

Bumpass, Larry L., R. Kelly Raley, and James A. Sweet. 1995. "The Changing Character of Stepfamilies: Implications of Cohabitation and Nonmarital Childbearing." *Demography* 32:425–36.

"Child Abuse and Neglect National Statistics." 2000 (April). National Clearinghouse on Child Abuse and Neglect Information. 330 C St., SW, Washington, D.C. 20447.

Clark, Charles. 1996. "Marriage and Divorce." *CQ Researcher,* 6(18):409–32.

Cole, Charles L., Anna L. Cole, and Jessica G. Gandolfo. 2000. "Marriage Enrichment for Newlyweds: Models for Strengthening Marriages in the New Millennium." Poster presentation at the 62nd Annual Conference of the National Council on Family Relations. Minneapolis, MN. November 10–13.

Coltrane, Scott and Randall Collins. 2001. *Sociology of Marriage and the Family: Gender, Love, and Property* 5th ed. Belmont CA: Wadsworth Publishing Co.

Coontz, Stephanie. 2000. "Marriage: Then and Now." *Phi Kappa Phi Journal* 80: 10–15.

Curtin, S. C. and J. A. Martin. 2000. "Preliminary Data for 1999." *National Vital Statistics Reports,* Vol. 48, no. 14. Hyattsville, MD: National Center for Health Statistics.

Daley, S. 2000 (April 18). "French Couples Take Plunge that Falls Short of Marriage." *The New York Times,* pp. A1, A4.

Daro, Deborah. 1998. "Public Opinion and Behaviors Regarding Child Abuse Prevention: 1998 Survey." Chicago: National Committee to Prevent Child Abuse. http://www. childabuse.org/poll98.html

Demo, David H., Mark A. Fine, and Lawrence H. Ganong. 2000. "Divorce as a Family Stressor." In *Families & Change: Coping with Stressful Events and Transitions,* 2nd ed., eds. P. C. McKenry and S. J. Price, pp. 279–302. Thousand Oaks, CA: Sage Publications.

Demo, David H. 1992. "Parent-Child Relations: Assessing Recent Changes." *Journal of Marriage and the Family* 54:104–17.

_____. 1993. "The Relentless Search for Effects of Divorce: Forging New Trails or Tumbling Down the Beaten Path?" *Journal of Marriage and the Family* 55: 42–45.

DiLillo, D., G. C. Tremblay, and L. Peterson. 2000. "Linking Childhood Sexual Abuse and Abusive Parenting: The Mediating Role of Maternal Anger." *Child Abuse and Neglect* 24:767–79.

"Domestic Violence Fact Sheet." 1999 (January). Department of Health and Human Services, Administration for Children and Families. http://www.acf.dhhs.gov/p...pa/facts/domsvio.htm

Donovan, Patricia. 1999. "The 'Illegitimacy Bonus' and State Efforts to Reduce Out-of-Wedlock Births." *Family Planning Perspectives* 31(2): 94–97.

Drummond, Tammerlin. 2000. "Mom on Her Own." *Time* (August 28), pp. 54–55.

Edin, Kathryn. 2000. "What Do Low-Income Single Mothers Say about Marriage?" *Social Problems* 47(1): 112–33.

Edin, Kathryn, and Laura Lein. 1997. *Making Ends Meet: How Single Mothers Survive Welfare and Low-Wage Work.* New York: Russell Sage Foundation.

Edwards, Tamala M. 2000. "Flying Solo." *Time* (August 28). pp. 49–53.

Elliott, D. M., and J. Briere. 1992. "The Sexually Abused Boy: Problems in Manhood." *Medical Aspects of Human Sexuality* 26:68–71.

Eltahawy, Mona. 2000 (March 6). "Giving Wives a Way Out." *U.S. News & World Report* 128(9):35.

Emery, Robert E. 1999. "Postdivorce Family Life for Children: An Overview of Research and Some Implications for Policy." In *The Postdivorce Family: Children, Parenting, and Society,* eds. R. A. Thompson and P. R. Amato, pp. 3–27. Thousand Oaks, CA: Sage Publications.

Family Court Reform Council of America. 2000. "Parental Alienation Syndrome." 31441 Santa Margarita Parkway, Suite A184. Rancho Santa Margarita, CA 92688.

Finkelhor, D., G. Hotaling, I. A. Lewis, and C. Smith. 1990. "Sexual Abuse in a National Survey of Adult Men and Women: Prevalence, Characteristics, and Risk Factors." *Child Abuse and Neglect* 14:19–28.

Fisher, Bonnie S., Francis T. Cullen, and Michael G. Turner. 2000. *The Sexual Victimization of College Women.* National Institute of Justice and Bureau of Justice Statistics. Washington, D.C.: U.S. Department of Justice.

Flory, Heather. 2000 (Spring). "I Promise to Love, Honor, Obey...and Not Divorce You: Covenant Marriage and the Backlash Against No-Fault Divorce." *Family Law Quarterly* 34(1):133–148.

Forum on Child and Family Statistics. 2000. *America's Children: Key National Indicators of Well-Being, 2000.* http://www.ChildStats.gov

Gardner, Richard A. 1998. *The Parental Alienation Syndrome,* 2nd edition,

Cresskill NJ: Creative Therapeutics, Inc.

Gelles, Richard J. 2000. "Violence, Abuse, and Neglect in Families." In *Families & Change: Coping with Stressful Events and Transitions*, 2nd ed., eds. P. C. McKenry and S. J. Price, pp. 183–207. Thousand Oaks, CA: Sage Publications.

Gelles, Richard J. 1993. "Family Violence." In *Family Violence: Prevention and Treatment*, eds. Robert L. Hampton, Thomas P. Gullotta, Gerald R. Adams, Earl H. Potter III, and Roger P. Weissberg, pp. 1–24. Newbury Park, CA: Sage Publications.

Gelles, Richard J., and Jon R. Conte. 1991. "Domestic Violence and Sexual Abuse of Children: A Review of Research in the Eighties." In *Contemporary Families: Looking Forward, Looking Back*, ed. Alan Booth, pp. 327–40. Minneapolis: National Council on Family Relations.

Global Study of Family Values. 1998. The Gallup Organization. http://198.175.140.8/Special_Reports/family.htm

Harrington, Donna, and Howard Dubowitz. 1993. "What Can Be Done to Prevent Child Maltreatment?" In *Family Violence: Prevention and Treatment*, eds. Robert L. Hampton, Thomas P. Gullotta, Gerald R. Adams, Earl H. Potter III, and Roger P. Weissberg, pp. 258–80. Newbury Park, CA: Sage Publications.

Henry, Ronald K. 1999 (Spring). "Child Support at a Crossroads: When the Real World Intrudes Upon Academics and Advocates." *Family Law Quarterly* 33(1): 235–64.

Hewlett, Sylvia Ann and Cornel West. 1998. *The War Against Parents: What We Can Do for Beleaguered Moms and Dads*. Boston: Houghton Mifflin Company.

Hochschild, Arlie Russell. 1997. *The Time Bind: When Work Becomes Home and Home Becomes Work*. New York: Henry Holt and Company.

Hochschild, Arlie Russell. 1989. *The Second Shift: Working Parents and the Revolution at Home*. New York: Viking.

Hogan, D. P., R. Sun, and G. T. Cornwell. 2000. "Sexual and Fertility Behaviors of American Females Aged 15–19 Years: 1985,

1990, and 1995. *American Journal of Public Health* 90:1421–25.

"In the News." 1998. Family Violence Prevention Fund. http://www.igc.org/fund/materials/speakup/02_13_98.htm

Jasinski, J. L., L. M. Williams, and J. Siegel. 2000. "Childhood Physical and Sexual Abuse as Risk Factors for Heavy Drinking among African-American Women: A Prospective Study. *Child Abuse and Neglect* 24: 1061–1071.

Jekielek, Susan M. 1998. "Parental Conflict, Marital Disruption and Children's Emotional Well-Being." *Social Forces* 76:905–35.

Johnson, Michael P. 2001. "Patriarchal Terrorism and Common Couple Violence: Two Forms of Violence Against Women." In *Men and Masculinity: A Text Reader*, ed. T. F. Cohen, pp. 248–260. Belmont CA: Wadsworth.

Johnson, Michael P. and Kathleen Ferraro. 2001. "Research on Domestic Violence in the 1990s: Making Distinctions." In *Understanding Families Into the New Millennium: A Decade in Review*, ed. Robert M. Milardo, pp. 167–182. Minneapolis: National Council on Family Relations.

Jorgensen, Stephen R. 2000 (Nov. 11). "Adolescent Pregnancy Prevention: Prospects for 2000 and Beyond." Presidential Address at the National Council on Family Relations 62nd Annual Conference. Minneapolis, MN.

Kaufman, Joan, and Edward Zigler. 1992. "The Prevention of Child Maltreatment: Programming, Research, and Policy." In *Prevention of Child Maltreatment: Developmental and Ecological Perspectives*, ed. Diane J. Willis, E. Wayne Holden, and Mindy Rosenberg, pp. 269–95. New York: John Wiley & Sons.

Knox, David (with Kermit Leggett). 1998. *The Divorced Dad's Survival Book: How to Stay Connected with Your Kids*. New York: Insight Books.

Knutson, John F., and Mary Beth Selner. 1994. "Punitive Childhood Experiences Reported by Young Adults over a 10-Year Period." *Child Abuse and Neglect* 18:155–66.

Krug, Ronald S. 1989. "Adult Male Report of Childhood Sexual Abuse by Mothers: Case Description, Mo-

tivations, and Long-Term Consequences." *Child Abuse and Neglect* 13:111–19.

Lachs, Mark S., Christianna Williams, Shelley O'Brien, Leslie Hurst, and Ralph Horwitz. 1997. "Risk Factors for Reported Elder Abuse and Neglect: A Nine-Year Observational Cohort Study." *Gerontologist* 37:469–74.

Lanz, Jean B. 1995. "Psychological, Behavioral, and Social Characteristics Associated with Early Forced Sexual Intercourse among Pregnant Adolescents." *Journal of Interpersonal Violence* 10:188–200.

Leite, Randy W. and Patrick C. McKenry. 2000. "Aspects of Father Status and Post-Divorce Father Involvement with Children." Poster session at the National Council on Family Relations 62nd Annual Conference. Minneapolis, MN: Nov. 10–13.

Lewin, Tamar. 2000 (November 4). "Fears for Children's Well-Being Complicates a Debate Over Marriage." *The New York Times on the Web*. http://www.nytimes.com/2000/11/04/arts/04MARR.html

Lloyd, Sally A. 2000. "Intimate Violence: Paradoxes of Romance, Conflict, and Control." *National Forum* 80(4): 19–22.

Lloyd, Sally A. and Beth C. Emery. 2000. *The Dark Side of Courtship: Physical and Sexual Aggression*. Thousand Oaks CA: Sage Publications.

Lloyd, S. A., and B. C. Emery. 1993. "Abuse in the Family: An Ecological, Life-Cycle Perspective." In *Family Relations: Challenges for the Future*, ed. T. H. Brubaker, pp. 129–52. Newbury Park, CA: Sage Publications.

Luker, Kristin. 1996. *Dubious Conceptions: The Politics of Teenage Pregnancy*. Cambridge, MA: Harvard University Press.

Magdol, L., T. E. Moffitt, A. Caspi, and P. A. Silva. 1998. "Hitting Without a License: Testing Explanations for Differences in Partner Abuse between Young Adult Daters and Cohabitors." *Journal of Marriage and the Family* 60:41–55.

Marlow, L., and S. R. Sauber. 1990. *The Handbook of Divorce Mediation*. New York: Plenum.

Mindel, Charles H., Robert W. Habenstein, and Roosevelt Wright,

Jr. 1998. *Ethnic Families in America: Patterns and Variations.* Upper Saddle River, NJ: Prentice Hall.

Monson, C. M., G. R. Byrd, and J. Langhinrichsen-Rohling. 1996. "To Have and to Hold: Perceptions of Marital Rape." *Journal of Interpersonal Violence* 11:410–24.

National Center for Injury Prevention and Control. 2000. "Intimate Partner Violence Fact Sheet." National Center for Injury Prevention and Control. Mailstop K60, 4770 Buford Highway NE, Atlanta, GA 30341-3724.

National Coalition for the Homeless. 1999. NCH Fact Sheet #1. "Why Are People Homeless?" http://www.nationalhomeless.org/causes.html

National Parenting Association. 1996. *What Will Parents Vote For?: Findings of the First National Survey of Parent Priorities.* New York: Author.

Nelson, B. S. and K. S. Wampler. 2000. "Systemic Effects of Trauma in Clinic Couples: An Exploratory Study of Secondary Trauma Resulting from Childhood Abuse." *Journal of Marriage and Family Counseling* 26:171–184.

Nielsen, L. 1999. "College Aged Students with Divorced Parents: Facts and Fiction." *College Student Journal* 33:543–572.

Nock, Steven L. 1995. "Commitment and Dependency in Marriage. *Journal of Marriage and the Family* 57: 503–514.

Parker, Marcie, R. Edward Bergmark, Mark Attridge, and Jude Miller-Burke. 2000. "Domestic Violence and its Effect on Children." *National Council on Family Relations Report* 45(4):F6–F7.

Pasley, Kay and Carmelle Minton. 2001. "Generative Fathering After Divorce and Remarriage: Beyond the 'Disappearing Dad.'" In *Men and Masculinity: A Text Reader*, ed. T. F. Cohen, pp. 239–248. Belmont CA: Wadsworth.

Peterson, Karen S. 1997. "States Flirt with Ways to Reduce Divorce Rate." *USA Today*, April 10, D1–2.

Peterson, Richard R. 1996. "A Reevaluation of the Economic Consequences of Divorce." *American Sociological Review* 61:528–536.

Popenoe, David. 1993. "Point of View: Scholars Should Worry about the Disintegration of the American Family." *Chronicle of Higher Education*, April 14, A48.

Popenoe, David. 1996. *Life without Father.* New York: Free Press.

Rennison, Callie M. and Sarah Welchans. 2000. "Intimate Partner Violence." U.S. Department of Justice. Office of Justice Programs. Washington, D.C.: Bureau of Justice Statistics.

Resnick, Michael, Peter S. Bearman, Robert W. Blum, Karl E. Bauman, Kathleen M. Harris, Jo Jones, Joyce Tabor, Trish Beubring, Renee E. Sieving, Marcia Shew, Marjore Ireland, Linda H. Berringer, and J. Richard Udry. 1997 (September 10). "Protecting Adolescents from Harm." *Journal of the American Medical Association* 278(10):823–32.

Russell, D. E. 1990. *Rape in Marriage.* Bloomington, IN: Indiana University Press.

Schacht, Thomas E. 2000. "Protection Strategies to Protect Professionals and Families Involved in High-Conflict Divorce." *UALR Law Review* 22(3):565–592.

Scott, K. L. and D. A. Wolfe. 2000. "Change Among Batterers: Examining Men's Success Stories." *Journal of Interpersonal Violence* 15:827–842.

Shapiro, Joseph P., and Joannie M. Schrof. 1995. "Honor Thy Children." *U.S. News and World Report*, February 27, 39–49.

Singh, Susheela and Jacqueline E. Darroch. 2000. "Adolescent Pregnancy and Childbearing: Levels and Trends in Developed Countries." *Family Planning Perspectives* 32(1):14–23.

Spiegel, D. 2000. "Suffer the Children: Long-Term Effects of Sexual Abuse." *Society* 37:18–20.

Stanley, Scott M., Howard J. Markman, Michelle St. Peters, and B. Douglas Leber. 1995. "Strengthening Marriage and Preventing Divorce: New Directions in Prevention Research." *Family Relations* 44: 392–401.

Statistical Abstract of the United States: 2000. 2000. 120th ed. U.S. Bureau of the Census. Washington, D.C.: U.S. Government Printing Office.

Statistical Abstract of the United States: 1999. 1999. 119th ed. U.S. Bureau of the Census. Washington, D.C.: U.S. Government Printing Office.

Stein, Theodore J. 1993. "Legal Perspectives on Family Violence against Children." In *Family Violence: Prevention and Treatment*, eds. Robert L. Hampton, Thomas P. Gullotta, Gerald R. Adams, Earl H. Potter III, and Roger P. Weissberg, pp. 179–97. Newbury Park, CA: Sage Publications.

Stets, J. E. and M. A. Straus. 1989. "The Marriage as a Hitting License: A Comparison of Assaults in Dating, Cohabiting, and Married Couples." In *Violence in Dating Relationships*, eds. M. A. Pirog-Good and J. E. Stets. New York: Greenwood Press. pp. 33–52.

Stock, J. L., M. A. Bell, D. K. Boyer, and F. A. Connell. 1997. "Adolescent Pregnancy and Sexual Risk-Taking among Sexually Abused Girls." *Family Planning Perspectives* 29:200–203.

Straus, Murray. 2000. "Corporal Punishment and Primary Prevention of Physical Abuse. *Child Abuse and Neglect* 24:1109–1114.

Straus, Murray A., David B. Sugarman, and Jean Giles-Sims. 1997. "Spanking by Parents and Subsequent Antisocial Behavior of Children." *Archives of Pediatric Adolescent Medicine* 151:761–67.

Thakkar, R. R., P. M. Gutierrez, C. L. Kuczen, and T. R. McCanne. 2000. "History of Physical and/or Sexual Abuse, and Current Suicidality in College Women." *Child Abuse and Neglect* 24:1345–1354.

Thompson, Ross A. and Jennifer M. Wyatt. 1999. "Values, Policy, and Research on Divorce." In *The Postdivorce Family: Children, Parenting, and Society*, eds. R. A. Thompson and P. R. Amato, pp. 191–232. Thousand Oaks, CA: Sage Publications.

Thompson, Ross A. and Paul R. Amato. 1999. "The Postdivorce Family: An Introduction to the Issues." In *The Postdivorce Family: Children, Parenting, and Society*, eds. R. A. Thompson and P. R. Amato, pp. xi–xxiii. Thousand Oaks, CA: Sage Publications.

United Nations Development Programme. 2000. *Human Development Report 2000.* Cary, North Carolina: Oxford University Press.

U.S. Bureau of the Census. 2000 (Sept.). "Money Income in the U.S." *Current Population Reports.*

Washington, D.C.: U.S. Government Printing Office.

U.S. Bureau of the Census. 1998. "Poverty in the United States." *Current Population Reports* P60-201. Washington, D.C.: U.S. Government Printing Office.

U.S. Department of Health and Human Services. 2000 (June 17). "HHS Fatherhood Initiative." http://www.hhs.gov/news/press/2000pres/20000617.html

U.S. Department of Justice. 1998 (March 16). "Murder by Intimates Declined 36 Percent Since 1976, Decrease Greater for Male than for Female Victims." Washington, D.C. http://www.ojp.usdoj.gov/bjs/pub/press/vi.pr

Ventura, Stephanie J. and Christine A. Bachrach. 2000 (Oct. 18). "Nonmarital Childbearing in the United States, 1940–99." *National Vital Statistics Report* 48(16).

Ventura, Stephanie J., Sally C. Curtin, and T. J. Mathews. 2000 (April 24). "Variations in Teenage Birth Rates, 1991–1998: National and State Trends." *National Vital Statistics Report* 48(6).

Viano, C. Emilio. 1992. "Violence among Intimates: Major Issues and Approaches." In *Intimate Violence: Interdisciplinary Perspectives*, ed. C. E. Viano, pp. 3–12. Washington, D.C.: Hemisphere.

Waite, L. and M. Gallagher. 2000. *The Case for Marriage: Why Married People are Happier, Healthier and Better off Financially*. New York: Doubleday.

Walker, Alexis J. 2001. "Refracted Knowledge: Viewing Families Through the Prism of Social Science." In *Understanding Families Into the New Millennium: A Decade in Review*, ed. Robert M. Milardo, pp. 52–65. Minneapolis: National Council on Family Relations.

Whiffen, V. E., J. M. Thompson, and J. A. Aube. 2000. "Mediators of the Link between Childhood Sexual Abuse and Adult Depressive Symptoms." *Journal of Interpersonal Violence* 15:1100–1120.

Willis, Diane J., E. Wayne Holden, and Mindy Rosenberg. 1992. "Child Maltreatment Prevention: Introduction and Historical Overview." In *Prevention of Child Maltreatment: Developmental and Ecological Perspectives*, ed. Diane J.

Willis, E. Wayne Holden, and Mindy Rosenberg, pp. 1–14. New York: John Wiley & Sons.

Chapter 6

AARP (American Association of Retired Persons). 2000. "Baby Boomers Envision their Retirement: An AARP Segmentation Analysis. Executive Summary Part I. http://research.aarp.org/econ/boomer_seg_1.html

Anetzberger, Georgia J., Jill E. Korbin, and Craig Austin. 1994. "Alcoholism and Elder Abuse." *Journal of Interpersonal Violence* 9:184–93.

AOA (Administration on Aging). 2001. "The Older Americans Act." http://www.aoa.gov/may2001/factsheets/OAA.htm

AOA (Administration on Aging). 2000a. "The Growth of America's Older Population." Fact Sheet. http://www.aoa.dhhs.gov/may2000/factsheets/growth.htm

_____. 2000b. "A Profile of Older Americans" http://www.aoa.gov/STATS/profile/default.html

_____. 2000c. "Demographic Changes." http://www.aoa.gov/stats/aging21/demography.html

_____. 2000d. "A Diverse Aging Population." http://www.aoa.dhhs.gov/may2000/factsheets/diverse.htm

_____. 2000e. "Older Women" http://www.aoa.dhhs.gov/may2000/factsheets/olderwomen.html

_____. 2000f. "Employment and the Older Worker." http://www.aoa.dhhs.gov/factsheets/employolderworker.html

_____. 2000g. "AOA Annual Profile of Older Americans Shows Drop in Poverty Rate." http://www.aoa.dhhs.gov/pr/pr2000/OAprofile.html

_____. 2000h. "The Elderly Nutrition Program." http://www.aoa.dhhs.gov/factsheets/enp.htm

AP (Associated Press). 2000. "Scientists Hunt Earliest Symptoms in Race against Alzheimer's Epidemic." CNN.com. July 9. http://www.cnn.com/2000/HEALTH/aging/07/09/bc.alzheimers.ap/index.htm

Arluke, Arnold, and Jack Levin. 1990. "'Second Childhood': Old Age in Popular Culture." In *Readings on Social Problems*, ed. W. Feigelman, pp.

261–65. Fort Worth: Holt, Rinehart and Winston.

Atchley, Robert C. 2000. *Social Forces and Aging*. Belmont, CA: Wadsworth.

Begley, Sharon. 2000. "The Stereotype Trap." *Newsweek*. November 6, pp. 66–68.

Bergmann, Barbara R. 1999. "A 'Help for Working Parents' Program Can Reduce Child Poverty." *Brown University Child and Adolescent Behavior Letter* 15(2):1.

Boudreau, Francois A. 1993. "Elder Abuse." In *Family Violence: Prevention and Treatment*, eds. R. L. Hampton, T. P. Gullota, G. R. Adams, E. H. Potter III, and R. P. Weissberg, pp. 142–58. Newbury Park, CA: Sage Publications.

Brazzini, D. G., W. D. McIntosh, S. M. Smith, S. Cook, and C. Harris. 1997. "The Aging Woman in Popular Film: Underrepresented, Unattractive, Unfriendly, and Unintelligent." *Sex Roles* 36:531–43.

Brooks-Gunn, Jeanne, and Greg Duncan. 1997. "The Effects of Poverty on Children." *Future of Children* 7(2):55–70.

Carpenter, MacKenzie and Ginny Kopas. 1999. "Casualties of Custody Wars: Special Report." *Pittsburgh Post-Gazette*. http://www.post-gazette.com/custody

CASA (Center on Addiction and Substance Abuse). 1999. "Back to School, 1999." The CASA National Survey of American Attitudes on Substance Abuse: Teens and their Parents." http://www.casacolumbia.org/usr_doc17645.pdf

CBS/New York Times Poll. 2000. "Priorities." September 9-11. http://www.pollingreport.com/priorit1/htm

CDF (Children's Defense Fund). 2000a. "Fair Start FAQs." http://www.childrensdefense.org/fairstart_faq.html

_____. 2000b. "The Children's Defense Fund has Assessed the State of America's Child..." CDF Press Release. http://www.childrensdefense.org/release00324.htm

_____. 2000c. "Where America Stands." http://www.childrensdefense.org/facts_america.html

_____. 2000d. "Child Care Now!" http://www.childrensdefense.org/childcare/cc_polls.html

CNN. 2001. "13 Year Old Convicted of First Degree Murder in Florida Wrestling Death." http://www.CNN.com/2001/LAW/01/25/wrestling.death.02

Clinton, William J. 2000, "Statement by the President." Office of the Press Secretary. The White House. Washington, D.C. November 13.

Costa, Dora L. 2000. "A Century of Retirement." *TIAA-CREF Participant.* November, pp.12–13.

Cowgill, Donald, and Lowell Holmes. 1972. *Aging and Modernization.* New York: Appleton-Century-Crofts.

Cummings, Elaine, and William Henry. 1961. *Growing Old: The Process of Disengagement.* New York: Basic Books.

DeAngelis, Tori. 1997. "Elderly May Be Less Depressed Than the Young." *APA Monitor,* October. http://www.apa.org/monitor/oct97/elderly.html

Duncan, Greg, W. Jean Yeung, Jeanne Brooks-Gunn, and Judith Smith. 1998. "How Much Does Childhood Poverty Affect the Life Chance of Children?" *American Sociological Review* 63:402–23.

Ennis, Dave. 2000. "Survey Finds Retirement Plans Often Include a Job." *Morning Star.* November 27, 1D, 3D.

FBI (Federal Bureau of Investigation). 2000. *Crime in the United States, 1999.* Uniform Crime Reports. Washington, D.C.: U.S. Government Printing Office

Fields, Jason, and Kristin Smith. 1998. "Poverty, Family Structure, and Child Well-Being." Population Division. Washington, D.C.: U.S. Bureau of Census.

FTC (Federal Trade Commission). 2000. "FTC Releases Report on the Marketing of Violent Entertainment to Children." Federal Trade Commission. http://wwwftc.gov/opa/2000/09/youthviol.htm

Gallup Poll. 2000. "Children and Violence." http://www.gallup.com/poll/indicators/indchild_violence.asp

Goldberg, Beverly. 2000. *Age Works.* New York: Free Press.

Harris, Diana K. 1990. *Sociology of Aging.* New York: Harper & Row.

Harris, Kathleen, and Jeremy Marmer. 1996. "Poverty, Paternal Involvement and Adolescent Well-Being." *Journal of Family Issues* 17(5):614–40.

HCFA (Health Care Financing Administration). 2000. "Medicare." http://www.hcfa.gov/medicare/medicare/htm

HHS (Health and Human Services). 1999. "Children's Health Insurance Program National Back-to-School Kick Off." HHS News, September 22. http://www.hcfa.gov/init/9909922wh.htm

Hooyman, Nancy R. and H. Asuman Kiyak. 1999. *Social Gerontology: A Multidisciplinary Perspective,* 2nd ed. Boston: Allyn and Bacon.

Jacobson, Linda. 2000. "Children's Early Needs Seen as Going Unmet." *Washington Post.* October 3. http://washingtonpost.com/wp-dyn/articles/A4014-2000Oct5.html

Leeman, Sue. 2000. "Report Details Child Poverty." http:www.excite.com/news/ap/000612/09/us-unicef-child-poverty/printstory-1

Livni, Ephrat. 2000. "Exercise, the Anti-Drug." September 21. http://abcnews.go.com/sections/living/Daily/News/depression_elderly000921.html

Matras, Judah. 1990. *Dependency, Obligations, and Entitlements: A New Sociology of Aging, the Life Course, and the Elderly.* Englewood Cliffs, NJ: Prentice-Hall.

Miner, Sonia, John Logan, and Glenna Spitze. 1993. "Predicting Frequency of Senior Center Attendance." *The Gerontologist* 33:650–57.

NCSC (National Council of Senior Citizens). 2000a. "Our Issues: Poverty." http://www.ncscinc.org/issues/poverty.htm

———. 2000b. "Affordable Housing." http://www.ncscinc.org/issues/affordhouse.htm

Newman, Cathy. 2000. "Older, Healthier and Wealthier." *Washington Post.* August 10, A03.

NIH (National Institute of Health). 2000a. "Nation's Children Gain in Many Areas." NIH News Release. July 13. http://www.nih.gov/news/pr/jul2000

———. 2000b. "Well Being Improves for most Older People, But Not for All, New Federal Report Says." NIH News Release. August 10. http://www.nih.gov/ma/news/pr/2000/0810.htm

NIMH (National Institute of Mental Health). 1999. "Older Americans:

Depression and Suicide Facts." http://www.nimh.nig.gov/publicat/edlerlydepsuicide.ctm

Peterson, Peter. 2000. "Gray Dawn." In *Social Problems of the Modern World,* ed. Frances Moulder , pp. 126–133. Belmont, CA: Wadsworth.

Reio, Thomas G. and Joanne Sanders-Reto. 1999. "Combating Workplace Ageism." *Adult Learning* 11:10–13.

Riley, Matilda White. 1987. "On the Significance of Age in Sociology." *American Sociological Review* 52 (February):1–14.

Riley, Matilda W., and John W. Riley. 1992. "The Lives of Older People and Changing Social Roles." In *Issues in Society,* eds. Hugh Lena, William Helmreich, and William McCord, pp. 220– 31. New York: McGraw-Hill.

Russakoff, Dale. 2000. "Report Paints Brighter Picture of Children's Health." *Washington Post.* July 14, A01.

Schieber, Sylvester. 2000. "The Global Aging Crisis." *Electric Perspectives* 25: 18–28.

Seeman, Teresa E., and Nancy Adler. 1998. "Older Americans: Who Will They Be?" *National Forum,* Spring, 22–25.

"Snapshots of the Elderly." 2000. *Washington Post.* http://www.washingtonpost.com/wp-srv/health/images/elderly.html

Statistical Abstract of the United States 2000, 120th ed. 2000. U.S. Bureau of the Census. Washington, D.C.: U.S. Government Printing Office.

Thurow, Lester C. 1996. "The Birth of a Revolutionary Class." *The New York Times Magazine,* May 19, 46–47.

Torres-Gil, Fernando. 1990. "Seniors React to Medicare Catastrophic Bill: Equity or Selfishness?" *Journal of Aging and Social Policy* 2(1):1–8.

Uhlenberg, Peter. 2000. "Integration of Young and Old." *The Gerontologist* 40:276–279.

UNICEF (United Nations Children's Fund). 1994. "The Progress of Nations." United Nations.

———. 1998. "The First Nearly Universally Ratified Human Rights Treaty in History." *Status.* Washington, D.C.: UNICEF. http://www.unicef.org/crc/status.html

———. 2000. "Convention on the Rights of the Child: FAQ." UNICEF. http://www.uniccf.org/crc/html

U.S. Senate, 1999. "Children, Violence and the Media." Federal Trade Commission Report to Senate Committee on the Judiciary. http:///.senate.gov/~judiciary/mediavio.htm

Weiss, Gregory L. and Lynn Lonnquist. 2000, "The Sociology of Health, Healing, and Illness," 3rd ed. Upper Saddle River, NJ: Prentice Hall.

Weissberg, Roger P., and Carol Kuster. 1997. "Introduction and Overview: Let's Make Healthy Children 2010 a National Priority." In *Enhancing Children's Well-Being*, eds. Roger Weissberg, Thomas Gullotta, Robert Hampton, Bruce Ryan, and Gerald Adams, pp. 1–16. Thousand Oaks, CA: Sage Publications.

Yamaguchi, Mari. 2000. "Japan is Fastest Growing Graying Country." *The Washington Post*. May 30, A2.

Chapter 7

American Council on Education and American Association of University Professors. 2000. *Does Diversity Make a Difference? Three Research Studies on Diversity in College Classrooms*. Washington, D.C.: American Council on Education and American Association of University Professors.

American Council on Education and University of California. 2000. *The American Freshman: National Norms for Fall 2000*. Los Angeles Higher Education Research Institute.

Beeman, Mark, Geeta Chowdhry, and Karmen Todd. 2000. "Educating Students About Affirmative Action: An Analysis of University Sociology Texts." *Teaching Sociology* 28(2):98–115.

Bobo, Lawrence, and James R. Kluegel. 1993. "Opposition to Race-Targeting: Self Interest, Stratification Ideology or Racial Attitudes?" *American Sociological Review* 58(4):443–64.

Children Now. 2000a. *Fall Colors I: How Diverse is the 1999-2000 Prime-Time Season?* http://www.childrennow.org

———. 2000b. *Fall Colors II: Exploring the Quality of Diverse Portrayals on Prime-Time Television*. http://www.childrennow.org

Children Now and the National Hispanic Foundation for the Arts. 2000. *Fall Colors III: Latinowood and TV: Prime Time for a Reality Check*. http://www.childrennow.org

Cohen, Mark Nathan. 1998. "Culture, Not Race, Explains Human Diversity." *Chronicle of Higher Education* 44(32): B4–B5.

Conley, Dalton. 1999. *Being Black, Living in the Red: Race, Wealth, and Social Policy in America*. Berkeley: University of California Press.

Current Population Survey, U.S. Bureau of the Census. 1997 (March). "Country of Origin and Year of Entry into the U.S. of the Foreign Born, by Citizenship Status: March 1997." http://www.bls.census.gov/cps/pub/1997/for_born.htm

Darity, William, Jr. 2000 (December 1). "Give Affirmative Action Time to Act." *The Chronicle of Higher Education*. p. B18.

Dees, Morris. 2000 (Dec. 28). Personal correspondence. Morris Dees, co-founder of the Southern Poverty Law Center. 400 Washington Avenue, Montgomery, AL 36104.

EEOC Press Release (Equal Employment Opportunity Commission). 2000 (October 30). "EEOC Settles Racial Harassment Suit for $249,000 Against Florida Citrus Grower Sun Ag, Inc." http://www.eeoc.gov/press/10-3000.html

Etzioni, Amitai. 1997. "New Issues: Rethinking Race." *The Public Perspective* (June–July):39–40. http://www.ropercenter.uconn.edu/pubper/pdf/!84b.htm

Federal Bureau of Investigation. 2001. *Hate Crime Statistics 1999*. http://www.fbi.gov/

Gaertner, Samuel L. and John F. Dovidio. 2000. *Reducing Intergroup Bias: The Common Ingroup Identity Model*. Philadelphia: Taylor & Francis Group.

Gallup/CNN/USA Today Poll. 2000 (January). http://www.pollingreport.com/race.htm

Gardyn, Rebecca and John Fetto. 2000 (June). "Demographics…It's All the Rage!" *American Demographics*. http://www.demographics.com/publications/htm

Glynn, Patrick. 1998. "Racial Reconciliation: Can Religion Work Where Politics Has Failed?" *American Behavioral Scientist* 41:834–41.

Goldstein, Joseph. 1999 (January). "Sunbeams." *The Sun* 277:48.

Goodnough, Abby. 2001(January 11). "New York City Is Short-Changed in School Aid, State Judge Rules." *The New York Times on the Web*. http://www.nytimes.com/2001/01/11/nyregion/11SCHO.html

Grieco, Elizabeth M. and Rachel C. Cassidy. 2001 (March). "Overview of Race and Hispanic Origin: Census 2000 Brief." U.S. Census Bureau. www.census.gov/prod/2001pubs/cenbr01-1.pdf

Guillebeau, Christopher. 1999. "Affirmative Action in a Global Perspective: The Cases of South Africa and Brazil." *Sociological Spectrum* 19(4): 443–465.

Guinier, Lani. 1998. Interview with Paula Zahn. CBS Evening News, July 18.

Gurin, Patricia. 1999 (Spring). "New Research on the Benefits of Diversity in College and Beyond: An Empirical Analysis." *Diversity Digest*: 5–15. Washington, D.C.: Association of American Colleges and Universities.

Halton, Beau. 1998 (March 26). "City's Housing Bias Called 'Abysmal.'" http://www.jacksonville.c...98/met_2blhousi.html

"Hate on Campus." 2000 (Spring). *Intelligence Report* 98:6–15.

Healey, Joseph F. 1997. *Race, Ethnicity, and Gender in the United States: Inequality, Group Conflict, and Power*. Thousand Oaks, CA: Pine Forge Press.

Hill, Mark E. 2000. "Color Differences in the Socioeconomic Status of African American Men: Results of a Longitudinal Study." *Social Forces* 78(4):1437–1460.

Hodgkinson, Harold L. 1995. "What Should We Call People?: Race, Class, and the Census for 2000." *Phi Delta Kappa*, October, 173–79.

Holzer, Harry and David Neumark. 2000. "Assessing Affirmative Action." *Journal of Economic Literature* 38(3): 483–568.

hooks, bell. 2000. *Where We Stand: Class Matters*. New York: Routledge.

Humphreys, Debra. 2000 (Fall). "National Survey Finds Diversity Requirements Common Around the Country." *Diversity Digest*. http://www.diversityweb.org/Digest/F00/survey.html

Humphreys, Debra. 1999. "Diversity and the College Curriculum: How Colleges & Universities Are Preparing Students for a Changing World." *DiversityWeb.* http://www.inform.umd.edu/EdRes/Topic/Di...Leadersguide/CT/curriculum_briefing.html

Immigration and Naturalization Service. 2001. "General Naturalization Requirements." http://www.ins.usdoj.gov/natz/general.html

"Intelligence Briefs." 2000 (September). *SPLC Report*, 30(3):3.

Intelligence Report. 1998 (Winter, issue no. 89). Montgomery, Ala.: Southern Poverty Law Center.

Jackson, Janine. 1999. "Affirmative Action Coverage Ignores Women—and Discrimination." *Extra!* (January/February). http://www.fair.org/extra/9901/affirmative-action.html

Jensen, Derrick. 2001 (April). "Saving the Indigenous Soul: An Interview with Martin Prechtel." *The Sun*, Issue 304:4–15.

Kaplan, David E. and Lucian Kim. 2000 (September 25). "Nazism's New Global Threat." *U.S. News Online.* http://www.usnews.com/usnews/issue/000925/nazi.htm

Keita, S. O. Y., and Rick A. Kittles. 1997. "The Persistence of Racial Thinking and the Myth of Racial Divergence." *American Anthropologist* 99(3):534–44.

King, Joyce E. 2000 (Fall). "A Moral Choice." *Teaching Tolerance*, pp. 14–15.

Kleg, Milton. 1993. *Hate, Prejudice and Racism*. Albany: State University of New York Press.

Kozol, Jonathan. 1991. *Savage Inequalities: Children in America's Schools*. New York: Crown.

Kumovich, Robert M. 1999 (August). "Conflict, Religious Identity, and Ethnic Intolerance in Croatia." In *Sociological Abstracts*. Abstracts of papers presented at the 94th Annual Meeting of the American Sociological Association, Chicago.

Landau, Elaine. 1993. *The White Power Movement: America's Racist Hate Groups*. Brookfield, CT: Millbrook Press.

Lawrence, Sandra M. 1997. "Beyond Race Awareness: White Racial Identity and Multicultural Teaching." *Journal of Teacher Education* 48(2):108–17.

Leggon, Cheryl B. 1999. "Introduction: Race and Ethnicity—A Global Perspective." *Sociological Spectrum* 19(4): 381–385.

Levin, Jack, and Jack McDevitt. 1995. "Landmark Study Reveals Hate Crimes Vary Significantly by Offender Motivation." *Klanwatch Intelligence Report*, August, 7–9.

Liu, Marian. 2000 (Aug. 16). "Asian Americans Divided Over Web's 'Mr. Wong.' *San Francisco Chronicle*. http://www.sfgate.com/cgi-bin/article.cgi?file=chronicle/archive/2000/08/16/DD35504.DTL

Lofthus, Kai R. 1998. "Swedish Biz Decries Racist Music." *Billboard*, January 24:71, 73.

Lollock, Lisa. 2001 (March). "The Foreign-Born Population in the United States: March 2000." *Current Population Reports* P20-534. Washington D.C.: U.S. Bureau of the Census.

Ludwig, Jack. 2000 (February 28). "Perceptions of Black and White Americans Continue to Diverge Widely on Issues of Race Relations in the U.S." Gallup Organization, Poll Releases. http://www.gallup.com/poll/releases/pr000228.asp

Marger, Martin N. 2000. *Race and Ethnic Relations: American and Global Perspectives*, 5th ed. Belmont, CA: Wadsworth.

Massey, Douglas, and Nancy Denton. 1993. *American Apartheid: Segregation and the Making of an American Underclass*. Cambridge, MA: Harvard University Press.

McLemore, S. Dale, Harriet D. Romo, and Susan Gonzalez Baker. 2001. *Racial and Ethnic Relations in America*, 6th ed. Needham Heights, MA: Allyn & Bacon.

Mishel, Lawrence, Jared Bernstein, and John Schmitt. 1999. *The State of Working America 1998–99*. Ithaca, NY: Cornell University Press.

Molnar, Stephen. 1983. *Human Variation: Races, Types, and Ethnic Groups*, 2nd ed. Englewood Cliffs, NJ: Prentice-Hall.

NAACP Press Release. 2001 (January 10). "NAACP National Civil Rights Groups File Florida Voting Rights Lawsuit to Eliminate Unfair Voting Practices." National Association for the Advancement of Colored People. www.NAACP.org

NAACP Press Release. 2000 (November 11). "NAACP Voting Irregularities Public Hearing." National Association for the Advancement of Colored People. www.NAACP.org

Nash, Manning. 1962. "Race and the Ideology of Race." *Current Anthropology* 3:258–88.

National Coalition on Black Civic Participation Press Release. 2000 (November 10). "National Coalition's Efforts Lead to Upsurge in Black Voter Turnout." http://www.bigvote.org/

Newburger, Eric C. and Andrea Curry. 2000 (March). "Educational Attainment in the United States." *Current Population Reports* p. 20-528. Washington, D.C.: U.S. Census Bureau.

Niemonen, Jack. 1999. "Deconstructing Cultural Pluralism." *Sociological Spectrum* 19(4):401–419.

Oliver, Melvin and Thomas Shapiro. 1997. *Black Wealth/White Wealth: A New Perspective on Racial Inequality*. New York: Taylor & Francis Group.

Parsons, Sharon, William Simmons, Frankie Shinhoster, and John Kilburn. 1999. "A Test of the Grapevine: An Empirical Examination of Conspiracy Theories Among African-Americans." *Sociological Spectrum* 19(2):201–222.

Population Reference Bureau. 1999. "World Population: More Than Just Numbers." Washington, D.C.: Population Reference Bureau.

Purdum, Todd S. 2001 (March 29). "California Census Confirms Whites Are in Minority." *The New York Times on the Web.* http://www.nytimes.com/2001/03/30/national/30CALI.html

Race Relations Reporter. 1999 (Oct. 15). Volume VII No. 8. New York: CH II Publishers, Inc. 200 West 57th St., New York, NY 10019.

Ramirez, Roberto R. 2000 (February). "The Hispanic Population in the United States." *Current Population Reports* p. 20-527. Washington D.C.: U.S. Census Bureau.

Reid, Evelyn. 1995. "Waiting to Excel: Biraciality in the Classroom." In *Educating for Diversity: An Anthology of Multicultural Voices*, ed. Carl A. Grant, pp. 263–73. Needham Heights, MA: Allyn & Bacon.

Rivera, Ismael. 2000 (August 17). "The Affirmative Action Debate." American Association for Affirmative Action. http://www.affirmativeaction.org/alerts/aadebate.html

Schaefer, Richard T. 1998. *Racial and Ethnic Groups*, 7th ed. New York: HarperCollins.

Schmitt, Eric. 2001 (April 4). "Analysis of Census Finds Segregation Along with Diversity." *The New York Times on the Web*. http://www.nytimes.com/2001/04/04/national/04CENS.html

Schmitt, Eric. 2001 (March 31). "Blacks Split on Disclosing Multiracial Roots." *The New York Times on the Web*. http://www.nytimes.com/2001/03/31/national/31RACE.html

Schuman, Howard and Maria Krysan. 1999. "A Historical Note on Whites' Beliefs About Racial Inequality." *American Sociological Review*, 64:847–855.

Schuman, Howard, Charlotte Steeh, Lawrence Bobo, and Maria Krysan. 1997. *Racial Attitudes in America: Trends and Interpretations*. Cambridge, MA: Harvard University Press.

Shipler, David K. 1998 (March 15). "Subtle vs. Overt Racism." *Washington Spectator* 24(6):1–3.

Snyder, Dean. 2000 (March 10). "Religious Leaders Report to Clinton on Racial Justice Work." United Methodist News Service. http://umns.umc.org/00/mar/136.htm

Statistical Abstract of the United States: 2000, 120th ed. U.S. Bureau of the Census. Washington, D.C.: U.S. Government Printing Office.

"Study Finds Benefits From Immigration." 1997 (May 18). *Minneapolis Star Tribune*, p. 4A.

Teaching Tolerance. 2000 (Fall). "Hear & Now." p. 5.

Turner, Margery Austin and Felicity Skidmore. 1999. *Mortgage Lending Discrimination: A Review of Existing Evidence*. Washington, D.C.: The Urban Institute.

United for a Fair Economy. 1999 (September 30). Press Release, "The Racial Wealth Gap: Left Out of the Boom." http://www.ufenet.org/press/racial_wealth_gap.html

Urban Institute. 2000 (Sept. 2). "Press Release." http://www.urban.org/news/press/CP_000911.html

U.S. Department of Labor. Undated. "Affirmative Action at OFCCP." http://www.dol.gov/dol/esa/public/regs/compliance/ofccp/what_is.htm

U.S. Department of Labor. 1999 (December). "Facts on Executive Order 11246 Affirmative Action." http://www.dol.gov/dol/esa/public/ofcp_org.htm

Wheeler, Michael L. 1994. *Diversity Training: A Research Report*. New York: The Conference Board.

Williams, David R., James S. Jackson, Tony N. Brown, Myriam Torres, Tyrone A. Forman, and Kendrick Brown. 1999. "Traditional and Contemporary Prejudice and Urban Whites' Support for Affirmative Action and Government Help." *Social Problems* 46(4): 503–527.

Williams, Eddie N., and Milton D. Morris. 1993. "Racism and Our Future." In *Race in America: The Struggle for Equality*, eds. Herbert Hill and James E. Jones Jr., pp. 417–24. Madison: University of Wisconsin Press.

Wilson, William J. 1987. *The Truly Disadvantaged: The Inner City, the Underclass and Public Policy*. Chicago: University of Chicago Press.

"The Year of Hate." 2001 (Spring). *Intelligence Report* Issue 101. http://www.splcenter.org/intelligenceproject/ip-index.html

Zack, Naomi. 1998. *Thinking about Race*. Belmont, CA: Wadsworth Publishing Co.

Zinn, Howard. 1993. "Columbus and the Doctrine of Discovery." In *Systemic Crisis: Problems in Society, Politics, and World Order*, ed. William D. Perdue, pp. 351–57. Fort Worth: Harcourt Brace Jovanovich.

Chapter 8

ACE (American Council on Education). 2001. "Making the Case for Affirmative Action in Higher Education." http://www.acenet.edu/bookstore/descriptions/making_the_case/threats/home.html

Anderson, John, and Molly Moore. 1998. "The Burden of Womanhood." In *Global Issues 98/99*, ed. Robert Jackson, pp. 170–175. Guilford, CT: Dushkin/McGraw-Hill.

Anderson, Margaret L. 1997. *Thinking about Women*. 4th ed. New York: MacMillan.

Americas. 2000. "Issues of Gender." May/June, 52.

Austin, Jonathan D. 2000. "U.N. Report: Women's Unequal Treatment Hurts Economies." CNN.com. September 20. http://www.cnn.com/2000/world/europe/09/20.un.population.report

Baker, Robin, Gary Kriger, and Pamela Riley. 1996. "Time, Dirt and Money: The Effects of Gender, Gender Ideology, and Type of Earner Marriage on Time, Household Task, and Economic Satisfaction among Couples with Children." *Journal of Social Behavior and Personality* 11:161–77.

Bannon, Lisa. 2000. "Why Girls and Boys Get Different Toys." *The Wall Street Journal*, February 14, B1.

Basow, Susan A. 1992. *Gender: Stereotypes and Roles*, 3rd ed. Pacific Grove, CA: Brooks/Cole.

Begley, Sharon. 2000. "The Stereotype Trap." *Newsweek*, November 6, 66–68.

Beutel, Ann M., and Margaret Mooney Marini. 1995. "Gender and Values." *American Sociological Review* 60:436–48.

Bianchi, Susanne M., Melissa A. Milkie, Liana C. Sayer, and John Robinson. 2000. "Is Anyone Doing the Housework? Trends in the Gender Division of Household Labor." *Social Forces* 79:191–228.

Bittman, Michael and Judy Wajcman. 2000. "The Rush Hour: The Character of Leisure Time and Gender Equity." *Social Forces* 79:165–189.

BLS (Bureau of Labor Statistics). 2000. *Report on the Youth Labor Force*. U.S. Department of Labor. Washington, D.C.

Burger, Jerry M. and Cecilia H. Solano. 1994. "Changes in Desire for Control over Time: Gender Differences in a Ten-Year Longitudinal Study." *Sex Roles* 31:465–72.

Cejka, Mary Ann and Alice Eagly. 1999. "Gender Stereotypic images of Occupations Correspond to the Sex Segregation of Employment." *Personality and Social Psychology Bulletin* 25: 413–423.

Chavez, Linda. 2000. *The Color Bind*. Berkeley: University of California Press.

Cianni, Mary and Beverly Romberger. 1997. "Life in the Corporation: A Multi-Method Study of the Experiences of Male and Female Asian, Black, Hispanic and White Employees." *Gender, Work and Organization* 4:116–29.

Civil Rights Monitor. 2000. "Sexual Harassment Decisions, Supreme Court 1997–1998 Term." http:///civirights.org/crlibrary/monitor/winter_spring1999

Cohen, Dan. 1995. "Female Kicker Practices on Football's Second String." *The Chronicle Online.* April 12. http://chronicle.duke.edu/chronicle/1995/01/12/merer.html

Cohen, Theodore. 2001. *Men and Masculinity.* Belmont, CA: Wadsworth.

Evans, Lorraine and Kimberly Davies. 2000. "No Sissy Boys Here." *Sex Roles* (Feb): 255–271.

Faludi, Susan. 1991. *Backlash. The Undeclared War against American Women.* New York: Crown Publishers.

Fitzgerald, Louise F. and Sandra L. Shullman. 1993. "Sexual Harassment: A Research Analysis and Agenda for the '90s." *Journal of Vocational Behavior* 40:5–27.

Fitzpatrick, Catherine. 2000. "Modern Image of Masculinity Changes with Rise of New Celebrities." *Detroit News,* June 24. http://detnews.com/2000/religion/0006/24

Hochschild, Arlie. 1989. *The Second Shift: Working Patterns and the Revolution at Home.* New York: Viking Penguin.

IWRP (International Women's Right's Project). 2000. "The First CEDAW Impact Study." http://www.yorku.ca/iwrp/cedawReport

Kalb, Claudia. 2000. "What Boys Really Want." *Newsweek,* July 2. http://www.msnbc.com/news/428301.asp

Kenworthy, Lane and Melissa Malami. 1999. "Gender Inequality in Political Representation: A Worldwide Comparative Analysis." *Social Forces* 78:235–269.

Kilbourne, Barbara S., Georg Farkas, Kurt Beron, Dorothea Weir, and Paula England. 1994. "Returns to Skill, Compensating Differentials, and Gender Bias: Effects of Occupational Characteristics on the Wages of White Women and Men." *American Journal of Sociology* 100:689–719.

Klein, Matthew. 1998. "Women's Trip to the Top." *American Demographics,* February, 22.

_____. 1997. "Blue Janes." *American Demographics,* August, 23.

Kopelman, Lotetta M. 1994. "Female Circumcision/Genital Mutilation and Ethical Relativism." *Second Opinion* 20:55–71.

Leeman, Sue. 2000. "The More things Change..." September 20. http://abcnews.go.com/sections/living/Daily/News/women_unreport00920.html

Leo, John. 1997. "Fairness? Promises, Promises." *U.S. News and World Report* 123(4):18.

Long, J. Scott, Paul D. Allison, and Robert McGinnis. 1993. "Rank Advancement in Academic Careers: Sex Differences and the Effects of Productivity." *American Sociological Review* 58:703–22.

Lorber, Judith. 1998. "Night to his Day." In *Reading Between the Lines,* eds. Amanda Konradi and Martha Schmidt, pp. 213–20. Mountain View, CA: Mayfield Publishing.

Marini, Margaret Mooney, and Pi-Ling Fan. 1997. "The Gender Gap in Earnings at Career Entry." *American Sociological Review* 62:588–604.

Martin, Patricia Yancey. 1992. "Gender, Interaction, and Inequality in Organizations." In *Gender, Interaction, and Inequality,* ed. Cecilia Ridgeway, pp. 208–31. New York: Springer-Verlag.

McCammon, Susan, David Knox, and Caroline Schacht. 1998. *Making Choices in Sexuality.* Pacific Grove, CA: Brooks/Cole Publishing Co.

Mensch, Barbara and Cynthia Lloyd. 1997. "Gender Differences in the Schooling Experiences of Adolescents in Low-Income Countries: The Case of Kenya." Policy Research Working Paper no. 95. New York: Population Council.

Moen, Phyllis and Yan Yu. 2000. "Effective Work/Life Strategies: Working Couples, Working Conditions, Gender and Life Quality." *Social Problems* 47:291–326.

Morin, Richard and Megan Rosenfeld. 2000. "The Politics of Fatigue." In *Annual Editions: Social Problems,* ed. Kurt Finsterbusch, pp. 152–154. Guilford, CT: Dushkin/McGraw-Hill.

Nichols-Casebolt, Ann and Judy Krysik. 1997. "The Economic Well-Being of Never and Ever-Married Mother Families." *Journal of Social Service Research* 23(1):19–40.

Olson, Josephine E., Irene H. Frieze, and Ellen G. Detlefsen. 1990. "Having It All? Combining Work and Family in a Male and a Female Profession." *Sex Roles* 23:515–34.

Parker, Kathleen. 2000. "It's Time for Women to Get Angry but Not at Men." *Greensboro News Record,* March 14.

Pollock, William. 2000a. *Real Boys' Voices.* New York: Random House.

Pollock, William. 2000b. "The Columbine Syndrome." *National Forum* 80: 39–42.

Population Reference Bureau. 1999. "World Population: More than Just Numbers." Washington, D.C. http://www.prb.org

Purcell, Piper and Lara Stewart. 1990. "Dick and Jane in 1989." *Sex Roles* 22:177–85.

Rabin, Sarah. 2000. "Feminists Take CEDAW Into Our Own Hands." National Organization for Women. http://63111.42.146/cgs/gs_article.asp?ArticleD=1863

Reid, Pamela T. and Lillian Comas-Diaz. 1990. "Gender and Ethnicity: Perspectives on Dual Status." *Sex Roles* 22:397–408.

Reskin, Barbara and Debra McBrier. 2000. "Why not Ascription? Organizations' Employment of Male and Female Managers." *American Sociological Review* 65:210–233.

Robinson, John P., and Suzanne Bianchi. 1997. "The Children's Hours." *American Demographics,* December, 1–6.

Rosenberg, Janet, Harry Perlstadt, and William Phillips. 1997. "Now That We Are Here: Discrimination, Disparagement, and Harassment at Work and the Experience of Women Lawyers." In *Workplace/Women's Place,* ed. Dana Dunn, pp. 247–59. Los Angeles: Roxbury.

Rubenstein, Carin. 1990. "A Brave New World." *New Woman* 20(10):158–64.

Saad, Lydia. 2000. "Most Working Women Deny Gender Discrimination in Their Pay." Gallup Poll. http://www.gallup.poll/releases/pr000207.asp

Sachs, Susan. 2000. "In Iran, More Women Leaving Nest for University." *The New York Times on the Web.* July 22. http://www10.nytimes.com/library/world/mideast

Sadker, Myra and David Sadker. 1990. "Confronting Sexism in the College Classroom." In *Gender in the Classroom: Power and Pedagogy*, eds. S. L. Gabriel and I. Smithson, pp. 176–87. Chicago: University of Illinois Press.

Sapiro, Virginia. 1994. *Women in American Society*. Mountain View, CA: Mayfield.

Schneider, Margaret, and Susan Phillips. 1997. "A Qualitative Study of Sexual Harassment of Female Doctors by Patients." *Social Science and Medicine* 45:669–76.

Schroeder, K. A., L. L. Blood, and D. Maluso. 1993. "Gender Differences and Similarities between Male and Female Undergraduate Students regarding Expectations for Career and Family Roles." *College Student Journal* 27:237–49.

Schwalbe, Michael. 1996. *Unlocking the Iron Cage: The Men's Movement, Gender Politics, and American Culture*. New York: Oxford University Press.

Sheehan, Molly. 2000. "Women Slowly Gain Ground in Politics." In *Vital Signs: The Environmental Trends That Are Shaping Our Future*, ed. Linda Starke. pp. 152–153. New York: W.W. Norton Company.

Shepard, Paul. 2000. "Education-gender Divide Growing with Black Wealth, Study Finds." *Boston Globe*, July 26, A5.

Signorielli, Nancy. 1998. "Reflections of Girls in the Media: A Content Analysis Across Six Media." Overview. http://childrennow.org/media/mc97/ReflectSummary.html

Smolken, Rachael. 2000. "Girls SAT Scores Still Lag Boys." *Post Gazette*. http://www.post-gazette.com/headlines/20000830sat2.asp

Sommers, Christina. 2000. *The War Against Boys*. New York: Simon and Schuster.

Statistical Abstract of the United States: 2000, 120th ed. U.S. Bureau of the Census. Washington, D.C.: U.S. Government Printing Office.

Statistical Abstract of the United States: 1999, 119th ed. U.S. Bureau of the Census. Washington, D.C.: U.S. Government Printing Office.

Suggs, Welch. 2000. "Duke U. Discriminated against Female Football Player, Jury Finds." *Chronicle of Higher Education*. October 13. http://

nersp.nerdc.ufl.edu/~lombardi/his01/Duke_kick_chron.html

Tam, Tony. 1997. "Sex Segregation and Occupational Gender Inequality in the United States: Devaluation or Specialized Training?" *American Journal of Sociology* 102(6):1652–92.

Tannen, Deborah. 1990. *You Just Don't Understand: Women and Men in Conversation*. New York: Ballantine Books.

Tomaskovic-Devey, Donald. 1993. "The Gender and Race Composition of Jobs and the Male/Female, White/Black Pay Gap." *Social Forces* 72(1):45–76.

Thomas, Karen. 1999. "Equality is for the Daughters." *USA Today*, February 17, 7A.

United Nations. 2000a. *The World's Women 2000: Trends and Statistics*. New York: United Nations Statistics Division.

———. 2000b. "Strengthening Women's Economic Capacity." United Nations Developing Fund for Women. http://www.unifem.undp.org/economic.htm

———. 2000c. "Convention on the Elimination of all Forms of Discrimination Against Women." http://un.org/womenwatch/daw/cedaw/

U.S. Census. 2000a. "Educational Attainment in the U.S." *Current Population Reports*. March. Washington, D.C.: U.S. Department of Commerce, Economics and Statistics Administration.

———. 2000b. "Women in the U.S.: A Profile." *Current Population Reports*. March. Washington, D.C.: U.S. Department of Commerce, Economics and Statistics Administration.

———. 2000c. "Women's History Month: Census Bureau Facts for Features." http://wwwcensusgov/Press-Release/www/2000

———. 1999. "Money Income in the United States:1998." http://wwwcensus.gov/prod/99pubs/p60-206.pdfpdf

Van Willigen, Marieke and Patricia Drentea. 1997. "Benefits of Equitable Relationships: The Impact of Sense of Failure, Household Division of Labor, and Decision-Making Power on Social Support." Presented at the American Socio-

logical Association, Toronto, Canada, August.

WHO (World Health Organization). 2001. "Prevalence Rates for FGM." http://www.who.int/frh-whd/FGM

Williams, Christine L. 1995. *Still a Man's World: Men Who Do Women's Work*. Berkeley: University of California Press.

Williams, John E. and Deborah L. Best. 1990. *Sex and Psyche: Gender and Self Viewed Cross-Culturally*. London: Sage Publications.

Wilmot, Alyssa. 1999. "First National Love Your Body Day a Big Success." Press Release. *National Organization for Women Newsletter* (Winter). http://www.now.org

WIN (Women's International Network) News. 2000. "Reports from around the World." Autumn, 50–58.

Winfield, Nicole. 2000. "Activists give U.S. mixed Rating for Efforts on Gender." *Boston Globe* June 8, A21.

Witt, S. D. 1996. "Traditional or Androgynous: An Analysis to Determine Gender Role Orientation of Basal Readers." *Child Study Journal* 26:303–318.

World Bank. 2001. "Engendering Development." *World Bank Policy Research Report*. http://www.worldbank.org/gender/pur/newsummary.htm

"The Year in Hate." 2001 (Spring). *Intelligence Report* 101:34–35.

Yamaguchi, Mari. 2000. "Female Government Workers Face Harassment." http://news.excite.com/news/ap/001227/05/int.Japan

Yoder, Janice D. and Patricia Aniakudo. 1997 "Outsiders within the Firehouse: Subordination and Difference in the Social Interactions of African American Women Firefighters." *Gender and Society* 11(3):324–41.

Yumiko, Ehara. "Feminism's Growing Pains." *Japan Quarterly* 47:41–48.

Zimmerman, Marc A., Laurel Copeland, Jean Shope, and T.E. Dielman. 1997. "A Longitudinal Study of Self-Esteem: Implications for Adolescent Development." *Journal of Youth and Adolescence* 26 (2):117–41.

Chapter 9

"1998 in Review." 1999. *Out*, January, p. 6.

"ACLU Fact Sheet: Overview of Lesbian and Gay Parenting, Adoption and Foster Care." 1999. American Civil Liberties Union. http://www.aclu.org/issues/gay/parent.html

"Adoption." 2000. Human Rights Campaign FamilyNet. http://Familynet.hrc.org/

Alsdorf, Matt. 2001 (March 16). "Portugal Grants Rights to Gay Couples." PlanetOut.com. http://www.planetout.com/news/article-print.html?2001/03/16/2

Bailey, Robert W. 1999. *Out and Voting II: The Gay, Lesbian, and Bisexual Vote in Congressional Elections, 1990–1998.* Washington, D.C.: The Policy Institute of the National Gay and Lesbian Task Force.

Bayer, Ronald. 1987. *Homosexuality and American Psychiatry: The Politics of Diagnosis,* 2nd ed. Princeton, NJ: Princeton University Press.

Besen, Wayne. 2000. "Introduction." In *Feeling Free: Personal Stories: How Love and Self Acceptance Saved Us from "Ex-Gay" Ministries.* Human Rights Campaign. p. 7. Washington, D.C.: Human Rights Campaign Foundation.

Black, Dan; Gary Gates, Seth Sanders, and Lowell Taylor. 2000 (May). "Demographics of the Gay and Lesbian Population in the United States: Evidence from Available Systematic Data Sources." *Demography* 37(2):139–154.

Brannock, J. C., and B. E. Chapman. 1990. "Negative Sexual Experiences with Men among Heterosexual Women and Lesbians." *Journal of Homosexuality* 19:105–10.

"Brazilian Killers Sentenced." 2001 (February 15). PlanetOut.com. http://www.planetout.com/news/article-print.html?2001/02/15/1

Bullough, Vern L. 2000 (July–Sept.). "Transgenderism and the Concept of Gender." *The International Journal of Transgenderism.* www.symposion.com

Butler, Amy C. 2000. "Trends in Same-gender Sexual Partnering, 1988–1998." *The Journal of Sex Research* 37(4):333–343.

Button, James W., Barbara A. Rienzo, and Kenneth D. Wald. 1997. *Private Lives, Public Conflicts: Battles over Gay Rights in American Communities.* Washington, D.C.: CQ Press.

Chase, Bob. 2000. "NEA President Bob Chase's Historic Speech from 2000 GLSEN Conference." http://www.glsen.org/templates/resources/record.html?section=14&record=255

"Constitutional Protection." 1999. GayLawNet. http://www.nexus.net.au/~dba/news.html#top

"Custody and Visitation." 2000. Human Rights Campaign FamilyNet. http://Familynet.hrc.org.

De Cecco, John P., and D. A. Parker. 1995. "The Biology of Homosexuality: Sexual Orientation or Sexual Preference? *Journal of Homosexuality* 28:1–28.

D'Emilio, John. 1990. "The Campus Environment for Gay and Lesbian Life." *Academe* 76(1):16–19.

Docll, R. G. 1995. "Sexuality in the Brain." *Journal of Homosexuality* 28:345–56.

Dozetos, Barbara. 2001 (March 7). "School Shooter Taunted as 'Gay'." PlanetOut.com. http://www.planetout.com/news/article-print.html?2001/03/07/1

Drinkwater, Gregg. 2001 (March 30). "Netherlands to Celebrate First Gay Marriages." PlanetOut.com http://www.planetout.com/news/article-print.html?2001/03/2

Durkheim, Emile. 1993. "The Normal and the Pathological." Originally published in *The Rules of Sociological Method, 1938.* In *Social Deviance,* ed. Henry N. Pontell, pp. 33–63. Englewood Cliffs, NJ: Prentice-Hall.

Elliot, David. 2001 (March 23). "Arkansas Court Strikes Down Sodomy Law." National Gay and Lesbian Task Force. http://www.ngltf.org/news/printed.cfm?releaseID=377

Esterberg, K. 1997. *Lesbian and Bisexual Identities: Constructing Communities, Constructing Selves.* Philadelphia: Temple University Press.

Ettelbrick, Paula. 2000. "Domestic Partnerships: Domestic Partner Benefits for State Employees." National Gay and Lesbian Task Force Policy Institute. www.ngltf.org/pi/dp.bstate.htm

Faulkner, Anne H., and Kevin Cranston. 1998. "Correlates of Same-Sex Sexual Behavior in a Random Sample of Massachusetts High School Students." *Journal of Public Health* 88 (February):262–66.

Firestein, B. A. 1996. "Bisexuality as Paradigm Shift: Transforming Our Disciplines." In *Bisexuality: The Psychology and Politics of an Invisible Minority,* ed. B. A. Firestein, pp. 263–291. Thousand Oaks, CA: Sage.

Fone, Byrne. 2000. *Homophobia: A History.* New York: Henry Holt and Company.

Frank, Barney. 1997. Foreword to *Private Lives, Public Conflicts: Battles over Gay Rights in American Communities,* by J. W. Button, B. A. Rienzo, and K. D. Wald. Washington D.C.: CQ Press.

Frank, David John and Elizabeth H. McEneaney. 1999. "The Individualization of Society and the Liberalization of State Policies on Same-Sex Relations, 1984–1995." *Social Forces* 77(3):911–44.

Franklin, Karen. 2000. "Antigay Behaviors Among Young Adults." *Journal of Interpersonal Violence* 15(4):339–362.

Franklin, Sarah. 1993. "Essentialism, Which Essentialism? Some Implications of Reproductive and Genetic Techno-Science." *Journal of Homosexuality* 24:27–39.

Freedman, Estelle B., and John D'Emilio. 1990. "Problems Encountered in Writing the History of Sexuality: Sources, Theory, and Interpretation." *Journal of Sex Research* 27:481–95.

Gallup Organization. 2000. "Gallup Poll Topics: A-Z." www.gallup.com//poll/indicators/indhomosexual.asp

Garnets, L., G. M. Herek, and B. Levy. 1990. "Violence and Victimization of Lesbians and Gay Men: Mental Health Consequences." *Journal of Interpersonal Violence* 5:366–83.

Gay and Lesbian International Lobby. 2000. "Recognition of Gay and Lesbian Partnerships in Europe." www.steff.suite.dk/partner.htm.

GLSEN's National School Climate Survey: Lesbian Gay, Bisexual and Transgender Students and Their Experiences in School. 1999. The Gay, Lesbian and Straight Education Network. www.glscn.org/pages/sections/news/natlnews/1999/sep/survey

Goode, Erica E., and Betsy Wagner. 1993. "Intimate Friendships." *U.S. News and World Report,* July 5, 49–52.

Greenhouse, Linda. 1998. "Gay Rights Case Fails in Bid for Supreme Court Hearing." *The New York Times*, January 13, 15, late edition, East Coast.

"A Historic Victory: Civil Unions for Same-Sex Couples. What Does it Mean?" *NCLR Newsletter* (National Center for Lesbian Rights). 2000 (Fall), pp. 3, 10. www.nclrights.org/index.html

Homophobia 101: Teaching Respect for All. 2000. The Gay, Lesbian, and Straight Education Network. www.glsen.org/

Human Rights Campaign. 2000a. *Feeling Free: Personal Stories: How Love and Self-Acceptance Saved Us from "Ex-Gay" Ministries.* Washington, D.C.: Human Rights Campaign Foundation.

——. 2000b. *The State of the Workplace for Lesbian, Gay, Bisexual and Transgendered Americans, 2000.* Washington, D.C.: Human Rights Campaign.

——. 2000c. "1999 FBI Hate Crime Statistics." www.hrc.org/

Human Rights Watch. 2001. *World Report 2001.* http://www.hrw.org

International Gay and Lesbian Human Rights Commission. 1999. "Antidiscrimination Legislation." www.iglhrc.org/news/factsheets/990604-antidis.html.

International Lesbian and Gay Association. 1999. "World Legal Survey 1999." www.ilga.org

"Jail, Death Sentences in Africa." 2001 (February 21). PlanetOut.com. http://www.planetout.com/news/articleprint.html?2001/02/21/2

Kinsey, A. C., Pomeroy, W. B., and Martin, C. E. 1948. *Sexual Behavior in the Human Male.* Philadelphia: W. B. Saunders.

Kinsey, A. C., Pomeroy, W. B., Martin, C. E. and Gebhard, P. H. 1953. *Sexual Behavior in the Human Female.* Philadelphia: W. B. Saunders.

Kirkpatrick, R. C. 2000. "The Evolution of Human Sexual Behavior." *Current Anthropology* 41(3):385.

Kite, M. E., and B. E. Whitley, Jr. 1996. "Sex Differences in Attitudes toward Homosexual Persons, Behavior and Civil Rights: A Meta-analysis." *Personality and Social Psychology Bulletin* 22:336–52.

Klassen, Albert D., Colin J. Williams, and Eugene E. Levitt. 1989. *Sex and Morality in the United States.*

Middletown, CT: Wesleyan University Press.

Lambda Legal Defense and Education Fund. 2000a. "Students and Salt Lake City School Board End Feud Over Gay-Supportive Clubs." Press Release, Oct. 6, 2000. www.lambdalegal.org

——. 2000b. *Student Advocacy for University Anti-Bias Policies that Include Sexual Orientation.* Publications July 6, 2000. www.lambdalegal.org

Landis, Dan. 1999 (Feb. 17). "Mississippi Supreme Court Made a Tragic Mistake in Denying Custody to Gay Father, Experts Say." American Civil Liberties Union, News. www.aclu.org

LAWbriefs. 2000 (Fall). "Recent Developments in Sexual Orientation and Gender Identity Law." Vol. 3, No. 3.

Lever, Janet. 1994. "The 1994 *Advocate* Survey of Sexuality and Relationships: The Men." *The Advocate,* August 23, 16–24.

Louderback, L. A., and B. E. Whitley. 1997. "Perceived Erotic Value of Homosexuality and Sex-Role Attitudes as Mediators of Sex Differences in Heterosexual College Students' Attitudes toward Lesbians and Gay Men." *Journal of Sex Research* 34:175–82.

Mathison, Carla. 1998. "The Invisible Minority: Preparing Teachers to Meet the Needs of Gay and Lesbian Youth." *Journal of Teacher Education* 49:151–55.

Michael, Robert T., John H. Gagnon, Edward O. Laumann, and Gina Kolata. 1994. *Sex in America: A Definitive Survey.* Boston: Little, Brown.

Mohr, Richard D. 1995. "Anti-Gay Stereotypes." In *Race, Class, and Gender in the United States,* 3rd ed., ed. P. S. Rothenberg, pp. 402–8. New York: St. Martin's Press.

Moore, David W. 1993. "Public Polarized on Gay Issue." *Gallup Poll Monthly* 331 (April):30–34.

The National Coalition of Anti-Violence Programs. 2000. *Anti-Lesbian, Gay, Bisexual and Transgender Violence in 1999.* New York: The New York City Gay & Lesbian Anti-Violence Project. 240 West 35th St., Suite 2000. New York, NY 10001.

National Gay and Lesbian Task Force. 2001a. "Specific Anti-Same-Sex Marriage Laws in the U.S.-January

2001." http://www.ngltf.org/downloads/marriagemap0201.pdf

——. 2001b. "Hate Crime Laws in the U.S.-January 2001." http://www.ngltf.org/downloads/hatemap0101.pdf

"NCLR" Wins Equal Tax Benefits for Non-biological Lesbian Mother." 2000 (Fall). NCLR Newsletter (National Center for Lesbian Rights). pp. 1, 10. www.nclrights.org/index.html

Nugent, Robert, and Jeannine Gramick. 1989. "Homosexuality: Protestant, Catholic, and Jewish Issues: A Fishbone Tale." *Journal of Homosexuality* 18:7–46.

Parker, Laura and Guillermo X. Garcia. 2000 (Oct.10). *USA Today*, pp. 1A, 2A.

Patterson, Charlotte J. 2001. "Family Relationships of Lesbians and Gay Men." In *Understanding Families Into the New Millennium: A Decade in Review*, ed. Robert M. Milardo, pp. 271–88. Minneapolis, MN: National Council on Family Relations.

Paul, J. P. 1996. "Bisexuality: Exploring/Exploding the Boundaries." In *The Lives of Lesbians, Gays, and Bisexuals: Children to Adults*, eds. R. Savin-Williams and K. M. Cohen , pp. 436–61. Fort Worth: Harcourt Brace.

Pillard, Richard C., and J. Michael Bailey. 1998. "Human Sexuality Has a Heritable Component." *Human Biology* 70 (April):347–65.

Platt, Leah. 2001. "Not Your Father's High School Club." *The American Prospect* 12(1):A37–A39.

"Post-Election Analysis." 2000. Human Rights Campaign. www.hrc.org

Price, Jammie, and Michael G. Dalecki. 1998. "The Social Basis of Homophobia: An Empirical Illustration." *Sociological Spectrum* 18:143–59.

Ricks, Thomas E. 2000 (July 22). "Pentagon Vows to Enforce 'Don't Ask.'" *Washington Post*. p. A01.

Rosin, Hanna, and Richard Morin. 1999 (January 11). "In One Area, Americans Still Draw a Line on Acceptability." *The Washington Post National Weekly Edition* 16(11):8.

Sanday, Peggy. R. 1995. "Pulling Train." In *Race, Class, and Gender in the United States*, 3d ed., ed. P. S. Rothenberg, pp. 396–402. New York: St. Martin's Press.

Schellenberg, E. Glenn, Jessie Hirt, and Alan Sears. 1999. "Attitudes Toward Homosexuals Among Students at a Canadian University." *Sex Roles* 40(1/2): 139–52.

SIECUS (Sexuality Information and Education Council of the United States). 2000. "Fact Sheets: Sexual Orientation and Identity." 130 West 42nd St. Suite 350. NYC, NY 10036-7802.

Simon, A. 1995. "Some Correlates of Individuals' Attitudes toward Lesbians." *Journal of Homosexuality* 29:89–103.

"Sodomy Fact Sheet: A Global Overview." 2000. The International Gay and Lesbian Human Rights Commission. 1360 Mission Street, San Francisco, CA 94103.

Sullivan, A. 1997. "The Conservative Case." In *Same-Sex Marriage: Pro and Con*, ed. A. Sullivan, pp. 146–54. New York: Vintage Books.

Thompson, Cooper. 1995. "A New Vision of Masculinity." In *Race, Class, and Gender in the United States*, 3d ed., ed. P.S. Rothenberg, pp. 475–81. New York: St. Martin's Press.

The United Methodist Church and Homosexuality. 1999. http://religioustolerance.org/hom_umc.htm

Wilcox, Clyde and Robin Wolpert. 2000. "Gay Rights in the Public Sphere: Public Opinion on Gay and Lesbian Equality." In *The Politics of Gay Rights*, eds. Craig A. Rimmerman, Kenneth D. Wald, and Clyde Wilcox, pp. 409–432. Chicago: University of Chicago Press.

Yang, Alan. 1999. *From Wrongs to Rights 1973 to 1999: Public Opinion on Gay and Lesbian Americans Moves Toward Equality*. New York: The Policy Institute of the National Gay and Lesbian Task Force.

Zernike, Kate. 2000 (August 29). "Scouts' Successful Ban on Gays Is Followed by Loss in Support." *The New York Times on the Web*. pp. 1–5. www.nytimes.com/library/national/082900scouts-contribute.html

Chapter 10

Albelda, Randy, and Chris Tilly. 1997. *Glass Ceilings and Bottomless Pits: Women's Work, Women's Poverty*. Boston, MA: South End Press.

Alex-Assensoh, Yvette. 1995. "Myths about Race and the Underclass." *Urban Affairs Review* 31:3–19.

"America's Poorest People Have No Place to Go." 2000 (February 1). *The Washington Spectator* 26(3):1–3.

Anderson, Sarah, John Cavanagh, Chuck Collins, Chris Hartman, and Felice Yeskel. 2000. *Executive Excess: Seventh Annual CEO Compensation Survey*. Boston: Institute for Policy Studies and United for a Fair Economy.

Barlett, Donald L., and James B. Steele. 1998. "The Empire of the Pigs." *Time*, November 30, 52–64.

Becker, Elizabeth. 2001 (February 26). "Millions Eligible for Food Stamps Aren't Applying." *The New York Times on the Web*. http://www.nytimes.com/2001/02/26/national/26FOOD.html

Briggs, Vernon M. Jr. 1998. "American-Style Capitalism and Income Disparity: The Challenge of Social Anarchy." *Journal of Economic Issues* 32(2):473–81.

Brown, Lester R. 2001. "Eradicating Hunger: A Growing Challenge." In *State of the World 2001*, eds. Lester R. Brown, Christopher Flavin, and Hilary French, pp. 43–62. New York: W.W. Norton & Co.

Children's Defense Fund. 2000. *The High Cost of Child Care Puts Quality Child Care Out of Reach for Many Families*. Washington, D.C.: Children's Defense Fund.

Children's Defense Fund and the National Coalition for the Homeless. 1998. *Welfare to What: Early Findings on Family Hardship and Well-Being*. Washington, D.C.: Children's Defense Fund.

Chossudovsky, Michel. 1998. "Global Poverty in the Late 20th Century." *Journal of International Affairs* 52(1):293–303.

Conley, Dalton. 2001. "Capital for College: Parental Assets and Postsecondary Schooling." *Sociology of Education* 74(January):59–72.

Corcoran, Mary, and Terry Adams. 1997. "Race, Sex, and the Intergenerational Transmission of Poverty." In *Consequences of Growing Up Poor*, eds. Greg J. Duncan and Jeanne Brooks-Gunn, pp. 461–517. New York: Russell Sage Foundation.

Cracker, David A. and Toby Linden, eds. 1998. *Ethics of Consumption: The Good Life, Justice, and Global Stewardship*. Lanhan MD: Rowman & Littlefield.

Dalaker, Joseph and Bernadette D. Proctor. 2000. *Poverty in the United States: 1999*. Current Population Reports P60–210. U.S. Census Bureau. Washington, D.C.: U.S. Government Printing Office.

Davis, Kingsley, and Wilbert Moore. 1945. "Some Principles of Stratification." *American Sociological Review* 10:242–49.

Deen, Thalif. 2000. "NGOs Call For UN Poverty Eradication Fund." Global Policy Forum. http://www.globalpolicy.org/msummit/millenni/millfor3.htm

Deng, Francis M. 1998. "The Cow and the Thing Called 'What': Dinka Cultural Perspectives on Wealth and Poverty." *Journal of International Affairs* 52(1):101–15.

Duncan, Greg J., and Jeanne Brooks-Gunn. 1997. "Income Effects across the Life Span: Integration and Interpretation." In *Consequences of Growing Up Poor*, eds. Greg J. Duncan and Jeanne Brooks-Gunn, pp. 596–610. New York: Russell Sage Foundation.

Duncan, Greg J. and P. Lindsay Chase-Lansdale. 2001. "Welfare Reform and Child Well-Being." Paper presented at the Blank/Haskins conference, "The New World of Welfare Reform." Washington, D.C., February 1–2.

Economic Policy Institute. 2000. "Issue Guide to the Minimum Wage." http://www.epinet.org/Issues-guides/minwage/minwagefaq.html

Flavin, Christopher. 2001. "Rich Planet, Poor Planet." In *State of the World 2001*, eds. Lester R. Brown, Christopher Flavin, and Hilary French. Worldwatch Institute. New York: W.W. Norton & Co.

Friedman, Pamela. 2000 (April). "The Earned Income Tax Credit." *Issue Note* 4(4). http://www.welfareinfo.org/friedmanapril.htm

Gans, Herbert J. 1972. "The Positive Functions of Poverty." *American Journal of Sociology* 78 (September): 275–388.

Global Poverty Report. 2000. G8 Okinawa Summit, July 2000.

Hill, Lewis E. 1998. "The Institutional Economics of Poverty: An Inquiry into the Causes and Effects of Poverty." *Journal of Economic Issues* 32(2):279–86.

hooks, bell. 2000. *Where We Stand: Class Matters.* New York: Routledge.

Hope VI Fact Sheet. 2000. U.S. Department of Housing and Urban Development. Rockville, MD.

HUD (U.S. Department of Housing and Urban Development). 2000a (October). "New Facts About Households Assisted by HUD's Housing Programs." *Recent Research Results,* Rockville, MD.

———. 2000b (April)."Gun-Related Violence: The Costs to Public Housing Communities." *Recent Research Results.* pp. 1–2. Rockville, MD.

Human Development Report 1997. 1997. United Nations Development Programme. New York: Oxford University Press.

Independent Sector. 1999. "Giving and Volunteering in the United States: Findings from a National Survey." http://www.independentsector.org/ GrandV/s_keyf.htm

Johnson, Nicholas, and Ed Lazere. 1998. "Rising Number of States Offer EITCs." Center on Budget and Policy Priorities. http://www.cbpp. org/9-14-98sfp.htm

Kennedy, Bruce P., Ichiro Kawachi, Roberta Glass, and Deborah Prothrow-Stith. 1998. "Income Distribution, Socioeconomic Status, and Self-Rated Health in the U.S.: Multilevel Analysis." *British Medical Journal* 317(7163):917–21.

Knickerbocker, Brad. 2000. "Nongovernmental Organizations are Fighting and Winning Social, Political Battles." Global Policy Forum. http://www.globalpolicy.org/ngos/ 00role.htm

Kraut, Karen, Scott Klinger, and Chuck Collins. 2000. *Choosing the High Road: Businesses that Pay a Living Wage and Prosper.* Boston: United for a Fair Economy.

Levitan, Sar A., Garth L. Mangum, and Stephen L. Mangum. 1998. *Programs in Aid of the Poor.* 7th ed. Baltimore: Johns Hopkins University Press.

Lewis, Oscar. 1966. "The Culture of Poverty." *Scientific American* 2(5):19–25.

Lewis, Oscar. 1998. "The Culture of Poverty: Resolving Common Social Problems." *Society* 35(2):7–10.

Luker, Kristin. 1996. *Dubious Conceptions: The Politics of Teenage Pregnancy.* Cambridge, MA: Harvard University Press.

Mann, Judy. 2000 (May 15). "Demonstrators at the Barricades Aren't Very Subtle, But They Sometimes Win." *The Washington Spectator,* 26(10):1–3.

Massey, D. S. 1991. "American Apartheid: Segregation and the Making of the American Underclass." *American Journal of Sociology* 96:329–57.

Mayer, Susan E. 1997a. *What Money Can't Buy: Family Income and Children's Life Chances.* Cambridge, MA: Harvard University Press.

Mayer, Susan E. 1997b. "Trends in the Economic Well-Being and Life Chances of America's Children." In *Consequences of Growing Up Poor,* eds. Greg J. Duncan and Jeanne Brooks-Gunn, pp. 49–69. New York: Russell Sage Foundation.

McIntyre, Robert and T.D. Coo Nguyen. 2000 (October). "Corporate Income Taxes in the 1990s." Institute on Taxation and Economic Policy. 1311 L Street, NW. Washington, D.C. 20005.

Mead, L. 1992. *The New Politics of Poverty: The Non-Working Poor in America.* New York: Basic Books.

Michel, Sonya. 1998. "Childcare and Welfare (In)justice." *Feminist Studies* 24:44–54.

Mishel, Lawrence, Jared Bernstein, and John Schmitt. 2001. *The State of Working America 2000/2001.* Ithaca, NY: Cornell University Press.

Narayan, Deepa. 2000. *Voices of the Poor: Can Anyone Hear Us?* New York: Oxford University Press.

National Law Center on Homelessness and Poverty. 2000 (December). "Myths and Facts About Homelessness." http://www.nlchp.org/ myths.htm

National Law Center on Homelessness and Poverty in America. 2000. "Homelessness and Poverty in America." http://www.nlchp.org/ h&pusa.htm

Newman, Katherine S. 1999. *No Shame in My Game: The Working Poor in the Inner City.* New York: Alfred

A. Knopf and The Russell Sage Foundation.

Office of Management and Budget. 2001. "A Citizens Guide to the Federal Budget." Washington, D.C.: U.S. Government Printing Office.

Parenti, Michael. 1998. "The Super Rich Are Out of Sight." *Dollars and Sense* 217(May–June):36–37.

Paul, James A. 2000. "NGOs and Global Policy-Making." Global Policy Forum. http://www. globalpolicy.org/ngos/analysis/ anal00htm

Pressman, Steven. 1998. "The Gender Poverty Gap in Developed Countries: Causes and Cures." *The Social Science Journal* 35(2):275–87.

Seccombe, Karen. 2001. "Families in Poverty in the 1990s: Trends, Causes, Consequences, and Lessons Learned." In *Understanding Families Into the New Millennium: A Decade in Review,* ed. Robert M. Milardo, pp. 313–332. Minneapolis, MN: National Council on Family Relations.

Speth, James Gustave. 1998. "Poverty: A Denial of Human Rights." *Journal of International Affairs* 52(1):277–86.

Stegman, Michael A., Roberto G. Quercia, and George McCarthy. 2000. "Housing America's Working Families." *New Century Housing* 1(1):1–48.

Stensel, Dean and Stephen Moore. 1997. "Federal Aid to Dependent Corporations." Cato Institute. http://www.cato.org/pubs/ wtpapers/corporatewelfare.htm

Streeten, Paul. 1998. "Beyond the Six Veils: Conceptualizing and Measuring Poverty." *Journal of International Affairs* 52(1):1–8.

Sweeney, Eileen P. 2000 (February 29). "Recent Studies Indicate that Many Parents Who are Current or Former Welfare Recipients Have Disabilities or Other Medical Conditions." Washington, D.C.: Center on Budget and Policy Priorities.

United Nations. 1997. *Report on the World Social Situation, 1997.* New York: United Nations.

United Nations Development Programme. 2000. *Human Development Report 2000.* New York: Oxford University Press.

_____. 1997. *Human Development Report 1997*. New York: Oxford University Press.

U.S. Census Bureau. 2000. "Historical Poverty Tables-People." http://www.census.gov/hhes/poverty/histpov/perindex.html

U.S. Department of Agriculture. 2000. "Food Stamps." http://www.fns.usda.gov/fsp/MENU/faqs/faqs.htm

U.S. Department of Health and Human Services. 2000. *Temporary Assistance to Needy Families (TANF) Program, Third Annual Report to Congress*. http://www.acf.dhhs.gov/programs/opre/annual3.doc

Van Kempen, Eva T. 1997. "Poverty Pockets and Life Chances: On the Role of Place in Shaping Social Inequality." *American Behavioural Scientist* 41(3):430–50.

Wilson, William J. 1996. *When Work Disappears: The World of the New Urban Poor*. New York: Alfred A Knopf.

Wilson, William J. 1987. *The Truly Disadvantaged: The Inner City, the Underclass, and Public Policy*. Chicago: University of Chicago Press.

World Bank. 2001. *World Development Report: Attacking Poverty, 2000/2001*. Herndon, VA: World Bank and Oxford University Press.

Wren, Christopher S. 2001 (January 9). "U.N. Report Maps Hunger 'Hot Spots.'" *The New York Times on the Web*. http://www.nytimes.com/2001/01/01/09/world/09HUNG.html

Chapter 11

AFL-CIO. 2000 (April). *Death on the Job: The Toll of Neglect*, 9th ed. http://www.aflcio.org/safety/infodth.htm

Ambrose, Soren. 1998. "The Case against the IMF." *Campaign for Human Rights Newsletter* no. 12. http://www.summersault.co...wsletter/news12.html

Anderson, Sarah. 2001. "Seven Years Under NAFTA." Institute for Policy Studies. Washington, D.C. www.ips-dc.org

Barlett, Donald L. and James B. Steele. 1998. "Corporate Welfare: First in a Series." *Time* 152(19):36–39.

Bassi, Laurie J. and Jens Ludwig. 2000 (January). "School-to-Work Programs in the United States: A Multi-Firm Case Study of Training,

Benefits, and Costs." *Industrial and Labor Relations Review* 53(2):219.

Bavendam, James. 2000. "Managing Job Satisfaction." Special Reports, Volume 6:1–2. Bavendam Research Incorporated. www.bavendam.com

Bell, Daniel. 1973. *The Coming of Post-Industrial Society*. New York: Basic Books.

Bello, Walden. 2001 (Jan./Feb.). "Lilliputians Rising: 2000: The Year of Global Protest Against Corporate Globalization." *Multinational Monitor* 22(No. 1 & 2). http://www.essential.org/monitor/mm2001/01jan-feb/

Benjamin, Medea. 1998. "What's Fair about Fair Labor Association (FLA)?" *Sweatshop Watch*. http://www.sweatshopwatch.org/swatch/headlines/1998/org/swatch/headlines/1998/gex_fla.html

"Big Business for Reform." 2000 (November). *Multinational Monitor* 21(11). http://www.essential.org/monitor/mm2000/00november/toc.html

Bond, James T., Ellen Galinsky, and Jennifer E. Swanberg. 1997. *The 1997 National Study of the Changing Workforce*. New York: Families and Work Institute.

Bowles, Diane O. 2000. "Growth in Telework." Paper presented at the symposium Telework and the New Workplace of the 21st Century, Xavier University, New Orleans, October 16, 2000. U.S. Department of Labor. http://www.dol.gov/dol/asp/public/telework.htm

Bronfenbrenner, Kate. 2000 (December). "Raw Power: Plant-Closing Threats and the Threat to Union By Organizing." *Multinational Monitor* 21(12). http://www.essential.org/monitor/mm2000/00december/toc.html

Bureau of Labor Statistics. 2001. "Unemployment Rates in Nine Countries, 1990-2000." ftp://ftp.bls.gov/pub/special.requests/ForeignLabor/flsjec.txt

Bureau of Labor Statistics. 2000a. "Workplace Injuries and Illnesses in 1999." http://stats.bls.gov/oshhome.htm

Bureau of Labor Statistics. 2000b. "National Census of Fatal Occupational Injuries, 1999." Washington D.C.: U.S. Department of Labor.

Bureau of Labor Statistics. 2000c. "Employment Characteristics of Families in 1999." http://stats.bls.gov/news.release/famee.nr0htm

Cantor, David, Jane Waldfogel, Jeffrey Kerwin, Mareena McKinley Wright, Kerry Levin, John Rauch, Tracey Hagerty, and Martha Stapelton Kudela. 2001. "Balancing the Needs of Families and Employers: The Family and Medical Leave Surveys, 2000 Update." U.S. Department of Labor. http://www.dol.gov/dol/asp/public/fmla/main2000.htm

Caston, Richard J. 1998. *Life in a Business-Oriented Society: A Sociological Perspective*. Boston: Allyn and Bacon.

Coalition on Human Needs. 2000. "The Need for the Jobs Agenda." http://www.chn.org/jobsagenda/

Common Cause. 2001 (February 7). "National Parties Raise Record $463 Million in Soft Money During 1999-2000 Election Cycle." *Common Cause News*. http://commoncause.org

Common Cause. 2000. "Top Soft-Money Donors: January 1, 1999 through December 31, 1999." http:www.commoncause.org/laundromat/topdonors99_new.htm

Danaher, Kevin. 1998. "Are Workers Waking Up?" Global Exchange: Education for Action. http://www.globalexchange.org/education/econnomy/laborday.html

Dorman, Peter. 2001. "Child Labour in the Developed Economies." Geneva: International Labour Office.

Durkheim, Emile. [1893] 1966. *On the Division of Labor in Society*, trans. G. Simpson. New York: Free Press.

Economic Policy Institute. 2001 (February 2). "Jobs Picture: Slowing Economy Catches Up to Labor Market." http://www.epinet.org/webfeatures/economicindicators/jobspict.html

Eitzen, Stanley, and Maxine Baca Zinn, eds. 1990. *The Reshaping of America: Social Consequences of the Changing Economy*. Englewood Cliffs, NJ: Prentice-Hall.

"Employment Situation Summary." 2001 (February 2). Bureau of Labor Statistics. http://stats.bls.gov/news.release/empsit.nr0htm

Epstein, Gerald, Julie Graham, and Jessica Nembhard, eds. 1993. "Third World Socialism and the Demise of COMECON." In *Creating a New World Economy: Forces of Change and Plans of Action*, pp. 405–20. Philadelphia: Temple University Press.

Frederick, James and Nancy Lessin. 2000 (November). "Blame the Worker: The Rise of Behavioral-Based Safety Programs." *Multinational Monitor* 21(11). http://www.essential.org/monitor/mm2000/00november/toc.html

Freeman, Richard B. and Joel Rogers. 1999. *What Workers Want*. Ithaca, NY: Cornell University Press.

Galinsky, Ellen, and James T. Bond. 1998. *The 1998 Business Work-Life Study*. New York: Families and Work Institute.

Galinsky, Ellen, James E. Riesbeck, Fran S. Rodgers, and Faith A. Wohl. 1993. "Business Economics and the Work-Family Response." In *Work-Family Needs: Leading Corporations Respond*, pp. 51–54. New York: The Conference Board.

"The Garment Industry." 2001. Sweatshop Watch. http://www.sweatshopwatch.org/swatch/industry/

"Global Labor Repression." 2000 (November). Behind the Lines. *Multinational Monitor* 21(11). http://www.essential.org/monitor/mm2000/00november/toc.html

Global March against Child Labor. 1998. "Global March against Child Labor." children@globalmarch-us.org

Gordon, David M. 1996. *Fat and Mean: The Corporate Squeeze of Working Americans and the Myth of Managerial "Downsizing."* New York: The Free Press.

Granny D Home Page. 2000. http://www.grannyd.com/

Greenhouse, Steven. 2001 (January 21). "Unions Hit Lowest Point in 6 Decades." *The New York Times on the Web*. http://nytimes.com

Greenhouse, Steven. 2000 (January 26). "Anti-Sweatshop Movement Is Achieving Gains Overseas." *The New York Times on the Web*. http://nytimes.com

Hargis, Michael J. 2001 (Jan./Feb.). "Bangladesh: Garment Workers Burned to Death." Industrial

Worker #1630, 98(1). http://parsons..ww.org/~iw/jan2001/stories/intl.html

Harris, Stew. 2001 (Jan./Feb.). "Students Against Sweatshops." *Multinational Monitor* 22 (1 & 2) http://www.essential.org/monitor/mm2001/01jan-feb/

Hewlett, Sylvia Ann, and Cornell West. 1998. *The War against Parents: What We Can Do for America's Beleaguered Moms and Dads*. Boston: Houghton Mifflin Company.

Hochschild, Arlie Russell. 1997. *The Time Bind: When Work Becomes Home and Home Becomes Work*. New York: Henry Holt and Company.

Human Rights Watch. 2001. *World Report 2001*. www.hrw.org/

Human Rights Watch. 2000a. *Fingers to the Bone: United States Failure to Protect Child Farmworkers*. Human Rights Watch. 350 Fifth Ave. 34th Floor. NYC, NY 10118-3299.

Human Rights Watch. 2000b. Unfair Advantage: Workers' Freedom of Association in the United States Under International Human Rights Standards. http://www.hrw.org/reports/2000/uslabor

International Labour Organization. 2001. *World Employment Report 2001: Life at Work in the Information Economy*. Geneva: International Labour Organization.

International Labour Organization. 2000. "Statistics: Revealing a Hidden Tragedy." www.ilo.org/public/...ish/standards/idec/simpoc/stats/4stt.htm

International Labor Rights Fund. 1997. "Congress Acts to Ban Imports Made with Child Labor." www.laborrights.org

Kenworthy, Lane. 1995. *In Search of National Economic Success*. Thousand Oaks, CA: Sage Publications.

Koch, Kathy. 1998. "High-Tech Labor Shortage." *CQ Researcher* 8(16):361–84.

Kruse, Douglas L. and Douglas Mahony. 2000 (October). "Illegal Child Labor in the United States: Prevalence and Characteristics." *Industrial and Labor Relations* 54(1):17.

"Labor's 'Female Friendly' Agenda." 1998. *Labor Relations Bulletin* no. 690, 2.

Lenski, Gerard, and J. Lenski. 1987. *Human Societies: An Introduction to*

Macrosociology, 5th ed. New York: McGraw-Hill.

Leonard, Bill. 1996 (July). "From School to Work: Partnerships Smooth the Transition." *HR Magazine* (Society for Human Resource Management). http://www.shrm.org/hrmag...articles/0796cov.htm

Levitan, Sar A., Garth L. Mangum, and Stephen L. Mangum. 1998. *Programs in Aid of the Poor*, 7th ed. Baltimore: Johns Hopkins University Press.

Lovell, Vicky and Hedieh Rahmanou. 2000 (November). "Paid Family and Medical Leave: Essential Support for Working Women and Men." Institute for Women's Policy Research Publication #A124. http://www.iwpr.org/

Mishel, Lawrence, Jared Bernstein, and John Schmitt. 2001. *The State of Working America, 2000-2001. Economic Policy Institute Series*. Ithaca, NY: Cornell University Press.

Mokhiber, Russell, and Robert Weissman. 2001 (January 4). "The Corporate Conservative Administration." Focus on the Corporation. www.essential.org

Mokhiber, Russell, and Robert Weissman. 1998. "Focus on the Corporation." *Multinational Monitor*. http://www.essential.org/...ocus/focus.9806.html

Multinational Monitor. 2000 (November). "Editorial: What is Society Willing to Spend on Human Beings?" *Multinational Monitor* 21(11). http://www.essential.org/monitor/mm2000/00november/

National Labor Committee. 2001 (January 16). "Nightmare at J.C. Penney Contractor." http://www.nlcnet.org/news/jcpenney_contractor.htm

National Safety Council. 1997. *Accident Facts 1997 Edition*. Itasca, IL: National Safety Council.

"New OSHA Policy Relieves Employees." 1998. *Labor Relations Bulletin* no. 687, 8.

Parker, David L. (with Lee Engfer and Robert Conrow). 1998. *Stolen Dreams: Portraits of Working Children*. Minneapolis: Lerner Publications Company.

PBS. 2001. *Trade Secrets: A Bill Moyers Report*. http://www.pbs.org/tradesecrets/

Report on the World Social Situation. 1997. New York: United Nations.

Ross, Catherine E., and Marylyn P. Wright. 1998. "Women's Work, Men's Work, and the Sense of Control." *Work and Occupations* 25(3):333–55.

Sack, Kevin. 2001 (January 4). "Judge Finds Labor Law Broken at Meat-Packing Plant." *The New York Times on the Web.* http://www.nytimes.com

Silvers, Jonathan. 1996. "Child Labor in Pakistan." *Atlantic Monthly* 277(2):79–92.

Simmons, Wendy W. 2001. "Despite Recent Wave of Corporate Layoffs, Most Americans Not Worried About Losing Their Job." Gallup News Service. http://www.gallup.com/poll/releases/pr010131.asp

"S.O.S. Stress at Work: Costs of Workplace Stress are Rising, with Depression Increasingly Common." 2000 (December). *World of Work* (Magazine of the International Labour Organization), No. 37.

Statistical Abstract of the United States: 2000, 120th ed. U.S. Bureau of the Census. Washington D.C.: U.S. Government Printing Office.

"Stay Put, Judge Tells Co." 2001 (Jan./Feb.). Behind the Lines. *Multinational Monitor* 22(1 & 2). http://www.essential.org/monitor/mm2001/01jan-feb/corp3.html

Thurow, Lester. 1996. *The Future of Capitalism: How Today's Economic Forces Shape Tomorrow's World.* New York: Morrow.

"Underage and Unprotected: Child Labor in Egypt's Cotton Fields." 2001 (January). *Human Rights Watch* 13(1).

UNICEF. 2000. *The Progress of Nations 2000.* New York: United Nations.

UNICEF. 1997. *State of the World's Children 1997.* New York: United Nations.

U.S. Department of Labor. 2001 (Jan. 16). "OSHA Statement: Statement of Assistant Secretary Charles N. Jeffress on Effective Date of OSHA Ergonomic Standard." http://www.osha.gov/media/statements/cjcrgo011601.html

U.S. Department of Labor. 1999. *Futurework: Trends and Challenges for Work in the 21st Century.* http://www.dol.gov/dol/asp/public/futurework/report/

U.S. Department of Labor. 1995. *By the Sweat and Toil of Children. Vol. 2, The Use of Child Labor in U.S. Agricultural Imports and Forced and Bonded Child Labor.* Washington, D.C.: U.S. Department of Labor, Bureau of International Labor Affairs.

Went, Robert. 2000. *Globalization: Neoliberal Challenge, Radical Responses.* Sterling, VA: Pluto Press.

Western, Bruce. 1995. "A Comparative Study of Working Class Disorganization: Union Decline in Eighteen Advanced Capitalist Countries." *American Sociological Review* 60:179–201.

Wright, Carter. 2001 (Jan./Feb.). "A Clean Sweep: Justice for Janitors." *Multinational Monitor* 22(1 & 2). http://www.essential.org/monitor/mm2001/01jan-feb/corp3.html

Chapter 12

ABCNews. 2001. "Could It Happen Here?" ABCNews.com. March 13. http://www.abcnews.go.com/sections/GMA/GoodMo.../GMA_School_violence

Aberle-Grasse, Melissa. 2000. "The Washington Study Service-Learning Year of Eastern Mennonite University: Reflections on 23 Years of Service Learning." *American Behavioral Scientist* 43:848–857.

Armas, Genaro. 2000. "More Americans Making the Grade." Associated Press, September 15. ABCNEWS.com. http://www.abcnews.go.com/sections/us/DailyNews

Ascher, Carol, Norm Fruchter, and Robert Berne. 1997. *Hard Lessons: Public Schools and Privatization.* New York: Twentieth Century Fund.

Associated Press. 1998. "Education Becomes Major Political Issue as Candidates Listen to Voters." *The New York Times,* September 20, A5.

Baker, David P. and Deborah P. Jones. 1993. "Creating Gender Equality: Cross-National Gender Stratification and Mathematical Performance." *Sociology of Education* 66:91–103.

Bankston, Carl and Stephen Caldas. 1997. "The American School Dilemma: Race and Scholastic Performance." *Sociological Quarterly* 38(3):423–29.

Bush, George W. 2001. "No Child Left Behind." http//www.first.gov

Bushweller, Kevin. 1995. "Turning Our Backs on Boys." *Education Digest,* January, 9–12.

CDC (Centers for Disease Control). 2000. "Facts about Violence Among Youth and Violence in School." National Center for Injury Prevention and Control Home Page. http://www.cdc.org

CDF (Children's Defense Fund). 2000. "Key Facts about Education." http://www.childrensdefense.org/keyfacts_education

Call, Kathleen, Lorie Grabowski, Jeylan Mortimer, Katherine Nash, and Chaimun Lee. 1997. "Impoverished Youth and the Attainment Process." Presented at the annual meeting of the American Sociological Association, Toronto, Canada, August.

Cochran-Smith, Marilyn. 2000. "Teacher Education at the Turn of the Century." *Journal of Teacher Education* 51:163–165.

Cohen, Warren. 1998. "Vouchers for Good and Ill." *U.S. News and World Report,* April 27, 46.

Coleman, James S., J. E. Campbell, L. Hobson, J. McPartland, A. Mood, F. Weinfield, and R. York. 1966. *Equality of Educational Opportunity.* Washington, D.C.: U.S. Government Printing Office.

Conciatore, Jacqueline. 2000. "Study Shows That More Than Half of American Colleges Now Have Diversity Requirements." *Black Issues in Higher Education* 17:22.

Congressional Budget Office. 2000. "Options to Expand Federal Health, Retirement, and Educational Activities: Preface." http://www.cbo.gov

Consumer Reports. 2000. "Reading, Writing and . . . Buying." In *Crisis in American Institutions,* eds. Jerome Skolnick and Elliott Currie, pp. 382–386. Boston: Allyn and Bacon.

Digest of Education Statistics. 2001. Chapters 1–6. National Center for Educational Statistics. http://nces.ed.gov/pubs2001/digest

Durham, Gisele. 2000. "U.S. Teens called Liars, Cheats." Associated Press. http://news.excite.com/news/ap/001016/o8/morality

ED Initiatives. 1998. "Helping All Children Reach High Standards."

September 25. Department of Education. http://www.ed.gov/pubs/EDInitiatives/98/98-09-25.html

Elam, Stanley M., Lowell C. Rose, and Alec M. Gallup. 1994. "The 26th Annual Phi Delta Kappa/Gallup Poll of the Public's Attitudes toward the Public Schools." *Phi Delta Kappan*, September, 41–56.

Embry, Dennis, Daniel Flannery, T. Alexander-Vazsonvi, Kenneth Powell, and Henry Atha. 1996. "PeaceBuilders: A Theoretically Driven School Based Model for Early Violence Prevention." *American Journal of Preventive Medicine* 12(5):91–100.

Evans, Lorraine and Kimberly Davies. 2000. "No Sissy Boys Here." *Sex Roles* (Feb): 255–271.

Fact Sheet. 2000. "2000 Head Start Fact Sheet." Research and Statistics. Department of Health and Human Services. Washington, D.C.

Finsterbusch, Kurt. 1999. *Taking Sides: Clashing Views on Controversial Social Issues*. Guilford, CT: Dushkin/McGraw-Hill.

Fletcher, Robert S. 1943. *History of Oberlin College to the Civil War*. Oberlin, OH: Oberlin College Press.

Flexner, Eleanor. 1972. *Century of Struggle: The Women's Rights Movement in the United States*. New York: Atheneum.

Gallup Organization. 2001. "Education Reform: The Public's Opinion." http://www.gallup.com/poll/releases

Goldberg, Carey. 1999. "After Girls Get Attention, Focus is on Boys' Woes." In *Themes of the Times: New York Times*. Upper Saddle River, NJ: Prentice-Hall, p. 6.

Harris Poll. 1998. "Question 415." Study No. 5818418: April. Chapel Hill, N.C.: Institute for Research in Social Sciences.

Jencks, Christopher and Meredith Phillips. 1998. "America's Next Achievement Test: Closing the Black-White Test Score Gap." *The American Prospect*, September/October, 44–53.

Kanter, Rosabeth Moss. 1972. "The Organization Child: Experience Management in a Nursery School." *Sociology of Education* 45:186–211.

Kozol, Jonathan. 1991. *Savage Inequalities: Children in America's Schools*. New York: Crown Publishers.

Lareau, Annette. 1989. *Home Advantage: Social Class and Parental Intervention in Elementary Education*. Philadelphia: Falmer Press.

Legislative Action. 1998. "Public Law 105-277." http://www.house.gov/eeo/testindex.htm

Leo, John. 1998. "Dumbing Down Teachers." *U.S. News and World Report*, August 3, 15.

Levinson, Arlene. 2000. "Study Evaluates Higher Education." Excite.News. http://www.excite.com/news/ap/00130/08/grading

Literacy. 2000. "Basic Literacy." Literacy Volunteers of America. http://www.literacyvolunteers.org/about/basic

Merton, Robert K. 1968. *Social Theory and Social Structure*. New York: Free Press.

Mollison, Andrew. 2001. "Boom in Charter Schools Continues Despite Mixed Results." *The Atlanta Journal Constitution*. September 24. http://www.accessatlanta.com/partners/ajc/epaper

Muller, Chandra and Katherine Schiller. 2000. "Leveling the Playing Field?" *Sociology of Education* 73:196–218.

Murnane, Richard J. 1994. "Education and the Well-Being of the Next Generation." In *Confronting Poverty: Prescriptions for Change*, eds. Sheldon H. Danziger, Gary D. Sandefur, and Daniel H. Weinberg, pp. 289–307. New York: Russell Sage Foundation.

NAAL (National Assessment of Adult Literacy). 2000. "FAQ." http://nces.ed.gov/naal/faq

Natriello, Gary. 1995. "Dropouts: Definitions, Causes, Consequences, and Remedies." In *Transforming Schools*, eds. Peter W. Cookson Jr. and Barbara Schneider, pp. 107–28. New York: Garland Publishing Co.

NCES (National Center for Education Statistics). 2000a. "Family Reading." http://www.nces.ed.gov/fastfacts

———. 2000b. "The Release of National Assessment of Education Progress (NAEP) 1999 Trends in Academic Progress." http://www.nces.ed.gov/commissioner/remarks/2000

———. 2000c. "Teacher Trends." http://www.nces.ed.gov/fastfacts

(The) New York Times. 2000. "Charter School Accountability." December 28, A22.

Noddings, Nel. 1995. "A Morally Defensible Mission for Schools in the 21st Century." *Phi Delta Kappan*, January, 365–68.

Orfield, G., and S. E. Eaton. 1997. *Dismantling Desegregation: The Quiet Reversal of Brown v. Bd. of Education*. New York: New Press.

Pinkerton, Jim. 1998. "Here's How to Pass While Schools Fail." *USA Today*, April 23, A11.

Pollock, William. 2000. *Real Boys' Voices*. New York: Random House.

Population Reference Bureau. 1999. "World Population: More Than Just Numbers." http://www.prb.org

Rand Institute. 1999. "Ethnic Gaps in Higher Education are Growing." News Release. http://www.rand.org/hot/Press/educationgap

Ray, Carol A. and Roslyn A. Mickelson. 1993. "Restructuring Students for Restructured Work: The Economy, School Reform, and Non–College-Bound Youths." *Sociology of Education* 66:1–20.

Rodriquez, Richard. 1990. "Searching for Roots in a Changing World." In *Social Problems Today*, ed. James M. Henslin, pp. 202–13. Englewood Cliffs, NJ: Prentice-Hall.

Roscigno, Vincent. 1998. "Race and the Reproduction of Educational Disadvantage." *Social Forces* 76(3):1033–60.

Rosenthal, Robert and Lenore Jacobson. 1968. *Pygmalion in the Classroom: Teacher Expectations and Pupils' Intellectual Development*. New York: Holt, Rinehart & Winston.

Rumberger, Russell W. 1987. "High School Dropouts: A Review of Issues and Evidence." *Review of Educational Research* 57:101–21.

Saporito, Salvatore and Annette Lareau. 1999. "School Selection as a Process: The Multiple Dimensions of Race in Framing Educational Choice." *Social Problems* 46:418–439.

Shanker, Albert. 1996. "Mythical Choice and Real Standards." In *Reducing Poverty in America*, ed. Michael Darby, pp. 154–72. Thousand Oaks, CA: Sage.

Silber, John. 1998. "The Correct Answer: Too Many Can't Teach." *The News & Observer*, July 8, A2.

Sommers, Christina. 2000. *The War Against Boys*. New York: Simon and Schuster.

State of our Nation's Youth. 2001. Alexandria, VA: Horatio Alger Association of Distinguished Americans. http://www.horatioalger.com

Statistical Abstract of the United States: 1999, 119th ed. U.S. Bureau of the Census. Washington, D.C.: U.S. Government Printing Office.

Stellin, Susan. 2001. "Shift for Education Technology." *The New York Times on the Web*, January 10. http://nytimes.com/2001/01/10/technology

Summary Report. 2001. "Building Their Futures." http://www2.acf.dhhs.gov/programs/hsb/EHS

Teachman, Jay D., Kathleen Paasch, and Karen Carver. 1997. "Social Capital and the Generation of Human Capital." *Social Forces* 75(4):1343–59.

Toch, Thomas. 1998. "Education Bazaar." *U.S. News and World Report*, April 27, 35–46.

United Nations Population Fund. 1999. "Campaign Issues: Facing the Facts." Face to Face. http://www.facecampaign.org

U.S. Census. 2000a. "Educational Attainment in the U.S." *Current Population Reports. March*. Washington, D.C.: U.S. Department of Commerce, Economics and Statistics Administration.

———. 2000b. "Women in the U.S.: A Profile." *Current Population Reports*. March. Washington, D.C.: U.S. Department of Commerce, Economics and Statistics Administration.

U.S. Department of Education. 2000. "The Baby Boom Echo: No End in Sight: Figure 2." http://www.ed.gov/pubs/bbecho00/figure2

U.S. Department of Justice. 2000. "Crimes in the Nation's Schools Declined in the 1990s According to Departments of Justice and Education." http://www.ojp.usdoj.gov/bjs/pub/press

U.S. Newswire. 2000a. "Fact Sheet" Building a Stronger Global Partnership." http://www.usnewswire.com/topnews/current_releases

———. 2000b. "NSBA Report Says Charter Schools Not Meeting Promises." http://www.usnewswire.com/topnews/current_releases

Walsh, Kenneth T. and Ben Wildavsky. 2001. "Bush Bounds Out of Starting Gate..." *U.S. News and World Report*, February 5, 17.

Washington Post. 2001. "School Vouchers." *Political News*. http://www.washingtonpost.com

Waterman, Alan. S., ed. 1997. *Service-Learning: Applications from the Research*. Mahwah, NJ: Lawrence Erlbaum Associates Publishers.

Webb, Julie. 1989. "The Outcomes of Home-Based Education: Employment and Other Issues." *Educational Review* 41:121–33.

Wilgoren, Jodi. 2001. "Calls for Change in the Scheduling of the School Day." *The New York Times on the Web*. January 10. http://www.nytimes.com/2001/01/10/nyregion/10scho.html

Weiner, Rebecca. 2000. "Industry Group's Education Study Draws Conclusions and Critics." *The New York Times on the Web*. http://www.nytimes.com/library/tect/00/08/cyber/education

Winters, Rebecca. "From Home to Harvard." *Time*. September 11. http://www.time.com/time/magazine/articles

Youth Indicators. 1996. U.S. Department of Education. National Center for Education Statistics: International Assessment of Educational Progress. Indicator 38. http://nces.ed.gov/pubs/yi/y9638a.html

Zernike, Kate. 2001. "Gap Between Best and Worst Widens on U.S. Reading Test." *The New York Times on the Web*. April 7. http://www.nytimes.com/2001/04/07/national

Zlotkowski, E. 1996. "Opportunity for all: linking service-learning and business education." *Journal of Business Ethics* 1:5–19.

Chapter 13

American Public Transportation Association. 2001 (January 31). "1995–2000 Ridership on the Nation's Public Transportation Systems Has Grown By Over 20%." APTANet. http://www.apta.com/

Anderson, Elijah. 1992. *Streetwise: Race, Class, and Change in an Urban Community*. Chicago: University of Chicago Press.

Archer, Dennis W. 1998. "The Lesson of Detroit: Never Underestimate a City." *Vital Speeches* 64(11):340–43.

Associated Press. 2000 (June 13). "Wolfpack Attacks." ABCNews.com. http://www.abcnews.go.com/sections/us/DailyNews/centralpark000613.html

"The Bridge to the 21st Century Leads to Gridlock in and around Decaying Cities." 1997. *The Washington Spectator* 23(12). The Public Concern Foundation, Inc.

Centers for Disease Control and Prevention. 2000. *HIV/AIDS Surveillance Report*. Vol. 12(1).

Claffey, Mike. 2000 (July 25). "Suspects' Own Words Detail Rampage in Park." *Daily News Online*. http://www.nydailynews.com/2000-07-25/News_and_Views/Crime_File/a-74395.asp

Clark, David. 1998. "Interdependent Urbanization in an Urban World: An Historical Overview." *The Geographical Journal* 164(1):85–96.

CNN. 2000 (June 14). "More Women, Girls Tell of Attacks by Mob of Men in New York." CNN.com. http://www.cnn.com/2000/US/06/14/central.park.assault.01/

Cohen, Natalie. 2000. "Business Location Decision-Making and the Cities: Bringing Companies Back." Center on Urban and Metropolitan Policy, The Brookings Institution. Washington, D.C.

Coles, H. Brent. 2001. "Priorities for 'the New American City.'" Washington, D.C.: United States Conference of Mayors.

Cowherd, Phil. 2001. "What is the Business Case for Investing in Inner-City Neighborhoods?" *Public Management* 83(1): 12–14.

"Crisis in Low-Income Rental Housing." 1998. *America* 179(1):3.

Dalaker, Joseph and Bernadette D. Proctor. 2000. U.S. Census Bureau, Current Population Reports, Series P60-210, *Poverty in the United States: 1999*. Washington D.C.: U.S. Government Printing Office.

Dunlap, Riley E. and Lydia Saad. 2001 (April 16). "Only One in Four Americans Are Anxious About the Environment." Gallup News Service. http://www.gallup.com/poll/releases/pr010416.asp

Durning, Alan. 1996. *The City and the Car*. Northwest Environment Watch. Seattle: Sasquatch Books.

Federal Bureau of Investigation. 2000. *Uniform Crime Reports: Crime in the United States—1999*. Washington, D.C.: U.S. Government Printing Office.

Fisher, Claude. 1982. *To Dwell among Friends: Personal Networks in Town and City*. Chicago: University of Chicago Press.

Froehlich, Maryann. 1998. "Smart Growth: Why Local Governments Are Taking a New Approach to Managing Growth in Their Communities." *Public Management* 80(5):5–9.

Gans, Herbert. [1962] 1984. *The Urban Villagers*, 2nd ed. New York: Free Press (first edition published in 1962).

Geddes, Robert. 1997. "Metropolis Unbound: The Sprawling American City and the Search for Alternatives." *The American Prospect* 35 (November–December):40–46.

Greenberg, Michael, K. Tyler Miller, Karen Lowrie, and Henry Mayer. 2001. "Surveying the Land: Brownfields in Medium-Sized and Small Communities." *Public Management* 83(1): 18–23.

Heubusch, Kevin. 1998 (January). "Small Is Beautiful." *American Demographics*. http://www.demographics.com/publications/AD/98_ad/9801_ad/ad980130.htm

Jargowsky, Paul A. 1997. *Poverty and Place: Ghettos, Barrios, and the American City*. New York: Russell Sage Foundation.

Jensen, Derrick. 2001 (February). "Road to Ruin: An Interview with Jan Lundberg." *The Sun* 302: 4–13.

Johnson, William C. 1997. *Urban Planning and Politics*. Chicago: American Planning Association, Planners Press.

Kelbaugh, Douglas. 1997. *Common Place: Toward Neighborhood and Regional Design*. Seattle: University of Washington Press.

National Association of Home Builders. 1999. "Smart Growth: Building Better Places to Live, Work and Play." Washington, D.C.: National Association of Home Builders. http://www.smartgrowth.org/pdf/smart/pdf

National League of Cities. 1998 (January 21). "Cities Are Increasing Service Levels and Range of Services as Problem Solvers for Citizens; Most Local Leaders Optimistic about Conditions and Outlook; Concerned about Mandates; Annual NLC Survey Finds Many Cities Successfully Involved with Information Technology." http://www.nlc.org/pres-opn.htm

———. 2000a (February 28). "NCL Insta-Poll Finds Many Cities Establishing Youth Curfews: Viewed as 'Useful Tool.'" http://www.nlc.org/curfew

———. 2000b (July 25). "Fiscal Conditions Remain Strong in Most Cities But Show Signs of Leveling Off." http://www.nlc.org/2000fiscal.htm

Nelessen, Anton C. 1997. "The Computer Commuter: Neighborhood Transit for the 21st Century." Urban Design, Tele-communication and Travel Forecasting Conference. http://www.bts.gov/tmip/p...ip/udes/nelessen.htm

Newman, Katherine S. 1999. *No Shame in My Game: The Working Poor in the Inner City*. New York: Alfred A. Knopf, Inc., and the Russell Sage Foundation.

Orfield, Myron. 1997. *Metropolitics: A Regional Agenda for Community and Stability*. Washington, D.C.: Brookings Institution Press and Cambridge, MA: The Lincoln Institute of Land Policy.

Pelley, Janet. 1999 (January 1). "Building Smart-Growth Communities." *Environmental Science &Technology News* 33(1):28A–32A.

Perry, Marc. J. and Paul J. Mackun. 2001. "Population Change and Distribution: 1990 to 2000." U.S. Census Bureau. Washington, D.C.: U.S. Government Printing Office.

Rodriguez, Luis. 2000 (April). "Urban Renewal: The Resurrection of An Ex-Gang Member" (in an interview with Derrick Jensen). *The Sun*, pp. 4–13.

Schmitt, Eric. 2001 (April 3). "Analysis of Census Finds Segregation Along with Diversity." *The New York Times on the Web*. http://www/nytimes.com/2001/04/04/national/04CENS.html

Schmoke, Kurt L. 1998. "Ingredients for a Successful City: Variety Is the Spice of Urban Life." *Vital Speeches of the Day* 65(4):110–17.

Sheehan, Molly O'Meara. 2001a. "Reinventing Cities for People and the Planet." In *State of the World 2001*, ed. Linda Starke. New York: W.W. Norton & Company.

Sheehan, Molly O'Meara. 2001b. "Making Better Transportation Choices." In *State of the World 2001*, ed. Linda Starke, pp. 103–122. New York: W.W. Norton & Company.

Sheehan, Molly O'Meara. 2000. "Urban Population Continues to Rise." In *Vital Signs: The Environmental Trends That Are Shaping Our Future*, ed. Linda Starke, pp. 104–105. New York: W.W. Norton & Co.

Shevis, Jim. 1999 (October 15). "More Affluent Than Their Inner-City Neighbors, Suburbanites Still Have Growth Problems." *The Washington Spectator* 25(19):1–3).

Smart Growth Network. 1999 (January 14). "About the Smart Growth Network: Mission Statement and Principles." http://www.smartgrowth.org/

Statistical Abstract of the United States: 1993, 113th ed., U.S. Bureau of the Census. Washington, D.C.: U.S. Government Printing Office.

Statistical Abstract of the United States: 1994, 114th ed., U.S. Bureau of the Census. Washington, D.C.: U.S. Government Printing Office.

Statistical Abstract of the United States: 2000, 120th ed., U.S. Bureau of the Census. Washington, D.C.: U.S. Government Printing Office.

"Technology Smooths the Ride for Santa Ana Commuters." 1998. *American City & County* 113(1):17–18.

Tittle, Charles. 1989. "Influences on Urbanism: A Test of Predictions from Three Perspectives." *Social Problems* 36(3):270–88.

Union of Concerned Scientists. 1999. "The Hidden Costs of Transportation." http://www.ucsusa.org/transportation/hidden.html

United Nations. 1994. "Programme of Action." The United Nations International Conference on Population and Development (ICPD). Cairo, Egypt. September 5–13, 1994.

United Nations Population Fund. 1999. *The State of World Population 1999*. New York: United Nations.

_____. 1996. *The State of the World Population Report 1996*. New York: United Nations.

United States Conference of Mayors. 2001a. *Traffic Congestion and Rail Investment*. Washington, D.C.: Global Strategy Group.

_____. 2001b (January 17). "300 Mayors, in Nation's Capital Release Groundbreaking Nationwide Poll Showing Strong Public Support for Passenger Rail Investment." Washington, D.C.: United States Conference of Mayors.

_____. 2000a (December). *A Status Report on Hunger and Homelessness in America's Cities*. Washington, D.C.: The United States Conference of Mayors.

_____. 2000b. *Recycling America's Land: A National Report on Brownfields Redevelopment*. Washington, D.C.: The United States Conference of Mayors.

_____. 2000c. *Examining Skills Shortages in America's Cities: Impact, City Responses, and Business Perspectives*. Washington, D.C.: The United States Conference of Mayors.

"Urbanization Trend Seen as Accelerating." 2001 (March–April). *Popline* 23:3.

U.S. Census Bureau. 2001. "Ranking Tables for Metropolitan Areas: 1990 and 2000." http://www.census.gov/population/www/cen2000/phc.t3.html

U.S. Department of Housing and Urban Development. 2001 (January 10). "Congress Approves 9 New EZs, 40 Renewal Communities, and More Money for Round II EZs." *EZ/EC News Flash*. http://www.hud.gov/

_____. 2000. *The State of the Cities 2000*. Washington, D.C.: U.S. Government Printing Office.

_____. 1999. *Now is the Time: Places Left Behind in the New Economy*. Washington, D.C.: U.S. Government Printing Office.

Warren, Roxanne. 1998. *The Urban Oasis: Guideways and Greenways in the Human Environment*. New York: McGraw-Hill.

Wilson, William Julius. 1996. *When Work Disappears: The World of the New Urban Poor*. New York: Alfred A. Knopf.

Wirth, Louis. 1938. "Urbanism as a Way of Life." *American Journal of Sociology* 44:8–20.

Wolff, Kurt H. 1978. *The Sociology of George Simmel*. Toronto: Free Press.

World Health Organization and United Nations Joint Programme on HIV/AIDS. 1998. "Report on the Global HIV/AIDS Epidemic— June 1998." http://www.who.int/emc-hiv/global_report/data/globrep-e.pdf

Chapter 14

Abramovitz, Janet N. 2000. "Paper Recycling Remains Strong." In *Vital Signs 2000*, eds. Lester R. Brown, Michael Renner, and Brian Halweil, pp.132–133. New York: W.W. Norton & Co.

Agbese, Pita Ogaba. 1995. "Nigeria's Environment: Crises, Consequences, and Responses." In *Environmental Policies in the Third World: A Comparative Analysis*, eds. O. P. Dwivedi and Dhirendra K. Vajpeyi, pp. 125–44. Westport, CT: Greenwood Press.

Alternative Energy Institute. 2001 (May 1). "Bush/Cheney Miss the Boat on Energy Crisis." *Alternative Energy Institute Bulletin* #23. www.altenergy.org/news/news.html

American Council on Education and University of California. 2000. "The American Freshman: National Norms for Fall 2000." Los Angeles Higher Education Research Institute: American Council on Education and University of California.

Associated Press. 2000 (January 7). "DaimlerChrysler Corporation Quits Global Climate Coalition." *Environmental News Network*. http://www.enn.com/news/

Bergman, Lester V. 1998 (December). "Cataract Development: It's Cumulative." *Environmental Health Perspectives* 106(12). http://ehpnet1.niehs.nih.gov/docs/1998/06-12/forum.html

Black, Harvey Karl. 1999 (February). "Complex Cleanup." *Environmental Health Perspectives* 107(2). http://ehpnet1.niehs.nih.gov/docs/1999/07-2/focus-abs.html

Boland, Reed, Sudhakar Rao, and George Zeidenstein. 1994. "Honoring Human Rights in Population Policies: From Declaration to Ac-

tion." In *Population Policies Reconsidered: Health, Empowerment, and Rights*, eds. Gita Sen, Adrienne Germain, and Lincoln C. Chen, pp. 89–105. Boston: Harvard School of Public Health.

Bongaarts, John, and Susan Cotts Watkins. 1996. "Social Interactions and Contemporary Fertility Transitions." *Population and Development Review* 22(4):639–82.

Brown, Lester R. 1995. "The State of the World's Natural Environment." In *Seeing Ourselves: Classic, Contemporary, and Cross-Cultural Readings in Sociology*, 3rd ed., eds. John J. Macionis and Nijole V. Benokraitis, pp. 411–16. Englewood Cliffs, NJ: Prentice-Hall.

_____. 1998a. "The Future of Growth." In *State of the World 1998*, eds. Lester R. Brown, Christopher Flavin, and Hilary French, pp. 3–20. New York: W.W. Norton & Co.

_____. 1998b. "Overview: New Records, New Stresses." In *Vital Signs 1998*, eds. Lester R. Brown, Michael Renner, and Christopher Flavin, pp. 15–24. New York: W.W. Norton & Co.

Brown, Lester R., and Jennifer Mitchell. 1998. "Building a New Economy." In *State of the World 1998*, eds. Lester R. Brown, Christopher Flavin, and Hilary French, pp. 168–187. New York: W.W. Norton & Co.

Brown, Lester R., Christopher Flavin, and Hilary French. 1998. Foreword. In *State of the World 1998*, eds. Lester R. Brown, Christopher Flavin, and Hilary French, pp. xvii–xix. New York: W.W. Norton & Co.

Bruce, Nigel, Rogelio Perez-Padilla, and Rachel Albalak. 2000. "Indoor Air Pollution in Developing Countries: A Major Environmental and Public Health Challenge." *Bulletin of the World Health Organization*, 78(9):1078–1092.

Bullard, Robert D. 2000. *Dumping in Dixie: Race, Class, and Environmental Quality*, 3rd ed. Boulder CO: Westview Press.

Bullard, Robert D., and Glenn S. Johnson. 1997. "Just Transportation." In *Just Transportation: Dismantling Race and Class Barriers to Mobility*, eds. Robert D. Bullard and

Glenn S. Johnson, pp. 1–21. Stony Creek, CT: New Society Publishers.

Buss, Dale. 2001. "Green Cars." *American Demographics* (January).

Carlson, Darren K. 2001 (January 23). "Scientists Deliver Serious Warning About Effects of Global Warming." *Gallup News Service*, The Gallup Organization. http://www.gallup.com/poll/releases/pr010123b.asp

Catley-Carlson, Margaret, and Judith A. M. Outlaw. 1998. "Poverty and Population Issues: Clarifying the Connections." *Journal of International Affairs* 52(1):233–43.

"China Phasing Out One-Child Policy." 2000 (May-June). *Popline* Vol. 22: 1.

Cincotta, Richard P. and Robert Engelman. 2000. *Human Population and the Future of Biological Diversity.* Washington D.C.: Population Action International.

Cooper, Mary H. 1998. "Population and the Environment." *The CQ Researcher* 8(26):601–24.

"Corporate Spotlight." 2001 (March/April). *Adbusters* 34: 38.

Cray, Charlie. 2001. "Taking on Toxics I: Stopping POPs." *Multinational Monitor* 22(1 & 2).

DaVanzo, Julie, David M. Adamson, Nancy Belden, and Sally Patterson. 2000. *How Americans View World Population Issues: A Survey of Public Opinion.* Santa Monica, CA: Rand Corporation.

DeCourcy, Michael/Public Agenda. 2001. "Earth in Balance." *American Demographics* (January).

"The Delicate Balance." 1994. *The National Center for Environmental Health Strategies*, 5(3–4). 1100 Rural Avenue, Voorhees, NJ 08043.

Denson, Bryan. 2000. "Shadowy Saboteurs." *The IRE Journal* (Investigative Reporters and Editors, Inc.) 23 (May/June): 12–14.

DesJardins, Andrea. 1997. "Sweet Poison: What Your Nose Can't Tell You about the Dangers of Perfume." http://members.aol.com/enviroknow/perfume/sweet–poison.htm

Dionis, Joanna. 1999. "Handle with Care." *Mother Jones*, January–February, 25.

Edwards, Bob, and Anthony Ladd. 1998. "Where the Hogs Are '97: Environmental Justice and Farm Loss in North Carolina, 1980–1997." Paper presented at the 2nd National Black Land Loss Summit in Tillery, NC, February 1998.

———. 2000. "Environmental Justice, Swine Production and Farm Loss in North Carolina." *Sociological Spectrum* 20(3):263–290.

Engelman, Robert, Richard P. Cincotta, Bonnie Dye, Tom Gardner-Outlaw, and Jennifer Wisnewski. 2000. *People in the Balance: Population and Natural Resources at the Turn of the Millennium.* Washington D.C.: Population Action International.

"EPA Cracks Down on Diesel Trucks." 2000 (September). *Environmental Defense* XXXI(3): 2.

Fisher, Brandy E. 1998 (December). "Scents and Sensitivity." *Environmental Health Perspectives* 106(12). http://ehpnet1.niehs.nih.gov/docs/1998/106-12/focus-abs.html

———. 1999 (January). "Focus: Most Unwanted." *Environmental Health Perspectives* 107(1). http://ehpnet1.niehs.nih.gov/docs/1999/107-1/focus-abs.html

Flavin, Christopher. 2000. "Wind Power Booms." In *Vital Signs 2000*, eds. Lester R. Brown, Michael Renner, and Brian Halweil, pp. 56–57. New York: W.W. Norton & Co.

French, Hilary. 2000. "Environmental Treaties Gain Ground." In *Vital Signs 2000*, eds. Lester R. Brown, Michael Renner, and Brian Halweil, pp.134–5. New York: W.W. Norton & Co.

Fridleifsson, Ingvar B. 2000. "Globeglance: Energy 2000." *United Nations Chronicle Online Edition* XXXVII(2):1–4.

Gallup, Alec, and Lydia Saad. 1997. "Public Concerned, Not Alarmed about Global Warming." *Gallup Poll Archives*. The Gallup Organization. http://198.175.140.8/poll%5Farchives/1997/971202.htm

Gardner, Gary. 1998. "Sanitation Access Lacking." In *Vital Signs 1998*, eds. Lester R. Brown, Michael Renner, and Christopher Flavin, pp. 70–1. New York: W.W. Norton & Co.

"Global Corporations Join Us to Reduce Greenhouse Gas Emissions." 2000 (November). *Environmental Defense* XXXI(4):5.

Greenhouse, Linda. 2001 (February 28). "E.P.A.'s Authority on Air Rules Wins Supreme Court's Backing." *The New York Times on the Web.* http://www.nytimes.com/2001/02/28/national/

Hager, Nicky and Bob Burton. 2000. *Secrets and Lies: The Anatomy of an Anti-Environmental PR Campaign.* Monroe, ME: Common Courage Press.

Halweil, Brian. 2000. "World Population Passes 6 Billion." In *Vital Signs 2000*, eds. Lester R. Brown, Michael Renner, and Brian Halweil, pp. 98–99. New York: W.W. Norton & Co.

Holcombe, Randall G. 1995. *Public Policy and the Quality of Life: Market Incentives versus Government Planning.* Westport, CT: Greenwood Press.

Hunter, Lori M. 2001. *The Environmental Implications of Population Dynamics.* Santa Monica, CA: Rand Corporation.

Intergovernmental Panel on Climate Change. 2001a. *Climate Change 2001: The Scientific Basis.* United Nations Environmental Programme and the World Meteorological Organization. www.ipcc.ch

———. 2001b. *Climate Change 2001: Impacts, Adaptation, and Vulnerability.* United Nations Environmental Programme and the World Meteorological Organization. www.ipcc.ch

———. 2000. *Land Use, Land-Use Change, and Forestry.* www.ipcc.ch

Jan, George P. 1995. "Environmental Protection in China." *Environmental Policies in the Third World: A Comparative Analysis*, eds. O. P. Dwivedi and Dhirendra K. Vajpeyi, pp. 71–84. Westport, CT: Greenwood Press.

"Japanese Toy Firm Offers Employees Fertility Incentives." 2000 (May-June). *Popline* 22: 2.

Jensen, Derrick. 1999. "The War on Truth: The Secret Battle for the American Mind: An Interview with John Stauber." *The Sun* 279(March):6–15.

Jensen, Derrick. 2001. "A Weakened World Cannot Forgive Us: An Interview with Kathleen Dean Moore." *The Sun* 303(March):13.

Kaplan, Sheila and Jim Morris. 2000. "Kids at Risk." *U.S. News & World Report*, June 19:47–53.

Karliner, Joshua. 1997. *The Corporate Planet: Ecology and Politics in the Age of Globalization.* Boulder, CO: Sierra Club Books.

_____. 1998. "Corporate Greenwashing." *Green Guide* 58 (August):1–3.

_____. 2001 (March 29). "The Global Warming President." *CorpWatch.* http:www.corpwatch.org/climate/updates/2001/jkarliner1.html

Kemps, Dominic. 1998. "Deaths, Diseases Traced to Environment." *Popline* 20 (May–June): 3.

Koenig, Dieter. 1995. "Sustainable Development: Linking Global Environmental Change to Technology Cooperation." *Environmental Policies in the Third World: A Comparative Analysis,* eds. O. P. Dwivedi and Dhirendra K. Vajpeyi, pp. 1–21. Westport, CT: Greenwood Press.

Lenssen, Nicholas. 2000. "Nuclear Power Rises Slightly." In *Vital Signs 2000,* eds. Lester R. Brown, Michael Renner, and Brian Halweil, pp.54–55. New York: W.W. Norton & Co.

Leutwyler, Kristin. 2001. "The Poor Face More Environmental Hazards." *Scientific American* (January 8). wysiwyg://110/http://www.sciam.com/news/010801/3.html

Lindauer, Wendy. 1999 (September). "Fact Sheet: Sick Building Syndrome." Environmental Health Center. http://www.nsc.org/ehc/indoor/sbs.htm

Lipke, David J. 2001 (January). "Good for Whom?" *American Demographics.*

Livernash, Robert, and Eric Rodenburg. 1998. "Population Change, Resources, and the Environment." *Population Bulletin* 53(1):1–36.

Mason, Karen Oppenheim. 1997. "Explaining Fertility Transition." *Demography* 34(4):443–54.

McGinn, Anne Platt. 2000. "Endocrine Disrupters Raise Concern." In *Vital Signs 2000,* eds. Lester R. Brown, Michael Renner, and Brian Halweil, pp. 130–131. New York: W.W. Norton & Co.

McMichael, Anthony J., Kirk R. Smith, and Carlos F. Corvalan. 2000. "The Sustainability Transition: A New Challenge." *Bulletin of the World Health Organization* 78(9):1067.

Mead, Leila. 1998. "Radioactive Wastelands." *The Green Guide* 53(April 14):1–3.

Miller, Norman. 2000 (May). "Rains of Terror." *Geographical* 75(5):90.

Mitchell, Jennifer D. 1998. "Before the Next Doubling." *World Watch* 11(1):21–29.

Mokhiber, Russell and Robert Weissman. 2000 (October 26). "National Breast Cancer Industry Month." *Focus on the Corporation.* http://www.corporatepredators.org

National Assessment Synthesis Team. 2000. *Climate Change Impacts on the United States: The Potential Consequences of Climate Variability and Change.* Washington D.C.: U.S. Global Change Research Program.

National Environmental Education and Training Foundation and Roper Starch Worldwide. 1999. *1999 NEETF/Roper Report Card.* Washington, D.C.: National Environmental Education and Training Foundation.

"New Rules for Feedlots." 1998. *Environmental Health Perspectives* 106(12). http://ehpnet1.niehs.nih.gov/docs/1998/106-12/forum.html

"News & Views." 2001 (February 28). "Powell, Whitman Join Opposition to Bush Global Gag Rule." Population Action International. http://www.populationaction.org/news/eboard_022801.htm

"Nukes Rebuked." 2000 (July 1). *The Washington Spectator* 26(13):4.

"Online Activism Lives Up to Its Promise." 2001 (March). *Environmental Defense* XXXII(2):8.

PBS. 2001. "Trade Secrets: A Moyers Report." www.pbs.org/tradesecrets/program/program.html

Perry, Marc J. and Paul J. Mackun. 2001 (April). "Population Change and Distribution." U.S. Census Bureau. Washington, D.C.: U.S. Government Printing Office.

Pimentel, David, and Anthony Greiner. 1997. "Environmental and Socio-Economic Costs of Pesticide Use." In *Techniques for Reducing Pesticide Use,* ed. D. Pimentel, pp. 50–78. New York: John Wiley & Sons.

Pimentel, David, Maria Tort, Linda D'Anna, Anne Krawic, Joshua Berger, Jessica Rossman, Fridah Mugo, Nancy Doon, Michael Shriberg, Erica Howard, Susan Lee, and Jonathan Talbot. 1998. "Ecology of Increasing Disease: Population Growth and Environmental

Degradation." *BioScience* 48(October):817–27.

Population Institute. 2000. *World Population Awareness Week 2000: Saving Women's Lives—A Guide to Action.* Washington, D.C.: The Population Institute.

_____. 1999. *World Population Awareness Week: The Year of 6 Billion.* Washington, D.C.: The Population Institute.

_____. 1998. "1998 World Population Overview and Outlook 1999." http://www.populationinstitute.org/overview98.html

Population Reference Bureau. 2000a (June). "2000 World Population Data Sheet." http://www.prb.org/prb/pubs/wpds2000/

_____. 2000b. "How Does Family Planning Influence Women's Lives?" *MEASURE Communication: Reports: Women 2000 Policy Briefs.* Washington, D.C.: Population Reference Bureau.

_____. 2000c. "Is Education the Best Contraceptive?" *MEASURE Communication: Reports: Women 2000 Policy Briefs.* Washington, D.C.: Population Reference Bureau.

"A Prescription for Reducing the Damage Caused by Dams." 2001 (March). *Environmental Defense* XXXII(2).

"President Bush's Tax and Budget Bumbles Label Him Commander in Cheap." 2001 (May 1). *The Washington Spectator* 27(9):1–3.

Reese, April. 2001 (February). "Africa's Struggle with Desertification." Population Reference Bureau. http://www.prb.org/regions/africa/africa_desertification.html

Renner, Michael. 2000. "Vehicle Production Increases." In *Vital Signs 2000,* eds. Lester R. Brown, Michael Renner, and Brian Halweil, pp. 86–87. New York: W.W. Norton & Co.

_____. 1996. *Fighting for Survival: Environmental Decline, Social Conflict, and the New Age of Insecurity.* New York: W.W. Norton & Co.

Roodman, David Malin. 2000. "Environmental Tax Shifts Multiplying." In *Vital Signs 2000,* eds. Lester R. Brown, Michael Renner, and Brian Halweil, pp. 138–9. New York: W.W. Norton & Co.

"San Francisco Bans Pesticides." 1997. *Green Guide* 35(February 7):1.

Scholand, Michael 2000. "Compact Fluorescents Light Up the Globe." In *Vital Signs 2000*, eds. Lester R. Brown, Michael Renner, and Brian Halweil, pp. 60–61. New York: W.W. Norton & Co.

"Senators Seek Gag Rule Overturn." 2001 (March–April). *Popline* 23:1–2.

Smith, Velma, John Coequyt, and Richard Wiles. 2000. *Clean Water Report Card*. Washington, D.C.: Environmental Working Group.

Statistical Abstract of the United States:1995, 115th ed. U.S. Bureau of the Census. Washington, D.C.: U.S. Government Printing Office.

Statistical Abstract of the United States: 1998, 118th ed. U.S. Bureau of the Census. Washington, D.C.: U.S Government Printing Office.

Statistical Abstract of the United States: 2000, 120th ed. U.S. Bureau of the Census. Washington, D.C.: U.S. Government Printing Office.

Stephens, Sharon. 1998. "Reflections on Environmental Justice: Children as Victims and Actors." In *Environmental Victims*, ed. Christopher Williams, pp. 48–71. London: Earthscan Publications.

Stiefel, Chana. 2000. (November 27). "Gaping Ozone Hole." *Science World* 57(6):5.

Stover, Dawn. 1995. "The Nuclear Legacy." *Popular Science*, August, 52–83.

Switzer, Jacqueline Vaughn. 1997. *Green Backlash: The History and Politics of Environmental Opposition in the U.S.* Boulder, CO: Lynne Rienner Publishers.

Tuxill, John. 1998. *Losing Strands in the Web of Life: Vertebrate Declines and the Conservation of Biological Diversity.* Worldwatch Paper 141. Washington, D.C.: Worldwatch Institute.

United Nations Population Division. 2001. *World Population Prospects: The 2000 Revision.* New York: United Nations.

United Nations Population Fund. 2000. *The State of the World Population 2000.* New York: United Nations.

_____. 1997. *1997 State of the World Population.* New York: United Nations.

U.S. Department of Health and Human Services. 2001 (January). *Ninth Report on Carcinogens.* Washington, D.C.: Public Health Service.

Vajpeyi, Dhirendra K. 1995. "External Factors Influencing Environmental Policymaking: Role of Multilateral Development Aid Agencies." *Environmental Policies in the Third World: A Comparative Analysis,* eds. O. P. Dwivedi and Dhirendra K. Vajpeyi, pp. 24–45. Westport, CT: Greenwood Press.

"Water Wars Forecast if Solutions Not Found." 1999 (January 1). Environment News Service. http://ens.lycos.com/ens/jan99/1999L-01-01-02.html

Weiner, Tim. 2001 (January 5). "Terrific News in Mexico City: Air Is Sometimes Breathable." *The New York Times on the Web.* http://www.nytimes.com/2001/01/05/world/05MEXI.html

World Resources Institute. 2000. *World Resources 2000–2001: People and Ecosystems: the Fraying Web of Life.* Washington D.C.: World Resources Institute.

_____. 1998. *Climate, Biodiversity, and Forests: Issues and Opportunities Emerging from the Kyoto Protocol.* Baltimore, MD: World Resource.

Wood, Stanley, Kate Sebastian, and Sara J. Scherr. 2000. *Agroecosystems: Pilot Analysis of Global Ecosystems.* Washington D.C.: International Food Policy Research Institute and World Resources Institute.

Zabin, L. S., and K. Kiragu. 1998. "The Health Consequences of Adolescent Sexual and Fertility Behavior in Sub-Saharan Africa." *Studies in Family Planning* 2(June 29):210–32.

Zwingle, Erla. 1998. "Women and Population." *National Geographic,* October, 35–55.

Chapter 15

Addison, John T., Douglas Fox, and Christopher Ruhm. 2000. "Technology, Trade Sensitivity, and Labor Displacement." *Southern Economic Journal* 66:682–699.

American Association of University Women. 2000. "Tech-Savvy: Educating Girls in the New Computer Age." http://www.aauw.org/2000/techsavvybd

Arnold, Stephen. 2001. "Internet Users at Risk." *Searcher* 9:24.

Bell, Daniel. 1973. *The Coming of Post-Industrial Society: A Venture in Social Forecasting.* New York: Basic Books.

Beniger, James R. 1993. "The Control Revolution." In *Technology and the Future*, ed. Albert H. Teich, pp. 40–65. New York: St. Martin's Press.

Boles, Margaret, and Brenda Sunoo. 1998. "Do Your Employees Suffer from Technophobia?" *Workforce* 77(1):21.

Brand, Bob. 2000. "DotComGuy." September 1. http://thebee.com/bweb

Brasher, Philip. 2000. "Government Probes Biotech Corn Allegations." Excite.News. http://news.excite.com/news/ap/000918

Brin, David. 1998. *The Transparent Society: Will Technology Force Us to Choose between Privacy and Freedom?* Reading, MA: Addison Wesley.

Buchanan, Allen, Dan Brock, Norman Daniels, and Daniel Wikler. 2000. *From Chance to Choice: Genetics and Justice.* Cambridge, New York: Cambridge University Press.

Bush, Corlann G. 1993. "Women and the Assessment of Technology." In *Technology and the Future*, ed. Albert H. Teich, pp. 192–214. New York: St. Martin's Press.

Carey, Patricia M. 1998. "Sticking It Out in the Sticks." *Home Office Computing* 16:64–69.

Ceruzzi, Paul. 1993. "An Unforeseen Revolution: Computers and Expectations, 1935–1985." In *Technology and the Future*, ed. Albert H. Teich, pp. 160–74. New York: St. Martin's Press.

Clarke, Adele E. 1990. "Controversy and the Development of Reproductive Sciences." *Social Problems* 37(1):18–37.

Cohen, Adam. 2001. "E-bay's Bid to Conquer All." *Time*, February 5, 48.

Conrad, Peter. 1997. "Public Eyes and Private Genes: Historical Frames, New Constructions, and Social Problems." *Social Problems* 44:139–54.

CyberAtlas. 2000a. "European Internet Use Still Behind the U.S." http://cyberatlas.internet.com/big_pictu...phics/article/0,1323,5911_351591,00.html

_____. 2000b. "Demographics of the Net Getting Older." http://cyberatlas.internet.com/big_picture

DotComGuy, Inc. 2001. http://dotcomguy.com

Dresser, Rebecca. 2000. "Regulating Assisted Reproduction." *The Hastings Center Report* 30:26–27.

Durkheim, Emile. [1925] 1973. *Moral Education*. New York: Free Press.

Ehrenfeld, David. 1998. "A Techno-Pox Upon the Land." *Harper's*, October, 13–17.

Eibert, Mark D. 1998. "Clone Wars." *Reason* 30(2):52–54.

Fix, Janet L. 1994. "Automation Makes Bank Branches a Liability." *USA Today*, November 28, B1.

Gallup Poll. 2000a. "Americans Say the Internet Makes Their Lives Better." http://www.gallup.com/poll/releases/pr000223
———. 2000b. "Abortion Issues." http://www.gallup.com/poll/indicators

Gibson, Brad. 2001. "PC Data: Retail Mac Sales Down." *MacCentral Online*. http://maccentral.macworld.com/news

GIP (Global Internet Project). 1998. "The Workplace." http://www.gip.org/gip2g.html

Glendinning, Chellis. 1990. *When Technology Wounds: The Human Consequences of Progress*. New York: William Morrow.

Global Reach. 2000. "Global Internet Statistics." http://www.glreach.com/globstats/index

Goodman, Paul. 1993. "Can Technology Be Humane?" In *Technology and the Future*, ed. Albert H. Teich, pp. 239–55. New York: St. Martin's Press.

Gottlieb, Scott. 2000. "Abortion Pill is Approved for Sale in United States." *British Medical Journal* 321:851.

Guernsey, Lisa. 2000. "You've Got Inappropriate Mail." *The New York Times*, April 5, C1.

Hafner, Katie. 1999. "Horse and Blender, Car and Crockpot." In *Themes of the Times: N.Y. Times*, p. 1. Upper Saddle River, NJ: Prentice Hall.

Hancock, LynNell. 1995. "The Haves and the Have-Nots." *Newsweek*, February 27, 50–53.

Hayes, Frank. 1998. "Age Bias an IT Reality." *Computerworld* 32(46):12.

Headcount. 1998. "Profile of U.S. Users." http://www.headcount.com/globalsource/profile/profile/index.htm?choice=the_us&id-144

Hormats, Robert D. 2001. "Asian Connection." *Across the Board* 38:47–50.

Hosmer, Ellen. 1986. "High Tech Hazards: Chipping Away at Workers' Health." *Multinational Monitor* 7 (January 31):1–5.

IFR (International Federation of Robotics). 1997. "1997 Key Data for the World Robot Market." http://www.ifr.org

Johnson, Jim. 1988. "Mixing Humans and Nonhumans Together: The Sociology of a Door-Closer." *Social Problems* 35:298–310.

Johnson, Margaret. 2000. "Committee Approves Watered-down Anti-Hacking Bill." October 6. http:///www2.infoworld.com/articles/hn/xml/00/10/06

Kahn, Brian. 1993. "Information Technology and Information Infrastructure." In *Empowering Technology: Implementing a U.S. Strategy*, ed. Lewis M. Branscomb, pp. 135–66. Cambridge, MA: MIT Press.

Kahn, A. 1997. "Clone Mammals...Clone Man." *Nature*, March 13, 119.

Kaplan, Carl S. 2000. "The Year in Technology Law." December 22, http://www.nytimes.com/2000/12/22/technology/22CYBERLAW

Kelly, Jason. 1997. "Technophobia." Atlanta Business Chronicle, August 18. http://www.amcity.com/atlanta/stories/1997/08/18/focus1.html

Klein, Matthew. 1998. "From Luxury to Necessity." *American Demographics* 20(8):8–12.

Kluger, J. 1997. "Will We Follow the Sheep?" *Time*, March 10, 70–72.

Kuhn, Thomas. 1973. *The Structure of Scientific Revolutions*. Chicago: Chicago University Press.

Lemonick, Michael, and Dick Thompson. 1999. "Racing to Map Our DNA." *Time Daily*, 153:1–6. http://www.time.com

Levy, Pierre. 1997. "Cyberculture in Question: A Critique of the Critique." *Revue-du-Mauss* 9:111–26.

Levy, Steven. 1995. "TechnoMania." *Newsweek*, February 27, 25–29.

Lewin, Tamar. 2001. "New State Laws Tackle Familiar National Issues." *The New York Times*, January 1, A7.

Lewin, Tamar. 1998. "Serious Gender Gap Remains in Technology." *N.Y. Times News Service*. October 18.

Lohr, Steve. 1999. "Bold Vision Propels Phone-TV Mergers Media Convergence." In *Themes of the Times: N.Y. Times*, p. 8. Upper Saddle River, NJ:Prentice Hall.

Macklin, Ruth. 1991. "Artificial Means of Reproduction and Our Understanding of the Family." *Hastings Center Report*, January/February, 5–11.

Markoff, John. 2000. "Report Questions a Number in Microsoft Trial." *The New York Times on the Web*. August 28. http://www.nytimes.com/library/tech/00/08

McCormick, S. J. Richard A. 1994. "Blastomere Separation." *Hastings Center Report*, March/April, 14–16.

McDermott, John. 1993. "Technology: The Opiate of the Intellectuals." In *Technology and the Future*, ed. Albert H. Teich, pp. 89–107. New York: St. Martin's Press.

McFarling, Usha L. 1998. "Bioethicists Warn Human Cloning Will Be Difficult to Stop." *Raleigh News and Observer*, November 18, A5.

Meeks, Brock N. 2000. "Growth of Internet Use Slows." MSNBC.com. September 21. February 12. http://www.msnbc.com/news/464681

Mehlman, Maxwell H., and Jeffery R. Botkin. 1998. *Access to the Genome: The Challenge to Equality*. Washington, D.C.: Georgetown University Press.

Merton, Robert K. 1973. "The Normative Structure of Science." In *The Sociology of Science*, ed. Robert K. Merton. Chicago: University of Chicago Press.

Mesthene, Emmanuel G. 1993. "The Role of Technology in Society." In *Technology and the Future*, ed. Albert H. Teich, pp. 73–88. New York: St. Martin's Press.

NIST (National Institute of Standards and Technology) 2000. "NIST's Fast Facts." http://www.nist.gov

NSF (National Science Foundation). 2000. "Pocket Data Book." Division of Science Resources Studies. Arlington, VA. NSF-00328.

Negroponte, Nicholas. 1995. "Nicholas Negroponte: The Multimedia Today Interview." *Multimedia Today*, July–September, 86–88.

Ogburn, William F. 1957. "Cultural Lag as Theory." *Sociology and Social Research* 41:167–74.

Papadakis, Maria. 2000. "Complex Picture of Computer Use in Home Emerges." National Science Foundation. March 31. NSF000-314.

Pascal, Zachary G. 1996. "The Outlook: High Tech Explains Widening Wage Gap." *Wall Street Journal*, April 22, A1.

Perrolle, Judith A. 1990. "Computers and Capitalism." In *Social Problems Today*, ed. James M. Henslin, pp. 336–42. Englewood Cliffs, NJ: Prentice-Hall.

Petersen, Melody. 2000. "U.S. to Keep a Closer Watch on Genetically Altered Crops." *The New York Times on the Web*, May 4. http://www. nytimes.com/library/national/science/health

Pollack, Andrew. 2000. "Nations Agree on Safety Rules for Biotech Food." *The New York Times*, January 30. http://www.nytimes.com/library/national/science

Postman, Neil. 1992. *Technopoly: The Surrender of Culture to Technology*. New York: Alfred A. Knopf.

Potter, Ned. 2001. "First Reading: Scientists Detail Human Genetic Code." ABCNews.com. February 11. http://www.abcnews.go.com. sections/wnt/Daily/News/

Powers, Richard. 1998. "Too Many Breakthroughs." Op-Ed. *The New York Times*, November 19, 35.

Quick, Rebecca. 1998. "Technology: Pieces of the Puzzle—Not So Private Lives: Will We Have Any Secrets in the Future?" *Wall Street Journal*, November 13, R27.

Rabino, Isaac. 1998. "The Biotech Future." *American Scientist* 86(2):110–12.

Regalado, Antonio. 2000. "The Great Gene Grab." *Technology Review* 103:48–55.

Reuters. 2000. "U.S. Unlikely to Follow Britain's Human Clone Lead." August 17. http://www.nytimes. com/library/national/science

Rifkin, Jeremy, 1996. *The End of Work: The Decline of the Global Labor Force and the Dawn of Post-Market Era*. Berkeley, CA: Putnam.

Rosenberg, Jim. 1998. "Troubles and Technologies." *Editor and Publisher* 131(6):4.

Sampat, Payal. 2000. "Internet Use Accelerates." In *Vital Signs: The Environmental Trends That Are Shaping Our Future*, ed. Linda Starke, pp.

94–94. New York: W.W. Norton Company.

Schaefer, Naomi. 2001. "The Coming Internet Privacy Scrum." *The American Enterprise* 12: 50–51.

Schwartz, John. 2001. "Wassail beats Lucres." *The New York Times*, January 8, C4.

Shand, Hope. 1998. "An Owner's Guide." *Mother Jones*, May/June, 46.

Statistical Abstract of the United States: 2000, 120th ed. U.S. Bureau of the Census. Washington, D.C.: U.S. Government Printing Office.

Stenger, Richard. 2000. "Does Dot-ComGuy Live in e-utopia or a Publicity Hut?" January 25. http://www.cnn.com/2000/TECH/computimg?01?25/dotcomguy

Tanaka, Jennifer. 2000. "An Extreme Reaction." *Newsweek*, September 25, p. 75.

Toffler, Alvin. 1970. *Future Shock*. New York: Random House.

U.S. Department of Commerce News. 1999. "Computer Use Up Sharply..." Press Release. http://www.census.gov/Press-Release/www/1999

U.S. Department of Justice. 1999. "1999 Report in Cyberstalking: A New Challenge for Law Enforcement and Industry." Washington, D.C.: Department of Justice.

Weinberg, Alvin. 1966. "Can Technology Replace Social Engineering?" *University of Chicago Magazine* 59 (October):6–10.

Welter, Cole H. 1997. "Technological Segregation: A Peek Through the Looking Glass at the Rich and Poor in an Information Age." *Arts Education Policy Review* 99(2):1–6.

Whine, Michael. 1997. "The Far Right on the Internet." In *The Governance of Cyberspace*, ed. Brian D. Loader, pp. 209–27. London: Routledge.

White, Lawrence. 2000. "Colleges must Protect Privacy in the Digital Age." *The Chronicle of Higher Education*, June 30, B5–6.

Winner, Langdon. 1993. "Artifact/Ideas as Political Culture." In *Technology and the Future*, ed. Albert H. Teich, pp. 283–94. New York: St. Martin's Press.

Witt, Louise. 2000. "Why Would Anyone Listen to a Guy Named Dot-ComGuy?" *Business Week Online*, August 4. http://www.businessweek. com

World Employment Report. 2001. "Digital Divide is Wide and Getting Wider." Geneva: International Labor Organization. http://www.ilo. org/public/english/bureau/inf/pkits/wer2001

Zehr, Mary Ann. 2000. "Study Finds Disparity in Internet Use." *Education Week*, May 24. http://www. edweek.org

Chapter 16

Albright, Madeleine. 2000. "Time to Renew Faith in the Nuclear Nonproliferation Treaty." March 7. U.S. Department of State: International Information Programs.

American Jewish Committee. 2000. "2000 Annual Survey of Jewish Opinion." American Jewish Committee. http://www.ajc.org/pre/survey2000

Amnesty International. 1995. *Human Rights Are Women's Right*. New York: Amnesty International USA.

AP (Associated Press). 2001. "Accused Spy's Sarasota Parents, Neighbors Shocked Over Arrest." February 23. http://www.naplesnews.com

Arms Control Association. 2000. "Start III at a Glance." September. http:www.armscontrol.org/FACTS

Barkan, Steven, and Lynne Snowden. 2001. *Collective Violence*. Boston: Allyn and Bacon.

BICC (Bonn International Center for Conversion). 1998. "Chapter Six." *Conversion Survey, 1998*. Bonn, Germany: BICC.

Brown, Seyom. 1994. *The Causes and Prevention of War*. New York: St. Martin's Press.

Card, Claudia. 1997. "Addendum to 'Rape as a Weapon of War.'" *Hypatia* 12:216–18.

Carneiro, Robert L. 1994. "War and Peace: Alternating Realities in Human History." In *Studying War: Anthropological Perspectives*, eds. S. P. Reyna and R. E. Downs, pp. 3–27. Langhorne, PA: Gordon & Breach Science Publishers.

Cauffman, Elizabeth, Shirley Feldman, Jaime Waterman, and Hans Steiner. 1998. "Posttraumatic Stress Disorder among Female Juvenile Offenders." *Journal of the American Academy of Child and Adolescent Psychiatry* 37:1209–17.

CCFR (Chicago Council on Foreign Relations). 2000. *American Public Opinion and U.S. Foreign Policy 1999*, ed. John E. Reilly. Chicago: Chicago Council on Foreign Relations.

Center for Defense Information. 2000. "Highlight of the FY 2001 Request." http://www.cdi.org/issues/usmi

CNN. 2001a. "Rape War Crime Verdict Welcomed." February 23. http://www.cnn.com/2001/WORLD/europe/

———. 2001b. "Libyan Bomber Sentenced to Life." January 31. http://europe.cnn.com/2001/LAW/

Cohen, Ronald. 1986. "War and Peace Proneness in Pre- and Postindustrial States." In *Peace and War: Cross-Cultural Perspectives*, eds. M. L. Foster and R. A. Rubinstein, pp. 253–67. New Brunswick, NJ: Transaction Books.

Conflict. 2000. "Ethiopia Eritrea Conflict." http://www.synapse.net/~acdi20/oua3/internet_reso/contempor/

Cooney, Mark. 1997. "From Warre to Tyranny: Lethal Conflict and the State." *American Sociological Review* 62:316–38.

Crock, Stan. 2001. "The Coming Firefight over Defense Spending." *Business Week Online*, January 19. http://www.businessweek.com/bwdaily/dnflash

Dixon, William J. 1994. "Democracy and the Peaceful Settlement of International Conflict." *American Political Science Review* 88(1):14–32.

Doyle, Michael. 1986. "Liberalism and World Politics." *American Political Science Review* 80 (December):1151–69.

Dudley, Steven. 2000. "Children of War Fill Colombia Slums." *The Washington Post*, August 8, A22.

Editorial. 2000. "Foreign Conservationists Under Siege." *The New York Times*. April 1, A14.

Enders, Walter, and Todd Sandler. 1993. "The Effectiveness of Anti-Terrorism Policies: A Vector-Autoregression-Intervention Analysis." *American Political Science Review* 87(4):829–44.

Funke, Odelia. 1994. "National Security and the Environment." In *Environmental Policy in the 1990s: Toward a New Agenda*, 2nd ed., eds. Norman J. Vig and Michael E.

Kraft, pp. 323–45. Washington, D.C.: Congressional Quarterly, Inc.

Garrett, Laurie. 2001. "The Nightmare of Bioterrorism." *Foreign Affairs* 80:76.

Gentry, John. 1998. "Military Force in an Age of National Cowardice." *The Washington Quarterly* 21:179–92.

Gioseffi, Daniela. 1993. *Introduction to On Prejudice: A Global Perspective*, ed. Daniela Gioseffi, pp. xi–1. New York: Anchor Books, Doubleday.

Gullo, Karen. 2001. "FBI Agent Accused of Russia Spying." February 20. http://news.excite.com/news/ap

Hayman, Peter, and Douglas Scaturo. 1993. "Psychological Debriefing of Returning Military Personnel: A Protocol for Post-Combat Intervention." *Journal of Social Behavior and Personality* 8(5):117–30.

Hawaii, State of. 2000. "Dual-Use Technologies." http://www.state.hi.us/dbedt/ert/key

Hooks, Gregory, and Leonard E. Bloomquist. 1992. "The Legacy of World War II for Regional Growth and Decline: The Effects of Wartime Investments on U.S. Manufacturing, 1947–72." *Social Forces* 71(2):303–37.

HRW (Human Rights Watch). 2001. "The Democratic Republic of Congo." http://www.hrw.org/wr2k1/africa/drc

Inglesby, Thomas, Donald Henderson, John Bartlett, Michael Ascher, et al. 1999. "Anthrax as a Biological weapon: Medical and Public Health Management." *Journal of the American Medical Association* 281: 1735–1745.

INTERPOL. 1998. "Frequently Asked Questions about Terrorism." http://www.kenpubs.co.uk/INTERPOL.COM/English/faq

Klingman, Avigdor, and Zehava Goldstein. 1994. "Adolescents' Response to Unconventional War Threat Prior to the Gulf War." *Death Studies* 18:75–82.

Kyl, Jon, and Morton Halperin. 1997. "Q: Is the White House's Nuclear-Arms Policy on the Wrong Track?" *Insight on the News* 42:24–28.

Lamont, Beth. 2001. "The New Mandate for UN Peacekeeping. " *The Humanist* 61:39–41.

Landmine. 2000. The Landmine Site. "About Landmines." http://www.thelandminesite.com

Lewis, Bernard. 1990. "The Roots of Islamic Rage." *The Atlantic*, September, 47–60.

Lifsher, Marc. 1999. "Critics Assail Plan to Dispose of Spray Cans." *Wall Street Journal*. June 16, CA1.

Lloyd, John. 1998. "The Dream of Global Justice." *New Statesman* 127:28–30.

Lorch, Donatella, and Preeston Mendenhall. 2000. "A War's Hidden Tragedy." *Newsweek*. http://msnbc.com/news

MacIntyre, Ben. 2001. "Bush Prepares to Slash the U.S. Nuclear Arsenal." *The Times*. February 10. http://www.thetimes.co.uk.article

Margasak, Larry. 2001. "FBI Agent Accused of Spying." February 21. http://dailynews.yahoo.com/h/ap

Miller, Susan. 1993. "A Human Horror Story." *Newsweek*, December 27, 17.

Moaddel, Mansoor. 1994. "Political Conflict in the World Economy: A Cross-National Analysis of Modernization and World-System Theories." *American Sociological Review* 59 (April): 276–303.

MPRI (Military Professional Resources, Inc.). 2001. "About MPRI." http://www.mpri.com/channels/about

Myers-Brown, Karen, Kathleen Walker, and Judith A. Myers-Walls. 2000. " Children's Reactions to International Conflict: A Cross-Cultural Analysis." Presented at the National Council of Family Relations. Minneapolis, Minnesota. November 20.

Mylvaganam, Senthil. 1998. "The LTTE: A Regional Problem or a Global Threat?" *Crime and Justice International* 14:1–2.

Myre, Greg. 2000. "Mediators seek Mideast Calm, Summit." Excite News. http://news.excite.com/news/ap/001012/12/new

Nelson, Murry R. 1999. "An Alternative Medium of Social Education—the 'Horrors of War' Picture Cards." *The Social Studies* 88:100–108.

Novac, Andrei. 1998. "Traumatic Disorders—Contemporary Directions." *The Western Journal of Medicine* 169:40–42.

Office of Management and Budget. 2001. "A Citizen's Guide to the Federal Budget." Office of Management and Budget. Washington, D.C.

Paul, Annie Murphy. 1998. "Psychology's Own Peace Corps." *Psychology Today* 31:56–60.

Perdue, William Dan. 1993. *Systemic Crisis: Problems in Society, Politics and World Order*. Fort Worth, TX: Harcourt Brace Jovanovich.

Perry, William J., and John M. Shalikashvili. 2000. "The U.S. Military: Still the Best by Far." *The Washington Post* August 9. http://www.washingtonpost.com/ac2/wp-dyn

Pfefferbaum, Betty. 1997. "Posttraumatic Stress Disorder in Children: A Review of the Past 10 Years." *Journal of the American Academy of Child and Adolescent Psychiatry* 36:1503–12.

Porter, Bruce D. 1994. *War and the Rise of the State: The Military Foundations of Modern Politics*. New York: Free Press.

Renner, Michael. 2000a. "Number of Wars on Upswing." In *Vital Signs: The Environmental Trends That Are Shaping Our Future*, ed. Linda Starke, pp. 110–111. New York: W.W. Norton Company.

———. 2000b. "Peacekeeping Expenditures Turn Up." In *Vital Signs: The Environmental Trends That Are Shaping Our Future*, ed. Linda Starke, pp. 112–113. New York: W.W. Norton Company.

———. 1993a. "Environmental Dimensions of Disarmament and Conversion." In *Real Security: Converting the Defense Economy and Building Peace*, eds. Kevin J. Cassidy and Gregory A. Bischak, pp. 88–132. Albany: State University of New York Press.

———. 1993b. "National Insecurity." In *Systematic Crisis: Problems in Society, Politics and World Order*, ed. William D. Perdue, pp. 136–41. Fort Worth, TX: Harcourt Brace Jovanovich.

Rosenberg, Tina. 2000. "The Unbearable Memories of a U.N. Peacekeeper." *The New York Times*, 4, 14. October 8.

Scheff, Thomas. 1994. *Bloody Revenge*. Boulder, CO: Westview Press.

Shanahan, John J. 1995. "Director's Letter." *The Defense Monitor* 24(6)8. Washington, D.C.: Center for Defense Information.

Shalikashvili, John M. 2001. "The Test Ban Solution." *The Washington Post*. January 6. http://www.clw.org/pub/clw/coalition.opedshali

SIPRI (Stockholm International Peace Research Institute). 2000. *SIPRI Yearbook 2000: Armaments, Disarmament and International Security*. Oxford: Oxford University Press.

Starr, J. R., and D. C. Stoll. 1989. "U.S. Foreign Policy on Water Resources in the Middle East." Washington, D.C.: The Center for Strategic and International Studies.

Strobel, Warren, David Kaplan, Richard Newman, Kevin Whitelaw and Thomas Grose. 2001. "A War in the Shadows." *U.S. News and World Report* 130:22.

Statistical Abstract of the United States: 2000, 120th ed. U.S. Bureau of the Census. Washington, D.C.: U.S. Government Printing Office.

Sutker, Patricia B., Madeline Uddo, Kevin Brailey, and Albert N. Allain, Jr. 1993. "War-Zone Trauma and Stress-Related Symptoms in Operation Desert Shield/Storm (ODS) Returnees." *Journal of Social Issues* 49(4):33–50.

Sweet, Alec Stone, and Thomas L. Brunell. 1998. "Constructing a Supranational Constitution: Dispute Resolution and Governance in the European Community." *American Political Science Review* 92:63–82.

UCS (Union of Concerned Scientists). 1998. "Comprehensive Test Ban Treaty." http://www.ucsusa.org/arms/ctbt.top.html

USIS (United States Information Source). 2000. "Patterns of Global Terrorism." http://www.usis.usemb.se/terror/rpt1999/review.html

Vesely, Milan. 2001. "UN Peacekeepers: Warriors or Victims?" *African Business* 261: 8–10.

Weida, William J. 1998. "The Hidden Cost of Our Nuclear Arsenal." The Brookings Institute. http:www.brook.edu/fp/projects/nucwcost/weida

World Commission on Environment and Development (Brundtland Commission). 1990. *Our Common Future*. New York: Oxford University Press.

Zakaria, Fareed. 2000. "The New Twilight Struggle." *Newsweek*. October 23. http://www.msnbc.com/news

Zimmerman, Tim. 1997. "Just When You Thought You Were Safe...Could a False Alarm Still Start a Nuclear War?" *U.S. News and World Report*, February 10, 38–40.

Epilogue

Alvarado, Diana. 2000. "Student Activism Today." *Diversity Digest*. http:www.inform.umd.edu/DiversityWeb/Digest/sm99/activism.html.

Babbie, Earl. 1994. *The Sociological Spirit: Critical Essays in a Critical Science*. Belmont, CA: Wadsworth.

Chinni, Dante, Marc Peyser, John Leland, Annetta Miller, Tom Morganthau, Peter Annin, and Pat Wingert. 1995. "Everyday Heroes." *Newsweek*, May 29, 26–39.

Lamielle, Mary. 1995. Personal communication. National Center for Environmental Health Strategies. 1100 Rural Avenue, Voorhees, NJ 08043.

Loeb, Paul Rogat. 1995. "The Choice to Care." *Teaching Tolerance*, Spring, 38–43.

Statistical Abstract of the United States 2000, 120 ed. U.S. Bureau of the Census. Washington, D.C.: U.S. Government Printing Office.

Name Index

Subject Index

Photo Credits

Chapter 1: 3: © AP/Wide World Photos; **4:** © Michael Okoniewski/Liaison Agency; **22:** © AP/Wide World Photos

Chapter 2: 34: © AFP/CORBIS; **40:** Courtesy of Lacy Hilliard; **47:** Courtesy of *World Health Magazine*, World Health Organization; **56:** © Mary Kate Denny/PhotoEdit; **59:** © Mark Richards/PhotoEdit.

Chapter 3: 66: © Georges Merillon/Liaison Agency; **73:** © John Kobal Foundation/Hulton/Archive; **80:** Partnership for a Drug-Free America®; **83:** AP/Wide World Photos.

Chapter 4: 98: © Stephen Shames/Matrix; **102:** © William Campbell/Sygma; **105:** © A. Ramey/PhotoEdit; **112:** © Billy E. Barnes/PhotoEdit.

Chapter 5: 124: © Elizabeth Crews/Stock Boston; **131:** © Carol Geddes/Sonoma Image; **137:** © American Amusement Machine Association; **142:** © Myrleen Ferguson/PhotoEdit.

Chapter 6: 162: © Mark Allan/Alpha/Globe Photos; **169:** © Tony Freeman/PhotoEdit; **174:** ©Tom Miner/The Image Works; **180:** © Frank Fournier/Contact Press Images

Chapter 7: 187: © AP/Wide World Photos; **192:** © Brad Markel/Liaison Agency; **201:** © Gamma Presse Images/Liaison Agency; **211:** © Mark Richards/PhotoEdit; **215 (all images):** Southern Poverty Law Center.

Chapter 8: 225: © Rick Berkowitz/Index Stock Imagery; **241:** Brown Brothers; **244 (left):** © AP/Wide World Photos; **244 (right):** © Bob Sacha.

Chapter 9: 249: © AP/Wide World Photos; **251:** © AP/Wide World Photos; **254:** Courtesy of Tracey St. Pierre; **260:** © AP/Wide World Photos; **271:** Courtesy of Tammy Baldwin.

Chapter 10: 299: © Elena Rooraid/PhotoEdit.; **304:** U. S. Department of Agriculture; **305:** © AP/Wide World Photos.

Chapter 11: 330: Jim Bourg/Liaison Agency; **339:** © AP/Wide World Photos; **341:** © AP/Wide World Photos; **347:** U. S. Bureau of Labor; **350:** © Reuters/Win McNamee/Archive Photos; **351:** © Reuters/Hulton/Archive

Chapter 12: 357: © Karen Kasmauski/Woodfin Camp & Associates; **360 (left):** © James D. Wilson/Woodfin Camp & Associates; **360 (right):** © Dan Habib/Impact Visuals; **365:** © Michael Newman/PhotoEdit; **373:** © Reuters/Kevin Lamarque/Hulton/Archive

Chapter 13: 398: © AP/Wide World Photos; **403:** © AP/Wide World Photos; **405:** © PhotoEdit.

Chapter 14: 419: © Jeff Greenberg/PhotoEdit; **425:** © Jeff Greenberg/PhotoEdit; **431:** © A. Ramey/PhotoEdit; **436:** Courtesy of Caroline Schacht.

Chapter 15: 457: (left): © Corbis; **457 (right):** © Douglas Mason/Woodfin Camp & Associates; **459:** © Robert Burke/Liaison Agency; **463:** © Adam Lubroth/STONE; **471:** © David Sams/Stock Boston.

Chapter 16: 483: © T. Hartwell/Sygma; **486:** © Bettmann/CORBIS; **496:** © S. Compoint/Sygma; **499:** © Jeff Greenberg/PhotoEdit.